Genetics and Evolution of Aquatic Organisms

Genetics and Evolution of Aquatic Organisms

Edited by

A.R. Beaumont

School of Ocean Sciences
University of Wales, Bangor
UK

CHAPMAN & HALL
London · Glasgow · New York · Tokyo · Melbourne · Madras

Published by Chapman & Hall, 2–6 Boundary Row, London SE1 8HN, UK

Chapman & Hall, 2–6 Boundary Row, London SE1 8HN, UK

Blackie Academic & Professional, Wester Cleddens Road, Bishopbriggs, Glasgow G64 2NZ, UK

Chapman & Hall GmbH, Pappelallee 3, 69469 Weinheim, Germany

Chapman & Hall Inc., One Penn Plaza, 41st Floor, New York NY10019, USA

Chapman & Hall Japan, Thomson Publishing Japan, Hirakawacho Nemoto Building, 6F, 1-7-11 Hirakawa-cho, Chiyoda-ku, Tokyo 102, Japan

Chapman & Hall Australia, Thomas Nelson Australia, 102 Dodds Street, South Melbourne, Victoria 3205, Australia

Chapman & Hall India, R. Seshadri, 32 Second Main Road, CIT East, Madras 600 035, India

First edition 1994

© 1994 A.R. Beaumont

Typeset in Times 9/11 pt by EXPO Holdings, Malaysia
Printed in Great Britain by Cambridge University Press

ISBN 0 412 49370 5

A catalogue record for this book is available from the British Library

Library of Congress Cataloging-in-Publication data

Genetics and evolution of aquatic organisms/edited by A.R. Beaumont.
  – 1st ed.
    p. cm.
  Includes bibliographical references and index.
  ISBN 0–412–49370–5 (acid-free paper)
  1. Animal genetics–Congresses. 2. Aquatic animals–Genetics–Congresses. 3. Evolution (Biology)–Congresses. 4. Aquatic animals–Evolution–Congresses. I. Beaumont, A.R. (Andy R.)
QH432.G44 1994   93–33230
591.1′5–dc20   CIP

Contents

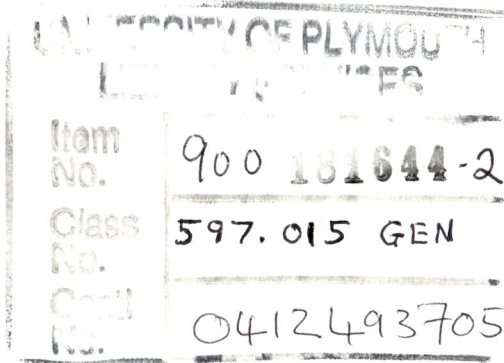

Contributors

Giuliana Allegrucci   Dipartimento di Biologia, Università di Roma 'Tor Vergata', Via E. Carnevale, 00173 Roma, Italy

John C. Avise   Department of Genetics, University of Georgia, Athens, GA 30602, USA

Janina Baršienė   Institute of Ecology, Akademijos 2, 2600 Vilnius, Lithuania

Bruno Battaglia   Department of Biology, University of Padova, Via Trieste 75, 35121 Padova, Italy

Giorgio Bavestrello   Istituto di Zoologia dell'Università di Genova, Via Balbi 5, I-16126 Genova, Italy

John A. Beardmore   School of Biological Sciences, University of Wales, Swansea, Singleton Park, Swansea SA2 8PP, UK

Andy R. Beaumont   School of Ocean Sciences, University of Wales, Bangor, Menai Bridge, Gwynedd LL59 5EY, UK

Khalid Belkhir   Laboratoire Génome et Populations, URA CNRS 1493, Université des Sciences et Techniques du Languedoc, CC 064, 34095 Montpellier Cedex 05, France

John A.H. Benzie   Australian Institute of Marine Science, PMB No. 3, Townsville Queensland 4810, Australia

Philippe Borsa   Laboratoire Génome et Populations, URA CNRS 1493, Université des Sciences et Techniques du Languedoc, CC 064, 34095 Montpellier Cedex 05, France (Present address: ORSTOM, BP A5, Noumea, New Caledonia)

Edwin Bourget   GIROQ, Département de Biologie, Université Laval, Ste-Foy, Québec G1K 7P4, Canada

Giancarlo Campesan   Istituto di Biologia del Mare, CNR, Riviera Sette Martiri 1364/A, 30122 Venezia, Italy

Mike Cantrell   Department of Pre-entry Science, University of Botswana, Private Bag 0022, Gaborone, Botswana

Gary R. Carvalho   Marine, Environmental and Evolutionary Research Group, School of Biological Sciences, University of Wales, Swansea, Singleton Park, Swansea SA2 8PP, UK

Manola Castelli   Laboratoire de Démographie des Poissons et d'Aquaculture, Service d'Ethologie et Aquarium, Faculté des Sciences, Université de Liège, 8 B Chemin de la Justice, 4500 Tihange, Belgium

Stefano Cataudella   Dipartimento di Biologia, Università di Roma 'Tor Vergata', Via E. Carnevale, 00173 Roma, Italy

Robert W. Chapman   Department of Biology and Institute for Coastal and Marine Resources, East Carolina University, Greenville, North Carolina 27858, USA

Marina Cobolli   Dipartimento di Biologia Animale e dell'Uomo, Università 'La Sapienza', Viale dell'Università 32, 00185 Roma, Italy

Denis Couvet   CEPE, route de Mende, BP 5051, 34033 Montpellier Cedex, France

John Davenport   School of Ocean Sciences, University of Wales, Bangor, Menai Bridge, Gwynedd LL59 5EY, UK (Present address: University Marine Biological Station, Millport, Isle of Cumbrae KA28 0EG, UK)

William S. Davidson   Department of Biochemistry, Memorial University, St John's, Newfoundland A1B 3X9, Canada

Amanda J. Day   Plymouth Marine Laboratory, Prospect Place, West Hoe, Plymouth PL1 3DH, UK

Elvira De Matthaeis   Dipartimento di Biologia Animale e dell'Uomo, Università 'La Sapienza', Viale dell' Università, 32, 00185 Roma, Italy

Thierry De Meeüs   URA CNRS 698, USTL, Pl. Eugène Bataillon, 34095 Montpellier Cedex 05, France

Evelyne Derelle   Observatoire Océanologique, URA CNRS 117, F-66650 Banyuls-sur-Mer, France

David R. Dixon   Plymouth Marine Laboratory, Citadel Hill, Plymouth PL1 2PB, UK

Brandon Eleby   Department of Biology, East Carolina University, Greenville, North Carolina 27858, USA

Jean-Pierre Féral   Observatoire Océanologique, URA CNRS 117, F-66650 Banyuls-sur-Mer, France

Andrew Ferguson   Division of Environmental and Evolutionary Biology, School of Biology and Biochemistry, The Queen's University of Belfast, Belfast BT7 1NN, UK

Carlo Fortunato   Dipartimento di Biologia, Università di Roma 'Tor Vergata', Via E. Carnevale, 00173 Roma, Italy

Patrick M. Gaffney   College of Marine Studies, University of Delaware, Lewes, Delaware 19958, USA

René Galzin   Laboratoire d'Ichthyoécologie Tropicale et Méditerranéenne, URA CNRS 1453, Université de Perpignan, 66860 Perpignan Cedex, France

Peter E. Gibbs   Plymouth Marine Laboratory Citadel Hill, Plymouth PL1 2PB UK

Chris J. Gliddon   School of Biological Sciences, University of Wales, Bangor, Gwynedd LL57 2UW, UK

Elizabeth M. Gosling   Regional Technical College, Dublin Road, Galway, Ireland

Jerome Goudet   School of Biological Sciences, University of Wales, Bangor, Gwynedd LL57 2UW, UK

Sheila E. Hartley Department of Biological and Molecular Sciences, University of Stirling, Stirling FK9 4LA, UK

Dennis Hedgecock Bodega Marine Laboratory, University of California, Davis, Bodega Bay, CA 94923, USA

Kathrin Hoare School of Ocean Sciences, University of Wales, Bangor, Menai Bridge, Gwynedd LL59 5EY, UK

Eric R. Holm GIROQ Département de Biologie, Université Laval, Ste-Foy, Québec G1K 7PA, Canada

Herman Hummel Centre for Estuarine and Coastal Ecology, Netherlands Institute of Ecology, Vierstraat 28, 4401 EA Yerseke The Netherlands

Mohammad G. Hussain Institute of Aquaculture, University of Stirling, Stirling FK9 4LA, UK (Present address: Fisheries Research Institute, Brackishwater Station, Paikgacha, Khulna-9280, Bangladesh)

David J. Innes Department of Biology, Memorial University of Newfoundland, St John's, Newfoundland A1B 3X9, Canada

Philippe Jarne Laboratoire de Génétique et Environnement, URA CNRS 327, Université des Sciences et Techniques du Languedoc, CC 064, 34095 Montpellier Cedex 05, France

Knut Eirik Jørstad Institute of Marine Research, Division of Aquaculture, P.O. Box 1872, Nordnes, N-5024 Bergen, Norway

Ravil Kamaltynov Institute of Limnology, Siberian Branch of the Russian Academy of Sciences, P.O. Box 4199, 664033 Irkutsk, Russia

Yuri P. Kartavtsev Institute of Marine Biology, Vladivostok 690041, Russia

Alexey S. Kondrashov Laboratory of Genetics, University of Wisconsin, Madison WI 53706, USA

Bruno Ladrón de Guevara Universidad Católica del Norte, Depto de Biología Marina, Casilla 117, Coquimbo, Chile

Daphna Lavee Department of Genetics, Alexander Silberman Institute of Life Sciences, The Hebrew University of Jerusalem, Jerusalem 91904, Israel

Marco Mattoccia Dipartimento di Biologia, Università 'Tor Vergata', Via di Passo Lombardo 430, 00100 Roma, Italy

Brendan J. McAndrew Institute of Aquaculture, University of Stirling, Stirling, FK9 4LA, UK

Luc De Meester Laboratory of Animal Ecology, University of Ghent, KL Ledeganckstraat 35, 9000 Ghent, Belgium

Axel Meyer Department of Ecology and Evolution, State University of New York, Stony Brook, New York 11794-5245, USA

Paraskeva Michailova Institute of Zoology, Bulgarian Academy of Sciences, 1 Rouski Boul., Sofia 1000, Bulgaria

Consuelo Montero University of Padova, Department of Biology, Via Trieste 75, 35121 Padova, Italy

Sununguko W. Mpoloka Department of Biology, University of Botswana, Private Bag 0022, Gaborone, Botswana

Jakob Müller Department of Population Biology, Zoological Institute, University of Mainz, Saarstraße 21, 6500 Mainz, Germany

Gunnar Nævdal Department of Fisheries and Marine Biology, University of Bergen, Bergen High-Technology Center, N-5020 Bergen, Norway

Claudio Palma Universidad de la Serena, Depto de Biología, La Serena, Chile

Juan J. Pasantes Department of Genetics, University of La Coruña, 15071 La Coruña, Spain

Philip L. Pascoe Plymouth Marine Laboratory, Citadel Hill, Plymouth PL1 2PB, UK

Tomaso Patarnello Department of Biology, University of Padova, Via Trieste 75, 35121 Padova, Italy

John C. Patton LGL Genetics, 1410 Cavitt Street, Bryan, Texas 77801 USA

Ole I. Paulsen Institute of Marine Research, Division of Aquaculture, P.O. Box 1872, Nordnes, N-5024 Bergen, Norway

David J. Penman Institute of Aquaculture, University of Stirling, Stirling FK9 4LA, UK

Véronique Perrot Zoologisches Institut Basel, Rheinsprung 9, 4051 Basel, Switzerland

Hervé Philippe Laboratoire de Biologie Cellulaire 4, Université Paris-Sud, URA CNRS 1134, F-91405 Orsay Cedex, France

Serge Planes Laboratoire d'Ichthyoécologie Tropicale et Méditerranéenne, URA CNRS 1453, Université de Perpignan, 66860 Perpignan Cedex, France

Grant H. Pogson Department of Biology, Dalhousie University, Halifax, Nova Scotia, B3H 431 Canada (Present address: Department of Zoology, University of Cambridge, Downing St., Cambridge CB2 3EJ, UK)

Paulo A., Prodöhl Division of Environmental and Evolutionary Biology, School of Biology and Biochemistry, The Queen's University of Belfast, Belfast BT7 1NN, UK

Sophie Richerd Laboratoire de Génétique et Evolution des Populations Végétales, URA CNRS 1185, GDR CNRS-IFREMER 1002, bat. SN2, Université Lille 1, 59655 Villeneuve d'Ascq Cedex, France

Bruce J. Riddoch Department of Biology, University of Botswana, Private Bag 0022, Gaborone, Botswana

Uzi Ritte Department of Genetics, Alexander Silberman Institute of Life Sciences, The Hebrew University of Jerusalem, Jerusalem 91904, Israel

John S. Ryland School of Biological Sciences, University of Wales, Swansea, Singleton Park, Swansea SA2 8PP, UK

Pierangela Saccoccio Dipartimento di Biologia Animale e dell'Uomo, Università 'La Sapienza', Viale dell'Università 32, 00185 Roma, Italy

Heikki Salemaa Department of Zoology and Division of Ecology, P.O. Box 17 (Rautatiekatu 13), SF-00014 University of Helsinki, Finland

Michele Sarà   Instituto di Zoologia dell'Università di Genova, Via Balbi 5, I-16126 Genova, Italy

Valerio Sbordoni   Dipartimento di Biologia, Università di Roma 'Tor Vergata', Via E. Carnevale, 00173 Roma, Italy

Felicita Scapini   Dipartimento di Biologia e Genetica, Università di Firenze, Via Romana 17, 50125 Firenze, Italy

Kim T. Scribner   Alaska Fish and Wildlife Research Center, US Fish and Wildlife Service, 1011 E. Tudor Road, Anchorage AL 99503, USA

Alfred Seitz   Department of Population Biology, Zoological Institute, University of Mainz, Saarstraße 21, 6500 Mainz, Germany

Paul Shaw   School of Biological Sciences, University of Wales, Swansea, Singleton Park, Swansea SA2 8PP, UK

Edson P. da Silva   Departamento de Biologia Geral, Universidade Federal Fluminense, Morro do Valonguinho s/n, 24.000 Niterói-RJ, Brazil

Doreen R. Singleton   Department of Biology, Memorial University of Newfoundland, St John's, Newfoundland A1B 3X9, Canada

David O.F. Skibinski   Molecular Biology Research Group, School of Biological Sciences, University of Wales, Swansea, Singleton Park, Swansea SA2 8PP, UK

Panom Sodsuk   Institute of Aquaculture, University of Stirling, Stirling FK9 4LA, UK

Antonio M. Solé-Cava   Departamento de Genética, Universidade Federal do Rio de Janeiro, Bloco A-CCS, Ilha do Fundao, 21.941 Rio de Janeiro-RJ, Brazil

Michele Stenico   Department of Biology, University of Padova, Via Trieste 75, 35121 Padova, Italy

John B. Taggart   Division of Environmental and Evolutionary Biology, School of Biology and Biochemistry, The Queen's University of Belfast, Belfast BT7 1NN, UK

Catherine Thiriot-Quiévreux   Laboratoire d'Océanographie Biochimique et Ecologie, Observatoire Océanologique, Université P. et M. Curie-CNRS-INSU, 06230 Villefranche-sur-Mer, France

Solveig Thorkildsen   Department of Fisheries and Marine Biology, University of Bergen, Bergen High-Technology Center, N-5020 Bergen, Norway

Myriam Valero   Laboratoire de Génétique et Evolution des Populations Végétales, URA CNRS 1185, GDR CNRS-IFREMER 1002, bat. SN2, Université Lille 1, 59655 Villeneuve d'Ascq Cedex, France

Eric Verspoor   SOAFD Marine Laboratory, P.O. Box 101, Aberdeen AB9 8DB, UK

Frederico Winkler   Universidad Católica del Norte, Depto de Biología Marina, Casilla 117, Coquimbo, Chile

Eleftherios Zouros   Department of Biology, Dalhousie University, Halifax, Nova Scotia B3H 4J1, Canada, and Department of Biology, University of Crete, Iraklion, Crete, Greece

Preface

It is now 15 years since the publication of Battaglia and Beardmore's (1978) volume on the genetics of marine organisms which presented the proceedings of a conference held in Venice the previous year. At that time marine biologists and geneticists were exploring many aspects of the increasing body of data which was being generated through electrophoretic studies. The main genetic themes were the measurement of genetic variation, its relationship to environmental variation, its maintenance by selection and its use in taxonomy. Since then, the technical advances in the study of genetics and the increasing dominance of neutral theory as the explanation for genetic variation at both the protein and DNA level have brought about major changes in approach. It was partly a recognition of these changes which instigated discussions at the 23rd European Marine Biology Symposium at Swansea and led to the idea of holding an international meeting on the genetics of aquatic organisms. Following my offer to organize such a meeting at Bangor, an Organizing Committee of Catherine Thiriot-Quiévreux, Tomaso Patarnello and Herman Hummel was set up at a Workshop on the Genetics of Marine and Estuarine Organisms at Yerseke, The Netherlands.

This book contains articles and papers selected from those presented at the Genetics and Evolution of Aquatic Organisms Conference held at Bangor on September 10–16, 1992. Each chapter (except Chapter 2) is opened by a review section covering a general topic area and this is then followed by a number of experience sections. Chapters 1 and 2 cover the uses of isozyme data in understanding genetic structure at the population level and taxonomy at the species level while Chapter 3 deals with heterozygosity, heterosis and adaptation. Chapters 4 and 5 introduce modern DNA technology and discuss its uses in the study of genetics and evolution of aquatic animals. Clonal animals and the role they play in the study of genetics and evolution are covered in Chapter 6, while the more traditional, but nonetheless fundamental, area of chromosomal genetics is the subject of Chapter 7. Genetics in relation to pollution and aquaculture are explored in Chapters 8 and 9. Clearly there is bound to be considerable overlap between these various themes but I hope that the book does nevertheless present a coherent overall approach.

My particular thanks are due to Catherine Thiriot-Quiévreux, Tomaso Patarnello and Herman Hummel, whose help and support were invaluable at every stage in the organization of the conference. I also thank Joan Lewis and Janice Chapman for expert secretarial assistance. All these, and others, including Jenny Fairbrother, Craig Wilding, Phil Papageorgiou and Kate Hoare, provided invaluable assistance during the conference itself. We are indebted to the National Environment Research Council and the FAR programme of the Commission of the European Communities for partial funding of the conference.

Every attempt has been made to ensure the highest scientific quality of the contributions and I would like to thank all those who gave of their time to referee papers for this book.

My special thanks are due to Kate Hoare, whose support and encouragement have kept me going throughout this enterprise.

References

Battaglia, B. and Beardmore, J.A. (1978) *Marine Organisms, Genetics, Ecology and Evolution*, Plenum Press, London.

A.R. Beaumont
Menai Bridge

1

Speciation and wide-scale genetic differentiation

1.1 SPECIATION AND SPECIES CONCEPTS IN THE MARINE ENVIRONMENT

Elizabeth M. Gosling

Abstract

The first part of this section presents a historical review of different species concepts, starting in the fourth century BC with the morphological species concept of Aristotle; proceeding on to the twentieth century, when this concept was replaced with the biological species concept of Dobzhansky and Mayr; and finishing with recent proposed alternatives to the biological species concept – the phylogenetic concept of Cracraft (1983), the recognition concept of Paterson (1985) and the cohesion concept of Templeton (1989). The advantages and limitations of each concept are discussed.

The second part of the section deals with speciation and species concepts in the marine environment. In order to test the operational usefulness or otherwise of the above described concepts, three different 'species situations' have been chosen: Cerastoderma (Cardium) edule *and* C. glaucum, *two cockle species, to illustrate a fairly straightforward situation, where to date there is no major systematic controversy;* Mytilus, *the blue mussel, to illustrate systematic problems in a genus whose members are morphologically and genetically very similar and where there is clear evidence of widespread hybridization; and* Lasaea, *to illustrate systematic difficulties in a brooding genus with sexual and asexual modes of reproduction. The importance of a multidisciplinary approach in arriving at systematic decisions is highlighted, as are the operational difficulties in attempting to apply a universal species concept to the complexity of variation patterns in nature.*

1.1.1 Introduction

The science of taxonomy started with Aristotle, who, writing in the fourth century BC, saw nature as organized gradually from lifeless matter through complex forms of plant and animal life. In classifying organisms into groups Aristotle adopted a purely phenetic approach: 'It is by resemblance of the shapes of their parts, or of their whole body, that the groups are marked off from each other'. Once major groups were identified he proceeded to

Genetics and Evolution of Aquatic Organisms. Edited by A.R. Beaumont.
Published in 1994 by Chapman & Hall, 2–6 Boundary Row, London SE1 8HN.
ISBN 0-412-49370-5

subdivide these further into smaller units, and the process was repeated until further subdivision was no longer possible. His formalization of individual kinds (species) and of collective groups (genera) certainly laid the groundwork for the more perceptive and elaborate classification systems which were to follow. The existence of discrete clusters of things that can be called species has, therefore, long been recognized. However, the concept of what a species is has changed many times in the course of history.

From the time of Aristotle and right up to the time of Linnaeus in the eighteenth century the prevailing species concept was a purely morphological one, species being defined as a collection of similar individuals sharing in the same essence or *eidos*. Thus, species were seen as well-defined units of nature, constant through time, and sharply delimited from each other. However, at the end of the eighteenth century the naturalist, George Buffon, while still supporting the concept of constant, well-delimited species, wrote the following:

> We should regard two animals as belonging to the same species if, by means of copulation, they can perpetuate themselves and preserve the likeness of the species; and we should regard them as belonging to different species if they are incapable of producing progeny by the same means.

and

> The species is a constant succession of similar individuals that can reproduce together.

(Buffon, 1749)

Buffon, and other naturalists after him, came tantalizingly close to a biological definition of species. But none of them took the seemingly small step of defining species in terms of a reproductively isolated assemblage of populations.

It was not until the 1920s, 1930s and 1940s that a biological species concept began to be applied consistently by scientists such as Rensch, Dobzhansky, Huxley and Mayr. The main tenets of this concept were that species should be envisaged not as types but as populations (or groups of populations), and should be defined by distinctness (i.e. the reproductive gap) rather than by difference (Mayr, 1982, p. 272). Several definitions appeared over the next few decades, the most oft-quoted one being Mayr's (1970) definition – 'Species are groups of interbreeding natural populations that are reproductively isolated from other such groups'. An ancestral species becomes transformed into two descendent species – usually through geographic isolation (see below) – when an array of populations able to interbreed becomes segregated into two reproductively isolated arrays. Reproductive isolation is used, therefore, as the fundamental criterion to define species because reproductive isolation allows gene pools to evolve independently of each other.

But how is the integrity of species being maintained? Dobzhansky (1937) coined the term 'isolating mechanism' for any agent that hinders the interbreeding of groups of individuals. Later Mayr (1963, p. 91) was more specific and defined isolating mechanisms as 'biological properties of individuals which prevent the interbreeding of populations that are actually or potentially sympatric'. Detailed descriptions of the various types of isolating mechanisms can be found in any elementary evolutionary text. Basically, they fall into two main categories: prezygotic barriers, which impede or prevent hybridization of members of different populations; and postzygotic barriers, which reduce the viability or fertility of hybrids that have arisen. Most species pairs are kept apart by a combination of several isolating mechanisms. Over the years there has been considerable discussion as to the origin and development of reproductive isolation mechanisms (RIMs). It is now generally accepted that, in allopatric situations, they arise as accidental byproducts of allopatric divergence, selection only serving to perfect them should recently formed species become sympatric (see Butlin (1989) for detailed discussion).

It is probably opportune at this point to very briefly describe the main ways in which speciation can occur. Elaborate classification systems abound in the literature and interested readers are referred to Mayr (1970), Dobzhansky *et al.* (1977), White (1978) and Grant (1981).

In the origin of species the gene pool of a population is severed from other populations of the parent species. The splinter population can then follow its own evolutionary course as changes in gene frequencies caused by selection, genetic drift and mutations occur undiluted by gene flow from other populations. The initial block to gene flow may be a geographic barrier that physically isolates the population. How formidable this barrier must be to keep allopatric populations apart depends on the dispersal ability of the organisms concerned. Alternatively, a new species may arise within the range of the parent population, i.e. without geographic isolation. This mode of speciation, termed sympatric, implies strong disruptive selection on a polymorphic population, since the process of divergence is constantly opposed by interbreeding between the divergent polymorphic types. Both allopatric and sympatric speciation involve the gradual divergence of populations, the development of RIMs as byproducts of genetic divergence usually requiring many thousands of generations. However, a third type of speciation, quantum speciation, is a rapid process and involves major chromosomal changes, such as translocations or polyploidy, in one or more individuals of a population. These individuals are then reproductively isolated from the parental population and, therefore, constitute a new species. It is now quite generally accepted that geographic (allopatric) speciation is the almost exclusive mode of speciation among animals, and most likely the prevailing one in plants. On the other hand, it would appear from a survey of the literature that sympatric and quantum speciation are mostly confined to plants (see Lynch (1989) for discussion).

1.1.2 Species concepts: an evaluation

The biological species concept (see above) has been the dominant one in the evolutionary literature since the 1940s. The concept, more often referred to nowadays as the isolation species concept (ISC), regards species 'as a reproductive community, a gene pool and a genetic system' (White, 1978), and because of this the concept is potentially useful in analysing speciation from a population genetics perspective. However, in recent years the ISC has come under increasing fire, particularly from plant systematists, but also from zoologists (Sokal and Crovello, 1970; Cronquist, 1978; Eldredge and Cracraft, 1980; Grant, 1981; Mishler and Donoghue, 1982; Donoghue, 1985; Paterson, 1985; Masters *et al.*, 1987; McKitrick and Zink, 1988; Cracraft, 1989; Templeton, 1989). The ISC has caused confusion primarily because it rests on the erroneous premise that there is a close correspondence between breeding groups and morphologically and ecologically discrete units, through gene flow, or the lack of it. For example, a lack of concordance is illustrated by sibling species, semispecies, polytypic species and polyploids – all common situations in higher plants. Many, therefore, find the ISC to be non-operational and impractical. In theory, while the ISC delimits species on the basis of non-interbreeding, in practice the fact that there is so little information on the breeding capabilities of the vast majority of species means that the basic definition of the species is of necessity phenetic. In other words, the morphological species concept (MSC) is essentially the default taxonomic base (Guiry, 1992).

Many find difficulties with the ISC in relation to the speciation process. The difficulties mainly arise from the fact that the ISC is defined in terms of RIMs. But these are irrelevant as isolating barriers during speciation – they cannot function as isolating barriers in allopatry

(Templeton, 1989; see also below). They are byproducts, not causative agents, of the gradual genetic divergence that follows isolation. Several authors have gone so far as to suggest that reproductive isolation should not be a part of species concepts (Nelson and Platnick, 1981; Cracraft, 1983; McKitrick and Zink, 1988).

The ISC also faces criticism from taxonomists interested in phylogenetic reconstruction. Species should consistently be single evolutionary units (McKitrick and Zink, 1988). However, according to the ISC, populations are assembled into groups on the basis of interbreeding but organisms that can interbreed are not necessarily closely related genealogically, i.e. they may be reproductively compatible but may not yield taxa that are monophyletic. A monophyletic group is one that contains all and only the descendants of a particular common ancestor (Donoghue, 1985).

The ISC also runs into difficulties regarding hybridization between allopatric taxa in a zone of contact. Strict application of the ISC would be to unite these taxa into a single biological species. But this is not usually done since there is some arbitrary level of hybridization which is permissible before such taxa are considered conspecific. In spite of hybridization many plant and animal species maintain their morphological, ecological and genetic integrity on either side of the hybrid zone. In plants there are many examples of hybridizing groups of species – defined as a syngameon by Grant (1981) – which behave like biological species but differ from them in terms of their complex internal structure. Syngameons are by no means limited to plants but have been reported in mammalian and invertebrate species (Templeton, 1989).

Finally, since the ISC is applicable only to sexually reproducing organisms, large portions of the organic world are excluded.

In recent years alternatives to the ISC have been proposed. Probably the one that has gained most acceptance is the phylogenetic species concept (PSC) of Cracraft (1983). A species according to the PSC is 'an irreducible (basal) cluster of organisms, diagnosably distinct from other such clusters, and within which there is a parental pattern of ancestry and descent' (Cracraft, 1989); therefore, the organisms included in any given species constitute a lineage, i.e. they are monophyletic. This definition is very similar to the 'evolutionary' species definition of Simpson (1951). These concepts were coined to give evolution more weight and to introduce a time element into species concepts. Species status is mainly decided on the basis of phenotypic cohesion within a group of organisms versus phenotypic discontinuity between groups. Therefore, the PSC emphasizes the most general aspect of taxonomic diversification, namely differentiation, some of which results in reproductive isolation and some of which does not (Cracraft, 1989). The PSC is, therefore, close to the operational species definition used by most practising taxonomists.The emphasis on differentiation (morphological, ecological and genetic) rather than on reproductive isolation – produced by the process of differentiation – means that the PSC can be applied to both sexual and asexual organisms, to allopatric and sympatric populations and to living and extinct groups. It is not surprising, in the light of the foregoing, that the PSC has been enthusiastically adopted by evolutionary biologists, taxonomists and palaeontologists. However, one difficulty with the PSC is deciding which traits are the important ones to use in defining species. Also, since the PSC deals only with the manifestation of cohesion rather than the evolutionary mechanisms responsible for cohesion it does not provide an adequate framework for integrating population genetic factors into the species concept (Templeton, 1989).

Another concept, the recognition species concept (RSC) of Paterson (1985), defines a species as a group of organisms that share a common fertilization system. This system, termed the specific-mate-recognition system (SMRS), includes all those features by which organisms recognize each other as mates. Excluded from the definition are all barriers to gene flow which act after fertilization, e.g. zygotic and hybrid inviability and hybrid sterility. Paterson sees the SMRS as the product of directional selection acting to maximize reproduction among members of a population, and a side-effect of the accompanying genetic change is sometimes the erection of an isolating barrier. Although the RSC may be more biologically meaningful for allopatric species and populations than the ISC, the emphasis of both concepts is on reproduction (the former emphasizing mechanisms that facilitate reproduction between members of a population, the latter emphasizing mechanisms that prevent reproduction between different populations). Therefore, the RSC is subject to much the same criticisms as the ISC (Paterson, 1985; Masters *et al.*, 1987; Raubenheimer and Crowe, 1987; Templeton, 1989; Coyne, Orr and Futuyma, 1988; Mayr, 1988; Masters and Spencer, 1989; Kimbel, 1991).

More recently, Templeton (1989) has argued that rather than placing sole emphasis on reproduction we should concentrate on those processes which cause groups of organisms to be similar to one another. The cohesion species concept (CSC) of Templeton defines species as the most inclusive population of individuals sharing a common developmental genetic system, physiology and ecology. The focus of this concept is on cohesion mechanisms (Templeton, 1989, Table 2) such as gene flow, stabilizing selection and ecological factors which maintain species homogeneity. The main advantages of the CSC are that, like the PSC (but unlike the ISC and RSC), it includes asexual taxa and the members of syngameons, and facilitates the study of the speciation process in that it deals with a broader array of evolutionary forces than any of the concepts discussed so far. However, the CSC suffers from the same operational difficulties as the ISC and RSC (Endler, 1989).

A summary of the different species concepts together with their limitations is presented in Table 1.1.1.

1.1.3 Species concepts in the marine environment

In a recent literature search on species concepts, going back to 1985, I found only one out of a total of close to 180 publications (Guiry, 1991) dealing with species concepts in the marine environment. It is obvious, therefore, that marine biologists have not been to the forefront in the debate on species concepts. This is unexpected in view of the extent of the marine biotope and species richness therein. One possible reason for the lack of involvement of marine biologists could be that information on speciation in the sea is only beginning to emerge. The best information on speciation – gleaned largely by the application of biochemical genetic techniques (Murphy *et al.*, 1990) – has been on those taxa with life histories very similar to those of terrestrial species, i.e. those with low dispersal and small population sizes. However, there is little information to date on speciation in high-dispersal taxa. These, while rare on land, are a dominant component of marine fauna. Such species have free-swimming adults or planktonic larvae that can drift for considerable distances before settlement. Typically, geographic ranges of such species are vast, populations are large and fecundities number in the millions of eggs per female (review: Palumbi, 1992). Marine bivalves are a good example of such high dispersal taxa.

In this section I will examine speciation and the use of species concepts in three groups of marine bivalves. The first example, the cockle species pair, *Cerastoderma (Cardium) edule* (L.) and *Cerastoderma (Cardium) glaucum* (Lmk.), was chosen to illustrate a fairly straightforward situation where to date there appears to be no major systematic controversy (apart from the recent change of the genus name from *Cardium* to *Cerastoderma*). The second example, the blue mussel *Mytilus*, was chosen to illustrate systematic problems in a genus whose members are morphologically and genetically very similar, and where there is widespread hybridization between taxa. The third was chosen to illustrate systematic

Table 1.1.1 Description and limitations of different species concepts

Concept	Limitations
Isolation species concept (Mayr, 1963) Species defined as groups of interbreeding natural populations that are reproductively isolated from other such groups. The genetic integrity of the species is maintained through reproductive isolating mechanisms.	Erroneous implication that there is a close correspondence between breeding groups and morphological and ecological units. Concept is operationally difficult: often have to adopt a phenetic approach to delimiting species especially in because reproductive isolation is usually difficult to ascertain, the case of allopatric taxa. Organisms included in any given species do not necessarily constitute a lineage, i.e. they are not monophyletic. Species are defined in terms of isolating mechanisms. But these are not mechanisms as such, but rather byproducts of evolutionary divergence. Hybridization is not compatible with this concept. The concept cannot be applied to asexual or fossil taxa.
Phylogenetic species concept (Cracraft, 1983) A species is defined as a group of individuals that share a common evolutionary fate through time (i.e. it is monophyletic) and is diagnosable on the basis of one or more genetically determined traits. Emphasis is on *differentiation* (morphological, ecological and genetic), rather than on *reproductive isolation*, produced by the process of differentiation. Therefore, this concept can be applied to both sexual and asexual organisms, and to living and extinct groups.	Difficult to decide which traits are the important ones in defining species. Deals only with the manifestation of cohesion rather than the evolutionary mechanisms responsible for cohesion; therefore, it does not provide an adequate framework for integrating population genetic factors into the species concept.
Recognition species concept (Paterson, 1985) A species is defined as the most inclusive group of individual biparental organisms which share a common fertilization system. This concept is viewed by many as the opposite side of the coin to the isolation species concept, i.e. selection operates to secure syngamy rather than to hinder fertilization.	Because of the emphasis on reproduction this concept suffers from the same limitations as the isolation species concept.
Cohesion species concept (Templeton, 1989) A cohesion species is defined as the most inclusive group of organisms whose range of phenotypic variation is limited by genetically and environmentally based cohesion mechanisms e.g. gene flow, genetic drift and natural selection. Because gene flow is not the only factor defining the boundaries of an evolutionary lineage this concept can be applied to syngameons and asexual taxa.	Suffers from the same operational difficulties as the isolation and recognition species concepts.

difficulties in the cosmopolitan brooding genus *Lasaea*, whose members exhibit sexual and asexual modes of reproduction.

For each of the three species situations I have tried to adopt a multidisciplinary approach in the delineation of species. Ideally such an approach should include data on (White, 1978):

1. geographic distribution, present and past;
2. morphology;
3. ecology, e.g. habitat, food, temperature, salinity, substrate preferences;
4. physiology and biochemistry;
5. reproductive cycle, behavioural or chemical isolating mechanisms;
6. genetics, to include data on allozymic, nuclear (n) and mitochondrial (mt) DNA polymorphisms, cytogenetics (chromosome number, morphology, karyotypes), immunology, hybridization, if any in nature, information on artificial hybridization.

I can say without fear of contradiction that in an assessment of a particular species situation there are few, if any, cases where all of this information has been collated.

(a)   Systematics of the sympatric taxa Cerastoderma (Cardium) edule *and* C. glaucum

Most of the work on the taxonomy of the genus *Cerastoderma (=Cardium)* in European waters has been carried out in Denmark by Brock, who has recently published a synthesis of her work (Brock, 1991), and interested readers are referred to this for details and references.

The geographic distribution of *C. edule* extends from the Barents Sea, southwards along the Atlantic coasts of Europe to western Africa. *C. glaucum*, with a more southerly distribution, occurs in the Mediterranean, Black and Caspian Seas, but also in the Baltic. Sympatric populations of the two species are frequent in British, Danish, Swedish and Norwegian waters (Brock, 1979).

It is believed that the ancestor of the modern *Cerastoderma (Cardium)* complex existed about 50 mya and that *C. edule*, a direct descendant, evolved in the Atlantic. *C. glaucum* probably evolved at a later date in the late Miocene in the Mediterranean Sea, and when the Gibraltar Straits reopened in the early Pliocene, about 4 mya, *C. glaucum* spread out of the Mediterranean Sea into the areas it now occupies today. This picture of the evolution of these two species is supported by evidence from nuclear DNA sequence analysis (Brock and Christiansen, 1989).

Allopatric populations of *C. edule* and *C. glaucum* are easily differentiated on the basis of shell characteristics, and although sympatric populations exhibit convergent character displacement for shell characters Brock (1978) has identified a morphological character which unequivocally separates the two species in sympatry. The two taxa occupy similar but overlapping niches but, of the two, *C. glaucum* is the more euryhaline and eurythermic species (Brock, 1991). There is no evidence of interspecific food partitioning but small differences in, for example, growth rate, oxygen consumption and parasitic infection have been reported (Brock, 1991). In sympatry, although the two taxa appear to have similar spawning periods, evidence from electrophoretic investigations on wild populations (Brock, 1978; Gosling, 1980) and from hybridization studies on laboratory populations (Kingston, 1973) indicates that cross-fertilization does not occur. Brock (1991) has suggested that perhaps some sort of prezygotic RIM, a type of gamete mate-recognition system, operates in sympatry. This is plausible in view of recent evidence in echinoderm and gastropod species for gamete incompatibility in interspecific crosses, caused by species-specific surface proteins (Palumbi, 1992).

Surprisingly, there have been few reports of electrophoretic variation within the genus. Those that have appeared have had as their main objective the search for enzyme loci which could be used in species identification (Jelnes, Petersen and Russell, 1971; Brock, 1978; Gosling, 1980). Brock (1978) found consistent differences at two malate dehydrogenase *(Mdh)* loci between allopatric and sympatric *C. edule* and *C. glaucum*; both loci are monomorphic but exhibit different electrophoretic mobilities in the two species. Brock (1979, 1980a, b, 1982, 1987) has repeatedly used these loci to identify the two species prior to other types of investigations (morphological, ecological, DNA analysis).

Although mtDNA analysis has revealed extensive homology between the two species (Brock and Christiansen, 1989), *C. edule* can consistently be separated from *C. glaucum* on the basis of one non-shared DNA fragment.

Using crossed immunoelectrophoresis Brock (1987) reported seven species-specific antigen–antibody reactions between *C. edule* and *C. glaucum*. Results from other species (see Gosling (1992a) for references) suggest that a minimum of one or two antigen differences can be expected in species pair comparisons.

To date, although the karyotype of *C. edule* has been investigated (Koulman and Wolff, 1977; Insua and Thiriot-Quiévreux, 1992), no comparisons have yet been made between *C. edule* and *C. glaucum.*

So what is the specific status of *C. edule* and *C. glaucum*? The two are reproductively isolated from each other and, therefore, by ISC or RSC criteria are good species. Although there is no evidence that each constitutes a separate lineage, i.e. is monophyletic, the degree of differentiation between the two taxa indicates that they are at least diagnosably distinct, thus satisfying the other main requirement of the PSC for species recognition. A corollary of such differentiation is that powerful mechanisms must be actively maintaining genetic and phenotypic cohesion within each taxon; therefore, by CSC criteria they are also good species.

(b)   Mytilus *systematics*

The systematics of this genus have been the subject of considerable discussion for well over a hundred years. Much of the controversy stems from the fact that, until relatively recently (as with other bivalve taxa) *Mytilus* systematics have been based solely on morphological shell characteristics. The development of various genetic techniques, in particular allozyme analysis, used in conjunction with morphometric analysis, has helped to clarify the systematic status of various taxa within the genus (McDonald, Seed and Koehn, 1991).

Most attention has focused on the systematic status of *Mytilus edulis* L. and the Mediterranean mussel, *M. galloprovincialis* Lmk., but a third mussel type, the recently discovered *M. trossulus* Gould, has now entered the fray. Morphological differences between the three taxa are small with no single character being clearly diagnostic (reviews: Gosling, 1984; 1992a,b; Gardner, 1992; Seed, 1992). Using a combination of characters – some of which necessitated the use of a microscope – McDonald, Seed and Koehn (1991) achieved a good separation of the different taxa. However, taxa were first identified using allozyme markers and samples were analysed only from areas of non-overlap between taxa in order to remove the confounding effect of hybrids.

Several enzyme loci have proved useful in differentiating the different taxa within the genus. The best discriminatory loci are mannose phosphate isomerase *(Mpi)*, peptidase-II *(Aap)* and glucose phosphate isomerase *(Gpi)*, although none of these is truly diagnostic. Using information from these and other loci, McDonald, Seed and Koehn (1991) have mapped the distribution of the three taxa on a global scale. *M. edulis* is widely distributed in the northern hemisphere; it occurs in European waters extending from the White Sea in Russia as far south as the Atlantic coast of southern France; it is also found in Iceland and on the east coast of North America and in South America. *M. galloprovincialis* is found in the

Mediterranean Sea and on the Atlantic coasts of Spain, Portugal and France, and in the British Isles as far north as the Shetland and Orkney Islands; it also occurs in Southern California, Japan, on the east China coast as far north as Korea, and in Tasmania, New Zealand and Australia. *M. trossulus* – found only in the northern hemisphere – inhabits the Baltic Sea, the east coast of Russia, the west coast of North America and the east coast of the Canadian Maritimes. Another form, *M. desolationis*, is found only in the Kerguelen Islands in the southern Indian Ocean, while *M. californianus* – for which there is no controversy regarding its specific status – is found on the west coast of the USA.

The genus is of relatively recent origin with apparently no records older than 2 000 000 years (Seed, 1976). *M. edulis* is believed to be the ancestral form, with *M. galloprovincialis* evolving in the Mediterranean Sea during its isolation from the Atlantic during one of the Pleistocene ice ages (cf. *Cerastoderma (Cardium) glaucum*). Outside western Europe, with the exception of some well-documented areas (Hong Kong, Japan and Southern Africa), the widespread occurrence of *M. galloprovincialis* in the southern hemisphere remains a dilemma. *M. trossulus*, confined to more northerly latitudes, is believed to be a zoo-geographic remnant of what was once (1–2 mya) a more widely distributed mussel (Koehn, 1991). Its presence in the low-salinity habitat of the Baltic Sea is of recent occurrence.

Everywhere that two of the above mussel forms come into geographic contact they hybridize and the size of the hybrid zone can be extensive – more than 1500 km in the case of *M. edulis/galloprovincialis* in western Europe. Results from electrophoretic analysis indicate that such zones are spatially complex, containing a mixture of pure, hybrid and introgressed individuals (Skibinski and Beardmore, 1979; Cousteau, Renaud and Delay, 1991; Väinölä and Hvilsom, 1991) – the degree of introgression varies depending on geographic location. This, together with results from artificial hybridization studies on *M. edulis* and *M. galloprovincialis,* indicates that there is little evidence of genetic incompatibility, although some differences in sperm morphology have been reported (references in Gosling, 1992a,b).

Results from cytological, immunological and mtDNA studies have also indicated that differences between the taxa are slight or non-existent (Gosling, 1992a; Blot, Legendre and Albert, 1990; Insua and Thiriot-Quiévreux, unpublished).

In summary, all the evidence to date indicates that *M. edulis*, *M. galloprovincialis* and *M. trossulus* (and *M. desolationis*) are very similar to one another, with no single morphological or genetic character being clearly diagnostic.

So what is the systematic status of the various forms within the genus? In areas of contact between two taxa the amount of hybridization (and in some cases introgression) taking place is probably too high by ISC or RSC standards for species recognition. By the same token there is little evidence that we are dealing with cohesive taxa and, therefore, also by CSC criteria these are not good species. While it could be argued that there are plenty of examples, both in plants and animals, of stable hybrid zones between what are generally recognized to be distinct species, such zones are invariably narrow, due to a stable balance between dispersal (gene flow) and selection (review: Barton and Hewitt, 1985). Deciding how much hybridization is permissible between taxa before they are considered conspecific is another operational difficulty of reproductive species concepts.

It is difficult to see how the PSC can be applied to the *Mytilus* situation. The emphasis of this concept is on diagnosability and monophyly. To determine whether individual taxa are monophyletic, i.e. constitute separate species, phylogenetic relationships are usually generated by cladistic analysis of a set of discrete character data derived from morphological or molecular data. The absence, to date, of discrete diagnostic characters, either at the morphological or genetic level, makes the application of the PSC a well-nigh impossible task.

The conclusion must be that, as they stand, the different forms of *Mytilus* do not merit full specific status. Whether they are in the process of differentiation (incipient sympatric speciation) or whether they are in the process of intergradation after a period in allopatry is impossible to say. Perhaps our efforts might be better served by directing more attention towards these processes rather than concentrating on the systematic status of forms within the genus; and accepting that since speciation is a gradual process, all populations will not be grouped into discrete species at any particular moment in history, and 'difficult' cases must be expected.

We are left then with the problem of what to call the different taxa. We could treat each taxon as a semispecies and apply the appropriate trinomial nomenclature (Väinölä and Hvilsom, 1991), or we could continue to refer to the individual taxa as *M. edulis*, *M. galloprovincialis* and *M. trossulus*, while at the same time recognizing that there is little evidence that full specific status is warranted. The latter is more attractive in that it is always simpler to retain the *status quo*, but also because, since there is little consensus on a definition of 'species', there is probably even less on a definition of 'semispecies'!

(c) Lasaea *systematics*

The genus *Lasaea* consists of minute (< 6 mm) hermaphroditic brooding clams that inhabit crevices on rocky intertidal shores. Keen (1938) listed more than 40 species in the genus but subsequently Ponder (1971), tending to echo an earlier view that the genus was monospecific, concluded that most of the nominal species were subspecies or ecotypes of the type species *L. rubra*. However, up to very recently several species have continued to be recognized within the genus: *L. rubra* on European and Mediterranean coasts, *L. subviridis* on northeastern Pacific coasts, *L. consanguinea* from the Kerguelen Islands in the southern Indian Ocean and *L. australis* from the western Pacific region. These have been identified on the basis of morphological characteristics such as shell shape, thickness, surface sculpture and colour and hinge shape and position (Bucquoy, Dautzenberg and Dollfus, 1892).

There are two distinctly different modes of reproduction and development within the genus. *L. australis* reproduces by cross-fertilization and releases its young as veliger larvae (Ó Foighil, 1988; Tyler-Walters and Crisp, 1989). In contrast, all other members of the genus release their young as crawl-away juveniles but, surprisingly, show a near-cosmopolitan geographic distribution (Ó Foighil, 1989). Results from allozyme studies show that populations with this developmental mode are composed of non-hybridizing frequently sympatric genetic strains; there is no evidence to date for cross-fertilization within this group (Crisp *et al.*, 1983; Crisp and Standen, 1988; Ó Foighil and Eernisse, 1988; Tyler-Walters and Crisp, 1989; Tyler-Walters and Davenport, 1990). In addition, all strains examined to date are polyploid but numbers of chromosomes vary depending on the strain: northeastern Pacific *Lasaea* are triploid ($n = 32$) with a variable number of supernumerary chromosomes; in Kerguelen strains a chromosome number of 100–120 was found but ploidy levels could not be determined; while in European strains the number of chromosomes varied greatly between 63 and 340 and could be grouped into different ploidy levels of 3, 4 and 5, with once again variable numbers of supernumerary chromosomes (Thiriot-Quiévreux *et al.*, 1988; Thiriot-Quiévreux, Insua Pombo and Albert, 1989; Ó Foighil and Thiriot-Quiévreux, 1991). In the sexual species *L. australis* a diploid number of $2n = 36$ has recently been reported (Thiriot-Quiévreux, 1992).

It has been quite difficult to sort out just how direct developers reproduce. Since they are simultaneous hermaphrodites with a minute male allocation, reproduction could be either by self-fertilization or by apomixis/parthenogenesis – the mitotic production of diploid eggs (Crisp *et al.*, 1983; McGrath and Ó Foighil, 1986; Ó Foighil, 1987; Crisp and Standen, 1988;

Ó Foighil and Eernisse, 1988; Tyler-Walters and Crisp, 1989). Recent results from egg–sperm interaction studies (Ó Foighil and Thiriot-Quiévreux, 1991) indicate that northeastern Pacific *Lasaea* are asexual and not self-fertile, as are undoubtedly direct development populations from other geographic areas (Ó Foighil, personal communication). Asexual *Lasaea* use their own sperm to trigger development – a process known as pseudogamy or autogynogenesis. Although pseudogamy has been reported in a wide variety of taxa (White, 1978), this is the first molluscan genus in which it has been detected.

Within individual regions, e.g. northeastern Pacific and Britain, populations of *Lasaea* are composed of many reproductively isolated strains or clones, morphologically similar but readily distinguishable by electrophoretic analysis (Crisp *et al.*, 1983); Crisp and Standen, 1988; Ó Foighil and Eernisse, 1988). In Britain the distribution of individual clones is strongly correlated with tidal level, and genetically determined physiological differences have been reported between clones (Tyler-Walters and Davenport, 1990).

In summary, the genus *Lasaea* encompasses, on the one hand, the sexual species *L. australis* with its rather restricted geographic distribution, and on the other, asexual *Lasaea* populations comprising an indefinitely large number of separate clones, the individual ranges of which are rather obscure at present.

So what is the specific status of these clones? Clearly, a strict application of the isolation or recognition species concepts would discount each clone as a separate species. Also, they appear to be outside the domain of the phylogenetic species concept. Ó Foighil and Thiriot-Quiévreux (1991) and Thiriot-Quiévreux (1992) postulated that the genus, which dates from the Eocene period (50 mya), was originally composed of cross-fertilizing herm-aphrodites with planktonic larval development – a condition retained by *L. australis* – and that the various clones appear to have arisen polyphyletically through multiple hybrid-izations between diploid (possibly Australian) parental species. The polyphyletic origin of these clones has been substantiated by very recent results from comparative analysis of nucleotide sequences of the mitochondrial cytochrome oxidase II gene (Ó Foighil and Smith, 1993).

This then leaves us with the CSC to consider. Since clones are sexually isolated, have separate fertilization systems and are genetically adapted to their own particular environmental conditions, then by CSC criteria each should be recognized as a separate species. Although this decision is probably taxonomically correct – and there is a certain amount of support for it (Crisp and Standen, 1988; Ó Foighil and Eernisse, 1988) – the task of detecting and naming hundreds, and perhaps even thousands, of new *Lasaea* species is, to say the least, a daunting one. A preferred approach might be to follow the recommendation of Echelle, Echelle and Middaugh (1989) and Echelle (1990) for gynogenetic fish of the genus *Menidia*. The situation in this genus is similar to that of *Lasaea* in that clones have arisen polyphyletically through multiple hybridizations between diploid parental species. Echelle and colleagues recommend that the clonal complex, formerly treated as a single species, *Menidia clarkhubbsi*, be referred to as the *Menidia clarkhubbsi* complex, indicating a collection of an as yet undetermined number of species. In the case of *Lasaea* I would recommend that the same system be adopted, referring to the cosmopolitan complex of *Lasaea* clones as the *Lasaea rubra* complex. The reason for choosing *L. rubra* is two-fold: it is the type species for the genus (Montagu 1803) and it was in *L. rubra* British populations that clones were first reported (Crisp *et al.*, 1983).

1.1.4 Concluding remarks

There is probably no other concept in biology that has generated so much discussion, and so much disagreement, as the species concept (see Mayr (1982) for historical review). As a result, myriads of species definitions have appeared in the literature, each with its own proponents and detractors. Much of the discussion has drifted so far from the operational identification of species into 'formidably esoteric philosophical dimensions that it is scarcely possible for an ordinary taxonomist to understand what the arguments are all about' (Smith, 1990).

Therefore, the aim of this section was to describe in simple terms four of the species concepts currently in use, highlighting their strengths and weaknesses. The intention was then to test the operational usefulness or otherwise of one or more concepts when applied to several 'species situations' in the marine environment. In each of these situations a multidisciplinary approach was taken, using all available published information to arrive at a taxonomic decision.

The examples chosen clearly demonstrate that no one species concept is inherently superior to another. The species concept chosen for a particular species situation will depend on whether one is dealing with sympatric versus allopatric taxa, sexual versus asexual organisms, and so on. In other words, a universal species concept probably does not exist. For a number of years now systematists, outside of the aquatic environment, have been calling for a more pluralistic outlook on species (Grant, 1981; Mishler and Donoghue, 1982; Donoghue, 1985; Endler, 1989), reasoning that since species situations are so diverse, then a variety of concepts would seem necessary and desirable to reflect this complexity. Let us too, as aquatic biologists, adopt a more pluralistic outlook on species; it is time for 'species' to suffer a fate similar to that of the classical concept of 'gene' (Mishler and Donoghue, 1982).

Acknowledgements

I wish to thank the following for helpful discussion: Michael Guiry, Ana Insua, David McGrath, Diarmaid Ó Foighil and Catherine Thiriot-Quiévreux.

References

Barton, N.H. and Hewitt, G.M. (1985) Analysis of hybrid zones. *Annu. Rev. Ecol. Syst.* **16**, 113–48.

Blot, M., Legendre, B. and Albert, P. (1990) Restriction fragment length polymorphism of mitochondrial DNA in subantarctic mussels. *J. Mar. Biol. Ecol.*, **141**, 79–86.

Brock, V. (1978) Morphological and biochemical criteria for the separation of *Cardium glaucum* (Bruguière) from *Cardium edule* (L.). *Ophelia*, **17**, 207–14.

Brock, V. (1979) Habitat selection of two congeneric bivalves, *Cardium edule* and *C. glaucum* in sympatric and allopatric populations. *Mar. Biol.*, **54**, 149–56.

Brock, V. (1980a) The geographical distribution of *Cerastoderma (Cardium) edule* (L.) and *C. lamarcki* (Reeve) in the Baltic and adjacent seas related to salinity and salinity fluctuations. *Ophelia*, **19**, 207–14.

Brock, V. (1980b) Evidence for niche differences in sympatric populations of *Cerastoderma edule* and *C. lamarcki*. *Mar. Ecol. Progr. Ser.*, **2**, 75–80.

Brock, V. (1982) Does displacement of spawning time occur in the sibling species *Cerastoderma edule* and *C. lamarcki*? *Mar. Biol.*, **67**, 33–38.

Brock, V. (1987) Genetic relations between the bivalves *Cardium (Cerastoderma) edule*, *Cardium lamarcki* and *Cardium glaucum*, studied by means of crossed immunoelectrophoresis. *Mar. Biol.*, **93**, 493–98.

Brock, V. (1991) *An Interdisciplinary Study of Evolution in the Cockles, Cardium (Cerastoderma) edule, C. glaucum, and C. lamarcki*, Vestjydsk Forlag, Denmark.

Brock, V. and Christiansen, G. (1989) Evolution of *Cardium (Cerastoderma) edule*, *C. lamarcki* and *C. glaucum*: studies of DNA-variation. *Mar. Biol.*, **102**, 505–11.

Bucquoy, E., Dautzenberg, P. and Dollfus, G. (1892) *Les Mollusques Marins du Roussillon*, T II, Facc. VI, *Pelecypoda*, Ballière et fils, Paris, pp. 221–72.

Buffon, G.L. (1749–1804) *Histoire Naturelle, Générale et Particulière*, 44 vols; Imprimerie Royale, puis Plassan, Paris.

Butlin, R. (1989) Reinforcement of premating isolation, in *Speciation and its Consequences* (eds D. Otte and J.A. Endler), Sinauer Associates, Inc., Sunderland, MA, pp. 158–79.

Cousteau, C., Renaud, F. and Delay, B. (1991) Genetic characterization of the hybridization between *Mytilus edulis* and *M. galloprovincialis* on the Atlantic coast of France. *Mar. Biol.*, **111**, 87–93.

Coyne, J.A., Orr, H.A. and Futuyma, D.J. (1988). Do we need a new species concept? *Syst. Zool.*, **37**(2), 190–200.

Cracraft, J. (1983) Species concepts and speciation analysis. *Curr. Ornithol.*, **1**, 159–87.

Cracraft, J. (1989) Speciation and its ontology: the empirical consequences of alternative species concepts for understanding patterns and processes of differentiation, in *Speciation and its Consequences* (eds D. Otte and J.A. Endler), Sinauer Associates, Inc., Sunderland, MA, pp. 28–59.

Cronquist, A. (1978) Once again, what is a species? in *Biosystematics in Agriculture* (ed. J.A. Romberger), Allanheld & Osmun, Montclair, New Jersey, pp. 3–20.

Crisp, D.J. and Standen, A. (1988) *Lasaea rubra* (Montagu) (Bivalvia: Erycinacea), an apomictic crevice-living bivalve with clones separated by tidal level preference. *J. Exp. Mar. Biol. Ecol.*, **117**, 27–45.

Crisp, D.J., Burfitt, A., Rodrigues K. and Budd, M.D. (1983) *Lasaea rubra*: an apomictic bivalve. *Mar. Biol. Lett.*, **4**, 127–36.

Dobzhansky, Th. (1937) *Genetics and the Origin of Species*, Columbia University Press, New York.

Dobzhansky, Th., Ayala, F.J., Stebbins, G.L. and Valentine, J.W. (1977) *Evolution*, W.H. Freeman, San Francisco.

Donoghue, M.J. (1985) a critique of the biological species concept and recommendations for a phylogenetic alternative. *The Bryologist*, **88**, 172–81.

Echelle, A. (1990) Nomenclature and non-Mendelian ('clonal') vertebrates. *Syst. Zool.*, **39**, 70–78.

Echelle, A.A., Echelle, A.F. and Middaugh, D.P. (1989) Evolutionary biology of the *Menidia clarkhubbsi* complex of unisexual fish (Atherinidae): origins, clonal diversity, and mode of reproduction, in *Evolution and Ecology of Unisexual Vertebrates* (eds R.M. Dawley and J.P. Bogart), Bulletin 466, New York State Museum, Albany, New York, pp. 144–52.

Eldredge, N. and Cracraft, J. (1980) *Phylogenetic Patterns and the Evolutionary Process*, Columbia University Press, New York.

Endler, J.A. (1989) Conceptual and other problems in speciation, in *Speciation and its Consequences* (eds D. Otte and J.A. Endler), Sinauer Associates, Inc., Sunderland, MA, pp. 625–48.

Gardner, J.P.A. (1992) *Mytilus galloprovincialis* (Lmk.) (Bivalvia, Mollusca): the taxonomic status of the Mediterranean mussel. *Ophelia*, **35**, 219–43.

Gosling, E.M. (1980) Gene frequency changes and adaptation in marine cockles. *Nature*, **286**, 601–2.

Gosling, E.M. (1984) The systematic status of *Mytilus galloprovincialis* in western Europe: a review. *Malacologia*, **25**(2), 551–68.

Gosling, E.M. (1992a) Genetics of *Mytilus*, in *The Mussel Mytilus* (ed. E.M. Gosling), Elsevier Press, Amsterdam, pp. 309–82.

Gosling, E.M. (1992b) Systematics and geographic distribution of *Mytilus*, in *The Mussel Mytilus* (ed. E.M. Gosling), Elsevier Press, Amsterdam, pp. 1–20.

Grant, V. (1981) *Plant Speciation*, Columbia University Press, New York.

Guiry, M.D. (1991) Crossability data and the species concept in marine red algae. *J. Phycol.* 27(3), p. 27.

Guiry, M.D. (1992) Species concepts in marine red algae. *Progr. Phycol. Res.*, **8**, 251–78.

Insua, A. and Thiriot-Quiévreux, C. (1992) Karyotypes of *Cerastoderma edule*, *Venerupis pullastra* and *V. rhomboides* (Bivalvia, Veneroida). *Aquat. Living Resour.*, **5**, 1–8.

Jelnes, J.E., Petersen, G.H. and Russell, P.J.C. (1971) Isoenzyme taxonomy applied on four species of *Cardium* from Danish and British waters with a short description of the distribution of the species (Bivalvia), *Ophelia*, **9**, 15–19.

Keen, A.M. (1938) New pelecypod species of the genera *Lasaea* and *Crassinella*. *Proc. Malacol. Soc. (Lond.)*, **23**, 18–32.

Kimbel, W.H. (1991) Species, species concepts and hominid evolution. *J. Hum. Evol.*, **20**, 355–71.

Kingston, P. (1973) Interspecific hybridization in Cardium. *Nature*, **243**, 360.

Koehn, R.K. (1991) The genetics and taxonomy of species in the genus *Mytilus*. *Aquaculture*, **94**, 125–45.

Koulman, J.G. and Wolff, W.J. (1977) The Mollusca of the estuarine region of the rivers Rhine, Meuse, and Scheldt in relation to the hydrography of the area. V. The Cardiidae. *Basteria*, **41**, 21–32.

Lynch, J.D. (1989) The gauge of speciation: on the frequencies of modes of speciation, in *Speciation and its Consequences* (eds D. Otte and J.A. Endler), Sinauer Associates, Inc., Sunderland, MA, pp. 527–53.

Masters, J.C. and Spencer, H.G. (1989) Why we need a new species concept. *Syst. Zool.*, **38**(3), 270–79.

Masters, J.C., Rayner, R.J., McKay, I.J. *et al.* (1987) The concept of species: recognition versus isolation. *South African J. Sci.*, **83**, 534–37.

Mayr, E. (1963) *Animal Species and Evolution*, Harvard University Press, Cambridge, MA.

Mayr, E. (1970) *Populations, Species and Evolution*, Harvard University Press, Cambridge, MA.

Mayr, E. (1982) *The Growth of Biological Thought. Diversity, Evolution and Inheritance*, Harvard University Press, Cambridge, MA.

Mayr, E. (1988) The why and how of species. *Biol. Philos.*, **3**, 431–41.

McDonald, J.H., Seed, R. and Koehn, R.K. (1991) Allozyme and morphometric characters of three species of *Mytilus* in the Northern and Southern hemispheres. *Mar. Biol.*, **111**, 323–35.

McGrath, D. and Ó Foighil, D. (1986) Population dynamics and reproduction of hermaphroditic *Lasaea rubra* (Montagu) (Bivalvia, Galeommatacea). *Ophelia*, **25**(3) 1, 209–19.

McKitrick, M.C. and Zink, R.M. (1988) Species concepts in ornithology. *Condor*, **90**, 1–14.

Mishler, B.D. and Donoghue, M.J. (1982) Species concepts: a case for pluralism. *Syst. Zool.*, **31**, 491–503.

Murphy, R.W., Sites, J.W., Buth, D.G. *et al.* (1990). Proteins I: isoenzyme electrophoresis, in *Molecular Systematics* (eds D.M. Hillis and C. Moritz), Sinauer Associates, Sunderland, MA, pp. 45–126.

Nelson, G. and Platnick, N. (1981) *Systematics and Biogeography: Cladistics and Vicariance*, Columbia University Press, New York.

Ó Foighil, D. (1987) Cytological evidence for self-fertilization in *Lasaea subviridis* (Galeommatacea: Bivalvia). *Int. J. Invert. Reprod. Devel.*, **12**, 83–90.

Ó Foighil, D. (1988) Random mating and planktotrophic larval development in the brooding hermaphrodite clam *Lasaea australis*. *The Veliger*, **31**, 214–21.

Ó Foighil, D. (1989) Planktotrophic larval development is associated with a restricted geographic range in *Lasaea*, a genus of brooding, hermaphroditic bivalves. *Mar. Biol.*, **103**, 349–58.

Ó Foighil, D. and Eernisse, D.J. (1988) Geographically widespread, non-hybridizing, sympatric strains of the hermaphroditic, brooding clam *Lasaea* in the northeastern Pacific Ocean. *Biol. Bull.*, **175**, 218–29.

Ó Foighil, D. and Smith, M.J. (1993) Molecular analysis of the evolution of asexuality in the cosmopolitan marine clam *Lasaea*. *Evolution* (in press).

Ó Foighil, D. and Thiriot-Quiévreux, C. (1991) Ploidy and pronuclear interaction in northeastern Pacific *Lasaea* clones (Mollusca: Bivalvia). *Biol. Bull.*, **181**, 222–31.

Palumbi, S.R. (1992) Marine speciation on a small planet. TREE, **4**, 114–18.

Paterson, H.E.H. (1985) The recognition concept of species. *Transvaal Mus. Monogr.*, **4**, 21–29.

Ponder, W.F. (1971) Some New Zealand and Subantarctic bivalves of the Cyamiacea and Leptonacea with descriptions of new taxa. *Records of the Dominion Museum (Wellington)*, **7**, 119–41.

Raubenheimer, D. and Crowe, T.M. (1987) The recognition species concept: is it really an alternative? *South African J. Sci.*, **83**, 530–34.

Seed, R. (1976) Ecology, in *Marine Mussels: their Ecology and Physiology* (ed. B.L. Bayne), Cambridge University Press, Cambridge, pp. 13–56.

Seed, R. (1992) Systematics, evolution and distribution of mussels belonging to the genus *Mytilus*: an overview. *Am. Malacol. Bull.*, **9**(2), 123–37.

Simpson, G.G. (1951) The species concept. *Evolution*, **5**, 285–98.

Skibinski, D.O.F. and Beardmore, J.A. (1979) A genetic study of intergradation between *Mytilus edulis* and *M. galloprovincialis*. *Experientia*, **35**, 1442–44.

Smith, H.A. (1990) The universal species concept. *Herpetologica*, **46**, 122–24.

Sokal, R.R. and Crovello, T.J. (1970) The biological species concept: a critical evaluation. *Am. Nat.*, **104**, 127–53.

Templeton, A.R. (1989) The meaning of species and speciation: a genetic perspective, in *Speciation and its Consequences* (eds D. Otte and J.A. Endler), Sinauer Associates, Inc., Sunderland, MA, pp. 3–27.

Thiriot-Quiévreux, C. (1992) Karyotype of *Lasaea australis*, a brooding bivalve species. *Aus. J. Mar. Freshwat. Res.*, **43**, 403–8.

Thiriot-Quiévreux, C., Soyer, J., de Bovée, F. *et al.* (1988) Unusual chromosome complement in the brooding bivalve *Lasaea consanguinea*. *Genetica*, **76**, 143–51.

Thiriot-Quiévreux, C., Insua Pombo, A.M. and Albert, P. (1989) Polyploidie chez un bivalve incubant *Lasaea rubra* (Montagu). *CR Acad. Sci.* (Paris), **308**, 115–200.

Tyler-Walters, H. and Crisp, D.J. (1989) The modes of reproduction in *Lasaea rubra* (Montagu) and *L. australis* (Lamarck): (Erycinidae; Bivalvia), in *Proceedings of the 23rd European Marine Biology Symposium*, Swansea (eds J.S. Ryland and P.A. Tyler), Olsen and Olsen, Denmark, pp. 299–308.

Tyler-Walters, H. and Davenport, J. (1990) The relationship between the distribution of genetically distinct inbred lines and upper lethal temperature in *Lasaea rubra*. *J. Mar. Biol. Assoc., UK*, **70**, 557–70.

Väinölä, R. and Hvilsom, M.M. (1991) Genetic divergence and a hybrid zone between Baltic and North Sea *Mytilus* populations (Mytilidae; Mollusca). *Biol. J. Linn. Soc.*, **43**, 127–48.
White, M.J.D. (1978) *Modes of Speciation*, W.H. Freeman, San Francisco.

1.2 GENETIC DIVERGENCE BETWEEN NATURAL POPULATIONS OF MEDITERRANEAN SANDHOPPERS (CRUSTACEA, AMPHIPODA)

Elvira De Matthaeis, Marina Cobolli, Marco Mattoccia, Pierangela Saccoccio and Felicita Scapini

Abstract

The genetic structure of two supralittoral talitrid species, Talitrus saltator *and* Talorchestia deshayesii, *was compared by analysing allozymic variation at 19 loci. The study was performed on 14 populations of the first species and five populations of the latter, collected from different localities along the western and the eastern coasts of the Italian peninsula. The levels of genetic differentiation within the two species were quite different. Within* T. deshayesii *the interpopulation genetic divergence was low (*D = 0.036*). Within* T. saltator *an average genetic distance of 0.4 was found between the western and eastern populations, while levels of genetic distance between populations of the same geographic group ranged from 0 to 0.1. The observed pattern of genetic variation could be the result of historical factors (Plio-Pleistocene climatic fluctuations) as well as of intrinsic ecological factors (plasticity of the species).*

1.2.1 Introduction

Studies on the genetic variation of species with a wide geographic range have provided useful data for estimating levels of genetic divergence between allopatric populations, enabling analysis of the role played by geographic barriers to gene flow (Burton, 1986; Hedgecock, 1986).

Talitrid amphipods appear to be particularly suitable for the evaluation of levels of genetic differentiation on both micro- and macrogeographic scales. Many species are abundant elements of the supralittoral strandline fauna of rocky, stony and sandy shores with wide geographic distributions, but nevertheless being endangered in many cases because of anthropogenic pressures on coastal environments.

Furthermore, talitrids living on sandy shores have been shown to have an innate sun compass which enables them to maintain their position in moist habitats at the water's edge (Pardi, 1960). This orientation behaviour has been shown to be genetically determined, with differences among natural populations related to the orientation of the seashore where they live (Scapini, Ugolini and Pardi, 1985). In particular, mass crosses between geographically distant populations of sandhoppers attributed to the single species *Talitrus saltator* revealed a partial or total disruption of this compass in offspring (Scapini and Fasinella, 1990).

From a systematic point of view *Talitrus* has been traditionally considered a monotypic genus with only the *saltator* species: it appears to hold a quite isolated position in the family Talitridae (Bousfield, 1982).

The typical form of the species has been described using British samples; a *mediterranea* form, smaller and more slender than the Atlantic form, has been recognized in the Mediterranean Sea and a *briani* form limited to the Adriatic Sea has been proposed by Ruffo since 1936 on the basis of some differences in the size of North Adriatic individuals collected from some localities near Venice (Ruffo, 1936).

The present study investigates the genetic structure of natural populations of *Talitrus saltator* in order to resolve the levels of genetic differentiation within the species. Five populations of *Talorchestia deshayesii* were also analysed with the aim of obtaining a comparable set of data on the genetic structure of a taxon considered to be closely related to *Talitrus*. Moreover, the two species show similar ecological and geographic trends, dwelling on sandy beaches from southern Scandinavia and the northeast Atlantic to the whole Mediterranean basin.

1.2.2 Materials and methods

The samples of *Talitrus saltator* (Montagu) were collected along the Ligurian, Tyrrhenian and Adriatic coasts of the Italian peninsula (Figure 1.2.1). The samples of *Talorchestia deshayesii* (Audouin) were collected along the Tyrrhenian, Adriatic and Ionian coasts of the Italian peninsula (Figure 1.2.1). Furthermore, two samples of the latter species were collected, one on the islet of Asinara (northwestern Sardinia) and one on the northern coast of Sicily (Figure 1.2.1).

The samples were transported alive to the laboratory and frozen at −80 °C.

Horizontal electrophoresis was performed on 12% starch gel (Connaught Laboratories) using crude homogenates of the whole body of adult specimens. Between 11 and 47 individuals from each population of the two species were analysed. Table 1.2.1 reports the list of the proteins investigated, the electrophoretic systems used, the staining techniques, the abbreviations and the Enzyme Commission numbers.

The following electrophoretic buffer systems were used: A = discontinuous Tris citrate, pH 8.6 (Poulik, 1957); B = Tris versene borate, pH 9.1 (Ayala *et al.*, 1972); B1 = as buffer B with NADP$^+$ added; C = continuous Tris citrate, pH 8.4 (modified from buffer C of Ayala *et al.* (1972)); D = Tris maleate, pH 7.4 (Brewer and Sing, 1970); E = continuous Tris citrate,

Table 1.2.1   A summary of the enzyme proteins investigated, enzyme codes, procedures used and loci detected in *Talitrus saltator* and *Talorchestia deshayesii* samples

Codes	Enzyme	E.C. No.	Buffer system	Staining techniques	Loci scored
ACPH	Acid phosphatase	3.1.3.2	A	Ayala *et al.* (1972)	*Acph-1*; *Acph-2*; *Acph-3*
ALDO	Aldolase	4.1.2.1.3	B1	Ayala *et al.* (1972)	*Aldo-1*; *Aldo-2*;
AO	Aldehyde oxidase	1.2.3.1	B	Ayala *et al.* (1974)	*Ao-1*
APH	Alkaline phosphatase	3.1.3.1	A	Ayala *et al.* (1972)	*Aph-1*; *Aph-2*
CA	Carbonic anhydrase	4.2.1.1	G	Brewer and Sing (1970)	*Ca-1*; *Ca-2*
EST	Esterase	3.1.1.1	A	Ayala *et al.* (1972)	*Est-1*; *Est-2*
GOT	Glutamic-oxalacetic transaminase	2.6.1.1	B1	Ayala, Valentine and Zumwalt (1975)	*Got-1*
MPI	Mannose phosphate isomerase	5.3.1.8	I	Harris and Hopkinson (1978)	*Mpi*
PEP	Peptidase	3.4.11	A	Ward and Beardmore (1977)	*Pep-1*; *Pep-2*; *Pep-3*
PGM	Phosphoglucose mutase	2.7.5.1	D	Brewer and Sing (1970)	*Pgm*
PHI	Phosphohexose isomerase	5.3.1.9	C	Brewer and Sing (1970)	*Phi*

See text for buffer systems.

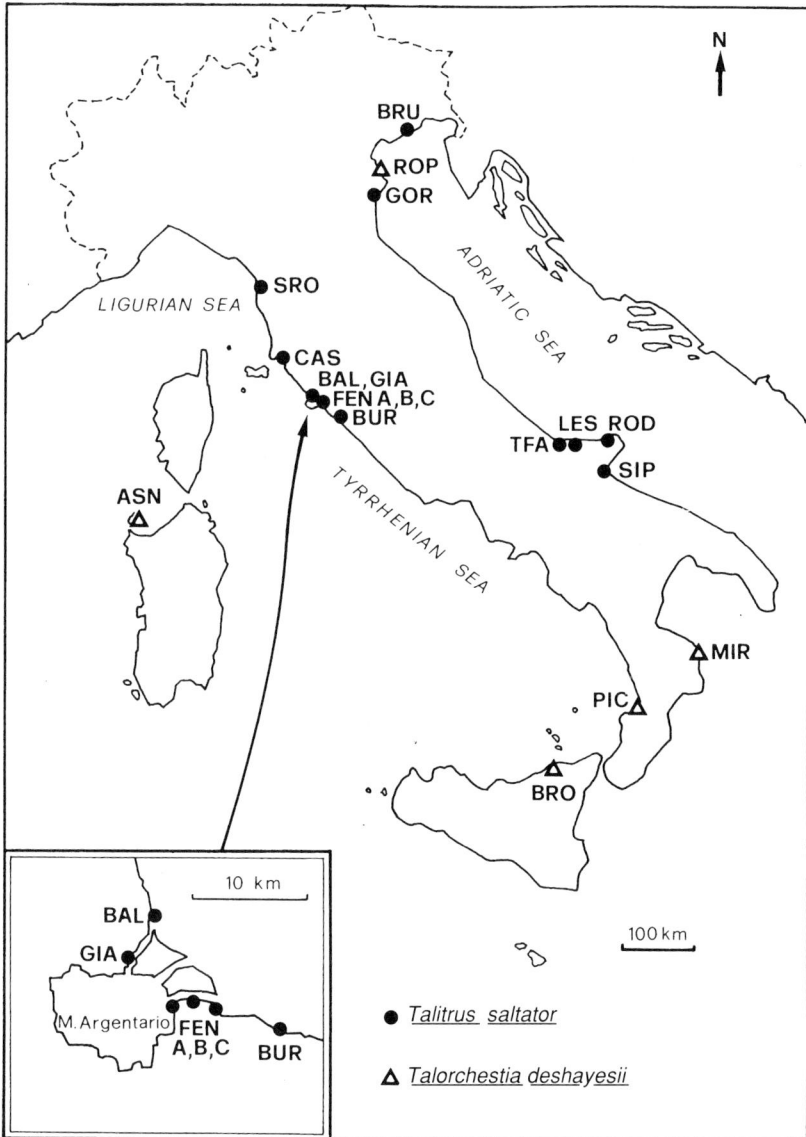

Figure 1.2.1 Sampling localities of *Talitrus saltator* (black circles) and *Talorchestia deshayesii* (open triangles): SRO = San Rossore; CAS = Castiglione della Pescaia; BAL = Bocca d'Albegna; GIA = Tombolo di Giannella; FENA, FENB, FENC = Tombolo di Feniglia; BUR = Burano; BRU = Brussa; GOR = Goro; TFA = Torre Fantine; LES = Marina di Lesina; ROD = Rodi Garganico; SIP = Siponto; ROP = Rosa Pineta; MIR = Mirto; BRO = Brolo; PIC = Pizzo Calabro; ASN = Isola dell'Asinara.

pH 8.0 (Ward and Beardmore, 1977); G = borate, pH 8 (Brewer and Sing, 1970); I = Tris phosphate, pH 8.3 (Harris and Hopkinson, 1978).

Data were analysed by the computer program BIOSYS-1 (Swofford and Selander, 1981). A matrix of Nei's (1978) genetic distance (*D*) was calculated for all populations sampled for this study. BIOSYS-1 was also used to construct an UPGMA dendrogram (Sneath and

Table 1.2.2 Allele frequencies at 13 loci in *Talitrus saltator* samples (from SRO to SIP) and in *Talorchestia deshayesii* samples (from ROP to ASN)

| | *Talitrus saltator* | | | | | | | | | | | | | | *Talorchestia deshayesii* | | | | |
| | Ligurian-Tyrrhenian populations | | | | | | | | Adriatic populations | | | | | | | | | | |
Locus	SRO	CAS	BAL	GIA	FENA	FENB	FENC	BUR	BRU	GOR	TFA	LES	ROD	SIP	ROP	MIR	BRO	PIC	ASN
Acph-1																			
(N)	20	30	30	14	19	16	22	29	19	15	34	47	20	42	12	15	11	13	39
A	0.200	1.000	1.000	1.000	1.000	1.000	1.000	0.034	1.000	1.000	1.000	1.000	1.000	1.000	1.000	1.000	1.000	1.000	1.000
B	0.800	0.000	0.000	0.000	0.000	0.000	0.000	0.966	0.000	0.000	0.000	0.000	0.000	0.000	0.000	0.000	0.000	0.000	0.000
Acph-3																			
(N)	18	30	30	14	19	16	22	29	15	15	34	47	20	42	12	11	11	13	39
B	0.000	0.000	0.000	0.000	0.000	0.000	0.000	0.000	0.000	0.000	0.000	0.000	0.000	0.000	1.000	0.727	1.000	1.000	1.000
C	0.000	0.000	0.000	0.000	0.000	0.000	0.000	0.000	1.000	1.000	1.000	1.000	1.000	1.000	0.000	0.000	0.000	0.000	0.000
D	1.000	1.000	1.000	1.000	0.974	0.969	1.000	0.966	0.000	0.000	0.000	0.000	0.000	0.000	0.000	0.000	0.000	0.000	0.000
E	0.000	0.000	0.000	0.000	0.026	0.031	0.000	0.034	0.000	0.000	0.000	0.000	0.000	0.000	0.000	0.000	0.000	0.000	0.000
F	0.000	0.000	0.000	0.000	0.000	0.000	0.000	0.000	0.000	0.000	0.000	0.000	0.000	0.000	0.000	0.273	0.000	0.000	0.000
Aldo-1																			
(N)	34	30	30	14	32	18	28	36	20	13	16	16	18	27	12	23	11	13	39
A	0.000	0.000	0.000	0.000	0.000	0.000	0.000	0.000	0.000	0.000	0.000	0.000	0.000	0.000	1.000	1.000	1.000	1.000	1.000
B	1.000	1.000	1.000	1.000	1.000	1.000	1.000	1.000	1.000	1.000	1.000	1.000	1.000	1.000	0.000	0.000	0.000	0.000	0.000
Aldo-2																			
(N)	34	30	30	14	32	18	28	36	20	13	16	16	18	27	12	23	11	13	39
A	0.000	0.000	0.000	0.000	0.000	0.000	0.000	0.000	0.000	0.000	0.000	0.000	0.000	0.000	1.000	1.000	1.000	1.000	1.000
B	1.000	1.000	1.000	1.000	1.000	1.000	1.000	1.000	1.000	1.000	1.000	1.000	1.000	1.000	0.000	0.000	0.000	0.000	0.000
Ao-1																			
(N)	24	30	30	14	32	35	28	28	26	18	34	22	40	19	12	18	11	13	39
A	0.000	0.000	0.000	0.000	0.000	0.000	0.000	0.000	0.000	0.000	0.000	0.000	0.000	0.000	1.000	1.000	1.000	1.000	1.000
B	1.000	1.000	1.000	1.000	1.000	1.000	1.000	1.000	0.000	0.000	0.000	0.000	0.000	0.000	0.000	0.000	0.000	0.000	0.000
C	0.000	0.000	0.000	0.000	0.000	0.000	0.000	0.000	1.000	1.000	1.000	1.000	1.000	1.000	0.000	0.000	0.000	0.000	0.000

Ca-1																			
(N)	34	30	35	32	14	35	32	28	21	18	34	47	46	39	12	29	11	23	11
C	1.000	1.000	1.000	1.000	1.000	1.000	1.000	1.000	1.000	1.000	1.000	1.000	1.000	1.000	0.000	0.000	0.000	0.000	0.000
D	0.000	0.000	0.000	0.000	0.000	0.000	0.000	0.000	0.000	0.000	0.000	0.000	0.000	0.000	1.000	1.000	1.000	1.000	1.000
Ca-2																			
(N)	34	30	35	32	14	35	32	28	21	18	29	19	19	28	12	29	11	23	11
B	1.000	1.000	1.000	1.000	1.000	1.000	1.000	1.000	1.000	1.000	1.000	1.000	1.000	1.000	0.000	0.000	0.000	0.000	0.000
C	0.000	0.000	0.000	0.000	0.000	0.000	0.000	0.000	0.000	0.000	0.000	0.000	0.000	0.000	0.000	0.000	0.000	0.000	0.000
D	0.000	0.000	0.000	0.000	0.000	0.000	0.000	0.000	0.000	0.000	0.000	0.000	0.000	0.000	1.000	1.000	1.000	1.000	1.000
Est-1																			
(N)	20	24	29	25	14	22	34	28	23	18	34	34	32	42	12	44	11	14	16
B	1.000	1.000	1.000	1.000	1.000	1.000	1.000	1.000	0.000	0.000	0.000	0.000	0.000	0.000	0.000	0.000	0.000	0.000	0.000
C	0.000	0.000	0.000	0.000	0.000	0.000	0.000	0.000	1.000	1.000	1.000	1.000	1.000	1.000	1.000	1.000	1.000	1.000	1.000
Est-2																			
(N)	20	24	29	25	14	22	34	28	23	18	34	34	32	42	12	44	11	14	16
B	1.000	1.000	1.000	1.000	1.000	1.000	1.000	1.000	0.000	0.000	0.000	0.000	0.000	0.000	0.000	1.000	1.000	1.000	1.000
C	0.000	0.000	0.000	0.000	0.000	0.000	0.000	0.000	1.000	1.000	1.000	1.000	1.000	1.000	1.000	0.000	0.000	0.000	0.000
Got-1																			
(N)	20	30	30	32	14	35	33	24	25	22	47	18	40	42	12	10	11	23	39
C	1.000	1.000	1.000	1.000	1.000	1.000	1.000	1.000	1.000	1.000	1.000	1.000	1.000	1.000	0.000	0.000	0.000	0.000	0.000
D	0.000	0.000	0.000	0.000	0.000	0.000	0.000	0.000	0.000	0.000	0.000	0.000	0.000	0.000	1.000	1.000	1.000	1.000	1.000
Mpi																			
(N)	24	24	30	9	22	22	22	36	13	14	18	15	31	21	12	27	6	3	11
A	0.083	0.042	0.000	0.056	0.091	0.091	0.045	0.097	0.615	0.736	0.833	0.900	0.806	0.333	0.000	0.000	0.000	0.000	0.000
B	0.896	0.917	0.967	0.889	0.864	0.886	0.909	0.819	0.385	0.214	0.167	0.100	0.194	0.667	0.000	0.000	0.000	0.000	0.000
C	0.000	0.000	0.000	0.000	0.000	0.000	0.000	0.000	0.000	0.000	0.000	0.000	0.000	0.000	1.000	0.981	0.083	0.833	1.000
D	0.021	0.021	0.000	0.000	0.023	0.023	0.000	0.000	0.000	0.000	0.000	0.000	0.000	0.000	0.000	0.000	0.000	0.000	0.000
E	0.000	0.000	0.000	0.000	0.000	0.000	0.000	0.000	0.000	0.000	0.000	0.000	0.000	0.000	0.000	0.019	0.917	0.167	0.000
G	0.000	0.021	0.033	0.056	0.023	0.000	0.045	0.083	0.000	0.000	0.000	0.000	0.000	0.000	0.000	0.000	0.000	0.000	0.000

Table 1.2.2 (contd)

| | Talitrus saltator | | | | | | | | | | | | | | Talorchestia deshayesii | | | | |
| | Ligurian-Tyrrhenian populations | | | | | | | | Adriatic populations | | | | | | | | | | |
Locus	SRO	CAS	BAL	GIA	FENA	FENB	FENC	BUR	BRU	GOR	TFA	LES	ROD	SIP	ROP	MIR	BRO	PIC	ASN
Pgm																			
(N)	27	24	30	5	12	19	11	41	19	12	18	34	20	20	10	31	5	4	15
A	0.278	0.021	0.033	0.000	0.042	0.000	0.000	0.000	0.000	0.000	0.000	0.000	0.000	0.000	0.000	0.000	0.000	0.000	0.000
C	0.574	0.125	0.183	0.200	0.083	0.026	0.136	0.232	0.000	0.000	0.000	0.000	0.025	0.000	0.000	0.000	0.000	0.000	0.000
E	0.000	0.000	0.000	0.000	0.000	0.000	0.000	0.000	0.000	0.000	0.000	0.000	0.000	0.000	0.000	0.532	0.000	0.500	0.000
H	0.148	0.854	0.767	0.800	0.792	0.974	0.864	0.768	0.895	0.875	1.000	1.000	0.975	0.375	1.000	0.387	0.900	0.500	1.000
J	0.000	0.000	0.017	0.000	0.042	0.000	0.000	0.000	0.105	0.125	0.000	0.000	0.000	0.625	0.000	0.000	0.000	0.000	0.000
K	0.000	0.000	0.000	0.000	0.042	0.000	0.000	0.000	0.000	0.000	0.000	0.000	0.000	0.000	0.000	0.032	0.000	0.000	0.000
L	0.000	0.000	0.000	0.000	0.000	0.000	0.000	0.000	0.000	0.000	0.000	0.000	0.000	0.000	0.000	0.048	0.100	0.000	0.000
Phi																			
(N)	18	17	21	9	10	13	11	24	22	11	22	29	19	14	6	5	8	9	16
E	0.000	0.118	0.000	0.000	0.000	0.000	0.000	0.021	0.000	0.000	0.000	0.000	0.000	0.000	0.000	0.000	0.000	0.000	0.000
F	0.028	0.441	0.571	0.611	0.550	0.769	0.727	0.667	0.000	0.045	0.000	0.000	0.000	0.000	0.000	0.000	0.000	0.222	0.406
G	0.972	0.441	0.429	0.389	0.450	0.231	0.273	0.312	0.932	0.955	1.000	1.000	1.000	0.964	1.000	0.900	1.000	0.722	0.281
H	0.000	0.000	0.000	0.000	0.000	0.000	0.000	0.000	0.000	0.000	0.000	0.000	0.000	0.000	0.000	0.100	0.000	0.056	0.000
J	0.000	0.000	0.000	0.000	0.000	0.000	0.000	0.000	0.068	0.000	0.000	0.000	0.000	0.036	0.000	0.000	0.000	0.000	0.000
K	0.000	0.000	0.000	0.000	0.000	0.000	0.000	0.000	0.000	0.000	0.000	0.000	0.000	0.000	0.000	0.000	0.000	0.000	0.313

N = number of individuals assayed.
See Figure 1.2.1 for population symbols.

Sokal, 1973). To measure population subdivision in *Talitrus saltator*, F-statistics were calculated according to Wright (1965). In *Talorchestia deshayesii*, F-statistics were not calculated because of the small number of individuals available in some populations at the four polymorphic loci.

1.2.3 Results

Nineteen presumptive gene loci from 11 enzyme systems were scorable in all populations. Designations of the alleles and the allele frequencies are detailed in Table 1.2.2. Electromorphs of *Acph-2*, *Aph-1*, *Aph-2*, *Pep-1*, *Pep-2* and *Pep-3* were monomorphic and fixed for the same allele in all population samples. These loci are not listed in Table 1.2.2.

(a) Genetic variation within Talitrus saltator

Differences in allele distribution among the populations sampled for this study are shown in Table 1.2.2.

The *Ao-1 B*, *Ca-2 B*, *Est-1 B* and *Est-2 B* alleles completely discriminated a western group, which included the Ligurian and Tyrrhenian populations, from an eastern group, which comprised the Adriatic populations.

Five loci were polymorphic at least in one population (*Acph-1*, *Acph-3*, *Mpi*, *Pgm*, *Phi*). The *Acph-1 B* allele was found with high frequency in the SRO population only. At the other four polymorphic loci the two groups of populations showed considerable differentiation. At the *Acph-3* locus, the alleles D and E characterized the Ligurian–Tyrrhenian populations, while the allele C was found in the Adriatic group. At the *Mpi* locus, the two groups differed in the frequency of the most common allele: allele A was found in high frequency in all Adriatic populations except SIP, while allele B characterized the Ligurian–Tyrrhenian group and it was found at a frequency of 0.667 in the SIP population. At the same locus two more alleles, D and G, were found at low frequency in the Ligurian–Tyrrhenian populations. The locus *Pgm* was polymorphic with five alleles in the Ligurian–Tyrrhenian group and with three alleles in the Adriatic one. Within the Adriatic group, the allele H characterized the TFA and LES populations; the allele H was found to be the most common in all populations except SRO and SIP. Allele C was found with a frequency of 0.574 in population SRO, while allele J was found with a frequency of 0.625 in population SIP. The locus *Phi* was polymorphic with three alleles in the Ligurian–Tyrrhenian group. Allele F was found to be the most common in all the Ligurian–Tyrrhenian populations except SRO and CAS. Allele G was found with a frequency of 0.972 in population SRO, while two alleles, F and G shared an identical high frequency of 0.441 in population CAS. At the same locus allele G was found to be the most common in the Adriatic populations.

The estimates of genetic variability are given in Table 1.2.3. All values indicated a low level of genetic polymorphism. The average expected heterozygosity was 0.055 in the Ligurian–Tyrrhenian group while it was 0.029 in the Adriatic one.

The results of the F-statistic analysis are shown in Table 1.2.4. The indices were calculated on all populations (Table 1.2.4(a)) and for the Ligurian–Tyrrhenian and Adriatic groups separately (Table 1.2.4(b) and 1.2.4(c) respectively) taking into account the polymorphic loci and the loci fixed for alternative alleles. Actually the F_{is} values were low and not significant (except for F_{is} at the *Phi* locus of the Ligurian–Tyrrhenian group), showing that the populations were basically in Hardy–Weinberg equilibrium. The F_{it} and F_{st} values for all populations were very high (average F_{is} and F_{st} = 0.778 and 0.812 respectively). The average F_{is} and F_{st} values for the Ligurian–Tyrrhenian group were 0.360 and 0.250 respectively. The average F_{is} and F_{st} values for the Adriatic group were 0.185 and 0.210 respectively.

Table 1.2.3   Estimates of genetic variability at 19 loci in 14 populations of *Talitrus saltator*

Population	Mean sample size per locus	Mean no of alleles per locus	P	H_o	H_e	H_i
Ligurian–Tyrrhenian						
SRO	19.1	1.3	21.1	0.048	0.065	0.030
	(2.2)	(0.2)		(0.026)	(0.038)	(0.006)
CAS	18.5	1.4	15.8	0.052	0.055	0.022
	(1.8)	(0.2)		(0.034)	(0.036)	(0.005)
BAL	21.0	1.3	15.8	0.041	0.050	0.022
	(2.1)	(0.2)		(0.026)	(0.032)	(0.005)
GIA	8.8	1.2	15.8	0.050	0.057	0.041
	(0.7)	(0.1)		(0.028)	(0.033)	(0.008)
FENA	18.3	1.5	21.1	0.050	0.063	0.021
	(2.2)	(0.3)		(0.026)	(0.034)	(0.006)
FENB	18.9	1.3	21.1	0.034	0.037	0.019
	(2.3)	(0.1)		(0.020)	(0.022)	(0.005)
FENC	18.5	1.2	15.8	0.033	0.044	0.017
	(2.3)	(0.1)		(0.019)	(0.026)	(0.005)
BUR	25.6	1.4	26.3	0.057	0.067	0.033
	(2.3)	(0.2)		(0.027)	(0.033)	(0.006)
Adriatic						
BRU	16.9	1.2	15.8	0.043	0.043	0.023
	(1.7)	(0.1)		(0.027)	(0.028)	(0.005)
GOR	11.2	1.2	15.8	0.032	0.035	0.023
	(1.1)	(0.1)		(0.023)	(0.022)	(0.005)
TFA	17.6	1.1	5.3	0.018	0.015	0.009
	(2.2)	(0.1)		(0.018)	(0.015)	(0.003)
LES	16.7	1.1	5.3	0.011	0.010	0.003
	(2.7)	(0.1)		(0.011)	(0.010)	(0.002)
ROD	22.5	1.1	10.5	0.016	0.019	0.010
	(2.7)	(0.1)		(0.014)	(0.017)	(0.003)
SIP	22.5	1.2	15.8	0.042	0.053	0.014
	(2.5)	(0.1)		(0.027)	(0.034)	(0.004)

P = frequency of polymorphic loci (those with a frequency of the most common allele ≤ 0.99); H_o = average observed heterozygosity; H_e = average heterozygosity (expected under Hardy–Weinberg equilibrium); H_i = average frequency of heterozygous loci per individual. (Standard errors in parentheses.) See Figure 1.2.1 for population symbols.

The genetic distances (*D*) between populations are shown in Table 1.2.5. The intragroup values of *D* varied from 0.001 to 0.1 in the Ligurian–Tyrrhenian group and from 0.002 to 0.039 in the Adriatic one. The intergroup values of *D* were quite different: *D* varied from 0.365 (FEN A-BRU) to 0.482 (BUR-LES).

(b)   Genetic variation within Talorchestia deshayesii

Among the 19 gene loci scored in the five populations of this species, four loci (*Acph-3*, *Mpi*, *Pgm* and *Phi*) were polymorphic in at least one population. The ROP population was monomorphic at all loci examined. Allele *F* at the *Acph-3* locus was only found in the MIR population with a frequency of 0.273. The *Mpi* locus was polymorphic with two alleles in the MIR, BRO and PIC populations. At this locus the BRO population differed from the other four populations, in possessing the allele *E* with a frequency of 0.917; this allele was not found in the ROP and ASN populations and it was found with a low frequency in the

Table 1.2.4 F-statistic indices (Wright, 1965) calculated: (a) for all the populations of *Talitrus saltator*; (b) for the Ligurian–Tyrrhenian populations; (c) for the Adriatic populations

(a)

Locus	F_{is}	F_{it}	F_{st}
Acph-1	0.028	0.940***	0.880***
Acph-3	-0.0097	0.970***	0.960***
Ao-1	0.000	1.000***	1.000***
Ca-2	0.000	1.000***	1.000***
Est-1	0.000	1.000***	1.000***
Est-2	0.000	1.000***	1.000***
Mpi	-0.023	0.440	0.460***
Pgm	0.116	0.440	0.260***
Phi	0.084	0.522***	0.450***

(b)

Locus	F_{is}	F_{it}	F_{st}
Acph-1	0.046	0.922***	0.861***
Acph-3	-0.015	0.200	0.017
Mpi	-0.06	-0.033	0.021
Pgm	0.147	0.323	0.158***
Phi	0.184**	0.389***	0.202**

(c)

Locus	F_{is}	F_{it}	F_{st}
Mpi	0.037	0.200**	0.185***
Pgm	0.058	0.540***	0.390***
Phi	-0.022	-0.186	0.040**

* $p \leq 0.05$, ** $p \leq 0.01$, *** $p \leq 0.001$ (Nei, 1977; Workman and Niswander, 1970).

other two populations. At the *Pgm* locus, allele *E* was only found in the MIR and PIC populations. At the *Phi* locus, the ASN population differed from the other four populations in possessing the private allele *K* at a frequency of 0.313 (Table 1.2.2).

The genetic distances varied from 0.008 (PIC-MIR) to 0.069 (ASN-BRO) (Table 1.2.5).

(c) Genetic differentiation between Talitrus *and* Talorchestia

Electromorphs of *Acph-3*, *Aldo-1*, *Aldo-2*, *Ao-1*, *Ca-1*, *Ca-2*, *Est-2*, *Got-1* and *Mpi* completely discriminated samples of *Talorchestia deshayesii* from *Talitrus saltator* (Table 1.2.2).

Genetic distances between *Talitrus* and *Talorchestia* varied from 0.634 to 0.864 (Table 1.2.5). In the dendrogram (Figure 1.2.2) which synthesizes the genetic relationships among all populations, the two species clustered at a value of *D* = 0.7.

1.2.4   Discussion

Analysing the intergeneric comparisons, populations belonging to *Talitrus* and *Talorchestia* show a relatively high level of genetic differentiation (Figure 1.2.2). Considering their taxonomic separation, an even larger genetic divergence might be expected. However, values of the same order of magnitude have been frequently reported in the literature in comparisons between species of different genera from many animal groups (Hedgecock, Tracey and Nelson, 1982; Thorpe, 1983).

Analysing the intraspecific comparisons, no clear geographic pattern of genetic differentiation is shown within *Talorchestia deshayesii*. The highest value of *D* (0.069) was observed between the two BRO and ASN populations. (Table 1.2.5), which are geographically distant, whereas a value of *D* = 0.047 resulted between BRO and PIC populations, which are the geographically closest populations (Figure 1.2.1). In some cases (i.e. polymorphic loci in the ASN population) these values of *D* are derived from few individuals, and thus these results might be somewhat premature. On the contrary, two quite different degrees of genetic differentiation emerge from the present study of *Talitrus saltator*. Values of genetic distance higher than 0.35 characterize comparisons between Ligurian–Tyrrhenian and Adriatic populations, while genetic distance values lower than 0.1 have been found in comparisons between populations within the same geographic group (Figure 1.2.2).

The situation described in this study is different from the findings of Bulnheim and Scholl (1986), who studied the genetic variation within *Talitrus saltator* and *Talorchestia deshayesii* from northern European and Atlantic locations. These authors found some differences in the distribution of alleles at the two polymorphic loci *Pgi* (equal to the *Phi* of the present work) and *Pgm*, while they found virtually no variation at the other 15 loci studied in the two species. In *Talorchestia deshayesii* this homogeneous genetic structure was found in the northern European and Atlantic samples (Bulnheim and Scholl, 1986) as well as in the Mediterranean ones (present work). In *Talitrus saltator* clinal variation has been found at the *Pgi* locus in the extra-Mediterranean samples (Bulnheim and Scholl, 1986) whereas among the Mediterranean populations the two geographic groups differed from each other in the frequency of the most common allele at the *Phi* locus.

The high genetic homogeneity of the extra-Mediterranean samples, collected along a wide geographic transect, strongly differs from the genetic differentiation found within the Mediterranean populations of *Talitrus saltator*. In order to look for comparable levels of genetic divergence in the extra-Mediterranean and in the Mediterranean groups we examined the genetic distance values within the Ligurian–Tyrrhenian and Adriatic populations separately. It should be noted that the genetic variation was slightly different within the two

Table 1.2.5 Matrix of genetic distance coefficients (Nei, 1978) between 19 populations of *Talitrus saltator* (from SRO to SIP) and of *Talorchestia deshayesii* (from ROP to ASN)

| | Talitrus saltator | | | | | | | | | | | | | | Talorchestia deshayesii | | | | |
| | Ligurian-Tyrrhenian populations | | | | | | | | Adriatic populations | | | | | | | | | | |
	SRO	CAS	BAL	GIA	FENA	FENB	FENC	BUR	BRU	GOR	TFA	LES	ROD	SIP	ROP	MIR	BRO	PIC	ASN
SRO	0.000																		
CAS	0.073	0.000																	
BAL	0.071	0.001	0.000																
GIA	0.075	0.002	0.000	0.000															
FENA	0.074	0.001	0.001	0.001	0.000														
FENB	0.100	0.005	0.005	0.003	0.004	0.000													
FENC	0.088	0.003	0.002	0.001	0.002	0.001	0.000												
BUR	0.043	0.057	0.055	0.054	0.056	0.056	0.054	0.000											
BRU	0.438	0.366	0.376	0.374	0.365	0.381	0.384	0.457	0.000										
GOR	0.453	0.381	0.391	0.388	0.378	0.394	0.398	0.470	0.002	0.000									
TFA	0.461	0.385	0.398	0.394	0.383	0.399	0.404	0.475	0.003	0.001	0.000								
LES	0.468	0.393	0.406	0401	0.390	0.406	0.411	0.482	0.005	0.002	0.000	0.000							
ROD	0.455	0.383	0.395	0.391	0.381	0.397	0.401	0.472	0.003	0.001	0.000	0.000	0.000						
SIP	0.417	0.375	0.379	0.382	0.372	0.398	0.394	0.468	0.020	0.025	0.035	0.039	0.033	0.000					
ROP	0.795	0.688	0.701	0.699	0.689	0.707	0.711	0.817	0.638	0.642	0.634	0.637	0.635	0.684	0.000				
MIR	0.786	0.719	0.726	0.727	0.716	0.744	0.741	0.848	0.674	0.676	0.676	0.679	0.675	0.687	0.022	0.000			
BRO	0.788	0.688	0.700	0.699	0.689	0.708	0.711	0.817	0.638	0.642	0.636	0.638	0.636	0.679	0.046	0.066	0.000		
PIC	0.804	0.705	0.710	0.708	0.701	0.716	0.716	0.824	0.680	0.682	0.683	0.685	0.682	0.701	0.018	0.008	0.047	0.000	
ASN	0.864	0.685	0.691	0.685	0.683	0.673	0.682	0.793	0.691	0.695	0.691	0.694	0.692	0.743	0.021	0.041	0.069	0.025	0.000

See Figure 1.2.1 for population symbols.

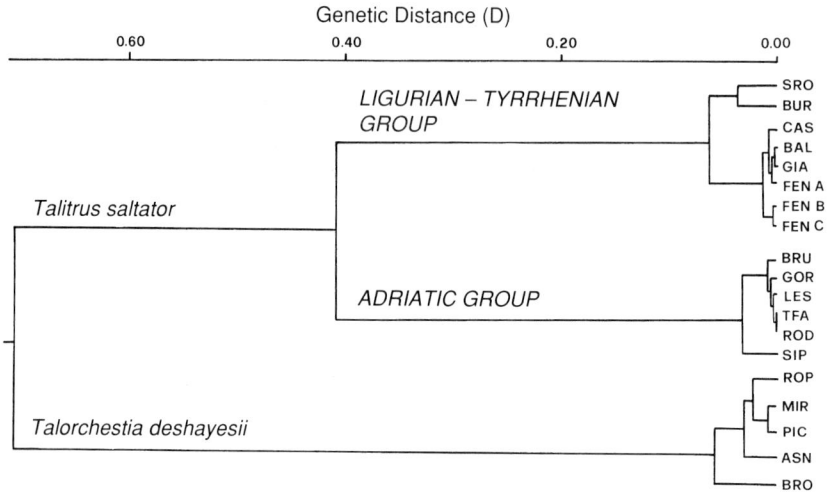

Figure 1.2.2 Dendrogram of *Talitrus saltator* and *Talorchestia deshayesii* populations based on UPGMA clustering (Sneath and Sokal, 1973) of the genetic distance data in Table 1.2.5. See Figure 1.2.1 for population symbols.

separately. It should be noted that the genetic variation was slightly different within the two geographic groups found in the Mediterranean.

Within the Ligurian–Tyrrhenian group the two populations farthest apart (SRO and BUR) grouped together in a subcluster contrary to their geographic distance (Figures 1.2.1 and 1.2.2). This outcome could be due to the presence of the *Acph-1 B* allele, which was only found in these two populations.

The genetic distances between the Adriatic populations are small (Table 1.2.5) in spite of their relatively great geographic separation (Figure 1.2.1). Only the SIP population exhibits a certain degree of genetic differentiation (average $D = 0.030$, Figure 1.2.2); it is also the Adriatic population with the highest expected heterozygosity (Table 1.2.3).

The western (Ligurian–Tyrrhenian) populations of sandhoppers cluster the eastern (Adriatic) ones at an average $D = 0.4$.

It should be noted that the values of genetic distance found between these two geographic groups of populations are higher than values recorded between sibling species of marine gammarids ($D = 0.201$) (Siegismund, Simonsen and Kolding, 1985) or between sibling species of the subterranean shrimp *Troglocaris*, for which a secondary sympatric occurrence has been postulated ($D = 0.209$) (Cobolli Sbordoni *et al.*, 1990). Moreover, Stewart and Griffiths (1992) found an average value of $D = 0.690$ between different species of freshwater crangonyctoids, and Scheepmaker (1990) found an average genetic distance of 0.501 between different species of freshwater gammarids: in both cases the evolutionary history of the group could be related to vicariant events. Furthermore, an average genetic distance of 0.39 has been found between a group of ten cave, phreatic and spring populations of *Niphargus longicaudatus* versus one spring population of *N. pasquinii* (Sbordoni, Cobolli Sbordoni and De Matthaeis, 1979).

On the ground of their morphological characters the Adriatic populations of the present study can be attributed to the *briani* form described by Ruffo (1936). Unfortunately sandhoppers have become extinct in the localities studied by Ruffo (1936) due to anthropogenic pressure in that region.

Allozyme data strongly support the morphological evidence of differentiation within the Mediterranean *Talitrus saltator*: the presence of fixed alternative alleles at several loci clearly discriminates the Adriatic populations from the others, providing evidence of the absolute lack of gene flow between the Adriatic and the Ligurian–Tyrrhenian populations. In addition the genetic subdivision into two distinct gene pools is outlined by the results of the *F*-statistics analysis (Table 1.2.4).

The results of the present study could explain the difficulties found in the experimental mass crosses between Ligurian–Tyrrhenian and Adriatic populations (Scapini and Fasinella, 1990). Two different gene pools, which normally never meet or exchange genes in the wild, were crossed in the laboratory. This observed 'outbreeding depression' probably resulted from the breaking up of coadapted gene complexes.

The ecology of dispersal of sandhoppers and/or the present pattern of the Italian coasts (Figure 1.2.1) as well as palaeogeographic evidence of the Italian peninsula during the Late Pliocene (Pasa, 1953) could help explain all these findings.

Talitrids lack larval stages, which are supposed to provide genetic homogeneity over wide geographic areas. According to Dahl (1946), their dispersal capacity over long distances would be generally low and related to coastal surface currents. It would be mainly passive and occasional, via animals attached to drifting wrack and wood. *Talorchestia* is more tolerant to aquatic immersion than *Talitrus* and therefore *Talorchestia* is likely to have higher plasticity than *Talitrus*. The relative lack of genetic variation observed within Mediterranean *Talorchestia deshayesii* can be partially explained on these grounds. However, talitrids are also able to move up to 200 m per night on sandy shores (Scapini *et al.*, 1992). Stretches of rocky shores and human settlements act as barriers to their dispersion. The slight differences observed in the *D* values within the Ligurian–Tyrrhenian and the Adriatic groups of *Talitrus saltator* can be due also to local constraints to active movements of individuals. However, the pattern of surface currents in the Mediterranean could explain the isolated condition of the Adriatic sea (Lacombe and Tchernia, 1972) which leads to the peculiarity of the faunistic composition of the Adriatic ecosystem (Bacescu, 1985; Tortonese, 1985). The genetic differentiation found between the two geographic groups of Mediterranean *Talitrus saltator* could be the result of the geographic isolation of the Adriatic sandhoppers.

Furthermore, Plio-Pleistocene climatic fluctuations could have played an important role in the past isolation of the Adriatic populations of many taxa (Bacescu, 1985). Possibly during this period the Adriatic populations of *Talitrus saltator* became separated physically from the Ligurian–Tyrrhenian group and began to diverge genetically. Moreover, the northern Adriatic coasts became available for colonization by sandhoppers only recently, after the last glaciation, due to the corresponding changes in sea level. This outcome can account for the levels of genetic similarity found between the Adriatic populations. A recent colonization could also account for the genetic homogeneity found by Bulnheim and Scholl (1986) between Atlantic and northern-European populations of *Talitrus saltator*.

Future research will concentrate on the analysis of the genetic structure of different populations from the whole geographic area of the *Talitrus saltator* complex. Ultrastructural morphological inspection of the characters usually utilized in the taxonomy of this group will also be accomplished with the aim of defining the distribution areas of the two electrophoretic entities.

Acknowledgements

The authors wish to thank Sandro Ruffo for stimulating this research and for his precious help in the identification of samples. Cinzia Boccali, Niccolò Falchi and Marco Oliverio kindly provided their technical assistance. Financial support for the project was provided by Ministero dell'Università e della Ricerca Scientifica (MURST 40%) and by the Consiglio Nazionale delle Ricerche (CNR), Comitato Ambiente). Some of the samples were collected during the cruises of the oceanographic ship 'Minerva' (CNR).

References

Ayala, F.J., Powell, J.R., Tracey, M.L., Mourao, C.A. and Perez-Salas, S. (1972) Enzyme variability in the *Drosophila willistoni* group. IV. Genetic variation in natural populations of *Drosophila willistoni*. *Genetics*, **70**, 113–139.

Ayala, F.J., Valentine, J.V., Barr, L.G. and Zumwalt, G.S. (1974) Genetic variability in a temperate intertidal phoronid, *Phoronopsis viridis*. *Biochem. Genet.*, **11**, 413–427.

Ayala, F.J., Valentine, J.V. and Zumwalt, G.S. (1975) An electrophoretic study of the Antarctic zooplankter *Euphasia superba*. *Limnol. Oceanogr.*, **20**, 635–640.

Bacescu, M. (1985) The effects of the geological and physico-chemical factors on the distribution of marine plants and animals in the Mediterranean, in *Mediterranean Marine Ecosystems* (eds M. Moraitou-Apostolopoulou and V. Kiortsis), Plenum Press, New York, pp. 195–212.

Bousfield, E.L. (1982) The amphipod superfamily Talitroidea in the north-eastern Pacific region. I. Family Talitridae: systematics and distributional ecology. *Natl. Mus. Canada Publ. Biol. Ocean.*, **11**, 1–73.

Brewer, G.J. and Sing, C.F. (1970) *An Introduction to Isozyme Techniques*, Academic Press, New York.

Bulnheim, H.P. and Scholl, A. (1986) Genetic differentiation between populations of *Talitrus saltator* and *Talorchestia deshayesii* (Crustacea: Amphipoda) from coastal areas of the north-western European continent. *Mar. Biol.*, **92**, 525–536.

Burton, R.S. (1986) Evolutionary consequences of restricted gene flow among natural populations of the copepod, *Tigriopus californicus*. *Bull. Mar. Sci.*, **39**, 526–535.

Cobolli Sbordoni, M., Mattoccia, M., La Rosa, P., De Matthaeis, E. and Sbordoni, V. (1990) Secondary sympatric occurrence of sibling species of subterranean shrimps in the Karst. *Int. J. Speleol.*, **19**, 9–27.

Dahl, E. (1946) The Amphipoda of the sound. Part I: Terrestrial Amphipoda. *Lunds Univ. Arssk. (Adv. 2)*, **42**, 1–53.

Harris, H. and Hopkinson, D.A. (1978) *Handbook of Enzyme Electrophoresis in Human Genetics*, Supplement, North-Holland Publishing Company, Amsterdam.

Hedgecock, D. (1986) Is gene flow from pelagic larval dispersal important in the adaptation and evolution of marine invertebrates? *Bull. Mar. Sci.*, **39**, 550–564.

Hedgecock, D., Tracey, M.L., and Nelson, K. (1982) Evolutionary divergence and speciation, in *The Biology of Crustacea*, Vol. 2 (ed. D.E. Bliss), Academic Press, New York, pp. 339–347.

Lacombe, H. and Tchernia, P. (1972) Caracteres hydrologiques et circulation des eaux en Mediterranee, in *The Mediterranean Sea* (ed. D.J. Stanley), Hutchinson & Ross, Stroudsburg, pp. 25–36.

Nei, M. (1977) F-statistics and analysis of gene diversity in subdivided populations. *Ann. Hum. Genet.* **41**, 225–233.

Nei, M. (1978) Estimation of average heterozygosity and genetic distance from a small number of individuals. *Genetics*, **89**, 583–590.

Pardi, L. (1960) Innate components in the solar orientation of littoral amphipods. *Cold Spring Harbor Symp. Quant. Biol.*, **25**, 394–401.

Pasa, A. (1953) Appunti geologici per la paleogeografia delle Puglie. *Mem. Biogeogr. Adriatica*, **2**, 175–286.

Poulik, M.D. (1957) Starch gel electrophoresis in discontinuous system of buffers. *Nature*, **180**, 1477–1479.

Ruffo, S. (1936) Studi sui Crostacei Anfipodi I. Contributo alla conoscenza degli Anfipodi dell'Adriatico. *Boll. Ist. Entomol. Univ. Stud. Bologna*, **9**, 23–32.

Sbordoni, V., Cobolli Sbordoni, M. and De Matthaeis, E. (1979) Divergenza genetica tra popolazioni e specie ipogee ed epigee di *Niphargus* (Crustacea, Amphipoda). *Lav. Soc. Ital. Biogeogr.*, (New Series), **4**, 1–23.

Scapini, F. and Fasinella, D. (1990) Genetic determination and plasticity in the sun orientation of natural populations of *Talitrus saltator*. *Mar. Biol.*, **107**, 141–145.

Scapini, F., Chelazzi, L., Colombini, I. and Fallaci, M. (1992) Surface activity, zonation and migrations of *Talitrus saltator* (Montagu, 1808) on a Mediterranean beach. *Mar. Biol.*, **112**, 573–581.

Scapini, F., Ugolini, A., and Pardi, L. (1985) Inheritance of solar direction finding in sandhoppers. II. Differences in arcuated coastlines. *J. Comp. Physiol.* (A), **156**, 729–735.

Scheepmaker, M. (1990) Genetic differentiation and estimated levels of gene flow in members of the *Gammarus pulex*-group (Crustacea, Amphipoda) in western Europe. *Bijdr. DierkD.*, **60**, 3–30.

Siegismund, H.R., Simonsen, V. and Kolding, S. (1985) Genetic studies of *Gammarus*. I. Genetic differentiation of local populations. *Hereditas*, **102**, 1–13.

Sneath, P.H.A. and Sokal, R.R. (1973) *Numerical Taxonomy*, Freeman, San Francisco.

Stewart, B.A. and Griffiths, C.L. (1992) A taxonomic reexamination of freshwater amphipods in the *Paramelita auricularius–P. crassicornis* complex, with descriptions of three additional species. *Crustaceana*, **62**, 166–192.

Swofford, D.L. and Selander, R.B. (1981) BIOSYS-1: a FORTRAN program for the comprehensive analysis of electrophoretic data in population genetics and systematics. *J. Hered.*, **72**, 281–283.

Thorpe, J.P. (1983) Enzyme variation, genetic distance and evolutionary divergence in relation to levels of taxonomic separation, in *Protein Polymorphism: Adaptive and Taxonomic Significance* (eds G.S. Oxford and D. Rollinson), Academic Press, New York, pp. 131–152.

Tortonese, E. (1985) Distribution and ecology of endemic elements in the Mediterranean fauna (fishes and echinoderms), in *Mediterranean Marine Ecosystems* (eds M. Moraitou-Apostolopoulou and V. Kiortsis), Plenum Press, New York, pp. 57–83.

Ward, R.D. and Beardmore, J.A. (1977) Protein variation in the plaice (*Pleuronectes platessa*). *Genet. Res.*, **30**, 45–62.

Workman, P.L. and Niswander, J.D. (1970) Population studies on southwestern Indian tribes. II. Local genetic differentiation in the Papago. *Am. J. Hum. Genet.*, **22**, 24–49.

Wright, S. (1965) The interpretation of population structure by F-statistics with special regard to system of mating. *Evolution*, **9**, 395–420.

1.3 A COMPARISON BETWEEN MORPHOLOGICAL AND GENETIC DATA IN TWO SPECIES OF *TETHYA* (PORIFERA, DEMOSPONGIAE)

Giorgio Bavestrello and Michele Sarà

Abstract

Genetic distances and divergence in the external morphology and spicular traits have been compared in four populations of Tethya aurantium *and five of* T. citrina *(Porifera, Demospongiae) taken from different Mediterranean and North Atlantic sites.* T. aurantium *is electrophoretically identical in all the examined populations while the populations of* T. citrina *show genetic variability. While the Tyrrhenian and Ionian populations of this species have a negligible genetic distance (D < 0.05), the populations of Limsky Canal (North Adriatic Sea) and Torbay (English Channel) are strongly separated from each other (D = 0.5) and from all other populations (D about 0.3). At the species level the morphological and electrophoretic data agree such that* T. aurantium *shows a greater homogeneity than* T. citrina. *Nevertheless, at the inter- or intrapopulational level, particularly in* T. citrina, *there is a considerable discrepancy between the spicular and genetic data. The genetic distances are more consistent with the differences in the external morphology and this fact emphasizes the taxonomic role of both these characters. The spicular variability is discussed in relation to the environmental and developmental factors that influence their growth and final shape and size.*

1.3.1 Introduction

Traditionally, sponge taxonomy has been based on the shape and size of the siliceous spicules and on their arrangement in the sponge skeleton while the external morphology was considered to be subject to considerable plasticity and to be unsuitable for species determination. In recent decades, with the field study of living specimens, a growing interest has developed in the diagnostic value of external characters, such as individual sponge shape and size, colour, consistency and surface texture (Barthel, Gutt and Tendal, 1991; Bavestrello and Sarà, 1992). On the other hand, observations of intraspecific variation of spicule shape and size (Jones, 1979, 1987) indicate that spicular traits should be critically evaluated in species determination.

The use of allozyme electrophoretic analysis may be a useful test to clarify, on a genetic basis, dubious taxonomic questions (Solé-Cava and Thorpe, 1986; Solé-Cava *et al.*, 1991a, b; Sarà *et al.*, 1989; Hooper *et al.*, 1990; Bavestrello and Sarà, 1992; Sarà, Bavestrello and Mensi, 1992). Some data indicate that the genetic divergence of sibling species is accompanied by some differences in the spicular morphometry. However, a more detailed analysis of morphological and electrophoretic data, at the inter- or intrapopulational level, is lacking in sponges. Studies of this kind have been carried out particularly on insects and vertebrates (e.g. Allegrucci *et al.*, 1992; Mensi *et al.*, 1992).

In this study we investigate consistency between morphological (individual size and colour, spicule size and shape) and allozymic differences in some electrophoretically surveyed populations of *Tethya aurantium* and *T. citrina* from the Mediterranean and North Atlantic.

1.3.2 Material and methods

This study was performed on several samples of *T. aurantium* and *T. citrina* collected in different localities of the Mediterranean and English Channel (Table 1.3.1).

The Stagnone of Marsala, Porto Cesareo and Porto Pozzo are coastal shallow lagoons with a salinity of 37%, similar to the open sea. Limsky Canal is an arm of the sea with a maximum depth of about 40 m. Punta Ala is a sheltered bay and Capo Caccia and Torbay are exposed seashores, the latter site being in the English Channel.

Table 1.3.1 Sampled localities for *Tethya aurantium* and *T. citrina*

Localities	T. aurantium	T. citrina
Lagoon of Porto Pozzo (East Sardinia, Tyrrhenian Sea)		X
Capo Caccia (West Sardinia, Sardinian Sea)	X	
Punta Ala (Tuscany, Tyrrhenian Sea)		X
Stagnone of Marsala (West Sicily, Sicily Canal)	X	X
Lagoon of Porto Cesareo (West Apulian, Ionian Sea)	X	
Limsky Canal (Istria, North Adriatic Sea)	X	X
Torbay (English Channel)		X

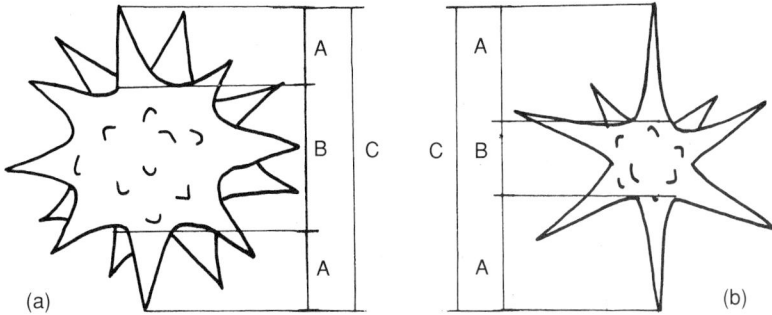

Figure 1.3.1 Semi-schematic drawing of the megasters of (a) *T. aurantium* and (b) *T. citrina*. A = ray length; B = centre diameter; C = total diameter.

Morphological analysis was conducted on 10 specimens of each population for each locality. Both external morphology (diameter, cortex colour and thickness) and spicular size and shape were taken into consideration. For the spicular analysis two slides were prepared from each specimen and the size of 50 megasters (Figure 1.3.1) on each slide was measured under a light microscope using a camera lucida and 'Graphtec' digitizer (Jones, 1987). The average diameter of the megasters and the ratio between the length of the ray and the diameter of the centre (R/C) (Sarà and Melone, 1965; Pulitzer-Finali, 1983) in each population were calculated and the differences tested with a *t*-test. From these data, the size frequency distributions were drawn and the differences tested by non-parametric Kolmogorov–Smirnov test (Smirnov, 1948).

These data were compared with Nei's (1972) genetic distances obtained from an electrophoretic analysis carried out previously on samples from the same populations (Sarà *et al.*, 1989; Sarà, Bavestrello and Mensi, 1992).

1.3.3   Results

(a)   Morphology

External morphology
All examined populations of *T. aurantium* show a remarkable homogeneity in their external morphology (Table 1.3.2). In fact, no significant differences are shown between the total average diameters of the specimens of different populations. Also, the cortex thickness (about 10% of the total diameter) and the cortex colour are very similar in the examined populations.

A different pattern is shown by *T. citrina*, a species more variable in its external morphology (Table 1.3.2). The specimen diameter is quite homogeneous in the Tyrrhenian and Ionic populations (about 20 mm) but this value shows a strong increase in the Torbay (30 mm) and particularly in the Limsky Canal (60 mm) specimens. The Mediterranean populations have a thin cortex (about 5% of the total diameter) but in the specimens from Torbay this structure is thicker (about 10% of the total diameter) and very similar to that of *T. aurantium*. Specimens of *T. citrina* from Torbay are also remarkable for their colour. In fact, 75% of specimens are yellow and the remaining 25% are orange, while all the Mediterranean specimens are yellowish-grey.

Table 1.3.2 External morphology of *Tethya aurantium* and *T. citrina* in the studied populations

		Total diameter (mm)	Cortex thickness (%)	Cortex colour
T. aurantium	Stagnone of Marsala	36.7 ± 6.2	11.34 ± 2.5	Orange
	Porto Cesareo	32.8 ± 8.9	10.8 ± 2.6	Orange
	Capo Caccia	32.3 ± 8.4	9.12 ± 2.9	Orange
	Limsky Canal	37.8 ± 10	11.45 ± 3.2	Orange
T. citrina	Stagnone of Marsala	21.6 ± 5.1	4.76 ± 1.4	Grey*
	Punta Ala	23.5 ± 4.6	5.12 ± 0.8	Grey
	Porto Pozzo	18.3 ± 3.9	5.08 ± 1.6	Grey
	Limsky Canal	61.8 ± 10.5	5.4 ± 1.8	Grey
	Torbay	32.5 ± 7.1	10.3 ± 2.6	Yellow–orange

The values of total individual diameter and the cortex thickness are the means ± SE. The cortex thickness is expressed as a percentage of the total diameter.
* Grey = yellowish-grey

Spicular morphometry

The measured spicular parameters are summarized in Table 1.3.3.

The frequency distributions of the total diameters of the megasters (Figure 1.3.2(a)) of *T. aurantium* from the Stagnone of Marsala and Limsky Canal completely overlap while those of Capo Caccia and Porto Cesareo are significantly different ($0.05 > P > 0.01$) from each other and from all other populations. The population of Porto Cesareo has a significantly lower ($P < 0.05$) R/C value than all other populations. No other populations showed significant differences (Figure 1.3.2(b)).

Analysis of the megaster diameters of *T. citrina* (Figure 1.3.3(a)) suggests two groups of populations: Punta Ala and Torbay versus Limsky Canal, Stagnone of Marsala and Porto Pozzo. A similar pattern is shown by the R/C size frequency distributions (Figure 1.3.3(b)) but, in this case, the distribution of the megasters of Marsala is grouped with those of Torbay and Punta Ala.

Table 1.3.3 Megaster morphometry of *Tethya aurantium* and *T. citrina* in the studied populations

		Total diameter (mm)	R/C
T. aurantium	Stagnone of Marsala	60.88 ± 15.3	0.49 ± 0.08
	Porto Cesareo	82.49 ± 19.7	0.44 ± 0.05
	Capo Caccia	67.54 ± 15.15	0.56 ± 0.05
	Limsky Canal	62.36 ± 11.34	0.54 ± 0.07
T. citrina	Stagnone of Marsala	81.97 ± 13.7	0.92 ± 0.12
	Punta Ala	52.73 ± 10.7	0.91 ± 0.1
	Porto Pozzo	78.33 ± 11.5	1.21 ± 0.09
	Limsky Canal	80.87 ± 12.9	1.16 ± 0.13
	Torbay	60.3 ± 14.9	0.79 ± 0.1

The values are the means ± SD. The R/C is the ratio between the length of the ray and the diameter of the centre of the spicule (see Figure 1.3.1).

(b)   Electrophoresis

The electrophoretic data were taken from two studies on the genetic variability of the European species of the genus *Tethya* (Sarà *et al.*, 1989; Sarà, Bavestrello and Mensi, 1992). Allele frequencies at the loci scored are summarized in Table 1.3.4. The four populations of *T. aurantium* had an identical electrophoretic pattern since all the loci scored were monomorphic. In *T. citrina* the Tyrrhenian and Ionic populations are electrophoretically

Table 1.3.4   Allele frequencies at up to 12 allozyme loci in *Tethya aurantium* and *T. citrina* populations from the Mediterranean and English Channel. All the studied populations of *T. aurantium* are electrophoretically identical

		T. citrina					*T. aurantium*
		Marsala	*Punta Ala*	*Porto Pozzo*	*Limsky Canal*	*Torbay*	*Mediterranean*
Got	a	0	0	0	0.7	0	0
	b	1	1	0.9	0.3	1	1
	c	0	0	0.1	0	0	0
6Pgd	a	1	0.55	1	1	1	0
	b	0	0.45	0	0	0	1
Hk	a	1	1	1	0	0.8	0
	b	0	0	0	0	0.1	0
	c	0	0	0	0.45	0	0
	d	0	0	0	0.55	0	0
	e	0	0	0	0	0	1
Mdh-1	a	0	0	0	0	0.58	0
	b	1	1	1	1	0.42	0
	c	0	0	0	0	0	1
Mdh-2	a	0	0	0	0		1
	b	1	1	1	1		0
Ak	a	1	1	1	1	1	0
	b	0	0	0	0	0	1
Es	a	0	0	0	0	0	1
	b	1	1	1	1	1	0
Idh	a	0	0	0	0	0.2	0
	b	0	0	0	0	0.8	0
	c	1	1	1	1	0	1
Fum	a	0	0	0	0	0	1
	b	1	1	1	1	0	0
	c	0	0	0	0	1	0
Pgi	a	0	0	0	0	0	1
	b	1	1	1	1	1	0
Mpi	a	0	0	0	0	0	1
	b	0	1	1	1	0.58	0
	c	0	0	0	0	0.42	0
Sod	a	0	0	0	1		0
	b	0	0	0	0		1
	c	0.78	0.9	1	0	0	0
	d	0.22	0.1	0	0	0	0

Data from Sarà *et al.* (1989) and Sarà, Bavestrello and Mensi (1992).
a–e are the designation for alleles at the allozyme loci.

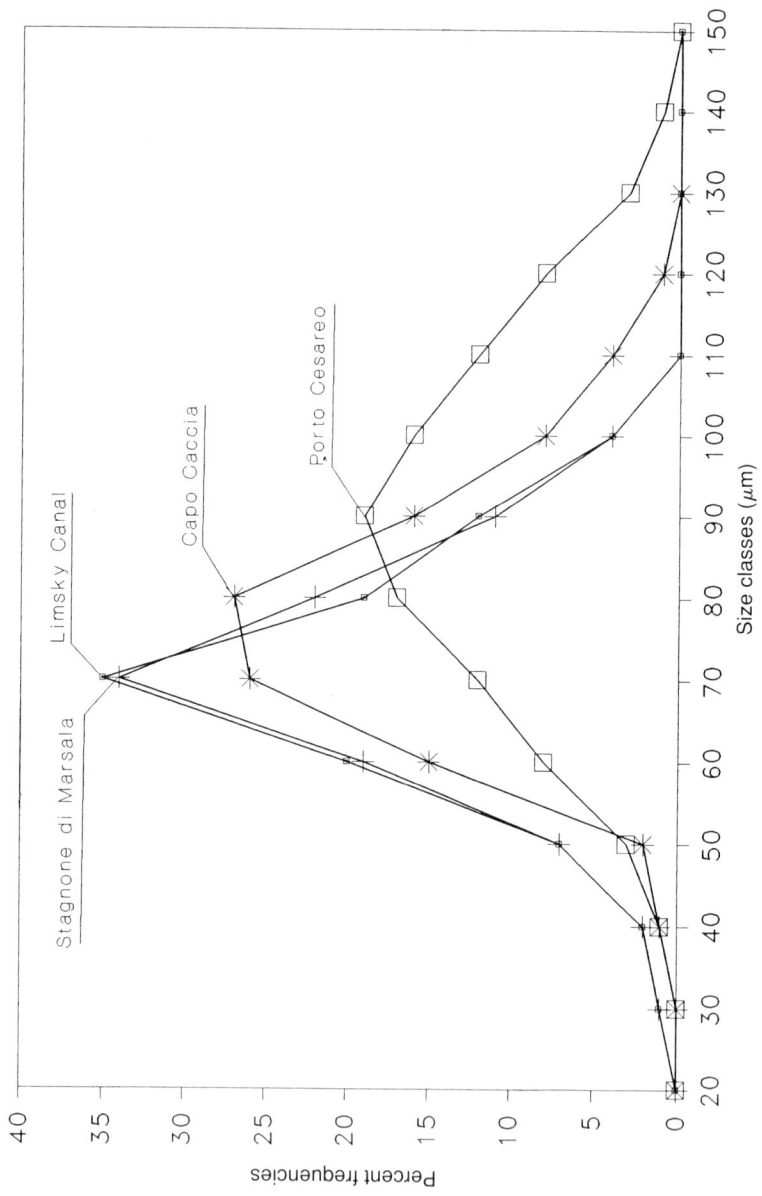

Figure 1.3.2 (a) *T. aurantium*. Frequency distributions of megaster diameters.

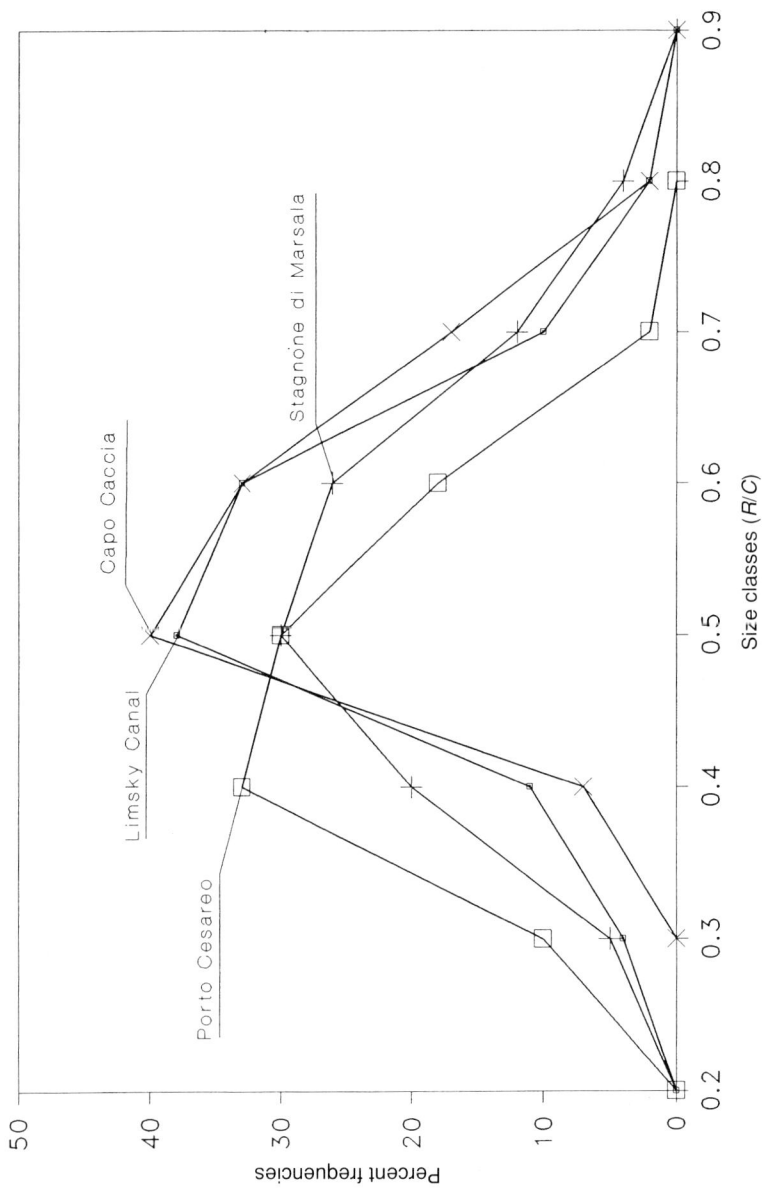

Figure 1.3.2 (b) *T. aurantium.* Frequency distributions of megaster *R/C*.

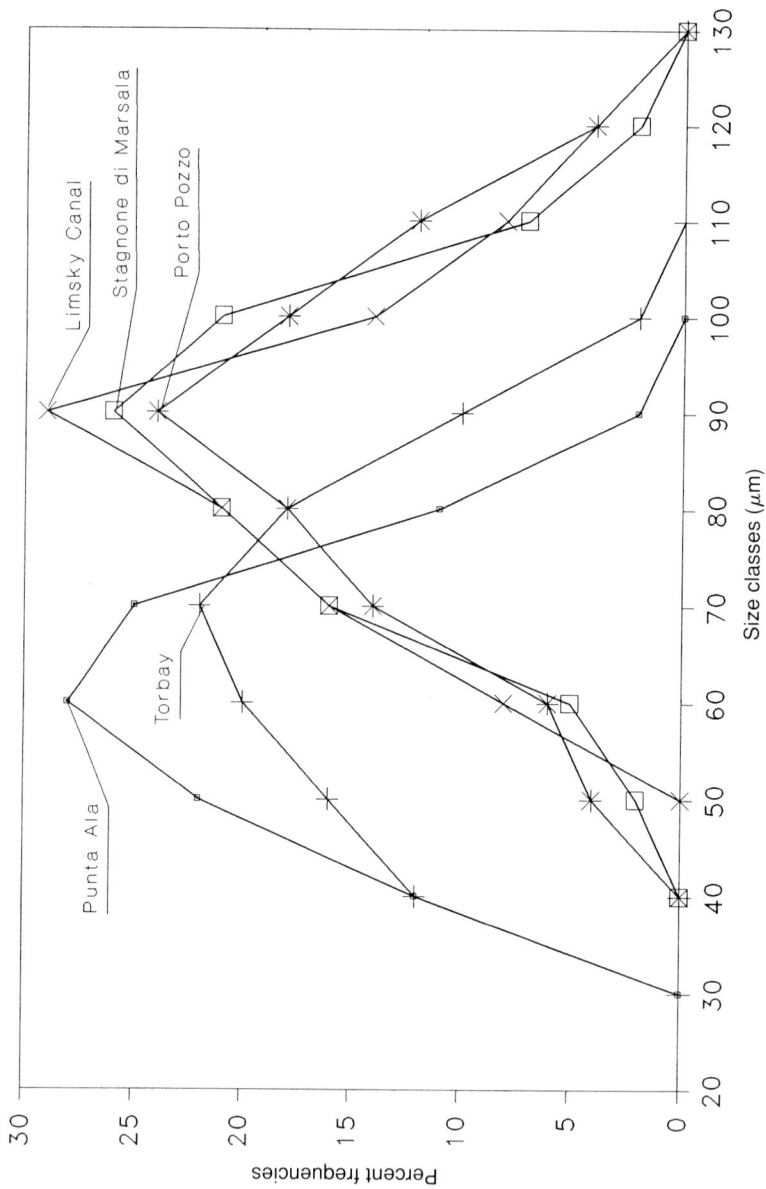

Figure 1.3.3 (a) *T. citrina.* Frequency distributions of megaster diameters.

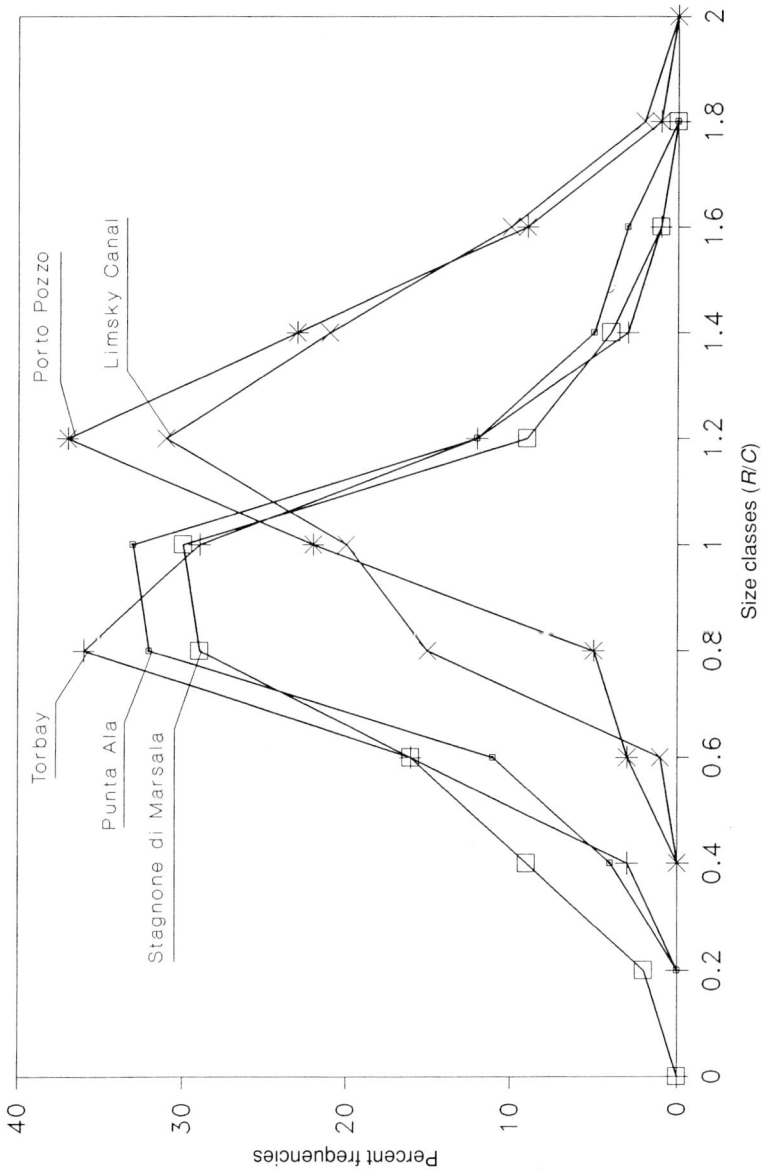

Figure 1.3.3 (b) *T. citrina.* Frequency distributions of megaster *R/C*.

similar ($D < 0.05$) with only a few polymorphic loci (8.3% in the Stagnone of Marsala and Porto Pozzo populations and 16.6% in the Punta Ala population) (Table 1.3.4). In contrast, the two populations of Limsky Canal and Torbay were separated from the others by the presence of strongly different allele frequencies at some polymorphic loci (Table 1.3.4) and diagnostic alleles at the *Idh* and *Fum* loci in the Torbay population and the *Hk* and *Sod* loci in the Limsky Canal population.

The Tyrrhenian and Ionian populations are a homogeneous group with genetic distances (Nei, 1972) less than 0.05 (Figure 1.3.4). The populations of the Limsky Canal and Torbay are strongly isolated from each other ($D = 0.5$) and also from the Tyrrhenian and Ionian populations ($D = 0.29$ and 0.35 respectively).

1.3.4   Discussion

The present data show that, at the species level, a general agreement exists between the morphological and electrophoretic data. In fact, the examined populations of *T. aurantium* were electrophoretically identical and were morphologically more similar both for individual size, cortex colour and thickness than those of *T. citrina*, which had different allozymic

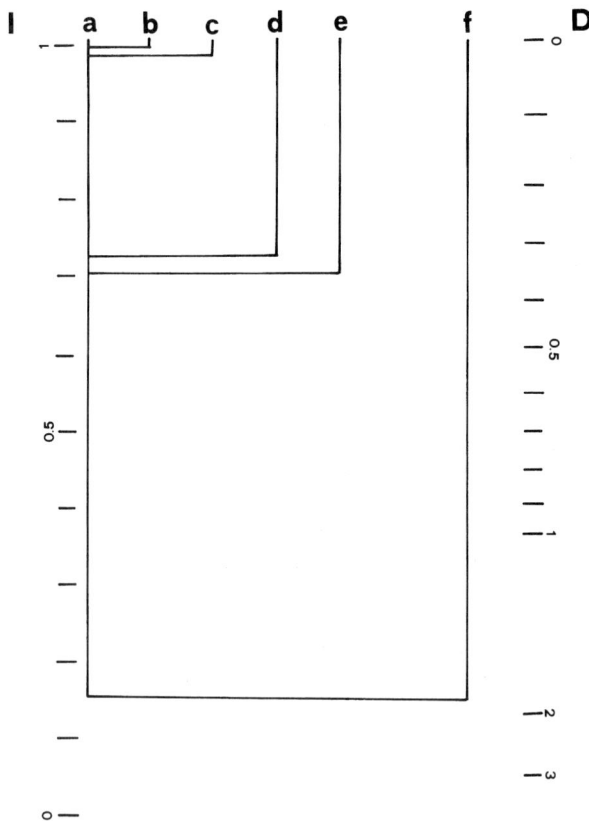

Figure 1.3.4 Genetic distances (D) and identities (I) among the populations of *T. citrina* (a = Stagnone of Marsala; b = Porto Pozzo; c = Punta Ala; d = Limsky Canal; e = Torbay) and *T. aurantium* (f = all populations).

patterns. This suggests a general correlation between the genotype and the variability of external morphology and spicular size and shape. But at the inter- and intrapopulational level, the correspondence between genetic and morphological data becomes more confused, particularly with regard to the following aspects.

Although all the tested samples of *T. aurantium* are electrophoretically identical, morphological differences were found by the morphometric analysis of megasters. While the size frequency distributions of the asters of the Limsky Canal and Stagnone of Marsala populations completely overlapped, those of Capo Caccia and Porto Cesareo were significantly different from all others, both in the average size and variance.

In *T. citrina* the two spicular parameters, size and shape (*R/C*), did not correlate with the electrophoretic distances. For example, the size frequency distributions of the megasters of the genetically distant populations of Limsky Canal and Stagnone of Marsala overlap. On the other hand, the megasters were very different between the populations of Punta Ala and Stagnone of Marsala that showed a negligible genetic distance.

The differences in the external morphology seem more linked to the electrophoretic banding pattern. In fact, the large specimen diameter in the population from the Limsky Canal, about double that in the other populations, agrees with its relative genetic isolation. The morphological differences (colour, size and cortex thickness) between the Torbay population and the Mediterranean ones were consistent with its high value of genetic distance. These data emphasize the taxonomic importance of the external morphology in the genus *Tethya*. However, some caution is needed in the use of megaster size and shape.

The discrepancy between the genetic distances and the morphometric parameters is a common phenomenon in different animal groups (Allegrucci *et al.*, 1992) and may be related to the influence of environmental cues on the genomic phenotypic expression (Smith-Gill, 1983). In sponges this phenomenon may influence both external (Fry, 1979) and spicular morphology. The spicules of *Haliclona rosea* and *H. elegans* show size variations during the seasonal cycle related to spiculogenetic activity (Jones, 1987). In *Hymeniacidon perleve* the average size of spicules is strongly related to the silica concentration in the seawater (Stone, 1970). Differences in the spicule size have also been detected along latitudinal or bathymetric gradients (Hartman, 1958; Hooper, 1992; Bavestrello, Bonito and Sarà, 1993) and Palumbi (1986) demonstrated an increase of spicule dimensions in shallow water sponges subjected to wave action.

A previous study on the megaster size in two species of *Tethya* (Bavestrello, Corriero and Sarà, 1992) indicates that the spicular plasticity is also influenced by sponge anatomy. In this genus, with a complex body organization, the megaster size increases from the medulla to the cortex.

In conclusion, in the genus *Tethya* the electrophoretic pattern is more in agreement with the external morphology than with the spicular features. This fact suggests a genetically constrained body organization of these sponges while the spicular traits may be particularly influenced by environmental cues.

The size and shape variability in sponge spicules is therefore regulated by a complex network of interactions between genetic, epigenetic and environmental factors. Sponge spicular variability, due to its simple quantitative nature, represents a useful feature for providing insights into the way in which genetic and extragenetic interactions may influence phenotypic traits.

Acknowledgement

This work has been partly supported by funds from CNR and MURST (40%).

References

Allegrucci, G., Caccone, A., Cesaroni, D. and Sbordoni, V. (1992) Evolutionary divergence in *Dolichopoda* cave crickets: a comparison of single copy DNA hybridization data with allozymes and morphometric distances. *J. Evol. Biol.*, **5**, 121–48.

Barthel, D., Gutt, J. and Tendal, O.S. (1991) New information on the biology of Antarctic deep-water sponges derived from underwater photography. *Mar. Ecol. Progr. Ser.*, **69**, 303–7.

Bavestrello, G. and Sarà, M. (1992) Morphological and genetic differences in ecologically distinct populations of *Petrosia ficiformis* (Porifera, Demospongiae). *Biol. J. Linn. Soc.*, **47**, 49–60.

Bavestrello, G., Bonito, M. and Sarà, M. (1993) Influence of depth on size variability of sponges spicules. *Scient. Mar.* (in press).

Bavestrello, G., Corriero, G. and Sarà, M. (1992) Differences between two sympatric species of *Tethya* (Porifera, Demospongiae) concerning the growth and final form of their megasters. *Zool. J. Linn. Soc.*, **104**, 81–7.

Fry, W.G. (1979) Taxonomy, the individual and the sponge, in *Biology and Systematics of Colonial Organisms* (eds G. Larwood and B.R. Rosen), Academic Press, London, pp. 49–80.

Hartman, W.D. (1958) Natural history of marine sponges of southern New England. *Bull. Peabody Mus. Nat. Hist.*, **12**, 1–155.

Hooper, J.A. (1992) Revision of the family Raspailliidae (Porifera, Demospongiae), with description of Australian species. *Invert. Taxon.*, **5**, 1179–418.

Hooper, J.A., Capon, R.J., Keenan, C.P. and Parry, D.L. (1990). Biochemical and morphological differences of two sympatric sibling species of *Clathria* (Porifera: Demospongiae: Microcionidae) from Northern Australia. *Invertebrate Taxonomy*, **4**, 123–48.

Jones, W.C. (1979) The microstructure and genesis of sponge biominerals, in *Biologie des Spongiaires* (eds C. Levi and N. Boury-Esnault), Colloque Internationaux du CNRS, Paris, pp. 425–48.

Jones, W.C. (1987) Seasonal variations in the skeleton and spicule dimensions of *Haliclona elegans* (Bowerbank) *sensu* Topsent (1887) from two sites in north Wales, in *European Contributions to the Taxonomy of Sponges* (ed. W.C. Jones) Litho Press Co., Middleton, pp. 109–29.

Mensi, P., Lattes, A., Macario, B. and Salvidio, S. (1992) Taxonomy and evolution of European brown frogs. *Zool. J. Linn. Soc.*, **104**, 293–311.

Nei, M. (1972) Genetic distance between populations. *Am. Nat.*, **106**, 283–92.

Palumbi, S.R. (1986) How body plans limit the acclimation: responses of a demosponge to wave force. *Ecology*, **67**, 208–14.

Pulitzer-Finali, G. (1983) A collection of Mediterranean Demospongiae (Porifera) with, in appendix, a list of Demospongiae hitherto recorded from the Mediterranean Sea. *Annal. Mus. Civic. Storia Natur. Genov.*, **84**, 445–621.

Sarà, M. and Melone, N. (1965) Una nuova specie del genere *Tethya*, *T. citrina* sp.n. dal Mediterraneo (Porifera, Demospongiae). *Atti Soc. Peloritana Sci. Fis. Matemat. Natur.*, **11**, 13–168.

Sarà, M., Bavestrello, G. and Mensi, P. (1992) Redescription of *Tethya norvegica* Bowerbank (Porifera, Demospongiae) with remarks on the genus *Tethya* in the North East Atlantic. *Zool. Scripta*, **21**, 211–16.

Sarà, M., Mensi, P., Manconi, R. and Balletto, E. (1989) Genetic variability in Mediterranean populations of *Tethya* (Porifera, Demospongiae), in *Reproduction, Genetics and Distributions of Marine Organisms* (eds J.S. Ryland and P.A. Tyler), Olsen & Olsen, Fredensborg, pp. 293–8.

Smirnov, N. (1948) Table for estimating the goodness of fit of empirical distributions. *Ann. Math. Statist.*, **19**, 279–81.

Smith-Gill, S.J. (1983) Developmental plasticity: developmental conversion vs. phenotypic modulation. *Am. Zool.*, **23**, 47–55.

Solé-Cava, A.M. and Thorpe, J.P. (1986) Genetic differentiation between morphotypes of the marine sponge *Suberites ficus* (Demospongiae, Hadromerida). *Mar. Biol.*, **93**, 247–53.

Solé-Cava, A.M., Thorpe, J.P. and Manconi, R. (1991) A new Mediterranean species of *Axinella* detected by biochemical genetic methods, in *Fossil and Recent Sponges* (eds J. Reitner and H. Keupp), Springer-Verlag, Berlin, pp. 313–21.

Solé-Cava, A.M., Klautau, M., Boury-Esnault, N., Borojevic, R. and Thorpe, J.P. (1991) Genetic evidence for cryptic speciation in allopatric populations of two cosmopolitan species of the calcareous sponge genus *Clathrina*. *Mar. Biol.*, **111**, 381–6.

Stone, A.R. (1970) Seasonal variations of spicule size in *Hymeniacidon perleve*. *J. Mar. Biol. Assoc. UK*, **50**, 343–8.

1.4 WIDE-SCALE GENETIC DIFFERENTIATION AMONG PINK SHRIMP
PANDALUS BOREALIS POPULATIONS

Yuri P. Kartavtsev

Abstract

Genetic differentiation among 22 samples of the pink shrimp Pandalus borealis *was investigated at four to five polymorphic allozyme loci (*sMdh, Fdhg, Gpi, Pgm, Pep-LT-3*) from populations in the Sea of Japan, the Bering Sea, the Gulf of Alaska, the Barents Sea and the Gulf of Saint Lawrence. There was statistically highly significant (P < 0.001) heterogeneity of allelic frequencies among samples from different sea basins for all five loci that were compared. Populations were genetically homogeneous within any sea or single area basin.*

Clustering, performed with UPGMA, factor and discriminant analyses, shows clearly differentiated groups for different sea basins but increased genetic similarity in the northern part of the area. The reasons for that and small genetic differentiation among many marine invertebrate populations are discussed.

1.4.1 Introduction

For any commercial species, like the pink shrimp *Pandalus borealis* Kroyer, the identification of multiple stocks, determination of the numbers and distributions of distinct stocks, and evaluation of levels of between-stock genetic differentiation are among the most important questions to investigate. Such information can be critical for the rational management of this resource. Despite the rich literature on the biology of the pink shrimp (e.g. Butler, 1964; Kuznetsov, 1964, Ivanov, 1964; Shumway *et al.*, 1985), little is known regarding the population genetic structure of this species.

As has been pointed out (Kartavtsev, Berenboim and Zgurovsky, 1991; Kartavtsev, Zgurovsky and Fedina, 1992a), the pink shrimp population structure from the Barents Sea has attracted considerable study. Both ecological and morphological differentiation of pink shrimp cohorts in the Barents Sea have been investigated (Bryazgin, 1970; Bryazgin and Rusanova, 1974; Berenboim, 1978, 1982; Teigsmark, 1983) along with larval dispersal (Lysyi, 1988). Recently, similar data sets have appeared for some regions of the Atlantic Ocean (Hopkins and Nilssen, 1990; Skuladottir, 1990; Savard and Parsons, 1990), and for the Far Eastern seas (Kartavtsev, Zgurovsky and Fedina, 1992b). Until recently the pink shrimp has not been the subject of population genetic research (Kartavtsev, Zgurovsky and Fedina, 1990, 1992a; Kartavtsev, Berenboim and Zgurovsky, 1991).

The main purpose of this section is to compare intra- and interpopulation genetic differentiation of the pink shrimp over the whole area studied.

1.4.2 Materials and methods

The shrimp were obtained by bottom trawling in the years from 1987 to 1991. Four samples were collected from the Sea of Japan, JS5–JS8 (Kartavtsev, Zgurovsky and Fedina, 1992a), six samples from the Bering Sea, BS1–BS4 (Kartavtsev, Zgurovsky and Fedina, 1992a) and BS16 and BS17 (Kartavtsev, Berenboim and Zgurovsky, 1991), six samples from the Gulf of Alaska, AK20–AK25 (Kartavtsev *et al.*, 1992), and five samples from the Barents Sea,

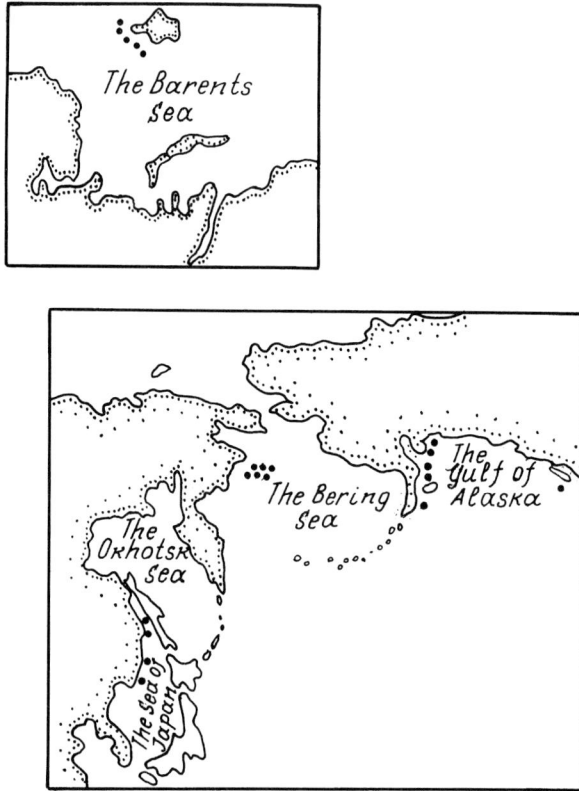

Figure 1.4.1 Location of the pink shrimp *Pandalus borealis* collections analysed in this study. Dots show sample sites. Collection from the Gulf of Saint Lawrence is not shown.

BR13–BR15 (Kartavtsev, Berenboim and Zgurovsky, 1991) and BR18 and BR19 (Kartavtsev *et al.*, 1992), which were all located near Spitzbergen Island. One sample was taken in the Gulf of Saint Lawrence in the Atlantic basin in 1991 (collection provided by Dr J.-M. Sevigny). Sample locations are shown in Figure 1.4.1. The total number of shrimp sampled in the 22 collections was 1976.

Electrophoresis of enzymes was carried out on starch gels (14–15%). Five enzymes were investigated: (1) malate dehydrogenase (sMDH, EC 1.1.1.37, locus *sMdh*); (2) formaldehyde dehydrogenase (FDHG, EC 1.2.1.1, *Fdh*); (3) glucose 6-phosphate isomerase (GPI, EC 5.3.1.9, *Gpi*); (4) phosphoglucomutase (PGM, EC 2.7.5.1, *Pgm*); (5) peptidase-leucyl-tyrosine (PEP-LT-3, EC —, *Pep-LT-3*). Details of electrophoresis and staining techniques are given elsewhere (Kartavtsev, Berenboim and Zgurovsky, 1991; Kartavtsev, Zgurovsky and Fedina, 1992a).

Statistical analyses were performed using BMDP (Dixon, 1977), NTSYS-pc (Rolf, 1988) and SPECSTAT (Kartavtsev and Soloviev, 1992). Analysis of the amount of among-sample relative differences and of the contributions of separate loci to the general picture of differentiation was based on the principal component and discriminant function methods (Afify and Azen, 1982), using both BMDP and NTSYS-pc. As major measurements of genetic similarity–dissimilarity, Nei's unbiased minimal distance D_m (Nei, 1978) and the similarity R_c metric (Zhivotovsky *et al.* 1989) were used.

1.4.3 Results and discussion

Genotypic and allelic frequencies were similar in males and females of the pink shrimp (Kartavtsev, Berenboim and Zgurovsky, 1991; Kartavtsev, Zgurovsky and Fedina, 1992a; Kartavtsev *et al.*, 1992). The absence of male–female differences allows us to pool sexes within samples for further genetic analysis. Individual collections exhibited no deviations from Hardy–Weinberg equilibrium genotypic frequencies among any of the loci studied (Kartavtsev, Berenboim and Zgurovsky, 1991; Kartavtsev, Zgurovsky and Fedina, 1992a; Kartavtsev *et al.*, 1992).

The allele frequency distributions in the area studied have one common theme: within single sea basins, genetic homogeneity is observed (with one exception for the *Fdhg* in the BR area that diminished in pooled χ^2), while between basins allelic frequencies are statistically heterogeneous (Table 1.4.1, $p < 0.001$). Similarly allelic frequencies are homogeneous at all five loci for the Gulf of Alaska samples investigated (Table 1.4.1) but

Table 1.4.1 The heterogeneity of frequencies of the most common alleles at five loci among the different sea-basin populations of the pink shrimp *Pandalus borealis*

Locus	Population unit	N	Mean frequency	χ^2	d.f.	Significance (p <)
sMdh	JS	440	0.873	5.89	5	0.5
	BS	469	0.579	9.58	5	0.1
	BR	396	0.496	1.90	4	0.5
	AK	596	0.997	4.80	5	0.1
	Total	1976	0.749	913.26	21	0.001
Fdhg	JS	307	0.860	6.35	4	0.5
	BS	457	0.980	9.97	5	0.1
	BR	396	0.980	9.91	4	0.05
	AK	596	0.972	3.91	5	0.5
	Total	1976	0.954	224.09	21	0.001
Gpi	JS	457	0.934	1.82	5	0.9
	BS	469	0.998	7.48	5	0.5
	BR	406	0.999	3.12	4	0.5
	AK	596	0.993	5.82	5	0.1
	Total	1976	0.984	174.56	21	0.001
Pgm	JS	452	0.706	4.01	5	0.5
	BS	482	0.917	2.56	5	0.5
	BR	396	0.995	4.15	4	0.1
	AK	596	0.930	6.06	5	0.1
	Total	1976	0.899	499.17	21	0.001
Pep-LT-3	BR	119	0.622	0.01	1	0.9
	AK	266	0.348	5.52	3	0.1
	Total	534	0.448	57.89	6	0.001
Total for loci sMdh, Fdhg, Gpi, Pgm						
	JS	–	–	18.07	19	0.9
	BS	–	–	29.59	20	0.1
	BR	–	–	19.08	16	0.1
	AK	–	–	21.04	20	0.1
	Total for all samples	–	–	1811.08	75	0.001

Chi-square (χ^2) comprises the results of the Workman–Niswander (1970) test for heterogeneity of most frequent alleles; d.f. = number of degrees of freedom; N = number of animals in the sample. Totals include data from the Gulf of Saint Lawrence.
JS = Sea of Japan; BS = Bering Sea; AK = Gulf of Alaska; BR = Barents Sea.

are significantly different at *sMdh* and over all four loci remaining, when compared to those in the geographically close Bering Sea samples: $\chi^2 = 608.47$, d.f. $= 11$, and $\chi^2 = 645.88$, d.f. $= 44$, $p < 0.001$ (Kartavtsev *et al.*, 1992).

There is no doubt about the heterogeneity of allelic frequencies for the whole area investigated among 22 samples: *sMdh*, $\chi^2 = 913.26$; *Fdhg*, $\chi^2 = 224.09$; *Gpi*, $\chi^2 = 174.56$; *Pgm*, $\chi^2 = 499.17$ (d.f. $= 21$, $p < 0.001$; Table 1.4.1). Similarly, at the *Pep-LT-3* locus, population heterogeneity was observed for the pooled sample set from the Gulf of Alaska, the Barents Sea and the Gulf of Saint Lawrence, $\chi^2 = 57.89$ (d.f. $= 6$, $p < 0.001$; Table 1.4.1), and in pairwise comparisons (Kartavtsev, Berenboim and Zgurovsky, 1991; Kartavtsev *et al.*, 1992). Cumulative χ^2 values for all loci show the same tendency (Table 1.4.1). These results are presented in Figure 1.4.2.

Relative genetic differences among populations can be quantified by cluster or factor analyses. The amount of relative genetic differentiation within and between the areas studied (Table 1.4.2) led us to expect good clustering with either method. The bivariate plot of the

Table 1.4.2 The estimations of population genetic differentiation in the pink shrimp *Pandalus borealis* from different regions

Locus	Population unit	D'_{st}	F'_{st}	$D_m \pm SE$	n	k
sMdh	JS	0.0009	0.0042	-0.0000 ± 0.0033	15	6
	BS	0.0031	0.0065	0.0029 ± 0.0095	15	6
	BR	0.0014	0.0029	-0.0022 ± 0.0057	10	5
	AK	0.0000	0.0049	-0.0000 ± 0.0000	15	6
	Total	0.0899	0.2254	0.0877 ± 0.0268	231	22
Fdhg	JS	0.0019	0.0084	0.0024 ± 0.0036	10	5
	BS	0.0002	0.0066	0.0002 ± 0.0000	15	6
	BR	0.0005	0.0121	0.0002 ± 0.0000	10	5
	AK	0.0002	0.0036	-0.0001 ± 0.0000	15	6
	Total	0.0028	0.0373	0.0023 ± 0.0018	231	22
Gpi	JS	0.0001	0.0012	-0.0008 ± 0.0001	15	6
	BS	0.0000	0.0050	0.0000 ± 0.0000	15	6
	BR	0.0000	0.0059	-0.0000 ± 0.0000	10	5
	AK	0.0001	0.0056	0.0000 ± 0.0000	15	6
	Total	0.0012	0.0389	0.0010 ± 0.0011	231	22
Pgm	JS	0.0016	0.0037	-0.0005 ± 0.0076	15	6
	BS	0.0004	0.0027	-0.0001 ± 0.0002	15	6
	BR	0.0001	0.0077	0.0000 ± 0.0000	10	5
	AK	0.0013	0.0097	0.0008 ± 0.0000	15	6
	Total	0.0260	0.1417	0.0248 ± 0.0143	231	22
Average for *sMdh*, *Fdhg*, *Gpi*, *Pgm*	JS	0.0010	0.0040	0.0000 ± 0.0037	55	–
	BS	0.0009	0.0052	0.0007 ± 0.0024	60	–
	BR	0.0005	0.0069	-0.0005 ± 0.0014	40	–
	AK	0.0004	0.0059	0.0000 ± 0.0000	60	–
	Within seas	0.0009	0.0044	0.0003 ± 0.0028	127	–
	Among seas	0.0092	0.0331	0.0168 ± 0.0093	533	–
	Total	0.0299	0.1108	0.0289 ± 0.0076	924	–

D'_{st} and F'_{st} are the measures of genetic differentiation among populations (Nei, 1987); $D_m =$ the unbiased minimal genetic distance (Nei, 1978); $n =$ the number of pairwise comparisons; $k =$ the number of samples (subpopulations). The standard errors (SE) are calculated from the variance matrices. For other abbreviations see Table 1.4.1.

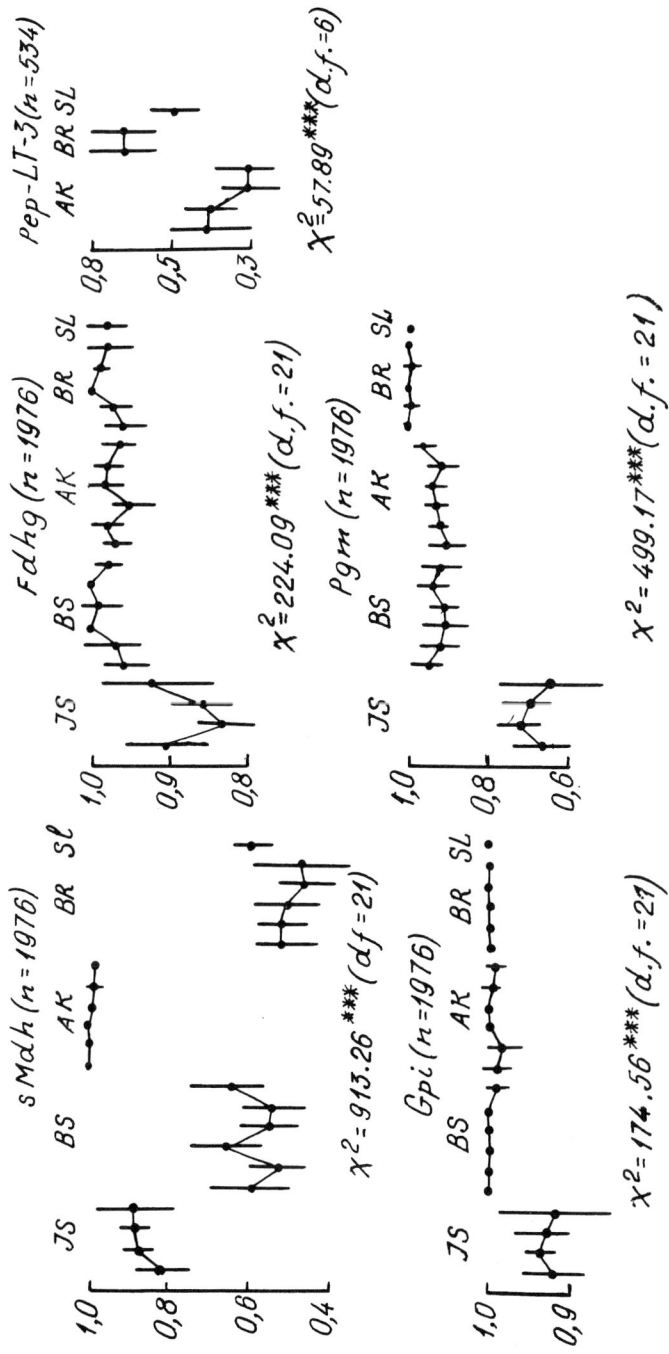

Figure 1.4.2 Allele frequency variability at five allozyme loci ($sMdh$, $Fdhg$, Gpi, Pgm, $Pep-LT-3$) among samples of the pink shrimp *Pandalus borealis* from the Sea of Japan (JS), the Bering Sea (BS), the Gulf of Alaska (AK), the Barents Sea (BR) and the Gulf of Saint Lawrence (SL). Allele frequencies are on ordinates. ***$p < 0.001$.

first two principal components (PC) for the original data matrix of genetic similarity R_c at four loci (*sMdh*, *Fdhg*, *Gpi*, *Pgm*, excluding *Pep-LT-3*) for 22 samples illustrates this result (Figure 1.4.3). Sharp differentiation among the pink shrimp populations from the Gulf of Alaska, from the Bering Sea and from the Barents Sea is seen at PC1 and less at PC2; on the other hand, the AK–BS–BR group, which is not greatly differentiated by PC2 scores, is clearly differentiated from the Sea of Japan population at PC2 (Figure 1.4.3). The sample from the Gulf of Saint Lawrence is very close to the BS and BR clusters. As has been shown before (Kartavtsev, Berenboim and Zgurovsky, 1991; Kartavtsev, Zgurovsky and Fedina, 1992a), no differences in allelic frequencies among the loci *sMdh*, *Fdhg*, *Gpi* and *Pgm* were evident when comparing samples from within a sea basin across different years. This characteristic of allele frequency variability together with strict interbasin differences (Figures 1.4.2 and 1.4.3) leads to biologically meaningful clustering of the samples. UPGMA on minimal distances (Nei, 1978) gave very similar clustering of the 22 populations (Kartavtsev *et al.*, 1992).

Similar results were obtained for 21 samples, excluding the Gulf of Saint Lawrence sample, by a factor analysis performed directly on the allele frequency data at four loci, as above, together with some 'environmental' variables like latitude, depth and sea basin (Kartavtsev *et al.*, 1992) with BMDP. A three-dimensional plot of the first three principal components is presented (Figure 1.4.4). The BMDP factor analysis version, unlike NTSYS,

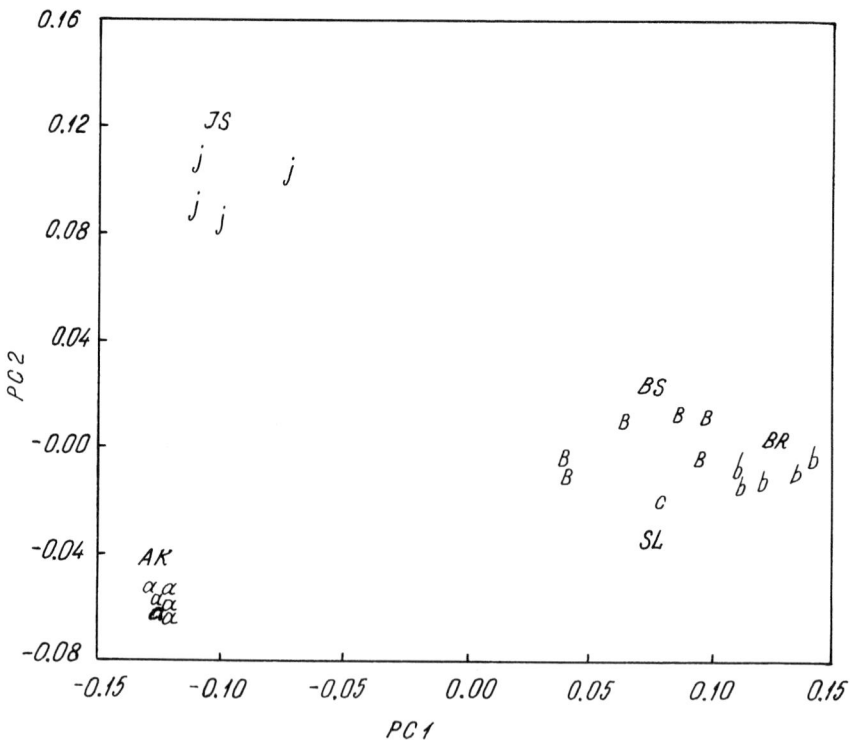

Figure 1.4.3 Bivariate plot of the first two principal components (PC) of the R_c matrix for four polymorphic allozyme loci (*sMdh*, *Fdhg*, *Gpi*, *Pgm*) of the pink shrimp *Pandalus borealis* from the Sea of Japan (JS), the Bering Sea (BS), the Gulf of Alaska (AK), the Barents Sea (BR) and the Gulf of Saint Lawrence (SL).

Table 1.4.3 Matrix of sample classification of the pink shrimp *Pandalus borealis* for three northern basins based on discriminant analysis of allele frequencies at four allozyme loci

Group	Percentage	Number of samples classified for group		
		BS	BR	AK
BS	100.0	6	0	0
BR	100.0	0	5	0
AK	100.0	0	0	6
Total	100.0	6	5	6

For abbreviations, see Table 1.4.1.

provides information on the contribution of variables to the PCs. As an example, we outline here only the contribution to PC1. The largest contributions to PC1, which again clearly differentiates the JS from AK–BS–BR, have the variables Pgm^4 (superscript denotes alleles), Gpi^1, Gpi^2, Pgm^3, $Fdhg^3$, $Fdhg^1$, latitude, $Fdhg^2$, and sea. The coefficients of correlation of these variables with PC1, which ranged proportionally according to the decrease of contribution of the variables to PC, were, respectively, –0.979, –0.966, 0.962, –0.834, 0.829, 0.788, 0.741 ($n = 21$, $p < 0.0005$). Comparison of Figures 1.4.3 and 1.4.4 also illustrates the increased power of the R_c metric in comparison with the euclidean metric to the BS and BR

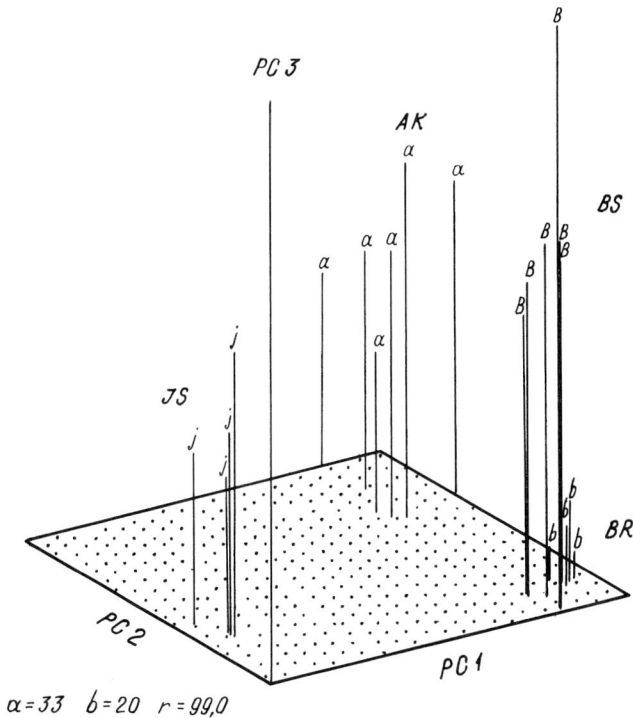

Figure 1.4.4 Three-dimensional plot of principal component (PC) scores for 21 samples of *Pandalus borealis* based on the allele frequency data for four loci and four environmental variables. On the axes: x = PC1, y = PC2, z = PC3. a, b = values of angle of rotation around z and x axes; r = relative distance from the operator or size of the figure. For other abbreviations, see Figure 1.4.3.

samples, which are distinguishable at PC1 in Figure 1.4.3 but are not differentiated in Figure 1.4.4.

All studied populations of the separate sea basins are unmistakably (100% accuracy) distinguishable from each other on the basis of cumulative genetic differences, with 100% correct assignment of population samples to their native basin in discriminant analysis. This is true even for the least differentiated shrimp cohorts from the BS, AK and BR regions (Table 1.4.3).

The results obtained here are in general accord with data on the biology of the species (Butler, 1964; Kuznetsov, 1964; Shumway *et al.*, 1985) and with the published data on population structure (Berenboim, 1978, 1982; Teigsmark, 1983; Lysyi, 1988; Kartavtsev, Berenboim and Zgurovsky, 1991; Kartavtsev, Zgurovsky and Fedina, 1992a). Dispersal of the planktonic larvae of the pink shrimp, which inhabit pelagic waters for 3–4 months (Barr, 1970; Shumway *et al.*, 1985), presumably allows massive migration exchange among cohorts over large distances (Berenboim, 1982; Shumway *et al.*, 1985; Lysyi, 1988). As a result, reasonably extensive gene flow probably occurs among local populations of the pink shrimp, which are naturally united by the circulation of currents within a gulf or sea basin. The genetic consequence of such a process is the formation of a large united gene pool and allele frequency homogeneity in the absence of differentiating natural selection.

Data on the homogeneity of allelic frequencies within the borders of large gulfs and sea basins contrast with the heterogeneity among different gulfs and seas both in pairwise comparisons and in pooled samples (Table 1.4.1; Kartavtsev, Berenboim and Zgurovsky, 1991; Kartavtsev, Zgurovsky and Fedina, 1992a; Kartavtsev *et al.*, 1992) and implies that the separate large gulfs as well as the separate seas are inhabited by different reproductive stocks. These stocks are genetically equivalent to Mendelian populations because of the unity of their gene pools (Table 1.4.1; Kartavtsev, Berenboim and Zgurovsky, 1991; Kartavtsev, Zgurovsky and Fedina, 1992a; Kartavtsev *et al.*, 1992), the Hardy–Weinberg equilibrium holding within them (Kartavtsev, Berenboim and Zgurovsky, 1991; Kartavtsev, Zgurovsky and Fedina, 1992a), and the temporal stability of allele frequencies (Kartavtsev, Berenboim and Zgurovsky, 1991; Kartavtsev, Zgurovsky and Fedina, 1992a). This conclusion certainly does not mean that in some other parts of the species area population structure will always be the same, but, given similar oceanic conditions, it is likely to be similar in most cases.

The duration of the planktonic larval stages should be maximal under conditions of low surface water temperatures in the northern part of the species range. Thus, 'northern' populations might have greater gene flow and consequently increased gene pool integrity in comparison with 'southern' populations. Furthermore, this may explain why the BS, BR and AK populations tend to be less differentiated among themselves than they are from the population in the Sea of Japan. Similarly, perhaps, the krill populations in two Antarctic seas show no differences in allelic frequencies at seven polymorphic loci (Schneppenheim and MacDonald, 1984). In both cases an alternative explanation based on the possible common adaptive nature of genetic similarity of 'specialists' at high latitudes is plausible (Valentine 1976). Support for the alternative hypothesis was presented for a narrower taxonomic group, a pair of krill species (Fevolden and Ayala, 1981). Data on the association of shrimp differentiation with latitude, as given above, are consistent with such an interpretation.

In summarizing our previous results we may conclude that: (1) there is statistically significant allele frequency heterogeneity at all studied polymorphic allozyme loci over the studied range; and (2) there is clear grouping of the samples from single sea basins into 'native' genetically homogeneous clusters and several local stocks therefore exist within the limits of the species range. I suggest the existence in the northern and northwestern Pacific of three or four local stocks of the pink shrimp, inhabiting the Sea of Japan, the Okhotsk Sea, the Bering Sea, the Gulf of Alaska and nearby waters. Values of F'_{st}, D'_{st} and D_m (Table 1.4.2) show the increase of genetic differentiation in the order: single local stock; different local stocks. The statistics used reveal only a relatively small amount of genetic divergence even for very distant populations of the pink shrimp, e.g. for shrimp from the Sea of Japan and from the Barents Sea, which contribute little to the total differentiation (Table 1.4.2). Similar results have been obtained for other groups of crustaceans, including prawns and lobsters (Hedgecock, Tracey and Nelson, 1982; Shaklee and Samollow, 1984; Chow and Fujio, 1985; Chow, Fujio and Nomura, 1985), and krill (Fevolden and Ayala, 1981; Schneppenheim and MacDonald, 1984).

These results can be partially plotted using as a measure of interpopulation differentiation the values of F_{st} (= G'_{st} or, better, F'_{st} (Nei, 1987)) (Figure 1.4.5). Obviously, as outlined by different authors (Hedgecock and Nelson, 1981; Hedgecock, Tracey and Nelson, 1982; Chow and Fujio, 1987), this plot demonstrates that living conditions or biological peculiarities or both can influence the level of genetic differentiation (the significance of at least some differences can be seen with the help of confidence intervals, 95%). It is obvious that freshwater forms are most differentiated (Figure 1.4.5). At the same time, however, a question remains whether the reason is connected with gene flow and population size decrease or with diversifying natural selection (Kartavtsev *et al.*, 1992).

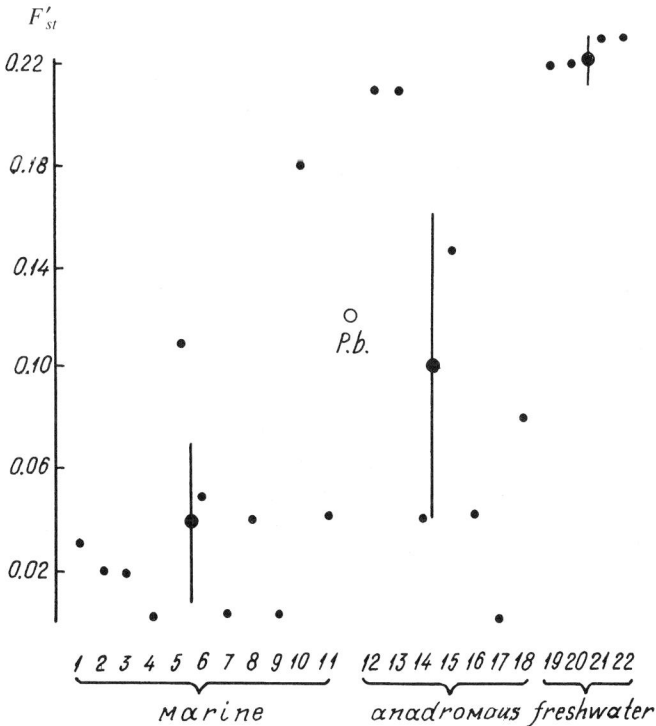

Figure 1.4.5 Plot of F'_{st} distribution among the higher crustaceans for three groups: marine (M), anadromous (A), and freshwater (F). The circle and the letters P.b. show the point for the pink shrimp. Dots with vertical lines show mean F'_{st} values with confidence intervals (95%) for the three groups M, A, F. Literature data, from Chow and Fujio (1987), were recalculated in the scale F'_{st} by the author.

The discreteness of local stocks as population units cannot be absolute and depends upon the stability of existing barriers to gene flow, and the stability of the adaptive population landscape. The value of gene flow or exchange of individuals (Nm) among populations of pink shrimp may be evaluated from F_{st} as $Nm = 5$ (Kartavtsev, Berenboim and Zgurovsky, 1991). If one takes into account the fact that generations overlap in pink shrimp, real exchange between some populations may be four to five or even six times less (depending on the number of generations sampled) and may not exceed one individual per generation. Under the approach suggested earlier (Slatkin 1985) and based on mean F'_{st} as the estimate of between-generation gene frequency variability for our four main loci (Table 1.4.2), we find that the average exchange rate for the 21 Pacific cohorts sampled (Nm) = 1.9/4 (assuming that at least four generations were considered), which is less than one individual per generation. For the local stocks from the Gulf of Alaska and the Bering Sea, the exchange rate could reach $Nm = 3.0/4 = 0.75$ (Kartavtsev *et al.*, 1992), which again is quite close to the previous estimates.

The lower heterozygosity of the Barents Sea shrimp (Kartavtsev, Berenboim and Zgurovsky, 1991) and decreased heterozygosity of both the Bering Sea and the Gulf of Alaska shrimp combined with their low genetic differentiation (see discussion above and data from Table 1.4.1 and Figures 1.4.3 and 1.4.4) agree both with Ayala and Valentine's (Valentine, 1976) adaptive specialist concept and with increased gene flow or a possible bottleneck homogenizing effect on populations in the northern area. Thus, more data on both additional gene markers and new south–north area coverage are needed.

Acknowledgements

I am very grateful to many people who support the pink shrimp investigation both conceptually, in data collecting and in discussion of the results. My special thanks go to Drs Konstantin Zgurovsky, Boris Berenboim, Douglas Eggers, Anthony Gharrett, James Shaklee and Jean-Marie Sevigny. I am grateful as well for partial financial support of the research from Russia state program 'Priority Directions in Genetics'.

References

Afify, A. and Azen, S. (1982) *Statistical Analysis. Computer Approach*, Mir Publ., Moscow (in Russian).

Barr, L. (1970) *Alaska Fishery Resources: the Shrimps. Fishery Leaflet 631* US Fish. Widl. Serv., Washington DC, pp. 1–10.

Berenboim, B.I. (1978) On the population differences in the shrimp *Pandalus borealis* from the Barents Sea. *Gidrobiol. Zh.*, **14**, 44–7 (in Russian).

Berenboim, B.I. (1982) Reproduction of the populations of the shrimp *Pandalus borealis* in the Barents Sea. *Okeanol.* **22**, 118–24 (in Russian).

Bryazgin, V.F. (1970) On the distribution and biology of the shrimp *Pandalus borealis* Kr. in open regions of the Barents Sea. *The Results of Economical Investigations of Fisheries in the Northern Basin. Murmansk*, **16** (2), 93–108 (in Russian).

Bryazgin, V.F. and Rusanova, M.N. (1974) Distribution patterns and population variability of *Pandalus borealis* Kr. in open regions of North-Eastern Atlantic, in *Hydrobiology and Biogeography of the Shelf of the Cold and Temperate Waters of World Oceans* (ed. A.N. Golikov), Nauka Publ., Leningrad, pp. 88–9.

Butler, T.H. (1964) Growth, reproduction and distribution of Pandalid shrimps in British Columbia. *J. Fish. Res. Bd. Can.*, **21**, 1403–52.

Chow, S. and Fujii, Y. (1985) Population genetics of the Palemonid shrimps (Decapoda: Crustacea). 1. Genetic variability and differentiation of local populations. *Tohoku J. Agr. Res.*, **36**, 93–108.

Chow, S. and Fujio, Y. (1987) Comparison of intraspecific genetic diversity levels among local populations in Decapod Crustacean species; with some references of genotypic diversity. *Nippon Suisan Gakkai.*, **53**, 5, 691–3.

Chow, S., Fujio, Y. and Nomura, T. (1985) Reproductive isolation and distinct population structure in two types of freshwater shrimp *Palaemon paucidens. Evolution*, **42**, 804–13.

Dixon, W.E. (ed.) (1982) *BMDP: Biomedical Computer Programs*, University of California Press, Los Angeles.

Fevolden, S.E. and Ayala, F.J. (1981) Enzyme polymorphism in Antarctic krill (Euphausiacea): genetic variation between populations and species. *Sarsia*, **66**, 167–81.

Hedgecock, D. and Nelson, K. (1981) Gene enzyme variation and adaptive strategies in the Crustacea, in *Genetics and Reproduction of Marine Organisms* (eds V. Kasyanov and A. Pudovkin), Publ. Far East Science Center, Vladivostok, pp. 105–29.

Hedgecock, D., Tracey, M. and Nelson, K. (1982) Genetics, in *The Biology of Crustacea* (ed. L.G. Abele), Academic Press, New York, pp. 283–403.

Hopkins, C.C.E. and Nilssen, E.M. (1990) Population parameters and life histories of the deep water prawn *Pandalus borealis* from different regions, *Abstr. Sci. Papers Posters, Shellfish Life History and Shellfishery Models Symposium*, Moncton, Canada, ICES, Palaegade 2-4, DK-1261, Copenhagen K, 85.

Ivanov, B.G. (1964) Results on the investigation of biology and distribution of shrimps in Pribilov Island region and in the Bering Sea. *VNIRO Proc.*, 49, 11–12.

Kartavtsev, Y.P., Berenboim, B.I. and Zgurovsky, K.A. (1991) Population genetic differentiation of the pink shrimp *Pandalus borealis* from the Barents and Bering seas. *J. Shellfish Res.* **10**, 333–9.

Kartavtsev, Y.P. and Soloviev, A.A. (1992) Programming mini complex SPECSTAT for the statistical analysis of allozyme population genetic data. *Genetica 28*, 194–7 (in Russian; package record is provided in English including description).

Kartavtsev, Y.P., Zgurovsky, K.A. and Fedina, Z.M. (1990) Analysis of the spacial structure of the pink shrimp (*Pandalus borealis*) from the Far Eastern seas using methods of population genetics and fenetics, *Abstr. Sci. Papers Posters, Shellfish Life History and Shellfishery Models Symposium*, Moncton, Canada, ICES, Palaegade 2-4, DK-1261, Copenhagen K, 61.

Kartavtsev, Y.P., Zgurovsky, K.A. and Fedina, Z.M. (1992a) Allozyme variability and differentiation of the pink shrimp *Pandalus borealis* from the Far-Eastern seas. *Genetica*, **28**, 110–22 (in Russian).

Kartavtsev, Y.P., Zgurovsky, K.A. and Fedina, Z.M. (1992b) Morphological variability of the pink shrimp *Pandalus borealis* from the Far Eastern seas and its relation to population structure of the species. *Russ. J. Mar. Biol.* (in Russian), **3–4**, 53–61.

Kartavtsev, Y P, Sitnikov, A.V., Komissarov, P.I. and Eggers, D. (1992) Allozyme variability and differentiation of the pink shrimp *Pandalus borealis* Kroyer from the Gulf of Alaska and the Bering Sea. *Genetica* (in Russian), **28**, 58–72.

Kuznetsov, V.V. (1964) *Biology of Mass and Common Crustacean Species in the Barents and White Seas*, Nauka Publ., Moscow–Leningrad (in Russian).

Lysyi, A. Yu. (1988) Larval ecology and population structure of the shrimp *Pandalus borealis* in the seas of Northern Europe. *Apatity, Kolsky Branch USSR Acad. Sci.*, **68**.

Nei, M. (1978) Estimation of average heterozygosity and genetic distance from a small number of individuals. *Genetics*, **89**, 583–90.

Nei, M. (1987) *Molecular Evolutionary Genetics*, Columbia University Press, New York.

Rolf, F.J. (1988) *NTSYS-pc: Numerical Taxonomy and Multivariate Analysis System*, Exeter Publ. Ltd, New York.

Savard, L. and Parsons, D.G. (1990) Geographic variation in the life history of *Pandalus borealis* from the Northwest Atlantic. *Abstr. Sci. Papers Posters, Shellfish Life History and Shellfishery Models Symposium*, Moncton, Canada, ICES, Palaegade 2-4, DK-1261, Copenhagen K, p. 23

Schneppenheim, R. and MacDonald, C.M. (1984) Genetic variation and population structure of krill (*Euphausia superba*) in the Atlantic sector of Antarctic waters and off the Antarctic Peninsula. *Polar Biol.*, **3**, 19–28.

Shaklee, J.B. and Samollow, P.B. (1984) Genetic variation and population structure in deep water snapper, *Pristipomoides filamentosus*, in the Hawaiian Archipelago. *Fish. Bull.*, **82**, 703–13.

Shumway, S.E., Perkins, H.C., Schick, D.F. and Stikney, A.P. (1985) Synopsis of biological data on the pink shrimp, *Pandalus borealis* Kroyer 1838. *FAO Fish. Synopsis*, **144**, 1–157.

Skuladottir, U. (1990) Defining stocks of *Pandalus borealis* off Northern Iceland using the maximum length and maturity of females as a measure, in *Abstr. Sci. Papers Posters, Shellfish Life History and Shellfishery Models Symposium*, Moncton, Canada, ICES, Palaegade 2-4, DK-1261, Copenhagen K, p. 75.

Slatkin, M. (1985) Gene flow in natural populations. *Annu. Rev. Ecol. Syst.*, **16**, 393–430.

Teigsmark, G. (1983) Populations of the deep-sea shrimp *Pandalus borealis* Kroyer in the Barents Sea. *Fisheridir. skr. Her. Havunders*, **17**, 377–430.

Valentine, J.W. (1976) Genetic strategies of adaptation, in *Molecular Evolution* (ed. F.J. Ayala), Sinauer Associates, Inc., Sunderland, Massachusetts, pp. 78–94.

Workman, P.Z. and Niswander, J.D. (1970) Population structure of south-western Indian tribes. II. Local genetic differentiation in Papago. *Am. J. Hum. Genet.*, 1, 24–9.

Zhivotovsky, L.A., Glubokovsky, M.K., Viktorovsky, R.M., *et al.* (1989) Genetic differentiation of the pink salmon. *Genetica*, **25**, 1261–74 (in Russian).

1.5 THE EVOLUTION OF GENETIC DIVERGENCE AT PROTEIN CODING LOCI AMONG ANADROMOUS AND NONANADROMOUS POPULATIONS OF ATLANTIC SALMON *SALMO SALAR*

Eric Verspoor

Abstract

*Interpopulation variation at 21 protein coding loci is compared between groups of anadromous and nonanadromous Atlantic salmon (*Salmo salar*) populations on the island of Newfoundland in eastern Canada. Levels of genetic divergence among nonanadromous populations, consistent with an absence of gene flow, is over 10 times higher than among anadromous populations, where limited gene flow is indicated. Overall, the pattern of allele frequency variation within and between the two population groups suggests divergence of the populations by genetic drift rather than natural selection. The pattern of genetic relatedness among the nonanadromous populations is consistent with their polyphyletic evolution during the current postglacial period. The genetic clustering of one nonanadromous population with European salmon is likely to be an artifact of genetic drift. However, the establishment of Atlantic salmon populations in Newfoundland from North American and European colonizers cannot be ruled out.*

1.5.1 Introduction

The Atlantic salmon *Salmo salar* L. has been the subject of considerable debate as regards the evolution of population variation in life-history type (e.g. Power, 1958; Behnke, 1972; Berg, 1985) as well as in frequencies of protein electromorphs (e.g. Payne, Child and Forrest, 1971; Verspoor, 1986; Verspoor and Jordan, 1989). Analyses of their covariation, which can provide insight into the evolutionary origins of population divergence, have been limited. The only studies carried out of the association of life-history and protein variation (Payne, 1974; Claytor and MacCrimmon, 1988) have been *ad hoc* and based on variation at only one or two loci.

Two basic population types exist in the Atlantic salmon with respect to life-history – anadromous and nonanadromous. In the former, egg and juvenile development occur in fresh water followed by migration to salt water for further growth and maturation; mature fish return to fresh water to spawn. In contrast, in nonanadromous populations, the whole life-cycle is completed in fresh water. The two population types can exist allopatrically (Scott and Crossman, 1964) and parapatrically (Vuorinen and Berg, 1989) as well as sympatrically (Verspoor and Cole, 1989).

Anadromous and nonanadromous populations were initially placed in two distinct sub-specific phyla, based largely on the life-history dichotomy itself – *S. salar salar* and *S. s. sebago*, respectively (Scott and Crossman, 1973). However, this phyletic dichotomy is now generally rejected based on meristic, morphometric and chromosomal data (Wilder, 1947;

Claytor and MacCrimmon, 1988). The current view is for a polyphyletic origin (Power, 1958; Behnke, 1972; Berg, 1985). Nonanadromous populations are believed to have evolved independently in many localities during the postglacial period from anadromous colonizers as a result of the development of physical barriers to migration to the sea such as impassable falls. This scenario is feasible given the existence in anadromous populations of life-history variation in both sexes in relation to anadromy (e.g. Lee and Power, 1976; O'Connell and Gibson, 1989).

A systematic assessment of genetic, rather than just phenotypic, divergence of anadromous and nonanadromous Atlantic salmon populations is lacking. So far, genetic studies have shown that major differences can occur between the two population types (Payne, 1974; Vuorinen and Berg, 1989; Palva, Lehvaslaiho and Palva, 1989), even in sympatry (Verspoor and Cole, 1989; Birt, Green and Davidson, 1991). The differences can be almost as large as those observed between North American and European populations. If the nonanadromous populations have evolved since the last glaciation, this raises the question as to how the differences evolved in such a short time given that continental divergence is likely to predate the last glaciation.

To gain insight into this question, an analysis is presented of the pattern of genetic divergence within and between groups of the two population types from the island of Newfoundland in eastern Canada. The analysis examines variation at loci coding for enzymatic proteins and assesses the relative importance of natural selection, gene flow, genetic drift and colonization history in population divergence.

1.5.2 Materials and methods

Atlantic salmon were sampled by electrofishing, netting or fixed traps from 12 anadromous and 12 nonanadromous populations in eastern Newfoundland in 1985 and 1986 (Table 1.5.1; Figure 1.5.1). Seven of the nonanadromous populations were lacustrine and five riverine. Sample sizes ranged from 20 to 107 and averaged 70 (SD = 22) and 55 (SD = 21) for the anadromous and nonanadromous groups, respectively.

Allozyme analysis of muscle and liver was carried out as previously described for 21 loci (20 structural and 1 regulatory – *AAT-1,2**, *AAT-3**, *CK-1**, *CK-2**, *GPI-1,2**, *GPI-3**, *IDDH-*

Table 1.5.1 Water bodies in Newfoundland with anadromous and nonanadromous populations of Atlantic salmon which were sampled and analysed for allozyme variation; numbers refer to Figure 1.5.1

Anadromous populations		Nonanadromous populations	
No.	*Water body*	*No.*	*Water body*
1	Indian River	13	Peter Stride Pond
2	Exploits River	14	Little Gull Lake
3	Campbellton River	15	Rodney Pond
4	Gander River	16	Big Triangle Pond
5	Ragged Harbour River	17	Fourth Pond
6	Middle Brook	18	Five Mile Pond West
7	NE Placentia River	19	Tors Cove Pond
8	SE Placentia River	20	North Aquaforte River
9	North Arm River	21	South Aquaforte River
10	Mobile River	22	Long Beach River
11	Biscay Bay River	23	Bristol Cove River
12	NE Trepassey River	24	St Shotts River

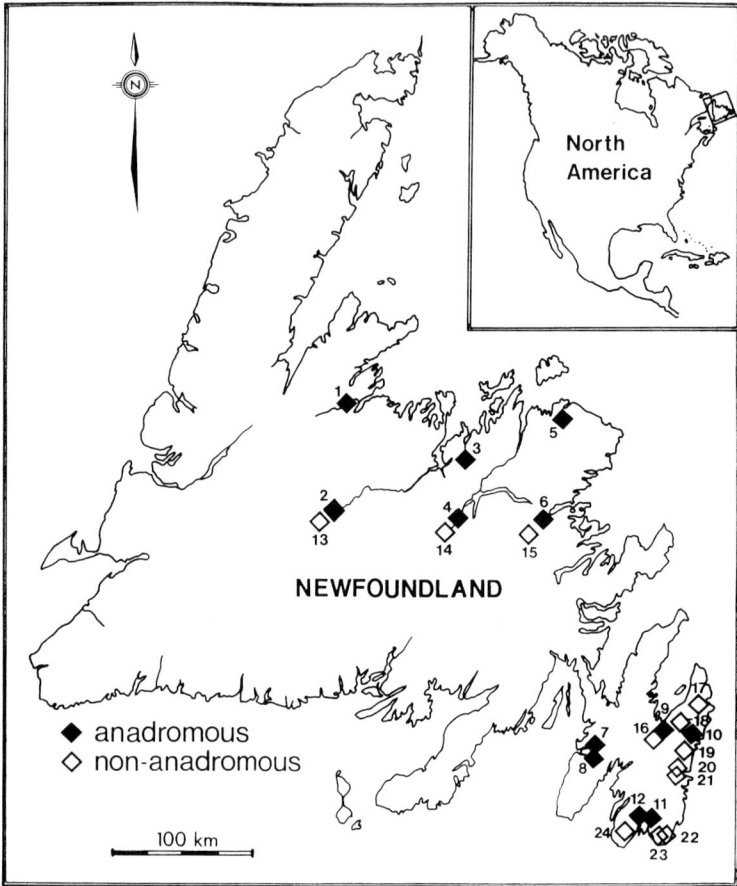

Figure 1.5.1 A map of Newfoundland showing the locations where samples of anadromous and nonanadromous Atlantic salmon were collected. Numbers refer to Table 1.5.1.

1, *IDDH-2*, *IDHP-3*, *LDH-4*, *MDH-1*, *MDH-2*, *MDH-3,4*, *MEP-2*, *PGM-1*, *PGM-2*, *SOD**, and *PGM-1r** (Verspoor, 1988a; Verspoor and Cole, 1989). Inheritance studies show variation at *AAT-3*, *IDDH-2*, *IDHP-3*, *MDH-3,4*, *MEP-2* and *PGM-1r* to be genetic (Johnson, 1984) and variants at all loci fit genetic models expected from the quaternary structure of their protein products. Family and population analyses show no linkage among the loci (Johnson, 1984; Verspoor, unpublished data). Allele and locus nomenclature follows Verspoor (1988a) and Verspoor, Fraser and Youngson (1991).

Data analyses were carried out using Biosys-1 Release 1.7 (Swofford and Selander, 1989) and Systat Version 4.0 (Wilkinson, 1988). Statistical methods are described in Sokal and Rohlf (1981). The relatedness of the Newfoundland populations to others in North America and Europe was tested using published and unpublished data for the rivers listed in Table 1.5.2. These span the species' range.

Table 1.5.2 List of river samples used to assess the genetic relatedness of the Newfoundland samples with North American and European salmon populations

North America	Ref.	Europe	Ref.
1 Saint John (New Brunswick)	Verspoor (1988a)	9 Sella (Spain)	Verspoor (unpublished)
2 Lahave (Nova Scotia)	Verspoor (1988a)	10 Elorn (France)	Verspoor (unpublished)
3 SW Miramichi (New Brunswick)	Verspoor (1988a)	11 Blackwater (Ireland)	Cross and Ward (1980)
4 York (Quebec)	Verspoor (unpublished)	12 Taw (England)	Verspoor (unpublished)
5 Baie Trinitie (Quebec)	Verspoor (unpublished)	13 Dee (Scotland)	Verspoor (unpublished)
6 Humber (Newfoundland – west)	Verspoor (unpublished)	14 Vididalsa (Iceland)	Verspoor (unpublished)
7 Eagle (Labrador – south)	Verspoor (unpublished)	15 Sokna (Norway–south)	Ståhl (1987)
8 Flowers (Labrador – north)	Verspoor (unpublished)	16 Namsen (Norway–central)	Vuorinen and Berg (1989)
		17 Alta (Norway – central)	Ståhl (1987)
		18 Torne (Sweden – Baltic)	Ståhl (1987)
		19 Lödge (Sweden – Baltic)	Ståhl (1987)
		20 Neva (Russia – Baltic)	Koljonen (1991)

1.5.3 Results

Departures of genotype proportions from Castle–Hardy–Weinberg equilibrium or linkage disequilibrium were present only in the Little Gull Lake sample. The deficiencies and disequilibrium observed at this location are attributable to the mixing of two genetically distinct sympatric populations, one anadromous and one nonanadromous. These populations have been described previously by Verspoor and Cole (1989). The data for these populations are taken from that analysis.

Comparing population types, the anadromous samples average more alleles per locus (A) and have more loci polymorphic (P) (Tables 1.5.3 and 1.5.4) than nonanadromous samples. However, excepting the Bristol Cove River, all loci in all populations are diallelic, so these measures are essentially the same. Th differences between the two groups are significant. (t-test, d.f. = 22. A: $t = 2.16$, $p = 0.042$. P: $t = 2.21$, $p = 0.038$). Eleven of 12 nonanadromous populations have one or more loci fixed for an allele, versus 6 of 12 for the anadromous group (Table 1.5.3). The variances of A and P in the two groups do not differ nor do their mean expected heterozygosities (H) (Mann-Whitney U-test: $U = 54.5$, $p = 0.312$). The variance of H is, however, significantly higher among the nonanadromous samples (Bartlett's test: $\chi^2 = 5.03$, $p = 0.025$). Neither A, P nor H are significantly correlated with sample size. (A: $r = 0.259$, P: $r = 0.213$, H: $r = -0.108$. $df = 22$)

Among the nonanadromous population samples, P shows a clear positive association with the degree of isolation of the population (as measured by distance from the sea or the presence of barriers to anadromous migration – Figure 1.5.2). An association also exists with H but this is weaker due to a high level of variation in this parameter among those populations which are 'landlocked', i.e. isolated above impassable falls.

Table 1.5.3 Allele frequency data for samples from Newfoundland populations of anadromous and nonanadromous Atlantic salmon (population numbers refer to Table 1.5.1; numbers of samples (*N*) in parentheses)

Locus						Population						
	1 (107)	2 (81)	3 (74)	4 (24)	5 (62)	6 (72)	7 (94)	8 (81)	9 (51)	10 (52)	11 (68)	12 (59)
*PGM-1r**												
a	0.387	0.333	0.403	0.418	0.336	0.373	0.399	0.471	0.313	0.340	0.437	0.451
b	0.613	0.667	0.597	0.582	0.664	0.627	0.601	0.529	0.687	0.660	0.563	0.549
*IDDH-2**												
100	0.025	0.043	0.054	0.110	0.031	0.042	0.000	0.119	0.127	0.000	0.029	0.000
−72	0.975	0.957	0.946	0.890	0.969	0.958	1.000	0.881	0.873	1.000	0.971	1.000
*MDH-3, 4**												
75/80	0.000	0.000	0.000	0.000	0.000	0.000	0.000	0.000	0.000	0.000	0.000	0.000
100	0.519	0.518	0.466	0.330	0.492	0.417	0.253	0.625	0.441	0.510	0.664	0.610
120	0.481	0.482	0.534	0.670	0.508	0.583	0.747	0.375	0.559	0.490	0.336	0.390
*AAT-3**												
100	0.196	0.102	0.108	0.213	0.094	0.042	0.016	0.187	0.147	0.221	0.329	0.000
50	0.804	0.898	0.892	0.787	0.906	0.958	0.984	0.813	0.853	0.779	0.671	1.000
*MEP-2**												
100	0.075	0.030	0.041	0.052	0.000	0.007	0.000	0.000	0.010	0.000	0.000	0.000
125	0.925	0.970	0.959	0.948	1.000	0.993	1.000	1.000	0.990	1.000	1.000	1.000
*GPI-1,2 **												
100	0.981	1.000	1.000	1.000	1.000	1.000	1.000	1.000	1.000	1.000	1.000	1.000
140	0.019	0.000	0.000	0.000	0.000	0.000	0.000	0.000	0.000	0.000	0.000	0.000
GPI-3												
100	1.000	1.000	0.986	1.000	0.984	1.000	1.000	1.000	1.000	1.000	1.000	1.000
90	0.000	0.000	0.014	0.000	0.016	0.000	0.000	0.000	0.000	0.000	0.000	0.000
*MDH-1**												
100	1.000	1.000	1.000	1.000	1.000	1.000	1.000	1.000	1.000	1.000	0.986	1.000
−50	0.000	0.000	0.000	0.000	0.000	0.000	0.000	0.000	0.000	0.000	0.014	0.000
IDHP-3												
100	1.000	1.000	1.000	1.000	1.000	0.972	0.984	0.988	1.000	1.000	0.977	1.000
116	0.000	0.000	0.000	0.000	0.000	0.028	0.016	0.012	0.000	0.000	0.023	0.000

13 (53)	14 (51)	15 (28)	16 (42)	17 (73)	18 (53)	19 (81)	20 (68)	21 (37)	22 (60)	23 (88)	24 (20)
0.194	0.000	0.534	0.154	0.000	00275	0.294	0.500	0.368	0.000	0.143	0.000
0.806	1.000	0.466	0.846	1.000	0.725	0.706	0.500	0.632	1.000	0.587	1.000
0.000	0.037	0.054	0.595	0.000	0.000	0.012	0.000	0.000	0.017	0.063	0.075
1.000	0.963	0.946	0.405	1.000	1.000	0.988	1.000	1.000	0.983	0.937	0.925
0.000	0.000	0.000	0.000	0.000	0.000	0.000	0.000	0.000	0.000	0.165	0.000
0.264	0.963	0.536	0.302	0.130	0.764	0.777	0.728	0.568	0.750	0.290	0.400
0.736	0.037	0.464	0.698	0.870	0.236	0.223	0.272	0.432	0.250	0.545	0.600
0.236	0.902	0.018	0.238	0.212	0.613	0.333	0.544	0.486	0.000	0.028	0.350
0.764	0.098	0.982	0.762	0.788	0.387	0.667	0.456	0.514	1.000	0.972	0.650
0.000	0.000	0.000	0.000	0.007	0.000	0.000	0.000	0.000	0.000	0.006	0.800
1.000	1.000	1.000	1.000	0.993	1.000	1.000	1.000	1.000	1.000	0.994	0.200
1.000	1.000	1.000	1.000	1.000	1.000	1.000	1.000	1.000	1.000	1.000	1.000
0.000	0.000	0.000	0.000	0.000	0.000	0.000	0.000	0.000	0.000	0.000	0.000
1.000	1.000	1.000	1.000	1.000	1.000	1.000	1.000	1.000	1.000	1.000	1.000
0.000	0.000	0.000	0.000	0.000	0.000	0.000	0.000	0.000	0.000	0.000	0.000
1.000	1.000	1.000	1.000	1.000	0.991	1.000	1.000	1.000	1.000	1.000	1.000
0.000	0.000	0.000	0.000	0.000	0.009	0.000	0.000	0.000	0.000	0.000	0.000
1.000	1.000	1.000	0.964	1.000	1.000	1.000	1.000	1.000	1.000	0.994	0.925
0.000	0.000	0.000	0.036	0.000	0.000	0.000	0.000	0.000	0.000	0.006	0.075

Table 1.5.4 Summary of levels of genetic variability in anadromous and nonanadromous Atlantic salmon populations in Newfoundland for nine variable enzyme loci

Population	A	P	H (SD)	Population	A	P	H (SD)
1	1.7	67	0.169 (0.069)	13	1.3	33	0.119 (0.060)
2	1.6	56	0.142 (0.066)	14	1.3	33	0.036 (0.021)
3	1.7	67	0.154 (0.067)	15	1.4	44	0.128 (0.072)
4	1.6	56	0.177 (0.068)	16	1.6	56	0.179 (0.069)
5	1.6	56	0.135 (0.067)	17	1.3	33	0.064 (0.042)
6	1.7	67	0.132 (0.067)	18	1.4	44	0.140 (0.069)
7	1.4	44	0.103 (0.063)	19	1.4	44	0.137 (0.067)
8	1.6	56	0.168 (0.070)	20	1.3	33	0.156 (0.079)
9	1.6	56	0.159 (0.067)	21	1.3	33	0.164 (0.082)
10	1.3	33	0.145 (0.074)	22	1.2	22	0.046 (0.042)
11	1.7	67	0.169 (0.074)	23	1.8	67	0.142 (0.077)
12	1.2	22	0.109 (0.072)	24	1.6	56	0.175 (0.068)
Mean	1.6	53.7	0.147	Mean	1.4	41.7	0.124

A = average number of alleles per locus; *P* = percentage of polymorphic loci – a locus is considered polymorphic if more than one allele was detected; *H* = heterozygosity, unbiased estimate (Nei, 1978). Figure in brackets is standard deviation.

The allele frequencies for the main polymorphic loci are heterogeneous among samples in both population groups. The only exception is *PGM-1r*[*] among anadromous samples (Tables 1.5.3, 1.5.5). Comparing the two groups, F_{st} values (Wright, 1978) are consistently higher for all loci (Table 1.5.5) and average values are over 10-fold greater. The greater allele frequency variation among the nonanadromous samples (Figure 1.5.3) is significant for all the main polymorphisms (Bartlett's test – Table 1.5.5). In contrast, allele frequency variation among loci within each group is similar. Differences in the nonanadromous group are not significant (Bartlett's test on arcsin-transformed data; $\chi^2 = 1.44$, $p = 0.837$). A significant difference for the

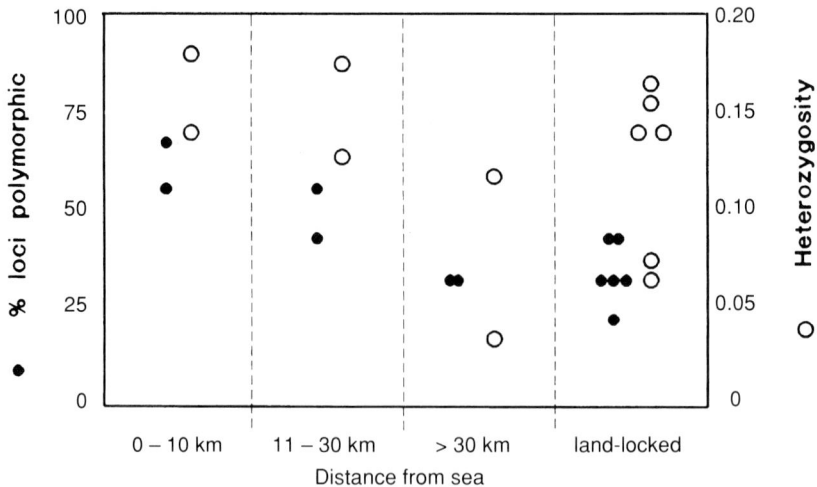

Figure 1.5.2 The relationship of the degree of isolation of nonanadromous Atlantic salmon populations and their percentage of polymorphic loci and level of heterozygosity. 'Landlocked' refers to populations which are isolated from the sea by virtue of an impassable waterfall.

Table 1.5.5 Results of tests of heterogeneity and F_{st} statistics, and comparison of means and variances for allele frequencies (*100) for anadromous and nonanadromous samples of Atlantic salmon

	Locus				
Parameter	*PGM-1r**	*IDDH-2**	*MDH-3,4**	*AAT-3**	*MEP-2**
Heterogeneity χ^2 (*p* value given)					
A	0.162	<0.0001	<0.0001	<0.0001	<0.0001
NA	<0.0001	<0.0001	<0.0001	<0.0001	<0.0001
F-statistics (F_{st})					
A	0.002	0.033	0.044	0.063	0.025
NA	0.202	0.380	0.231	0.296	0.766
Difference in variances of A and NA (Bartlett's Test: χ^2, d.f. = 1)					
p =	0.001	0.046	0.007	0.024	0.001
Difference between means of A and NA (Mann–Whitney *U*-test)					
p =	0.037	0.410	0.040	0.173	0.525

Comparison of means and variances of two groups based on arcsin-transformed allele frequencies. A = anadromous; NA = nonanadromous.

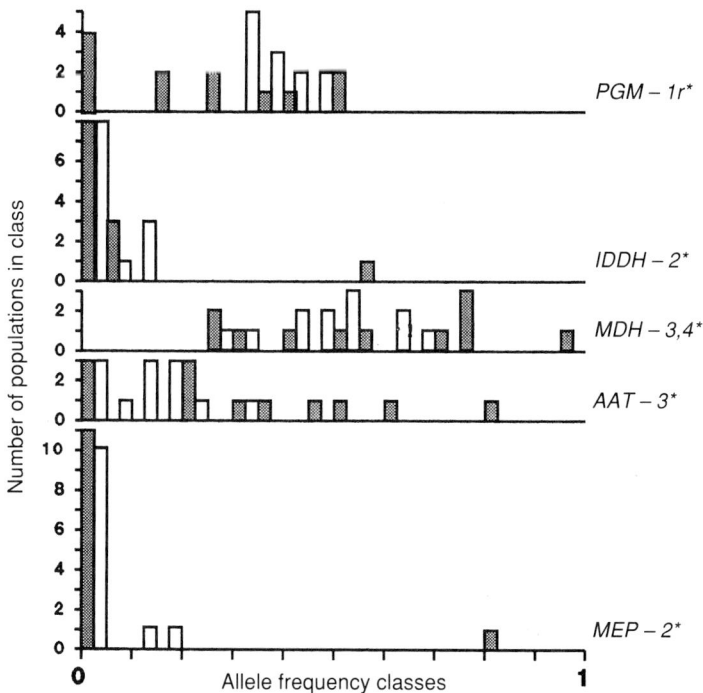

Figure 1.5.3 The distribution of frequencies of the *100 allele among anadromous and nonanadromous populations for the main polymorphic loci.

anadromous group (χ^2 = 12.99, p = 0.011) is due to the low level of variation detected at *PGM-1r*. Excluding *PGM-1r**, no heterogeneity among loci is detectable (χ^2 = 2.71, p = 0.438).

Differences in the mean allele frequencies of the two groups are weak or absent (Table 1.5.5). Two loci, *PGM-1r** and *AAT-3**, show marginally significant differences using the Mann–Whitney *U*-test, perhaps appropriately given the inequality of the variances of the two groups. However, the equivalent *t*-test shows only one significant result and, when adjusted for multiple *t*-tests (Sidak's correction for unplanned multiple tests), this disappears. Thus support for any differences in mean allele frequencies is weak.

Genetic distance analysis among the samples using Nei's (1978) unbiased *D* shows a clear clustering of anadromous Newfoundland populations with other such North American populations (Figure 1.5.4). Interspersed in this anadromous cluster are two of the nonanadromous populations. The other nonanadromous populations join either singly or in small groups, with the exception of Little Gull Lake. Little Gull Lake lumps with the anadromous populations from Europe in a separate cluster. The only 'pure' cluster of nonanadromous samples is composed of four – Five Mile Pond West, North Aquaforte River, South Aquaforte River and Tors Cove Pond – all from the same relatively small geographical area (Table 1.5.1; Figure 1.5.1), though each is in a different watershed and isolated above impassable falls. Clear clustering according to life-history type is absent and there is no association of anadromous and nonanadromous samples from the same watersheds. Genetic distances between anadromous and nonanadromous populations and among nonanadromous populations tend to be greater than those within the anadromous group.

The UPGMA clustering pattern for Nei's (1978) unbiased distance (*D*) produces essentially the same dendrogram using either the average or minimum measures. This pattern differs somewhat from that produced by a UPGMA analysis based on Cavalli-Sforza and Edwards (1967) arc distances (Figure 1.5.5). The main difference is the inclusion of a second nonanadromous sample (St Shott's River) with the European population cluster. Within each continental cluster there is also some marked reorganization.

Principal component analysis (Figure 1.5.6) shows a clear separation of North American and European anadromous population samples based on Factor 1. Factor 1 does not separate the anadromous and nonanadromous Newfoundland samples but the latter tend towards the European cluster and the Little Gull Lake and St Shotts samples rest approximately equidistant between the two continental groups. Factor 1 accounts for about 50% of the variance explained by the first four component factors. Factor 2 was not informative with regard to life-history or continent of origin but Factor 3 (Figure 1.5.6) provided for partial separation of the anadromous and nonanadromous North American samples. However, Factor 3 accounts for less than one-third of the variance explained by Factor 1. The loading for Factor 1 is primarily associated with variation at the main polymorphic loci *AAT-3**, *IDDH-2**, *MDH-3,4**, *MEP-2** and *PGM-1r**. Loadings for Factor 3 derive predominantly from variation at the *IDHP-3** and *MDH-1** loci, two of the minor polymorphisms.

1.5.4 Discussion

Present Atlantic salmon populations in Newfoundland are likely to have originated postglacially from recolonization in the light of the almost total cover of the island by the Wisconsin ice sheet (Scott and Crossman, 1964; Behnke, 1972; Rogerson, 1981). At the same time, its island status means that it is reasonable to assume that the colonizers were anadromous. Thus differences in the allozyme frequencies of nonanadromous and anadromous populations can only have evolved during the relatively short time which has elapsed since then, in the order of 10 000 years. This represents only a few thousand generations (based on a 4–9-year life-cycle) and means that new mutations can essentially be

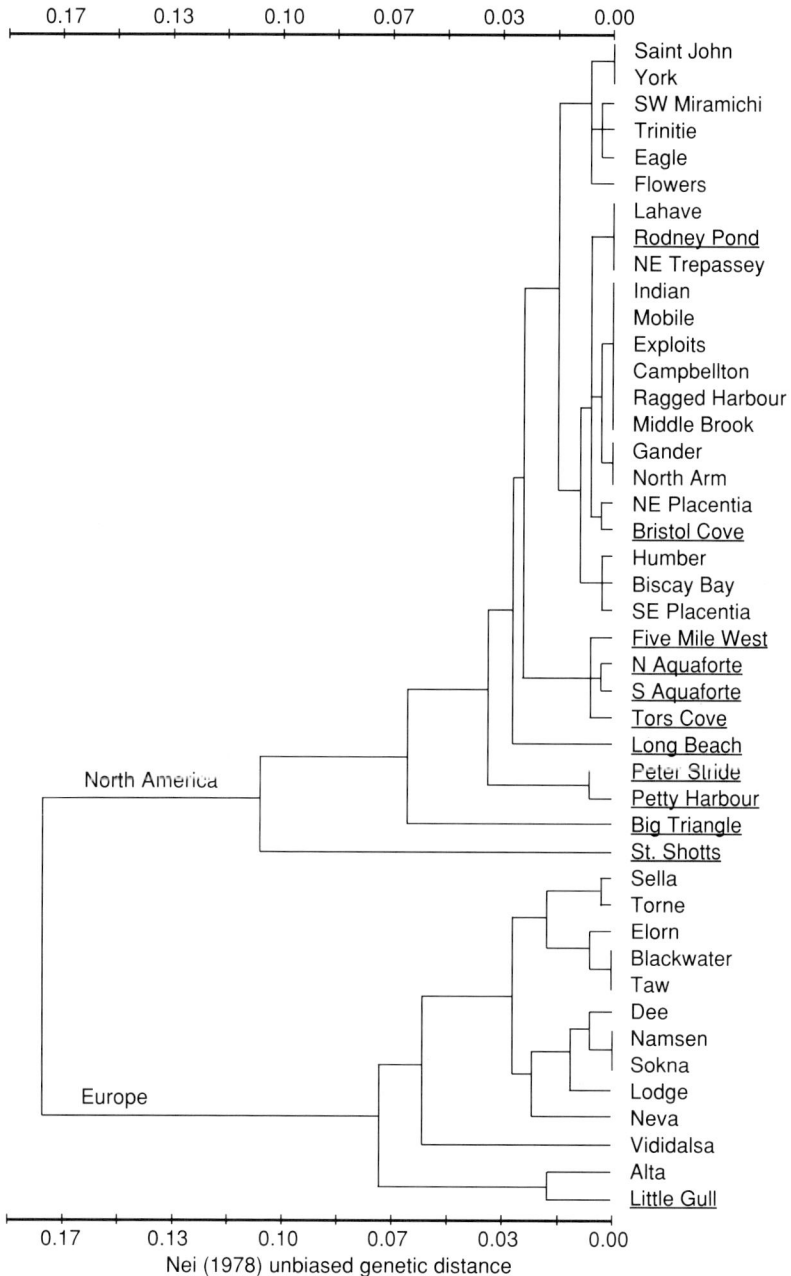

Figure 1.5.4 An UPGMA dendrogram of genetic association of anadromous and nonanadromous (underlined) Newfoundland Atlantic salmon populations and other anadromous populations from North America and Europe based on Nei's (1978) unbiased average genetic distance.

ruled out as a factor in both the life-history and allozyme divergence. This view is consistent with the observed lack of novel 'private' allelic variants in any of the population samples and of allozyme variation unique to Newfoundland when compared to populations elsewhere (Verspoor, 1988a,b).

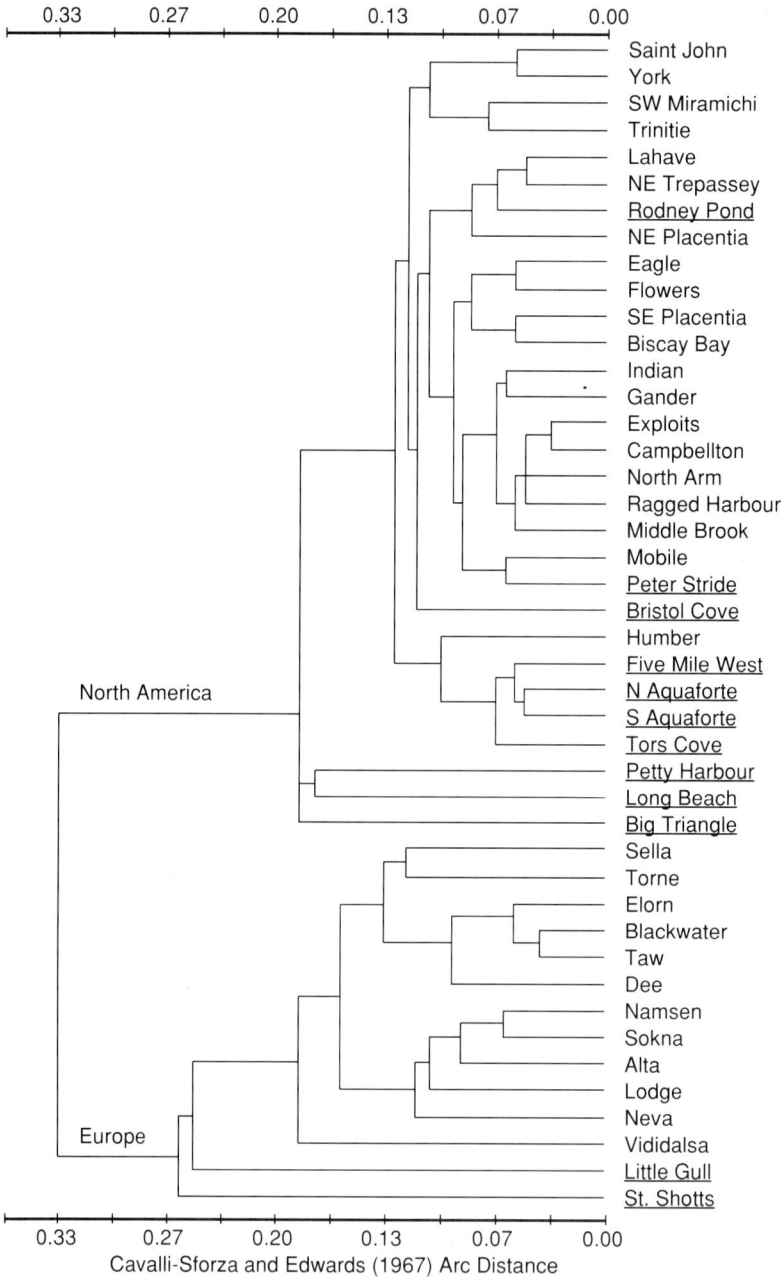

Figure 1.5.5 An UPGMA dendrogram of genetic association of anadromous and nonanadromous (underlined) Newfoundland Atlantic salmon populations and other anadromous populations from North America and Europe based on Cavalli-Sforza and Edwards (1967) arc distance.

The allozyme divergence among the anadromous and nonanadromous populations is greater than any so far reported among populations within any similar-sized region in the species range (Ståhl, 1987; Verspoor, 1988a; Koljonen, 1989; Crozier and Moffett, 1989; McElligot and Cross, 1991; Jordan *et al.*, 1992). In some cases, it approaches the level of

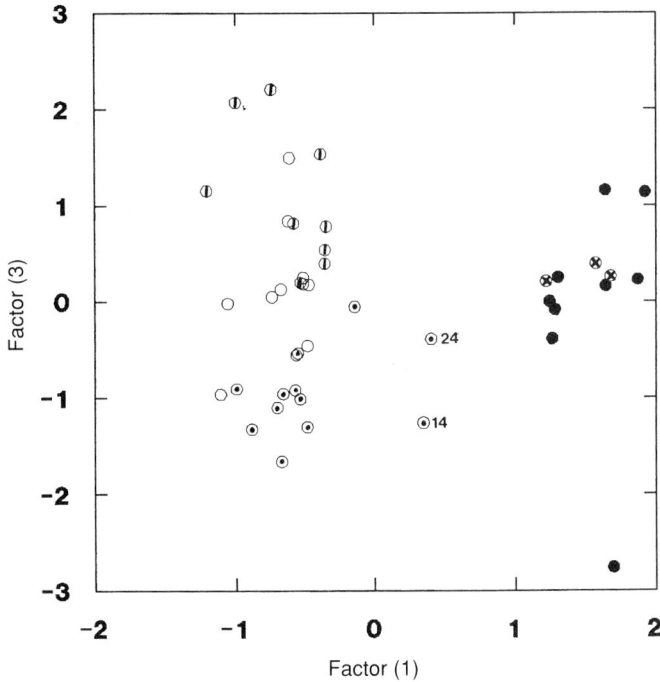

Figure 1.5.6 A principal component analysis plot for the two factors which show the greatest discrimination among Canadian (◑) and European (⊗ = Baltic and ● = Atlantic) Atlantic salmon populations and among anadromous (○) and nonanadromous (⊙) populations from Newfoundland. Analysis based on arcsin-transformed data.

divergence between North American and European salmon. It is also similar in magnitude to levels reported among sympatric anadromous and nonanadromous populations of the congeneric brown trout *S. trutta* (e.g. Ferguson and Taggart, 1991).

The greater genetic divergence among nonanadromous salmon populations compared to anadromous populations is in line with expectations based on general assessments of the effect of life-history on interpopulation variation (Gyllensten, 1985). The greater divergence is expected due to the increased physical isolation of freshwater habitats restricting gene flow and promoting genetic drift and facilitating differentiation by local selective pressures. Increased drift would cause the nonanadromous populations to have greater allele frequency variation, more frequent loss of alleles and lower heterozygosity (Kimura, 1983). Selection might be expected to cause the variances in allele frequencies among loci, or mean allele frequencies of the two population types to differ, or concordance of allele frequencies for the two types of populations where they cohabit the same river systems. However, with the possible exception of *PGM-1r**, the observations conform to the pattern expected under a model of genetic drift. Thus the extent of population isolation would appear to be the dominant factor in the evolution of genetic divergence among Newfoundland salmon populations.

Assuming effective neutrality of the allozyme variation, the level of gene flow among populations within each life-history group can be estimated using Wright's fixation index F_{st} (Wright, 1978) where $F_{st} = 1/(4N_e m + 1)$; N_e is the effective population size and m is the rate of genetically successful migration. An equal probability of exchange between all populations (island model) is also assumed, not unreasonably in the present case since there is no correlation of genetic and geographic distances among the populations. For the nonanadromous populations with an average F_{st} of 0.375 for the main polymorphic loci, the historical average migration rate (N_em) is 0.42 individuals per generation. N_em estimates between anadromous and nonanadromous populations when they are sympatric (Little Gull Lake–Gander River, N_em = 0.5) and parapatric (Big Triangle Pond–North Arm River, N_em = 0.4) are similar. These values are consistent with an absence of gene flow among the populations for most of their history. Among nonanadromous populations, this is expected given their physical isolation. However, the absence of gene flow in sympatry and parapatry, where physical barriers to interbreeding are lacking, raises some interesting questions given that the two forms can interbreed successfully under artificial conditions (Hutchings and Myers, 1985).

Anadromous populations with an average F_{st} of 0.033 have an N_em of 7.3, similar to values found in other studies of anadromous salmon from different rivers (e.g. Ståhl, 1981; Jordan *et al.*, unpublished data). This level of genetic exchange is significant but still insufficient to generate panmixis among the populations, as evidenced from their highly significant genetic heterogeneity. Thus, in spite of significant gene flow, the anadromous salmon in different watersheds still appear to be divided into essentially discrete reproductive populations.

The pattern of variation for *PGM-1r** suggests a possible involvement of selection in population divergence at this locus. Significantly lower allele frequency divergence than the other loci among the anadromous populations and possible differences between the two population groups in mean allele frequency were present. Consistent with this, *PGM-1r** has the lowest variance among loci in nonanadromous group. This is not significant but this may be because of the greater allele frequency variation in this group due to genetic drift. Interestingly, in the rainbow trout *Onchoryhnchus mykiss*, a similar polymorphism also appears to be adaptively relevant (Allendorf, Knudsen and Leary, 1983). In contrast, evidence for adaptive differentiation at *MEP-2**, found in other studies of Atlantic salmon (Verspoor and Jordan, 1989; Jordan, Youngson and Webb, 1990; Verspoor, Fraser and Youngson, 1991), was not found. However, this does not rule out selection. Drift may simply be obscuring selective effects or acting by moving populations among adaptive allele frequency equilibria determined by epistatic interactions between multiple loci (Wright, 1978).

Behnke (1980) has suggested that Atlantic salmon colonizing Newfoundland may have come from both European and North American refugia; salmon from the two continents share a common feeding area in the northwestern Atlantic and show major allele frequency differences (Ståhl, 1987; Verspoor, 1988b). If so, this could be a factor generating genetic diversity among salmon populations in Newfoundland if some populations were established predominantly or entirely by colonizers from one continental group or the other. If so the greatest effect would be on variation among the isolated nonanadromous populations. This dual colonization could also explain the clustering of nonanadromous populations with populations from Europe.

The analysis does not, however, support this as a major factor in current interpopulation variation. Firstly, none of the allozyme variants detected in Newfoundland are diagnostic for European populations. This in itself is not surprising as such allozyme variants are rare. Secondly, the clustering of the Little Gull Lake nonanadromous sample with European populations is associated with a low level of heterozygosity and a high level of fixation for common alleles which are shared by salmon from the two continental groups (Table 1.5.3; Stahl, 1987; Verspoor, 1988a,b). Given this, the population's 'European' affinity is most likely simply an extreme product of genetic drift.

The conclusion to which the analysis leads is that genetic drift is the major evolutionary force behind the genetic divergence among both anadromous and nonanadromous Atlantic salmon populations in Newfoundland. The relative importance of founder effects and bottle-necks, as opposed to chronic small population size, is unknown. Additionally, the data support a polyphyletic origin of nonanadromous populations from anadromous colonizers in the postglacial period as argued by Power (1958), Benhke (1972) and Berg (1985). This appears to be the case for other salmonid species as well (e.g. Ryman, 1983; Foote, Wood and Withler, 1989). Colonization of Newfoundland by both Canadian and European salmon would not appear to be a factor in population divergence. However, it does not rule out the possibility that European salmon were amongst the colonizers, a question which can only be answered by a detailed geographic study of continentally diagnostic variation at mitochondrial and nuclear gene loci.

Acknowledgements

The work was funded by a postdoctoral fellowship and contract from Fisheries and Oceans Canada, St John's, Newfoundland. I am indebted, in particular, to Dr John Pippy for his support of the work and to Professor Bryan Clarke and Dr David Parkin for comments on early drafts of the manuscript. Lloyd Cole and others kindly assisted with the collection of samples.

References

Allendorf, F.W., Knudsen, K.L. and Leary, R.F. (1983) Adaptive significance of differences in the tissue-specific expression of a phosphoglucomutase gene in rainbow trout. *Proc. Natl. Acad. Sci. USA*, **80**, 1397–400.

Behnke, R.J. (1972) The systematics of salmonid fishes of recently glaciated lakes. *Fish. Res. Bd Can.* **29**, 639–71.

Behnke, R.J. (1980) A systematic review of the genus *Salvelinus*, in *Charrs, Salmonid Fishes of the Genus Salvelinus* (ed. E.K. Balon), Dr W. Junk bv Publishers, The Hague, pp. 441–79.

Berg, O.K. (1985) The formation of non-anadromous populations of Atlantic salmon, *Salmo salar* L., in Europe. *J. Fish Biol.*, **27**, 805–15.

Birt, T.P., Green, J.M. and Davidson, W.S. (1991) Mitochondrial DNA variation reveals genetically distinct sympatric populations of anadromous and nonanadromous Atlantic salmon, *Salmo salar. Can. J. Fish. Aquat. Sci.*, **48**, 577–82.

Cavalli-Sforza, L.L. and Edwards, A.W.F. (1967) Phylogenetic analysis: models and estimation procedures. *Evolution*, **21**, 550–70.

Claytor, R.R. and MacCrimmon, H.R. (1988) Morphometric and meristic variability among North American Atlantic salmon (*Salmo salar*). *Can. J. Zool.*, **66**, 310–17.

Cross, T.F. and Ward, R.D. (1980) Protein variation and duplicate loci in the Atlantic salmon, *Salmo salar* L. *Genet. Res. Camb.*, **36**, 147–65.

Crozier, W.W. and Moffett, I.J.J. (1989) Amount and distribution of biochemical-genetic variation among wild populations and a hatchery stock of Atlantic salmon, *Salmo salar* L., from north-east Ireland. *J. Fish Biol.*, **35**, 665–77.

Ferguson, A. and Taggart, J.B. (1991) Genetic differentiation among the sympatric brown trout (*Salmo trutta*) populations of Lough Melvin, Ireland. *Biol. J. Linn. Soc.*, **43**, 221–37.

Foote, C.J., Wood, C.C. and Withler, R.E. (1989) Biochemical genetic comparison of Sockeye salmon and Kokanee, the anadromous and nonanadromous forms of *Oncorhynchus nerka. Can. J. Fish. Aquat. Sci.*, **46**, 149–58.

Gyllensten, U. (1985) The genetic structure of fish: differences in the intraspecific distribution of biochemical genetic variation between marine, anadromous and freshwater species. *J. Fish Biol.*, **26**, 691–9.

Hutchings, J.A. and Myers, R.A. (1985) Mating between anadromous and nonanadromous Atlantic salmon, *Salmo salar. Can. J. Zool.*, **63**, 2219–21.

Johnson, K. (1984) Protein variation in the Salmoninae: genetic interpretations of electrophoretic banding patterns, linkage associations among loci, and evolutionary relationships among species. PhD thesis, Pennsylvania State University, USA.

Jordan, W.C., Youngson, A.F. and Webb, J.H. (1990) Genetic variation at the malic enzyme-2 locus and age at maturity in sea-run Atlantic salmon (*Salmo salar*). *Can. J. Fish. Aquat. Sci.*, **47**, 1672–7.

Jordan, W.C., Youngson, A.F., Hay, D.W. and Ferguson, A. (1992) Genetic protein variation in natural populations of Atlantic salmon (*Salmo salar*) in Scotland: temporal and spatial variation. *Can. J. Fish. Aquat. Sci.*, **49**, 1863–72.

Kimura, M. (1983) *The Neutral Theory of Molecular Evolution*, Cambridge University Press, Cambridge.

Koljonen, M.-L. (1989) Electrophoretically detectable genetic variation in natural and hatchery stocks of Atlantic salmon in Finland. *Hereditas*, **110**, 23–35.

Lee, R.L.G. and Power, G. (1976) Atlantic salmon (*Salmo salar*) of the Leaf River, Ungava Bay, Canada. *J. Fish. Res. Bd Can.* **33**, 2616–21.

McElligott, E.A. and Cross, T.F. (1991) Protein variation in wild Atlantic salmon, with particular reference to southern Ireland. *J. Fish Biol.* **39** (Suppl. A), 35–42.

Nei, M. (1978) Estimation of average heterozygosity and genetic distance from a small number of individuals. *Genetics*, **89**, 583–90.

O'Connell, M.F. and Gibson, R.J. (1989) The maturation of anadromous female Atlantic salmon, *Salmo salar* L., stocked in a small pond in urban St. John's, Newfoundland, Canada. *J. Fish Biol.*, **34**, 937–46.

Palva, T.K., Lehvaslaiho, H. and Palva, E.T. (1989) Identification of anadromous and non-anadromous salmon stocks in Finland by mitochondrial DNA analysis. *Aquaculture*, **81**, 237–44.

Payne, R.H. (1974) Transferrin variation in North American populations of the Atlantic salmon *Salmo salar*. *J. Fish. Res. Bd Can.*, **31**, 1037–41.

Payne, R.H., Child, A.R. and Forrest, A. (1971) Geographic variation in the Atlantic salmon. *Nature*, **231**, 250–2.

Power, G (1958) The evolution of freshwater races of the Atlantic salmon (*Salmo salar* L.) in Eastern North America. *Arctic*, **11**, 86–91.

Rogerson, R.J. (1981) The tectonic evolution and surface morphology of Newfoundland, in *The Natural Environment of Newfoundland, Past and Present* (eds A.G. Macpherson and J.B. Macpherson), Memorial University Printing Services, St John's, Newfoundland, pp. 24–55.

Ryman, N. (1983) Patterns of distribution of biochemical genetic variation in salmonids: differences between species. *Aquaculture*, **33**, 1–21.

Scott, W.B. and Crossman, E.J. (1964) Fishes occurring in the freshwaters of insular Newfoundland. Life Sciences, Contribution No. 58, Royal Ontario Museum, University of Toronto.

Scott, W.B. and Crossman, E.J. (1973) Freshwater fishes of Canada. *Bull. 184, Fish. Res. Bd Can.*, Ottawa.

Sokal, R.R. and Rohlf, F.J. (1981). *Biometry*, W.H. Freeman and Company, San Francisco.

Ståhl, G. (1981) Genetic differentiation among natural populations of Atlantic salmon (*Salmo salar*) in Northern Sweden. *Ecol. Bull.*, **34**, 95–105.

Ståhl, G. (1987) Genetic population structure of Atlantic salmon, in *Population Genetics and its Applications to Fisheries Management* (eds N. Ryman and F. Utter), University of Washington Press, Seattle, pp. 121–40.

Swofford, D.L. and Selander, R.B. (1989) *Biosys-1: A Computer Program for the Analysis of Allelic Variation in Population Genetics and Biochemical Systematics – Release 1.7*, University of Illinois, Illinois Natural History Survey, Urbana.

Verspoor, E. (1986) Spatial correlations of transferrin allele frequencies in Atlantic salmon (*Salmo salar*) populations from North America. *Can. J. Fish. Aquat. Sci.*, **43**, 1074–8.

Verspoor, E. (1988a) Reduced genetic variability in first-generation hatchery populations of Atlantic salmon (*Salmo salar*). *Can. J. Fish. Aquat. Sci.*, **45**, 1686–90.

Verspoor, E. (1988b) Identification of stocks in the Atlantic salmon, in *Proceedings of the Symposium on Future Atlantic Salmon Management* (ed R.H. Stroud), Marine Recreational Fisheries Series, National Coalition for Marine Conservation, Inc., Savannah, Georgia, pp. 37–46.

Verspoor, E. and Cole, L.J. (1989) Genetically distinct sympatric populations of resident and anadromous Atlantic salmon, *Salmo salar*. *Can. J. Zool.*, **67**, 1453–61.

Verspoor, E., Fraser, N.H.C. and Youngson, A.F. (1991) Protein polymorphism in Atlantic salmon within a Scottish river: evidence for selection and estimates of gene flow between tributaries. *Aquaculture*, **98**, 217–230.

Verspoor, E. and Jordan, W.C. (1989) Genetic variation at the *Me-2* locus in the Atlantic salmon within and between rivers: evidence for its selective maintenance. *J. Fish Biol.*, 35 (Suppl. A), 205–13.

Vuorinen, J. and Berg, O.K. (1989) Genetic divergence of anadromous and nonanadromous Atlantic salmon (*Salmo salar*) in the river Namsen, Norway, *Can. J. Fish. Aquat. Sci.*, **46**, 406–409.

Wilder, D.G. (1947) A comparative study of the Atlantic salmon, *Salmo salar* Linneaus, and the lake salmon, *Salmo salar sebago* (Girard). *Can. J. Res. D*, **25**, 175–89.

Wilkinson, L. (1988) *Systat*, Systat, Inc., Evanston, Illinois.

Wright, S. (1978) *Evolution and Genetics of Populations IV: Variability within and among Natural Populations*, The University of Chicago Press, Chicago.

1.6 PATTERNS OF GENE FLOW IN THE GREAT BARRIER REEF AND CORAL SEA

John A.H. Benzie

Abstract

Surveys of allozyme variation in three species of giant clam, three species of sponge, one species of starfish, and a foraminiferan have revealed some consistent spatial patterns of genetic differentiation in populations from the Coral Sea. Gene flow within the highly connected Great Barrier Reef (GBR) system is extensive, but the data indicate significant differentiation of the Swain region within this system. Patterns observed in the western Coral Sea show marked differentiation of populations on a north–south axis that appears to be related to major current systems (the South Equatorial Current divergence). In contrast, greater connections were observed between the GBR and the Philippines, and between the Solomons, Vanuatu and the Philippines than directly between the GBR and the Solomons or Vanuatu. This is despite the strong flow of the South Equatorial Current from the eastern islands over a number of isolated reefs in the middle of the Coral Sea towards the GBR. It is suggested that these patterns of genetic differentiation reflect past episodes of genetic exchange rather than present-day gene flow.

1.6.1 Introduction

Information on the genetic structuring of tropical marine organisms, particularly those from coral reefs, is limited. Although the Great Barrier Reef (GBR) in Australia has been the subject of biological research for some time, population genetic work prior to the 1980s was limited to the study of a single population of giant clams (Ayala *et al.*, 1973) and its comparison with a geographically distant population in the Pacific (Campbell, Valentine and Ayala, 1975). There have been no systematic studies of the spatial patterns of genetic variation in coral reef species of the GBR until recently. The basic data that might allow an assessment of the levels of dispersal among reefs, or an assessment of the recent evolution of regional populations, are only now becoming available.

The Coral Sea has a rich diversity of reef types. The GBR, on the western margin of the Coral Sea, is a large complex of more than 2500 coral reefs developed on the continental shelf of the east coast of tropical Australia (Figure 1.6.1). The GBR stretches for more than 2000 km from Cape York near New Guinea to the Capricorn–Bunker group of reefs off southern Queensland. The island chains of the Solomons, Vanuatu and New Caledonia on the eastern rim of the Coral Sea have a range of lagoons and fringing reefs. A series of offshore reefs occurs on the submerged Queensland Plateau and on isolated sea-mounts between these island chains and Australia, and there are isolated reefs on the southern margin of the Coral Sea.

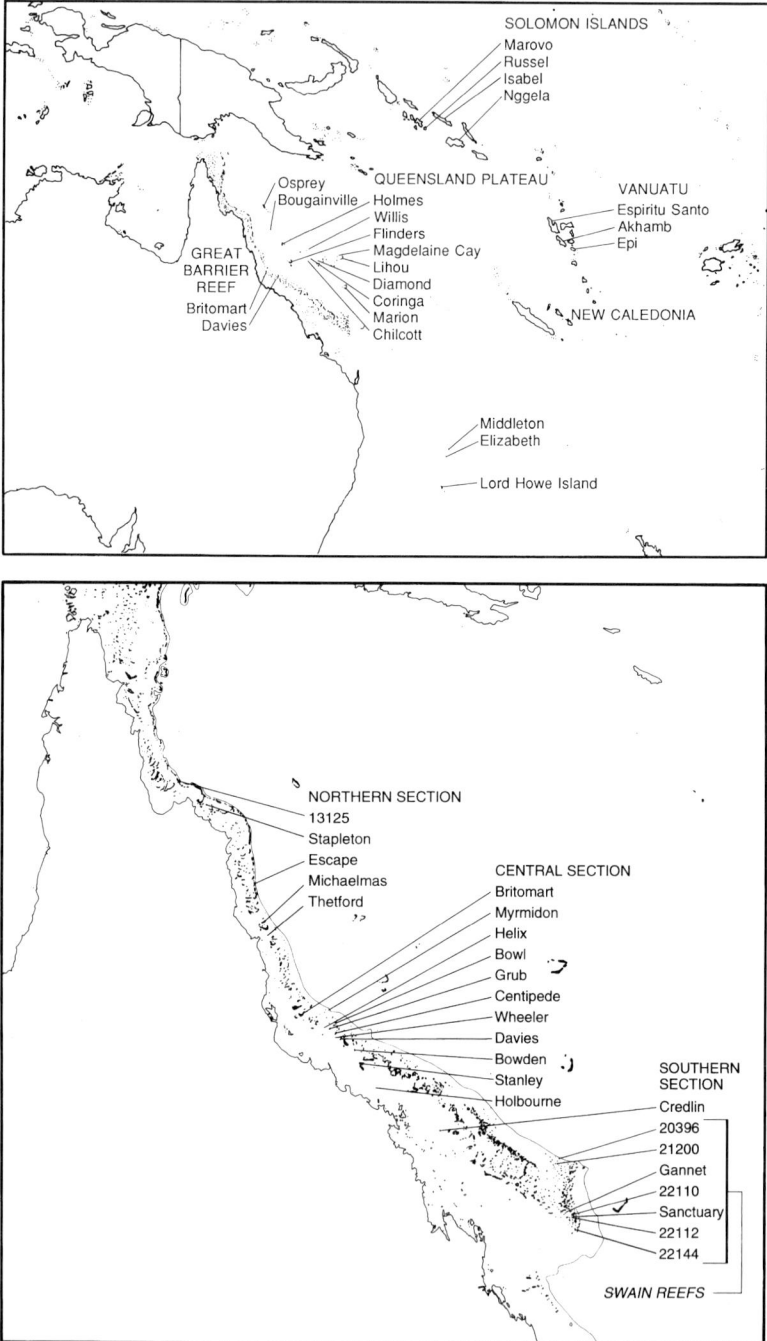

Figure 1.6.1 Map of the localities where the samples of reef organisms listed in the dendrograms in Figures 1.6.2 and 1.6.3 are placed in their geographic context. The localities where given species have been sampled can be obtained from the dendrograms in Figures 1.6.2 and 1.6.3.

Broadscale surveys throughout the region have now been conducted on crown-of-thorns starfish (*Acanthaster planci*) and on three species of giant clams (*Tridacna* spp.) for reasons specific to these organisms (Benzie, 1992, 1993). Genetic studies of crown-of-thorns starfish on the GBR were designed to investigate whether large increases in population size on certain reefs were the result of the dispersal of animals from a single source or whether the 'outbreak' populations were each independent events (Benzie and Stoddart, 1992a). The significant population differentiation among normal low-density or 'non-outbreak' populations, and the lack of genetic differentiation among the 'outbreak' populations, suggested that the 'outbreak' populations were indeed derived by dispersal from a single source. The surveys of giant clams were designed to estimate the genetic diversity in wild populations prior to the development of mariculture and restocking programmes in the Pacific. However, the surveys also formed part of a broader strategy of the Australian Institute of Marine Science to survey a number of taxa from a variety of taxonomic groups and with a variety of life-histories in order to investigate the population structure and evolution of reef species in the region. This has now included surveys of three sponge species and a foraminiferan in the western Coral Sea.

The aim of this section is to review recent work on the Coral Sea reefs which examines the spatial structure of wild populations on a variety of scales, to identify any consistent patterns of differentiation, and to discuss the implications these may have for the evolution of populations in the region.

1.6.2 Materials and methods

(a) Species surveyed

Extensive surveys of genetic variation over a variety of geographic scales in the GBR and/or the Coral Sea reefs have now been published for the crown-of-thorns starfish, *Acanthaster planci*, (Nash, Goddard and Lucas, 1988; Benzie, 1992; Benzie and Stoddart, 1989, 1992a,b), the giant clams *Tridacna gigas, T. derasa* and *T. maxima* (Benzie and Williams, 1992a,b; Macaranas *et al.*, 1992; Benzie, 1993) and the foraminiferan *Marginopora vertebralis* (Benzie and Pandolfi, 1991; Benzie, 1991). Information is also available for the sponges *Phyllospongia alcicornis, P. lamellosa* and *Carteriospongia sp.* (Benzie, Sandusky and Wilkinson, unpublished data).

(b) Geographic coverage

The GBR has been surveyed most intensively with data published from 6 to 13 populations in each of four species: *A. planci, T. gigas, T. derasa* and *T. maxima* (Table 1.6.1). Although samples were not always obtained from exactly the same reefs, comparable regional coverage was achieved over most of the GBR. Sample sizes for most reef populations were greater than 50 individuals, so providing a sound basis for establishing gene frequencies. Data for giant clams have also been published for populations from the Solomon Islands and pertinent comparisons made with Philippines populations (Macaranas *et al.*, 1992; Benzie, 1993). Information on *A. planci* obtained from Vanuatu (Benzie, 1990a) and from isolated reefs on the southern margins of the Coral Sea (Elizabeth and Middleton Reefs and Lord Howe Island), is compared here with data from the Philippines (Benzie, unpublished data). The isolated reefs in the western Coral Sea have been little studied but surveys of *T. maxima, M. vertebralis* and three species of dictyoceratid sponges were made during a cruise by the Akademik Oparin to the region in 1989. Sample sizes were limited to approximately 20 individuals in most of these populations (Table 1.6.2).

Table 1.6.1   Summary of population genetic statistics for samples from the GBR of three giant clam species, *T. gigas, T. maxima* and *T. derasa*, the starfish *A. planci*, and the foraminiferan *M. vertebralis*

	T. gigas	T. maxima	T. derasa	A. planci		M. vertebralis
				Non-outbreak	*Outbreak*	
Number of polymorphic loci screened	8	6	9	9	9	4
Number of populations screened	6	8	7	5	8	2
Number of individuals screened per population	66	68	52	48	81	128
Within-population variation						
Mean number of alleles per locus	2.0–2.4	4.0–5.0	3.2–3.7	2.1–3.1	2.5–3.4	1.5–3.0
Percentage of loci polymorphic	63–75	83–100	89–100	75–88	88–100	50–100
Direct count heterozygosity	0.22–0.25	0.36–0.41	0.41–0.46	0.21–0.35	0.25–0.28	0.10–0.15
Between-population variation						
Mean genetic distance between populations	0.001	0.005	0.013	0.013	0.001	0.005
F_{st} (all populations)	0.000[a]	0.003[a]	0.012[a]	0.018[*]	0.004[a]	0.006[a]
$N_e m$ (all populations)	Infinity	83.1	20.1	13.6	75.5	41.2
F_{st} (minus Swain)	0.000[a]	0.001[a]	0.007[a]	0.003[a]	0.004[a]	–
$N_e m$ (minus Swain)	Infinity	249.8	35.5	83.1	75.5	–
F_{st} (pooled Swain versus pooled others)	–	0.005[***]	0.013[***]	0.005[a]	0.002[a]	–
$N_e m$ (pooled Swain versus pooled others)	–	49.8	19.0	49.8	124.8	–
Larval life (days)	7–10	7–10	7–10	14–28	14–28	Unknown

[a] Not significant; [*] $0.5 > p > 0.01$; [***] $p < 0.001$. F_{st}, $N_e m$: see text.
Data for *A. planci* have been separated into outbreak and non-outbreak populations given the differences in dispersal and hence connectedness that might be expected in the two sets of populations. Data abstracted or calculated from Benzie (1993), Benzie and Williams (1992a,b), Macaranas *et al.* (1992) and Benzie and Stoddart (1992a).

(c)   Genetic markers and analysis

All the studies examined genetic variation using electrophoretically detectable protein variation at a number of polymorphic loci ranging from three to ten polymorphic systems (Tables 1.6.1 and 1.6.2). Details of methods for the starch gel and cellulose acetate electrophoresis used for *A. planci* are given in Nash, Goddard and Lucas (1988) and Benzie (1990b), those for the giant clams *T. gigas, T. derasa* and *T. maxima* are given in Benzie, Williams and Macaranas (1993) and those for the foraminiferan *M. vertebralis* are given in Benzie and Pandolfi (1991). Methods for detecting allozyme variation in sponges for the enzymes hexokinase (HK), malate dehydrogenase (MDH), esterase (EST) and glucose phosphate isomerase (GPI) have been developed by Benzie, Sandusky and Wilkinson (unpublished data). GPI, HK and MDH were run on cellulose acetate gels (Cellogel) for 2 h using a 0.01 M citrate phosphate buffer, pH 6.4 (see Benzie, Williams and Macaranas (1993) for details). EST and also MDH were run on 12% starch gels for 5 h using a Tris–EDTA–maleate pH 7.4 buffer (stock was 121.0 g Tris, 116.0 g maleic acid, 37.2 g EDTA

Table 1.6.2 Summary of population genetic statistics for samples from the western Coral Sea of three sponge species, *P. alcicornis*, *P. lamellosa* and *Carteriospongia sp.*, the giant clam *T. maxima*, and the foraminiferan *M. vertebralis*

	P. alcicornis	*P. lamellosa*	*Carteriospongia*	*T. maxima*	*M. vertebralis*
Number of polymorphic loci screened	5	4	3	6	4
Number of populations screened	4	5	4	10	12
Number of individuals screened per population	18	22	16	18	23
Within-population variation					
Mean number of alleles per locus	1.8–2.8	2.0–3.0	2.3–3.3	2.7–3.8	1.0–3.0
Percentage of loci polymorphic	60–100	50–100	67–100	67–100	0–100
Direct count heterozygosity	0.18–0.33	0.16–0.28	0.27–0.34	0.29–0.44	0.00–0.09
Between-population variation					
Mean genetic distance between populations	0.261	0.148	0.315	0.016	0.180
F_{st} (all populations)	0.292***	0.226***	0.330***	0.011[a]	0.345***
N_em (all populations)	0.61	0.86	0.51	22.5	0.24
F_{st} (average within regions)	0.194***	0.121***	0.176***	−0.002[a]	0.229***
N_em (average within regions)	1.04	1.82	1.17	Infinity	0.84
F_{st} (pooled north versus pooled south)	0.345***	0.146***	0.275***	0.019**	0.222***
N_em (pooled north versus pooled south	0.48	1.46	0.66	12.9	0.88
Larval life (days)	Unknown (< 3?)	Unknown (< 3?)	Unknown (< 3?)	7–10	Unknown

Data abstracted or calculated from Benzie (1991), Benzie and Williams (1992b) and from unpublished data (Benzie, Sandusky and Wilkinson). For abbreviations see Table 1.6.1.

(Na$_2$ salt), 20.0 g MgCl$_2$ and 50.0 g NaOH made up to 2 litres in distilled water; gel buffer, 5 ml stock made up to 250 ml with distilled water; electrode buffer, diluted one part stock plus four parts distilled water). Five polymorphic loci were assayed in *Phyllospongia alcicornis* (*Est, Gpi, Hk, Mdh-1* and *Mdh-2*), four in *P. lamellosa* (*Est, Gpi, Hk* and *Mdh-1*) and three in *Carteriospongia* sp. (*Gpi, Hk* and *Mdh-1*).

The analyses here have been restricted to comparisons of published population genetic statistics and any patterns of dispersal identified in the relevant papers. Where necessary, additional calculations of Weir and Cockerham's (1984) F_{st} (the standardized genetic variance among populations corrected for sample size), and N_em (the average number of migrants per generation), following equations in Waples (1987), were made of specific subsets of populations or pooled groups of populations to better examine patterns of gene exchange. In these cases an island model was assumed, where $N_em = [((1/F_{st})-1)/4]$.

1.6.3 Results

The available data fall into three groups: surveys of the GBR, surveys of the western Coral Sea reefs with some links to the GBR, and, for two taxa, additional information available from the islands on the eastern and/or southern fringe of the Coral Sea that allows comment on the genetic structure of these species over the whole region.

(a) GBR

Low levels of divergence among populations were observed in all the species studied. Genetic distances (Nei's (1978) D) among populations ranged from 0.001 to 0.013, but were usually less than 0.010 (Table 1.6.1). F_{st} ranged up to 0.018, but the difference from zero was usually not statistically significant, suggesting panmixis (random breeding) among populations throughout the GBR. High levels of gene flow of 13.6 to an infinite number of migrants per generation were inferred for all species throughout the GBR. The low levels of population differentiation observed were not the result of low levels of genetic variation. All species were highly polymorphic (63–100%) and had high levels of heterozygosity (0.21–0.46) and average numbers of alleles per locus (2.0–5.0).

Dendrograms illustrating the genetic relationships among populations in the GBR showed a consistent separation of the Swain reefs region in *T. maxima*, *T. derasa* and *A. planci* (Figure 1.6.2). When Swain populations were excluded from the data sets, F_{st} values were lower, and inferred levels of $N_e m$ higher, for all groups with the exception of the set of outbreak populations of *A. planci* (Table 1.6.1). Comparison of the pooled Swain populations with the pooled set of populations from the rest of the GBR demonstrated highly significant differentiation of *T. maxima* and *T. derasa* populations from the Swain region. Even where F_{st} was not significantly different from zero, $N_e m$ was much reduced between the Swain region and the rest of the GBR relative to the exchange within each area. $N_e m$ within the north–central GBR region was two to five times higher than that between regions in all cases except the set of outbreak populations of *A. planci*.

(b) Western Coral Sea reefs

The principal finding from surveys of the western Coral Sea reefs was that populations were differentiated along a north–south axis (Figure 1.6.3). The patterns were similar for the three sponge taxa, *P. alcicornis*, *P. lamellosa* and *Carteriospongia sp.*, and for the foraminiferan *M. vertebralis* in that the break in genetic continuity was generally coincident with the southern limit of the South Equatorial Current divergence (SEC divergence). The giant clam, *T. maxima*, showed little structure in the region of the SEC divergence, and displayed only two outlier populations of Lihou and Osprey. Although linked in the dendrogram, the outlier populations were widely separated in a principal coordinates plot, suggesting no particularly close relationship between these outliers to each other relative to the main cluster consisting of all the other populations. At the time *T. maxima* breeds, in the austral summer, the SEC divergence lies just south of Osprey (Figure 1.6.4), suggesting that oceanographic conditions might play a role in the genetic distinction of this population from all the others studied. Comparison of *M. vertebralis* and *T. maxima* populations from the GBR demonstrated no separation of the Queensland Plateau populations as a group from the GBR.

Genetic distances among populations were greater than in the GBR, reflecting the greater isolation of the populations concerned, and, in addition, a probable lower dispersal capacity of the sponges. Nei's (1978) D was 0.016 in *T. maxima*, about 2–3 times greater than genetic distances among populations of this species in the GBR, and was 0.180 in *M. vertebralis*, about 36 times greater than the genetic distance between two populations of this species in the GBR. Nei's (1978) D ranged from 0.180 to 0.315 among the sponge populations (Table 1.6.2). F_{st} values were also greater, and highly significant in all taxa except for the giant clam (F_{st} = 0.011), ranging from 0.226 to 0.345 in the sponges and foraminiferan. With the

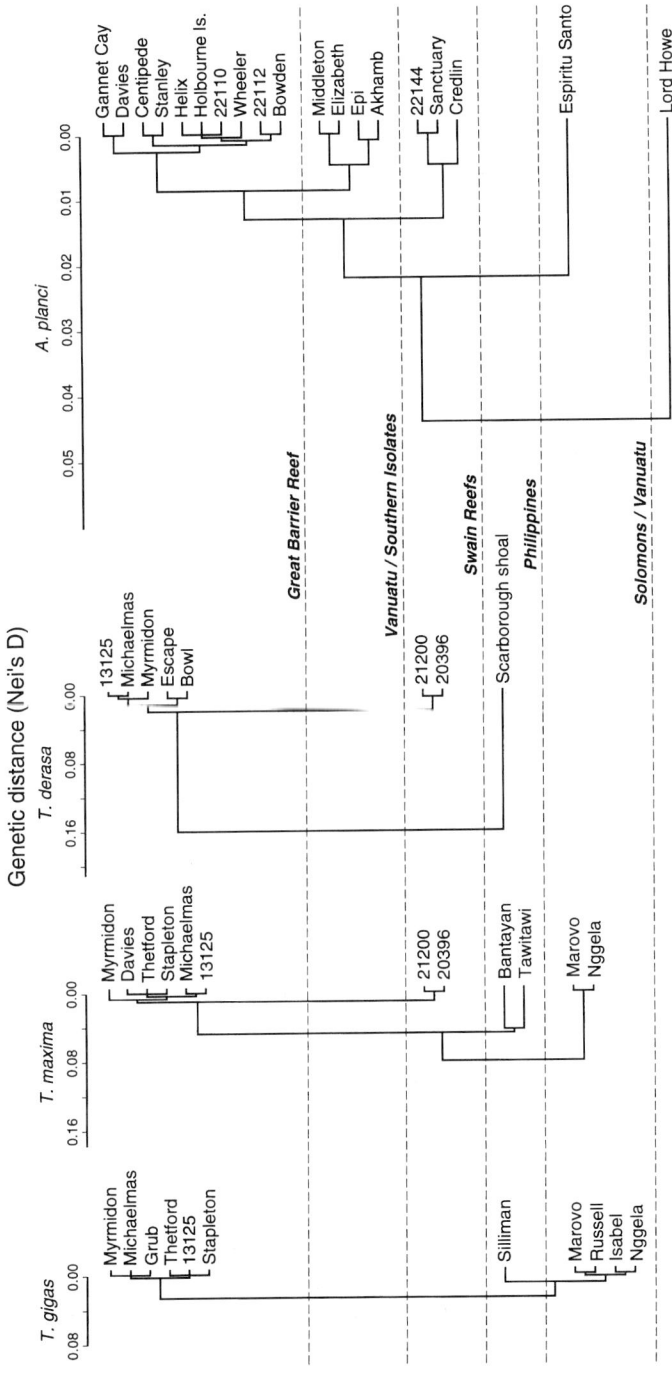

Figure 1.6.2 Dendrograms illustrating the similarity of the genetic relationships among populations from the Coral Sea of the giant clams *Tridacna gigas*, *T. maxima* and *T. derasa*, and the starfish *Acanthaster planci*. Data from the Philippines have also been included where possible to illustrate the closer relationships of the eastern and western Coral Sea reef populations to the Philippines than to each other. In each case Nei's (1978) unbiased genetic distances were clustered using the UPGMA method. Data from Benzie (1990a, 1993) and Benzie and Stoddart (1992a,b).

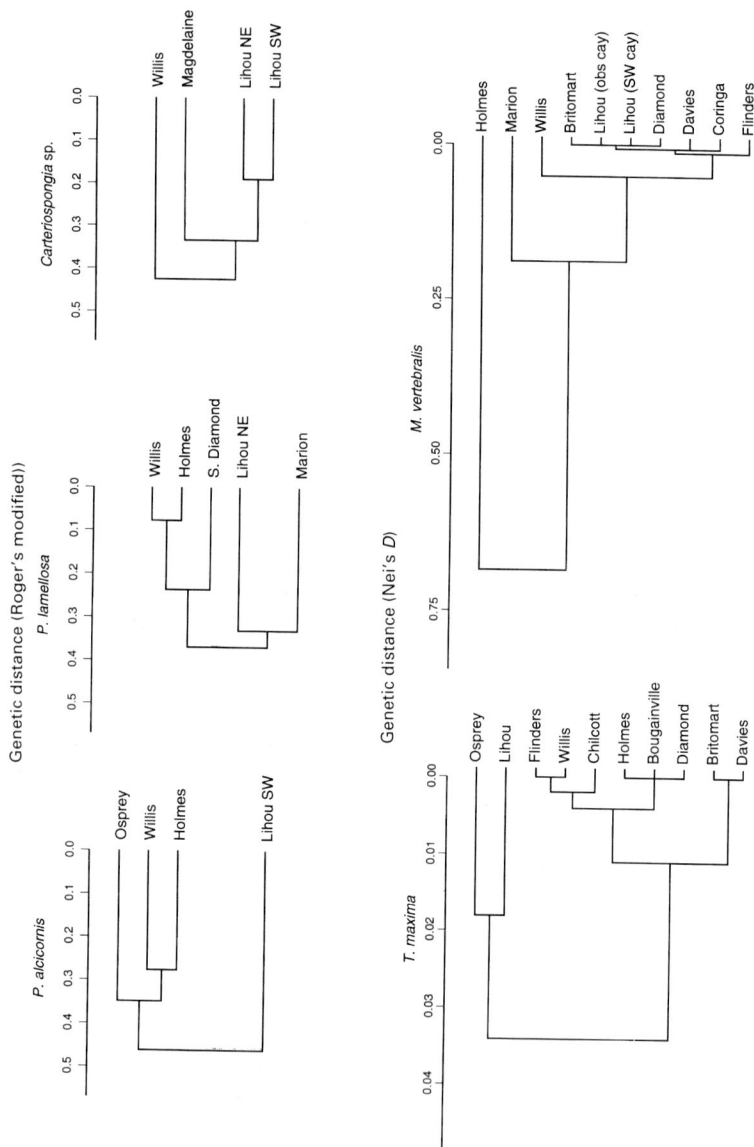

Figure 1.6.3 Dendrograms illustrating the north–south differentiation in the genetic relationships among populations of the sponges *P. alcicornis*, *P. lamellosa* and *Carteriospongia* sp., the giant clam *T. maxima* and the foraminiferan *Marginopora vertebralis*. In each case genetic distances were clustered using the UPGMA method. Data from Benzie (1991), Benzie and Williams (1992b) and Benzie *et al.* (unpublished data).

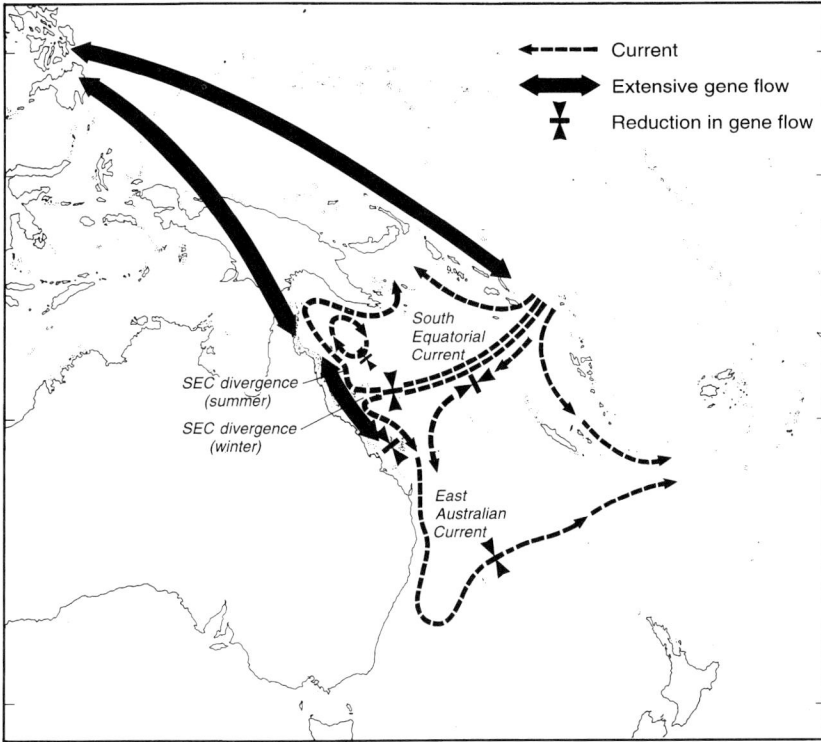

Figure 1.6.4 Map illustrating the general current flows in the Coral Sea, and the generalized patterns of genetic connectedness inferred from the surveys of eight species of reef invertebrates.

exception of *T. maxima*, average levels of gene flow were low ($N_em < 1.0$) among the western Coral Sea reef populations. However, all species showed higher gene flow within northern and southern regions than between regions (Table 1.6.2). Gene flow between regions was approximately half that within regions, with significant differentiation occurring between regions for the *T. maxima* populations. Polymorphism within populations ranged from 0 to 100% and levels of heterozygosity were high, 0.16–0.44, except in *M. vertebralis*, where they were 0.00–0.09. Similarly, the average numbers of alleles per locus was reasonably high, 1.8–3.3, except for *M. vertebralis*, where they were 1.0–3.0.

(c) Coral Sea

Data that span the Coral Sea are limited but suggest lower dispersal directly between the reefs on the eastern and western margins of the Coral Sea relative to that between either set of reefs and islands in the Indo-Pacific, and the occurrence of genetic isolates on the remote Lord Howe Island on the southern margin of the Coral Sea. The average number of migrants per generation (N_em) between the GBR and the islands on the eastern margin of the Coral Sea were less than expected on the basis of their geographic separation relative to the degree of exchange between the Philippines and the GBR, and between the Philippines and the Solomon Islands or Vanuatu, for both *A. planci* and *T. maxima* (Table 1.6.3; Figure 1.6.4). Indeed, the value of N_em for *T. maxima* across the Coral Sea was less than that between the

Table 1.6.3   Average genetic distance (Nei's *D*), F_{st} and N_em between populations of *A. planci* and *T. maxima* in the Coral Sea and the Philippines

	Genetic distance	F_{st}	N_em	N_em *expected*
T. maxima				
Philippines – eastern				
Coral Sea islands	0.077	0.065	3.6	–
Philippines – GBR	0.049	0.040	6.0	–
GBR – eastern				
Coral Sea islands	0.082	0.157	1.3	12.3
A. planci				
Philippines – eastern				
Coral Sea islands	0.012	0.022	11.1	–
Philippines – GBR	0.003	0.008	31.0	–
GBR – eastern				
Coral Sea islands	0.006	0.012	20.6	49.8
Lord Howe Island – GBR	0.038	0.106	2.1	55.3
Lord Howe Island – Vanuatu	0.038	0.088	2.6	46.9

Expected N_em values for exchange between the GBR to the eastern Coral Sea islands were calculated assuming the same average rate of increase in F_{st} per 1000 km as observed between the Philippines and the GBR and between the Philippines and the eastern Coral Sea reefs to occur over the distance of the Coral Sea and using this F_{st} to calculate N_em. The eastern Coral Sea islands were Vanuatu in the case of *A. planci* and the Solomon Islands in the case of *T. maxima*.

Philippines and either the GBR or the Solomon Islands. Similar patterns were reflected in the genetic distances among these populations (Table 1.6.3).

A. planci populations from Lord Howe Island were more differentiated genetically from those on the GBR and the intervening Elizabeth and Middleton reefs, or Vanuatu, than expected on the basis of their geographic separation (Figure 1.6.2; Table 1.6.3).

1.6.4   Discussion

Most of the Coral Sea region is little studied and much remains to be discovered about the evolution of its coral reef floras and faunas. The results of population genetic studies on a variety of organisms are, however, beginning to reveal consistent patterns of genetic structuring in the GBR and the Coral Sea reef populations. Some of these patterns appear to correlate well with present-day circulation patterns but others appear contrary to expectation, and raise a number of questions concerning the factors which might have given rise to them.

For example, all the studies published for the GBR concerned species which have relatively long larval phases (range 7–28 days) that provide the potential for dispersal over several hundred kilometres within the reef system. Similarly, they all breed within the austral summer, when net southerly currents (there are northerly flows at times) are found throughout the GBR from north of Cairns south to the Swain region. The low level of genetic divergence among populations in the GBR is therefore not unexpected. However, the differentiation of the Swain region is difficult to understand given the strong currents from the north and central areas towards the south. Gene flow is still high between the Swain region and the north–central region of the GBR, but a marked reduction in gene flow between the regions is apparent in most surveys. A dense packing of reefs just north of the Swain region diverts water, and tides, to the east and west of the GBR midline, so reducing the degree of direct current exchange and perhaps providing some explanation for the reduced genetic exchange.

As the overall level of population differentiation increases, these differences become stronger and statistically significant (e.g. compare *A. planci* with *T. maxima* with *T. derasa*). However, while the larval lifespan of the clams is less than that of the crown-of-thorns starfish and correlates with reduced gene flow, no marked differences between the clams is obvious that might account for differences in the levels of population differentiation between the clam species. The fact that high levels of gene flow can occur in the present day between these regions is demonstrated by the high levels of gene flow between the rest of the GBR and the Swain region for outbreak populations of crown-of-thorns starfish. The crown-of-thorns starfish have undergone two sets of population explosions in the last 30 years and large numbers are thought to have dispersed southwards through the GBR (Benzie and Stoddart, 1992a).

Similarly, the limited data from populations in the western Coral Sea suggest differentiation north and south of the SEC divergence in a variety of taxa. The correlation of differentiation perpendicular to this strong current might suggest that high gene flow should occur parallel to the current. However, despite the SEC flowing directly from the Solomon Islands to Australia, and a 10-day or 14-day planktonic lifespan for tridacnids and crown-of-thorns starfish respectively, gene flow between these populations is only 20–50% of that expected given the levels of gene flow along the GBR or Melanesian island chains to the Philippines.

One can only speculate that the strong genetic connections to the Indo-Pacific, which mirror biogeographic patterns of species distributions (Springer, 1982), reflect past dispersal events. Sea levels were reduced several times in the last million years and were some 150 m lower than today at the peak of the ice age. Island chains are postulated to have become solid land barriers and at least stronger filters, or deflectors, of ocean currents (Potts, 1985). The reduced gene flow across the Coral Sea may also reflect the lower probability of successful dispersal across wide stretches of ocean relative to 'island hopping' along the reefs associated with the island chains leading to the Philippines. Similarly, the genetic differentiation of *A. planci* populations on Lord Howe Island is not explained solely by the fact that currents do not pass directly to Lord Howe Island from the GBR but take a longer route through cooler regions (Hamilton, 1992) where *A. planci* larvae are unlikely to survive. Occasional jets of warm water reach Lord Howe Island directly from the reefs to the north and Benzie and Stoddart (1992b) have suggested that the genetic distinctions of the Lord Howe Island populations result in part from founder effects arising from the small number of individuals rarely recruited, citing as supporting evidence the skewed gene frequencies and reduced genetic variation of the Lord Howe Island population.

Populations of several species on the isolated reefs in the western Coral Sea, separated by hundreds of kilometres of deep water, show greater genetic differentiation as a result of their isolation, and a number showed marked gene frequency differences. Although all the sponges, and *T. maxima*, showed differences on a north–south axis, samples to the south of the SEC divergence were often restricted to Lihou. Lihou reef was a strong outlier in the *T. maxima* study, and it is possible that the patterns on the sponge dendrograms reflect the isolation of Lihou rather than a strong north–south differentiation. However, in the *T. maxima* study, the genetic difference of the Lihou population was equivalent to that expected on the basis of that reef population's geographic separation from the other populations (Benzie and Wiliams, 1992b), and Lihou was not an outlier in the *M. vertebralis* study (Benzie, 1991). Overall, the data for the sponges taken in conjunction with the clam and foraminiferan data support a north–south differentiation rather than Lihou as a genetic isolate.

Unknown, possibly stochastic, historical factors have been suggested to give rise to the marked differentiation of isolated reefs like Holmes and Marion for *M. vertebralis* and Lord

Howe Island for *A. planci* (Benzie, 1991; Benzie and Stoddart, 1992b). Equally intriguing is the close connection of *M. vertebralis* and *T. maxima* populations from the GBR and the Queensland Plateau, and the absence of any particularly close relationship among the Queensland Plateau reefs as a group separate to the GBR, despite the close connections that would have existed between them at times of low sea level.

The patterns of genetic differentiation of populations of marine organisms in the Coral Sea region have, therefore, provided interesting similarities with and differences from those expected on the basis of present-day current patterns. The lack of fit with these has raised more questions than it has answered, but suggests that many of the features of the patterns of gene exchange result from historical events rather than processes occurring today.

Acknowledgements

I thank the scientists and crew of the R.V. Akademik Oparin of the Pacific Institute of Bioorganic Chemistry of the Far East Branch of the USSR Academy of Sciences, Vladivostock for their assistance in collecting samples from the western Coral Sea, and C. Sandusky for her help in the sponge genetic analysis. This is contribution number 633 from the Australian Institute of Marine Science.

References

Ayala, F.J., Hedgecock, D., Zumwalt, G.S. and Valentine J.W. (1973) Genetic variation in *Tridacna maxima*, an ecological analog of some unsuccessful evolutionary lineages. *Evolution*, **27**, 177–91.

Benzie, J.A.H. (1990a) Genetic relationships of crown-of-thorns starfish, in *Vanuatu Marine Resources* (eds T.J. Done and K. Navin), Australian Institute of Marine Science, Townsville, pp. 114–18.

Benzie, J.A.H. (1990b) *Techniques for the Electrophoresis of Crown-of-thorns Starfish (Acanthaster planci) Enzymes*, Australian Institute of Marine Science, Townsville.

Benzie, J.A.H. (1991) The genetic relatedness of foraminiferan (*Marginopora vertebralis*) populations from reefs in the western Coral Sea and Great Barrier Reef. *Coral Reefs*, **10**, 29–36.

Benzie, J.A.H. (1992) Review of the genetics, dispersal and recruitment of crown-of-thorns starfish (*Acanthaster planci*). *Aust. J. Mar. Freshwat. Res.*, **43**, 597–610.

Benzie, J.A.H. (1993) The genetics of giant clams: an overview, in *Biology and Mariculture of Giant Clams* (ed. W.K. Fitt), Australian Centre for International Agricultural Research, Proceedings number 47, ACIAR, Canberra (in press).

Benzie, J.A.H. and Pandolfi, J.M. (1991) Allozyme variation in *Marginopora vertebralis* (Foraminifera: Miliolidae) from coral reef habitats in the Great Barrier Reef, Australia. *J. Foram. Res.*, **21**, 222–27.

Benzie, J.A.H. and Stoddart, J.A. (1989) Genetic approaches to ecological problems: crown-of-thorns starfish outbreaks, in *Proc. 6th Internat. Coral Reef Symp.*, (eds J.H. Choat, D. Barnes, M.A. Borowitzka *et al.*) Townsville **2**, 119–24.

Benzie, J.A.H. and Stoddart, J.A. (1992a) Genetic structure of outbreaking and non-outbreaking crown-of-thorns starfish (*Acanthaster planci*) populations on the Great Barrier Reef. *Mar. Biol.*, **112**, 119–30.

Benzie, J.A.H. and Stoddart, J.A. (1992b) The genetic structure of crown-of-thorns starfish (*Acanthaster planci*) in Australia. *Mar. Biol.*, **112**, 631–9.

Benzie, J.A.H. and Williams, S.T. (1992a) No genetic differentiation of giant clam (*Tridacna gigas*) populations in the Great Barrier Reef, Australia. *Mar. Biol.*, **113**, 373–77.

Benzie, J.A.H. and Williams, S.T. (1992b) Genetic structure of giant clam (*Tridacna maxima*) populations from reefs in the western Coral Sea. *Coral Reefs*, **11**, 135–41.

Benzie, J.A.H., Williams, S.T. and Macaranas, J. (1993) Allozyme electrophoretic methods for analysing genetic variation in giant clams (*Tridacnidae*). Australian Centre for International Agricultural Research, *Technical Report* **23**.

Campbell, C.A., Valentine, J.W. and Ayala, F.J. (1975) High genetic variability in a population of *Tridacna maxima* from the Great Barrier Reef. *Mar. Biol.*, **33**, 341–5.

Hamilton, L.J. (1992) Surface circulation in the Tasman and Coral Seas: climatological features derived from bathy-thermograph data. *Aust. J. Mar. Freshwat. Res.*, **43**, 793–822.

Macaranas, J.M., Ablan, C.A., Pante, Ma.J.R., Benzie, J.A.H. and Williams, S.T. (1992) Genetic structure of giant clam (*Tridacna derasa*) populations from reefs in the Indo-Pacific. *Mar. Biol.*, **113**, 231–8.

Nash, W.J., Goddard, M. and Lucas, J.S. (1988) Population genetic studies of the crown-of-thorns starfish *Acanthaster planci* (L.) in the Great Barrier Reef region. *Coral Reefs*, **7**, 11–18.

Nei, M. (1978) Estimation of average heterozygosity and genetic distance from a small number of individuals. *Genetics*, **89**, 583–90.

Potts, D.C. (1985) Sea-level fluctuations and speciation in Scleractinia. *Proc. 5th Internat. Coral Reef Cong.*, Centenne Museum-Ephe, Moorea, **4**, 127–32.

Springer, V.G. (1982) Pacific plate biogeography, with special reference to shorefishes. *Smithson. Cont. Zool.*, **367**, 1–181.

Waples, R.S. (1987) A multispecies approach to the analysis of gene flow in marine shore fishes. *Evolution*, **41**, 385–400.

Weir, B.S. and Cockerham, C.C. (1984) Estimating F-statistics for the analysis of population structure. *Evolution* **38**, 1358–70.

2

Gene flow and population structure

2.1 THE DIFFERENT LEVELS OF POPULATION STRUCTURING OF THE
DOGWHELK, *NUCELLA LAPILLUS*, ALONG THE SOUTH DEVON COAST

Jerome Goudet, Thierry De Meeüs, Amanda J. Day and Chris J. Gliddon

Abstract

Allozymic data on Nucella lapillus *obtained by A.J. Day from 15 sites along the south Devon coast are analyzed using a technique based on Wright's F-statistics. After appropriate corrections of F-statistics to eliminate sources of bias, we calculate F_{is} for different groupings of the 15 samples, and graphically represent its changes. Three different levels of population structuring can be discerned, and a significant statistical difference is obtained for the third level after withdrawal of one of the eight loci which is shown to behave differently from the others. We were unable, however, to identify the breeding unit, and suggest two hypotheses: (1) dogwhelks, at a small spatial scale, are best described by the isolation by distance model; (2) the sampling strategy was inappropriate. Suggestions are made about how to distinguish between these two hypotheses.*

2.1.1 Introduction

In studies of natural populations, it is often necessary to understand the underlying genetic structure of natural populations before testable hypotheses can be erected to explain the observed pattern of genetic variability. Two types of methods are employed: direct methods, based on the observation of the movements of animals, gametes or marker genes; and indirect methods, based on the observed spatial distribution of alleles, genotypes, chromosomal segments or phenotypic traits. Unfortunately, the direct methods lead to a very restricted view of gene flow in time and space since occasional long-range migration can be very difficult to detect and observed movements of animals or gametes may not lead to reproduction or fertilization. In contrast, indirect methods allow inferences to be made about the historical levels or patterns of gene flow that have given rise to the observed pattern of genetic variation (Slatkin, 1985). The latter methods are sensitive to rare events in the past that affect the present structure of the population. One of these indirect methods is based on

Genetics and Evolution of Aquatic Organisms. Edited by A.R. Beaumont.
Published in 1994 by Chapman & Hall, 2–6 Boundary Row, London SE1 8HN.
ISBN 0-412-49370-5

the *F*-statistics of Wright (1921, 1951). It partitions any deficit of heterozygotes into within-population and among-population components, allowing inferences to be made about the underlying levels of inbreeding and gene flow. Our purpose in this study is to derive and then use appropriate estimators of these *F*-statistics to extract the different levels of genetic structuring in a geographic study of the dogwhelk, *Nucella lapillus*, using data collected by Day (1990).

2.1.2   Materials and methods

(a)   Wright's F-statistics and estimators

When surveying a population using gel electrophoresis, one obtains both the genotypic and allelic frequency arrays of the sample from a single set of observations. The *F*-statistics are a set of tools that allow a comparison of the genotypic array of the population under investigation with that of a random mating population. This is done by measuring the ratio of the difference between observed and expected heterozygosity under Hardy–Weinberg equilibrium (HWE) to the expected heterozygosity:

$$F = \frac{H_{exp} - H_{obs}}{H_{exp}} = 1 - \frac{H_{obs}}{H_{exp}}$$

(2.1.1)

where *F* stands for fixation index. If individuals are more homozygous than expected under HWE, *F* will be positive with a maximum of 1, and if they are less, *F* will be negative, with a minimum of –1. If the sampled population consists of subsamples from different locations, the fixation index could be measured either at the level of the subsamples, when it is called F_{is}, where '*i*' stands for individuals and '*s*' for subpopulation, or at the level of the total sample, when it is called F_{it}, '*i*' for individuals and '*t*' for total population. If there is a discrepancy between these two fixation indices, it is due to unequal allelic frequencies over the subsamples. This is known as the Wahlund effect (Wahlund, 1928). Pooling together the subsamples for calculation of the expected heterozygosity will lead to a smaller value than if the expected heterozygosity was calculated in each subsample and then averaged. This source of heterozygote deficit is measured by F_{st}, '*s*' for subpopulation and '*t*' for total population.

These tools are often used by population biologists because they can easily be associated with the inbreeding coefficient (Wright, 1921) and the probability of identity by descent (Malécot, 1948). Wright (1943) also showed that the quantity F_{st} could lead to an estimate of gene flow if it is assumed that the studied population is structured as in an island model. The relation between F_{st} and gene flow is given by:

$$F_{st} \simeq \frac{1}{4N_e m_e + 1}$$

(2.1.2)

where N_e is the effective local subsample size and m_e is the effective migration rate. This relationship holds providing that m_e is small and N_e large. Similar developments have been applied to other models such as the stepping-stone model (Crow and Aoki, 1984) and the isolation by distance (IBD) model (Slatkin and Barton, 1989), but no well-defined answers emerged, although some insight has been gained into the effects of population structuring.

Although measuring gene flow by means of *F*-statistics seems a difficult task, identification of the different levels of structuring should be possible, providing that appropriate estimators are used. The main feature of a good estimator is that it will be independent of both sample sizes and the number of subpopulations sampled. To date, two groups of such estimators exist, one developed by Cockerham (1969, 1973; Weir and Cockerham, 1984) by means of hierarchical analysis of variance, and the other developed by Nei (1973, 1977; Nei and Chesser, 1983), that seek unbiased estimators of the different components of the F-statistics. Both groups of estimators give the same result for F_{is} when sample sizes are equal, but Weir and Cockerham's (1984) F_{is} deals explicitly with the problem of unequal sample sizes by weighting the F_{is} obtained from each sample by its sample size, whereas Nei and Chesser (1983) advocate the use of harmonic mean of sample size as the weighting factor. As the sizes of our samples vary widely from one sampled location to another (from 6 individuals to 39 individuals), and because the technique outlined below relies on unbiased estimates, we will use Weir and Cockerham's (1984) F_{is}.

(b) Unravelling the genetic structure

As previously mentioned, unbiased estimators will give the same result whatever the sample size, as long as the individuals constituting the sample belong to the same breeding group. However, if individuals belong to different breeding groups, the estimated statistic will be a combination of both the within-subpopulation heterozygote deficit due to the breeding behaviour of the species (e.g. inbreeding) and the among-subpopulation deficit due to the Wahlund effect. Assessing the level at which the population is structured is a prerequisite for discriminating between these two factors. Once determined, the population structure should then help to disentangle the different facets of selection acting on the species.

The principle of our analyses is based on the very interesting feature described by the dashed curve in Figure 2.1.1. This figure was obtained from a computer-generated stepping-stone model. Nets of increasing mesh size are superimposed on the genotypes obtained from the stepping-stone model, and, for each mesh size, F_{is} is estimated and shown on the figure. When the mesh is smaller or equal to the deme size, F_{is} remains the same, but starts increasing when the mesh size encompasses two or more demes. When investigating the breeding pattern of a species, the first step is to sample at the appropriate scale. If we have no prior knowledge of the reproductive system of the organism, we need to obtain samples from as restricted an area as possible, to ensure that these samples belong to the same breeding group. F_{is} can then be calculated, using a modification of Weir's (1990) program for the calculation of F-statistics.

For the next step, samples are pooled in pairs. The pooling is subjective but, usually, geographic distances form a good basis. Once pooled, the data are reprocessed using Weir's program to get a new estimate of F_{is} for the first pooling level. The procedure is repeated until all samples are pooled together, leading to the last estimate, F_{it} (F in Weir and Cockerham's (1984) notation). The different estimates are then plotted on a graph as in Figure 2.1.1, the x-axis representing the different levels of pooling and the y-axis the equivalent F_{is} value. If pooling did not merge together two different breeding groups, then F_{is} stays constant. However, if two or more different breeding units are pooled together, there will be an increase in the value of F_{is}, the magnitude of this increase being dependent on the amount of gene flow between the different breeding groups. This curve can be obtained for each allele, each locus, or any combination of them, providing that appropriate combined estimators are used (Weir and Cockerham (1984) describe such combined estimators). The only modifications needed to the input data file for the different levels of pooling are the labels of the different samples.

It is important to note that estimators are unbiased: the solid curve on Figure 2.1.1 represents the values of a biased F_{is} (estimated without any corrections). It can be seen from Figure 2.1.1 that, unless appropriate estimators are used (dashed line), the levels of structuring may stay hidden (solid line).

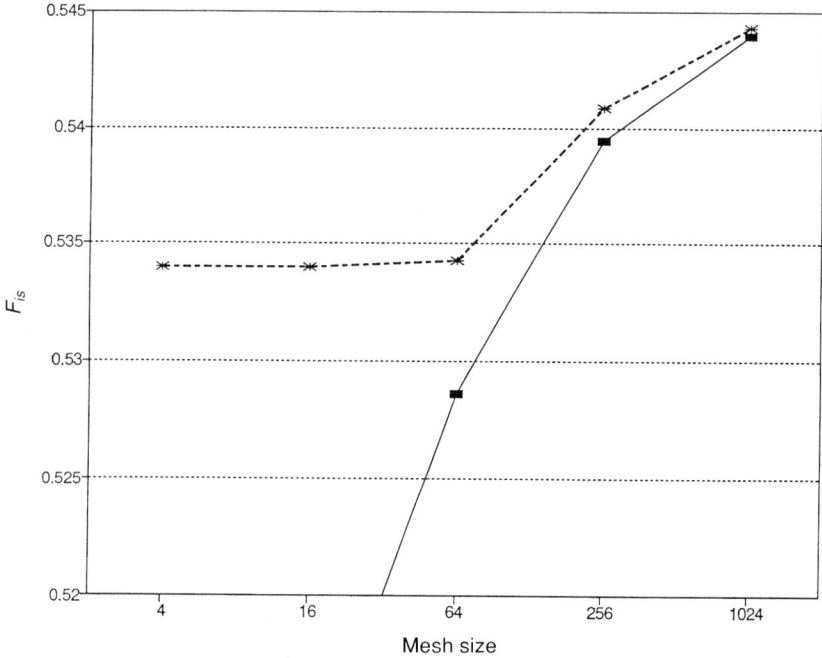

Figure 2.1.1 Change of F_{is} as a function of sampling mesh size. The computer-generated population was a two-dimensional stepping-stone model. The deme size was 64, the migration proportion was 20% and there was 70% selfing. Dashed line: F_{is} estimated using Weir and Cockerham (1984) stay constant when the samples are made of smaller units than the real deme size, but increase as soon as the sampling mesh is larger than the deme size. Solid line: same parameter set as for dashed line. The solid curve represents the change of Wright's F_{is}, from which it is impossible to detect the level of structuring.

Weir's (1990) program also provides bootstrap and jackknife confidence intervals of the statistics. These two techniques are based on recurrent resampling of the data, leading to non-parametric or distribution-free statistics, so that different estimates can be compared without the assumptions of normality or homoscedasticity of the variances. A bootstrap consists of producing a large number (1000) of replicated data sets by sampling, with replacement, from L loci if the survey consisted of L loci and re-estimating the different F for each new combination. Based on the assumption that the loci examined are a random sample of all possible loci, a confidence interval at 95% for the different F-statistics is obtained by taking the inner 95% estimates.

A jackknife consists of omitting each locus in turn and re-estimating the F. For L loci, one obtains L different estimates. The jackknife mean is the average of the L estimates, and the jackknife variance is their variance. With jackknifing, it is also possible to obtain an estimate per locus by jackknifing over sampled location. It would obviously be pointless to apply a bootstrap over sampled locations, at least as far as F_{st} and F_{it} are concerned.

(c)   The dogwhelk, Nucella lapillus

The dogwhelk is a widely distributed predatory intertidal gastropod, feeding on mussels and barnacles. It is ubiquitous on rocky substrates around the coasts of Great Britain and Ireland,

and extends from Iceland to Portugal (Berry and Crothers, 1974). Restriction in adult movement (a few metres only) (Hughes, 1972), associated with the absence of a dispersed planktonic stage, are likely to result in local differentiation of subpopulations (structuring).

Shell shape is different in exposed and sheltered sites: whelks from exposed sites have a thinner, shorter shell than whelks from sheltered sites, and a larger aperture that allows them to resist wave action. On the other hand, a thicker shell allows whelks from sheltered sites to resist the action of predatory crabs during their growth (Currey and Hughes, 1982).

Chromosome number has been found to vary between sheltered and exposed sites, with $2n = 26$ for exposed sites and $2n = 36$ for sheltered sites (but see section 7.2). This polymorphism is due to Robertsonian translocations (Bantock and Cockayne, 1975). However, when no chromosome number polymorphism can be found, the number of chromosomes is $2n = 26$ and this is the case in most of the British Isles apart from the English Channel (Bantock and Cockayne, 1975).

(d) The data

Allozyme data were obtained by Day (1990), on dogwhelks from 15 sites 50 m to 21 km apart in south Devon, southwestern England (Figure 2.1.2). Sites 1 to 5 (between Blackpool and Start Point) are very exposed to wave action and are quite distant from each other

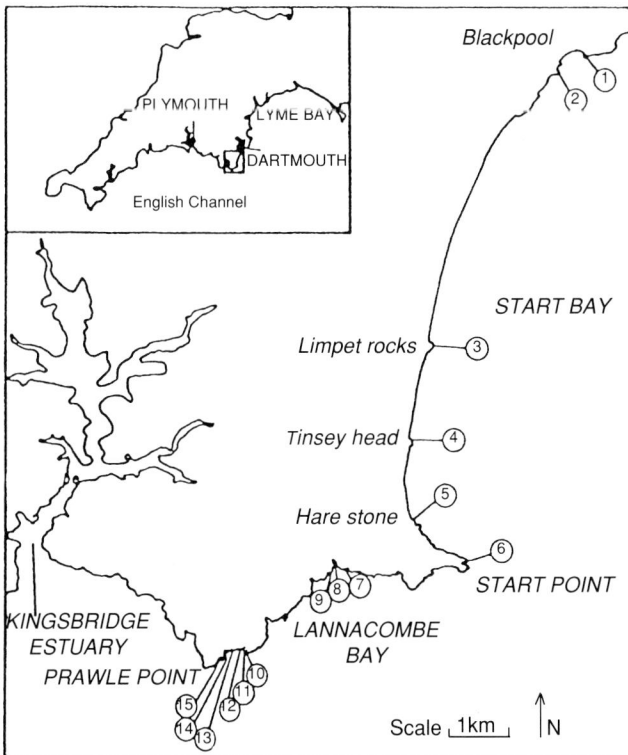

Figure 2.1.2 Location of the sampled sites for *N. lapillus* along the south Devon coastline.

(0.8–6.7 km). All sites with whelk populations were sampled. These populations were usually dense, with easily identified breeding aggregations in crevices. Each sample consisted of all the whelks within a single aggregation (Day, 1990). To the south of Start Point, the coastline is mainly sheltered. Ten sites were sampled along this strip of coast, one at Start Point itself (six individuals), three around Lannacombe Bay, and six to the east of Prawle Point. At both Prawle and Lannacombe, the samples came from a 0.5-km stretch of coastline, and the distance between samples was 50–150 m at the former and 150–300 m at the latter (Day, 1990). Whelks at these 10 sites were dispersed and no aggregations could be found; therefore, samples were taken within foraging areas less than 4 m^2 (following estimates of maximum distance travelled by whelks) (Hughes, 1972) in an attempt to ensure that individuals would be part of the same breeding unit. The size of the sampled areas in the sheltered sites meant that no more than 21 whelks were found at a single sample location (Day, 1990).

Samples were analyzed for allozyme variation at eight soluble enzyme loci: *Est-3*, *Lap-1*, *Lap-2*, *Mdh-1*, *Pep-1*, *Pep-2*, *Pgm-1* and *Pgm-2*. Nomenclature, electrophoresis buffers and staining methods follow those of Day and Bayne (1988) modified by Day (1990). Raw data are presented in Day (1990). All loci were polymorphic, and the number of alleles per locus ranged from two for *Pep-1*, *Pgm-1*, *Est-3* and *Pep-2* to four for *Pgm-2* and *Lap-2*. The interesting feature of this sample collection is that it is particularly well suited to the type of analysis described above: the sampling locations are restricted to a very small area, the distances between samples have been recorded, and the data consist of both allelic and genotypic frequencies, a crucial point if one is to use F_{is} to estimate levels of structuring. For all sites pooling strategy was:

1. all samples independently
2. 8–9, 10–11, 14–15 pooled together
3. 7–9, 10–12, 13–15 pooled together
4. 7–9, 10–15 pooled together
5. 1–2, 4–5, 7–9, 10–15 pooled together
6. 1–2, 3–5, 7–9, 10–15 pooled together
7. 1–5, 6–9, 10–15 pooled together
8. 1–5 and 6–15 pooled together
9. all sites together.

Each curve will consist, therefore, of nine data points. For the study of sheltered sites only, the pooling strategy was:

1. all sheltered samples independently
2. 8–9, 10–11, 14–15 pooled together
3. 8–9, 10–11, 12–13, 14–15 pooled together
4. 7–9, 10–12, 13–15 pooled together
5. 7–9, 10–15 pooled together
6. 6–9, 10–15 pooled together
7. all sheltered sites together.

An extra level of pooling between level 2 and 3 (8–9, 10–11, 12–13, 14–15) was added to check the effect of the somewhat subjective pooling strategy. There will therefore be seven data points.

In a previous analysis, Day (1990) found that high F_{st} values resulted, in the whole sample, from a high degree of heterogeneity from exposed (quite monomorphic) to sheltered (rather polymorphic) sites. Some evidence of a smaller scale of population structuring came from the analysis of some of the eight loci studied but without the opportunity of calculating confidence intervals or, consequently, the precise scale at which such structuring might take place.

2.1.3 Results

(a) Analysis per site and per locus

Frequencies of the most common allele per site and per locus are given in Table 2.1.1, together with the number of monomorphic sites and loci using the 95% criterion. The number of monomorphic loci per site gives an idea of the amount of variability present. For the exposed sites (1–5), an average of 60% of the loci within site are monomorphic, whereas this number falls to 25% for the sheltered sites (6–15). *Pgm-1* is the most monomorphic locus, with 11 out of 15 sites monomorphic. F_{is} values per locus and site are given in Table 2.1.2 together with the percentage contributed by each to the global F_{is} value (this contribution is the weighting factor for $F_{is_{il}}$, $N_i \Sigma_k p_k q_k$, where N_i is the sample size, and $\Sigma_k p_k q_k$ the sum over all alleles at the locus of the product pq). The exposed sites contribute 24.13% of the F_{is} value, whereas the Prawle Point sites account for 49.94%, and Lannacombe Bay 23.37%. Overall F_{is} as estimated using the Weir and Cockerham (1984) method is –0.0021 with a 95% confidence interval (CI) using bootstrapping over loci of [–0.10,0.08]. The jackknife over loci gives essentially the same CI. The overall F_{is} therefore leads us to infer that there is random mating within sites. F_{is} values per locus for all sites, together with 95% CI obtained from jackknifing over populations, and 95% CI obtained from bootstrapping over loci, are given in Figure 2.1.3. The jackknife CI provides information about the consistency of F_{is} estimates over populations, but is not independent of the level of variability found at the locus (if most of the sites are monomorphic at one locus, dropping these sites in turn will lead to the same jackknife estimate, therefore narrowing the CI). Indeed, *Pgm-1* has the smallest CI, but this is to be expected because of its lack of variability. *Lap-2* stands out, with an F_{is} value of –0.25, and a CI that does not overlap with other loci. If bootstrapping or jackknifing over loci CI is used instead of jackknifing over populations, *Lap 2* is still the locus which stands out (of the other loci, *Pep-1*, *Lap-1* and *Pgm-2* had means just outside of the range of the bootstrap CI).

Table 2.1.1 *N. lapillus*: frequency of the most common allele per locus and per site

				Loci					Fixed
	Pep-1	Lap-1	Pgm-1	Pgm-2	Lap-2	Pep-2	Mdh-1	Est-3	
Sites									
1	0.731	0.875	0.929	0.917	0.964	0.964	0.929	0.964	3
2	0.733	0.879	0.983	0.8	1	1	0.967	1	5
3	0.718	1	0.962	0.731	1	1	0.859	1	5
4	0.929	1	1	1	1	1	0.786	1	6
5	0.875	0.979	0.979	0.913	1	1	0.854	1	5
6	0.8	0.75	1	0.667	1	0.8	0.833	0.833	2
7	0.962	0.964	0.893	0.893	0.929	0.893	0.929	0.929	2
8	0.895	0.868	0.895	0.895	0.789	0.583	0.553	0.553	0
9	0.765	0.794	0.853	0.882	0.824	0.765	0.706	0.853	0
10	0.969	0.875	0.969	0.938	0.563	0.688	0.75	0.656	2
11	0.952	0.881	0.976	0.833	0.619	0.53	0.5	0.5	2
12	0.852	0.971	0.971	0.971	0.529	0.559	0.559	0.559	3
13	1	1	1	0.972	0.528	0.719	0.917	0.917	4
14	0.976	0.857	0.976	0.952	0.69	0.524	0.881	0.881	3
15	0.976	0.857	1	0.833	0.595	0.675	0.857	0.875	2
Fixed	5	6	11	3	6	5	1	5	

Extreme column and row represent the numbers of monomorphic sites and loci respectively using the 95% criterion. For site numbers see Figure 2.1.2.

Table 2.1.2 N. lapillus: F_{is} values per site and per locus

Sites		Pep-1	Lap-1	Pgm-1	Pgm-2	Loci Lap-2	Pep-2	Mdh-1	Est-3	Contribution of the site
	1	-0.335	-0.100	-0.046	-0.052			-0.020		3.17%
	2	0.039	-0.117		-0.234					6.50%
	3	0.378			0.295			-0.151		9.88%
	4	-0.006						-0.201		0.71%
	5	-0.122			-0.073			-0.148		3.87%
	6	-0.143	0.616	-0.084	0.333	-0.046	1.000	-0.111	-0.109	2.56%
	7			-0.093	0.307	0.079	-0.084	-0.046	-0.046	3.38%
	8	-0.094	0.335	-0.143	0.462	0.217	0.001	0.125	-0.146	10.75%
	9	0.372	-0.231		0.459		0.371	0.004	0.323	9.24%
	10		-0.154		-0.017	-0.241	-0.132	-0.132	0.063	7.88%
	11		0.341		-0.112	-0.393	-0.280	-0.024	-0.024	11.42%
	12	-0.143				-0.222	0.428	-0.044	-0.044	8.92%
	13		0.635			-0.464	0.256	-0.063	-0.067	4.81%
	14		-0.142		0.168	-0.201	0.069	-0.112	-0.112	8.12%
	15		0.141		0.128	-0.462	-0.230	-0.142	-0.118	8.79%
All sites		0.098		-0.061		-0.245	0.057	-0.060	-0.031	
Contribution of the locus		9.22%	12.96%	4.42%	13.25%	17.57%	12.11%	14.98%	15.50%	

Empty cells are for monomorphic sites. Extreme column and row represent the contributions to global F_{is} of loci and sites respectively. For site numbers see Figure 2.1.2.

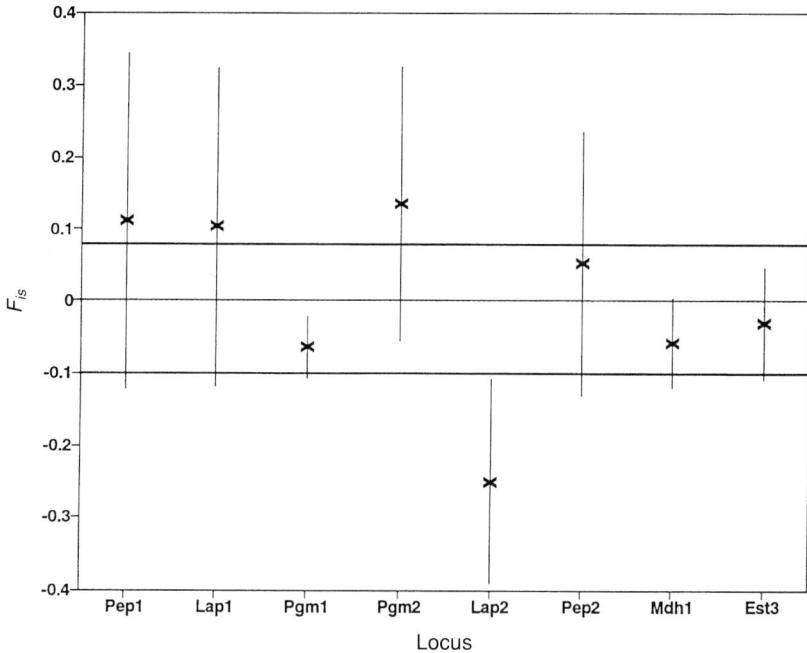

Figure 2.1.3 *N. lapillus* F_{is} values per locus. Vertical bars give the over-population jackknife 95% confidence intervals (CI) per locus. Thick horizontal lines correspond to bootstrap 95% CI.

The overall F_{st} is 0.3326, with bootstrapped 95% CI [0.16,0.44] and an equivalent jackknifed CI. This value of F_{st} is an indication of high structuring, and if the island model of population structure was assumed to be true, this would lead to an estimate of $N_e m_e$ using equation (2.1.2) of 0.5, that is, one effective migrant per site every other generation.

However, the variation in contribution to F_{is} from the three main areas (exposed (1–5), Lannacombe (7–9) and Prawle (10–15)) is an indication that these three groups should be re-analyzed (Table 2.1.3).

The F_{is} value (0.02) and CI [–0.11,0.09] for the exposed sites (1–5) are similar to the analysis encompassing all three areas. *Pep-1* gives as the highest deficit (0.11), and *Lap-1* and *Mdh-1* the greatest excess of heterozygotes (–0.10 and –0.13 respectively). *Pep-1* is the only locus that shows more variability in the exposed than in the sheltered area. The contribution to F_{is} came primarily from the loci *Pep-1* (34%), *Pgm-2* (28%) and *Mdh-1* (19%) (Figure 2.1.4 (a)).

The estimated F_{is} from the sheltered sites 10–15 at Prawle Point is –0.08, with 95% CI [–0.20, 0.05], again similar to the three areas together, although slightly more negative. *Lap-1* shows a large deficit of heterozygotes (0.20), and *Lap-2* an excess (–0.34). Contributions to F_{is} came primarily from *Lap-2* (24%), *Pep-2* (22%), *Mdh-1* and *Est-3* (16% each) (Figure 2.1.4 (b)).

The Lannacombe Bay sheltered sites 7–9 give unexpected results in that the estimated F_{is} is 0.09, but the 95% CI is [0.02, 0.18], implying that there is a heterozygote deficit in this area. This is also where allelic frequencies at four loci are most variable (allele 9 for *Lap-2*, *Est-3* and *Mdh-1*, and allele 11 for *Pep-2*) as shown in Day (1990, Figure 2). All loci contribute more or less the same amount to F_{is} (Figure 2.1.4 (c)) with *Pgm-2* showing a large deficit of heterozygotes (0.41).

Table 2.1.3 *N. lapillus:* F_{is} values per area and per locus

	Loci								
Area	*Pep-1*	*Lap-1*	*Pgm-1*	*Pgm-2*	*Lap-2*	*Pep-2*	*Mdh-1*	*Est-3*	*All*
Exposed	0.1095	-0.1044	-0.0245	0.07	-0.014	-0.0146	-0.1349	-0.0146	0.0201
Lannacombe Bay	0.1862	0.0185	-0.11	0.4173	0.1156	0.1194	0.0627	-0.0002	0.0935
Prawle Point	-0.0732	0.2015	-0.0008	0.0265	-0.3361	0.0075	-0.0769	-0.0357	-0.0792

Exposed, sites 1–5; Lannacombe Bay (sheltered), sites 7–9; Prawle Point (sheltered), sites 10–15.
For site numbers see Figure 2.1.2.
Site 6 not included due to small sample size.

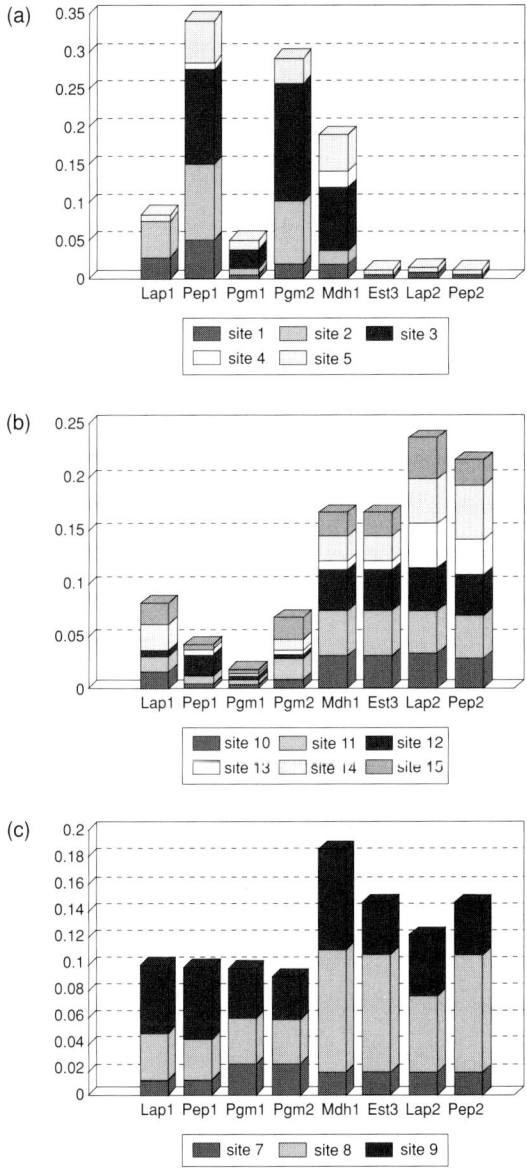

Figure 2.1.4 *N. lapillus* contribution of the different loci and sites to the per area F_{is}. (a) = Exposed sites, 1-5; (b) = Prawle point sites (sheltered), 10–15; (c) = Lannacombe Bay sites (sheltered), 7–9. For site numbers see Figure 2.1.2.

(b)   Recurrent pooling of sites

All sites

Results from the technique are shown in Figure 2.1.5(a). Pooling levels 2 and 3 led to a slight increase in the value of F_{is}, but the first major discontinuity occurs at pooling level 4, when all sheltered Prawle sites are grouped together. This could be considered as a first level of structuring, although confidence intervals of the mean (jackknifing over loci) between level 3 and 4 overlap. The next three pooling levels consist in grouping of exposed populations, and the F_{is} values obtained are constant. We have already mentioned that the exposed area was fairly monomorphic, and it is not surprising to see that these pooling stages do not provide any new information. If there is structuring in this area, the data set is unable to detect it. Level 7 also adds the Start Point samples (only six individuals) to those of Lannacombe without changing the F_{is} value.

The next discontinuity on the graph occurs between level 7 and 8, when pooling together Lannacombe Bay samples with Prawle Point samples. Here, the difference in F_{is} value is large enough for the two CI not to overlap. This is the second level of structuring. Pooling all samples together reveals the third level of structuring, highlighting the difference in genetic make-up of exposed and sheltered sites.

Sheltered sites only
Figure 2.1.5(b) describes the change of F_{is} in sheltered sites only. The graph shows essentially a similar behaviour to that for all sites (Figure 2.1.5(a)). The extra pooling level 3 (pooling of sites 12 and 13 together) leads to a slight increase in F_{is}, followed by a decrease when pooling 10–12 and 13–15. Although the difference is not significant, it suggests that 12 and 13, only 50 m apart, belong the different breeding units and confirms that the pooling strategy adopted is appropriate.

The observation that *Lap-2* is significantly different from all other loci (Figure 2.1.3) led to a re-analysis of sheltered sites excluding this locus. Results are given in Figure 2.1.5(c). The graph shows essentially the same behaviour, but confidence intervals of the mean are much narrower, and allow discrimination between pooling levels 4 and 5, the third level of structuring, which corresponds to the pooling together of all Prawle Point samples. The pooling of all sheltered sites (level 7) remains highly significant.

2.1.4 Discussion

The existence of structuring within the south Devon dogwhelk population is confirmed, and the discriminatory power of the analysis allowed the detection of three different levels of structuring.

1. Sheltered and exposed sites, which display differences in shell morphology and chromosome numbers typical of whelk populations (Crothers, 1975; Bantock and Cockayne, 1975; Berry 1983; but see section 7.2), also display wide differences in allozyme variation. Exposed sites are more monomorphic, and polymorphism is expressed at different loci from sheltered sites (Figure 2.1.4). Start Point seems to be a geographic barrier to gene flow (Day, 1990).
2. Lannacombe Bay and Prawle Point, separated by 3.3 km, correspond to two different populations with little genetic exchange.
3. Prawle Point itself is subdivided into two entities, with less gene flow between sites 10–12 and 13–15 than within these sites although sites 12 and 13 are separated by only 50 m. Withdrawal of *Lap-2* allows us to conclude that the sheltered sites of Prawle Point consist of two different breeding units, with less gene flow between sites 10–12 and 13–15 than within them. Other loci were withdrawn without altering the qualitative behaviour of the curve.

A possible fourth level of structure could exist but sites 7–9, 10–12 and 3–15 do not seem to be random breeding units, and the slight increase found between pooling levels 1, 2, 3 and 4 suggests that isolation by distance (IBD) occurs within these areas.

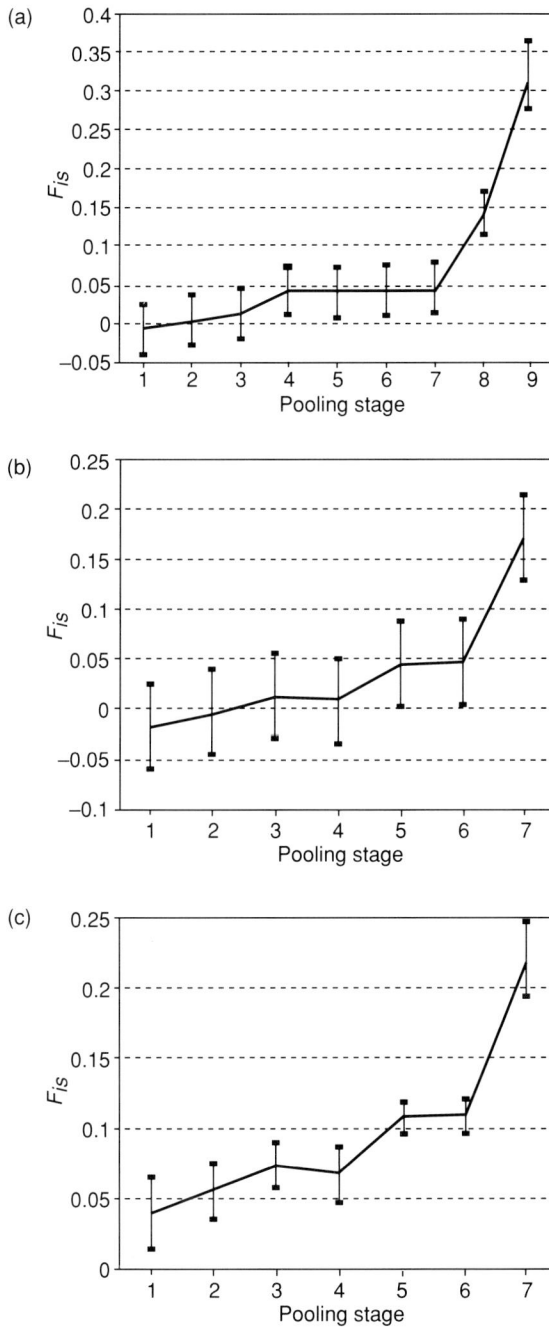

Figure 2.1.5 *N. lapillus* change in F_{is}. Pooling stages are described in the text. 95% CI of the mean are obtained by jackknifing over loci. (a) = All sites, all loci; (b) = change in F_{is} for sheltered sites only; (c) = change in F_{is} for sheltered sites only, excluding locus *Lap-2*.

A question could be raised as to how this technique compares with hierarchical *F*-statistics analyses, including more than one level. In a hierarchical analysis, one needs to decide *a priori* the different levels of structuring. In this research, no assumptions were made about how and where levels of structuring should be, other than pooling together close geographic locations. On the other hand, we cannot interpret the values of F_{is} obtained, because they correspond to a mixture of different factors affecting the genotypic structure of the population (even at the lowest level of structuring, there is still the suggestion that IBD is occurring, therefore preventing the inference of the magnitude of gene flow). Once the different levels of structuring are identified, it may prove useful to measure the respective magnitude of the levels of isolation by means of a hierarchical approach, but a problem still remains: by increasing the number of levels in an analysis of variance, one loses power, and may therefore be unable to partition effectively the different effects.

The significant heterozygote deficits obtained from the Lannacombe Bay area (sheltered) could be due either to disruptive selection and lowered fitness of the heterozygotes, or to the sampling strategy: whelks were necessarily collected from foraging sites rather than breeding sites as no aggregations were present. If more or less isolated breeding groups forage on the same site, the genetic analysis will display a lack of heterozygotes due to the Wahlund effect. It would be interesting to carry out experiments such as rearing individuals from each site and obtaining F_1 crosses. If the F_1 crosses have a lower fitness than the parental types, the first hypothesis will be confirmed.

Heterogeneity in allozyme polymorphism is striking, as shown in Figure 2.1.4. The contribution to F_{is} is very different between exposed and sheltered sites, and four clines of allelic frequencies coexist, with allelic change being largest at Lannacombe Bay, where the clines are steepest (Day, 1990). This could imply that investigation of the population structure by means of indirect methods using allozyme data is inappropriate. However, the present analysis seems to indicate that information can still be extracted, and is consistent with the finding of Hughes (1972) that movement in whelks estimated by direct methods is very restricted.

Using resampling CI, we have been able to show that *Lap-2* behaves differently from all other loci (Figure 2.1.3). In fact, this locus has been shown to be correlated to many ecological factors such as exposure to wave action, shell shape and chromosomal polymorphism (Day, 1990). Its exclusion led to a significant increase in F_{is} when pooling together all Prawle samples. Other loci have been shown to correlate with ecological factors, but their influence on the behaviour of the statistic is much smaller, and we could not distinguish them from the other loci.

It is necessary to identify 'badly behaved' loci before application of the technique. As Slatkin (1985) stated:

Estimates based on data from one or two loci should be suspect, but if estimates are based on data from numerous loci and there is consistency in the estimates using different methods, it is reasonable to have some confidence in the conclusions.

The identification of the breeding unit is a necessary prerequisite for any inference to be made about gene flow. The technique developed here allows the highlighting of variation in levels of gene flow. We were not able to find the precise limits of the random breeding units, and this could be for two reasons. Firstly, a random breeding unit does not exist in whelks, and the IBD model would be a more appropriate model of the genetic structure of populations of this organism. This is suggested by the slow increase of the curve in the first stages of pooling, a behaviour similar to that which is observed from IBD computer-generated data sets. Secondly, the sampling strategy, consisting of collecting whelks from foraging sites, is not appropriate, because different breeding units forage in the same site. To discriminate between these two hypotheses, it would be necessary to find whelk populations existing at higher densities than are found in south Devon and in which refuges are easily identified. Such sites exist all over Britain. One of them is on the island of Anglesey, North Wales. This site has other advantages: the distance between sheltered and exposed sites is much smaller than in Devon, not exceeding 300 m, and there is no numerical chromosomal polymorphism. A similar sampling of this site, coupled with ecological mark/recapture work and use of molecular markers to increase the number of variable loci, thereby minimizing the problem of badly behaved loci, may well shed more light on the population biology of this organism.

References

Bantock, C.R and Cockayne, W.C. (1975) Chromosomal polymorphism in *Nucella lapillus*. *Heredity,* **34**, 231–45.

Berry, R.J. (1983) Polymorphic shell banding in the dogwhelk, *Nucella lapillus* (Mollusca). *J. Zool,* **200**, 455–70

Berry, R.J. and Crothers, J.H (1974) Visible variation in dogwhelk, *Nucella lapillus*. *J. Zool,* **174**, 123–48.

Cockerham, C.C (1969) Variance of gene frequencies. *Evolution,* **23**, 72–84.

Cockerham, C.C (1973) Analysis of gene frequencies. *Genetics,* **74**, 679–700.

Crothers, J.H. (1975) On variation in *Nucella lapillus* (L): shell shape in population from the south coast of England. *Proc. Malacol. Soc, Lond.,* **41**, 489–98.

Crow, J and Aoki, K (1984) Group selection for a polygenic behavioural trait: estimating the degree of population subdivision. *Proc. Natl. Acad. Sci., USA,* **81**, 6073–7.

Currey, J.D. and Hughes, R.N (1982) Strength of the dogwhelk *Nucella lapillus* and the winkle *Littorina littorea* from different habitats. J. *Anim. Ecol.,* **51**, 47–56.

Day, A.J. (1990) Microgeographic variation in allozyme frequencies in relation to the degree of exposure to wave action in the dogwhelk *Nucella lapillus* (L.) (Prosobranchia: Muricacea). *Biol. J. Linn. Soc.,* **40**, 245–61.

Day, A.J and Bayne, B.L. (1988) Allozyme variation in populations of the dog-whelk *Nucella lapillus* (Prosobranchia: Murcacea) from the South West peninsula of England. *Mar. Biol.,* **99**, 93–100.

Hughes, R.N (1972) Annual production of 2 Nova Scotian populations of *Nucella lapillus* L. *Oecologia,* **8**, 356–70.

Malécot, G. (1948) *Les Mathématiques de l'Hérédité*, Masson, Paris.

Nei, M. (1973) Analysis of gene diversity in subdivided populations. *Proc. Natl. Acad. Sci. USA,* **70**, 3321–3.

Nei, M. (1977) F-statistics and analysis of gene diversity in subdivided populations. *Ann. Hum. Genet.* **41**, 225–33.

Nei, M. and Chesser, R.K. (1983) Estimation of fixation indices and gene diversities. *Ann. Hum. Genet.* **47**, 253–59.

Slatkin, M. (1985) Gene flow in natural populations. *Annu. Rev. Ecol. Syst.,* **16**, 393–430.

Slatkin, M. and Barton, N.H (1989). A comparison of three indirect methods of estimating average levels of gene flow. *Evolution,* **43**, 1349–68.

Wahlund, S. (1928). Zusammensetzung von population und korrelationsersheinung vom standpunkt der vererbungslehre aus betrachtet. *Hereditas,* **14**, 56–63.

Weir, B.S. (1990) *Genetic Data Analysis*, Sinauer Associates, Sunderland, Massachusetts.

Weir, B.S. and Cockerham, C.C. (1984) Estimating F-statistics for the analysis of population structure. *Evolution,* **38**, 1358–70.

Wright, S. (1921) Systems of mating. *Genetics,* **6**, 111–78.

Wright, S. (1943) Isolation by distance. *Genetics,* **28**, 114–38.

Wright, S. (1951) The genetical structure of populations. *Ann. Eugen.,* **15**, 323–54.

2.2   GENETIC VARIATION AND LOCALIZED GENE FLOW IN THE FAIRY SHRIMP, *BRANCHIPODOPSIS WOLFI*, IN TEMPORARY RAINWATER POOLS IN SOUTHEASTERN BOTSWANA

Bruce J. Riddoch, Sununguko W. Mpoloka and Mike Cantrell

Abstract

Local-scale population differentiation was studied in the anostracan Branchipodopsis wolfi, *which inhabits small temporary rainwater pools in southern Africa. Five polymorphic enzyme loci (Apk, Got-1, Got-2, Pgi and Pgm) were analysed from seven populations on a linear granite outcrop of about 100 m length. Single-locus F_{st}s showed significant genetic differentiation between populations for all five loci. The estimated number of migrants per generation (Nm), calculated from $\bar{F}_{st}$ by various methods, was between two and four. Jackknifing across loci using $\bar{F}_{st}$ and Nm produced lower jackknife estimates of Nm using the latter method. These estimates are unreliable because genetic distances (D_m) between populations indicated greater similarity between neighbouring pools so that the probability of dispersal between populations was not equal, which is a violation of the assumptions of the island model. Different agents of gene flow appear to be effective at different spatial scales and it is suggested that wind will cause observed similarities between neighbouring pools against a background of more random gene flow caused by vectors, which may become relatively more important over greater distances.*

2.2.1   Introduction

Electrophoretic studies of parthenogenetic and sexual pond-dwelling crustaceans have revealed considerable divergence in gene frequencies between local populations (Hebert, 1974; Hebert and Moran, 1980; Korpelainen, 1984; Boileau and Hebert, 1988, 1991; Chow, Fujio and Nomura, 1988; Boileau, Hebert and Schwartz, 1992). This has been explained by founder effects followed by restricted gene flow as a result of limited dispersal ability (Boileau and Hebert, 1991). The crustaceans in these studies spread by passive dispersal enhanced by various adaptations, including dormancy, gastrointestinal resistance and morphological accessories (Boileau, Hebert and Schwartz, 1992). Dormancy also makes these taxa highly resistant to extinction, as a consequence of environmental perturbations, and to evolutionary change, through the formation of a large dormant propagule pool (Hairston and De Stasio, 1988).

The amount of migration between local populations of pond-dwelling crustaceans, expressed as the number of migrants per generation (*Nm*), has been estimated indirectly from the level of gene frequency divergence ($\bar{F}_{st}$) assuming an island model (Wright, 1943), and these results suggest that migration rates are quite low (Boileau and Hebert, 1988; Boileau, Hebert and Schwartz 1992). However, at least two of the assumptions of the island model may not be valid in many cases. Firstly, dispersal amongst populations may not be equal or symmetrical, and secondly, the variance in gene frequencies may not be at equilibrium. Boileau, Hebert and Schwartz (1992) suggested that, for pond-dwelling organisms in physically compact regions, dispersal among populations may be equally probable. However, these authors demonstrated that their study populations may not be in equilibrium because when a small number of founders are capable of increasing population size very rapidly, gene frequencies become remarkably resistant to change from quite high numbers of dispersers. Thus the rate of decay of interpopulation variation in gene frequencies can be extremely slow; in their simulations several thousand generations. According to their simulations, lack of equilibrium could lead to underestimates of Nm by as much as two orders of magnitude. This was a particular problem in their study because their populations were less than 3000 years old. The majority of studies of genetic divergence and gene flow in pond-dwelling organisms have been conducted in northern temperate or Arctic regions with a recent glacial history, so most populations studied so far may be too young to have reached equilibrium.

Rock pools in the semi-arid subtropics are probably much older, the main cause of long-term temporal environmental heterogeneity having been a cyclical pattern of dry and wet phases (Thomas and Shaw, 1992). Consequently, the fauna of these ponds are more likely to have reached equilibrium.

The anostracan *Branchipodopsis wolfi* inhabits temporary rainwater pools in the semi-arid regions of southern Africa. The pools are dry throughout the dry season and dry out several times during the course of a single wet season. Consequently, the reproductive phase of the *B. wolfi* life-cycle must be complete within a few days. Anostracans are characterized by multi-year dormancy, which makes their populations extremely resistant to extinction.

In this section we examine interpopulation genetic variability at five loci in seven local populations on a single rock outcrop in southeastern Botswana. The aims are:

1. to extend the study of the genetic structure of populations of pond-dwelling organisms to the semi-arid subtropics;
2. to estimate the level of gene flow from $\bar{F}_{st}$ under the assumption that these populations are much older than those in previous studies of pond-dwelling crustaceans and, therefore, more likely to be at equilibrium, and that the compactness of the sample site should ensure equal probability of exchange of alleles;
3. to examine the assumption that the probability of dispersal between populations is equal at the local level.

2.2.2 Materials and methods

The sample site consists of a series of small temporary rainwater pools on a granite outcrop situated in semi-arid acacia savanna 7 km south of Gaborone ($24°43'$ S, $25°53'$ E) in southeastern Botswana. The mean duration of pools from filling to drying is approximately eight days (M. Cantrell, unpublished data). Rainfall is highly variable, averaging about 500 mm per annum, all of which falls in the summer months. The 1991–1992 wet season was particularly poor. Air temperatures frequently exceed $40\,°C$ in the summer, and soil surface temperatures can rise to $70\,°C$ (Arntzen and Veenendaal, 1986).

Seven pools were sampled on 20 February 1992; their relative positions are shown in Figure 2.2.1. Adult *B. wolfi* were collected from each pond using several sweeps with an aquarium net, which was then washed thoroughly before sampling another pond to eliminate the possibility of eggs being transferred between ponds.

Five polymorphic loci (frequency of commonest allele, $p < 0.95$) were scored: arginine phosphokinase (*Apk*, EC 2.7.3.3), glutamate-oxaloacetate transferase (*Got-1* and *Got-2*, EC 2.6.1.1), phosphoglucose isomerase (*Pgi*, EC 5.3.1.9) and phosphoglucomutase (*Pgm*, EC 2.7.5.1).

Electrophoresis was conducted on 12% starch gels using a Tris/citric acid buffer system (electrode buffer: 85 mM Tris, 75 mM citric acid, adjusted to pH 7.1 using 1 M Tris; gel: 27 mM Tris, 7.2 mM citric acid, pH 8.4).

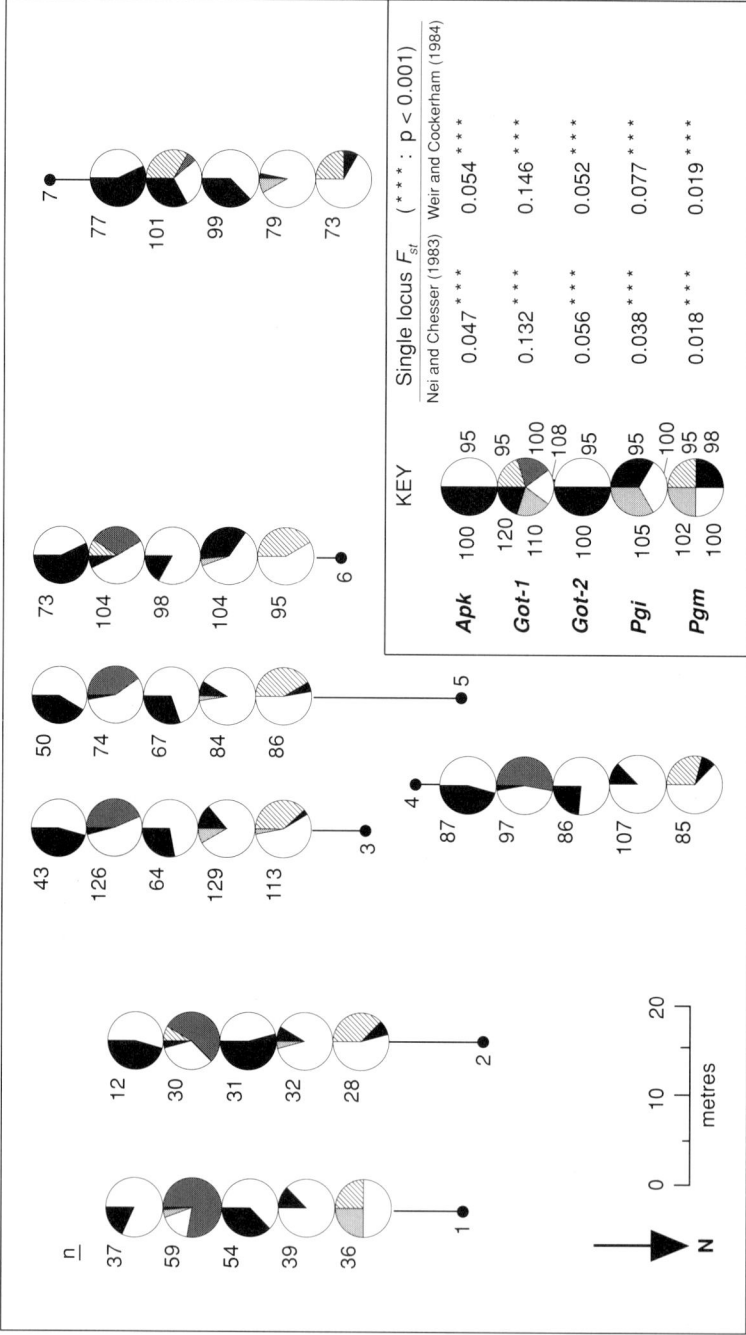

Figure 2.2.1 *B. wolfi*. The sample site showing the relative positions of the seven sample ponds (indicated by black spots), the allele frequencies of the five polymorphic loci in the seven study populations and the single-locus F_{st}s.

F_{st} and $\bar{F}_{st}$ were calculated using the methods of Nei and Chesser (1983), where $\bar{F}_{st}$ is equivalent to their unbiased G_{st}, and Weir and Cockerham (1984). $\bar{F}_{st}$ was used to estimate the number of migrants per generation (Nm) where:

$$Nm = \left(\frac{1}{\bar{F}_{st}} - 1\right)/4$$

The jackknife procedure recommended by Weir and Cockerham (1984) was used to derive a less biased estimate of Nm and to put confidence limits on it. F_{st} for each locus was evaluated using a χ^2 test (Workman and Niswander, 1970). The conditional average frequency method of Slatkin (1985a) was not used to estimate Nm because the number of 'private' alleles was too low and this method appears to be less reliable (Johnson, Clarke and Murray, 1988).

Nei's unbiased minimum genetic distance (D_m) (Nei, 1978) was calculated for each pair of populations using only the five polymorphic loci, and a dendrogram was constructed using the UPGMA method (Sneath and Sokal, 1973). Preliminary studies showed that some enzymes had more than one zone of staining and it was unclear whether these represented different loci or post-transcription modifications of a single locus. Such potentially mono-morphic loci were not included in the analysis and consequently genetic distances presented in this study are only used for comparisons between study populations.

2.2.3 Results

Allele frequency data are presented in Figure 2.2.1. χ^2-tests of F_{st} values (null hypothesis: $F_{st} = 0$) indicated significant heterogeneity ($p < 0.001$) between populations for all five polymorphic loci (Figure 2.2.1). The two methods of calculating F_{st} produced congruent results except for *Pgi*, where Weir and Cockerham's (1984) method gave a higher estimate. Variability was highest at *Got-1*. In population 7, alleles *Got-1*^{95} and *Got-1*^{120} were at relatively high frequencies, whereas the frequency of *Got-1*^{100} was markedly lower than for the other six populations. An additional allele, *Got-1*^{110}, was present in population 1, which was also characterized by *Pgm*^{102} at a frequency of 0.24.

The various estimates of Nm suggest that the number of dispersers exchanged per generation is somewhere between 2 and 4 (Table 2.2.1). Weir and Cockerham's (1984) method gave slightly lower estimates of Nm but the biggest change in Nm occurred if the jackknifed estimate was calculated from Nm rather than $\bar{F}_{st}$. Jackknifing from Nm produced a lower estimate by both Nei and Chesser's (1983) and Weir and Cockerham's (1984) methods.

The dendrogram of D_m (Figure 2.2.2) suggests that greater similarity exists between populations in close proximity to one another, although geographic distance is not the only factor influencing D_m. For example, the genetic distance between ponds 6 and 7 ($D_m = 0.057$) is greater than that between pond 7 and the more distant ponds 3, 4 and 5 (D_m ranges from 0.046 to 0.049).

2.2.4 Discussion

This study indicates that the high levels of genetic differentiation found between local populations of pond-dwelling crustaceans in Arctic and northern temperate species also occur in the semi-arid subtropics. The estimates of Nm for *B. wolfi* are similar to most of the values calculated by Boileau, Hebert and Schwartz (1992) for a range of species, which included two anostracans, *Artemiopsis stefanssoni* and *Branchinecta paludosa*. The estimates for *B. wolfi* are very similar to those for *A. stefanssoni* ($Nm = 3.1$) but markedly higher than those for *B. paludosa* ($Nm = 0.4$), which had the lowest estimated Nm of all the species in their study.

Table 2.2.1   Gene flow in populations of *B. wolfi*: estimates of *Nm* derived from $\bar{F}_{st}$

Estimate	$\bar{F}_{st}$	*Nm*
Nei and Chesser (1983)		
Direct	0.064	3.656
Jackknife ($\bar{F}_{st(N)}$)	0.065	3.596
	(0.045–0.086)	(2.671–5.368)
Jackknife (*Nm*)	–	2.866
		(1.127–4.605)
Weir and Cockerham (1984)		
Direct	0.074	3.124
Jackknife ($\bar{F}_{st(W)}$)	0.076	3.046
(0.051–0.100)	(0.051–0.100)	(2.239–4.625)
Jackknife (*Nm*)	–	2.386
		(0.849–3.922)

Values in parentheses are 95% confidence limits for jackknifed estimates. For details of calculation see text.

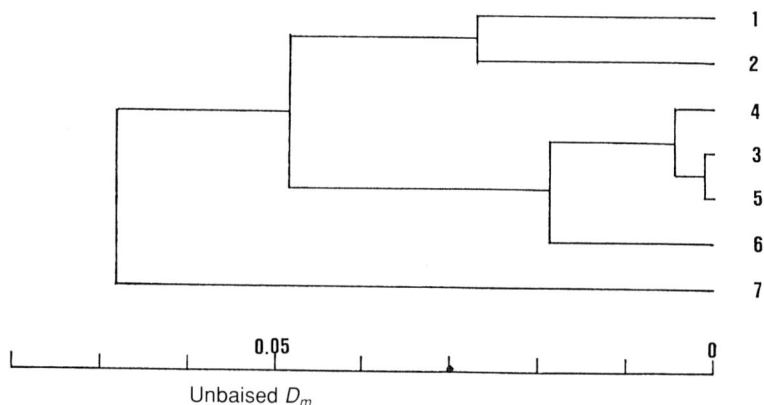

Figure 2.2.2 Dendrogram of the seven *B. wolfi* populations separated by Nei's unbiased minimum genetic distance calculated from the five polymorphic loci, and clustered by the UPGMA method.

The apparent similarities in estimated levels of gene flow between this study and most of the findings of Boileau, Hebert and Schwartz (1992) should be treated with some caution. If the assumption that our populations are old enough to have reached equilibrium is correct, either Arctic populations are not in equilibrium and experience higher levels of gene flow, or they are in equilibrium and experience similar levels of gene flow. An ecological comparison of the anostracan species might give some insight into this problem. If the Arctic populations are in equilibrium we might expect *B. paludosa* to have a lower dispersal capacity; this does not appear to be the case (Boileau, Hebert and Schwartz, 1992).

The slightly lower estimate of *Nm* using the method of Weir and Cockerham (1984) is in agreement with the findings of Waples (1987) for marine shore fishes. Goudet *et al.*

(Section 2.1) suggest that the method of Weir and Cockerham is more appropriate. However, in the case of pond-dwelling fauna, Boileau, Hebert and Schwartz (1992) have set the precedent by using Nei and Chesser's (1983) method with corrections for sampling biases (Nei, 1986); hence, for comparative studies this method may be more suitable.

It is clear from Table 2.2.1 that the jackknife estimate of Nm is strongly dependent on whether $\bar{F}_{st}$ or Nm values are jackknifed; this is because of the non-linear relationship between them. Both approaches have been used in published work: Nm by Johnson, Clarke and Murray (1988) and $\bar{F}_{st}$ by Benzie and Stoddart (1992). Intuitively we would support jackknifing Nm because it is the value we are assessing. However, the asymmetric confidence limits around estimates of Nm derived from jackknifing $\bar{F}_{st}$ make some evolutionary sense because the effect of different levels of gene flow on other evolutionary processes is probably not linear. The difference in the two estimates is of particular evolutionary significance in this study because the confidence limits on the Nm jackknifed estimates extend approximately to $Nm = 1$, the point below which genetic drift is not suppressed by gene flow (Slatkin, 1985b, 1987).

When considering the above estimation of gene flow it should be noted that at least one of the assumptions of the island model has been violated. We should take into account the spatial relationship between populations even on this local scale. Nm would be expected to be higher than our estimates amongst the central populations (3, 4 and 5) and lower amongst the more peripheral populations (Figure 2.2.1). To get a better idea of the processes involved in determining the population genetic structure of *B. wolfi*, we need to use approaches that incorporate spatial relationships between populations such as stepping-stone models (Slatkin, 1991, 1993) and spatial autocorrelation (Slatkin and Arter, 1991; Sokal and Jacquez, 1991). This would require a larger number of populations. However, genetic distance is not only a function of geographic distance; processes are involved which are not correlated with distance (distance-independent gene flow, geographically independent selection and genetic drift). With a larger data set we could try to assess the relative contribution of these to the overall pattern of genetic differentiation by filtering out the geographic component and then re-analysing the data. It may be possible to use the conceptual approach of time-series analysis where a time series is gradually degenerated to white noise (genetic drift) by the sequential filtering out of trends and pulses of various frequencies.

The results suggest that a combination of distance-dependent and distance-independent agents are acting on populations of *B. wolfi*. Possible agents of gene flow would be overflow, vectors and/or wind.

We would expect these agents of gene flow to be effective on different spatial scales. Dispersal via overflow channels will only occur between specific populations. The only ponds in this study connected by an overflow channel are 4 and 5 but there is no evidence that these are any more similar to each other than either is to pond 3. Dispersal by vectors and wind may both be important at this study site. We believe that wind would produce the spatial pattern observed at this site and may be the most important agent of localized gene flow in this semi-arid environment. Wind would probably be less influential in northern temperate and Arctic regions where sediment may be more compacted following pond drying and where ponds are flooded for considerably longer. Therefore, Boileau, Hebert and Schwartz's (1992) suggestion of equal probabilities of gene flow between populations in physically compact sites is probably valid for their sites. Dispersal by vectors may explain similarities between more widely separated populations. Clearly there is a need to sample more widely distributed populations of *B. wolfi* to see if there is a more random relationship between clusters of populations than within clusters.

It may also be possible to examine the effect of gene flow on the rate of adaptive evolution in *B. wolfi* through a comparative study of specific life-history parameters (e.g. maturation rate) in ponds of different duration and level of isolation. In this semi-arid environment the length of time a pond remains flooded will be a powerful selective force and it would be interesting to find out whether increased gene flow can constrain its effects.

Acknowledgements

We are very grateful to the University of Botswana for financial assistance and to Shameela Essack for research assistance.

References

Arntzen, J.W. and Veenendaal, E.M. (1986) *A Profile of Environment and Development in Botswana.* National Institute of Research and Documentation, University of Botswana, Gaborone.

Benzie, J.A.H. and Stoddart, J.A. (1992) Genetic structure of outbreaking and non-outbreaking crown-of thorns starfish (*Acanthaster planci*) populations on the Great Barrier Reef. *Mar. Biol.*, **112**, 119–30.

Boileau, M.G. and Hebert, P.D.N. (1988) Genetic differentiation of freshwater pond copepods at arctic sites. *Hydrobiologia* **167/168**, 393–400.

Boileau, M.G. and Hebert, P.D.N. (1991) Genetic consequences of passive dispersal in pond-dwelling copepods. *Evolution*, **45**, 721–33.

Boileau, M.G., Hebert, P.D.N. and Schwartz, S.S. (1992) Non-equilibrium gene frequency divergence: persistent founder effects in natural populations. *J. Evol. Biol.*, **5**, 25–39.

Chow, S., Fujio, Y. and Nomura, T. (1988) Reproductive isolation and distinct population structures in two types of the freshwater shrimp *Palaemon paucidens. Evolution*, **42**, 804–13.

Hairston, N.G. and De Stasio Jr, B.T. (1988) Rate of evolution slowed by a dormant propagule pool. *Nature*, **336**, 239–42.

Hebert P.D.N. (1974) Enzyme variability in natural populations of *Daphnia magna*. I. Population structure in East Anglia. *Evolution*, **28**, 546–56.

Hebert, P.D.N. and Moran, C. (1980) Enzyme variability in natural populations of *Daphnia carinata*. *Heredity*, **45**, 313–21.

Johnson, M.S., Clarke, B. and Murray, J. (1988) Discrepancies in the estimation of gene flow in *Partula*. *Genetics*, **120**, 233–8.

Korpelainen, H. (1984) Genic differentiation of *Daphnia magna* populations. *Hereditas*, **101**, 209–16.

Nei, M. (1978) Estimation of average heterozygosity and genetic distance from a small number of individuals. *Genetics*, **89**, 583–90.

Nei, M. (1986) Definition and estimation of fixation indices. *Evolution*, **40**, 643–5.

Nei, M. and Chesser, R.K. (1983) Estimation of fixation indices and gene diversities. *Ann. Human Genet.*, **47**, 253–9.

Slatkin, M. (1985a) Rare alleles as indicators of gene flow. *Evolution*, **39**, 53–65.

Slatkin, M. (1985b) Gene flow in natural populations. *Annu. Rev. Ecol. Syst.*, **16**, 393–430.

Slatkin, M. (1987) Gene flow and the geographic structure of natural populations. *Science*, **236**, 787–92.

Slatkin, M. (1991) Inbreeding coefficients and coalescence times. *Genet. Res.*, **58**, 167–75.

Slatkin, M. (1993) Isolation by distance in equilibrium and non-equilibrium populations. *Evolution*, **47**, 264–79.

Slatkin, M. and Arter, H.E. (1991) Spatial autocorrelation methods in population genetics. *Am. Nat.*, **138**, 499–517.

Sneath, P.H.A. and Sokal, R.R. (1973) *Numerical Taxonomy*, W.H. Freeman and Company, San Francisco.

Sokal, R.R. and Jacquez, G.M. (1991) Testing inferences about microevolutionary processes by means of spatial autocorrelation analysis. *Evolution*, **45**, 152–68.

Thomas, D.S.G and Shaw, P.A. (1992) *The Kalahari Environment*, Cambridge University Press, Cambridge.

Waples, R.S. (1987) A multispecies approach to the analysis of gene flow in marine shore fishes. *Evolution*, **41**, 385–400.

Weir, B.S. and Cockerham, C.C. (1984) Estimating *F*-statistics for the analysis of population structure. *Evolution*, **38**, 1358–70.

Workman, P.L. and Niswander, J.D. (1970) Population studies on southwestern Indian tribes. II. Local genetic differentiation in the Papago. *Am. J. Human Genet.*, **22**, 24–49.

Wright, S. (1943) Isolation by distance. *Genetics*, **28**, 114–38.

2.3 GENETIC STRUCTURE OF THE PALOURDE *RUDITAPES DECUSSATUS* L. IN THE MEDITERRANEAN

Philippe Borsa, Philippe Jarne, Khalid Belkhir and François Bonhomme

Abstract

The genetic structure of the bivalve Ruditapes decussatus *was studied at several spatial scales: over the whole Mediterranean, among lagoons in the same region (Languedoc), among sites in the same lagoon (Thau), and at a temporal scale, among year cohorts at a given site, in order to estimate levels and patterns of gene flow among populations. Genetic divergence at each level was expressed by means of Cockerham's coancestry coefficient (θ), estimated from genotypic data at 6–9 polymorphic enzyme loci. Significant divergence found between regions at the scale of the whole Mediterranean ($\theta \pm SD = 0.0149 \pm 0.0072$) indicated that gene flow between regions is limited. There was no evidence for heterogeneity within a region ($\theta = 0.0015 \pm 0.0011$), nor within a lagoon ($\theta = 0.0006 \pm 0.0005$). However, the analysis of divergence between year cohorts at a given site revealed discrepancies among loci in single-locus θ values, indicating, perhaps, short-term selection or genetic drift.*

2.3.1 Introduction

The planktonic larval stage has been interpreted as a means of dispersal and of maintenance of the homogeneity of the gene pool of a species over its distribution area (e.g. Scheltema, 1971). However, high potential dispersal does not necessarily imply gene flow. Genetic surveys have shown that gene flow between populations of marine invertebrates with a planktonic larval stage may be limited. Significant genetic heterogeneities among habitats (Koehn, Turano and Mitton, 1973), steep clines (Tracey *et al.*, 1975; Koehn *et al.*, 1984; McDonald and Siebenaller, 1989), and the maintenance of hybrid populations over generations (Skibinski, Beardmore and Cross, 1983) in spite of larval dispersal, have been reported. At the microgeographic level, patchiness in allele frequencies which may be due to drift or selection in swarms of larvae or early spat that were recruited by pulses has also been reported (Johnson and Black, 1982, 1984a,b; Watts, Johnson and Black, 1990). Thus, differentiation between local populations might be the rule rather than the exception (Burton, 1983; Hedgecock, 1986) and one is left with an apparently complex, multiple-scale array of population genetic patterns in marine invertebrates.

Here we report the results of a study of the genetic structure of *Ruditapes decussatus*, a bivalve of Mediterranean lagoons which has a planktotrophic, planktonic larval stage. The analysis was performed at several scales, from microgeographic to zoogeographic. Our objective was to infer levels of gene flow among populations or subpopulations from the estimation and comparison of genetic heterogeneities at each scale (i.e. temporal, microgeographic, regional and over the whole Mediterranean) in order to determine the boundaries of a population in *R. decussatus*. This should enhance our understanding of differentiation and speciation in marine invertebrates, a topic still subject to some debate (e.g. Palumbi, 1992; section 1.1). The present study includes a re-analysis of genotypic data from earlier surveys (Worms and Pasteur, 1982; Jarne, Berrebi and Guelorget, 1988), and extends the data of Borsa (1990) and Borsa, Zainuri and Delay (1991).

2.3.2 Materials and methods

(a) Zoogeographic and biological features of R. decussatus

The area of distribution of *R. decussatus* extends from the North Sea to the coasts of Senegal and along the coasts of the whole Mediterranean (Black Sea excepted), reaching as far east as the Red Sea (Mars, 1966; Fischer-Piette and Métivier, 1971). Its preferential habitats in the Mediterranean are coastal lagoons (Mars, 1966) and shallow-water, protected bays (H. Massé and O. Guelorget, personal communication). These habitats occur discretely all around the Mediterranean and are subject to anoxic crises which may cause local extinctions (Amanieu *et al.*, 1975; Borsa, Jousselin and Delay, 1992).

The duration of the planktonic larval stage of *R. decussatus* is 8–10 days at 24–25 °C (P. Borsa and A. Diter, unpublished data from hatchery experiments involving *R. decussatus* from Thau). This is approximately the temperature at which reproduction is known to occur in peri-Mediterranean populations of the species (Vilela, 1949; Gallois, 1977; Breber, 1980; Borsa and Millet, 1992). Since such temperatures occur in summer along the coasts of the Mediterranean (Robinson, 1973), and with coastal currents of a few centimetres per second, the distance that a larva is potentially able to cross ranges from about 10 to 100 km.

(b) Collection of samples

Surveys of genetic variation at electrophoretic loci of some of the samples have been reported earlier as indicated below. The samples studied here originated from the lagoons of Prévost, France (Worms and Pasteur, 1982; D. Monti and S. Salvidio, unpublished data), Ebro (Spain), Faro (South Portugal), Bizerte (Tunisia), Temsah (Egypt) (Jarne, Berrebi and Guelorget, 1988) and Thau, France (D. Monti and S. Salvidio, unpublished data; Borsa, 1990; Borsa, Zainuri and Delay, 1991; Borsa, Jousselin and Delay, 1992; P. Borsa, unpublished data). The list of samples, with sampling locations and dates, sample sizes and loci investigated, is presented in Table 2.3.1.

(c) Allozyme electrophoresis

Electrophoretic analyses of all samples, including those of earlier surveys, were conducted in the same laboratory, using the same protocols (except where noted below), i.e. media, buffers, trays, running conditions and staining procedures (Pasteur *et al.*, 1987).

When enzymes were encoded by genes at several presumptive loci, these were numbered consecutively, from slow to fast as in Borsa and Thiriot-Quiévreux (1990). Note that the loci were numbered in the opposite direction by Worms and Pasteur (1982). Ten loci were scored in all: *Aat-1* (encoding the slower aspartate aminotransferase; EC 2.6.1.1), *Est-D* (4-methyl-umbelliferyl-specific esterase; EC 3.1.1.1), *Glo* (glyoxalase; EC 4.4.1.5), *Idh-1* and *Idh-2* (respectively, slower and faster isocitrate dehydrogenases; EC 1.1.1.42), *Lap-1* (slower leucine aminopeptidase; EC 3.4.11.1), *Mdh-1* and *Mdh-2* (respectively, slower and faster malate dehydrogenases; EC 1.1.1.37), *Me-1* (slower malic enzyme; EC 1.1.1.40) and *Pgm-1* (slower phosphoglucomutase; EC 2.7.5.1).

The identities of electromorphs from different samples were ascertained by side-by-side runs (for all enzymes except LAP and EST-D). The buffer systems used for LAP and EST-D enzymes were not the same for all samples. For both enzymes, the discontinuous Tris-citrate-borate LiOH, pH 7.0 (TCBL) system was used for all Thau samples except TA-TD (Table 2.3.1), whereas for all other samples continuous Tris-maleate, pH 7.4, was used for EST-D extracts, and continuous Tris-EDTA-borate, pH 8.6, for LAP extracts. The use of

Table 2.3.1 List of samples of *Ruditapes decussatus* referred to in the present study, with sampling locality, sampling date, sample size (*N*) and loci scored per sample (*) that were used for the comparisons between populations

Name	Locality	Date	N	Aat-1	Est-D	Glo	Idh-1	Idh-2	Lap-1	Mdh-1	Mdh-2	Me-1	Pgm-1
	Prévost (43°30'N, 3°53'E)												
P1[a]	Prévost	1982	47	*	*	*				*	*	*	*
P2[b]	Prévost	07 84	64		*	*			*	*	*	*	*
P3[b]	Prévost	08 84	52		*	*			*	*	*		*
P4[b]	Prévost	11 84	64		*	*	*		*	*	*		*
	Thau (43°20'N, 3°40'E)												
T1[c]	Le Barrou	12 87	105	*	*	*	*	*	*	*			*
T2	Le Barrou	02 88	156		*	*	*	*	*	*			*
T3	Le Barrou	07 88	81		*	*	*	*	*	*			*
T4	Le Barrou	09 88	97		*	*	*	*	*	*			*
T5	Le Barrou	12 88	92		*	*	*	*	*	*			*
T6	Le Barrou	06 89	96		*	*	*	*	*	*			*
T7	Le Barrou	08 89	103		*	*	*	*	*	*			*
T8[c]	Balaruc-Port	12 87	92	*	*	*	*	*	*	*			*
T9[c]	Balaruc-Z.I.	02 89	92		*	*	*	*	*	*			*
T9bis	Balaruc-Z.I.	02 89	119		*	*	*	*	*	*			*
T10[c,d]	Balaruc-Z.I.	06 87	27		*	*	*	*	*	*			*
T11[d]	Balaruc	07 87	60		*	*	*	*	*	*			*
T12[c]	Le Mourre	08 87	101	*	*	*	*	*	*	*			*
T13[c]	Balaruc-Port	07 87	30		*	*	*	*	*	*			*
T14	Le Mourre	07 87	59		*	*	*	*	*	*			*
T15[c]	Bouzigues	05 87	117	*	*	*	*	*	*	*			*
T16[c,d]	Mourre-Blanc près Mèze	07 87	56		*		*	*	*	*	*	*	*
T17[c]	Lido	08 87	74		*	*	*	*	*				*
T18	Le Barrou	11 89	84		*	*	*	*	*	*			*

Table 2.3.1 (*cont*) List of samples of *Rudiatapes decussatus* referred to in the present study, with sampling locality, sampling date, sample size (*N*) and loci scored per sample (*) that were used for the comparisons between populations

Name	Locality	Date	N	Aat-1	Est-D	Glo	Idh-1	Idh-2	Lap-1	Mdh-1	Mdh-2	Me-1	Pgm-1
T19	Plan de Mèze	02 88	28		*	*	*	*	*	*	*		*
T20	Balaruc-Port	09 88	62		*	*	*	*	*	*			*
T21	Lido	07 87	20		*	*	*	*	*	*	*		*
TA[b]	Conque de Mèze	07 84	39		*	*			*	*			*
TB[b]	Conque de Mèze	07 84	39		*	*			*	*			*
TC[b]	Conque de Mèze	08 84	18		*	*			*	*			*
TD[b]	Conque de Mèze	11 84	61		*	*			*	*			*
	Other localities												
	Faro[c], Portugal (37°30′N, 8°00′W)	04 85	27		*	*	*					*	*
	Temsah[e], Egypt (30°40′N, 32°15′E)	03 85	28		*	*	*			*		*	*
	Bizerte[e] Tunisia. (37°15′N, 9°49′E)	03 86	113		*	*	*			*		*	*
	Ebro[c], Spain (40°40′N, 0°55′E)	03 86	32		*	*	*			*		*	*

[a]See Worms and Pasteur (1982).
[b]Samples analysed by D. Monti and S. Salvidio.
[c]See Borsa, Zainuri and Delay (1991).
[d]See Borsa, Jousselin and Delay (1992).
[e]See Jarne, Berrebi and Guelorget (1988).

TCBL improved the resolution of patterns at the *LAP-1* and *EST-D* loci, but did not modify the migration ranks of the electromorphs. Because of variability in resolution among samples, some poolings of *EST-D* and *LAP-1* electromorphs were necessary (Table 2.3.2). For all other enzymes, buffer systems were the same as those used by Worms and Pasteur (1982) and Borsa and Thiriot-Quiévreux (1990).

(d) Coancestry coefficient

The coancestry coefficient θ, proposed by Cockerham (1969), equals F_{st} in S. Wright's notation. This parameter has the properties of being affected neither by the numbers of alleles observed per locus, the numbers of individuals sampled per population, nor by the number of populations (Weir and Cockerham, 1984), thus allowing comparisons among surveys with different sample structures, which is the case here.

Multiallelic monolocus, and multilocus, θ values were estimated respectively as weighted averages over alleles and over loci (Equation 10 of Weir and Cockerham (1984)). Under the hypothesis, derived from the neutral model, that each locus can be used separately to estimate θ, its variance was estimated from the set of all monolocus estimates using the jackknife procedure as advocated by Weir and Cockerham (1984). Each multilocus θ was then compared to zero by a one-tailed *t*-test (Sokal and Rohlf, 1969).

(e) Genetic distance

A distance measure based on the coancestry coefficient, $D = -\ln(1 - \theta)$, of Reynolds, Weir and Cockerham (1983) was estimated. This distance is designed to measure divergence between populations under the model of genetic drift, so mutation is not included, unlike with other measures of genetic distance usually employed (Weir, 1990). We assume this to be an appropriate use in the case of *R. decussatus* populations around the Mediterranean, since no private allele was present in any population at any locus (see Table 2.32 for Thau, and Jarne, Berrebi and Guelorget (1988) for other peri-Mediterranean populations).

Table 2.3.2 Electromorph frequencies at 10 enzyme loci in *R. decussatus* from Thau (French Mediterranean). Electromorph mobilities are relative to the mobility of the commonest (*100*). N = number of individuals scored for each locus

Locus	N	Electromorph frequencies
Aat-1	218	*030*: 0.002, *100*: 0.991, *220*: 0.007
Est-D[a]	1649	*060*: 0.173, *088*: 0.071, *100*: 0.471, *116*: 0.284
Glo	1665	*070*: 0.001, *100*: 0.889, *135*: 0.111
Idh-1	1721	*075*: 0.001, *100*: 0.902, *125*: 0.096, *142*: 0.001
Idh-2	1590	*064*: 0.001, *085*: 0.369, *100*: 0.625, *118*: 0.005
Lap-1[a]	1539	*050*: 0.012, *066*: 0.030, *083*: 0.002, *098*: 0.002, *100*: 0.488
		102: 0.004, *133*: 0.378, *135*: 0.011, *167*: 0.074, *169*: 0.000
Mdh-1	1857	*071*: 0.008, *085*: 0.128, *100*: 0.861, *130*: 0.002
Mdh-2	136	*100*: 0.993, *130*: 0.007
Me-1	24	*080*: 0.146, *100*: 0.854
Pgm-1	1772	*080*: 0.006, *100*: 0.632, *124*: 0.359, *140*: 0.002

[a]Poolings of electromorphs were made for comparing the Thau data with those of other lagoons: *Est-D* {*088*, *100*}; *Lap-1* {*050*, *066*, *083*}, {*098*, *100*, *102*}, {*135*, *137*}, {*167*, *169*}

(f)   Correspondence analysis (CA)

CA (Benzécri, 1973; Greenacre, 1984; Mathieu *et al.*, 1990) was performed on a contingency table, in which each sample was described by its gene counts at several loci. This analysis was made for comparing the populations at the regional scale, for which data from two neighbouring lagoons (e.g. Thau and Prévost) were available. Computations were run using the ECOLOGIX package by Roux (1982).

2.3.3   Results

All results of the computations of single-locus and multilocus θ at several scales, temporal (among year cohorts within a sample) and spatial (among sites within a lagoon, among lagoons within a region, and among regions in the periphery of the Mediterranean), are presented in Table 2.3.3. Globally, the larger single-locus θ values were observed at the scale of the whole Mediterranean and the lower at the very local scale, i.e. among year cohorts within a site or among sites within a lagoon. However, two exceptions were noted: large θ values were observed at locus *Idh-1* among presumptive year cohorts in the Bouzigues sample, and at locus *Glo* among year cohorts in the Balaruc-Z.I. sample. A comparison of the different sets of single-locus θ values at different scales was made by means of a Kruskal–Wallis test (Sokal and Rohlf, 1969). No significant difference appeared between the within-sample scale and that of the lagoon ($p = 0.375$), nor between the scale of the lagoon and the regional scale ($p = 0.674$). A significant difference was observed between the regional scale and that of the whole Mediterranean ($p = 0.014$). Multilocus θ values were not significantly different from zero, either among cohorts, among sites, or among lagoons at the regional scale. The value observed among lagoons at the scale of the Mediterranean was significant ($p < 0.05$). Thus, the data in Table 2.3.3 do not yield any indication of population structuring at any scale below that of the whole Mediterranean, at which the null hypothesis of homogeneity among populations could be rejected.

A multidimensional treatment (CA) of multiple-locus (*Est-D, Glo, Lap-1, Mdh-1* and *Pgm-1*) data was run for samples from Thau ($n = 13$) and from Prévost ($n = 4$) which were characterized at all five loci simultaneously, and were of relatively large sizes ($N \geq 47$) (Table 2.3.1). By using this approach, we expected to be able to point out possible multiple-locus associations that would allow a distinction, if any, between the populations of these two lagoons. We found that the variability among the 13 samples from Thau encompassed that of all four samples from Prévost (Figure 2.3.1).

The matrix of genetic distances based on the multilocus coancestry coefficient (D), between populations at the peri-Mediterranean scale is given in Table 2.3.4. Genetic distances ranged from 0.0000 (Thau/Ebro) to 0.0590 (Bizerte/Temsah). Only one value (Thau/Bizerte) could be considered significant by using a *t*-test as for multilocus θ (one-tailed $t = 2.83$; 5 d.f.; $p < 0.025$). This is probably due to the fact that sample sizes in this case were relatively important (see Table 2.3.1). The non-significance of the D values, all larger, between populations that are separated by larger geographic distances (Table 2.3.4) is likely to be related to their smaller sample sizes. Nevertheless, no significant correlation of genetic distance with geographic distance was found (Mantel's test for the association of parameters in a matrix with internal correlation (Manly, 1985); one-tailed; $p = 0.125$).

Table 2.3.3 Single-locus and multilocus values of coancestry coefficient (θ) (Weir and Cockerham, 1984) among populations of *R. decussatus* at different scales, temporal and spatial

Locus	(k)	Among cohorts within sample		Among sites within lagoon (Thau)	Regional (Thau versus Prévost)	Mediterranean
		Bouzigues[a]	*Balaruc Z.I.*			
Aat-1	(3)	–	–	0.0069	−0.0019	–
N		–	–	218	265	–
n		–	–	4	2	–
Est-D	(4)	−0.0070	−0.0013	0.0007	0.0041	0.0055
N		108	209	1649	1753	1827
n		3	2	8	2	5
Glo	(3)	−0.0107	0.0136	0.0029	−0.0013	0.0571
N		105	206	1665	1891	1850
n		3	2	9	2	5
Idh-1	(4)	0.0316	−0.0039	−0.0013	0.0071	0.0303
N		106	211	1721	1777	1922
n		3	2	8	2	5
Idh-2	(4)	0.0055	−0.0049	0.0011	–	–
N		107	211	1590	–	–
n		3	2	8	–	–
Lap-1	(4)	−0.0069	−0.0045	−0.0009	0.0004	–
N		110	211	1886	2113	–
n		3	2	9	2	–
Mdh-1	(4)	0.0039	−0.0039	0.0015	−0.0004	0.0287
N		107	211	1857	2084	2054
n		3	2	9	2	5
Mdh-2	(2)	–	–	−0.0057	−0.0029	–
N		–	–	136	263	–
n		–	–	3	2	–
Me-1	(2)	–	–	–	–	0.0113
N		–	–	–	–	212
n		–	–	–	–	5
Pgm-1	(4)	−0.0118	0.0068	0.0012	0.0000	−0.0001
N		110	212	1772	1998	1968
n		3	2	9	2	5
Multilocus		−0.0014	0.0001	0.0006	0.0015	0.0149[*]
SD (jackknife)		0.0070	0.0021	0.0005	0.0011	0.0072

[a]Source: Borsa, Zainuri and Delay (1991)

[*]$p < 0.05$; one-tailed *t*-test.

N = total number of individuals; n = number of samples; k = number of electromorphs per locus.

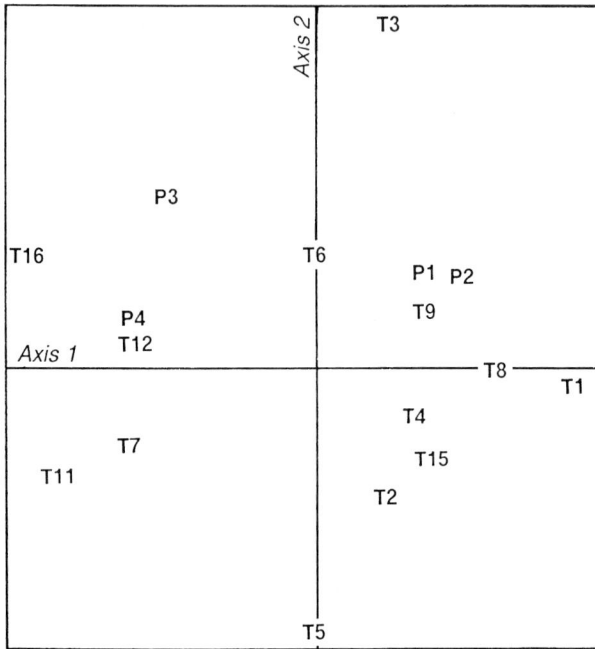

Figure 2.3.1 Correspondence analysis: projection on the plane defined by axes 1 and 2 (respectively, 28.4% and 21.4% of total inertia) of samples of *R. decussatus* from Thau (*n* = 13) and Prévost (*n* = 4), each defined by its genotypic frequencies at loci *Est-D*, *Glo*, *Lap-1*, *Mdh-1* and *Pgm-1*. Symbols for samples as in Table 2.3.1.

Table 2.3.4 Below diagonal: values of pairwise genetic distance among five peri-Mediterranean populations of *R. decussatus*. *D* = genetic distance based on multilocus coancestry coefficient (Reynolds, Weir and Cockerham 1983); *SD* = standard deviation (jackknife). Above diagonal: approximate geographic distances (km) measured following summer coastal surface currents (after Nielsen, 1912)

		Thau	*Ebro*	*Faro*	*Bizerte*	*Temsah*
Thau		–	400	1500	2800	5500
Ebro	*D*	0.0000	–	1100	2400	5100
	SD	0.0041				
Faro	*D*	0.0099	0.0002	–	1800	4500
	SD	0.0107	0.0030			
Bizerte	*D*	0.0181	0.0030	0.0084	–	2700
	SD	0.0064	0.0058	0.0110		
Temsah	*D*	0.0233	0.0336	0.0421	0.0590	–
	SD	0.0188	0.0359	0.0425	0.0553	

2.3.4 Discussion

The present study, which was based on the estimation of coancestry coefficients between populations at different scales, spatial and temporal, provided an insight into the genetic structure of *R. decussatus*.

Genetic differences were found between populations of *R. decussatus* from different regions of the Mediterranean, separated by an effective distance of several thousand kilometres. At that scale, no correlation between genetic distance and geographic distance was observed, which is consistent with an island model of population structure (Wright, 1969). The estimates of coancestry coefficients at the scale of the Mediterranean were homogeneous among loci. Since gene flow is expected to affect all loci to the same extent (Slatkin, 1985), homogeneity among single-locus θ values is consistent with the expectation of differentiation having arisen solely through genetic drift.

No heterogeneity could be detected at the within-region scale, i.e. between neighbouring lagoons. Therefore, all individuals at a regional scale should be considered as belonging to the same, genetically homogeneous population. In other words, high gene flow occurs, or has occurred recently, at that scale. The geographic limits of the population could even be extended to neighbouring regions: there was no indication for divergence, albeit small, between Thau and Ebro, both located along the northwestern coast of the western Mediterranean, about 400 km apart. Gene flow among lagoons at the regional scale is possible through larval transport (with the conditions that larvae are released from one lagoon and brought by currents into the next lagoon, where they are recruited and develop to the reproductive stage for participating as parents to the next generation), and may be enhanced by the recolonization of habitats left vacant after an extinction. These conditions for massive migration might not be met very often. However, under the current models of population genetic structure, an average of more than one successful migrant per generation is sufficient to prevent alternative alleles being fixed in geographically separated local populations (Slatkin, 1985).

The comparison of year cohorts within a local population revealed some heterogeneity among loci: large θ values were found at one locus in each of the two samples for which separate year-cohort data were available. Discrepancies among single-locus coancestry coefficient values suggest that selection or possibly genetic drift affects a particular enzyme locus, or a locus which is in gametic disequilibrium with it. In the face of gene flow, such differentiation, if real, can only be short-term.

A recent report based on a comparison of results from different studies of genetic structure of the American oyster, using nuclear RFLPs, mitochondrial DNAs and allozyme genes (Karl and Avise, 1992), stressed the presumption that allozyme genes are subjected to balancing selection, at least in bivalve molluscs, a hypothesis also consistent with the reported correlations of allozymic heterozygosity with reproductive performance or survival (e.g. Zouros *et al.*, 1983). This could be true in *R. decussatus*, in which correlations of allozymic heterozygosity with survival in stress conditions have also been reported (Borsa, 1990; Borsa, Jousselin and Delay, 1992). In order to get a more complete picture of the genetic structure of Mediterranean *R. decussatus*, and to eliminate doubts as to the existence of composition that is masked by balancing selection at allozyme loci, as cautioned by Karl and Avise (1992), this study should be completed using other genetic markers.

Acknowledgements

We are grateful to Dr N. Pasteur for access to raw electrophoretic data for all Prévost samples and to Drs M. Amanieu, J.F. Dallas, B. Delay and C. Thiriot-Quiévreux for corrections or comments on earlier versions of this paper. Dr O. Guelorget collected the samples from Ebro, Faro, Bizerte and Temsah. The costs of the electrophoretic analyses were supported by CNRS/PIREN Ecothau (Amanieu *et al.*, 1989), IFREMER contract no. 84/443 and ATP 'Bases biologiques de l'aquaculture'. P.B. benefitted from a PhD fellowship (no. 86/207) allocated by the French Ministère de la Recherche.

References

Amanieu, M., Baleux, B., Guelorget, O. and Michel, P. (1975) Etude biologique et hydrologique d'une crise dystrophique (malaïgue) dans l'Etang du Prévost à Palavas (Hérault). *Vie Milieu*, **25**, 175–204.

Amanieu, M., Legendre, P., Troussellier, M. and Frisoni, G.-F. (1989) Le programme Ecothau: théorie écologique et base de la modélisation. *Oceanol. Acta*, **12**, 189–99.

Benzécri, J.-P. (1973) *L'analyse des Données*, Dunod, Paris.

Borsa, P. (1990) *Génétique des populations de bivalves en milieu lagunaire: la palourde dans l'Etang de Thau (Méditerranée)*. Thèse de Doctorat de l'Université Paris 6, Paris.

Borsa, P., Jousselin, Y. and Delay, B. (1992) Relationships between allozymic heterozygosity, body size, and survival to natural anoxic stress in the palourde *Ruditapes decussatus* L. (Bivalvia: Veneridae). *J. Exp. Mar. Biol. Ecol.*, **155**, 169–81.

Borsa, P. and Millet, B. (1992) Recruitment of the clam *Ruditapes decussatus* in the lagoon of Thau, Mediterranean. *Est., Coast. Shelf Sci.*, **35**, 289–300.

Borsa, P. and Thiriot-Quiévreux, C. (1990) Karyological and allozymic characterization of *Ruditapes philippinarum, R. aureus* and *R. decussatus* (Bivalvia: Veneridae). *Aquaculture*, **90**, 209–27.

Borsa, P., Zainuri, M. and Delay, B. (1991) Heterozygote deficiency and population structure in the bivalve *Ruditapes decussatus* L. *Heredity*, **66**, 1–8.

Breber, P. (1980) Annual gonadal cycle in the carpet-shell clam *Venerupis decussata* in Venice Lagoon, Italy. *Proc. Natl. Shellfish. Assoc.*, **70**, 31–5.

Burton, R.S. (1983) Protein polymorphisms and genetic differentiation in marine invertebrate populations. *Mar. Biol. Letts.*, **4**, 193–206.

Cockerham, C.C. (1969) Variance of gene frequencies. *Evolution*, **23**, 72–84.

Fischer-Piette, E. and Métivier, B. (1971) Révision des Tapetinae (Mollusques Bivalves). *Mém. Mus. Natl. Hist. Nat. (Paris)*, **71**, 1–106.

Gallois, D. (1977) Sur la reproduction des palourdes, *Venerupis decussata* (Linné), et des clovisses, *Venerupis aurea* (Gmelin) de l'Etang de Thau. *Vie Milieu*, **27**, 233–55.

Greenacre, M.J. (1984) *Theory and Application of Correspondence Analysis*, Academic Press, London.

Hedgecock, D. (1986) Is gene flow from pelagic larval dispersal important in the adaptation and evolution of marine invertebrates? *Bull. Mar. Sci.*, **39**, 550–64.

Jarne, P., Berrebi, P. and Guelorget O. (1988) Variabilité génétique et morphométrique de cinq populations de la palourde *Ruditapes decussatus* (mollusque, bivalve). *Oceanol. Acta*, **11**, 401–7.

Johnson, M.S. and Black, R. (1982) Chaotic genetic patchiness in an intertidal limpet, *Siphonaria* sp. *Mar. Biol.*, **70**, 157–64.

Johnson, M.S. and Black, R. (1984a) The Wahlund effect and the geographical scale of variation in the intertidal limpet *Siphonaria* sp. *Mar. Biol.*, **79**, 295–302.

Johnson, M.S. and Black, R. (1984b) Pattern beneath the chaos: the effect of recruitment on the genetic patchiness in an intertidal limpet. *Evolution*, **38**, 1371–83.

Karl, S.A. and Avise, J.C. (1992) Balancing selection at allozyme loci in oysters: implications from nuclear RFLPs. *Science*, **256**, 100–2.

Koehn, R.K., Turano, F.J. and Mitton, J.B. (1973) Population genetics of marine pelecypods. II. Genetic differences in microhabitats of *Modiolus demissus*. *Evolution*, **27**, 100–105.

Koehn, R.K., Hall J.G., Innes, D.J. and Zera, A.J. (1984) Genetic differentiation of *Mytilus edulis* in eastern North America. *Mar. Biol.*, **79**, 117–26.

Manly, B.F.J. (1985) *The Statistics of Natural Selection on Animal Populations*, Chapman and Hall, London.

Mars, P. (1966) Recherches sur quelques étangs du littoral méditerranéen français et sur leurs faunes malacologiques. *Vie Milieu*, **suppl. 20**, 1–359.

Mathieu, E., Autem, M., Roux, M. and Bonhomme, F. (1990) Epreuves de validation dans l'analyse de structures génétiques multivariées: comment tester l'équilibre panmictique? *Rev. Statist. Appl.*, **1**, 47–66.

McDonald, J.H. and Siebenaller, J.F. (1989) Similar geographic variation at the *Lap* locus in the mussels *Mytilus trossulus* and *M. edulis*. *Evolution*, **43**, 228–31.

Nielsen, J.N. (1912) Hydrography of the Mediterranean and adjacent waters. *Danish Oceanogr. Exp. 1908–1910 Rep.*, **1**, 55–69.

Palumbi, S.R. (1992) Marine speciation on a small planet. *TREE*, **7**, 114–18.

Pasteur, N., Pasteur, G., Bonhomme, F., Catalan, J. and Britton-Davidian, J. (1987) *Manuel Technique de Génétique par Électrophorèse des Protéines*, Lavoisier, Paris.

Reynolds, J., Weir, B.S. and Cockerham, C.C. (1983) Estimation of the coancestry coefficient: basis for a short-term genetic distance. *Genetics*, **105**, 767–79.

Robinson, M.K. (1973) Monthly sea surface and subsurface temperature and depth of the top of the thermocline. Mediterranean, Black and Red Seas. *Fleet Numerical Weather Central, Monterey, Technical Note*, **73-3**, 1–15.

Roux, M. (1982) *Logiciel ECOLOGIX, 1ᵉ version*, Centre National Universitaire Sud de Calcul, Montpellier.

Scheltema, R.S. (1971) Larval dispersal as a means of genetic exchange between geographically separated populations of shallow-water marine gastropods. *Biol. Bull.*, **140**, 284–322.

Skibinski, D.O.F., Beardmore J.A. and Cross, T.F. (1983) Aspects of population genetics of *Mytilus* (Mytilidae, Mollusca) in the British Isles. *Biol. J. Linn. Soc.*, **19**, 137–83.

Slatkin, M. (1985) Gene flow in natural populations. *Annu. Rev. Ecol. Syst.*, **16**, 393–430.

Sokal, R.R. and Rohlf, F.J. (1969) *Biometry*, Freeman and Co., San Francisco.

Tracey, M.L., Nelson, K., Hedgecock, D., Shleser, R.A. and Pressick, M.L. (1975) Biochemical genetics of lobsters: genetic variation and the structure of American lobster (*Homarus americanus*) populations. *J. Fish. Res. Bd Can.*, **32**, 2091–101.

Vilela, H. (1949) Quelques données sommaires sur l'écologie de *Tapes decussatus* (L.). *Rapp. P.–v. Réun. Cons. Int. Explor. Mer*, **128**, 60–2.

Watts, R.J., Johnson, M.S. and Black, R. (1990) Effects of recruitment on the genetic patchiness in the urchin *Echinometra matthaei* in Western Australia. *Mar. Biol.*, **105**, 145–51.

Weir, B.S. (1990) *Genetic Data Analysis*, Sinauer, Sunderland.

Weir, B.S. and Cockerham, C.C. (1984) Estimating *F*-statistics for the analysis of population structure. *Evolution*, **38**, 1358–70.

Worms, J. and Pasteur, N. (1982) Polymorphisme biochimique de la palourde, *Venerupis decussata*, de l'étang du Prévost (France). *Oceanol. Acta*, **5**, 395–7.

Wright, S. (1969) *Evolution and the Genetics of Populations*, Vol. 2, *The Theory of Gene Frequencies*, University of Chicago Press, Chicago.

Zouros, E., Singh, S.M., Foltz, D.W. and Mallet, A.L. (1983) Post-settlement viability in the American oyster (*Crassostrea virginica*): an overdominant phenotype. *Genet. Res. Camb.*, **41**, 259–70.

2.4 GEOGRAPHIC STRUCTURE AND GENE FLOW IN THE MANINI (CONVICT SURGEONFISH, *ACANTHURUS TRIOSTEGUS*) IN THE SOUTH-CENTRAL PACIFIC

Serge Planes, Philippe Borsa, René Galzin and François Bonhomme

Abstract

The geographic structure of the manini, Acanthurus triostegus, *a coral reef surgeonfish of Polynesia, was investigated using allozyme markers. Four samples, from Bora-Bora (Society Archipelago), Moorea (Society), Takapoto (Western Tuamotu) and Moruroa (Eastern Tuamotu), were studied at 10 polymorphic loci, and the data analysed using* F*-statistics and permutation tests. Single-locus and multiple-locus genotypic proportions in each sample did not differ from Hardy–Weinberg expectations, whereas significant differences were observed between populations. No significant increase of genetic distance with geographic distance was observed. The estimates of gene flow between pairs of populations from different archipelagos, calculated under Wright's infinite island model, were homogeneous (average Nm ranging from 3.4 to 7.3). Moderate amounts of gene flow (i.e. effective migration) were thus inferred. These few successful migrants at each generation do not contribute much to recruitment. Hence, from the reef ecologist's point of view, island populations can be* *considered as primarily self-seeding, despite prior expectations of large oceanic transport in reef fish such as* A. triostegus, *inferred from their high passive dispersal capacities. In contrast, this moderate level of gene flow is enough to prevent the fixation of different alleles in the different archipelagos.*

2.4.1   Introduction

Most coral reef fishes have a pelagic larval phase lasting 10 days or more (Brothers, Williams and Sale, 1983; Victor, 1986; Wellington and Victor, 1989). Because of this, and because recruitment of larvae is very small relative to the great fecundity of adults (Talbot, Russell and Anderson, 1978; Sale, 1980; Williams, 1980; Victor, 1986), the prevailing view among reef fish ecologists is that larvae disperse widely and that populations of different reefs are interconnected by larval exchange (Sale, 1980; Mapstone and Fowler, 1988). However, some ecological data may also suggest that self-recruitment of reefs is a major process (Johannes, 1978; Sammarco and Andrews, 1988), enhanced by the occurrence of hydrodynamic eddies favouring particle retention (Lobel and Robinson, 1986; Hamner and Wolanski, 1988; Black, Moran and Hammond, 1991).

Genetic data available so far do not provide a clear-cut answer. Negligible differentiation has been found among samples of the damselfish *Stegastes fasciolatus* collected from reef islands of the Hawaiian archipelago up to 2500 km apart (Shaklee, 1984) and among populations of the Hawaiian snapper *Pristimoides filamentosus* (Shaklee and Samollow, 1984). In contrast, a high degree of differentiation has been reported between reef populations of the anemonefish *Amphiprion clarkii* (Bell, Moyer and Numachi, 1982) and of the damselfish *Stegastes partitus* (Lacson *et al.*, 1989) separated by a distance short enough to be within the potential dispersal range of larvae.

While the above ecological problem focuses on the importance of larval migration between populations and subsequent allorecruitment on reefs, the estimates of gene flow inferred from population structure are expected to give an answer concerning effective migration, which is another concept. The genetic impact of allorecruitment operates at an evolutionary scale, not at the ecological scale. A small amount of gene flow, mediated by the successful recruitment of a small proportion of foreign larvae (i.e. small amounts of allorecruitment), can suffice to prevent differentiation between islands. As stressed by Slatkin (1987), the comparison of gene flow data with knowledge of life-history (dispersal capacity in particular) may prove a fruitful approach in population biology studies. The comparison of effective migration rate with dispersal capacity is also an important issue in the current debate about patterns of structure and processes of regulation in coral reef fish assemblages, and more generally in supply-side ecology (Underwood and Fairweather, 1989; Ayre, 1990).

The purpose of the present study was to provide preliminary information on the genetic structure and gene flow of Polynesian populations of the manini (convict surgeonfish, *Acanthurus triostegus*), a reef fish present throughout the Pacific, whose biology has been investigated (Randall, 1961a,b) and whose larval stage is estimated as being relatively long (up to 70 days). We thereby intended to test self-recruitment versus allorecruitment in coral reef fish from the southern Pacific.

2.4.2   Materials and methods

(a)   Sampling

Acanthurus triostegus L. is one of the most abundant surgeonfishes in the reef habitat of the Pacific. Thirty-two to 36 individuals were caught by spear-fishing between February and

April 1990 in each of four islands of French Polynesia (Figure 2.4.1): Moorea (Society Archipelago), Bora-Bora (Society), Takapoto (Western Tuamotu) and Moruroa (Eastern Tuamotu). Moorea and Bora-Bora are two oceanic islands 250 km apart. Takapoto is a closed atoll, 650 km from Moorea. Moruroa is an open atoll, 1200 km from Moorea. Fish were collected inside the lagoon, except in Takapoto, where collection was on the oceanic side of the reef in order to compare with the other samples, all from an open habitat. The fish were dissected on site. Samples of tissue (liver, eyes and about 2 g of dorsal muscle) were kept in liquid nitrogen until storage at –80 °C in the laboratory.

(b) Enzyme electrophoresis

The samples of tissue were homogenized at 0–4 °C in an equal volume of Tris-EDTA-NADP, pH 6.8 buffer (Pasteur *et al.*, 1988) using an Ultraturrax homogenizer. The homogenates were centrifuged at 22 000*g* for 30 min at 4 °C and the supernatant, used as a source of soluble enzymes, was stored for a few weeks to a few months at –80 °C until electrophoresis. Horizontal starch gel electrophoresis and subsequent enzyme staining were performed according to Pasteur *et al.* (1988). The enzyme systems amenable to interpretation are listed in Table 2.4.1.

(c) Data analysis

Single-locus *F*-statistics were estimated using the parameters F, f and θ of Weir and Cockerham (1984), which in Wright's notation correspond to F_{it}, F_{is} and F_{st}, respectively. Their standard deviations over loci were estimated using a jackknife procedure. Weighted averages of F, f and θ (Weir and Cockerham, 1984) were then compared to zero using a *t*-test (Sokal and Rohlf, 1981). Genetic distances based on the coancestry coefficient (θ) were computed as $D = -\ln(1 - \theta)$ (Reynolds, Weir and Cockerham, 1983). The correlation of genetic distance with geographic distance was tested using Mantel's test of the association of two parameters in data matrices with internal correlation (Manly, 1985). Dixon's test for detecting outliers in a normal population (Sokal and Rohlf, 1981, p. 413) was used on the sets of single-locus f and θ. The reason for discarding outliers was that consistency is necessary for considering the information at different loci as replicates. The occurrence of a locus presenting large differences between populations (large θ values), among loci at which differences are consistently smaller, can be interpreted in terms of selection on this or a closely linked locus (Slatkin, 1987). Where Mendelian inheritance of allozyme phenotypes has not been formally established, as is the case here, it is also wise to discard those loci at which within-population phenotypic frequencies obviously depart from Mendelian expectations for codominant alleles and/or interpopulation differences are not consistent with those at other loci.

Estimates of gene flow were calculated from the estimates of coancestry coefficients assuming an infinite island model at equilibrium as $Nm = (1 - \theta)/4\theta$ (Wright, 1969).

Additionally, departure from panmixis was assessed using a multiple-locus resampling test (Mathieu *et al.*, 1990), which is expected to point out genotypic disequilibria *sensu* Weir (1990). For this, a dispersion index for the natural sample was computed and compared to the distribution obtained for the collection of randomized replicates (Mathieu *et al.*, 1990). This test, based on permutations on contingency tables of individual X multiple-locus genotype, allows evaluation at the same time of both inter- and intralocus departures from the panmictic equilibrium.

F-statistics, Mantel's test and permutation tests were performed using the FST, MANTEL and PERMU procedures of the GENETIX package (Bonhomme *et al.*, 1993).

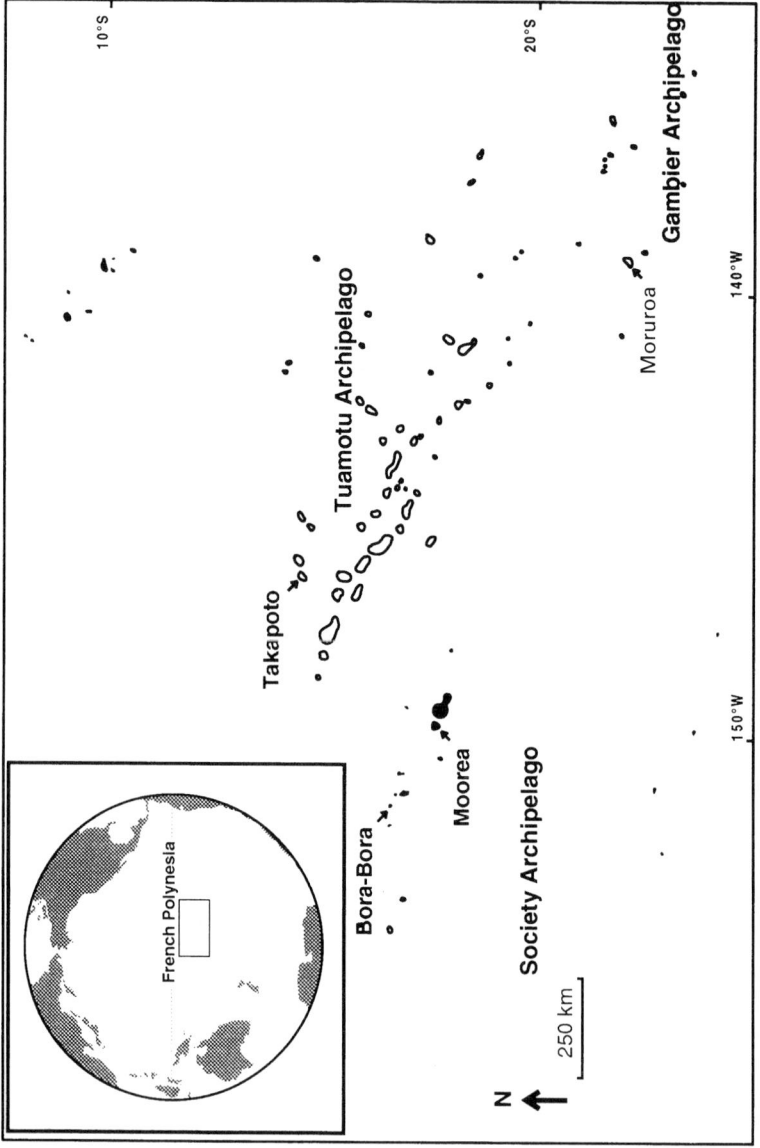

Figure 2.4.1 Islands of the south-central Pacific sampled for *Acanthurus triostegus*.

Table 2.4.1 Enzyme systems and presumptive loci scored in *Acanthurus triostegus*

Enzyme system	Tissue	Buffer	Locus	Polymorphism
Adenosine deaminase	Muscle	C	*Ada*	+
Adenylate kinase	Eye	D	*Ak*	–
Alcohol dehydrogenase	Liver	A	*Adh*	+
Aspartate aminotransferase	Liver	C	*Aat-1*	+
	Liver	C	*Aat-2*	+
	Muscle	C	*Aat-3*	–
Creatine kinase	Liver	C	*Ck*	–
Fumarase	Muscle	A	*Fum*	–
Glucose phosphate isomerase	Eye	A	*Gpi-1*	+
	Eye	A	*Gpi-2*	+
α-Glycerophosphate dehydrogenase	Muscle	D	*αGpd*	–
Glyoxalase-I	Liver	A	*Glo*	–
Guanine deaminase	Liver	C	*Gda*	+
Hexose phosphate dehydrogenase	Liver	A	*Hpd*	+
Isocitrate dehydrogenase	Liver	C	*Idh-L*	–
	Muscle	C	*Idh-M*	–
Lactate dehydrogenase	Eye	A	*Ldh-1*	–
	Eye	A	*Ldh-2*	–
	Eye	A	*Ldh-3*	–
Leucine aminopeptidase	Muscle	D	*Lap*	–
Malate dehydrogenase	Eye	D	*Mdh-1*	–
	Eye	D	*Mdh-2*	+
Malic enzyme	Muscle	D	*Me-1*	–
	Muscle	D	*Me-2*	–
Mannose phosphate isomerase	Muscle	C	*Mpi*	–
Phosphoglucomutase	Muscle	C	*Pgm*	+
6-Phosphogluconate dehydrogenase	Eye	C	*6-Pgd*	–
Sorbitol dehydrogenase	Liver	B	*Sdh*	+
Superoxide dismutase	Liver	A	*Sod*	–

+ = Locus polymorphic (frequency of the commonest electromorph < 0.95 in at least one population); – = sample monomorphism; A = discontinuous Tris-citrate-borate, pH 8.7 buffer; B = discontinuous Tris-HCl, pH 8.5 buffer; C = continuous Tris-citrate, pH 6.7 buffer; D = continuous Tris-citrate, pH 8.0 buffer. Loci *Aat-1* and *Aat-3* were scored on samples other than those of the present study (Planes, 1992).

2.4.3 Results

The estimates of single-locus F-statistics from genotype counts (see Appendix) are given in Table 2.4.2. Assuming that N has a normal distribution under the null hypothesis of Hardy–Weinberg equilibrium (Brown, 1970), no single-locus f value was significantly different from zero ($N = 33$; $p > 0.10$ at each locus). Neither could multilocus f values be considered different from zero. Therefore, agreement with Hardy–Weinberg proportions in each population was not rejected.

Differences in genetic composition between samples, expressed as large θ values, were observed. Assuming a normal distribution, the average multilocus θ value (0.0439) could be considered significantly different from zero (Student's t-test; 8 d.f.; $p = 0.038$). The *Gda* locus was discarded for the above calculations because the θ value at this locus departed significantly from the distribution of all other single-locus θs (Dixon's test: $p = 0.02$).

Permutation tests on matrices of genotypic data from each population did not result in the detection of multiple-locus genotypic disequilibria whereas significant departure from random expectations was evident for the pool of all samples (Figure 2.4.2).

Pairwise coancestry coefficients (θ) and genetic distances (D) between populations are presented in Table 2.4.3. D values did not show a clear increase with geographic distance

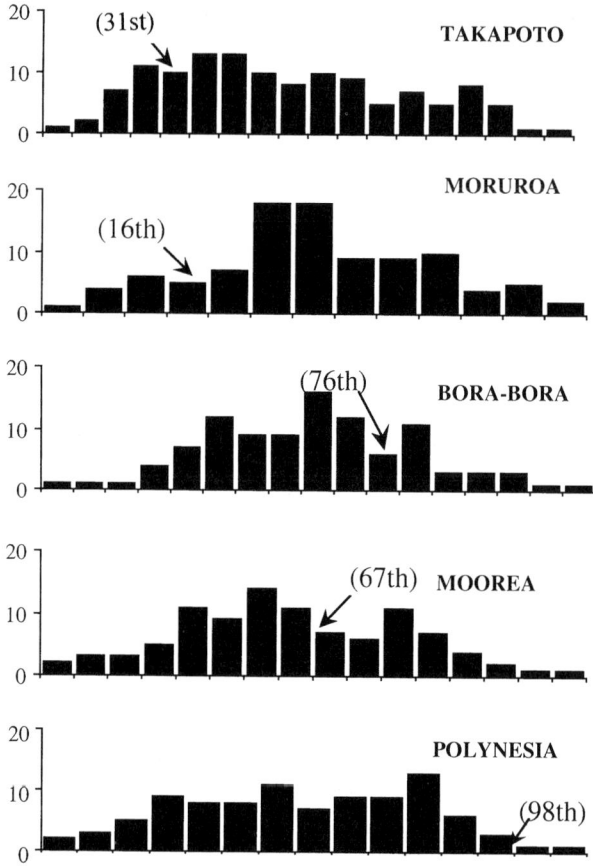

Figure 2.4.2 Distributions of dispersion indices from permutation tests (100 random permutations) on the matrices of individual X multiple-locus genotypes for each of four populations of *Acanthurus triostegus* and on the matrix of pooled data. Abscissa: percentiles of total range. Ordinate: frequencies. Arrow: dispersion index value of the wild sample (rank in brackets) among the random distribution.

Table 2.4.2 *F*-statistics on French Polynesian *Acanthurus triostegus* data

Locus	f	θ	F
Ada	0.1630	0.0250	0.1839
Adh	−0.1082	0.0247	−0.0808
Aat-2	0.0314	0.0855	0.1142
Gpi-1	0.1739	−0.0070	0.1681
Gpi-2	0.0207	0.0005	0.0212
Gda	−0.0216	0.3140[a]	0.2992
Hpd	0.1879	0.1275	0.2915
Mdh-2	−0.2483	−0.0019	−0.2506
Pgm	−0.0577	0.0275	−0.0286
Sdh	−0.0190	0.0481	0.0300
Multilocus (all 10 loci)	0.0101	0.0771[b]	0.0864
SD	0.0547	0.0361	0.0663
Multilocus (*Gda* excluded)	0.0132	0.0439[c]	0.0565
SD	0.0600	0.0173	0.0678

[a]Value detected as outlier ($p = 0.02$); [b]$p = 0.064$; [c]$p = 0.038$.
SD = standard deviation (jackknife).

Table 2.4.3 Estimates of genetic distance (*D*) and gene flow (*Nm*, with *SD* confidence intervals) derived from coancestry coefficients (average values over the nine loci of Table 2.4.2) between pairs of *Acanthurus triostegus* populations

Pairwise comparison	Geographic distance (km)	$\theta \pm SD$	D	Nm	
Bora-Bora/Moorea	210	0.0031 ± 0.0100	0.0031	80.4	(18.8, ∞)
Bora-Bora/Takapoto	755	0.0686 ± 0.0557	0.0711	3.4	(1.8, 19.2)
Bora-Bora/Moruroa	1450	0.0480 ± 0.0283	0.0492	5.0	(3.0, 12.4)
Moorea/Takapoto	665	0.0329 ± 0.0224	0.0335	7.3	(4.3, 23.6)
Moorea/Moruroa	1200	0.0423 ± 0.0245	0.0432	5.7	(3.5, 13.8)
Takapoto/Moruroa	1035	0.0629 ± 0.0300	0.0650	3.7	(2.4, 7.4)

(Mantel's test; $g = 1.25$; $p = 0.083$). Hence, consistency of these results with the expectations of an island model was not rejected. Estimates of gene flow between pairs of populations from different archipelagos, calculated for an island model at equilibrium (*Nm*; Table 2.4.3.), ranged from 3.4 to 7.3 migrants per generation, an array of values which can be considered as quite homogeneous. In contrast, the estimate of migration between Bora-Bora and Moorea (same archipelago) was $Nm = 80.4$.

2.4.4 Discussion

The present genetic study of *Acanthurus triostegus* in French Polynesia, the first one in coral reef fishes in this region of the Pacific, yielded results of both biogeographic and ecological interest.

Island populations at the scale of the whole of French Polynesia were significantly different from each other, whereas populations were not found to be internally structured, and hence could be considered as separate panmictic units. The model of one single population for the whole region, which could be expected from the passive dispersal capacities of *A. triostegus*, has therefore to be rejected.

The lack of correlation, on our limited sampling, between genetic distance and geographic distance indicates that the patterns observed may fit an island model, although the genetic distance estimate available for islands within the same archipelago (i.e. Moorea and Bora-Bora) was smaller. Further data encompassing a large number of samples from island within archipelagos should provide more precise information on the genetic structure of *A. triostegus*. These should also allow the testing of some current models of population genetic structure for reef fishes (e.g. island model versus two-dimensional stepping-stone model).

From a population genetic perspective, the values of effective migration rates were moderately large, i.e. small enough to allow significant divergence of gene frequencies between islands, although large enough (> 1) to counteract the effect of genetic drift if one assumes an infinite island model. From the ecologist's point of view, the meaning of these results is that unexpectedly low effective migration (see Section 2.4.1) occurs between islands, even within the same archipelago. Even though larval transport between islands cannot be determined directly from gene flow data on samples of recruited individuals (many larvae arriving on a reef are not necessarily recruited), the present results show that most of the population of recruits is contributed by self-recruitment in the reef fish *A. triostegus*.

As stressed by Le Fèvre and Bourget (1992), larval dispersal is determined not only by the duration of the larval stage, but also by hydrodynamic conditions and larval behaviour. Larvae of *A. triostegus* have been captured at relatively large depths; 50 m in Hawaii (Randall, 1961b) and more than 20 m in the Great Barrier Reef lagoon (Leis, 1991), depths at which wind-induced Ekman surface currents are much attenuated and retention is therefore favoured. Because of obvious sampling constraints, little more is known about the larval behaviour of *A. triostegus*, but there is some indication that larvae of other fish species are able to resist passive transport offshore by selective vertical movements (e.g. Rijnsdorp, Van Stralen and Van der Veer, 1985). The manini, like the majority of the coral reef species, disperses its eggs and larvae toward the ocean. Adult manini congregate in the pass of the reef for spawning (Randall, 1961*b*), where currents are stronger and supposedly export the eggs offshore. Such behaviour would be expected to impede larval homing, so active larval behaviour favouring inshore movement seems likely.

Up to now, gene flow data have only answered the question of how much allorecruitment occurs in *A. triostegus*. Investigations on the genetic structure of larval populations before settlement, now accessible through the refinement of sampling methods such as networks of larval traps (Doherty, 1987), and the advent of miniaturized techniques such as PCR, should contribute to a better understanding of the pre-recruitment stage in reef fish.

We showed that equilibrium-effective migration rates do not support the hypothesis of massive larval dispersal and inter-island transportation in the manini surgeonfish in the south-central Pacific. Even within the same archipelago, the absolute number of successful migrants is very small and does not contribute significantly to the population of the recruited larvae, which thus results almost entirely from self-recruitment. One can then ask why a long larval stage is

2.4.5   Appendix. Genotype numbers at 10 loci in four populations of *Acanthurus triostegus*: Takapoto, Moruroa, Bora-Bora and Moorea

Locus	Genotype	Takapoto $N = 32$	Moruroa $N = 36$	Bora-Bora $N = 32$	Moorea $N = 32$
Ada	090/090	9	7	4	3
	100/060	1	1	0	2
	100/090	13	13	13	9
	100/100	9	15	15	18
Adh	050/050	5	5	10	11
	100/050	18	20	19	15
	100/100	9	11	3	6
Aat-2	100/070	1	0	0	0
	100/100	14	5	18	16
	140/100	13	19	11	13
	140/140	4	12	3	3
Gpi-1	070/070	4	2	2	1
	100/070	9	9	8	9
	100/100	19	24	22	22
	120/100	0	1	0	0
Gpi-2	100/070	2	0	0	0
	100/100	27	32	31	27
	120/100	3	4	1	5
Gda	100/100	8	34	3	20
	140/100	19	2	15	9
	140/140	5	0	14	3
Hpd	100/100	2	9	10	7
	120/100	4	14	17	11
	120/120	26	13	5	14
Mdh-2	050/050	5	3	2	4
	100/050	14	27	19	20
	100/100	13	6	11	8
Pgm	100/060	0	2	6	2
	100/100	32	34	26	30
Sdh	020/020	6	15	8	6
	100/020	14	19	16	16
	100/100	12	2	8	10

N = sample size.

maintained over an evolutionary timescale in the manini, if there are no short-term advantages to maintaining a pelagic phase with dispersal capacities. In our opinion, the classical extinction–recolonization hypothesis of metapopulation models (Levins, 1971; Olivieri, Couvet and Gouyon, 1990) is not adequate for explaining the maintenance of a long larval stage in reef fish such as the manini, since the reef environment of oceanic islands, most of them being millions of years old, appears to have been stable over large periods of time, and the presence of coral reef communities on the reef's outer slope is not known to have been interrupted. Further ecological or genetic hypotheses are therefore needed to address the question.

Acknowledgements

We thank Drs K. Belkhir, J.F. Dallas and G. Kotulas for discussions and comments, anonymous reviewers for comments on earlier manuscripts, and also J. Algret, G. Berrebi, P. Berrebi, A.S. Cabanban, M. Calvas, R. Carossi, B. Delesalle, J.-P. Gattuso, G. Grosjean, C. Jardin, Y. Michalakis, T. Pambrun, C. Payri, B. Salvat and the Australian Institute of Marine Science. We acknowledge support from Mr Ducousso and Mr Bablet of the Service Mixte de Contrôle Biologique (Convention EPHE/SMCB no. 230), from P. Cabral of the Etablissement pour la Valorisation des Activités Aquacoles et Maritimes and from the Délégation à l'Environnement de Polynésie Française (convention Bora-Bora no. 89-1893).

References

Ayre, D.J. (1990) Population subdivision in Australian temperate marine invertebrates: larval connections versus historical factors. *Aust. J. Ecol.*, **15**, 403–11.

Bell, L.J., Moyer, J.T. and Numachi, K.I. (1982) Morphological and genetic variation in Japanese populations of the anemonefish *Amphiprion clarkii*. *Mar. Biol.*, **72**, 99–108.

Black, K.P., Moran, P.J. and Hammond, L.S. (1991) Numerical models show coral reefs can be self seeding. *Mar. Ecol. Prog. Ser.*, **74**, 1–11.

Bonhomme, F., Belkhir, K., Mathieu, E. and Roux, M. (1993) GENETIX – *Logiciel d'analyse des données en génétique des populations*, version 0.0, Université des Sciences et Techniques du Languedoc, Montpellier.

Brothers, E.B., Williams, D. McB. and Sale, P.F. (1983) Length of larval life in twelve families of fishes at One Tree Lagoon, Great Barrier Reef, Australia. *Mar. Biol.*, **76**, 319–24.

Brown, A.H.D. (1970) The estimation of Wright's fixation index from genotypic frequencies. *Genetica*, **41**, 399–406.

Doherty, P.J. (1987) Light-trap: selective but useful device for quantifying the distributions and abundances of larval fishes. *Bull. Mar. Sci.*, **41**, 423–31.

Hamner, W.P. and Wolanski, E. (1988) Hydrodynamic forcing functions and biological processes on coral reefs: a status review, in *Proceedings of the 6th International Coral Reef Symposium* (eds H.J. Choat, D. Barnes, M.A. Borowitzka *et al.*,), Townsville, Vol. 1, pp. 103–113.

Johannes, R.E. (1978) Reproductive strategies of coastal marine fishes in the tropics. *Environ. Biol. Fish.*, **3**, 741–60.

Lacson, J.M., Riccardi, V.M., Calhoun, S.W. and Morizot, D.C. (1989) Genetic differentiation of bicolor damselfish (*Eupomacentrus partitus*) populations on the Florida Keys. *Mar. Biol.*, **103**, 445–51.

Le Fèvre, J. and Bourget, E. (1992) Hydrodynamics and behaviour: transport processes in marine invertebrate larvae. *TREE*, **7**, 288–9.

Leis, J.M. (1991) Vertical distribution of fish larvae in the Great Barrier Reef Lagoon, Australia. *Mar. Biol.*, **109**, 157–66.

Levins, R. (1971) *Evolution in Changing Environments*, 2nd edn, Princeton University Press, Princeton.

Lobel, P.S. and Robinson, A.R. (1986) Transport and entrapment of fish larvae by mesoscale eddies and currents in Hawaiian waters. *Deep Sea Res.*, **33**, 483–500.

Manly, B.F.J. (1985) *The Statistics of Natural Selection on Animal Populations*, Chapman and Hall, London.

Mapstone, B.D. and Fowler, A.J. (1988) Recruitment and the structure of assemblages of fish on coral reefs. *TREE*, **3**, 72–7.

Mathieu, E., Autem, M., Roux, M. and Bonhomme, F. (1990) Epreuves de validation dans l'analyse de structures génétiques multivariées: comment tester l'équilibre panmictique? *Rev. Statist. Appl.*, **1**, 47–66.

Olivieri, I., Couvet, D. and Gouyon, P.-H. (1990) The genetics of transient populations: research at the metapopulation level. *TREE*, **5**, 207–10.

Pasteur, N., Pasteur, G., Bonhomme, F., Catalan, J. and Britton-Davidian, J. (1988) *Practical Isozyme Genetics*, Ellis Horwood, Chichester.

Planes, S. (1992) *Les échelles spatiales de dispersion des larves de poissons récifaux en Polynésie française: influence sur la différenciation géographique des populations*. PhD thesis, Université P.-&-M. Curie, Paris.

Randall, J.E. (1961a) Observations on the spawning of surgeonfishes (Acanthuridae) in the Society islands. *Copeia*, **2**, 237–8.

Randall, J.E. (1961b) A contribution to the biology of the convict surgeonfish of the Hawaiian islands, *Acanthurus triostegus sandwicensis*. *Pacific Sci.*, **15**, 215–72.

Reynolds, J., Weir, B.S. and Cockerham, C.C. (1983) Estimation of the coancestry coefficient: basis for a short-term genetic distance. *Genetics*, **105**, 767–79.

Rijnsdorp, A.D., Van Stralen, M. and Van der Veer, H.W. (1985) Selective tidal transport of North Sea plaice larvae *Pleuronectes platessa* in coastal nursery areas. *Trans. Am. Fish. Soc.*, **114**, 461–70.

Sale, P.F. (1980) The ecology of fishes on coral reefs. *Oceanogr. Mar. Biol. Annu. Rev.*, **18**, 367–421.

Sammarco, P.W. and Andrews, J.C. (1988) Localized dispersal and recruitment in Great Barrier Reef corals: the Helix experiment. *Science*, **239**, 1422–4.

Shaklee, J.B. (1984) Genetic variation and population structure in the damselfish *Stegastes fasciolatus* throughout the Hawaiian archipelago. *Copeia*, **3**, 629–40.

Shaklee, J.B. and Samollow, P.B. (1984) Genetic variation and population structure in a deepwater snapper, *Pristipomoides filamentosus* in the Hawaiian archipelago. *Fish. Bull.*, **82**, 703–13.

Slatkin, M. (1987) Gene flow and the geographic structure of natural populations. *Science*, **236**, 787–92.

Sokal, R.R. and Rohlf, F.J. (1981) *Biometry*, 2nd edn, Freeman and Co., San Francisco.

Talbot, F.H., Russell, B.C. and Anderson, G.R.V. (1978) Coral reef communities: unstable, high diversity systems? *Ecol. Monogr.*, **48**, 425–40.

Underwood, A.J. and Fairweather, P.G. (1989) Supply-side ecology and benthic marine assemblages. *TREE*, **4**, 16–20.

Victor, B.C. (1986) Planktonic larval stage duration of one hundred species of Pacific and Atlantic wrasses (family Labridae). *Mar. Biol.*, **90**, 317–26.

Weir, B.S. (1990) *Genetic Data Analysis*, Sinauer Associates, Sunderland.

Weir, B.S. and Cockerham, C.C. (1984) Estimating F-statistics for the analysis of population structure. *Evolution*, **38**, 1358–70.

Wellington, G.M. and Victor, B.C. (1989) Planktonic larval duration of one hundred species of Pacific and Atlantic damselfishes (Pomacentridae). *Mar. Biol.*, **101**, 557–67.

Williams, D. McB. (1980) The dynamics of the Pomacentrid community on small patch reefs in One Tree Lagoon, Great Barrier Reef. *Bull. Mar. Sci.*, **30**, 159–70.

Wright, S. (1969) *Evolution and the Genetics of Populations*, Vol. 2, *The Theory of Gene Frequencies*, University of Chicago Press, Chicago.

2.5   DOES VARIANCE IN REPRODUCTIVE SUCCESS LIMIT EFFECTIVE POPULATION SIZES OF MARINE ORGANISMS?

Dennis Hedgecock

Abstract

Substantial variation in reproductive success is made possible by the great fecundity and high early mortality of many marine organisms. A small minority of individuals can replace the entire population in each generation by a sweepstakes-chance matching of reproductive activity with oceanographic conditions conducive to spawning, fertilization, larval survival and successful recruitment. Most individuals may fail to match reproductive activity with permissive oceanographic conditions, and the resulting population variance in reproductive success may be orders of magnitude larger that the binomial or Poisson variance of theoretically ideal populations.

This hypothesis, which accounts for widespread, large discrepancies between effective and actual population numbers and for local differentiation despite high gene flow, predicts (1) temporal genetic change owing to random drift and (2) lower genetic diversity in cohorts of new recruits than exists in the spawning population. The first prediction is supported here by observations of genetic drift that yield at least a 10^5-fold discrepancy between estimated effective and actual population numbers for a semi-isolated population of Pacific oysters Crassostrea gigas. *The second prediction can be tested by future comparisons of genetic diversities among adults, larvae and juveniles in this population, using enzymatic amplification of mitochondrial DNA.*

2.5.1 Introduction

For organisms with very high fecundities and high mortalities of early life stages (type III survivorship curves), there exists the potential for a large variance in the number of offspring that individuals contribute to the next generation of reproducing adults. To the extent that this potential is realized, population genetic theory tells us that effective population sizes of these species may be substantially smaller than actual population sizes, with important consequences for adaptation and evolution. At equilibrium, the variance effective number of a dioecious population, as a function of breeding population number, N, and variance in the number of offspring per parent, V_k is $N_e = (4N - 4)/(V_k + 2)$ (Crow and Kimura, 1970; Crow and Denniston, 1988). Binomial or Poisson V_k, which approximates to variance in offspring number for most terrestrial animals, yields a ratio of N_e/N that is nearly 1.0. However, in species with type III survivorship, V_k may be much greater than binomial or Poisson V_k, and N_e/N much less than 1.0.

Reproduction of marine animals is mediated by spatially and temporally varying oceanographic processes affecting gonad maturation, spawning, fertilization, survival of larval broods and recruitment. Marine animals with a pelagic larval phase are generally highly fecund, but studies of artificial reproduction have revealed great variability in gamete quantity, quality and competence (Lannan, 1980a,b; Lannan, Robinson and Breese, 1980; Muranaka and Lannan, 1984; Gharrett and Shirley, 1985; Withler, 1988). Fertilization success of externally fertilizing species in the field has been shown to be reduced drastically by dilution of sperm or by low density or aggregation of spawning adults (Denny and Shibata, 1990; Levitan, Sewell and Chia, 1992). Thus, the potential for variance in reproductive success exists prior to fertilization.

The larval phase likewise offers ample opportunities for differential reproductive success. Marine animals with feeding pelagic larval stages have classic type III survivorship. Daily mortality of northern anchovy larvae, for example, averaged 0.18 for the years 1954–1984, which means that on average 95% of annual larval production died before recruiting to juvenile schools (Peterman and Bradford, 1987). Variation in larval mortality among years typically results in density-independent relationships between breeding population size and strength of recruitment for marine animals (Blaxter and Hunter, 1982; Gaines and Roughgarden, 1985). Variation in recruitment among years is caused chiefly by variation in climate, either indirectly through effects on food availability (Hjort, 1914; Lasker, 1975, 1978; Barber and Chavez, 1983; Peterman and Bradford, 1987) or directly through physical transport of larvae away from nursery areas or suitable settling sites (Iles and Sinclair, 1982; Parrish, Nelson and Bakun, 1981; Roughgarden, Gaines and Possingham, 1988).

As can now be appreciated from satellite imagery (Roughgarden, Gaines and Possingham, 1988; Roughgarden, Running and Matson, 1991), oceanographic processes and conditions that affect the reproduction of marine animal life vary not only among years but also within and among seasons and over mesoscale distances (tens to hundreds of kilometres). Temporal and spatial oceanographic variability has been correlated broadly with community structure (Parrish, Nelson and Bakun, 1981) and more narrowly with overall or regional recruitment success for a variety of taxa (Roughgarden, Gaines and Possingham, 1988; Ebert and Russell, 1988). Nevertheless, the extent to which variability of the marine environment might also enhance variance in offspring numbers among conspecific individuals has not been considered.

This section explores the hypothesis that, in each generation, a small minority of individuals replaces the entire population by a sweepstakes-chance matching of reproductive activity with oceanographic conditions conducive to spawning, fertilization, larval development and recruitment. Most individuals, given a small chance of matching reproductive activity to oceanographic windows of opportunity, fail to reproduce. The resulting population variance in offspring number, which balances the sweepstakes reproductive success of the minority with the reproductive failure of the majority, may be orders of magnitude larger than binomial or Poisson variance.

This hypothesis makes two testable predictions. First, random genetic drift resulting from the low N_e/N ratios imposed by large V_k ought to be measurable in natural populations of marine animals. Indeed, in the absence of migration, temporal genetic change within local populations should yield estimates of variance effective population numbers that are substantially smaller than actual population numbers. Second, specific cohorts of new recruits, to the extent that they represent the output of a reproductively successful minority, should have less genetic diversity than the total adult population. The first prediction is tested here with allozyme data for a well-studied, semi-isolated population of Pacific oyster *Crassostrea gigas* inhabiting Dabob Bay, Washington. I present evidence that temporal variance in allozyme frequencies in this oyster population is attributable to genetic drift and that the estimated variance effective population number is at least five orders of magnitude less than average annual harvest. Published data for other benthic marine species, including the American oyster, sea urchins and limpets, indicate that such temporal genetic change is not unusual.

2.5.2   Materials and methods

(a)   The Dabob Bay population of Pacific oysters

Two features of Dabob Bay that make it attractive as a study site for testing the hypothesis of large V_k are its relatively simple oceanography and a long-term record of larval oyster production. Dabob Bay, including Quilcene Bay, is a temperate fjord having very little tidal exchange with Hood Canal, from which it is partially isolated by an underwater sill, and negligible estuarine circulation, being fed only by relatively small streams (Figure 2.5.1) (Schaefer, 1938; McGary and Lincoln, 1977; Frost, 1988, personal communication). As a result of these factors, and in the absence of turbulent vertical mixing by sustained winds, a layer of warm water develops at the surface of the bay during the summer (Schaefer, 1938). Retention of this warm-water layer within the bay allows oysters to spawn and to produce commercially useful quantities of recruiting juveniles (spat) in 7 out of 10 years (Westley, 1956); annual spatfall intensity in Dabob Bay ranged over three orders of magnitude from 1952 to 1978 (Packer, 1980). In order to forecast spatfall in Dabob Bay, the Washington State Department of Fisheries (WDF) has monitored, for over 30 years, the occurrence, abundance and growth of larvae, and the environmental factors affecting larval growth and recruitment – water temperature, depth of thermocline, and wind speed and direction (Packer and Mathews, 1980; Packer, 1980). These data indicate that the Dabob Bay population of Pacific oysters is semi-isolated (i.e. the source of larval recruits is predominantly local and rarely immigrant) and that the origin, distribution and fate of larval cohorts may be determined with some accuracy.

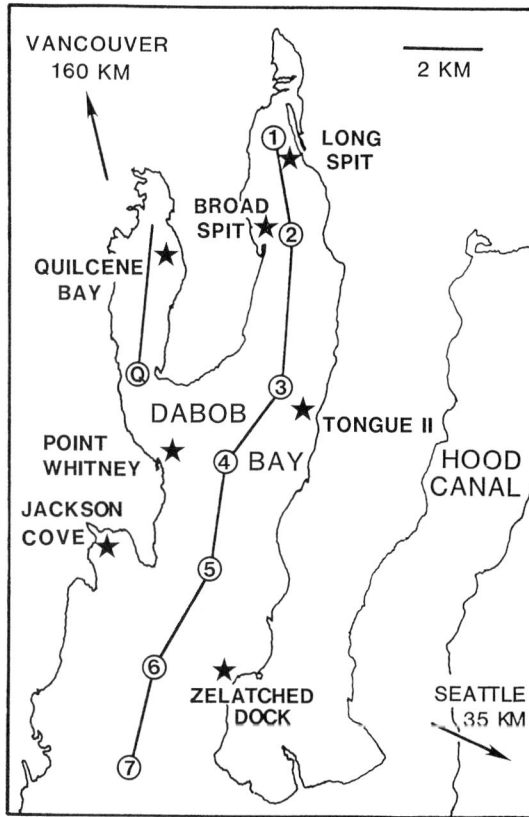

Figure 2.5.1 Map of Dabob and Quilcene Bays showing features of the Washington State Department of Fisheries sampling programme for *C. gigas* larvae. Circles are bathythermograph (BT) recording stations, lines are plankton sample transects and filled stars are sites where test cultch are deployed to measure spatfall intensity. The 1985 and 1990 year class samples of *C. gigas* were obtained near the north end of the Quilcene Bay plankton sampling line.

(b) Allozyme studies of Dabob Bay oysters

Data on allozyme variation are available for Dabob Bay oysters sampled as newly settled spat on oyster-shell cultch in 1971 and 1972 (Buroker, 1975) (six loci) and in 1985 (Hedgecock and Sly, 1990) (14 loci). Juvenile oysters representing the 1990 year class of Dabob Bay natural spat set were removed from larger, 2-year-old oysters produced by two controlled mating experiments in 1989 and grown in Quilcene Bay until February and May of 1991 (Hedgecock, Cooper and Hershberger, 1991). Data on genotypes at 19 polymorphic loci were obtained from 60 individuals of the 1990 year class. Electrophoretic methods, buffers, nomenclature and estimation of genetic parameters were as described by Hedgecock and Sly (1990). Alkaline phosphatase (APH) and mannose phosphate isomerase (MPI) enzymes were not resolved. Eight proteins that were resolved in addition to those reported by Hedgecock and Sly (1990), their genetic symbols, and the electrophoretic buffers used to resolve them were: adenylate kinase (*Adk*, buffer C); glyceraldehyde-3-phosphate dehydrogenase (*G-3-pdh*, buffer E); isocitrate dehydrogenase (two loci, *Idh-1*, *Idh-2*, buffer E); muscle protein (*Mp-1*, buffers D, E); 6-phosphogluconate dehydrogenase (*6pgdh*, buffer C); and phosphoglucomutase (*Pgm*, buffer E).

These four allozyme data sets allow estimation of genetic change in the Dabob Bay population over six intervals of time, 1971–1972, 1971–1985, 1972–1985, 1971–1990, 1972–1990 and 1985–1990, which are assumed to represent 1, 7, 6, 9.5, 9 and 2 elapsed generations, respectively. Estimates of effective population number were not qualitatively altered by assuming mean generation lengths ranging from 2 to 2.5 years. Mean temporal variances, F_k, in allelic frequencies, and estimates of mean effective population numbers, N_k, over these intervals were calculated according to Pollak (1983), using Nei and Tajima's (1981) estimator for loci with two alleles, as described by Hedgecock and Sly (1990) and Hedgecock, Chow and Waples (1992). Ninety-five per cent confidence limits around mean N_k were calculated according to Waples (1989).

If random genetic drift is the only cause of temporal change in allelic frequencies, $nF_k/E(F_k)$ is distributed as χ^2 with degrees of freedom, n, equal to the number of loci (Lewontin and Krakauer, 1973, 1975; Nei and Tajima, 1981; Waples, 1989). Although the distribution and expected value of F_k are unknown, simulations in which sample mean $\bar{F}_k$ is substituted for $E(F_k)$ show reasonable agreement with χ^2 distributions (Nei and Tajima, 1981; Waples, 1989). χ^2 probability plots (Gnanadesikan, 1977) and Kolmogorov–Smirnov goodness-of-fit tests, made with SYSTAT (Wilkinson, 1987), were used to assess whether variances of F_ks among loci are stochastic. F_ks for the 1971–1990 ($n = 6$), 1972–1990 ($n = 6$), and 1985–1990 ($n = 11$) intervals were first transformed to $nF_k/\bar{F}_k$. These values were then plotted against expected χ^2 values obtained from the inverse χ^2 function, XIF(α, d.f. = n), where α is the cumulative χ^2 probability for each F_k associated with its rank r in the distribution, $(r - 0.5)/n$. Plots should be linear if the distribution of $nF_k/\bar{F}_k$ is indeed χ^2.

2.5.3 Results

Allozyme frequencies for a sample of the 1990 year class of Pacific oyster spat from Dabob Bay are given in the Appendix. Of the 19 loci studied, 17 (89.5%) are polymorphic. Mean sample size per locus is 59.3 ± 0.3, mean number of alleles per locus is 3.8 ± 0.4, and mean observed and expected (Nei, 1978) heterozygosities are 0.323 ± 0.050 and 0.333 ± 0.051, respectively. There are no significant deviations of phenotypic frequencies from those expected under codominant Mendelian inheritance and Hardy–Weinberg–Castle equilibrium.

Estimates of mean temporal variance of allelic frequencies, F_k, and effective population number, N_k, have been calculated over six intervals (Table 2.5.1). The numbers of loci in

Table 2.5.1   Mean temporal variances in allelic frequencies, F_k, and estimated effective population numbers, N_k, over six intervals of time in a population of Pacific oysters, *Crassostrea gigas*, from Dabob Bay, Washington

t	l	F_k	Sampling variance	Actual variance	lCL	N_k	uCL
1	6	0.0234	0.0114	0.0120	13.4	41.7	218.8
2	11	0.0192	0.0172	0.0020	63.6	511.6	∞
6	5	0.0237	0.0148	0.0089	68.0	336.7	∞
7	5	0.0206	0.0136	0.0070	93.0	501.9	∞
9	6	0.0340	0.0141	0.0199	77.1	226.0	804.7
9.5	6	0.0252	0.0139	0.0113	115.0	418.7	9293

Interval length in generations is t, l is the number of loci studied, sampling variance is the harmonic mean of sample sizes per locus in the two populations compared, actual variance is F_k minus sampling variance, and lCL and uCL are the lower and upper 95% confidence limits on N_k.

comparisons involving samples from 1971 and 1972 are constrained to five or six (Buroker, 1975); comparisons of the 1985 and 1990 year classes, on the other hand, are based on 11 loci. Temporal variance, F_k, comprises both actual temporal change and random binomial sampling error; the latter component is $1/2S_0 + 1/2S_1$, where S_0 and S_1 are the harmonic means over loci of the number of individuals sampled at times $t = 0$ and $t = 1$, respectively, weighted by the number of independent alleles at each locus. These sampling variances, which range from 0.0114 to 0.0172, are subtracted from F_k to obtain an estimate of the actual temporal variance in allelic frequencies. The reciprocal of twice actual temporal variance, multiplied by the number of generations over which this change accumulated, provides an estimate of mean N_k (Pollak, 1983). Estimated mean effective population numbers for the Dabob Bay population range from 41.7, for the interval between 1971 and 1972, to 511.6, for the interval from 1985 to 1990. Mean N_k over the five intervals spanning two or more generations is 399 ± 54.

Probability plots of temporal variances for loci studied in comparisons of each of the 1971, 1972 and 1985 samples with the 1990 sample are linear (Figure 2.5.2), with no significant deviations from the corresponding χ^2 distributions (Kolmogorov–Smirnov tests all non-significant).

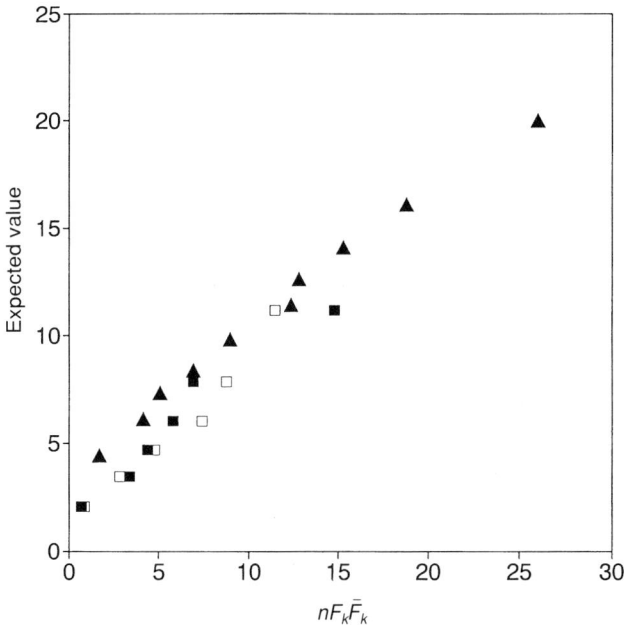

Figure 2.5.2 Probability plots of transformed variances for loci studied in temporal comparisons between the 1990 Quilcene Bay sample and each of the samples from 1971 (filled squares), 1972 (empty squares) and 1985 (filled triangles). The variances, F_k, are transformed by $nF_k/\bar{F}_k$, where n is the number of loci, and plotted against corresponding values from the χ^2 distribution (see text). Straight lines indicate no deviation from χ^2 distributions; variance of F_k among loci is stochastic, as expected if random genetic drift is the only cause of temporal genetic change.

2.5.4   Discussion

Part of the rationale for the hypothesis that large variance in reproductive success limits effective population numbers of marine organisms is that this hypothesis resolves two major paradoxes concerning the population genetics of marine organisms. The first paradox concerns widespread, large discrepancies between the abundances or actual population sizes of marine animal species and their long-term effective population sizes as estimated from their total genetic diversities. The second paradox concerns observations of slight but statistically significant genetic differences among local populations of species that have high rates of gene flow via dispersing pelagic larvae.

Often missed in the debate over whether balancing selection or drift of selectively neutral amino acid substitutions maintains protein polymorphism is the paradox that most species do not have enough genetic variation under either mechanism, given their enormous population numbers (Hedgecock, Tracey and Nelson, 1982; Nei and Graur, 1984; Nei, 1987; Avise, Ball and Arnold, 1988). The great abundance of many marine animals is well documented. Yet various authors, working on different taxa and employing different biochemical methods, have reported intraspecific genetic diversities implying discrepancies of 10^2–10^5 between abundance and effective population size (Hedgecock, Tracey and Nelson, 1982; Nei and Graur, 1984; Graves, Ferris and Dixon, 1984; Gyllensten, 1985; Vawter and Brown, 1986; Avise *et al.*, 1986; Gyllensten and Wilson, 1987; Avise, Ball and Arnold, 1988; Becker *et al.*, 1988; Reeb and Avise, 1990; Palumbi and Wilson, 1990). Two specific examples are cited for illustration. Seventy-eight million pounds of the American lobster *Homarus americanus* are landed annually (Williams, 1986), yet N_e for this crustacean with a harvest body weight of about one pound and a mean generation length of about seven years is calculated from allozyme data and neutral theory to be about 10^4 (Hedgecock, Tracey and Nelson, 1982). The number of American oysters harvested annually is estimated to be about 1.5×10^{10} (Food and Agriculture Organization, 1986: annual harvest of 3×10^5 metric tons meat weight, at 20 g meat weight per individual). Yet mitochondrial DNA sequence diversities within Gulf of Mexico and Atlantic populations imply effective female population sizes of 125 000 and 75 000, respectively (Reeb and Avise, 1990), and mean allozyme heterozygosities for these same populations imply effective sizes of 2.9×10^5 and 2.4×10^5, respectively (calculated from data of Buroker (1983a), excluding samples from the Florida Peninsula).

There are few explanations for such discrepancies. The use of neutral theory to estimate abundance from genetic diversity may be wrong; but, as Nei and Graur (1984) point out, if N_e is close to N, selection must drastically reduce, not enhance, intraspecific genetic variability while permitting interspecific gene substitutions to occur. A model of purifying selection fails to improve the fit of theory to data (Nei and Graur, 1984; Nei, 1987). This leaves only two population genetic factors that might account for low N_e/N ratios – widespread population bottlenecks or large V_k. Bottleneck population size must be very small, however, certainly less than 100 and more like 10, and the rate of population increase low in order to achieve the observed reductions in N_e (Nei, Maruyama and Chakraborty, 1975; Chakraborty and Nei, 1977). While population collapses certainly occur in the sea (e.g. Lessios, Robertson and Cubit, 1985), it is difficult to accept and even more difficult to test that essentially all marine animal lineages have suffered such catastrophes. V_ks of the order of 10^2–10^5, on the other hand, would provide an ecologically relevant and testable explanation of low N_e/N ratios.

The second paradox concerning the genetics of marine animals that disperse by means of planktotrophic larvae is the occurrence of significant local or microscale population structure despite high gene flow (Burton, 1983; Hedgecock, 1982, 1986). While similarity of allelic frequencies throughout the ranges of such species is generally consistent with high gene flow, genetic differentiation on a local scale requires either differential survival of genotypes after recruitment (e.g. Koehn, Newell and Immermann, 1980) or temporal variation in the genetic composition of recruits. Johnson and colleagues favour the latter explanation for limpets (Johnson and Black, 1981, 1984) and sea urchins (Watts, Johnson and Black, 1990); these studies show that 'chaotic' genetic variation among local benthic adult populations is a byproduct of temporal variation in the genetic composition of recruits. Moreover, temporal variance in allelic frequencies is exceeded by spatial variance only on scales of several hundreds to thousands of kilometres. Although Johnson and Black (1984) believe that this temporal variance arises from temporally and spatially varying selection on larvae, an alternative explanation is sampling variance, owing to sweepstakes reproductive success of a small fraction of the population.

(a) Random genetic drift in natural populations of oysters

Analysis of temporal genetic change is a powerful means of measuring random genetic drift, estimating effective population numbers, and testing population genetic hypotheses. The method is particularly robust over intervals of two to ten generations and when N_e is truly finite (Waples, 1989), and has proved illuminating in the study of isolated, hatchery-propagated stocks of fish and shellfish (Waples and Teel, 1990; Hedgecock and Sly, 1990; Hedgecock, Chow and Waples, 1992). Application of the temporal method to natural populations now appears useful in testing the hypothesis that variance in reproductive success limits effective population numbers of many marine animals.

Temporal genetic analysis is sensitive to the numbers of loci and individuals sampled, as illustrated by some results of the present study. The upper 95% confidence limits (CL) on N_k for three of the intervals studied are infinite (Table 2.5.1), which in the absence of the remaining data would undermine confidence in the point estimates of average N_e for the Dabob Bay oyster population. Two of these cases, comparisons of 1971 and 1972 with 1985, appear to be based on too few loci (only five). If, for example, these same five loci are used for the comparison of 1971 to 1972, F_k is 0.0192 and N_k is 62.8, with a lower CL of 15.0 and an upper CL of infinity (cf. Table 2.5.1). Thus, the additional information provided by one locus (*Mp-1*) makes the difference between infinite and finite upper CLs for the 1971–1972 interval and for the 1971–1990 and 1972–1990 intervals as well. The 1985–1990 comparison, for which 11 loci were studied, also has an infinite upper CL. In this case, however, numbers of individuals in the two samples were only 59 and 60, so that sampling variance was 28% higher than the average sampling variance in other comparisons. Another reason for the higher N_k and infinite upper CL in this comparison might also be that N_e during this interval was actually larger; WDF reports indicate substantial spatfall in 1985, 1986, 1988 and 1990.

The magnitudes of genetic change measured over intervals of from 1 to 19 years in this large, semi-isolated population of oysters are remarkable. Mean temporal variance in allozyme frequencies exceeds binomial sampling error in all six intervals. Moreover, distributions of temporal variance across loci conform to χ^2 distributions (Figure 2.5.2), suggesting that temporal changes were caused by random genetic drift of allozyme frequencies in a finite population. Resulting estimates of effective population number over intervals of two or more generations average only about 400. N_e in this population is thus estimated to be at least five or six orders of magnitude less than an annual harvest from Dabob Bay of 10^7 or 10^8 oysters (R. Burge, personal communication).

Random genetic drift has also been inferred for natural populations of the American oyster *Crassostrea virginica* (Hedgecock, Chow and Waples, 1992). Allozyme data have been reported for four populations sampled 4 years apart (compare Buroker (1983a) with Vrijenhoek, Ford and Haskin (1990) for Long Island, Delaware Bay and James River; and Buroker (1983b) with Paynter and DiMichele (1990) for upper Chesapeake Bay). Variance of allelic frequencies between these four sets of temporal samples is substantial (Table 2.5.2). Distributions of temporal variances among loci fit χ^2 distributions (not shown), implying that these temporal changes represent random genetic drift.

Table 2.5.2   Temporal genetic variance over two generations in four populations of the American oyster *Crassostrea virginica*

Locality	l	F_k	Sampling variance	Actual variance	lCL	N_k	uCL
Long Island	6	0.0158	0.0162	−0.0004	62.3	∞	∞
Delaware Bay	6	0.0424	0.0127	0.0296	13.8	33.8	79.4
Chesapeake Bay	4	0.0974	0.0304	0.0670	4.5	14.9	48.2
James River	6	0.0433	0.0100	0.0333	13.5	30.0	60.8

After Hedgecock, Chow and Waples (1992). Column headings as in Table 2.5.1.

For the Delaware Bay, Chesapeake Bay and James River samples, temporal variance corrected for sampling error, which is a normalized variance directly comparable to Wright's measure of spatial variance F_{st}, is as large or larger than allele-frequency variance throughout the range of Reeb and Avise's (1990) Atlantic mitochondrial lineage ($F_{st} = 0.029$; calculated from data of Buroker (1983a)). Actual temporal variance in Chesapeake Bay is nearly twice as large as variance from Canada to Texas ($F_{st} = 0.039$) (Buroker, 1983a). Thus, temporal changes cannot be explained by immigration from other localities and imply random genetic drift in finite local populations maintained by larval retention (cf. Hedgecock, 1982). Partial isolation of major estuarine populations would help explain the evolution of local physiological races (Loosanoff and Nomejko, 1951; Hedgecock and Okazaki, 1984). Interestingly, there is no temporal genetic change at the Milford, Connecticut, site, owing perhaps to greater gene flow in the more oceanic Long Island Sound. F_ks for Delaware Bay and Chesapeake Bay populations yield surprisingly small estimates of N_e with finite 95% upper confidence limits (Table 2.5.2). To what extent these small effective population numbers reflect recent devastation of these oyster populations by disease (Ford and Haskin, 1988; Krantz, 1992) is uncertain.

These observations of random genetic drift in natural populations of oysters confirm a major prediction of the hypothesis that variance in reproductive success in many marine animals may be large enough to limit effective population numbers to fractions of actual abundances. Temporal genetic change can be measured in these local populations and the temporal variances yield estimates of N_e that are at least 10^5-fold smaller than actual population numbers. These discrepancies are of magnitude similar to those found in comparisons of species abundances and long-term N_e estimated from total genetic diversity. Thus, the hypothesis of large variance in reproductive success, which was generated in part to explain discrepancies between species abundances and population genetic equilibria on evolutionary timescales, makes predictions that have been confirmed by observations of temporal change on ecological timescales.

A second prediction of the hypothesis is for lower genetic diversity in particular cohorts of larvae or newly recruited juveniles than exists in the spawning adult stock. This prediction may be verified in the future by detailed comparisons of genetic diversities among adults, larvae and juveniles. Because mitochondrial DNA appears to be predominantly maternally inherited in animals, polymorphisms in this genome are ideal genetic markers for studies of larval broods. Advances in molecular biology, in particular the development of enzymatic amplification of DNA by the polymerase chain reaction (PCR), now make possible population genetic studies of marine larvae (Banks, Waters and Hedgecock, 1993), which have not generally been amenable to allozyme analysis. Dabob Bay is an ideal locality to carry out detailed genetic studies of oyster larvae, because temporally well-separated larval cohorts can be readily identified in plankton samples during a spawning season.

(b)   Broader significance of the hypothesis

To the extent that large variance in reproductive success in marine animals is mediated by oceanographic conditions and processes, there is a strong and direct linkage between population genetics and oceanography. This linkage must be forged if we are to understand broader questions about marine populations, such as their responses to global climate change (Incze and Walsh, 1991). Within populations, large V_k might make population responses to selection pressures more complex and indeterminate than is presently appreciated by modellers of population dynamics. On the other hand, adaptive divergence among populations with the potential for gene exchange via dispersing pelagic larvae might be facilitated by a coupling of large V_k with mechanisms of larval retention, as perhaps illustrated by the evolution of physiological races of American oysters along the eastern US seaboard. At the operational level, detailed studies of genetic diversities within and between cohorts of larvae might, for example, provide useful information on the spatial and temporal dimensions of windows of oceanographic conditions conducive to reproduction and recruitment.

Variance in reproductive success can also affect population structure because of temporal and spatial variation in the genetic composition of recruiting larvae. The temporal genetic instability of marine populations should be taken into account in studies of physiological genetics and ecology. Large V_k might have a role in two controversial observations of bivalve population structure – excesses of homozygosity, particularly in recruits, and correlations of fitness with heterozygosity.

Finally, speciation in the sea may be more understandable if effective numbers of marine organisms are orders of magnitude smaller than abundance and if marine species are therefore subject to shifting-balance evolutionary processes.

2.5.5   Appendix. Allelic frequencies for 19 protein coding loci in a sample of 1990 year class Pacific oysters *Crassostrea gigas* from Dabob Bay, Washington

Allele	*Locus and sample size*									
	Aat 60	*Acon-1* 60	*Acon-2* 60	*Adk* 59	*Dia* 59	*G3pdh* 58	*Gpi* 58	*Dap* 60	*Idh-1* 55	*Idh-2* 60
A	0.617	0.633	0.858	0.797	0.559	0.707	0.922	0.358	0.964	0.758
B	0.000	0.175	0.067	0.110	0.314	0.293	0.000	0.508	0.000	0.017
C	0.108	0.183	0.008	0.059	0.000	0.000	0.017	0.117	0.027	0.142
D	0.275	0.008	0.000	0.025	0.102	0.000	0.000	0.008	0.009	0.000
E	0.000	0.000	0.067	0.008	0.025	0.000	0.034	0.000	0.000	0.042
F	0.000	0.000	0.000	0.000	0.000	0.000	0.017	0.000	0.000	0.000
G	0.000	0.000	0.000	0.000	0.000	0.000	0.000	0.008	0.000	0.042
H	0.000	0.000	0.000	0.000	0.000	0.000	0.009	0.000	0.000	0.000
H_{exp}	0.537	0.539	0.256	0.352	0.583	0.418	0.149	0.604	0.071	0.404
H_{obs}	0.550	0.500	0.200	0.356	0.576	0.345	0.155	0.533	0.073	0.433

Allele	*Locus and sample size*								
	Lap-2 60	*Tap-2* 60	*Mdh-1* 60	*Mdh-2* 60	*6Pgdh* 59	*Pgm* 60	*Mp-1* 58	*Sod-1* 60	*Sod-2* 60
A	0.775	0.717	0.992	1.000	0.780	0.550	0.353	1.000	0.967
B	0.000	0.175	0.000	0.000	0.000	0.100	0.647	0.000	0.000
C	0.108	0.092	0.008	0.000	0.059	0.117	0.000	0.000	0.033
D	0.017	0.000	0.000	0.000	0.093	0.108	0.000	0.000	0.000
E	0.008	0.017	0.000	0.000	0.034	0.042	0.000	0.000	0.000
F	0.083	0.000	0.000	0.000	0.025	0.025	0.000	0.000	0.000
G	0.008	0.000	0.000	0.000	0.000	0.025	0.000	0.000	0.000
H	0.000	0.000	0.000	0.000	0.008	0.033	0.000	0.000	0.000
H_{exp}	0.383	0.451	0.017	0.000	0.381	0.664	0.461	0.000	0.065
H_{obs}	0.317	0.450	0.017	0.000	0.390	0.683	0.500	0.000	0.067

Acknowledgements

I thank Dan McGoldrick for help with the electrophoretic analyses and Dr Elmarie Hutchinson and two anonymous reviewers for comments on the manuscript. Electrophoretic analysis of the 1990 year class was supported by the University of California Agricultural Experiment Station, Project N CA-D*-XXX-5099-H.

References

Avise, J.C., Ball, R.M. and Arnold, J. (1988) Current versus historical population sizes in vertebrate species with high gene flow: a comparison based on mitochondrial DNA lineages and inbreeding theory for neutral mutations, *Mol. Biol. Evol.*, **5**, 331–44.

Avise, J.C., Helfman, G.S., Saunders N.C. and Hales, L.S. (1986) Mitochondrial DNA differentiation in North Atlantic eels: population genetic consequences of an unusual life history pattern. *Proc. Natl. Acad. Sci. USA*, **83**, 4350–4.

Banks, M.A., Waters, C. and Hedgecock, D. (1993) Discrimination between closely related Pacific oyster species (*Crassostrea*) via mitochondrial DNA sequences coding for large subunit rRNA. *Mar. Mol. Biol. Biotechnol*, **2**, 129–36..

Barber, R.T. and Chavez, F.P. (1983) Ocean variability in relation to living resources during the 1982–83 El Nino. *Nature*, **319**, 279–85.

Becker, I.I., Grant, W.S., Kirby, R. and Robb, F.T. (1988) Evolutionary divergence between sympatric species of southern African hakes, *Merluccius capensis* and *M. paradoxus*. II. Restriction enzymes analysis of mitochondrial DNA. *Heredity*, **61**, 21–30.

Blaxter, J.H.S. and Hunter, J.R. (1982) The biology of clupeoid fishes. *Adv. Mar. Biol.*, **20**, 1–223.

Buroker, N.E. (1975) A survey of protein variation in populations of the Pacific oyster, *Crassostrea gigas*. MS dissertation, Univ. Washington, Seattle.

Buroker, N.E. (1983a) Population genetics of the American oyster *Crassostrea virginica* along the Atlantic coast and the Gulf of Mexico. *Mar. Biol.*, **75**, 99–112.

Buroker, N.E. (1983b) Genetic differentiation and population structure of the American oyster *Crassostrea virginica* (Gmelin) in Chesapeake Bay. *J. Shellfish. Res.*, **3**, 153–67.

Burton, R.S. (1983) Protein polymorphisms and genetic differentiation of marine invertebrate populations. *Mar. Biol. Lett.*, **4**, 193–206.

Chakraborty, R. and Nei, M. (1977) Bottleneck effects on average heterozygosity and genetic distance with the stepwise mutation model. *Evolution*, **31**, 347–56.

Crow, J.F. and Denniston, C. (1988) Inbreeding and variance effective population numbers. *Evolution*, **42**, 482–95.

Crow, J.F. and Kimura, M. (1970) *Introduction to Population Genetics Theory*, Harper and Row, New York.

Denny, M.W. and Shibata, M.F. (1990) Consequences of surf-zone turbulence for settlement and external fertilization. *Am. Natur.*, **134**, 859–89.

Ebert, T.A. and Russell, M.P. (1988) Latitudinal variation in size structure of the west coast purple sea urchin: a correlation with headlands. *Limnol. Oceanogr.*, **33**, 286–94.

Food and Agriculture Organization (1986) *Yearbook of Fishery Statistics, Catches and Landings*, Vol. 62, FAO, Rome.

Ford, S.E. and Haskin, H.H. (1988) Management strategies for MSX (*Haplosporidium nelsoni*) disease in eastern oysters, in *Disease Processes in Marine Bivalves* (ed. W.S. Fisher), American Fisheries Society, Publ. No. 18, Bethesda, pp. 249–56.

Frost, B.W. (1988) Variability and possible adaptive significance of diel vertical migration in *Calanus pacificus*, a marine planktonic copepod. *Bull. Mar. Sci.*, **43**, 675–94.

Gaines, S.D. and Roughgarden, J. (1985) Larval settlement rate: a leading determinant of structure in an ecological community of the marine intertidal zone. *Proc. Natl. Acad. Sci. USA*, **82**, 3707–11.

Gharrett, A.J. and Shirley, S.M. (1985) A genetic examination of spawning methodology in a salmon hatchery. *Aquaculture*, **47**, 245–56.

Gnanadesikan, R. (1977) *Methods for Statistical Data Analysis of Multivariate Observations*, J. Wiley and Sons, New York.

Graves, J.E., Ferris, S.D. and Dizon, A.E. (1984) Close genetic similarity of Atlantic and Pacific skipjack tuna (*Katsuwonis pelamis*) demonstrated with restriction endonuclease analysis of mitochondrial DNA. *Mar. Biol.*, **79**, 315–19.

Gyllensten, U. (1985) The genetic structure of fish: differences in the intraspecific distribution of biochemical genetic variation between marine, anadromous, and freshwater species. *J. Fish. Biol.*, **26**, 691–9.

Gyllensten, U. and Wilson, A.C. (1987) Mitochondrial DNA of salmonids, in *Population Genetics and Fishery Management* (eds N. Ryman and F. Utter), Washington University Press, Seattle, pp. 301–17.

Hedgecock, D. (1982) Genetical consequences of larval retention: theoretical and methodological aspects, in *Estuarine Comparisons* (ed. V.S. Kennedy), Academic Press, New York, pp. 553–68.

Hedgecock, D. (1986) Is gene flow from pelagic larval dispersal important in the adaptation and evolution of marine invertebrates? *Bull. Mar. Sci.*, **39**, 550–64.

Hedgecock, D., Chow, V. and Waples, R.E. (1992) Effective population numbers of shellfish broodstocks estimated from temporal variance in allelic frequencies. *Aquaculture*, **108**, 215–32.

Hedgecock, D., Cooper, K. and Hershberger, W. (1991) Genetic and environmental components of variance in harvest body size among pedigreed Pacific oysters *Crassostrea gigas* from controlled crosses. **J.** *Shellfish. Res.*, **10**, 516.

Hedgecock, D. and Okazaki, N.B. (1984) Genetic diversity within and between populations of American oysters (*Crassostrea*). *Malacologia*, **25**, 535–49.

Hedgecock, D. and Sly, F.L. (1990) Genetic drift and effective population sizes of hatchery-propagated stocks of the Pacific oyster *Crassostrea gigas*. *Aquaculture*, **88**, 21–38.

Hedgecock, D., Tracey, M.L. and Nelson, K. (1982) Genetics, in The *Biology of Crustacea*, vol. 2 (ed. L.G. Abele), Academic Press, New York, pp. 297–403.

Hjort, J. (1914) Fluctuations in the great fisheries of northern Europe viewed in the light of biological research. *Rapp. Procs-Verb. Reun., Cons. Int. Explor. Mer*, **20**, 1–228.

Iles, T.D. and Sinclair, M. (1982) Atlantic herring: stock discreteness and abundance. *Science*, **215**, 627–33.

Incze, L.S. and Walsh, P.J. (1991) *GLOBEC Workshop on Biotechnology Applications to Field Studies of Zooplankton*, GLOBEC Report Number 3, Joint Oceanographic Institutions, Inc., Washington.

Johnson, M.S. and Black, R. (1982) Chaotic genetic patchiness in an intertidal limpet, *Siphonaria* sp. *Mar. Biol.*, **70**, 157–64.

Johnson, M.S. and Black, R. (1984) Pattern beneath the chaos: the effect of recruitment on genetic patchiness in an intertidal limpet. *Evolution*, **38**, 1371–83.

Krantz, G.E. (1992) *Maryland Oyster Population Status Report 1991 Fall Survey*, CBRM-OX-92-1, Chesapeake Bay Research and Monitoring Division, Cooperative Oxford Laboratory, Oxford, MD.

Koehn, R.K., Newell, R.I.E. and Immermann, F. (1980) Maintenance of an aminopeptidase allele frequency cline by natural selection. *Proc. Natl. Acad. Sci., USA*, **77**, 5385–9.

Lannan, J.E. (1980a) Broodstock management of *Crassostrea gigas*: I. Genetic variation in survival in the larval rearing system. *Aquaculture*, **21**, 323–36.

Lannan, J.E. (1980b) Broodstock management of *Crassostrea gigas*: III. Selective breeding for improved larval survival. *Aquaculture*, **21**, 346–52.

Lannan, J.E., Robinson A. and Breese, W.P. (1980) Broodstock management of *Crassostrea gigas*: II. Broodstock conditioning to maximize larval survival. *Aquaculture*, **21**, 337–45.

Lasker, R. (1975) Field criteria for survival of anchovy larvae: the relation between inshore chlorophyll maximum layers and successful first feeding. *Fish. Bull., US*, **73**, 453–62.

Lasker, R. (1978) The relation between oceanographic conditions and larval anchovy food in the California Current: identification of factors contributing to recruitment failure. *Rapp. P.-V. Reun. Cons. Int. Explor. Mer*, **173**, 212–30.

Lessios, H.A., Robertson, D.R. and Cubit, J.D. (1985) Spread of *Diadema* mass mortality through the Caribbean. *Science*, **226**, 335–7.

Levitan, D.R., Sewell, M.A. and Chia, F.-S. (1992) How distribution and abundance influence fertilization success in the sea urchin *Stronglocentrotus franciscanus*. *Ecology*, **73**, 248–54.

Lewontin, R.C. and Krakauer, J. (1973) Distribution of gene frequency as a test of the theory of the selective neutrality of polymorphisms. *Genetics*, **74**, 175–95.

Lewontin, R.C. and Krakauer, J. (1975) Testing the heterogeneity of F values. *Genetics*, **80**, 397–8.

Loosanoff, V.L. and Nomejko, C.A. (1951) Existence of physiologically different races of oysters, *Crassostrea virginica*. *Biol. Bull.*, **101**, 151–6.

McGary, N. and Lincoln, J.H. (1977) *Tide Prints: Surface Tidal Currents in Puget Sound*, Washington Sea Grant Publ., University of Washington Press, Seattle.

Muranaka, M.S. and Lannan, J.E. (1984) Broodstock management of *Crassostrea gigas*: environmental influences on broodstock conditioning. *Aquaculture*, **39**, 217–28.

Nei, M. (1978) Estimation of average heterozygosity and genetic distance from a small number of individuals. *Genetics*, **89**, 583–590.

Nei, M. (1987) *Molecular Evolutionary Genetics*, Columbia University Press, New York.

Nei, M. and Graur, D. (1984) Extent of protein polymorphism and the neutral mutation theory. *Evol. Biol.*, **17**, 73–118.

Nei, M., Maruyama, T. and Chakraborty, R. (1975) The bottleneck effect and genetic variability in populations. *Evolution*, **29**, 1–10.

Nei, M. and Tajima, F. (1981) Genetic drift and estimation of effective population size. *Genetics*, **98**, 625–40.

Packer, J.F. (1980) Prediction of Pacific oyster spatfall intensity in Dabob Bay. *Wash. St. Dept. Fish., Prog. Rep.*, 126, pp. 1–80.

Packer, J.F. and Mathews, S.B. (1980) Vertical movement behaviour, growth, and prediction of spatfall for Pacific oyster larvae in Dabob Bay. *Wash. St. Dept. Fish., Prog. Rep.*, 97.

Palumbi, S.R. and Wilson, A.C. (1990) Mitochondrial DNA diversity in the sea urchins *Stronglyocentrotus purpuratus* and *S. droebachiensis*. *Evolution*, **44**, 403–15.

Parrish, R.D., Nelson, C.S. and Bakun, A. (1981) Transport mechanisms and reproductive success of fishes in the California Current. *Biol. Oceanogr.*, **1**, 175–203.

Paynter, K.T. and DiMichele, L. (1990) Growth rate of tray-cultured oysters (*Crassostrea virginica* Gmelin) in Chesapeake Bay. *Aquaculture*, **87**, 289–97.

Peterman, R.M. and Bradford, M.J. (1987) Wind speed and mortality rate of a marine fish, the northern anchovy (*Engraulis mordax*). *Science*, **235**, 354–6.

Pollak, E. (1983) A new method for estimating the effective population size from allele frequency changes. *Genetics*, **104**, 531–48.

Reeb, C.A. and Avise, J.C. (1990) A genetic discontinuity in a continuously distributed species: mitochondrial DNA in the American oyster, *Crassostrea virginica*. *Genetics*, **124**, 397–406.

Roughgarden, J., Gaines, S. and Possingham, H. (1988) Recruitment dynamics in complex life cycles. *Science*, **241**, 1460–6.

Roughgarden, J., Running, S.W. and Matson, P.A. (1991) What does remote sensing do for ecology? *Ecology*, **72**, 1918–22.

Schaefer, M.B. (1938) Preliminary observations on the reproduction of the Japanese common oyster, *Ostrea gigas*, in Quilcene Bay, Washington. *Wash. Dept. Fish. Biol. Rep.* 36E.

Vawter, L. and Brown, W.M. (1986) Nuclear and mitochondrial DNA comparisons reveal extreme rate variation in the molecular clock. *Science*, **234**, 194–6.

Vrijenhoek, R.C., Ford, S.E. and Haskin, H.H. (1990). Maintenance of heterozygosity during selective breeding of oysters for resistance to MSX disease. *J. Hered.*, **81**, 418 -23.

Waples, R.S. (1989) A generalized approach for estimating effective population size from temporal changes in allele frequency. *Genetics*, **121**, 379–91.

Waples, R.S. and Teel, D.J. (1990) Conservation genetics of Pacific salmon I. Temporal changes in allele frequency. *Conserv. Biol.*, **4**, 144–56.

Watts, R.J., Johnson, M.S. and Black, R. (1990) Effects of recruitment on genetic patchiness in the urchin *Echinometra mathaei* in Western Australia. *Mar. Biol.*, **105**, 145–51.

Westley, R.E. (1956) Retention of Pacific oyster larvae in an inlet with stratified waters. *Wash. St. Dept. Fish., Res. Paper* 1 (4), pp. 25–31.

Wilkinson, L. (1987) *SYSTAT: The System for Statistics*, SYSTAT Inc. Evanston, Il.

Williams, A.B. (1986) Lobsters – identification, world distribution and U.S. trade. *Mar. Fish. Rev.*, **48** (2), 1–36.

Withler, R.E. (1988) Genetic consequences of fertilizing chinook salmon (*Onchorhyncus tsawytscha*) eggs with pooled milt. *Aquaculture*, **68**, 15–25.

3

Heterozygosity, heterosis and adaptation

3.1 THE PRESENT STATUS OF THE RELATIONSHIP BETWEEN HETEROZYGOSITY AND HETEROSIS

Eleftherios Zouros and Grant H. Pogson

Abstract

The general observation that the rates of divergence of most macromolecules roughly follow a model of linear dependence on time has resulted in the de facto *acceptance of the neutral theory of molecular evolution. As a consequence, questions concerning both the significance and the mechanisms of maintenance of molecular variation no longer have a wide appeal among population geneticists. The dichotomy forced by neutral theory concerning molecular polymorphisms is whether they are transient or balanced. As a result of neutral theory's success in interpreting patterns of molecular divergence in historical terms, proponents of the school of balanced polymorphisms have also adopted a historical approach which claims to show that, at least for some polymorphisms, present patterns are incompatible with neutrality. Again, this has entailed a shift away from attempts to explain polymorphisms through studying the relationship between genotype and phenotype.*

Neutral theory's explicit recognition of constraints on the evolution of molecules, and the growing realization that there are limits to how much we can learn about selection from studying its footprints on DNA sequences, should force biologists to rediscover the phenotype. A positive correlation between the degree of individual heterozygosity and fitness-related traits is expected under a wide range of models that can be grouped into two classes; those treating the observed genetic variants as agents of the correlation and those treating these variants as markers of genetic conditions that are responsible for the correlation. We refer to the first class collectively as the selection hypothesis and to the second as the associative overdominance hypothesis. This conceptually fundamental distinction has proved difficult to test experimentally. Direct attempts to relate the effect of heterozygosity at a locus to the biochemical functions of its enzyme products have in most cases supported the view of direct allozyme involvement, but only rarely provided support for overdominance. On the other hand, the parallel observation of strong deficiencies of heterozygotes in most studies reporting a heterozygosity–growth correlation has provided evidence favouring the associative overdominance hypothesis.

Genetics and Evolution of Aquatic Organisms. Edited by A.R. Beaumont.
Published in 1994 by Chapman & Hall, 2–6 Boundary Row, London SE1 8HN.
ISBN 0-412-49370-5

The ability to score genetic variation at the recognition sites of restriction endonucleases and the discovery that the genomes of all higher organisms contain multiple repeats of various kinds has enabled new ways of measuring multilocus heterozygosity. Heterozygosity at these anonymous and most likely functionless loci should not correlate with growth (or any other fitness-related trait) under the selection hypothesis, but should under the associative overdominance hypothesis because these markers are as capable of measuring inbreeding levels as allozymes. We have applied this approach to a cohort of juvenile scallops (Placopecten magellanicus) *in which there was a correlation between allozyme heterozygosity and growth rate. No such correlation was observed for the anonymous DNA markers. Thus, a basic prediction of the associative overdominance hypothesis was not upheld. Since repeatability is especially important in this type of study, this conclusion cannot be considered as established before similar results are obtained from other studies.*

3.1.1 DNA sequences and the eclipse of the phenotype

The general observation that the rates of divergence of most macromolecules (DNA or proteins) roughly follow the model of linear dependence on time has resulted in a wide acceptance of the neutral theory of molecular evolution. Molecular polymorphisms, according to this view, represent transient states in the life of the species, caused by neutral mutations on their stochastic course to fixation or extinction (Kimura, 1983). The alternative view that polymorphisms are governed by natural selection predicts that polymorphisms will either have shorter lives than neutral theory predicts (ending with the fixation of favoured alleles), or will survive for considerably longer times as stable polymorphisms.

The availability of molecular data in the form of homologous DNA sequences from the same or different species offers, in principle, the possibility to test these predictions of the two theories (Kaplan, Darden and Hudson, 1988; Kreitman and Hudson, 1991). This approach has been applied so far to a few systems with mixed results. It appears that in the *Drosophila melanogaster* subgroup the *Adh* locus contains higher amounts of polymorphism than the neutral theory predicts, that the *D. melanogaster* fast/slow polymorphism is maintained by some form of balancing selection, and that the number of interspecific differences in fixed replacements is higher than allowed by neutral theory (Kreitman, 1983; Kreitman and Aguade, 1986; Kreitman and Hudson, 1991; McDonald and Kreitman, 1991) (but see Graur and Li (1991) and Whittam and Nei (1991) for a different view). Selective maintenance of molecular polymorphisms leading to high levels of antigenic diversity has been demonstrated in the complementary-determining regions of V_H genes (Tanaka and Nei, 1989) and the major histocompatibility complex (Hughes, Ohta and Nei, 1990). The rich *Xdh* polymorphism of *D. pseudoobscura*, on the other hand, appears to be compatible with the theory of neutral evolution (Riley, Hallas and Lewontin, 1989; Riles, Kaplan and Venille, 1992).

Inferring natural selection from its footprints on DNA sequences is free of the complications arising from the relationship between genotype and phenotype. In this respect it resembles molecular taxonomy. However, it is arguable whether intraspecific variation in DNA sequences can tell us as much about the forces that shape molecular polymorphisms as interspecific molecular differences can tell us about phylogenies. The smaller the timescale, the more difficult it becomes to extract information concerning evolutionary events from DNA sequences. A well-known example of this difficulty is the current debate over what the variation of the human mtDNA molecule can tell us about the recent history of the species (Maddison, 1991; Templeton, 1992).

3.1.2 The need to rediscover the phenotype

The success of the neutral theory of molecular evolution in revolutionizing the study of phylogeny must be attributed not only to the large timescales involved, but also to the ability of these studies to focus on parts of molecules that are free of selective constraints. The neutral theory's recognition of such constraints is a recognition of the role that natural selection can play in molecular evolution. Evolutionary biologists interested in explaining the origin and fate of molecular polymorphisms have still to define a set of rules for a rigorous *a priori* distinction between constrained and unconstrained parts of the molecule. Indeed, there is a growing suspicion that there might be a large and shifting area of overlap between the two. It is remarkable that allozymes, the polymorphisms that acted as a catalyst for the development of the neutral theory of evolution, remain stubbornly in the grey area claimed by both neutralists and selectionists.

The task of mapping the domains of neutrality and selection on the molecule cannot be accomplished without reference to some phenotypic aspect of the organism, either bio-chemical, morphological or behavioural. In principle, proponents of the neutral theory are not against studies searching for relationships between molecular polymorphisms and phenotype. They are not, however, particularly fond of such studies because of their innate bias against the neutral theory of molecular evolution. Such studies use neutrality as the null hypothesis, never to be proven, only rejected. A direct and clear relationship between an enzyme genotype and some phenotypic measure would be considered sufficient evidence that the enzyme polymorphism is under selection; failure to see a relationship would not eliminate the hypothesis of selection. The converse is also true. Selection can only be proven, never rejected. This is only one of the many problems in the adaptationist's approach discussed by Gould and Lewontin (1979).

Correlations between the degree of heterozygosity and fitness traits provide a connection between genotype and phenotype about which both selection and neutrality provide specific predictions and therefore can be tested on equal terms. This is because the selectionist's explanation of the correlation, and allozyme heterozygosity is *per se* responsible for the relationship, is in principle not restricted to a select set of enzyme loci, to a specific set of organisms, or to specific conditions of the organism's environment. Any restriction of the domain of applicability of the explanation would amount to a weakening of the hypothesis. This would be acceptable only if these caveats arise from our understanding of the mechanism that underlies the correlation, and not from attempts to account for the absence of the correlation when it is not observed. It follows that the selective explanation for the correlation is falsifiable. The neutralist's explanation for the correlation is that it results from the association between allozyme homozygosity and homozygosity for recessive deleterious genes. The causes of this association should affect uniformly all types of polymorphisms. If allozyme and bona fide neutral polymorphisms (e.g. a restriction site polymorphism residing within an intron) are shown to produce an equally strong correlation, then the neutralist's explanation must be accepted.

3.1.3 The relationship between heterozygosity and heterosis: present status

A widely held misconception about correlations between multilocus heterozygosity and fitness-related characters is that they establish overdominance. In fact, overdominance is only one of the many mechanisms that may produce such relationships. This topic has been discussed at length elsewhere (Zouros and Foltz, 1987) and has been the subject of some controversy (Koehn, 1990; Zouros, 1990). Zouros and Mallet (1989) have recognized six hypotheses that can explain the heterozygosity–fitness correlation: overdominance, multiple-locus dominance, balanced enzyme pathways, null alleles, chromosomal loss and associative overdominance. Recently, Chakraborty (1989) has proposed a new hypothesis, that of molecular imprinting. Viewed over the timespan of a single generation, molecular imprinting is formally equivalent to the hypothesis of null alleles, as it assumes that an individual inherits a silenced gene from one of its parents. The dynamics of gene frequency change over many generations will, however, differ from a true null allele and, depending on its rate, imprinting can produce higher levels of apparent homozygosity excess and stronger associations between the apparent degree of heterozygosity and phenotypic traits. At present, we know virtually nothing about molecular imprinting in the organisms in which heterozygosity/fitness correlations are most commonly observed, notably marine molluscs and conifers. Therefore, it is difficult to comment on the plausibility of this hypothesis as a general explanation for heterozygosity relationships.

The associative overdominance explanation is quite different from all the others, including imprinting, in that it assumes that allozymes are neutral markers of effects caused by other genes with which the allozyme loci are in a state of linkage association (linkage disequilibrium). This association may be caused by inbreeding, random drift or population admixture. A certain allele at a certain locus is expected to be in linkage disequilibrium with a deleterious gene only for a limited number of generations. However, because such linkages are continuously generated with other alleles at the same, or at different, loci, an assay that pools genotypes into heterozygosity classes irrespective of alleles or loci is likely to produce a positive correlation between the degree of allozyme homozygosity and homozygosity for one or more deleterious recessives. Because of the power of the associative overdominance hypothesis in accounting for heterozygosity relationships, it has been the subject of intense debate. Several reviews of the evidence (Mitton and Grant, 1984; Zouros and Foltz, 1987; Gaffney, 1990) suggest that there are no strong reasons, at present, for rejecting the associative overdominance hypothesis.

Subsequent studies have not altered this conclusion. Gentili and Beaumont (1988) reported a significant correlation between growth and heterozygosity in mussels grown at high density. Expression of the correlation under stressful conditions was also reported in the clam, *Mulinia lateralis* (Scott and Koehn, 1990), and the palourde, *Ruditapes decussatus* (Borsa, Jousselin and Delay, 1992). Stress has not, however, appeared to have been a factor in a similar correlation seen by Zouros, Romero-Dorey and Mallet (1988) in the mussel, *Mytilus edulis*, or by Alvarez *et al.* (1989) in the oyster, *Ostrea edulis*. Several studies (e.g. Patarnello, Bisol and Battaglia, 1989; Torrissen, 1991; DiMichele and Powers, 1991) have reported higher performances for single-locus heterozygotes for such characters as survival, growth rate and developmental time. On the other hand, lack of a correlation between multilocus heterozygosity and growth was described in natural populations of mussels by Skibinski and Roderick (1989) and by Gosling (1989), and in laboratory crosses by Beaumont (1991). In a comprehensive study, Houle (1989) examined size, development rate and fluctuating asymmetry in relation to heterozygosity at 11 enzyme loci in *Drosophila melanogaster*, but failed to observe a relationship with any of these characters. Booth, Woodruff and Gould (1990) obtained similar results with land snails of the genus *Cerion*.

Interest in the heterozygosity–growth correlation was strengthened by the demonstration of an inverse relationship between the degree of heterozygosity and metabolic activity. The first observations involved correlations of heterozygosity with rates of oxygen consumption (Koehn and Shumway, 1982; Garton, Koehn and Scott, 1984; Diehl *et al.*, 1985), but subsequent studies have provided evidence for higher rates of protein turnover among less heterozygous individuals (Hawkins, Bayne and Day, 1986; Hawkins *et al.*, 1989). Danzmann, Ferguson and Allendorf (1988) also observed a negative correlation between oxygen consumption and heterozygosity in rainbow trout, *Oncorhynchus mykiss*, thus extending these findings to a different group of species. One interpretation of these findings is that individuals with multiple allozyme heterozygosity use less energy for enzyme biosynthesis, and convert a higher proportion of available energy in growth, reproduction or other activities related to food acquisition or predator avoidance; a general scheme of how this may result in correlations of heterozygosity with fitness-related traits in species with different life-history strategies can be found in Volckaert and Zouros (1989). These attempts to link genetic with physiological parameters are necessary to further our understanding of the variation in performance among conspecific individuals. They do not, however, address the question of whether allozymes act as the agents or as markers of the genetic basis of this variation. In the extreme case of a perfect association between allozyme homozygosity and homozygosity for recessive deleterious genes, it is possible to imagine how a higher protein turnover can be the manifestation of a cumulative homozygosity for such genes. Convincing evidence for the energetic advantage of heterozygosity would come only with the demonstration that heterozygosity at a specific enzyme locus results in a lower turnover rate of enzyme coded by that locus.

An implicit assumption of hypotheses assigning a direct role of allozymes to correlations between heterozygosity and fitness characters is that, depending on their function, some loci should contribute more strongly to the correlation than others. In contrast, associative overdominance predicts that the expected contribution of individual loci is more or less the same, since inbreeding, random drift or population mixing are expected to affect all enzyme loci in the same manner. Theoretical work and empirical studies (e.g. Gillespie and Langley, 1974; Kacser and Burns, 1981) support the view that enzyme activities in heterozygous individuals are intermediate relative to the activities of the corresponding homozygotes. Given this, studies demonstrating that specific enzyme activities of heterozygous individuals exceed those of homozygotes may signify the presence of genuine overdominance. Pogson (1991) and Sarver, Katoh and Foltz (1992) have recently produced such evidence in marine molluscs. The three most common heterozygotes at the phosphoglucomutase-2 locus in the Pacific oyster, *Crassostrea gigas*, exhibit significantly higher enzyme activities than homozygotes (Pogson, 1991). The implication that these higher activities may affect growth rate is supported by the observations that *Pgm-2* heterozygotes are on average larger than homozygotes (Fujio, 1982), and by the effect of this enzyme activity variation on tissue glycogen levels (Pogson unpublished data). The octopine dehydrogenase (*Odh*) and leucine aminopeptidase (*Lap*) loci are highly polymorphic in the mussel *Mytilus trossulus* and the oyster *Crassostrea virginica*, respectively. Sarver, Katoh and Foltz (1992) measured *in vitro* specific activities in a large number of individuals at both loci and found in both cases that the mean levels of heterozygotes exceeded homozygotes.

Another approach to establishing a direct role for allozymes in heterozygosity relationships is to show consistent differences among enzyme loci in their contribution to the correlation. Koehn, Diehl and Scott (1988) reported a strong correlation between size and heterozygosity at 15 enzyme loci in the coot clam, *Mulinia lateralis*. These loci were subsequently characterized according to their metabolic function and ranked by their effect on the heterozygosity–growth correlation. Enzymes involved in protein catabolism, glycolysis or the metabolic pathways feeding into glycolysis (pre-glycolytic) had a much stronger effect on the relationship than others. In addition to being incompatible with associative overdominance, this observation provides further support for the hypothesis that allozyme heterozygosity leads to higher energetic efficiency. This intriguing observation has not been rigorously tested in other animals, although it does not appear to hold in the earthworm, *Eisenia foetida* (Diehl, 1988). Interestingly, the *M. lateralis* study by Koehn, Diehl and Scott (1988) also provides one of the best pieces of evidence in favour of the associative overdominance hypothesis. The follow-up analysis on the same data by Gaffney *et al.* (1990)

produced a strong negative relationship between the effect of a locus on the heterozygosity–growth correlation and the magnitude of heterozygote deficiency at that locus. This correlation is not expected under the hypothesis of direct allozyme involvement, but is a prediction of the most common version of the associative overdominance explanation, i.e. that the association between heterozygosity and the fitness-related character results from the correlation between allozyme homozygosity and homozygosity for deleterious genes. Population subdivision coupled with selection favouring animals with non-functional enzyme alleles (either because of null alleles, molecular imprinting or chromosomal loss) can explain additional features of the data, such as an excess of multiply homozygous and multiply heterozygous genotypes (Gaffney *et al.*, 1990), but associative overdominance cannot be completely excluded as a cause of the heterozygosity–size relationship.

3.1.4   Other approaches to explaining the heterozygosity–fitness correlation

It would not be exaggeration to suggest that the connection between heterozygosity and fitness is not understood today any better than when originally proposed as a hypothesis by Lerner (1954), or when it was first empirically demonstrated in plants (Schaal and Levin, 1976) and marine molluscs (Singh and Zouros, 1978). There is nothing exceptional about this. Other important issues in biology, such as the general question of selection versus neutrality in regard to molecular polymorphisms, are in the same state of uncertainty. Like the selection versus neutrality controversy, the search for correlations involving multiple-locus heterozygosity has also yielded important insights into other issues, such as population structure and the physiological energetics of growth. Research on heterozygosity will continue, but what direction should it take?

The question of how general the correlation is remains to be determined. Houle's (1989) conclusion is that it does not occur in large outbreeding populations and that marine molluscs are an exception to this rule (the common occurrence of the correlations in plants is not an exception because these correlations are best explained by inbreeding). Gaffney (1990) observes that there has not been a biased reporting of positive correlations in molluscs and concludes that, at least in this group of organisms, the phenomenon is real and demands an explanation. Recent technical developments in molecular genetics have enabled new possibilities for testing the various explanations of the correlation. In this section we present the results from one such attempt.

The genome of most eukaryotes is interspersed with repetitive DNA sequences whose copy number per 'locus' is highly variable. There are two major classes of these loci. Minisatellites consist of a repeating core unit of 10 to 70 base pairs (bp) in length. These repeating units form arrays ranging from 0.1 to 29 kilobases (kb) long and follow Mendelian inheritance (Jeffreys, Wilson and Thein, 1985). In addition to polymorphism in the number of repeats per array, nucleotide sequence may also vary among individual repeat units (Jarman and Wells, 1989). Microsatellites consist of smaller units, 2–10 bp each, that are also tandemly repeated in polymorphic size arrays (Tautz, 1989). These polymorphisms, also known under the collective acronym VNTRs (variable number of tandem repeats), can be scored either by the detection of differently sized restriction fragments after digestion of genomic DNA with restriction endonucleases, or by direct DNA sequencing. Polymorphism due to the simple gain or loss of a restriction site can also be scored in the same manner. An indispensable requirement for the use of these polymorphisms in heterozygosity studies is the ability to identify codominant alleles of individual 'loci'. Profiles of multiple bands, even though unique to each individual, cannot be profitably used in population studies unless it is possible to distinguish allelic from non-allelic bands.

Under the associative overdominance hypothesis, heterozygosity at minisatellite or microsatellite loci, or for restriction sites in noncoding DNA, should have the same effect on the correlation as heterozygosity at allozyme loci. In contrast, the hypothesis assigning a direct role to allozymes would predict that no correlation would be produced if heterozygosity was scored for either restriction site or VNTR polymorphisms. These predictions are not without some qualifications. First, it may be claimed that VNTR polymorphisms mark long segments of the genome devoid of functional genes. If this were the case, there would be little opportunity for linkage disequilibrium to exist between VNTR alleles and deleterious genes. Second, VNTRs are known to have much higher mutation rates than allozymes (Jeffreys *et al.*, 1985). This would tend to reduce both fixation indices and linkage disequilibrium values, the two main causes of associative overdominance. Third, VNTR alleles of the same size may in fact be different when scored for internal polymorphisms within repeating units (Jeffreys *et al.*, 1991). At present, we do not know whether VNTR or allozyme homozygosity provides a better index of identity by descent. Since homozygosity by descent is an essential assumption of the associative overdominance hypothesis, these limitations would make it more difficult to obtain evidence for this hypothesis. The competing hypothesis of direct allozyme involvement makes a clear distinction between heterozygosity for allozymes (which must affect the fitness character) and heterozygosity for non-allozyme markers (which must not) and is, therefore, not affected by these limitations.

We have applied this approach to a cohort of the deep-sea scallop, *Placopecten magellanicus*. The adductor muscle was excised from 222 juveniles collected randomly from wild spat. Shell height was measured and part of the muscle was used for allozyme electrophoresis and part for the extraction of genomic DNA. Single-locus VNTR and restriction site polymorphisms were obtained by screening random probes from a scallop adductor muscle cDNA library against genomic DNA digested with various restriction endonucleases (see section 4.3). In total, we scored genotypes at seven polymorphic enzyme loci, two general proteins of unknown function, five single-locus VNTR polymorphisms, and three restriction site polymorphisms. Details will be presented elsewhere (Pogson and Zouros, in preparation), but the main results are presented in Table 3.1.1.

The correlation between the degree of heterozygosity and shell height was not significant when all 17 loci were used (Table 3.1.1). When the loci were divided into two classes, those with known enzyme functions and those with unknown or no function (we refer to these as marker loci), a clear difference emerged: the correlation was positive and significant for the seven allozymes, but negative and non-significant for the 10 marker loci. A *t*-test comparing the correlation coefficients of these two classes gave $t = 1.90$, $P = 0.059$. For a more direct test of the hypothesis that allozymes are more likely to produce a significant positive correlation than a random set of loci, we calculated the correlation coefficients of all 19 448 combinations of seven loci drawn from the set of 17 and asked how many produced a result that equalled or surpassed that of the seven allozymes. One thousand and sixty-one, or 5.46% of the total, did. Being at the borderline of the conventional level of significance, these results do not reject the associative overdominance hypothesis but provide, nevertheless, negative evidence against it.

Examination of the effects of individual loci on the correlation showed a large amount of variation within each of the two classes, the allozymes and the markers. Variations in single enzyme locus effects are expected from previous studies and can be attributed to the functional role of the enzyme, as shown by Koehn, Diehl and Scott (1988). Among the seven allozymes we have scored, five belonged to the class that, according to Koehn, Diehl and Scott (1988), should have an effect on the correlation and two to the class that should not. The latter two loci ranked first and sixth in their effect on the correlation, so that at face value our results do not support the functionality hypothesis suggested by these authors. It

Table 3.1.1   Results of analyses on the relationships between heterozygosity and shell height in juvenile scallops using all 17 loci in total and when grouped in 'enzyme' or 'marker' classes; see text for details

Comparison	Regression statistics				
	Intercept	Slope	N	r	P
Shell height versus degree of heterozygosity at all 17 loci	52.40	+0.252	222	0.071	0.293
Shell height versus degree of heterozygosity at 7 enzyme loci	52.16	+0.811	222	0.148	0.028
Shell height versus degree of heterozygosity at 10 marker loci	55.10	−0.154	222	−0.032	0.626
Standardized effect of heterozygosity[a] at all 17 loci versus heterozygote deficiency (D)	−0.033	−0.083	17	−0.045	0.876
Standardized effect of heterozygosity[a] at 7 enzyme loci versus heterozygote deficiency (D)	−0.045	−0.102	7	−0.033	0.947
Standardized effect of heterozygosity[a] at 10 marker loci versus heterozygote deficiency (D)	−0.023	+0.108	10	0.084	0.816

[a](Mean heterozygote shell height − mean homozygote shell height)/Mean overall shell height.

must be noted, however, that the number of enzyme loci scored is far too low to allow a meaningful test of this hypothesis.

It is more difficult to explain the variation in the effects of the 10 marker loci, a few of which had a greater contribution to the correlation than several of the allozyme loci. We have mentioned above a number of reasons why VNTR polymorphisms may not be the best candidates for testing the associative overdominance hypothesis. The first reason does not apply to our study because, by using cDNA probes as the means of detecting this type of polymorphism, only transcribed parts of the genome were assayed. However, because VNTR polymorphisms may behave differently than the other loci (both allozyme and marker), we excluded them and repeated the analysis on the remaining 12 loci. The results are shown in Table 3.1.2. Removal of the VNTR markers had only a minor effect on the overall relationship between heterozygosity and size. The correlation between shell height and the five non-VNTR marker loci became completely flat, indicating that the VNTR loci were largely responsible for the negative relationship shown by the marker loci in Table 3.1.1.

Many studies that have reported positive correlations between growth and heterozygosity have also described significant deficiencies of heterozygotes (Zouros, 1987). We have already noted that in the extensive study on *Mulinia lateralis* the strong heterozygosity–growth correlation is accompanied by a significant negative relationship between the effect

Table 3.1.2 Results of analyses on the relationships between heterozygosity and shell height in juvenile scallops when the VNTR loci are excluded and the remaining 12 loci grouped into 'enzyme' or 'marker' classes; see text for details

Comparison	Regression statistics				
	Intercept	Slope	N	r	P
Shell height versus degree of heterozygosity at 12 non-VNTR loci	52.56	+0.441	222	0.110	0.105
Shell height versus degree of heterozygosity at 7 enzyme loci	52.16	+0.811	222	0.148	0.028
Shell height versus degree of heterozygosity at 5 non-VNTR marker loci	54.33	+0.002	222	0.010	0.996
Standardized effect of heterozygosity[a] at 12 non-VNTR loci versus heterozygote deficiency (D)	−0.025	−0.135	12	−0.055	0.858
Standardized effect of heterozygosity[a] at 7 enzyme loci versus heterozygote deficiency (D)	−0.045	−0.102	7	−0.033	0.947
Standardized effect of heterozygosity[a] at 5 non-VNTR loci versus heterozygote deficiency (D)	−0.003	+0.191	5	0.205	0.740

[a](Mean heterozygote shell height – mean homozygote shell height)/Mean overall shell height.

of a locus on the growth correlation and the magnitude of heterozygote deficiency at that locus (Gaffney *et al.*, 1990). At 13 of the 17 loci in our study the number of homozygous individuals was larger than expected from Hardy–Weinberg expectations, but the deviation was significant for only one enzyme locus (*Pgm*). There was no relationship between the degree of heterozygote deficiency and the relative contribution of individual loci to the heterozygosity–growth correlation (see Table 3.1.1 and 3.1.2). This result applies whether we consider the entire set of 17 loci, the 12 non-VNTR loci, or the 7 allozyme loci. Given the fact that there was not a significant deficiency of heterozygotes in our population sample, however, this lack of a correlation is expected and cannot be used as evidence against the hypothesis of associative overdominance.

The study whose main findings we have presented here was designed to produce direct evidence for either the associative overdominance hypothesis or for the competing hypothesis that fitness-related characters are affected only by allozyme heterozygosity. We have performed two analyses on our data. In the first, we examined all 17 loci as a set and asked whether allozyme loci produce a better correlation between an individual's degree of heterozygosity and the individual's size than a random set of loci. The answer was positive, a result that contradicts the associative overdominance hypothesis. In the second, we excluded VNTR loci on the grounds that they are not expected to be in linkage disequilibrium with deleterious recessives. The result from this second analysis was essentially the same as the first. The other two sets (VNTR, and non-VNTR and non-enzyme) produced either a negative relationship, or no relationship at all. We conclude that the study failed to provide direct support for one of the main predictions of the associative overdominance hypothesis and should be interpreted as providing indirect evidence for direct involvement of allozyme loci in a heterozygosity–size correlation. Given the history of the heterozygosity–fitness character debate, this conclusion should not be considered as established before other studies from different laboratories and involving different species produce similar results.

Acknowledgements

The work reported here was supported by an NSERC operating grant to E.Z., a Killam Postdoctoral Fellowship to G.H.P., and by the Marine Gene Probe Laboratory of Dalhousie University. We would like to thank an anonymous reviewer for providing constructive comments on the manuscript. We also thank K. Mesa for providing technical assistance in the scoring of heterozygosity at the DNA level.

References

Alvarez, G., Zapata, C., Amaro, R. and Guerra, A. (1989) Multi-locus heterozygosity at protein loci and fitness in the European oyster, *Ostrea edulis L. Heredity*, **63**, 359–72.

Beaumont, A.R. (1991) Genetic studies of laboratory reared mussels, *Mytilus edulis*: heterozygote deficiencies, heterozygosity and growth. *Biol. J. Linn. Soc.*, **44**, 273–85.

Booth, C.L., Woodruff, D.S. and Gould, S.J. (1990) Lack of significant associations between allozyme heterozygosity and phenotypic traits in the land snail *Cerion. Evolution*, **44**, 210–13.

Borsa, P., Jousselin, Y. and Delay, B. (1992) Relationships between allozymic heterozygosity, body size, and survival to natural anoxic stress in the palourde, *Ruditapes decussatus* L. (Bivalvia: Veneridae). *J. Exp. Mar. Biol. Ecol.*, **155**, 169–81.

Chakraborty, R. (1989) Can molecular imprinting explain heterozygote deficiency and hybrid vigor? *Genetics*, **122**, 113–17.

Danzmann, R.G., Ferguson, M.M. and Allendorf, F.W. (1988) Heterozygosity and components of fitness in a strain of rainbow trout. *Biol. J. Linn. Soc.*, **33**, 285–304.

Diehl, W.J. (1988) Genetics of carbohydrate metabolism and growth in *Eisenia foetida* (Oligochaeta: Lumbricidae). *Heredity*, **61**, 379–8.

Diehl, W.J., Gaffney, P.M., McDonald, J.H. and Koehn, R.K (1985) Relationship between weight-standardized oxygen consumption and multiple-locus heterozygosity in the mussel, *Mytilus edulis*, in *Proceedings of the 19th European Marine Biology Symposium* (ed. P.E. Gibbs), Cambridge University Press, Cambridge, pp. 529–34.

DiMichele, L. and Powers, D.A. (1991) Allozyme variation, developmental rate, and differential mortality in the teleost *Fundulus heteroclitus. Physiol. Zool.*, **64**(6), 1426–43.

Fujio, Y. (1982) A correlation of heterozygosity and growth rate in the Pacific oyster, *Crassostrea gigas. Tohoku J. Agric. Res.*, **33**, 66–75.

Gaffney, P.M (1990) Enzyme heterozygosity, growth rate, and viability in *Mytilus edulis*: another look. *Evolution*, **44**, 204–10.

Gaffney, P.M., Scott, T.M., Koehn, R.K. and Diehl, W.J. (1990) Interrelationships of heterozygosity, growth rate and heterozygote deficiencies in the coot clam, *Mulinia lateralis. Genetics*, **124**, 687–99.

Garton, D.W., Koehn, R.K. and Scott, T.M. (1984) Multiple-locus heterozygosity and the physiological genetics of growth in the coot clam, *Mulinia lateralis*, from a natural population. *Genetics*, **108**, 445–55.

Gentili, M.R. and Beaumont, A.R. (1988) Environmental stress, heterozygosity and growth rate in *Mytilus edulis. J. Exp. Mar. Biol. Ecol.*, **120**, 145–53.

Gillespie, J.H. and Langley, C.H. (1974) A general model to account for enzyme variation in natural populations. *Genetics*, **76**, 837–48.

Gosling, E.M. (1989) Genetic heterozygosity and growth rate in a cohort of *Mytilus edulis* from the Irish coast. *Mar. Biol.*, **100**, 211–15.

Gould, S.J. and Lewontin, R.C. (1979) The spandrels of San Marcos and the Panglossian paradigm. A critique of the adaptationist programme. *Proc. R. Soc. Lond. B.*, **205**, 581–98.

Graur, D. and Li, W.-H. (1991) Neutral mutation hypothesis test. *Nature*, **354**, 114–15.

Hawkins, A.J.S., Bayne, B.L. and Day, A.J. (1986) Protein turnover, physiological energetics and heterozygosity in the blue mussel, *Mytilus edulis*: the basis of variable age-specific growth. *Proc. R. Soc. Lond. B.*, **229**, 161–76.

Hawkins, A.J.S., Bayne, B.L., Day, A.J., Rusin, J. and Worrall, C.M. (1989) Genotype-dependent interrelations between energy metabolism, protein metabolism and fitness, in *Reproduction, Genetics, and the Distributions of Marine Organisms. 23rd European Marine Biology Symposium* (eds J.S. Ryland and P.A. Tyler), Olsen and Olsen, Fredensborg, Denmark, pp. 283–92.

Houle, D. (1989) Allozyme-associated heterosis in *Drosophila melanogaster. Genetics*, **123**, 789–801.

Hughes, A.L., Ohta, T. and Nei, M. (1990) Positive Darwinian selection promotes charge profile diversity in the antigen-binding cleft of class 1 major-histocompatibility-complex molecules. *Mol. Biol. Evol.*, **7**, 515–24.

Jarman, A.P. and Wells, R.A. (1989) Hypervariable minisatellites: recombinators or innocent bystanders? *TIG*, **5**, 367–71.

Jeffreys, A.J., Wilson, V. and Thein, S.L. (1985) Individual-specific 'fingerprints' of human DNA. *Nature*, **316**, 76–9.

Jeffreys, A.J., MacLeod, A., Tamaki, K., Neil, D.L. and Monckton, D.G. (1991) Minisatellite repeat coding as a digital approach to DNA typing. *Nature* **354**, 204–9.

Kacser, H. and Burns, J.A. (1981) The molecular basis of dominance. *Genetics*, **97**, 639–66.

Kaplan, N.L., Darden, T. and Hudson, R.R. (1988) The coalescent process in models with selection. *Genetics*, **120**, 819–29.

Kimura, M. (1983) *The Neutral Theory of Molecular Evolution*, Cambridge University Press, Cambridge.

Koehn, R.K. (1990) Heterozygosity and growth in marine bivalves: comments on the paper by Zouros, Romero-Dorey, and Mallet (1988). *Evolution*, **44**, 213–16.

Koehn, R.K., Diehl, W.J. and Scott, T.M. (1988) The differential contribution by individual enzymes of glycolysis and protein catabolism to the relationship between heterozygosity and growth rate in the coot clam, *Mulinia lateralis. Genetics*, **118**, 121–30.

Koehn, R.K. and Shumway, S.E. (1982) A genetic/physiological explanation for differential growth rate among individuals of the American oyster, *Crassostrea virginica* (Gmelin). *Mar. Biol. Lett.*, **3**, 35–42.

Kreitman, M. (1983) Nucleotide polymorphism at the alcohol dehydrogenase locus of *Drosophila melanogaster. Nature*, **304**, 412–17.

Kreitman, M. and Aguade, M. (1986) Excess polymorphism at the Adh locus in *Drosophila melanogaster. Genetics*, **114**, 93–110.

Kreitman, M. and Hudson, R.R. (1991) Inferring the evolutionary histories of the Adh and Adh-dup loci in *Drosophila melanogaster* from patterns of polymorphism and divergence. *Genetics*, **127**, 565–82.

Lerner, I.M. (1954) *Genetic Homeostasis*, John Wiley, New York.

Maddison, D.R. (1991) African origin of human mitochondrial DNA reexamined. *Syst. Zool.*, **40**, 355–63.

McDonald, J.H. and Kreitman, M. (1991) Adaptive protein evolution at the Adh locus in *Drosophila. Nature*, **351**, 652–4.

Mitton, J.B. and Grant, M.C. (1984) Associations among protein heterozygosity, growth rate, and developmental homeostasis. *Annu. Rev. Ecol. Syst.*, **15**, 479–99.

Patarnello, T., Bisol, P.M. and Battaglia, B. (1989) Studies on differential fitness of PGI genotypes with regard to temperature in *Gammarus insensibilis* (Crustacea: Amphipoda). *Mar. Biol.*, **102**, 355–9.

Pogson, G.H. (1991) Expression of overdominance for specific activity at the phosphoglucomutase-2 locus in the Pacific oyster, *Crassostrea gigas. Genetics*, **128**, 133–41.

Riley, M.A., Hallas, M.E. and Lewontin, R.C. (1989) Distinguishing the forces controlling genetic variation at the Xdh locus in *Drosophila pseudoobscura. Genetics*, **123**, 359–69.

Riley, M.A., Kaplan, S.R. and Veuille, M. (1992) Nucleotide polymorphism at the xanthine dehydrogenase locus in *Drosophila pseudoobscura. Mol. Biol. Evol.*, **9**, 56–69.

Sarver, S.K., Katoh, M. and Foltz, D.W. (1992) Apparent overdominance of enzyme specific activity in two marine bivalves. *Genetica*, **85**, 231–9.

Schaal, B.A. and Levin, D.A. (1976) The demographic genetics of *Liatris cylindracea* Michx. (Compositae). *Am. Nat.*, **110**, 191–206.

Scott, T.M. and Koehn, R.K. (1990) The effect of environmental stress on the relationship between heterozygosity and growth rate in the coot clam *Mulinia lateralis* (Say). *J. Exp. Mar. Biol. Ecol.*, **135**, 109–16.

Singh, S.M. and Zouros, E. (1978) Genetic variation associated with growth rate in the American oyster (*Crassostrea virginica*). *Evolution*, **32**, 342–53.

Skibinski, D.O.F. and Roderick, E.E. (1989) Heterozygosity and growth in transplanted mussels. *Mar. Biol.*, **102**, 73–84.

Tanaka, T. and Nei, M. (1989) Positive Darwinian selection observed at the variable-region genes of immunoglobulins. *Mol. Biol. Evol.*, **6**, 447–59.

Tautz, D. (1989) Hypervariability of simple sequences as a source for polymorphic DNA markers. *Nucleic Acids Res.*, **17**, 6463–71.

Templeton, A.R. (1992) Human origins and analysis of mitochondrial DNA sequences. *Science*, **255**, 737.

Torrissen, K.R. (1991) Genetic variation in growth rate of Atlantic salmon with different trypsin-like isozyme patterns. *Aquaculture*, **93**, 299–312.

Volckaert, F. and Zouros, E. (1989) Allozyme and physiological variation in the scallop *Placopecten magellanicus* (Gmelin), and a general model for the effects of heterozygosity on fitness in marine molluscs. *Mar. Biol.*, **103**, 51–61.

Whittam, T.S. and Nei, N. (1991) Neutral mutation hypothesis test. *Nature*, **354**, 115–16.

Zouros, E. (1987) On the relation between heterozygosity and heterosis: an evaluation of the evidence from marine mollusks. *Isozymes: Curr. Top. Biol. Med. Res.*, **15**, 255–79.

Zouros, E. (1990) Heterozygosity and growth in marine bivalves: response to Koehn's comments. *Evolution*, **44**, 216–18.

Zouros, E. and Foltz, D.W. (1987) The use of allelic isozyme variation for the study of heterosis. *Isozymes: Curr. Top. Biol. Med. Res.*, **13**, 1–59.

Zouros, E. and Mallet, A.L. (1989) Genetic explanations of the growth/ heterozygosity correlation in marine mollusks, in *Reproduction, Genetics and Distributions of Marine Organisms: 23rd European Marine Biology Symposium* (eds J.S. Ryland and P.A. Tyler), Olsen and Olsen, Fredensborg, Denmark, pp. 317–24.

Zouros, E., Romero-Dorey, M. and Mallet, A.L. (1988) Heterozygosity and growth in marine bivalves: further data and possible explanations. *Evolution*, **42**, 1332–41.

3.2   HETEROSIS AND HETEROZYGOTE DEFICIENCIES IN MARINE BIVALVES: MORE LIGHT?

Patrick M. Gaffney

Abstract

Positive associations between allozyme heterozygosity and fitness correlates such as growth rate and viability are frequently reported in marine bivalves. Typically, only loci that exhibit deficiencies of heterozygotes relative to Hardy–Weinberg equilibrium expectations are found to show such allozyme-associated heterosis (AAH). Hypotheses offered to explain these phenomena together or separately include inbreeding, scoring bias, selection, null alleles, aneuploidy and population mixing. Previous analysis of a large-scale population survey (multiple loci, numerous individuals scored) of the dwarf surfclam Mulinia lateralis *found that none of these factors was sufficient as a sole explanation of the observed patterns of heterozygote deficiencies and AAH.*

Two approaches may shed further light on these enigmatic phenomena. First, virtually all studies to date have focused largely on single-locus analyses and phenotypic associations with multilocus heterozygosity. Multilocus analyses of genotypic disequilibria may provide more resolving power than single-locus tests. I examine here three large data sets from marine bivalves, using the complete two-locus disequilibrium statistics of Weir and Cockerham. Secondly, preliminary results from laboratory crosses of M. lateralis *are assessed to evaluate the roles of null alleles, spontaneous aneuploidy, genomic imprinting and selection as sources of heterozygote deficiencies and AAH.*

3.2.1 Introduction

Positive associations between allozyme heterozygosity and correlates of fitness such as growth rate, viability and fecundity are frequently reported for marine bivalves (review: Zouros, 1987), although often comparably designed and analysed studies do not detect such associations (Gaffney, 1990). It is also common for allozyme surveys of marine bivalve populations to find heterozygote deficiencies, i.e. fewer heterozygous individuals than expected under Hardy–Weinberg equilibrium (review: Zouros and Foltz, 1984; Beaumont, 1991). A connection between allozyme-associated heterosis (AAH) and heterozygote deficiencies was first noted by Zouros (1987) and subsequently confirmed in the dwarf surfclam *Mulinia lateralis* (Gaffney *et al.*, 1990), but is not always found (Figure 3.2.1).

A number of hypotheses have been advanced to explain one or both of these phenomena: inbreeding, scoring bias, null alleles, spontaneous aneuploidy, genomic imprinting (selective

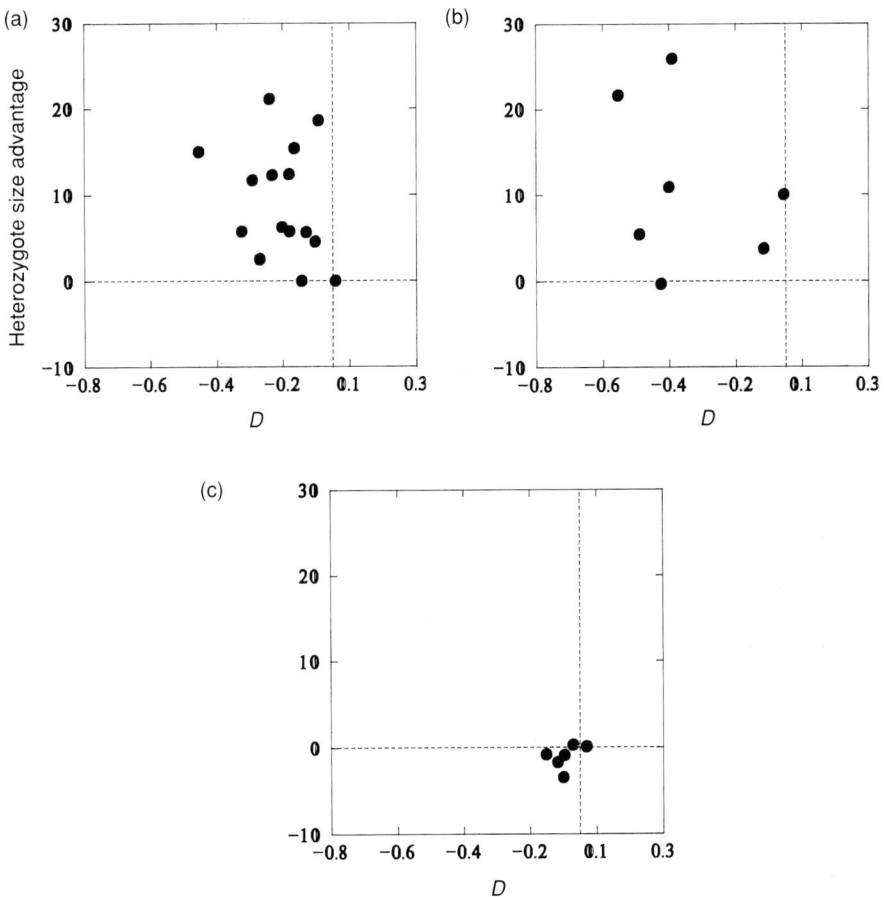

Figure 3.2.1 Locus-specific relative size advantage of heterozygotes in natural populations of marine bivalves in relation to locus-specific heterozygote deficiencies (*D*; see text). Size advantage is defined as 100 × (median size of heterozygotes – median size of homozygotes)/(median size of homozygotes). (a) *Mulinia lateralis*, size measured as total wet weight; (b) *Crassostrea virginica*, size measured as total wet weight; (c) *Argopecten irradians*, size measured as wet tissue weight, excluding shell.

inactivation of maternal or paternal alleles), selection for and against heterozygotes, and population mixing. Most studies have attempted to test these hypotheses against survey data by applying a series of single-locus tests, with a limited examination of multilocus patterns, chiefly in the form of multilocus heterozygosity. Using this approach, Gaffney *et al.* (1990) found that no single factor appeared capable of explaining patterns observed in a large-scale study of *M. lateralis*.

If no single factor appears capable of explaining both phenomena and their inter-relationship, two approaches offer some hope. The first is the application of new statistical methods to extract more information from the conventional multilocus data set obtained from population surveys. The second, arguably more powerful, is to make controlled crosses in the laboratory. The analysis of pair matings offers several advantages: (1) the genetic interpretation of zymograms may be confirmed and the potential for scoring bias assessed; (2) null alleles and genomic imprinting will be revealed by aberrant genotypes or systematic deviations from expected genotypic proportions; (3) genotypic fitnesses (viabilities, growth rates) may be estimated for siblings grown in a common environment; (4) population phenomena, i.e. inbreeding and the Wahlund effect, are eliminated. Beaumont (1991) has demonstrated the utility of this approach in *Mytilus edulis*, although the number of pair matings examined was not large enough to allow confident extrapolation to natural popu-lations. I discuss here the further analysis of three large-scale allozyme surveys of marine bivalves, and present preliminary data from a series of pair matings in the dwarf surfclam.

3.2.2   Materials and Methods

(a)   Two-locus disequilibria

Data from three large-scale allozyme surveys of marine bivalves were examined. The oyster data set consisted of $N \approx 1800$ year-old oysters (*Crassostrea virginica*) from a natural cohort scored at seven enzyme loci, as originally reported by Zouros, Singh and Miles (1980). The dwarf surfclam (*M. lateralis*) data set consisted of several thousand animals from a single year class, scored at 15 loci (Koehn, Diehl and Scott, 1988; Gaffney *et al.*, 1990). The scallop data set consisted of 1700 individuals from a single year class of bay scallops (*Argopecten irradians*) scored at six loci (M. Krause, unpublished data). For the calculation of two-locus disequilibria in *M. lateralis*, only individuals with complete genotypic data at all loci were used ($N = 1906$).

The complete two-locus disequilibrium coefficients of Weir and Cockerham (1989) comprise the following: (1) D_A and D_B, which measure departures from Hardy–Weinberg equilibrium at loci A and B, respectively; (2) Δ_{AB}, a composite linkage disequilibrium coefficient applicable to situations where coupling and repulsion double heterozygotes cannot be distinguished, e.g. allozyme surveys; (3) D_{AAB} and D_{ABB}, two trigenic disequilibria that measure the associations between triples of genes after the associations between all appropriate pairs of genes have been removed; and (4) Δ_{AABB}, a composite quadrigenic coefficient. Calculation and testing of two-locus disequilibrium statistics were done for the three data sets using the LD86 Fortran program of Weir (1990). Each locus was converted to a two-allele system by collapsing all alleles other than the most common one into a second synthetic allele. For each data set, significance tests were done using a sequential Bonferroni procedure (Rice, 1989) to adjust for the number of tests made for each coefficient. For example, of the 105 tests of Δ_{AB} in *M. lateralis*, the smallest *P*-value exceeded $0.05/105 = 0.0005$, so none was considered significant.

The conventional measure of single-locus heterozygote deficiencies, D = (observed number of heterozygotes/expected number of heterozygotes) -1, was used for graphical purposes (Figure 3.2.1).

(b)   Laboratory crosses

Pair matings of *M. lateralis* were made by combining eggs and sperm from individuals induced by thermal stimulation to spawn in separate tubes. Parents were frozen at –80 °C immediately after spawning. Progeny were reared in individual culture containers for 60–120 days using standard bivalve culture methods (Ludwig and Gaffney, 1991) and then frozen for electrophoretic analysis.

Allozyme genotypes of parents and progeny were determined by standard starch gel electrophoresis for 11 loci. Buffers and staining systems for dipeptidases (AP, EC 3.4.1.2), β-galactosidase (βGAL, EC 3.2.1.23), glucose phosphate isomerase (GPI, EC 5.3.1.9), mannose phosphate isomerase (MPI, EC 5.3.1.8), 6-phosphogluconate dehydrogenase (PGD, EC 1.1.1.44) and strombine dehydrogenase (SDH, EC 1.5.1-) were described in Koehn, Diehl and Scott (1988). Phosphoglucomutase (PGM, EC 5.4.2.2) was resolved using the method of Pogson (1989). Esterase (EST, EC 3.1.1.1) was run on a 0.24 M Tris-citrate pH 7.0 buffer system and stained using a fluorescent substrate (4-methylumbelliferyl acetate). Octopine dehydrogenase (ODH, EC 1.5.1.11) was run on a LiOH system and stained according to Dando *et al.* (1981).

Genotypic distributions in progeny were compared to Mendelian expectations to detect the presence of null alleles, genomic imprinting, selection, and *de novo* nulls arising from mutation or spontaneous aneuploidy. Approximately 50 families have been scored to date, but genotypic proportions have been examined in only 14.

3.2.3   Results

There were no significant two-locus disequilibria in any of the data sets, although departures from Hardy–Weinberg equilibrium, invariably in the form of heterozygote deficiencies, were common (Table 3.2.1).

In the pair matings, null alleles proved to be frequent at several loci (Table 3.2.2). These estimates are conservative, as some matings (e.g. $AA \times AA$) will not reveal the presence of null alleles. For all loci with the exception of *Pgd*, the observed frequencies were considerably higher than the average null frequency of 2×10^{-3} in *Drosophila melanogaster* (Langley *et al.*, 1981) and 3×10^{-3} in pine trees (Allendorf, Knudsen and Blake, 1982). They were also much higher than frequencies estimated in the original survey on the basis of two-banded staining patterns in the dimeric enzymes coded by the *Ap-2* and *Gpi* loci (Gaffney *et al.*, 1990), suggesting some scoring bias in routine population surveys.

Because several families were segregating for null alleles, we could in many cases distinguish null heterozygotes and null homozygotes from progeny carrying two active (wild-type) alleles. This allowed a comparison of growth rate of wild-type versus null-carrying clams. By comparing observed proportions with Mendelian expectations, the

Table 3.2.1   Complete two-locus disequilibria in bivalve populations; number of significant tests (numerator) and total number of tests made (denominator)

	D_A	Δ_{AB}	D_{AAB}	D_{ABB}	Δ_{AABB}
C. virginica	6/7	0/21	0/21	0/21	0/21
M. lateralis	13/15	0/105	0/101	0/101	0/95
A. irradians	2/6	0/15	0/15	0/15	0/15

Some coefficients could not be tested because variances were too small (Weir, 1990). D_A = departure from Hardy–Weinberg equilibrium at locus A; Δ_{AB} = composite two-locus linkage disequilibrium; D_{AAB} and D_{ABB} = trigenic disequilibria; Δ_{AABB} = composite quadrigenic disequilibrium (see text).

Table 3.2.2   Frequencies of null alleles in *M. lateralis* observed by progeny testing

Locus	N_s	N_x (max)		p_x	(max)	D_{exp}	(max)	D_{obs}
Ap-1	132	14	(17)	0.106	(0.129)	−0.192	(−0.228)	−0.242
Ap-2	132	6	(7)	0.045	(0.053)	−0.087	(−0.101)	−0.182
Ap-3	88	4	(6)	0.045	(0.068)	−0.087	(−0.128)	−0.405
βGal	108	7	(8)	0.065	(0.074)	−0.122	(−0.138)	−0.275
Est-2	132	1	(28)	0.008	(0.212)	−0.015	(−0.350)	−
Gpi	148	1	(10)	0.007	(0.068)	−0.013	(−0.127)	−0.053
Mpi	106	1	(4)	0.009	(0.038)	−0.019	(−0.073)	−0.220
Odh	32	4	(5)	0.125	(0.156)	−0.222	(−0.270)	−
Pgd	128	0	(6)	0.000	(0.047)	0.000	(−0.090)	−0.094
Pgm-1	116	2	(20)	0.017	(0.172)	−0.034	(−0.294)	−0.131
Sdh	116	1	(16)	0.009	(0.138)	−0.017	(−0.242)	−

N_s = number of alleles (parental chromosomes) examined; N_x = number of null alleles identified (maximum number possible given in parentheses); p_x = null allele frequency in parents; D_{exp} = heterozygote deficiency expected under Hardy–Weinberg equilibrium with null allele frequency p_x; D_{obs} = heterozygote deficiencies reported by Koehn, Diehl and Scott (1988), calculated as (observed number of heterozygotes/expected number of heterozygotes) − 1.

viability of null-bearing clams could be estimated. Although the error in these estimates is fairly large, since the number of individuals sampled in each family was typically on the order of 20–150, there were many families in which the null heterozygotes were larger than wild-type heterozygotes, and also many cases in which null heterozygotes appeared more viable (Figure 3.2.2). Overall, null heterozygotes appeared to be somewhat less viable, except at *Ap-1* and perhaps *Odh*.

Aside from segregating null alleles, genotypic proportions in progeny analysed to date (14 families) were in accord with Mendelian expectations. There was no evidence of

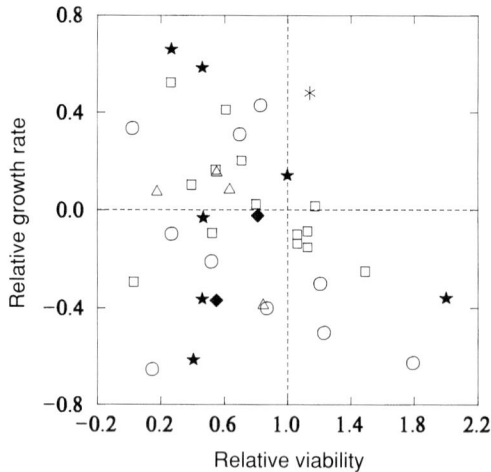

Figure 3.2.2 Relative viabilities and growth rates of null heterozygotes in *Mulinia lateralis*. Each point represents a within-family comparison of null heterozygotes at a given locus to heterozygous siblings with two active (wild-type) alleles. Growth rate differences are expressed as the standardized shell length of null heterozygotes minus the standardized shell length of wild-type heterozygotes. Viabilities of null heterozygotes were estimated by conventional cross-product ratios and are expressed relative to wild-type heterozygote viability of 1. Open circles = *Odh*; open squares = *Ap-1*; open triangles = *Ap-2*; filled stars = *Ap-3*; filled diamonds = *Gpi*; asterisk = *Est-2*.

genomic imprinting, i.e. failure of an active allele observed in a parent to be expressed in its progeny. There also appeared to be few viability differences associated with allozyme genotypes. Five significant departures from Mendelian proportions were detected, of which three involved apparent viability differences among heterozygous genotypes, and two (*βGal* and *Ap-2*) showed greater apparent viability of heterozygotes compared to homozygotes. Growth rates of heterozygotes and homozygotes were not significantly different, except for two cases in which heterozygotes were significantly larger than homozygotes (*Odh* and *Ap-1*). Aberrant genotypes were observed not infrequently at several loci. These may be the result of spontaneous aneuploidy or *de novo* null mutations, but this is not certain until the potential for scoring errors or contamination of the cultures has been carefully evaluated.

Scoring artifacts were occasionally noted at some loci (*Ap-3*, *βGal*, *Mpi*). For example, progeny of an *Ap-3 AA* male and an *Ap-3 CC* female were blindly scored as *BB* rather than *AC*, because the heterodimer stained much more intensely than either homodimer. Heterozygotes for two alleles of similar mobility were sometimes indistinct, and scored as homozygotes for one or the other allele.

3.2.4 Discussion

The absence of significant two-locus disequilibria in any of the data sets suggests that the commonly observed single-locus heterozygote deficits are not likely to be due to the mixing of substantially differentiated subpopulations (Wahlund effect), unless these happen to show no covariance of allelic frequencies (Nei and Li, 1973). However, without information on allelic frequencies in larval subpopulations which combine at settlement, the contribution of the Wahlund effect to single-locus deficiencies remains unclear. The absence of quadrigenic disequilibria suggests that inbreeding is negligible in natural populations of these species, even in the bay scallop, a facultative selfer (Castagna, 1975; Wilbur and Gaffney, 1991).

The next step in the analysis of multilocus data sets such as these must focus on aspects of the multilocus heterozygosity distribution (MLHD). A simple test involves comparing the observed MLHD with that expected on the basis of single-locus heterozygosities (Gaffney *et al.*, 1990). No departure of observed from expected MLHD is seen for the oyster or scallop data, but the pattern in *M. lateralis* is clearly deviant (Figure 3.2.3). Although the cause of this deviant pattern, which can be viewed as a multilocus extension of the two-locus quadrigenic disequilibrium, was suggested by Charlesworth (1991) to be inbreeding, there was no evidence of quadrigenic disequilibria in the two-locus analysis. Further understanding of this discrepancy awaits the development of multiallelic models that treat the combined effects of inbreeding and selection on MLHD.

In the pair crosses, there was no evidence for genomic imprinting. This is consistent with other breeding studies of bivalves, and is to be expected if the evolutionary rationale for genomic imprinting (Moore and Haig, 1991) is correct.

The observation of scoring artifacts at several loci underscores the need to verify the genetic interpretation of zymograms through breeding studies. Routine surveys rarely employ high-resolution staining methods (e.g. sequential electrophoresis), and in order to sample numerous loci may include poorly resolved enzyme systems.

The most surprising finding from our preliminary analysis of controlled laboratory crosses is the high frequency of null alleles at several loci in *M. lateralis* collected in the wild. These frequencies are sufficient to explain large heterozygote deficiencies, but are considerably higher than expected for even mildly deleterious alleles at mutation–selection equilibrium.

Null alleles are often detected in bivalve breeding studies. In *C. virginica*, segregating null alleles have been found at *Lap-2* and *Mpi* (Foltz, 1986) and *Gpi* (P.M. Gaffney and S.K. Allen unpublished data). In *M. edulis*, null alleles were observed at *Lap-1* and *Lap-2*

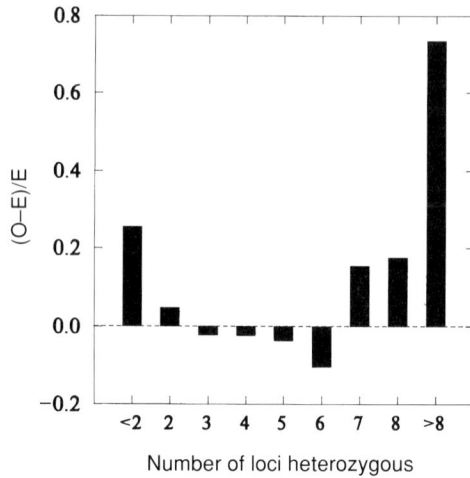

Figure 3.2.3 Significant ($P < 0.025$) deviation of observed MLHD from the distribution expected on the basis of observed single-locus heterozygosities in *Mulinia lateralis*, using the collapsed (two-allele) data set.

(Mallet *et al.*, 1985) and *Gpi* and *Odh* (K. Hoare and A.R. Beaumont unpublished data). In *C. gigas*, a null allele was observed at *Lap-2* (Guo and Gaffney, 1993). On the basis of enzyme activity assays, Katoh and Foltz (1988) estimated the frequency of *Lap-2* nulls in *C. virginica* to be 0.0465. Collectively, these reports suggest that null allele frequencies in bivalves may be considerably higher than in pine trees and *Drosophila*.

It is noteworthy that the loci exhibiting high null allele frequencies in *M. lateralis* (*Odh* and several aminopeptidase loci) show strong overlap with the above list. These loci often show large heterozygote deficiencies in allozyme surveys of bivalves. This could mean that null alleles at these loci are not invariably deleterious, either because the loss of enzymatic activity is compensated for by other loci with similar function (e.g. peptidases, 'pyruvate reductases'), or because low enzymatic activity is favoured under certain conditions. In other organisms, null alleles may reach high frequencies or even fixation at loci that are members of multiple isozyme families, e.g. lactate dehydrogenase in salmonids (Allendorf, Ståhl and Ryman, 1984) and amylase in isopods (Oxford, 1986). The data presented here indicate that null alleles may indeed play a key role in generating heterozygote deficiencies in bivalves; whether they contribute significantly to AAH remains to be determined.

Acknowledgements

I thank Michelle Moore, Leslie Vrolijk and Kristin Churchill for their able assistance in clam culture and electrophoresis, and Dr E. Zouros and Maureen Krause for the use of their oyster and scallop data sets. Dr Bruce Weir kindly provided his program LD86.FOR for calculating and testing two-locus disequilibrium coefficients. This work was supported in part by National Science Foundation grant BSR 90-06896 to P.M.G.

References

Allendorf, F.W., Knudsen, K.L. and Blake, G.M. (1982) Frequencies of null alleles at enzyme loci in natural populations of ponderosa and red pine. *Genetics*, **100**, 497–504.

Allendorf, F.W., Ståhl, G. and Ryman, N. (1984) Silencing of duplicate genes: a null allele polymorphism for lactate dehydrogenase in brown trout (*Salmo trutta*). *Mol. Biol. Evol.*, **1**, 238–48.

Beaumont, A.R. (1991) Genetic studies of laboratory reared mussels, *Mytilus edulis*: heterozygote deficiencies, heterozygosity and growth. *Biol. Linn. Soc.*, **44**, 273–85.

Castagna, M. (1975) Culture of the bay scallop, *Argopecten irradians*, in Virginia. *Mar. Fish. Rev.*, **37**, 19–24.

Charlesworth, D. (1991) The apparent selection on neutral marker loci in partially inbreeding populations. *Genet. Res.*, **57**, 159–75.

Dando, P.R., Storey, K.B., Hochachka, P.W. and Storey, J.M. (1981) Multiple dehydrogenases in marine molluscs: electrophoretic analysis of alanopine dehydrogenase, strombine dehydrogenase, octopine dehydrogenase and lactate dehydrogenase. *Mar. Biol. Lett.*, **2**, 249–57.

Foltz, D.W. (1986) Null alleles as a possible cause of heterozygote deficiencies in the oyster *Crassostrea virginica* and other bivalves. *Evolution*, **40**, 869–70.

Gaffney, P.M (1990) Enzyme heterozygosity, growth rate, and viability in *Mytilus edulis*: another look. *Evolution*, **44**, 204–10.

Gaffney, P.M., Scott, T.M., Koehn, R.K. and Diehl, W.J. (1990) Interrelationships of heterozygosity, growth rate and heterozygote deficiencies in the coot clam, *Mulinia lateralis*. *Genetics*, **124**, 687–99.

Guo, X. and Gaffney, P.M. (1993) Artificial gynogenesis in the Pacific oyster, *Crassostrea gigas*: II. Allozyme inheritance and early growth. *J. Hered.*, **84**, 311–5.

Katoh, M. and Foltz, D.W. (1988) Determination of null allele frequency at an allozyme locus in a natural oyster population. *J. Shellfish Res.* **7**, 203 (abstract only).

Koehn, R.K., Diehl, W.J. and Scott, T.M. (1988) The differential contribution by individual enzymes of glycolysis and protein catabolism to the relationship between heterozygosity and growth rate in the coot clam, *Mulinia lateralis*. *Genetics*, **118**, 121–30.

Langley, C.H., Voelker, R.A., Leigh Brown, A.J. *et al.* (1981) Null allele frequencies at allozyme loci in natural populations of *Drosophila melanogaster*. *Genetics*, **99**, 151–6.

Ludwig, A.N. and Gaffney, P.M. (1991) Quantitative genetics of growth in the dwarf surfclam *Mulinia lateralis* (Say, 1822) *J. Shellfish Res.*, **10**, 451–3.

Mallet, A.L., Zouros, E., Gartner-Kepkay, K.E., Freeman, K.R. and Dickie, L.M. (1985) Larval viability and heterozygote deficiency in populations of marine bivalves: evidence from pair matings of mussels. *Mar. Biol.*, **87**, 165–72.

Moore, T. and Haig, D. (1991) Genomic imprinting in mammalian development: a parental tug-of-war. *TIG*, **7**, 45–9.

Nei, M. and Li, W.-S. (1973) Linkage disequilibrium in subdivided populations. *Genetics*, **75**, 213–19.

Oxford, G.S. (1986) Multiple amylase loci in *Asellus* (Crustacea: Isopoda): genetics and linkage. *Heredity*, **56**, 105–10.

Pogson, G.H. (1989) Biochemical characterization of genotypes at the phosphoglucomutase-2 locus in the Pacific oyster, *Crassostrea gigas*. *Biochem. Genet.*, **27**, 571–89.

Rice, W.R. (1989) Analyzing tables of statistical tests. *Evolution*, **43**, 223–5.

Weir, B.S. (1990) *Genetic Data Analysis*, Sinauer Associates, Inc., Sunderland, Massachusetts.

Weir, B.S. and Cockerham, C.C. (1989) Complete characterization of disequilibrium at two loci, in *Mathematical Evolutionary Theory* (ed M.W. Feldman), Princeton University Press, Princeton, New Jersey, pp. 86–110

Wilbur, A.E. and Gaffney, P.M. (1991) Self-fertilization in the bay scallop, *Argopecten irradians*. *J. Shellfish. Res.*, **10**, 274 (abstract only).

Zouros, E. (1987) On the relationship between heterozygosity and heterosis: an evaluation of the evidence from marine mollusks. *Isozymes: Curr. Top. Biol. Med. Res.*, **15**, 255–70.

Zouros, E. and Foltz, D.W. (1984) Possible explanations for heterozygote deficiency in bivalve molluscs. *Malacologia*, **25**, 583–91.

Zouros, E., Singh, S.M. and Miles, H.E. (1980) Growth rate in oysters: an overdominant phenotype and its possible explanations. *Evolution*, **34**, 856–67.

3.3　SETTLEMENT BEHAVIOUR AND MICROGEOGRAPHIC POPULATION STRUCTURE IN THE BARNACLE *SEMIBALANUS BALANOIDES*

Eric R. Holm and Edwin Bourget

Abstract

At St Andrews, New Brunswick, Semibalanus balanoides *sampled in 1988 from crevices and exposed surfaces revealed significant microgeographic variation in mannose-6-phosphate isomerase (*Mpi*) genotypes. Previous experiments demonstrated interpopulation variation in habitat choice in* Semibalanus; *larvae from the Gulf of St Lawrence preferred crevices and those from the Atlantic coast exposed surfaces. We hypothesized that the population structure observed at St Andrews may be due to intrapopulation variation in settlement behaviour, and proposed four criteria which distinguish population structure due to settlement behaviour from that due to other processes. We used these criteria to determine if habitat choice generated spatial variation in* Mpi *genotypes at St Andrews. Spatial and/or temporal variation observed in the pattern of differentiation between habitats in 1991 and 1992 suggested that settlement behaviour produces no microgeographic population structure at the* Mpi *locus.*

Our study showed no population structure due to habitat choice. Nonetheless, distribution of electrophoretic markers can be a valuable indicator of intrapopulation variation in settlement behaviour. Further investigation of microgeographic population structure in sessile marine invertebrates is necessary to determine how often such spatial variation occurs, and the role settlement behaviour plays in its formation.

3.3.1　Introduction

The evolution of settlement behaviour in larvae of sessile marine invertebrates requires that variation in behaviour has an additive genetic component. While larval habitat choice has been well documented in the laboratory and field (review: Crisp, 1984; Strathmann, Branscomb and Vedder, 1981; Wethey, 1986; Raimondi, 1988; Johnson and Strathmann, 1989; Jensen and Morse, 1990; Roberts *et al.*, 1991; Walters, 1992), we have little knowledge as to its genetic basis or evolution. Heritability of settlement behaviour has been estimated under controlled conditions (MacKay and Doyle, 1978; Holm, 1990), but the extent to which genetic variation in behaviour is expressed in nature is unknown.

Spatial variation due to habitat choice, in both electrophoretic and morphological markers, has been observed in insects, molluscs and crustaceans (review: Jaenike and Holt, 1991), but apparently not among larvae of sessile marine invertebrates. Cases where settlement generates population genetic structure can indicate intrapopulation variation in habitat choice behaviour, and may therefore be useful in studying the genetic basis and evolution of these behaviours in the field. Other processes, however, such as reduced dispersal (Grosberg, 1987) and/or 'chaotic' spatio-temporal variation in the settlement of genetically differentiated cohorts of larvae (Johnson and Black, 1982, 1984), may also contribute to population structure at settlement. For settlement behaviour to cause microgeographic population structure, and thus suggest variability in larval habitat choice, spatial variation in allele or genotype frequencies should: (1) arise at settlement, (2) reflect the spatial distribution of potential settlement cues, (3) be consistent within and across settlement cohorts/generations, and, if the behaviour is adaptive, (4) persist into the adult stage. Population structure due to settlement of differentiated cohorts of larvae is unlikely to be consistent across cohorts (Johnson and Black, 1982, 1984), or to correspond to the distribution of cues. At the small spatial scales relevant here (< 10 m), population structure is also not likely to be caused by restricted dispersal in species which release larvae with prolonged development in the plankton.

We used the criteria proposed above to examine the role of larval habitat choice in producing spatial variation in genotype frequencies at the mannose-6-phosphate isomerase (*Mpi*) locus in the barnacle *Semibalanus balanoides*. In the Gulf of St Lawrence (Capucins, Québec), cyprid larvae of *S. balanoides* settle mainly in crevices, while at St Andrews (New Brunswick, Atlantic coast) larvae settle in crevices and on exposed, horizontal surfaces, but prefer the latter (Chabot and Bourget, 1988). These two populations are also differentiated at the *Mpi* locus, with allele *Mpi²* occurring at higher frequencies on the Atlantic coast than in the Gulf (Bourget *et al.*, 1989; Martel, 1990). Re-analysis of Martel's (1990) data from St Andrews (Table 3.3.1) revealed small but significant differences in *Mpi* genotype frequencies between barnacle samples from crevices and exposed surfaces ($G_{adj} = 6.79$, d.f. = 2, $p < 0.05$). Microgeographic population structure paralleled that observed on the regional scale – the frequency of *Mpi²* was lower in crevice samples than in samples from exposed surfaces. This pattern in *Mpi* genotypes at St Andrews may have been caused by intrapopulation variation in larval responses to settlement cues associated with crevices and exposed surfaces. We report here on the results of sampling carried out at St Andrews in 1991 and 1992 to determine if the spatial variation observed in Martel's (1990) data was still present, and, if so, whether the patterns were generated by settlement behaviour.

3.3.2   Materials and methods

On 29 May 1991 cobbles were collected from the intertidal zone at Minister Island (MI), St Andrews, their vertical and horizontal surfaces were marked, and they were transported to Université Laval for processing. Recruits of *S. balanoides* (age < 1 month) were removed from the vertical and horizontal sides of the cobbles and the whole animal, including shell, was homogenized in 5–10 μl of grinding buffer (50 mM Tris-HCl, 1 mM MgCl$_2$, 1 mM DTT, 50% v/v glycerol, adjusted to pH 7.5). Samples were stored at –20 °C until electrophoresis.

In May 1992 we returned to St Andrews to examine local-scale and microgeographic population structure. Ten crevices and adjacent exposed, horizontal surfaces were sampled at Big Bay (BB), near Chamcook, New Brunswick, and near the Huntsman Marine Science Center (HMSC, as sampled by Martel (1990) in St Andrews). Attached, unmetamorphosed cyprid larvae, newly metamorphosed recruits (age < 1–2 days) and adults were sampled from each crevice-exposed surface pair at BB. Only adults (age < 1 year) were sampled at HMSC, as settlement density there was very low. Animals were processed in the field. Large adults were homogenized in 50 or 100 μl, small adults in 8 μl, and larvae and new recruits in 3.5 μl of ice-cold buffer. Samples were frozen until transport to Université Laval for electrophoresis.

Presumptive genetic variation at the *Mpi* (EC 5.3.1.8) locus was resolved by isoelectric focusing of polyacrylamide gels (124 × 258 × 0.3 mm; 1.7 ml 30% w/v acrylamide + 1.6% w/v bis-acrylamide, 2 ml 50% v/v glycerol, 1.7 ml distilled water, 375 μl Pharmalyte pH 2.5–5, 375 μl Pharmalyte pH 5–6, 30 μl 0.5% DTT, 7.5 μl TEMED, 75 μl 10% ammonium persulphate). DL-glutamic acid (0.04 M) served as the anode electrolyte, and 0.1 M NaOH as the cathode electrolyte. Gels were prefocused for 5 min at 5 W before application of samples. Sample extract, 3.5 μl, was applied at the cathode, and the gel was focused for 40 min at 5 W and then for 40 min at 10 W. Gels were stained with a solution of 4.3 ml distilled water, 9.0 ml 0.5 M Tris-HCl pH 8, 0.3 ml 50 mM D-mannose-6-phosphate, 0.18 ml 30 mM NADP, 4.2 units glucose-6-phosphate dehydrogenase, 30 units phosphoglucose isomerase, 0.6 ml

Table 3.3.1 *Mpi* genotype and allele frequencies for *Semibalanus balanoides* from all sampling sites, surfaces and dates

Date	Site	Surface	n	Genotype frequencies					G	Allele frequencies				D_{23}
				13	22	23	33	34		1	2	3	4	
1988	HMSC	Exposed	75	0.013	0.240	0.373	0.373	–		0.007	0.427	0.567	–	+0.056*
		Crevice	69	0.014	0.101	0.565	0.319	–	6.79*	0.007	0.384	0.609	–	–0.049
1991	MI	Horizontal	78	0.038	0.218	0.346	0.397	–		0.019	0.391	0.590	–	+0.058*
		Vertical	85	0.012	0.118	0.447	0.424	–	4.60	0.006	0.341	0.653	–	+0.002
1992	BB	Horizontal	99	–	0.131	0.354	0.505	0.01		–	0.308	0.687	0.005	+0.035
		Vertical	100	–	0.120	0.440	0.440	–	1.56	–	0.340	0.660	–	+0.004
1992	HMSC	Horizontal	81	0.012	0.111	0.395	0.481	–		0.006	0.309	0.685	–	+0.014
		Vertical	75	0.027	0.133	0.467	0.373	–	1.90	0.013	0.367	0.620	–	–0.006

Sites: HMSC = Huntsman Marine Science Center; MI = Minister Island; BB = Big Bay. n = sample size; G = G-test statistic for comparison of genotype frequencies between surfaces (d.f. = 2 in all cases); D_{23} = disequilibrium coefficient (Weir, 1990) describing deviations from Hardy–Weinberg equilibrium for alleles 2 and 3. *$p < 0.05$. Data from HMSC in 1988 are re-analysed from Martel (1990).

15 mM MTT, 0.6 ml 5 mM PMS, and 15 ml 2% agar. Bands observed on staining were labelled according to their distance from the cathode, with the closest band as allele 1. All samples were run against controls of known genotype.

Deviations of genotype frequencies from Hardy–Weinberg equilibrium were tested following the disequilibrium coefficient approach of Weir (1990). We tested for variation in genotype frequencies between surfaces using G-tests for independence (Sokal and Rohlf, 1981). To ensure that all expected cell frequencies were > 5, the rare and next most common genotypes were pooled before analysis of genotype frequencies by G-test.

3.3.3 Results

Isoelectric focusing revealed four alleles at the *Mpi* locus, with two common alleles. Rare alleles always occurred as heterozygotes. Banding patterns were consistent with previously reported (Murphy *et al.*, 1990) quaternary structures for MPI.

We observed slight, but non-significant, variation in *Mpi* genotype frequencies between recruits sampled from the horizontal and vertical sides of cobbles in 1991 (Table 3.3.1, $G = 4.60$, d.f. $= 2$, $0.1 < p < 0.25$). Frequency of the Mpi^2 allele was marginally higher on horizontal than on vertical surfaces, as in 1988. Cyprid larvae and newly metamorphosed recruits collected in 1992 failed to stain strongly enough to allow analysis of *Mpi* variation. The results presented for BB are therefore based on the sample of adults. At both the BB and HMSC sites we again found no significant differences between samples from within (vertical) or outside (horizontal) crevices (Table 3.3.1).

Samples from each surface orientation were generally in Hardy–Weinberg equilibrium (D_{23} values, Table 3.3.1). Two disequilibrium coefficients were significant; barnacles taken from exposed surfaces in 1988 and from horizontal surfaces of cobbles in 1991 showed a deficit of heterozygotes (Table 3.3.1, $D_{23} = 0.056$ and 0.058, respectively; $p < 0.05$ for both). Although only two of the deviations were significant, a trend in the values was apparent. Samples from exposed surfaces consistently exhibited larger heterozygote deficits than those from vertical surfaces (Table 3.3.1).

3.3.4 Discussion

At St Andrews in 1988, *Semibalanus balanoides* sampled from within crevices and from adjacent, exposed habitats differed in their genotype frequencies at the *Mpi* locus. Allele frequencies in crevice samples more closely resembled the population at Capucins (Gulf of St Lawrence), where larvae preferentially settle in crevices (Chabot and Bourget, 1988) than did samples from exposed surfaces. Based on these data we hypothesized that the microgeographic population structure observed at St Andrews may have been due to intrapopulation variation (possibly genetic in nature) in settlement behaviour, and proposed four criteria with which to test the hypothesis. We did not obtain *Mpi* genotype frequencies for newly settled *S. balanoides*; however, it is still possible to determine whether adaptive settlement behaviour influences this system in the hypothesized way.

It seems likely that larval habitat choice does not produce microgeographic population structure in *S. balanoides* at St Andrews. Variation at the *Mpi* locus in 1988 corresponded to a potential settlement cue and was present in the adult stage. In 1991 and 1992, however, no significant population structure relative to surface orientation was detected, either in recruits or adults. These results demonstrate that spatial variation in *Mpi* genotype frequencies is not consistent within or across cohorts. The analysis carried out in 1988 does not allow us to determine whether the pattern observed at that time was due to a recruitment process like that observed by Johnson and Black (1982, 1984) or post-settlement selection.

Systems in which microgeographic population structure is produced by settlement behaviour can provide valuable 'natural experiments' for the study of genetic control and evolution of larval habitat choice. It is difficult to predict how frequently the effects of behaviour may be observed; most studies of sessile invertebrates at appropriate spatial scales have shown no population structure or variation inconsistent with a hypothesis of habitat choice (e.g. Koehn, Turano and Mitton, 1973; Tracey, Bellet and Gravem, 1975; Flowerdew and Crisp, 1976; but see Achituv and Mizrahi, 1987). Description of microgeographic population structure in organisms and environments where such spatial variation might be expected, and continued study of organisms that exhibit small-scale variation (e.g. Achituv and Mizrahi, 1987) in order to distinguish the effects of settlement behaviour from other processes (review: Grosberg, 1991), will improve our understanding of the evolution of settlement behaviour in marine invertebrate larvae.

Acknowledgements

We wish to thank the staff of the Huntsman Marine Science Center for the use of their facilities in 1991 and 1992. C. Pearce assisted with the processing of samples in 1991. N. Martel graciously answered all questions on her 1988 research, and provided the preserved gels used in re-analysis of the data. Comments from B. Boudreau, V. Brock and three anonymous reviewers substantially improved earlier drafts of this manuscript. These studies were supported by grants from NSERC and FCAR to E. Bourget, and are a contribution to the programme of GIROQ (Group interuniversitaire de recherches océanographiques du Québec).

References

Achituv, Y. and Mizrahi, L. (1987) Allozyme differences between tidal levels in *Tetraclita squamosa* Pilsbry from the Red Sea. *J. Exp. Mar. Biol. Ecol.*, **108**, 181–9.

Bourget, E., Martel, N., Lapointe, L. and Bussières, D. (1989) Behavioural, morphological and genetic changes in some North Atlantic populations of the barnacle *Semibalanus balanoides*, in *Evolutionary Biogeography of the Marine Algae of the North Atlantic* (eds D.J. Garbary and G.R. South), NATO ASI Series, Vol. G22, Springer-Verlag, Berlin, pp. 87–106.

Chabot, R. and Bourget, E. (1988) Influence of substratum heterogeneity and settled barnacle density on the settlement of cypris larvae. *Mar. Biol.*, **97**, 45– 56.

Crisp, D.J. (1984) Overview of research on marine invertebrate larvae, 1940–1980, in *Marine Biodeterioration: An Interdisciplinary Study* (eds J.D. Costlow and R.C. Tipper), Naval Institute Press, Annapolis, Maryland, pp. 103–26.

Flowerdew, M.W. and Crisp, D.J. (1976) Allelic esterase isozymes, their variation with season, position on the shore and stage of development in the cirripede *Balanus balanoides*. *Mar. Biol.*, **35**, 319–25.

Grosberg, R.K. (1987) Limited dispersal and proximity-dependent mating success in the colonial ascidian *Botryllus schlosseri*. *Evolution*, **41**, 372–84.

Grosberg, R.K. (1991) Sperm-mediated gene flow and the genetic structure of a population of the colonial ascidian *Botryllus schlosseri*. *Evolution*, **45**, 130–42.

Holm, E.R. (1990) Attachment behaviour in the barnacle *Balanus amphitrite amphitrite* (Darwin): genetic and environmental effects. *J. Exp. Mar. Biol. Ecol.*, **135**, 85–98.

Jaenike, J. and Holt, R.D. (1991) Genetic variation for habitat preference: evidence and explanations. *Am. Nat.*, **137**, S67–90.

Jensen, R.A. and Morse, D.E. (1990) Chemically induced metamorphosis of polychaete larvae in both the laboratory and ocean environment. *J. Chem. Ecol.*, **16**, 911–30.

Johnson, L.E. and Strathmann, R.R. (1989) Settling barnacle larvae avoid substrata previously occupied by a mobile predator. *J. Exp. Mar. Biol. Ecol.*, **128**, 87–103.

Johnson, M.S. and Black, R.B. (1982) Chaotic genetic patchiness in an intertidal limpet, *Siphonaria* sp. *Mar. Biol.*, **70**, 157–64.

Johnson, M.S. and Black, R.B. (1984) Pattern beneath the chaos: the effect of recruitment on genetic patchiness in an intertidal limpet. *Evolution*, **38**, 1371–83.

Koehn, R.K., Turano, F.J. and Mitton, J.B. (1973) Population genetics of marine pelecypods. II. Genetic differences in microhabitats of *Modiolus demissus*. *Evolution*, **27**, 100–105.

MacKay, T.F.C. and Doyle, R.W. (1978) An ecological genetic analysis of the settling behaviour of a marine polychaete: I. Probability of settlement and gregarious behaviour. *Heredity*, **40**, 1–12.

Martel, N. (1990) Différenciation génétique des cirripèdes (*Semibalanus balanoides*) du golfe du Saint-Laurent et de la côte atlantique canadienne. MS Thesis, Université Laval, Ste-Foy, Québec.

Murphy, R.W., Sites, J.W. Jr., Buth, D.G. and Haufler, C.H. (1990) Proteins I: Isozyme electrophoresis, in *Molecular Systematics* (eds D.M. Hillis and C. Moritz), Sinauer, Sunderland, Massachusetts, pp. 45–126.

Raimondi, P.T. (1988) Settlement cues and determination of the vertical limit of an intertidal barnacle. *Ecology*, **69**, 400–407.

Roberts, D., Rittschof, D., Holm, E. and Schmidt, A.R. (1991) Factors influencing initial larval settlement: temporal, spatial and surface molecular components. *J. Exp. Mar. Biol. Ecol.*, **150**, 203–21.

Sokal, R.R. and Rohlf, F.J. (1981) *Biometry*, Freeman, San Francisco, California.

Strathmann, R.R., Branscomb, E.S. and Vedder, K. (1981) Fatal errors in set as a cost of dispersal and the influence of intertidal flora on set of barnacles. *Oecologia*, **48**, 13–18.

Tracey, M.L., Bellet, N.F. and Gravem, C.D. (1975) Excess allozyme homozygosity and breeding population structure in the mussel *Mytilus californianus*. *Mar. Biol.*, **32**, 303–11.

Walters, L.J. (1992) Field settlement locations on subtidal marine hard substrata: is active larval exploration involved? *Limnol. Oceanogr.*, **37**, 1101–1107.

Weir, B.S. (1990) *Genetic Data Analysis*, Sinauer, Sunderland, Massachusetts.

Wethey, D.S. (1986) Ranking of settlement cues by barnacle larvae: influence of surface contour. *Bull. Mar. Sci.*, **39**, 393–400.

3.4   GENETIC VARIATION AND POPULATION STRUCTURE IN THE TROPICAL MARINE BIVALVE *ANOMALOCARDIA BRASILIANA* (GMELIN) (VENERIDAE)

Edson P. da Silva and Antonio M. Solé-Cava

Abstract

In this work we studied the levels of gene variation and population structuring of the euryhaline cockle, Anomalocardia brasiliana. *This species is exploited commercially in Brazil, and has a large geographical and ecological distribution in the Caribbean and along the eastern coast of South America, where it can be found both in metahaline (salinity of up to 65 parts per 1000) and estuarine environments. Since adaptive clines of gene frequencies with salinity for the leucine aminopeptidase (*Lap*) locus have been reported in some estuarine bivalves, we tested whether the same phenomenon could be observed in a species living in a wider range of salinities. Horizontal starch gel electrophoresis was used to analyse 10 gene loci in 790 individuals of* A. brasiliana *from seven localities along 2000 km of Brazilian coast. Observed heterozygosities were high (*H_o = 0.19 to 0.26*) and levels of population structuring were low (*F_{st} = 0.07, *smallest gene identity between localities = 0.92). Several samples presented heterozygote deficiencies (*F_{is} = 0.15; *p < 0.01). We observed a negative correlation between the (arc sin of the square root transformed) frequency of allele C of the* Lap-1 *locus in* A. brasiliana *and water salinity (*R^2 = 96.3%; *p < 0.003). This cline was more marked in hypohaline than in metahaline environments. This may reflect the effect of different selection regimes, related to different types of osmoregulation, acting on the animals living in hypotonic or hypertonic conditions.*

3.4.1   Introduction

Anomalocardia brasiliana is a tropical marine bivalve that lives in sandy and muddy bottoms in the intertidal zone (Rios, 1970; Schaeffer-Novelli, 1976) in sheltered waters of bays and lagoons from the Caribbean Sea to Uruguay (Warmke and Abbot, 1961). *A. brasiliana* can reach very high population densities, and it is consumed and exploited commercially. It is dioecious and reproduces sexually through external fertilization (Narchi, 1976; Schaeffer-Novelli, 1976). The larvae produced can remain in the plankton up to four weeks until settlement (Narchi, 1976).

A very interesting aspect of the biology of *A. brasiliana* is its broad euryhalinity (Leonel, 1981; Leonel, Lunetta and Salomão, 1982; Leonel, Magalhães and Lunetta, 1983). Populations of this species are common both in metahaline (up to 65 ppt salinity) and hypohaline (down to 20 ppt) environments (Schaeffer-Novelli, 1976; Mayr *et al.*, 1989; Silva and Fernandes, 1990).

One of the rare examples of an adaptive cline for a molecular marker in natural populations is the correlation usually observed between salinity and gene frequencies at the leucine aminopeptidase (*Lap*) locus of some euryhaline species of the mussel genus *Mytilus* (review: Koehn, 1983). This cline has been observed in salinities ranging from about 18 ppt up to 35 ppt, and its adaptive role was later confirmed in the laboratory (Beaumont *et al.*, 1988, 1989). In this section we demonstrate the presence of a significant ($p < 0.003$) *Lap* cline in *A. brasiliana*. This cline was strong in hypohaline conditions, but very weak in salinities above that of normal oceanic seawater (34 ppt).

Allozyme electrophoresis has been widely used to estimate levels of gene variation and population structuring in several groups of marine invertebrates, notably in molluscs (Berger, 1983; Burton, 1983; Gartner-Kepkay *et al.*, 1983; Koehn, 1985; Allendorf, Ryman and Utter, 1987; Smith, 1988). However, in spite of its economic importance, nothing is known about the genetics of the Brazilian cockle *A. brasiliana*. This section represents the first survey of gene variation and population structuring in this species.

3.4.2 Materials and methods

A total of 790 individuals were collected, during March 1991, from the intertidal zone of seven localities along 2000 km of Brazilian coast (Table 3.4.1 and Figure 3.4.1). Salinities at the collection points varied from 27 ppt to 68 ppt. The samples were transported on ice to the laboratory, where they were kept at -20 °C for up to 3 weeks, until electrophoresis.

Horizontal gel electrophoresis was performed by standard methods using 12.5% starch gels (Harris and Hopkinson, 1978; Murphy *et al.*, 1990). The gels were stained for 39 enzyme systems, of which seven gave useful results, interpreted as the expression of 10 gene loci (Table 3.4.2). The buffer system used was a discontinuous lithium hydroxide, pH 8.1 (Selander *et al.*, 1971).

Table 3.4.1 *Anomalocardia brasiliana* collection sites

Locality	Abbreviation	Latitude (S)	Longitude (W)	Salinity (ppt)	Distance from Bahia (km)
1. Bahia	BA	12°55′	38°29′	32.5	0
2. Boca do Canal	BC	22°53′	42°00′	35.0	1410
3. Palmeiras	PP	22°52′	42°03′	45.0	1420
4. Praia Seca	PS	22°54′	42°20′	65.0	1430
5. Charitas	CH	22°58′	43°10′	32.0–36.9	1550
6. Catalão	CA	22°50′	43°15′	27.0–32.0	1570
7. Paraná	PR	25°30′	48°33′	28.0–31.0	1950

ppt = parts per thousand.

Table 3.4.2 Allele frequencies at 10 gene loci in *Anomalocardia brasiliana* from sites on the coast of Brazil

Locus	BA	BC	PP	PS	CH	CA	PR
				Sampling site			
αEst-1							
N	26	52	56	49	9	40	40
a	0.154	0.192	0.116	0.092	0.000	0.275	0.375
b	0.423	0.433	0.482	0.408	0.556	0.488	0.400
c	0.423	0.375	0.402	0.500	0.444	0.237	0.225
αEst-2							
N	34	26	22	26	9	41	39
a	0.147	0.019	0.091	0.096	0.278	0.159	0.218
b	0.574	0.885	0.795	0.885	0.667	0.720	0.667
c	0.279	0.096	0.114	0.019	0.056	0.122	0.115
Gdh							
N	23	4	4	9	2	7	12
a	0.978	1.000	1.000	1.000	1.000	1.000	1.000
b	0.022	0.000	0.000	0.000	0.000	0.000	0.000
Lap-1							
N	40	123	143	129	9	131	106
a	0.000	0.004	0.017	0.019	0.056	0.053	0.019
b	0.988	0.972	0.969	0.977	0.944	0.916	0.948
c	0.013	0.020	0.010	0.004	0.000	0.023	0.033
d	0.000	0.004	0.003	0.000	0.000	0.008	0.000
Lap-2							
N	40	160	165	169	9	137	110
a	0.100	0.103	0.155	0.109	0.111	0.069	0.086
b	0.837	0.847	0.809	0.843	0.722	0.777	0.823
c	0.063	0.047	0.033	0.044	0.167	0.117	0.077
d	0.000	0.003	0.003	0.003	0.000	0.036	0.014
Mpi							
N	9	75	73	71	9	70	69
a	0.389	0.747	0.699	0.803	0.889	0.900	0.957
b	0.611	0.213	0.212	0.169	0.056	0.086	0.043
c	0.000	0.040	0.089	0.028	0.056	0.014	0.000
Pep-1							
N	11	37	45	64	9	21	19
a	1.000	1.000	1.000	1.000	1.000	1.000	1.000
Pep-2							
N	11	14	20	14	9	31	20
a	0.000	0.214	0.225	0.143	0.389	0.323	0.275
b	0.955	0.750	0.750	0.857	0.556	0.645	0.725
c	0.045	0.036	0.025	0.000	0.056	0.032	0.000
Pgi							
N	10	120	112	109	9	105	104
a	0.050	0.179	0.214	0.179	0.056	0.152	0.048
b	0.550	0.404	0.446	0.436	0.111	0.419	0.322
c	0.400	0.417	0.339	0.385	0.833	0.429	0.630
Sod							
N	24	71	71	83	9	86	50
a	0.042	0.000	0.000	0.000	0.000	0.000	0.010
b	0.958	1.000	0.993	1.000	1.000	0.994	0.970
c	0.000	0.000	0.007	0.000	0.000	0.006	0.020

N = sample size. Abbreviations for sampling sites as in Table 3.4.1. α *Est* = esterase; *Gdh* = glucose dehydrogenase; *Lap* = leucine amino peptidase; *Mpi* = mannose phosphate isomerase; *Pep* = peptidase; *Pgm* = phosphoglucomutase; *Sod* = superoxide dismutase.

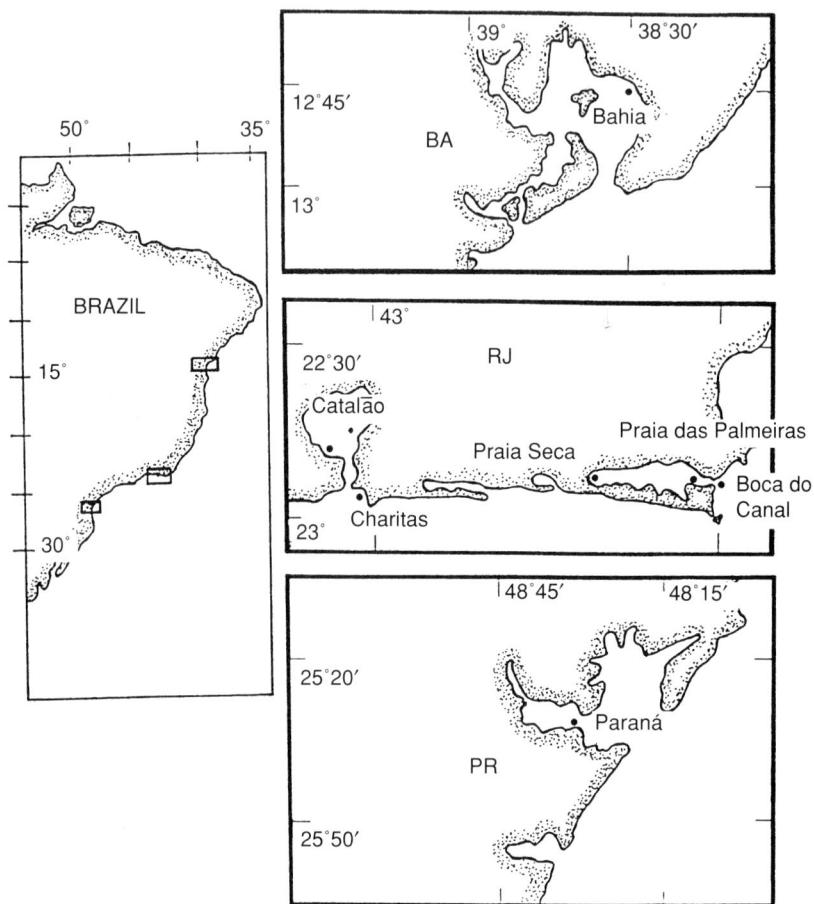

Figure 3.4.1 Collection sites of *Anomalocardia brasiliana*.

Levels of gene variation were estimated at the population level through the mean number of observed heterozygotes (H_o) and the mean number of Hardy–Weinberg expected heterozygotes (H_e) per locus (Nei, 1978, 1987). Interpopulational gene variation was estimated with Wright's fixation indices (Wright, 1951; Nei, 1987) and the use of pairwise gene identities (Nei, 1978). Heterogeneity in allele frequencies between localities was tested through χ^2 contingency table analysis with the pooling of rare alleles (Sokal and Rohlf, 1981). The presence of clines for the *Lap* locus with salinity was tested by regression analysis with linear and exponential models, using arcsin transformed allele frequencies (Sokal and Rohlf, 1981).

Gene frequencies, expected heterozygosities, departures from Hardy–Weinberg equilibrium, gene identity and fixation indices were calculated using the BIOSYS program (Swofford and Selander, 1981).

3.4.3 Results

Allele frequencies at the ten enzyme loci that could be genetically interpreted are shown in Table 3.4.2. Observed heterozygosities were high (H_o = 0.19 to 0.26), and levels of

Table 3.4.3 *F*-statistics for nine enzyme loci in *Anomalocardia brasiliana*

Locus	F_{is}	F_{it}	F_{st}
αEst-1	0.181	0.213	0.040
αEst-2	0.065	0.119	0.058
Gdh	−0.022	−0.003	0.019
Lap-1	0.184	0.195	0.013
Lap-2	0.085	0.097	0.013
Mpi	0.389	0.496	0.176
Pep-2	0.190	0.250	0.074
Pgi	0.052	0.128	0.080
Sod	−0.031	−0.009	0.021
Mean	0.148	0.207	0.069

Table 3.4.4 Pairwise unbiased gene identities (*I*, Nei (1978); above diagonal) and unbiased genetic distances (*D*, Nei (1978); below diagonal) between sampling sites of *Anomalocardia brasiliana*

Sampling site	1	2	3	4	5	6	7
1 Bahia	–	0.969	0.974	0.967	0.920	0.946	0.939
2 Boca do Canal	0.031	–	1.000	1.000	0.972	0.993	0.984
3 Palmeiras	0.026	0.000	–	0.999	0.970	0.992	0.978
4 Praia Seca	0.033	0.000	0.001	–	0.969	0.986	0.977
5 Charitas	0.083	0.028	0.030	0.031	–	0.982	0.986
6 Catalão	0.056	0.007	0.008	0.014	0.018	–	0.997
7 Paraná	0.063	0.016	0.022	0.023	0.014	0.003	–

population structuring, inferred from Wright's (1951) fixation and Nei's (1978) pairwise identity indices, were low to moderate (Tables 3.4.3 and 3.4.4 and Figure 3.4.2). Four loci presented significant (Fisher's exact test; $p < 0.01$) departures from Hardy–Weinberg equilibrium in different populations, due to heterozygote deficiencies. However, because of the high number of tests performed (53), it was important to discover whether those tests might

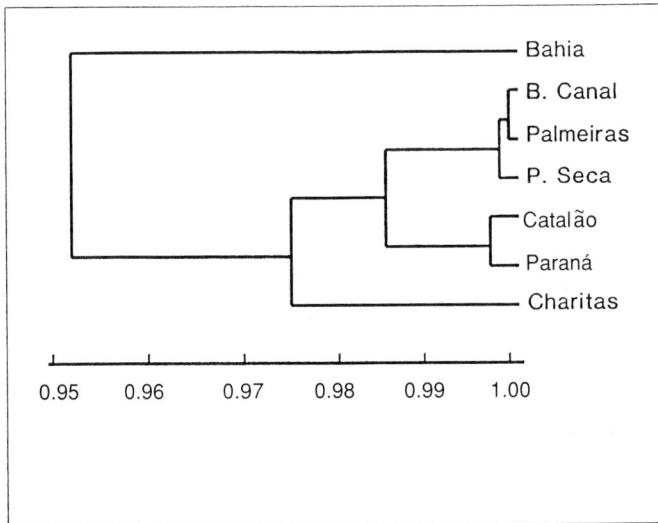

Figure 3.4.2 UPGMA phenogram of unbiased genetic identities (*I*, Nei, 1978) between samples of *Anomalocardia brasiliana*.

$$\text{ArcSin } \sqrt{f_{\text{lap-1-c}}} = e^{(2.97(\pm0.25) - 0.025(\pm0.006)\text{ppt S})}$$

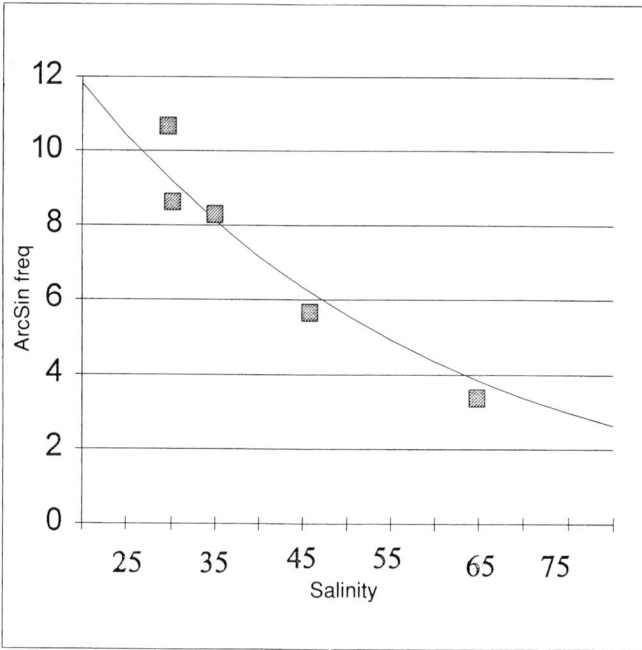

Figure 3.4.3 Regression analysis of the frequency of the allele *Lap-1^c* of *Anomalocardia brasiliana* in relation to salinity (ppt).

have given 'significant' results by chance alone (Lessios, 1992). Indeed, when a sequential Bonferroni analysis (Rice, 1989) was performed, none of the tests were significant at the 5% level.

A significant heterogeneity in allele frequencies ($\chi^2 = 26.04$; $p < 0.01$) was observed between localities for the two *Lap* loci. A cline was observed for allele 'c' from the *Lap-1* locus with salinity (Table 3.4.2). Arcsin ($\sqrt{f_x}$) transformed frequencies of *Lap-1^c* were significantly correlated with salinity, according to an exponential model (Figure 3.4.3); regression analysis $r^2 = 0.963$; $p < 0.003$.

3.4.4   Discussion

The most important observation in this section is a cline in allele frequencies at the *Lap* locus in a natural population of a bivalve mollusc living over a very wide range of salinities (from 27 to 65 ppt). Significant heterogeneities ($p < 0.01$) were observed for the allele frequencies of the two *Lap* loci. No clear pattern of variation could be found for any of the alleles of the *Lap-2* locus, but a clear (eight-fold) salinity-related cline was found for the *Lap-1^c* allele. A regression analysis of frequencies of this allele with salinity showed a very good fit to an exponential model ($r^2 = 96.3\%$; $p < 0.003$; Figure 3.4.3). The exponential regression fitted our data best because the relationship between salinity and gene frequency was most noticeable over the low-salinity part of the range, becoming approximately asymptotic to the *x*-axis at higher salinities. This seems to indicate that, if natural selection is responsible for the maintenance of this cline, it is acting more effectively at salinities less than 34 ppt. This conclusion is supported by physiological studies on four species of the mussel *Modiolus* which osmoregulate at low salinities, but do not control their cell volume at high salinities, simply losing water to the medium (Pierce, 1971). The effect of the *Lap* genotype on ammonia excretion in the gastropod *Thais haemastoma* also varies according to salinity, because this enzyme is related to the rate of nitrogen loss. This gives more scope for natural selection to act on the *Lap* locus at lower salinities (Garton and Berg, 1989) and could be an additional cause of the result observed here.

Another important observation from the genetic data is the relatively low level of genetic differentiation (mean $F_{st} = 0.07$) between samples from sites up to 2000 km apart. The actual importance of larval dispersal to gene flow has recently become the focus of a heated debate (Hedgecock, 1986; Olson and McPherson, 1987). This was fuelled principally by the discovery of very high levels of population structuring in many species with long-lived planktotrophic larvae, which indicated that the potential for gene flow is not always realized in marine invertebrates (Todd, Havenhand and Thorpe, 1988; Solé-Cava and Thorpe, 1992). The low level of population structuring observed in *A. brasiliana* suggests that the planktonic dispersion of *A. brasiliana* may be effective in minimizing the differentiation of allele frequencies over long distances. The level of F_{st} found in the population of *A. brasiliana* was similar to that reported for other species with presumed extensive dispersal, such as some fish (Waples, 1987), and marine invertebrates with long-lived planktonic larvae such as the starfish *Acanthaster planci* ($F_{st} = 0.070$) (Nishida and Lucas, 1988), the sea anemone *B. caissarum* ($F_{st} = 0.042$) (Russo, Solé-Cava and Thorpe, unpublished data) and the sea-urchin *Echinometra mathei* ($F_{st} = 0.013$) (Watts, Johnson and Black, 1990).

Following table-wide sequential Bonferroni analysis no significant deviations from the Hardy–Weinberg model were evident but at seven of the nine loci F_{is} was positive. Furthermore, mean F_{is} was 0.148 and it is fair to say that overall heterozygote deficiencies were observed in *A. brasiliana* at the seven localities studied. Heterozygote deficiencies are very common among marine invertebrates, especially in bivalve molluscs (Zouros and Foltz, 1984; Singh and Green, 1985; Volkaert and Zouros, 1989). This phenomenon could be due to inbreeding, bias during gel interpretation, the presence of null alleles, the Wahlund effect or natural selection. For a detailed discussion of this phenomenon see Volkaert and Zouros (1989), and Borsa, Zainuri and Delay (1991).

The high heterozygosities found in *A. brasiliana* were similar to those found in other molluscs (Beaumont and Beveridge, 1984; Nevo, 1988). The question of why bivalves have such high levels of genetic variation does not have an easy answer: several models, trying to relate levels of gene variation in many organisms to deterministic or stochastic forces, have been proposed in recent years (reviews: Nevo, 1978, 1988; Nei, 1987). However, because of the complexity and variability of the phenomena involved, it may be unwise to choose any one particular model to explain all cases observed (Beaumont and Beveridge, 1984; Solé-Cava and Thorpe, 1991). We can try to explain the high heterozygosity levels of *A. brasiliana* in two ways. First, one could argue that, since *A. brasiliana* is an intertidal organism, it must support variation in time (tidal cycle) and in space (vertical distribution on the beach). The levels of gene variation would then increase because of overdominance or balanced selection acting on the isozyme or closely linked loci (Levins, 1968; Levinton, 1975; Nevo, 1978; Noy, Lavie and Nevo, 1987; Solé-Cava and Thorpe, 1991). Second, a neutralist alternative explanation for the high levels of heterozygosity observed would be the random accumulation of neutral mutations in the population, which is expected to be directly proportional to its effective size (Ohta, 1973; Kimura, 1983; Solé-Cava and Thorpe, 1991). *A. brasiliana* probably has very high effective population sizes, since it can be found in high densities along the Brazilian, Caribbean and Uruguayan coasts. The neutralist explanation would seem, thus, to be sufficient to explain the heterozygosity levels found in this species.

Acknowledgements

The authors thank Silvana Brocanello and Evandro Pereira da Silva for helping with the collection of samples, and Claudia Russo for invaluable help in all stages of this work. This work was funded by grants from the Conselho Nacional de Pesquisas, CNPq, and the Coordenação de Aperfeiçoamento de Pessoal de Ensino Superior, CAPES, Brazil.

References

Allendorf, F., Ryman, N. and Utter, F. (1987) Genetics and fishery management: past, present and future, in *Population Genetics and Fishery Management* (eds N. Ryman and F. Utter), Washington Sea Grant Program, University of Washington Press, Washington, pp. 1–19.

Beaumont, A.R. and Beveridge, C.M. (1984) Electrophoretic survey of genetic variation in *Pecten maximus*, *Chlamys opercularis*, *C. varia* and *C. distorta* from the Irish Sea. *Mar. Biol.*, **81**, 299–306.

Beaumont, A.R., Beveridge, C.M., Barnett, E.A., Budd, M.D. and Smyth-Chamosa, M.D. (1988) Genetic studies of laboratory reared *Mytilus edulis*. I – Genotype specific selection in relation to salinity. *Heredity*, **61**, 389–400.

Beaumont, A.R., Beveridge, C.M., Barnett, E.A. and Budd, M.D. (1989) Genetic studies of laboratory reared *Mytilus edulis*. II – Selection at the leucine aminopeptidase (LAP) locus. *Heredity*, **62**, 169–76.

Berger, E.M. (1983) Population genetics of marine gastropods and bivalves, in *The Mollusca, Volume 6: Ecology* (ed. W.D. Russel-Hunter), Academic Press, London, pp. 563–96.

Borsa, P., Zainuri, M. and Delay, B. (1991) Heterozygote deficiency and population structure in the bivalve *Ruditapes decussatus*. *Heredity*, **66**, 1–8.

Burton, R.S. (1983) Protein polymorphisms and genetic differentiation of marine invertebrate populations. *Mar. Biol. Lett.*, **4**, 193–206.

Gartner-Kepkay, K.E., Zouros, E., Dickie, L.M. and Freeman, K.R. (1983) Genetic differentiation in the face of gene flow: a study of mussel populations from a single Nova Scotian embayment. *Can. J. Fish. Aquat. Sci.*, **40**, 443–51.

Garton, D.W. and Berg, D.J. (1989) Genetic variation at the *Lap* locus and ammonia excretion following salinity transfer in an estuarine snail. *Comp. Biochem. Physiol.*, **92A**, 71–4.

Harris, H. and Hopkinson, D.A. (1978) *Handbook of Enzyme Electrophoresis in Human Genetics*, North Holland, Amsterdam.

Hedgecock, D. (1986) Is gene flow from pelagic larval dispersal important in the adaptation and evolution of marine invertebrates? *Bull. Mar. Sci.*, **39**, 550–64.

Kimura, M. (1983) *The Neutral Theory of Molecular Evolution*, Cambridge University Press, Cambridge.

Koehn, R.K. (1983) The biochemical and physiological bases of aminopeptidase-I (LAP) polymorphism in *Mytilus edulis*. *Proc. Acad. Biol. Colloq.*, **41**, 117–46.

Koehn, R.K. (1985) Adaptive aspects of biochemical and physiological variability, in *Proceedings of the 19th European Marine Biology Symposium* (ed. P.E. Gibbs), Cambridge University Press, London, pp. 425–41.

Leonel, R.M.V. (1981) Influência da salinidade sobre algumas respostas osmóticas de *Anomalocardia brasiliana* (Gmelin, 1791) (Mollusca: Bivalvia), PhD Thesis, Instituto de Biociências, Universidade de São Paulo.

Leonel, R.M.V., Lunetta, J.E. and Salomão, L.C. (1982) Effects of variation in salinity on the osmotic regulation of *Anomalocardia brasiliana* (Gmelin, 1791) (Mollusca: Bivalvia). *Brazilian J. Med. Biol. Res.*, **15**, 335.

Leonel, R.M.V., Magalhães, A.R.M. and Lunetta, J.E. (1983) Sobrevivência de *Anomalocardia brasiliana* (Gmelin, 1791) (Mollusca: Bivalvia), em diferentes salinidades. *Boletim Fisiol. Anim.*, **7**, 63–72.

Lessios, H.A. (1992) Testing electrophoretic data for agreement with Hardy–Weinberg expectations. *Mar. Biol.*, **112**, 517–23.

Levins, R. (1968) *Evolution in Changing Environments*, Princeton University Press, Princeton, New Jersey.

Levinton, J.S. (1975) Levels of genetic polymorphism at two enzyme encoding loci in eight species of the genus *Macoma* (Mollusca: Bivalvia). *Mar. Biol.*, **33**, 41–7.

Mayr, L.M., Tenenbaum, D.R., Villac, M.C. *et al.* (1989) Hydrobiological characterization of Guanabara Bay, in *Coastlines of Brazil (Coastal Zone 89)* (eds C. Neves and O.T. Morgan), American Society of Civil Engineers, New York, pp. 124–38.

Murphy, R.W., Sites J.W. Jr., Buth, D.G. and Haufler, C.H. (1990) Proteins I: Isozyme electrophoresis, in *Molecular Systematics*, (eds D.M. Hillis and C. Moritz), Sinauer Associates, Sunderland, Massachusetts, pp. 45–126.

Narchi, W. (1976) Ciclo anual da gametogenese de *Anomalocardia brasiliana* (Gmelin, 1791) (Mollusca: Bivalvia). *Boletim Zool. Univ. S. Paulo*, **1**, 331–49.

Nei, M. (1978) Estimation of average heterozygosity and genetic distance from a small number of individuals. *Genetics*, **89**, 583–90.

Nei, M. (1987) *Molecular Evolutionary Genetics*, Columbia University Press, New York.

Nevo, E. (1978) Genetic variation in natural populations: patterns and theory. *Theor. Pop. Biol.*, **13**, 121–77.

Nevo, E. (1988) Natural selection in action: the interface of ecology and genetics in adaptation and speciation at the molecular and organismal levels, in *The Zoogeography of Israel* (eds Y. Yom-Tov and E. Tchernov), Dr W. Junk Publishers, Dordrecht, The Netherlands, pp. 411–38.

Nishida, M. and Lucas, J.S. (1988) Genetic differences between geographic populations of the crown-of-thorns starfish throughout the Pacific region. *Mar. Biol.*, **98**, 359–68.

Noy, B., Lavie, B. and Nevo, E. (1987) The niche-width variation hypothesis revisited: genetic diversity in the marine gastropods *Littorina punctata* (Gmelin) and *L. neritoides* (L.). *J. Exp. Mar. Biol. Ecol.*, **109**, 109–16.

Ohta, T. (1973) Slightly deleterious mutant substitutions in evolution. *Nature*, **246**, 96–8.

Olson, R.R. and McPherson, R. (1987) Potential vs. realized larval dispersal: fish predation on larvae of the ascidian *Lissoclinum patella* (Gottschaldt). *J. Exp. Mar. Biol. Ecol.*, **110**, 245–56.

Pierce, S.K. (1971) Volume regulation and valve movements by marine mussels. *Comp. Biochem. Physiol.*, **39A**, 103–17.

Rice, W.R. (1989) Analysing tables of statistical tests. *Evolution*, **43**, 223–5.

Rios, E.C. (1970) *Coastal Brazilian Seashells*, Fundação Universidade do Rio Grande, Rio Grande.

Schaeffer-Novelli, Y. (1976) Alguns aspectos ecológicos e análise da população de *Anomalocardia brasiliana* (Gmelin, 1791) Mollusca-Bivalvia, na praia do Saco da Ribeira, Ubatuba, Estado de São Paulo. PhD Thesis, Instituto de Biociências, Universidade de São Paulo, São Paulo.

Selander, R.K., Smith, M.H., Yang, S.Y., Johnson, W.E. and Gentry, J.R. (1971) Biochemical polymorphism and systematics in the genus *Peromyscus*. I – Variation in the old-field mouse (*Peromyscus polionotus*). *Stud. Genet. VI – Univ. Texas Publ. 7103*, 49–70.

Silva, E.P. and Fernandes, F.C. (1990) O bentos das praias arenosas da Lagoa de Araruama, RJ, in *II Simpósio de Ecossistemas da Costa Sul-Sudeste do Brasil: Estrutura, Função e Manejo* (ed S. Watanabe) 71, 210–223 Academia de Ciências do Estado de São Paulo, São Paulo.

Singh, S.M. and Green, R.H. (1985) Excess of allozyme homozygosity in marine molluscs and its possible biological significance. *Malacologia*, **25**, 569–81.

Smith, P. (1988) Biochemical-genetic variation in the green-lipped mussel *Perna canaliculus* around New Zealand and possible implications for mussel farming. *N.Z. J. Mar. Fresh. Res.*, **22**, 85–90.

Sokal, R.R. and Rohlf, F.J. (1981) *Biometry*, Freeman, San Francisco.

Solé-Cava, A.M. and Thorpe, J.P. (1991) High levels of genetic variation in natural populations of marine lower invertebrates. *Biol. J. Linn. Soc.*, **44**, 65–80.

Solé-Cava, A.M. and Thorpe, J.P. (1992) Genetic evidence of reproductive isolation between sympatric and allopatric populations of the various morphs of the common intertidal sea anemone *Actinia* (Anthozoa: Actiniaria) in Britain. *Mar. Biol.*, **112**, 243–52.

Swofford, D.L. and Selander, R.B. (1981) BIOSYS-1: A FORTRAN program for the comprehensive analysis of electrophoretic data in population genetics and systematics *J. Hered.*, **72**, 281–3.

Todd, CD., Havenhand, J.N. and Thorpe, J.P. (1988) Genetic differentiation, pelagic larval transport and gene flow between local populations of the intertidal marine mollusc *Adalaria proxima* (Alder & Hancock). *Funct. Ecol.*, **2**, 441–51.

Volckaert, F. and Zouros, E. (1989) Allozyme and physiological variation in the scallop *Placopecten magellanicus* and a general model for the effects of heterozygosity on fitness in marine molluscs. *Mar. Biol.*, **103**, 51–61.

Waples, R.S. (1987) A multispecies approach to the analysis of gene flow in marine shore fishes. *Evolution*, **41**, 385–400.

Warmke, G.L. and Abbot, R.T. (1961) *Caribbean Seashells*, Livingston, Pennsylvania.
Watts, R.J., Johnson, M.S. and Black, R. (1990) Effects of recruitment on genetic patchiness in the urchin *Echinometra mathaei* in western Australia. *Mar. Biol.*, **105**, 145–52.
Wright, S. (1951) The genetical structure of populations. *Ann. Eugenics*, **15**, 323–54.
Zouros, E. and Foltz, D.W. (1984) Possible explanation of heterozygote deficiency in bivalve molluscs. *Malacologia*, **25**, 583–91.

3.5 GENETIC VARIABILITY AND SUCCESS IN COLONIZATION IN TWO INTERTIDAL MUSSELS

Daphna Lavee and Uzi Ritte

Abstract

Electrophoretic variability was examined in two Red Sea, intertidal mussels, Brachidontes variabilis *(BV) and* Modiolus auriculatus *(MA). After the opening of the Suez Canal, BV, but not MA, colonized the Mediterranean. Both species were found to have a high level of variability, suggesting that extensive genetic variability by itself is not an attribute that can distinguish the colonizer from the non-colonizer. The major reason for the difference between the two species seems to be ecological – BV is more r-selected, and thus less sensitive to the competitive interactions from the Mediterranean mussel,* Mytilaster minimus. *A possible genetic contribution to successful colonization may be seen in 'Venice', a Red Sea site which both species have colonized recently. In contrast to the 'Venice' population of the colonizer BV, which maintained a high level of electrophoretic variability, the variability of the 'Venice' population of the non-colonizer MA was reduced.*

3.5.1 Introduction

The Red Sea and the Mediterranean have been separated for about 5 000 000 years, since the Miocene (Ekman, 1967). The opening of the Suez Canal, in 1869, connected these two seas (Por, 1977), and created an opportunity for immigration and colonization.

So far, more than 250 species have used this opportunity, mostly Red Sea species colonizing the Mediterranean (Spanier and Galil, 1991). These colonizing species, which unintentionally participated in 'some of the most important field experiments ever carried out' (Crawley, 1986), can be compared to their closely related non-colonizing species, in order to find out whether there are general attributes that distinguish between them (Baker and Stebbins, 1965; Safriel and Ritte, 1980; Parsons, 1984). This section considers genetic attributes, and presents the analysis of electrophoretic variability in samples of two similar Red Sea mytilid mussels, of which only one colonized the Mediterranean. In addition, colonizer populations in the target area were compared with those of the source area, in order to study the differences that may have occurred during this process.

3.5.2 Materials and methods

(a) Species

The colonizer is *Brachidontes variabilis* (BV), and the non-colonizer is *Modiolus auriculatus* (MA). Both are Indo-Pacific species, occupying the rocky intertidal zone (Barash and Danin, 1973). In the Red Sea, they are sympatric. BV is found mainly in the midlittoral (zonation terminology after Stephenson and Stephenson, 1972), is rare in the upper midlittoral (UML), is sometimes quite common in the lower midlittoral (LML) and infralittoral fringe (ILF), and is most abundant in the mid-midlittoral (MML) (Safriel, Felsenburg and Gilboa, 1980). MA is found nearly anywhere in the mid-midlittoral, often in the lower midlittoral (Safriel, Gilboa and Felsenburg, 1980), or in the infralittoral fringe.

MA is more common than BV, inhabiting a wide variety of habitats, including reef tables in the Gulf of Elat. Adult MA are about four times bigger than adult BV. Because of this difference, and because BV occupies a less stable environment in the Red Sea, it should be closer than MA to the *r* end of the *r–K* continuum.

MA is common in the Red Sea and in the Great Bitter Lake, and rare in the Little Bitter Lake and Lake Timsah (Safriel, Gilboa and Felsenburg, 1980). However, apart from a few empty shells, no living specimens of MA have been found in the Mediterranean (Safriel, Gilboa and Felsenburg, 1980). BV, on the other hand, now lives in both the Red Sea and the Mediterranean. In the Red Sea, it forms monospecific clusters, and its populations usually show a stable age distribution (Safriel, Felsenburg and Gilboa, 1980). In the Mediterranean it is found within the beds of the local mussel *Mytilaster minimus*, in densities that are 1/70 of the densities at which it occurs in the Red Sea (Safriel, Gilboa and Felsenburg, 1980). BV populations in the Mediterranean almost always grow exponentially, due to the continuous settlement of larvae in patches that had been cleared by storms (Sasson, 1981).

(b) Sampling

Sampling sites are shown in Figure 3.5.1. The criterion for sampling these species in the Red Sea was the presence of both species at the same site (except for site 6, which had only MA, and site 7, which had only BV). In the Mediterranean we selected the sites according to their distance from the Suez canal: minimal to maximal distance – sites 3, 2, 1.

Site 4 ('Venice'), in the north of the Gulf of Elat, is a small lagoon which was dug in the early 1960s. Its bottom is covered by granite stones; both species have invaded it and it is therefore an example of recent colonization.

Samples were collected over a 2-year period. Each site was sampled four times in the first year, and at least once during the second year (except for sites 3 and 9, which were only sampled once).

In the Red Sea, groups of adjacent MA and BV mussels were collected; in the Mediterranean, individual specimens of BV were collected from the beds of the local mussel *M. minimus*. Following collection, mussels were kept alive, in open containers, but were immediately frozen to −70 °C at the laboratory to await processing. Each sample included at least 30 specimens.

(c) Electrophoresis and data analysis

Electrophoresis was carried out on horizontal 10% starch gels following the procedure of Koehn, Milkman and Mitton (1976).

Altogether, 11 enzymes, representing 17 polymorphic and 3 monomorphic loci in BV, and 16 polymorphic loci in MA, were scored (Table 3.5.1). The frequencies of each allele and each genotype were determined separately for each sample, and together for all samples from the same site. Deviation from the Hardy–Weinberg distribution was computed as $D_t = (\Sigma H_o - \Sigma H_e)/\Sigma H_e$, where H_o is the observed and H_e is the expected number of heterozygotes at a locus (Koehn, Milkman and Mitton, 1976).

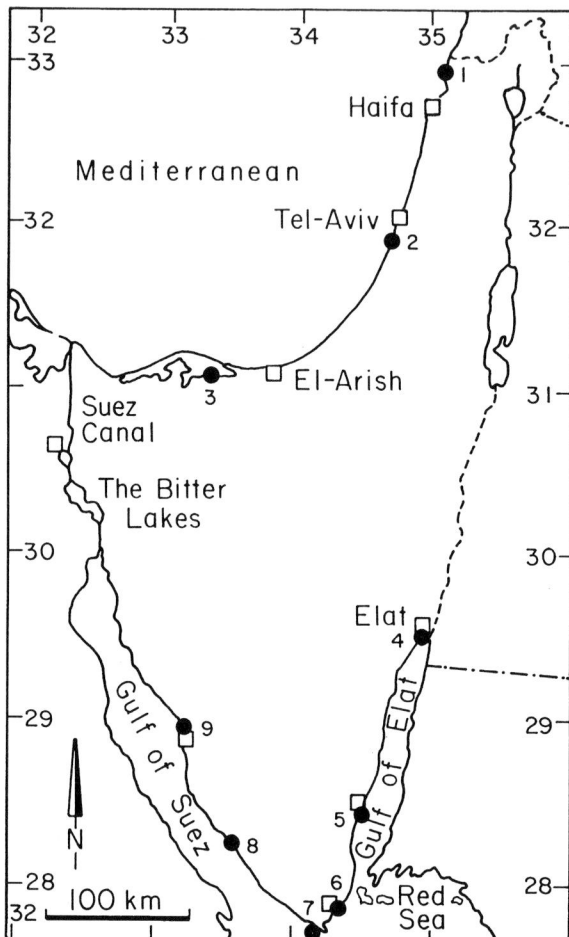

Figure 3.5.1 *B. variabilis* and *M. auriculatus*: map of the study area. Black circles indicate sampling sites. In the Mediterranean: 1 = Shave-Ziyyon (SZ); 2 = Palmahim (PL); 3 = Abu Tlul (TL). In the Red Sea: 4 = 'Venice' (VN); 5 = Dahab (DL); 6 = Marsa al Mauka (MG); 7 = Ras Muhammad (RM); 8 = Gebel Hammam Sidna Musa (AT); 9 = Abu-Zanima (AZ). Sites 1, 2, 4, 5, 6, 7 and 8 were repeatedly sampled.

Genetic distances were calculated between sites on pooled data for all collections at each site. Rogers' genetic distance was calculated using the PHYLAL2 program (Rogers, 1984, 1986).

The smallest space analysis (SSA) – a subprogram of the SPSS package (Levy, 1981) – calculates the smallest Euclidean space for a configuration of points. The data are similarity coefficients between samples, which are based on allele frequencies. The samples are scattered as points around a centre. The coordinates of each point designate its distance from the centre, and the program finds, for each locus, the sample that deviates most from it.

Table 3.5.1   Numbers of alleles at each locus in different sampling sites

Enzyme	Locus	B. variabilis				M. auriculatus		
		To	Re	VN	Me	To	Re	VN
Adenylate kinase (2.7.4.3)	Ak-1	7	7	5	7	6	6	6
	Ak-2	6	5	5	6			
	Ak-3	6	6	6	6			
Esterase (3.1.1.1)	Est-1	2	2	1	2	3	2	2
	Est-2	5	5	3	5	4	4	4
	Est-3	6	5	5	6	5	5	5
	Est-5	3	3	3	2	4	4	4
	Est-6	3	2	–	2	5	5	3
	Est-7	6	6	6	6	6	6	6
Glucokinase (2.7.1.2)	Gk	8	7	8	7	7	7	5
Isocitrate dehydrogenase (1.1.1.42)	Idh	6	6	6	5	7	7	6
Leucine aminopeptidase (3.4.1.1)	Lap	7	6	7	5	7	7	7
Malate dehydrogenase (1.1.1.37)	Mdh	*				5	5	5
'Fast' malate dehydrogenase	FMdh	5	6	5	4	6	6	4
Malic enzyme (1.1.1.40)	Me	4	3	3	6	4	4	4
Phosphoglucomutase (2.7.5.1)	Pgm	8	8	7	8	7	7	7
6-Phosphogluconate dehydrogenase (1.1.1.44)	6Pgd	7	6	6	7	8	8	8
Phosphohexose isomerase (5.3.1.9)	Phi	10	10	10	10	10	10	9

To = total number; *Re* = Red Sea without 'Venice'; *VN* = 'Venice'; *Me* = Mediterranean.
* Three monomorphic loci.

3.5.3   Results

Table 3.5.1 presents the number of alleles at each locus for all sites of the Red Sea (source area) except site 4, for site 4 ('Venice') in the Red Sea (site of recent colonization of both species), and in the Mediterranean (the area colonized by BV). In both species, at most of the loci, there were 2–3 relatively common alleles, with other alleles being infrequent.

The full data on allele frequencies at all sites for the different sampling dates have been presented elsewhere (Lavee, 1981) and are not included here. χ^2 contingency table tests on those data have demonstrated frequent examples of significant temporal (within-site) and geographic (between-site) variation in allele frequencies.

Samples were ranked, for each locus, according to their degree of similarity in allele frequencies. Usually, samples from the same site did not exhibit more similarity than samples from different sites, and samples from the same month in subsequent years were not more similar than samples from different months. The cluster of samples of BV, either from the Mediterranean (1, 2, 3) or the Red Sea (4–9), did not yield greater within-cluster similarity than the value of similarity within the cluster of the entire samples from both seas.

The SSA of BV showed that the most exceptional samples were from (1) August, (2) November and (3) October and in MA the samples from (1) November and January, (2) May and (3) March. It seems that in BV the major changes in allele frequencies occur during autumn, while in MA they occur in winter and spring.

Rogers' genetic distance, based on 13 polymorphic loci (Rogers, 1984, 1986), between the populations of MA was 0.309 (Figure 3.5.2(a)). The populations from the Gulf of Elat and from the Red Sea cluster separately from those in the Gulf of Suez. Most distant was the new lagoon population of 'Venice' (VN). The genetic distance between the BV populations was calculated twice: firstly using the seven polymorphic loci covering all seven populations

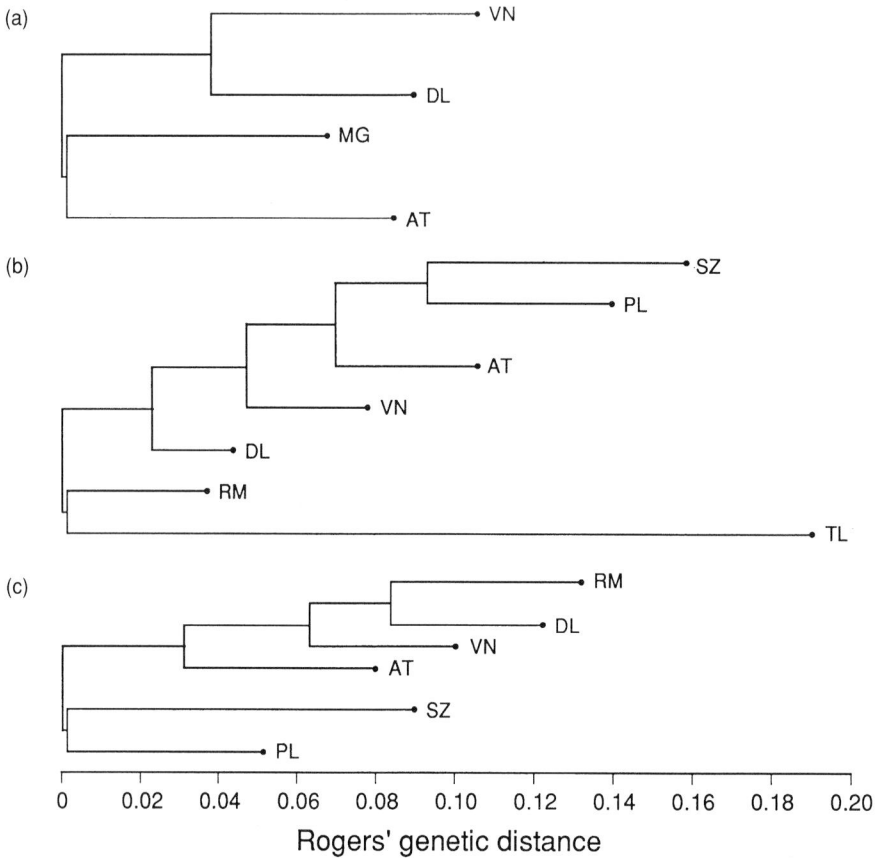

Figure 3.5.2 *B. variabilis* (BV) and *M. auriculatus* (MA): phylogenetic trees. (a) = MA − 13 polymorphic loci, four populations; (b) = BV − seven polymorphic loci, seven populations; (c) = BV − 13 polymorphic loci, six populations. Mediterranean sites; SZ = Shave-Ziyyon (site 1); PL = Palmahim (site 2); TL = Abu Tlul (site 3). Red Sea sites: VN = 'Venice' (site 4); DL = Dahab (site 5); MG = Marsa al Mauka (site 6); RM = Ras Muhammad (site 7); AT = Gebel Hammam Sidna Musa (site 8). Topology from TRETOP subroutine of PHYLAL2; optimization of ancestors and branch length by HAP subroutine.

(Figure 3.5.2(b)), and secondly using 13 polymorphic loci, covering only six populations (Figure 3.5.2(c)). The two calculations yielded different values, 0.512 and 0.392 respectively, but demonstrated a similarity in the relative distance between the populations of the Mediterranean, which diverged in both calculations, from the same ancestor. The TL population was very different.

At most loci (13 in BV and 10 in MA) D_t was negative, indicating a deficiency of heterozygotes (Table 3.5.2). In 37% of the samples of BV, and in 33% of the samples of MA, the deviation from Hardy–Weinberg equilibrium was significant at the 5% level in individual χ^2 goodness of fit tests. Application of Bonferroni-type tests to correct for type *I* error (Lessios, 1992) would reduce the number of significant results but there nevertheless remains a significant majority of negative D_t values.

Table 3.5.2 *B. variabilis* and *M. auriculatus*; mean deviation from Hardy–Weinberg equilibrium (D_t, see text) in samples from the Red Sea and the Mediterranean

Locus	*Brachidontas variabilis* $D_t \pm SD$	N_1	N_2	*Modiolus auriculatus* $D_t \pm SD$	N_1	N_2
Ak-1	-0.118 ± 0.256	(27)	916	0.121 ± 0.310	(14)	441
Ak-2	-0.050 ± 0.332	(13)	476			
Ak-3	0.050 ± 0.198	(12)	354			
Est-1	-0.186 ± 0.321	(7)	269	-0.287 ± 0.961	(4)	42
Est-2	-0.351 ± 0.306	(11)	304	-0.507 ± 0.285	(7)	168
Est-3	0.065 ± 0.259	(26)	1184	0.031 ± 0.268	(16)	638
Est-5	-0.454 ± 0.502	(5)	149	-0.310 ± 0.170	(2)	48
Est-6	-0.252 ± 0.541	(5)	119	-0.315 ± 0.092	(2)	41
Est-7	-0.190 ± 0.242	(8)	262	-0.060 ± 0.170	(5)	165
FMdh	-0.099 ± 0.324	(18)	520	0.304 ± 0.126	(9)	272
Gk	-0.203 ± 0.251	(34)	1037	0.103 ± 0.208	(18)	563
Idh	-0.144 ± 0.232	(36)	1956	-0.046 ± 0.287	(16)	529
Lap	-0.244 ± 0.258	(37)	1903	-0.141 ± 0.253	(18)	504
Mdh				-0.047 ± 0.568	(20)	936
Me	-0.459 ± 0.475	(21)	537	-0.489 ± 0.496	(14)	498
Pgm	-0.163 ± 0.239	(26)	1079	0.058 ± 0.171	(20)	921
Phi	0.098 ± 0.137	(39)	1944	0.039 ± 0.265	(22)	940
6Pgd	-0.350 ± 0.235	(37)	1951	-0.242 ± 0.354	(15)	478

SD = standard deviation, N_1 = number of samples, N_2 = number of individuals.

3.5.4 Discussion

High genetic variability might be considered a preadaptation for heterogeneous environments (Hedrick, Ginevan and Ewing, 1976). Since successful colonization often involves the occupation of new habitats, the genetic variability of the colonizer BV should thus be more extensive than that of the non-colonizer MA (Mayr, 1965; Safriel and Ritte, 1980). More recent studies (Gray, 1986; Barrett and Richardson, 1986) have not supported this suggestion – successful colonizers may differ from non-colonizers by genetically determined characters that are specifically responsible for the success of colonization (Parsons, 1983), but not in total genetic variability. As can be seen in the present study, MA did not colonize the Mediterranean, in spite of its extensive electrophoretic variability.

The non-involvement of high electrophoretic variability in general, and the possible involvement of specific genes, has been demonstrated in several other studies. No electrophoretic variability was found in the colonizing population of the pulmonate land snail *Rumina decollata* (a clonal species) in North America (Selander and Kaufman, 1973). In *Drosophila*, successful colonization may be due to a single gene: the colonizing species *D. melanogaster*, which is attracted to fermenting fruit, has active alleles at the alcohol dehydrogenase (*Adh*) locus, while the non-colonizing *D. inornata* and *D. hibisci* are *Adh*-null (Parsons, 1984). In Hawaiian *D. grimshawi* populations, the generalists (colonists) differ from the specialists (non-colonists) by the ability of the former to raise their larvae on a variety of plants, while the larvae of the latter are reared exclusively on the rotting bark of *Wikstroemia* (Carson and Ohta, 1981).

In both BV and MA, the average number of alleles per locus and the average values of heterozygosity were compared with those in other organisms (Nevo, Beiles and Ben-shlomo, 1984), and found to be high. MA (the non-colonizer) had even slightly higher values than BV (the colonizer).

Both species colonized 'Venice', the newly established lagoon site in the Red Sea, at approximately the same time. While the colonizer BV retained its overall level of electrophoretic variability (losing a few alleles at some loci but gaining new alleles at others), electrophoretic variability of the non-colonizer MA in 'Venice' was reduced. The inability of MA to maintain its high level of electrophoretic variability may be associated with its absence from the Mediterranean.

Differences in physiological endurance to high salinity between larvae of the two species are ruled out as factors which could prevent the colonization of MA in the Mediterranean: MA has been reported from the Bitter Lakes, and was even found to be common in the Great Bitter Lake (Safriel, Gilboa and Felsenburg, 1980).

The major reason for the success of BV, and the failure of MA, to colonize the Mediterranean seems to be ecological. BV occupies, in the Red Sea, a less stable habitat than MA. It is also a much smaller species than MA. In the Eastern Mediterranean, where density-independent selection pressures prevail, the more r-selected BV may have the ability to occupy vacant spaces, in competition with the local mussel, *Mytilaster minimus* (Sasson, 1981), while MA may not.

In both species, the significant differences in allele frequencies, both among samples from the same period in different sites, and among samples from the same site in different periods, may be due to the unique histories of recruitment, involving selective mortality of the planktonic larvae, as was shown by Johnson and Black (1984) for the intertidal limpet *Siphonaria jeanae*.

The almost uniform deficiency of heterozygotes may be due to the structure of the entire population, which is subdivided into small breeding units, adjacent to each other. Probably, most of the fertilization takes place between members of the same breeding unit. Each breeding unit is subjected to different selection pressures, favouring different alleles. As a consequence, differences in allele frequencies between breeding units arise, resulting in heterozygote deficiencies within the entire population due to the Wahlund effect (Gaffney *et al.*, 1990).

Two loci (*Est-3* and *Phi*), on the other hand, showed, in both species, an excess of heterozygotes, which was significant in some cases. This excess could be due to heterozygote advantage, associated with components of fitness, such as growth rate and variability (Zouros, Romero-Dorey and Mallet, 1988; Koehn, 1990; Gaffney, 1990).

The extensive electrophoretic variability of BV in the Mediterranean, including new alleles that were not found in the Red Sea samples, may indicate that the establishment of the Mediterranean populations did not involve a passage through a genetic bottleneck, and that a continuous supply of propagules is a possibility. The genetic distance between the two northern populations (SZ and PL) in the Mediterranean and the other populations (Figure 3.5.2) may therefore indicate the operation of different selection pressures rather than the effects of isolation by distance.

Acknowledgement

We thank D. Cohen for fruitful discussions, U.N. Safriel for his continuous support, J.S. Rogers for the supply of the PHYLAL2 program, E. Neufeld for help in the laboratory work, and two anonymous reviewers for their remarks on an earlier draft of this paper. This paper is publication no. 19 in the series 'Colonization of the Eastern Mediterranean by Red Sea Species immigrating through the Suez Canal', supported by grant no. 699 from the United States Israel Binational Science foundation (BSF), Jerusalem, Israel.

References

Baker, H.G. and Stebbins, G.L. (eds) (1965) *The Genetics of Colonizing Species*, Academic Press, New York.

Barash, Al. and Danin, Z. (1973) The Indo-Pacific species of Mollusca in the Mediterranean and notes on a collection from the Suez Canal. *Israel J. Zool.*, **21**, 301–74.

Barrett, S.C.H. and Richardson, B.J. (1986) Genetic attributes of invading species, in *Ecology of Biological Invasions* (eds R.H. Groves and J.J. Burdon). Cambridge University Press, Cambridge, pp. 21–33.

Carson, H.L. and Ohta, A.T. (1981) Origin of the genetic basis of colonizing ability, in *Evolution Today* (eds G.G.E. Scudder and J.L. Reveal), Hunt Institute for Botanical Documentation, Carnegie-Mellon University, Pittsburgh, pp. 365–70.

Crawley, M.J. (1986) The population biology of invaders, *Phil. Trans. R. Soc. Lond. Ser. B*, **314**, 711–31.

Ekman, S. (1967) *Zoogeography of the Sea*, Sidgewick and Jackson Ltd, London.

Gaffney, P.M. (1990) Enzyme heterozygosity, growth rate, and viability in *Mytilus edulis*: another look. *Evolution*, **44**, 204–10.

Gaffney, P.M., Scott, T.M., Koehn, R.K. and Diehl, W.J. (1990) Interrelationships of heterozygosity, growth rate and heterozygote deficiencies in the coot clam, *Mulinia lateralis*. *Genetics*, **124**, 683–99.

Gray, A.J. (1986) Do invading species have definable genetic characteristics? *Phil. Trans. R. Soc. Lond. Ser. B*, **314**, 655–74.

Hedrick, F.W., Ginevan, M.E. and Ewing, E.P. (1976) Genetic polymorphism in heterogeneous environments. *Annu. Rev. Ecol. Syst.*, **7**, 1–32.

Johnson, M.S. and Black, R. (1984) Pattern beneath the chaos: the effect of recruitment on genetic patchiness in an intertidal limpet. *Evolution*, **38**, 1371–83.

Koehn, R.K. (1990) Heterozygosity and growth in marine bivalves: comments on the paper by Zouros, Romero-Dorey, and Mallet (1988). *Evolution*, **44**, 213–6.

Koehn, R.K., Milkman, R. and Mitton, J.B. (1976) Population genetics of marine pelecypods. IV. Selection, migration and genetic differentiation in the blue mussel *Mytilus edulis*. *Evolution*, **30**, 2–32.

Lavee, D. (1981) Genetic variability in a colonizing system: comparison between the mussels *Brachiodontes variabilis* and *Modiolus auriculatus* (Bivalia: Mytilidae). PhD Thesis, The Hebrew University of Jerusalem, Jerusalem (in Hebrew)

Lessios, H.A. (1992) Testing electrophoretic data for agreement with Hardy–Weinberg expectations. *Mar. Biol.*, **112**, 517–23.

Levy, S. (1981) Lawful roles of facets in social theories, in *Multidimensional Data Representation: When and Why?* (ed. I. Borg), Mathesis Press, Ann Arbor, pp. 65–107.

Mayr, E. (1965) Summary, in *The Genetics of Colonizing Species* (eds H.G. Baker and G.L. Stebbins), Academic Press, New York and London, pp. 553–62.

Nevo, E., Beiles, A. and Ben-Shlomo, R. (1984) The evolutionary significance of genetic diversity: ecological, demographic and life-history correlates, in *Evolutionary Dynamics of Genetic Diversity (Lecture Notes in Biomathematics, Vol. 53)* (ed. G.S. Mani), Springer, Heidelberg, pp. 13–213.

Parsons, P.A. (1983) *The Evolutionary Biology of Colonizing Species*, Cambridge University Press, Cambridge.

Parsons, P.A. (1984) Colonizing species: a probe into evolution via the organism. *Endeavour, New Ser.*, **8**, 108–12.

Por, F.D. (1977) *Lessepsian Migration – The Influx of Red Sea Biota into the Mediterranean by Way of the Suez Canal*, Ecological Studies 23, Springer, Berlin.

Rogers, J.S. (1984) Deriving phylogenetic trees from allele frequencies. *Syst. Zool.*, **33**, 52–63

Rogers, J.S. (1986) Deriving phylogenetic trees from allele frequencies: a comparison of nine genetic distances. *Syst. Zool.*, **35**, 297–310.

Safriel, U.N., Felsenburg, T. and Gilboa, A. (1980) The distribution of *Brachidontes variabilis* (Krauss) along the Red Sea coasts of Sinai. *Argamon*, **7**, 31–44.

Safriel, U.N. Gilboa, A. and Felsenburg, T. (1980) Distribution of rocky intertidal mussels in the Red Sea Coasts of Sinai, the Suez Canal and the Mediterranean coast of Israel, with special reference to recent colonizers. *J. Biogeog.*, **7**, 39–62.

Safriel, U.N. and Ritte, U. (1980) Criteria for the identification of potential colonizers. *Biol. J. Linn. Soc.*, **13**, 287–97.

Sasson, Z. (1981) Interspecific competition in a colonizing system. MSc Thesis, The Hebrew University of Jerusalem, Jerusalem (in Hebrew).

Selander, R.K. and Kaufman, D.W. (1973) Self-fertilization and genetic structure in a colonizing land snail. *Proc. Natl. Acad. Sci. USA*, **70**, 1186–90.

Spanier, E. and Galil, B.S. (1991) Lessepsian migration: a continuous biogeographical process. *Endeavour, New Ser.*, **15**, 102–6.

Stephenson, T.A. and Stephenson, A. (1972) *Life Between Tidesmarks on Rocky Shores*, Freeman, San Francisco.

Zouros, E., Romero-Dorey, M. and Mallet, A.L. (1988) Heterozygosity and growth in marine bivalves: further data and possible explanations. *Evolution*, **42**, 1332–41.

4

DNA technology and genetics of aquatic invertebrates

4.1 THE POTENTIAL OF DNA TECHNIQUES IN THE POPULATION AND
EVOLUTIONARY GENETICS OF AQUATIC INVERTEBRATES

David O.F. Skibinski

Abstract

The techniques currently available for analysing genetic variation at the DNA level in natural populations are reviewed. Attention is focused on the application of these techniques in studies of the population genetics and taxonomy of aquatic invertebrates. It is concluded that DNA analysis has made little impact in this area compared with the technique of protein electrophoresis. This situation might change in future with further technical innovations based on the polymerase chain reaction.

4.1.1 Introduction

The discovery of effective methods of measuring genetic variation within and between species has always been an important technical objective in population genetics and taxonomy. Before the 1960s, techniques for the routine study of protein or DNA polymorphisms had not been developed. Attention was focused on heritable variation of diverse kinds. British ecological geneticists pursued snails, butterflies or plants displaying distinctive visible polymorphisms in natural populations. In the USA great effort was devoted to the study of visible mutations, lethals and chromosomal polymorphisms in *Drosophila*. These studies were highly informative in many ways, but failed to give a clear picture of the amount of genetic variation in natural populations. The results of experiments in quantitative genetics indicated that continuous characters such as body size respond readily to selection and sometimes have high heritability. Although a few experiments were undertaken to locate polygenes, it was never possible to determine in general whether the underlying genetic variation was due to a few highly polymorphic loci or many loci exhibiting low levels of variation. Thus deductions could not be made about overall genomic heterozygosity (Lewontin, 1974).

Genetics and Evolution of Aquatic Organisms. Edited by A.R. Beaumont.
Published in 1994 by Chapman & Hall, 2–6 Boundary Row, London SE1 8HN.
ISBN 0-412-49370-5

Since the 1960s the application of the technique of protein electrophoresis in the measurement of polymorphisms in natural populations led to advances in a number of important areas. It allowed genetic variation to be easily assayed at 20 or more independent loci and gave a clear indication of the levels of variation that exist in natural populations. This led to the refutation of the classical view of population genetic structure which held that populations have only low levels of variation for rather deleterious mutants held at low frequencies by the balance between mutation and selection (Lewontin, 1974). Early studies of protein variation were aimed at discovering whether or not the variation was of adaptive significance or was instead the result of mutation and genetic drift of neutral mutations. Two distinct approaches were used. In the first, patterns of variation in individual species or in pooled data from many species were compared with predictions derived from neutral theory or from considerations of the ecological circumstances of species. In the second approach individual polymorphisms were studied in detail in order to search for and discover the functional significance of alternative alleles at protein coding loci (Powers, 1990).

Protein electrophoresis has also been widely used in the study of the population genetic structure of natural populations and in the estimation of migration rates. It has been used in proving differences between commercial stocks, in providing strain- or stock-specific genetic markers, in resolving taxonomic controversies at the species level, and in the analysis of the structure and dynamics of hybrid zones. Finally, the technique has been applied successfully in systematics for the construction of evolutionary trees.

Protein electrophoresis was important because it allowed studies to be extended quickly across the whole taxonomic range of plants and animals. Relatively few organisms proved to be completely refractory to the technique. The main constraint is that a certain minimum amount of fresh tissue is required. The spat of *Mytilus* and similarly sized juvenile stages of other bivalves can be scored electrophoretically for some enzyme systems, but for others the small amount of tissue provides insufficient activity. Larvae would certainly tend to be too small. This frustrates the use of the technique in studies of dispersal and gene flow in aquatic invertebrates with pelagic larvae. These size constraints have been released by the development of DNA techniques based on the polymerase chain reaction.

Since the 1960s, the number of published studies of protein variation in population genetics and taxonomy has increased every year even during the last decade. This observation might seem to contradict an impression that the advances in and applications of DNA technology have led to a decrease in interest in allozyme techniques. What does seem to have happened is that the number of papers on protein electrophoresis in mainstream genetics journals has declined but the variety of journals publishing papers using the technique has widened. Such journals include those devoted specifically to the study of particular taxonomic groups.

4.1.2 DNA techniques and their applications

The development of recombinant DNA techniques in the 1970s gave population geneticists the ability to look directly at DNA sequences. The ability to join and clone foreign DNA in bacterial or phage vectors made the amplification and then sequencing of DNA possible. The discovery of restriction enzymes provided a method of rapid screening for DNA base sequence variation in populations.

Although DNA technology has had an enormous impact in biology generally, its impact in population genetics is possibly less than might be supposed. Compared with allozyme analysis a variety of different techniques need to be mastered. This helps to put DNA work beyond the reach of institutions and individuals lacking access to dedicated molecular biology laboratories, their equipment and personnel. Thus the increased use of allozyme analysis in population biology has not been matched in such a dramatic way by the increased use of DNA technology.

A dominant feature in published work on DNA variation is that the numbers of individuals that can be sampled is far less than in allozyme studies. There are a number of reasons for this. DNA extraction itself and the techniques for assaying variation are time-consuming. In the early days, to obtain DNA sequence information for a gene for a sample of individuals, it was necessary to clone DNA from each individual into a vector prior to sequencing. With nuclear DNA there was the additional problem of cloning both copies of the gene if nuclear genotypes were to be determined. The use of probes in restriction fragment length polymorphism (RFLP) analysis also involves a complex series of steps involving restriction digestion, electrophoresis, blotting, hybridization and visualization of bands (Sambrook, Fritsch and Maniatis, 1989). All these steps require optimization, and protocols may not always transfer effectively between species. It is perhaps easy for a trained molecular biologist working with human DNA with excellent support in a well-equipped and well-funded laboratory to develop a production line approach to the generation of RFLP data for large samples of individuals. It is less easy for a lone scientist working in an 'organism' – orientated laboratory.

The difficulties inherent in DNA technology have probably influenced the choice of problems which can be attacked. Population genetics requires data on large samples of individuals from each population in order to provide accurate estimates of allele and genotype frequencies. With DNA technology this is time-consuming and costly in terms of resources for most currently available approaches. Apart from work on humans, population genetics of nuclear DNA can be said to be advanced in only a few *Drosophila* species.

A major area of investigation in allozyme work in which aquatic invertebrates figure prominently is the functional study of polymorphisms at specific loci. Examples are the work on the *Lap* locus in *Mytilus* (Koehn and Hilbish, 1987) and the *Got* locus in *Tigriopus* (Burton and Feldman, 1983). These studies provide evidence of functional differences between alleles even though the precise nature of the selective forces operating is more uncertain. There is as yet no well-developed parallel for these functional studies in the DNA world. By contrast, DNA work in population genetics is dominated by a search for polymorphisms which can be used as markers in studies of population or species differentiation.

In studies of molecular evolution and taxonomy, interspecific variation is more important than intraspecific variation and much useful information can be obtained by comparing the sequence of a single copy of a gene obtained from each of several different species. Thus given the resources to obtain ten DNA sequences it would be more profitable to sequence the gene in ten different species than to sequence ten copies of it in a single species. It is therefore not surprising to find that DNA sequencing has been widely employed in systematics for the construction of phylogenetic trees, in the study of evolutionary rates, and in the comparative anatomy and evolution of specific molecules.

New techniques also lead on to new unenvisaged problems. For example, studies of mitochondrial DNA (mtDNA) initiated with the aim of surveying and characterizing geographic differentiation have revealed the interesting phenomena of length variation and heteroplasmy (e.g. Bermingham, Lamb and Avise, 1986). To account for length variation, emphasis might shift to the molecular characterization of individual mtDNA genomes through cloning, restriction mapping and sequencing (LaRoche *et al.*, 1990). Knowledge of the base sequence of an entire mtDNA genome would allow comparisons with other known mtDNA genomes and to consideration of higher order systematics through the analysis of gene order, e.g. in mussels (Hoffman, Boore and Brown, 1992), in sea urchins (Cantatore *et al.*, 1989), in sea stars (Smith *et al.*, 1990) and in the brine shrimp (Batuecas *et al.*, 1988). In this way population genetics could lose out to other areas of investigation.

An important new area of research opened up by DNA technology is the analysis of mtDNA. This molecule has many features which make it useful for population genetic analysis; see Avise (1991) and reviews cited therein. It is small and occurs in large amounts in cells, which facilitates analysis using techniques which do not depend on the availability of a hybridization probe. It is inherited maternally and reproduces asexually and thus can be used to define monophyletic groups of individuals. This makes it useful not only for systematic analysis but also for the study of population structure within species. Certain regions of the molecule evolve rapidly and are highly polymorphic. Thus mtDNA analysis has higher resolving power than allozyme techniques at lower levels of taxonomic divergence.

With mtDNA, differences between populations can be assessed statistically by considering the variation in genotype frequencies between populations as with allozyme data. However, mtDNA also offers the possibility of estimating the sequence divergence between genotypes, e.g. in sea urchins (Palumbi and Wilson, 1990), or of constructing phylogenetic trees of genotypes, e.g. in the horseshoe crab (Saunders, Kessler and Avise, 1986); for reviews see Avise *et al.* (1987) and Meyer (section 5.1). Thus mtDNA analysis can provide more detailed information than allozyme variation on the rates of evolution and evolutionary history. The way in which mtDNA analysis can extend the results of geographic surveys of allozyme variation is illustrated by recent work on the commercially important red rock lobster *Jasus edwardsii* (Ovenden, Brasher and White, 1992). A mtDNA analysis was undertaken to test the hypothesis that this species, which has a long-lived planktonic larva, should exhibit little or no population subdivision over a 4600-km geographic range between southern Australia and New Zealand. No evidence was found for significant genetic differentiation between populations. However, the study of a second species, *J. verreaux* (Brasher *et al.* 1992), revealed the existence of significant mtDNA differences between Australia and New Zealand. Though the two species have similar life-histories it was suggested that differentiation in *J. verreaux* may be related to the ability of the juvenile form to regulate dispersal through vertical movements in the water column. A similar experimental approach was used in an analysis of the systematic relationships of five lobster species, which previously had been defined using morphological criteria alone (Brasher, Ovenden and White, 1992). The mtDNA data were used for both a phenetic and cladistic analysis which gave similar clustering of the five species. Two of those previously named as distinct species were found to have closely similar mtDNA genotypes and it was suggested that they should be regarded as conspecific.

For the reasons given above, an examination of the literature reveals that mtDNA has been widely used in systematics, studies of molecular evolution and population genetics. By contrast, although nuclear DNA analysis has played a prominent role in studies of systematics and molecular evolution, it has to date played a relatively minor role in population genetics.

4.1.3   DNA studies in aquatic invertebrates: a taxonomic overview

A computer search of the literature since 1986 was carried out to identify DNA studies of aquatic invertebrates by using taxonomic keywords combined with variations of the keywords DNA and mitochondria. A total of 169 references were obtained. The majority of these were in non-genetic areas, e.g. biochemistry, physiology, neurobiology and molecular biology. When these were excluded, 63 references remained. Many of these were in areas such as molecular genetics, genetic toxicology or developmental genetics. These were excluded together with references devoted solely to molecular characterization of DNA within single species even if they included arguments relating to higher order taxonomy or molecular evolution. Twenty-seven papers remained which could be classified as falling within the area of population genetics and taxonomy. Examination of these papers resulted in the identification of new references which increased the total from 27 to 45 during the period surveyed. Of these 45, 26 are orientated towards population genetics in that they involve the analysis of intraspecific variation. The remaining 19 are orientated towards taxonomy or the comparison of DNA sequences in related species. Twenty-six of the papers consider mtDNA variation, 15 consider nuclear DNA variation and four consider both types of variation. The number of papers by year are 1986 (1 paper), 1987 (2), 1988 (1), 1989 (7), 1990 (10), 1991 (16) and 1992 (8) plus more expected. It can be concluded that DNA techniques have only recently made an impact in the population and evolutionary genetics of aquatic invertebrates, and this impact appears as yet to be quite small. It is interesting to contrast these figures with those for allozymes. In the allozyme database constructed by the author and colleagues (see Ward, Skibinski and Woodwark, 1992) the number of papers meeting the restrictive criteria of including 15 or more individuals scored for 15 or more loci, in two or more populations or species, is 103 for this group of organisms. Bearing in mind that the allozyme literature is still expanding, it will be some time before DNA studies catch up.

A taxonomic breakdown of the DNA papers is as follows. Amongst echinoderms there are eight papers on sea urchins (Degiorgi, 1988; Durica *et al.*, 1989; Barbieri *et al.*, 1990; Palumbi and Wilson, 1990; Smith, Boom and Raff, 1990; Degiorgi *et al.*, 1991; Palumbi and Kessing, 1991; Palumbi and Metz, 1991) and four papers on sea stars (Jacobs *et al.*, 1989; Smith *et al.*, 1989, 1990; Sainz and Cornudella, 1990). Amongst crustacea there are three papers on lobsters (Brasher *et al.*, 1992; Brasher, Ovenden and White, 1992; Ovenden, Brasher and White, 1992), three papers on crabs (Cunningham, Buss and Anderson, 1991; Cunningham, Blackstone and Buss, 1992; Neigel *et al.*, 1991), five papers on Daphnia (Crease, Stanton and Hebert 1989; Crease, Lynch and Spitze, 1990; Crease and Lynch, 1991; Stanton, Crease and Hebert 1991; Mort and Streit, 1992) and two papers on copepods, *Calanus* (Bucklin and Kann, 1991) and *Tisbe* (Steinbruck *et al.*, 1991). Amongst molluscs there are 11 papers on bivalves, comprising six papers on mussels (Edwards and Skibinski, 1987; Blot, Lendre and Albert, 1990; Fisher and Skibinski, 1990; Gardner and Skibinski, 1991; Hoeh, Blakley and Brown, 1991; Milyutina and Petrov, 1989), three papers on oysters (Reeb and Avise, 1990; Brown and Paynter, 1991; Karl and Avise, 1992), two papers on scallops (Snyder *et al.*, 1987; Gjetvaj, Cook and Zoures, 1992) and one paper on cockles (Brock and Christiansen, 1989). There are two papers on gastropods (Rollinson and Kane, 1991; Hauser *et al.*, 1992). Finally, there are three papers on corals (McMillan and Miller, 1989, 1990; McMillan *et al.*, 1991), and a single paper each on rotifers (Fu, Hirayama and Natsukari, 1991), the horseshoe crab *Limulus* (Saunders, Kessler and Avise, 1986) and a brachiopod *Terebratulina* (Cohen *et al.*, 1991).

The most widely studied groups of organisms thus seem to be bivalve molluscs, echinoderms, particularly sea urchins, and crustacea. However, while a majority of the echinoderm papers are orientated towards molecular evolution, the bivalve and crustacean papers are orientated more towards population genetics. It is likely that some references are omitted from this list, though it should give a fairly accurate picture at time of writing (July 1992). The dominance of crustacea and bivalves in this list is matched in the sample of allozyme studies referred to above which contain respectively 44 and 18 papers for these taxa. The sample of allozyme papers contains relatively more on gastropods (16 papers) and fewer on echinoderms (2 papers).

4.1.4   A survey of DNA techniques

This section is intended as a brief review for aquatic biologists who have little experience of DNA techniques but would like to apply these techniques in their own work. Where possible, reference is made to work on aquatic invertebrates. There are numerous excellent manuals available (e.g. Sambrook, Fritsch and Maniatis, 1989; Hillis and Moritz, 1990). Useful papers describing methods for rapid extraction of mtDNA are those of Lansman *et al.* (1981) and Chapman and Powers (1984). A variety of kits are available commercially for most techniques and these usually come with experimental protocols. Although kits might be relatively expensive, it is a false economy for a beginner to try to get by without, when time lost in the preparation of solutions and the correction of errors is accounted for.

(a)   DNA extraction

A wide variety of methods exist for the extraction of DNA from tissues. Nuclear DNA is usually prepared by treatment with a proteolytic enzyme and an anionic detergent such as SDS. Prior homogenization is not always necessary. DNA is extracted using the organic solvents phenol and chloroform and precipitated from aqueous solution with alcohol. Alternatively, impurities can be removed from the DNA preparation using high concentrations of salt or commercially available ion exchange systems.

Homogenization and differential centrifugation are usually used to prepare partially purified mitochondria. The preparation of supercoiled mtDNA from lysed mitochondria by caesium chloride density gradient centrifugation has been applied widely, e.g. in scallop (Gjetvaj, Cook and Zouros, 1992) and in oysters (Reeb and Avise, 1990). This method gives pure DNA but is time-consuming, and other more rapid techniques have been developed. For example, purified mitochondria can be lysed with SDS and DNA extracted with organic solvents, e.g. in mussels (Skibinski and Edwards, 1987). This gives mtDNA preparations which can be badly contaminated with nuclear DNA. Nuclear contamination can be reduced by lysing the mitochondria with an anionic detergent such as nonidet and removing chromosomal debris by centrifugation prior to DNA extraction, e.g. in lobsters (Brasher, Ovenden and White, 1992). Alternatively, the method of alkaline lysis can be used to selectively precipitate contaminating nuclear DNA (Palva and Palva, 1985). Mitochondrial DNA can also be purified away from nuclear DNA by excising mtDNA bands from agarose gels (Chapman and Powers, 1984). Problems which seem to be common with some organisms, e.g. molluscs, are shearing of DNA by nucleases or failure to get the DNA to cut with restriction enzymes. Shearing can usually be reduced by increasing the concentration of EDTA and avoiding prolonged incubations of homogenized tissue. Rollinson and Kane (1991) were forced to resort to an extreme measure in *Bulinus*. Nuclear DNA was extracted from tissue suspended in agarose plugs, a gentle technique commonly used for preparing very high molecular weight DNA for pulsed field electrophoresis. Problems with restriction cannot always be solved and might reflect the formation of covalent complexes between the DNA and impurities such as polysaccharides during extraction. Incubation with the cationic detergent CTAB prior to organic extraction has been useful in mussels (Fisher and Skibinski 1990), and Stine (1989) reported the removal of mucopolysaccharides in molluscs by centrifugation of homogenate through a sucrose step gradient. The addition of spermidine to restriction enzyme digestions has been recommended, e.g. in rotifers (King, 1989). Low yields of mtDNA seem to be a problem with gastropods, and might account for the paucity of studies in this group. Steinbruck *et al.* (1991) had difficulty in obtaining pure nuclear DNA from *Tisbe* species by phenol chloroform extraction and resorted to caesium chloride centrifugation. A general problem seems to be that the purity and yield of DNA obtained by different techniques vary considerably between tissues and organisms.

(b) Direct visualization of DNA bands

DNA separated on an agarose gel can be stained directly with a fluorescent stain such as ethidium bromide. Restricted nuclear DNA produces an uninformative smear, though distinct bands can sometimes be seen for high copy number repeated sequences. Such a repeated sequence was observed in coral species and the DNA recovered from the gel for cloning (McMillan and Miller, 1990). Mitochondrial DNA, being small and in high copy number, can be analysed by ethidium bromide staining of gels. This approach has been used successfully in a number of population surveys, e.g. in mussels (Edward and Skibinski, 1987) and in oysters (Brown and Paynter, 1991). Some nuclear DNA contamination can be tolerated but high yields of mtDNA are required because of the relatively low sensitivity. Radioactive end-labelling of digested mtDNA followed by visualization of bands by autoradiography allows greater sensitivity, but highly purified DNA is required. DNA purified by caesium chloride centrifugation is ideal and this approach has been prominent in studies of mtDNA variation, e.g. in sea urchins (Palumbi and Wilson, 1990). Some investigators have successfully applied the technique to mtDNA extracted directly from mtDNA pellets with phenol and chloroform, e.g. in lobsters (Brasher, Ovenden and White, 1992). DNA excised directly from gels can also be end-labelled effectively. DNA separated on acrylamide gels can be silver stained (Tegelstrom, 1986). This technique is more sensitive than ethidium bromide staining but higher nuclear backgrounds are obtained.

(c) Filter hybridization

The use of DNA as a probe in the hybridization of filter-bound DNA is one of the most valuable techniques in molecular genetics. Restricted DNA separated on an agarose gel is transferred to a nylon filter by capillary blotting or pressure and then exposed to radio-labelled probe DNA. Binding of the probe to homologous sequences on the filter is then visualized by autoradiography. The technique is several orders of magnitude more sensitive than ethidium bromide staining and thus releases the restriction imposed by limited availability of material in mtDNA analysis. The sensitivity is also sufficient for the detection of single copy nuclear DNA. Moreover, because of the specificity of the technique, the presence of non-homologous DNA causes no problems. Thus it is not necessary for nuclear contamination to be removed from mtDNA preparations. In general the technique is less successful with fragments of DNA less than a few hundred base pairs. Non-radioactive alternatives exist for the labelling of probes. These methods are based on enzyme-linked immunoassay and, though somewhat less sensitive and more time-consuming than radio-active techniques, are safer. Sensitivity is certainly sufficient for the analysis of mtDNA polymorphisms, e.g. in mussels (Fisher and Skibinski, 1990), and for analysis of single copy nuclear DNA, e.g. in stone crabs (Neigel *et al.*, 1991).

The essential requirement of the technique is the DNA probe. In mtDNA analysis, highly purified DNA can itself be used as a probe, if nuclear contamination which might lead to artifactual bands is eliminated. As a result, mtDNA cloned into a phage or plasmid vector is usually used as a probe. Heterologous probes derived from related species can also be used effectively. A disadvantage of filter hybridization is thus that an additional development phase is required to clone DNA from a species of interest. This is less of a problem if bulk preparations of mtDNA can be prepared than if the species under investigation yields little mtDNA, as is often the case with molluscs. In this circumstance a library must be created and screened with a heterologous mtDNA probe, and putative mtDNA clones validated carefully.

With nuclear DNA, homologous probes for DNA of known function are likely to be available only for species which are widely studied in molecular genetics. Aquatic invertebrates do not fall into this category. In this circumstance, heterologous probes from other species can be used to search for genetic markers for use in population genetics and taxonomy. For example, Rollinson and Kane (1991) used a heterologous cloned ribosomal RNA probe to study differences between four species of the planorbid freshwater snail *Bulinus*. Steinbruck *et al.* (1991) tested heterologous probes for six genes for markers for use in species identification and phylogenetics in *Tisbe*. Only three of the six probes gave interpretable banding patterns. An alternative approach is to clone random fragments of nuclear DNA from the species under investigation and use these to search for 'anonymous' restriction site or fragment size polymorphisms. The use of cloned nuclear DNA as probes is a key technique in many areas of genetics, including human population genetics, but has not been prominent in animal population genetics generally. However, examples of the application of the technique are provided by studies on the lesser snow goose (Quinn and White, 1987; Quinn *et al.*, 1987), the honeybee (Hall, 1986, 1990) and the striped bass (Wirgin and Maceda, 1991). In all these organisms allozyme studies had revealed low levels of variation, and it was hoped that the DNA approach would provide additional polymorphic markers. In the studies of bass and snow geese, large, 10–20-kb fragments of DNA were cloned into phage vectors; in the bee studies, smaller fragments were cloned into a plasmid vector. The clones were then screened using a total genomic DNA as a probe to identify those containing primarily unique sequence DNA. The presence of extensive repetitive DNA in the probes would lead to uninterpretable multibanded patterns or smearing on the autoradiograph. Thirty per cent of the bass and over 50% of the snow goose clones contained unique sequence DNA. The unique sequence clones were then used to search for polymorphisms. Of 17 probes tested in both these studies, all revealed polymorphisms in snow goose while only four revealed polymorphisms in bass, presumably reflecting a difference in level of genetic variation between the species. Nevertheless, useful polymorphisms were detected in all species. These were used for studying stock identification in the bass, introgression between African and European strains of honeybees, and maternity and paternity analysis in the snow goose, studies which would have been impossible with allozyme markers alone. There are a few examples of the application of the technique in aquatic invertebrates. Brock and Christiansen (1989) cloned fragments of nuclear DNA from the cockle *Cerastoderma* (*Cardium*) *edule* into a plasmid vector and used these to probe southern blots of restriction-digested nuclear DNA from *C. edule* and two related species. The clones were not screened for repetitive sequences and several gave smearing; however, three out of seven gave patterns that were useful in systematic analysis of the three species. Neigel *et al.* (1991) tested nine randomly chosen clones from a plasmid nuclear library of the stone crab *Menippe adia* in a search for DNA polymorphisms of value in population studies. Prescreening was not carried out and the results obtained suggested that single copy clones occur at sufficiently high frequency for this procedure to be unnecessary. A southern blot of restriction digestions of the nine clones was also probed sequentially with labelled genomic DNA from two related species of stone crab to compare and assess the relative abundance of the repeated sequences between species present within the majority of the clones.

Though there are some technical problems to be solved with the use of random nuclear DNA fragments as probes, they do not seem to be severe ones. It is thus surprising that the technique has not been used more widely in population genetics, as the number of potential markers is essentially unlimited. Possibly those individuals and laboratories with the expertise to carry out such work have found it more profitable to work in other areas.

A number of variant hybridization techniques have been applied in the assay of polymorphisms, primarily in human genetics. For example, if DNA sequence information is available for a polymorphic site, then short (15–20-bp) probes can be made which overlap the site. If there are two alleles, differing by a single base, two probes are made, one for each allele. If DNA from one homozygote is dotted directly onto a filter and hybridized with both probes under appropriate conditions of stringency, a signal will be obtained from the homologous probe only. The heterozygote will give a signal from both probes. Thus all three genotypes can be identified without using restriction enzymes, electrophoresis and conventional blotting (Saiki *et al.*, 1986). The technique is powerful if many markers are available, but considerable development work is required to identify polymorphic sites within regions of known sequence. The technique lends itself to automation and forensic applications, particularly if flanking primers can be used to amplify the target sequence from small DNA samples using the polymerase chain reaction (PCR). Kits are now available which allow the rapid cloning of cDNA from tissue samples. These could also be used for screening for polymorphisms. A problem might be that of complex banding patterns arising from the introns present in genomic DNA. By analogy with the studies of normal and mutant alleles associated with human genetic diseases such as cystic fibrosis, it is likely that cDNA clones will play a role if functional studies on DNA polymorphisms are to be pursued in future.

(d)   Polymerase chain reaction (PCR) techniques

The technique of PCR (Saiki *et al.*, 1988) has had a profound impact on molecular biology and has great potential as a tool for detecting polymorphisms. In PCR, two small oligonucleotide primers are used to amplify a region of DNA flanked by primers. The amplification takes place in a thermocycler and is mediated by a thermostable polymerase. Thus, in principle, if primers are available which flank a polymorphic region of the genome, this region can be amplified and the product restricted and analysed on ethidium-bromide-stained gels. This would be much more rapid than conventional RFLP analysis using hybridization and blotting techniques. The approach could be applied to both nuclear and mitochondrial DNA and could be carried out with minute amounts of starting tissue. Examples of the application of this technique are provided by the work of Karl, Bowen and Avise (1992) on the green turtle *Chelonia mydas* and Karl and Avise (1992) on the oyster *Crassostrea virginica*. In the study of *C. mydas*, random fragments of nuclear DNA in the 500–2000-bp range were cloned into a plasmid vector and screened to identify those containing only low or single copy DNA. Both ends of cloned inserts were then sequenced and primers synthesized so that the DNA sequence between the primers could be amplified by PCR in genomic DNA prepared from individual turtles. Of 15 inserts analysed, seven produced a single band of amplified DNA of the correct size. To identify polymorphisms, DNA was amplified using these seven primer pairs from a sample of nine turtles and restricted with a battery of 40 restriction enzymes. In the end, nine polymorphic restriction sites and three fragment size variants were identified and these were used in a geographic survey. A previous study of mtDNA variation had suggested high substructuring with respect to female lineages. The nuclear DNA study suggested moderate male-mediated gene flow between populations.

This appears to be a simple technique which can be applied to both nuclear and mtDNA. Small amounts of tissue can be used and the polymorphisms scored quickly on ethidium-bromide-stained gels. Some caution is required, however, as noted by Karl, Bowen and Avise (1992). The development stage was time-consuming and expensive. Fifteen primer pairs had to be synthesized, yet only seven worked well. The cloned pieces of DNA have to be much smaller than in the snow goose study because PCR will not amplify large pieces of DNA reliably. Thus the chance of detecting polymorphisms for a given clone is smaller. Also, when the cloned DNA is used as a hybridization probe, polymorphisms flanking the cloned regions can be detected. This is impossible with the PCR technique. Forty restriction enzymes had to be tested to identify the nine polymorphisms. Given that, in total, 8.4 kb of DNA in the seven inserts was screened, this means roughly one polymorphism found per kilobase per 40 enzymes tested. The same approach can be used for studying RFLPs in mtDNA. For example, Hall and Smith (1991) used the technique for distinguishing between African and European honeybees and documenting the spread of African honeybees. The authors note that the technique facilitated the analysis of large numbers of populations. The technique has also been applied to a 360-bp region of the mtDNA cytochrome *b* gene by Bucklin and Cann (1991) to provide markers of species identity in three species of the planktonic copepod *Calanus*. The authors remark that the technique may in future permit the estimation of gene flow in planktonic species populations. Steinbruck *et al.* (1991) also express enthusiasm for the technique after less successful application of heterologous probes in filter hybridization. A 1.8-kb fragment of the 18S RNA gene was amplified, and restriction patterns used to construct a phylogenetic tree of four *Tisbe* species. The development and use of 'universal' primers (Kocher, Thomas and Meyer, 1989), which can be used for amplifying DNA across diverse taxonomic groups, will avoid time-consuming development stages requiring cloning and sequencing.

(e) Analysis of repeated DNA sequences

The use of repeated sequences as hybridization probes for the production of DNA 'fingerprints' is now well established in population biology. Perhaps the biggest impact has occurred as the result of the cloning and analysis of human minisatellite sequences in which alternative alleles at a locus differ in the number of tandem duplications of a single short repeated sequence (Jeffreys, Wilson and Thein, 1985). The minisatellite loci are scattered throughout the genome and, though heterogeneous in base sequence, share a core sequence which is believed to enhance mutation rate in repeat number. When the core is used as a hybridization probe at low stringency it reveals a complex multilocus banding pattern, highly variable between individuals, arising from many multiallelic loci. When an entire minisatellite repeat is used as a probe at high stringency a single-locus pattern is revealed in which homozygotes have single bands, and heterozygotes two bands because of repeat number difference between alleles.

The cloned human minisatellites have been used as heterologous probes in a wide variety of species where they reveal complex multibanded patterns. For example, Hauser *et al.* (1992) used this approach to demonstrate that British populations of the freshwater snail *Potamopyrgus antipodarum* consist of three distinct parthenogenetic clones. Because of the high level of polymorphism, individual specific fingerprint patterns can be used for exclusion purposes in the study of parentage, and to this end have been used with great success in birds (Burke, 1989; Weatherhead and Montgomerie, 1991). Many different short DNA sequences also reveal complex DNA fingerprints when used as hybridization probes. Turner, Elder and Laughlin (1991) found that in the obligate selfer *Rivulus marmoratus* a GACA repeat probe gave a different pattern for almost every fish examined. The patterns were inherited stably and a high mutation rate was implicated as a cause of the diversity. Allozyme studies on the same fish species failed to detect clonal variation.

Multilocus fingerprint patterns suffer a number of disadvantages. In many organisms it appears difficult to obtain satisfactory fingerprints and the results are often inconsistent. The resolution of these problems can be demanding in terms of time and resources. Also, the multilocus patterns are not very amenable to genetic interpretation; bands cannot easily be assigned to loci. Thus the technique is of limited value in population genetics. This is, of course, not a limitation in the use of single-locus probes, where genotypes can be scored and allele frequencies estimated. The contrast between multilocus and single-locus probes is thus reminiscent of that between isozymes and allozymes. Bands attributable to single loci can sometimes be dissected out of multilocus patterns. Amos, Barrett and Dover (1991) used this approach successfully in a study of pod structure in pilot whales using human probes. A disadvantage of the single-locus approach is that minisatellite loci must be identified by screening a genomic library for each new species. The development effort can, however, be offset by the discovery of useful highly polymorphic loci. Minisatellite loci evolve rapidly and it is assumed that they will have limited use in phylogenetic studies. Some loci are monomorphic, others have tens of alleles. The very highly polymorphic loci may be of less value in studies of population differentiation, and especially species differences, than those with smaller numbers of alleles because of their rapid rate of evolution.

The applications of minisatellite variation were extended by Jeffries *et al.* (1991) by the development of digital DNA fingerprinting. Repeat units within a minisatellite allele were found to differ in base sequence, and by using PCR primers specific for different repeats, the internal variation of alleles could be characterized as a digital code. The codes were sufficiently variable to uniquely identify every individual in a population. It appears that there are regional differences in levels of polymorphisms along minisatellite alleles which may facilitate their use in the study of both population and species differentiation.

Microsatellites are permutations of simple repeat motifs such as dinucleotides up to a few hundred base pairs in length, dispersed throughout the genome but occurring usually in unique sequence DNA every few kilobases (Tautz, 1989). Microsatellite alleles vary in length, and high levels of polymorphism at these loci are thought to be due to replication slippage (Schlotterer and Tautz, 1992). Microsatellite loci are isolated by using repeat sequence probes to screen a genomic library. Clones are sequenced in order to make PCR primers flanking the repeat region. If one primer is radioactively labelled, alleles differing in length by only a single base pair can be detected on sequencing gels. Alternatively, the amplified products can be visualized by ethidium bromide staining. The technique is faster than blotting and hybridization, and loci with high heterozygosity can be obtained (Weber and May, 1989). Once flanking primers are developed, they may be useful in related species. For example, Schlotterer, Amos and Tautz (1991) discovered high conservation of microsatellite sequences in 11 whale species. Moore *et al.* (1991) observed that a majority of cattle primers worked successfully in sheep but not in humans. Because they occur so frequently within the genome, microsatellites provide an inexhaustible source of variation not only in population genetics but also in linkage studies (e.g. Love *et al.*, 1990; Hudson *et al.*, 1992). As with minisatellite variation the principal disadvantage appears to be that a development phase of cloning and sequencing may be required.

Repeated sequences have great potential for use as markers in population studies because of their high levels of polymorphism. However, they experience mutational processes more complex than the simple base substitutions which may well underly most allozyme variation. These processes include those responsible for concerted evolution (Dover, 1986), such as unequal crossing over, replication slippage and biased gene conversion, and may cause rapid evolution of DNA sequences and differentiation of populations in the face of gene flow. The observation that seven populations of the fish *Cyprinodon variatus* each possessed a different diagnostic variant of a 170-bp repeat sequence is of relevance in this context (Turner, Elder and Laughlin, 1991). Mutation rate does not appear in the frequently applied equations derived from neutral theory which relate migration rate to extent of geographic differentiation. It can be disregarded provided it is very much less than migration rate (Hartl and

Clark, 1989). However, the mutational properties and dynamics of repeated sequences may need to be considered carefully when this type of variation is used in studies of population structure. Crease and Lynch (1991), in a study of ribosomal DNA repeat variation within and between individuals and populations of *Daphnia pulex*, reported that the distribution of repeat types was not in accord with expectations based on theoretical models developed for sexual organisms. Thus it might be difficult to formulate generalizations about how this type of variation can be used to estimate population parameters, given the variety of reproductive modes of aquatic invertebrates.

An example of the use of repeated sequences in taxonomic analysis is provided by studies on the coral subgenus *Acrospora* (McMillan and Miller, 1989, 1990; McMillan *et al.*, 1991). Members of this group are widely distributed in the Indian, Pacific and Atlantic Oceans. The genus shows great phenotypic diversity, arising partly from environmentally induced growth form variations, and the methods of classical taxonomy based on morphology have led to the description of approximately 150 species. A highly repeated DNA sequence of about 120 bp is found throughout the *Acrospora* but is absent from related genera. In *A. formosa* it was discovered that this repeated sequence, which is about 120 bp long, accounts for 5% of the genome and occurs in several hundred thousand copies, tandemly repeated. Copies of the repeat were cloned for seven species and three techniques were used to obtain taxonomic information. First, nuclear DNA from each of the species was blotted onto a filter and hybridized with a cloned *A. formosa* repeat. The variation in the intensity of signal at high stringency was used as a measure of sequence similarity of the repeats between the species. Second, nuclear DNA from each species was digested with the restriction enzyme *Taq*I, separated on an agarose gel, southern blotted and hybridized as above. *Taq*I has either one or two cut sites within individual repeats, the number of cut sites varying between repeats both within and between species. As a result, the banding patterns differ between species and can again be used for taxonomic purposes. Third, the cloned repeats were sequenced and phylogenetic techniques were used to construct a tree. Sequence variation between repeats was observed both within and between species. The different methods of DNA analysis produced generally consistent results. The relationships between species based on the DNA data showed major differences from those based on gross morphology, but agreed more closely with those based on sperm morphology. An interesting feature of the sequence data is that the repeats contained a short conserved motif homologous with the chi sequence in *E. coli* which is a recombination signal. This sequence also occurs within the human minisatellite repeats (Jeffreys, Wilson and Thein, 1985) which suggests that unequal crossing over might be involved in the concerted evolution of repeats within species. The evolutionary dynamics of these repeats might complicate the application of methods of phylogenetic analysis which assume single base substitution in homologous sequences between species.

(f) DNA sequencing

Until the development of PCR, DNA sequencing required the cloning of the target DNA into a phage or plasmid vector. Amplification of the cloned DNA in the vector then provided the relatively large quantities of DNA required by sequencing protocols. The use of sequencing in population genetic analysis thus involved the cloning of the homologous DNA region from every individual sampled in a geographic survey. This time-consuming work has been carried out with great success in *Drosophila* (Kreitman and Hudson, 1991) but the resources required probably placed it beyond the reach of many laboratories. Thus sequencing work was orientated towards phylogenetic and evolutionary studies, where it was sufficient to clone and sequence the DNA regions from only one individual representing each species under study.

The development of PCR lead to the technique of direct sequencing, which bypasses the need for cloning vectors (Wrischnik *et al.*, 1987). In this technique PCR is used to provide sufficient amplified mtDNA to support a sequencing analysis using either one of the PCR primers or a third internal primer. The amplified DNA can be re-amplified using one primer in excess (Gyllensten and Ehrlich, 1988). This provides a single-strand template, which improves the quality of the sequencing results. The technique has been applied successfully to analysis of mtDNA regions in population and phylogenetic studies. For example, Bartlett and Davidson (1991) amplified and sequenced a 307-bp region of the cytochrome *b* gene from four tuna (*Thunnus*) species and discovered both intraspecific and interspecific variation. Species-specific markers of potential importance in enforcing fishing regulations were also identified. An advantage of the technique, a feature of PCR, is that it can be applied with very small tissue samples, e.g. those obtained from fin clippings or skin scrapings. A further advantage is that because only intact DNA can be sequenced the technique avoids potential artefacts in cloned ancient DNA induced by bacterial repair systems (Paabo and Wilson, 1988). Thomas *et al.* (1990) demonstrated its use in the analysis of museum specimens of three subspecies of kangaroo rat *Dipodomys panamintinus* in which DNA was extracted from dried skin. Identical phylogenies were obtained for modern and museum specimens, thus validating the technique.

Direct sequencing does not completely bypass the need for cloning vectors because sequence information is required to construct the primers. In many cases, however, 'universal' primers for highly conserved regions of the genome will work successfully with quite distantly related species (Kocher, Thomas and Meyer, 1989). To be useful in population genetic analysis it is necessary that these conserved primers flank a variable region. With a single primer pair the technique provides information on only a small region of the genome, and thus the data obtained will not be representative of the entire genome as in a conventional RFLP analysis. A possible limitation of the technique seems to be that it cannot always be applied in the population genetic analysis of nuclear DNA because of uncertainty in the reliable detection of heterozygotes on sequencing gels. This is unlikely to be a problem when the technique is used in phylogenetic analyses in which intraspecific variation is disregarded.

(g) DNA–DNA hybridization

In DNA–DNA hybridization, DNA of two species is annealed in solution and the nucleotide sequence similarity of the species is estimated from the melting profile of the hybrid DNA. Repetitive DNA which would obscure sequence similarity is first removed from each species as a rapidly annealing fraction. If the species are distantly related the base pairing in the hybrid DNA will be less accurate than if the species are closely related, and the DNA will melt at a lower temperature. A one degree difference in melting temperature corresponds to about a 1% difference in base sequence.

At first sight the technique seems ideal for phylogenetic analysis, for it incorporates the overall base sequence difference of the single copy DNA of the genomes of two species in a single measure. Unfortunately it seems that considerable expertise is needed to get the technique to work well; replicate experiments can often give different answers. The technique is most usefully applied to distantly related species; it is not sufficiently accurate for comparisons of intraspecific populations. DNA–DNA hybridization has been used with success in the systematics of birds (Sibley and Alquist, 1981) and primates (Sibley and Alquist, 1987) but because of the problems mentioned above has not been applied widely. However, some examples exist for aquatic invertebrates. The technique has been used in a phylogenetic study of mytilinids by Milyutina and Petrov (1989). Four species of *Mytilus* and *Crenomytilus grayanus* were compared with two more distantly related genera represented by *Modiolus modiolus* and *Septifer keenae*. The branching of the phylogenetic tree for the *Mytilus* species and *Modiolus* was consistent with that obtained from allozyme data. The results suggested variable rates of evolution in different parts of the tree, e.g. more rapid evolution in *Mytilus* than in *Crenomytilus*. Palumbi and Metz (1991) used both restriction site variation of mtDNA and DNA–DNA hybridization of nuclear DNA to study the phylogenetic relationships between four morphotypes of the sea urchin *Echinometra*. The branching pattern of the phylogenetic tree was identical with both techniques and both gave similar estimates of nucleotide sequence divergence between the morphotypes.

A technique which has a similar aim as solution hybridization is the use of conventional filter hybridization with the target DNA filter bound and labelled probe DNA in solution. Brock and Christiansen (1989) used total nuclear DNA from *Mytilus edulis* and *Mya arenaria* as hybridization probes of blots of restricted DNA from three related species of cockle. Multibanded patterns were obtained which could not be interpreted genetically but allowed conclusions to be drawn, on the basis of band sharing, about the systematic relationships of the three species. Because one DNA is filter bound and some sequences inaccessible to the probe, it is perhaps difficult to draw quantitative conclusions about nucleotide sequence divergence.

(h)   RAPD analysis

The random amplification of polymorphic DNA (RAPD analysis) (Williams *et al.*, 1990; Welsh and McClelland, 1990; Hadrys, Balick and Schierwater, 1992) for review of applications in population biology) is a recent PCR development. Genomic DNA is used as template in amplification reaction with a single short oligonucleotide primer which binds at random throughout the genome. A characteristic banding pattern which can be visualized on an ethidium-bromide-stained agarose gel will be produced as a result of amplification at all those sites in the genome where two primers are close enough together for PCR to work. Base sequence differences between individuals at the primer sites will result in polymorphisms for the presence or absence of bands. Ten-based-pair primers are optimal in producing interpretable patterns of up to 20 bands. Shorter primers produce multibanded patterns similar to multilocus DNA fingerprint patterns (Caetano-Anolles, Bassam and Gresshof, 1991). Different ten-base primers produce different patterns, and thus the number of possible markers is potentially unlimited.

The technique does not involve radioactivity, hybridization, restriction enzymes or sequencing, and because of its simplicity and rapidity has been hailed as the DNA equivalent of allozyme electrophoresis. Applications of the technique include its use to detect cultivar- and species-specific DNA markers in *Brassica* (Quiros *et al.*, 1991) and tomato *L. esculentum* (Klein-Lankhorst *et al.*, 1991). In both these studies markers were located to specific chromosomes by analysing chromosome addition or substitution lines. The technique has been used in the study of introgression in irises following the identification of species-specific markers (Arnold, Buckner and Robinson, 1991) and has been used to identify DNA markers closely linked to disease resistance genes in lettuce (Michelmore, Paran and Kesseli, 1991). Hadrys, Balick and Schierwater (1992) argue that the technique may have potential in phylogenetic analysis because of the possibility of obtaining specific markers, represented by fixed bands, at the level of the species and genus. The technique has the potential to be extended in many ways (Hadrys, Balick and Schierwater, 1992). The amplification reaction could be restricted before electrophoresis to identify RFLPs. RAPD bands could be purified from gels and used as hybridization probes for establishing band homologies or could be cloned and partially sequenced to generate band-specific primers for use in PCR-based RFLP analysis.

The RAPD banding pattern is superficially similar to a multilocus minisatellite pattern in that it is generated by a number of loci. There is no problem in assigning bands to loci, if co-migration is disregarded, as each band of different mobility will arise from a different region of the genome. Although heterozygotes may have more bands on average than homozygotes, the presence of a band is a dominant characteristic, and thus heterozygotes cannot be scored or allele frequencies estimated directly. This will create difficulties in population genetic analysis of outbreeding organisms, though the technique is well suited to the study of clonal organisms.

The major problem with RAPD analysis is that it is very sensitive to the PCR conditions employed and artifactual bands are easily generated. This has proved to be a problem in paternity analysis (Riedy, Hamilton and Aquadro, 1992), and might create difficulties in studies estimating similarity through band-sharing coefficients. The technique is perhaps most suited to organisms in which crosses can be easily made in the laboratory, which is a drawback for studies of most aquatic invertebrates.

(i)   Denaturing gradient gel electrophoresis (DGGE)

In the technique of DGGE (Myers, Maniatis and Lerman, 1987), PCR amplification of a polymorphic DNA region is followed by electrophoresis on a gel containing an increasing linear gradient of denaturing agents such as urea and formamide. The point in the gel at which the double-stranded DNA is denatured, and its electrophoretic mobility reduced, depends on the base sequence of the DNA and the presence of mispaired bases. Losekoot *et al.* (1990) applied the technique in the identification of β-thalassaemia mutations. The single base pair change from T to A which causes the difference between the normal *HbA* and mutant *HbS* alleles could be detected as a mobility difference following amplification and electrophoresis of a 300-bp fragment from the two homozygotes. Towards the end of the PCR amplification, two different heteroduplex molecules can be formed between DNA from different alleles. Mispairing in the heteroduplex causes melting at a lower concentration of denaturant, and hence heterozygotes produce four bands which can aid in their identification. Although DGGE can only be applied to small fragments, this may not be a disadvantage because the technique is not limited to the detection of base pair changes at restriction enzyme sites. It is also rapid because DNA bands can be visualized by ethidium bromide staining. However, development work is required to identify the optimal gradient conditions for each fragment and especially long PCR primers are needed in some applications of the technique. Moreover, it is possible that different base sequence changes will produce bands of identical mobility, and thus the resolution not be as high as with direct sequencing. If the alleles are cloned and sequenced, then information relating to the phylogeny of alleles may also be extracted. This seems a promising technique, though it has not yet been widely applied in population genetics. An example of its use is the study by Lessa (1992) of geographic variation in the frequency of three alleles for a β-globin intron in eight populations of pocket gophers *Thomomys bottae*.

In a related technique, single-strand conformation polymorphism (SCCP) (Orita *et al.*, 1989), PCR-amplified DNA fragments up to a few hundred base pairs in length are heated to dissociate them into single strands prior to electrophoresis on a non-denaturing poly-acrylamide gel. The two single strands of a homoduplex, differing in base sequence, will adopt slightly different conformations and thus migrate at different rates in the gel. Thus a homozygote will have two bands. In a variant homozygote with a single base substitution the two strands will have different conformations and the position of the two bands will be shifted. Heterozygotes will thus have four bands. Kirkpatrick (1992) successfully used this technique with silver staining to resolve genotypes for a three-allele polymorphism in a 339-bp intron region of the bovine growth hormone gene.

4.1.5   A case study of aquatic invertebrates: mtDNA analysis of marine bivalves

Marine bivalves have been widely studied using allozyme approaches, and thus it is not surprising that DNA analysis is also advanced in this group of organisms. Skibinski (1985) and Edwards and Skibinski (1987) used a small number of six-base restriction enzymes to survey mtDNA variation within the zone of intergradation of the mussels *Mytilus edulis* and *M. galloprovincialis*. Phenol chloroform extraction combined with ethidium bromide staining of agarose gels was used to assay variation. Because of nuclear contamination, a whole mtDNA genome from *Mytilus* was later cloned into the phage lambda vector EMBL3 (Skibinski and Edwards 1987) for use as a hybridization probe in Southern blots of restricted mtDNA (Blot, Lendre and Albert, 1990; Gardner and Skibinski, 1991). The results of this work showed that in contrast to the results obtained with many small animals, allopatric populations did not have unique mtDNA variants but instead were polymorphic for shared variants. Thus, as with allozyme loci, populations differed in the frequency of genotypes possessed by all populations and no perfectly diagnostic genotypes were found for the two forms of mussel. This situation demands the analysis of samples sufficiently large to accurately establish frequencies within populations. Some hybrid populations with intermediate allozyme frequencies had a disproportionately high frequency of those mtDNA genotypes predominating in *M. edulis*. This observation made possible some deductions about the direction of larval dispersal which would have been impossible with allozyme data alone. At the allozyme level, *M. galloprovincialis* from the Mediterranean and from Britain are similar to each other but distinct from British *M. edulis*. By contrast, British *M. galloprovincialis* and *M. edulis* have similar mtDNA genotype frequencies (Skibinski, Fisher and Edwards, unpublished data). This is consistent with the introgression of *M. edulis* mtDNA genotypes into *M. galloprovincialis* following its immigration from the Mediterranean.

Mitochondrial DNA prepared from whole body homogenates exhibits a high degree of heteroplasmy (Fisher and Skibinski, 1990; Hoeh, Blakley and Brown, 1991). Heteroplasmic individuals occurred at frequencies exceeding 50% in some populations. The most unusual feature of the heteroplasmy is that it is confined almost exclusively to males, which explains why it was not observed initially in preparations from female gonad (Fisher and Skibinski, 1990). In British populations the majority of heteroplasmic individuals possess two distinct genomes (F and M genomes) which exhibit over 10% sequence divergence, suggesting that the heteroplasmy might be ancient. The M genome has so far been observed only in heteroplasmic individuals. Paternal leakage of mtDNA has been favoured as an explanation for heteroplasmy and the use of PCR has now provided evidence for a low level of paternal transmission in mice and fruit flies (Gyllensten *et al.*, 1991; Kondo, Matsuura and Chigusa, 1992). Crosses have also been performed in *Mytilus*, from which direct evidence for paternal transmission has been obtained (Zouros *et al.*, 1992). The sex difference in *Mytilus* suggests that, following fertilization, the M genome is amplified selectively. Evidence for selective effects on mtDNA transmission and dynamics is growing (e.g. Matsuura, 1991) though the existence of sex-specific heteroplasmy has not so far been reported in other organisms. Recently over 80% of the F genome cloned by Skibinski and Edwards (1987) has been sequenced and the arrangement of the genes determined (Hoffman, Boore and Brown, 1992). The gene order is quite unlike that in any other known metazoan mtDNA, and the molecule is unique in having an additional tRNA gene.

A study which throws light on the forces influencing genetic variation has been carried out with the American oyster *Crassostrea virginica*. This was a population survey of 212 oysters collected from continuously distributed populations from Canada to Texas (Reeb and Avise, 1990). Mitochondrial DNA was purified using caesium chloride centrifugation and digested with 13 restriction enzymes with six-base recognition sites, and fragments were radioactively end-labelled. Bands were visualized by autoradiography following separation on agarose gels. The genetic data were used to construct a dendrogram in which the mtDNA genotypes fell into two distinct groups, one on Atlantic coasts and the other within the Gulf of Mexico, with a 'breakpoint' on the coast of Florida. It was suggested that vicariant processes in the past initiated separation of the two groups, which evolved in isolation, and that the pattern of ocean currents and ecological influences minimize gene flow despite the high dispersal capability of the organism. This hypothesis is supported by evidence of a breakpoint at the same locality in the horseshoe crab (Saunders, Kessler and Avise, 1986). The clear evidence of a breakpoint in oysters contrasts with the geographic homogeneity observed in allele frequencies at several allozyme loci. Reeb and Avise (1990) consider a number of hypotheses for the discrepancy between the allozyme and mtDNA results. One possibility is that balancing selection is maintaining allozyme frequencies at similar values throughout the species range. Another is that because the effective population size is greater for nuclear loci, allozyme frequencies drift apart more slowly than mtDNA genotype frequencies. The study was extended by Karl and Avise (1992) to include four nuclear DNA polymorphisms detected by restriction enzyme analysis of PCR-amplified DNA. The pattern of population differentiation for the nuclear DNA loci resembles that obtained with mtDNA in showing a breakpoint with sharp differentiation between the Atlantic coast and the Gulf of Mexico. The pattern contrasts sharply with that for allozyme loci, which are also coded for by nuclear genes. It was concluded that the allozyme and nuclear DNA loci are affected by different evolutionary forces. The explanation favoured is that while the nuclear DNA loci are neutral and free to diverge as a result of drift north and south of the breakpoint, the allozyme polymorphisms are maintained globally by balancing selection.

A similar observation has been made in another study of this species by Brown and Paynter (1991). *C. virginica* is important commercially but in recent years production has declined, possibly as a result of disease or overexploitation. Because of this, selection and crossbreeding programmes are underway. In order to assess the potential impact on levels of genetic variation of populations founded from small numbers of individuals, levels of mtDNA variation were compared in native populations with a line inbred for eight to ten generations and with an estimated effective population size of 15. Following an analysis of 166 oysters, very clear-cut mtDNA haplotype frequency differences were observed between the native and inbred oysters, even though allozyme differences between the populations were reported to be small. The favoured explanation invoked genetic drift and the smaller effective population size for mtDNA than for nuclear alleles.

As with the majority of higher animal taxa examined in population studies, the size of the mtDNA molecule in both *Crassostrea virginica* and *Mytilus edulis* is about 17 kb with occasional length variants. However, mtDNA genomes of increased length have been reported in many higher animals (Rand and Harrison, 1989), and are usually the result of tandem duplication of a repeated element. An unusually large mtDNA genome has been reported for the scallop *Placopecten magellanicus* (Snyder *et al.*, 1987; LaRoche *et al.*, 1990). The average size is about 34 kb but is highly variable as a result of a tandem duplication of a 1.45-kb repeated unit, which can occur in from two to eight copies per genome. Molecular characterization involved the restriction mapping and sequencing of the entire repeat. This information allowed some deductions to be made about the properties of the repeat, e.g. that it has the capability of producing DNA curvature and cruciform structures. It was concluded that unequal crossing over or replication errors might be responsible for the variation in repeat number. These studies led to a population survey of mtDNA variation in *P. magellanicus* and six other scallop species (Gjetvaj, Cook and Zouros, 1992). Homology between repeat units from different species was assessed by Southern blotting and hybridization with either cloned repeats or with DNA fragments excised from low melting point agarose gels and labelled directly by random primer synthesis. The survey indicated that one species, *Argopecten irradians*, was quite unlike *P. magellanicus* in having a mtDNA genome invariant in size while the remaining species all had unusually large genomes and displayed varying degrees of size polymorphism.

The analysis of mtDNA variation in bivalves has thus revealed a number of unusual phenomena such as size variation, heteroplasmy and paternal transmission, as well as providing new insights in conventional population genetics analysis. What is clear in this group is that the adoption of DNA techniques, for whatever reason, has revealed a range of new and interesting scientific problems.

4.1.6   Conclusions

An ideal DNA technique from the viewpoint of the experimental population geneticist is one which matches protein electrophoresis in terms of the number of individual animals which can be processed. There is the expectation that the technique would be highly sensitive in detecting genetic differences between individuals, and a bonus might be that it provides information on linked loci and the phylogeny of alleles or genotypes. Out of the wide variety of techniques available, the PCR approach used by Karl, Bowen and Avise (1992) looks very promising for an aquatic biologist entering the field without experience of DNA methods. Once polymorphic loci are detected they can be screened almost as rapidly as allozyme polymorphisms. Moreover, the possibility of analysing closely linked markers provides a powerful new dimension. Cloning DNA fragments and sequencing the ends is not now really difficult given the availability of excellent kits. In future the cost of synthesizing primers will no doubt decrease and the technique of PCR will become increasingly reliable, with the possibility of amplifying larger fragments and thus reducing the expense and effort required to identify polymorphisms. In general, the technique would appear to have high potential in the study of introgression between closely related species, and levels of genetic variation may well be higher in invertebrates than in turtles. If this optimism for the technique is at all justified then the days of blotting and hybridization are surely numbered.

The development of the RAPD technique has been hailed as the DNA equivalent of protein electrophoresis. However, there must be considerable doubts about its suitability for population genetic problems which depend on clear genetic interpretations. One possible extension is to combine RAPD analysis with restriction enzyme analysis to simulate the amplification of cloned and sequenced random nuclear DNA fragments. In interpopulation studies, large, consistently amplified species-specific bands would be ideal targets.

A major question mark hanging over the use of anonymous DNA markers in population studies is that the function and mutational properties of the underlying DNA regions are unknown. Allozyme loci, by contrast, are labelled with specific enzyme or protein functions and it is likely that alternative alleles have arisen through simple base substitutions. Thus they exhibit less complexity and variety. Nuclear DNA studies in the region of the *Adh* locus in *Drosophila* have demonstrated the value of combining DNA and allozyme approaches (Kreitman and Hudson, 1991) as has the study on oysters (Karl and Avise, 1992). It is to be hoped that this example is followed in other organisms so that patterns of variation in the growing body of DNA data can be set beside and considered in the light of the growing body of allozyme data.

Acknowledgements

I am grateful to colleagues and students in the School of Biological Sciences, University of Wales, Swansea, for discussing DNA techniques with me. I thank Mathew Woodwark for preparing the information on allozyme studies used in this section.

References

Amos, B., Barrett, J. and Dover, G.A. (1991) Breeding behavior of pilot whales revealed by DNA fingerprinting. *Heredity*, **67**, 49–55.

Arnold, M.L., Buckner, C.M. and Robinson, J.J. (1991) Pollen mediated introgression and hybrid speciation in Louisiana irises. *Proc. Natl. Acad. Sci.* USA, **88**, 1398–402.

Avise, J.C. (1991) 10 unorthodox perspectives on evolution prompted by comparative population genetic findings on mitochondrial DNA. *Annu. Rev. Genet.*, **25**, 45–69.

Avise J.C., Arnold, J., Ball, R.M. *et al.* (1987) Intraspecific phylogeography: the mitochondrial DNA bridge between population genetics and systematics. *Annu. Rev. Ecol. Syst.*, **1987**, 489–522.

Barbieri, R., Duro, G., Izzo, V. *et al.* (1990) Polymorphisms in the intergenic region of the sea-urchin *Paracentrotus lividus* ribosomal DNA. *Cell Biol. Int. Rep.*, **14**, 917–26.

Bartlett, S.E. and Davidson, W.S. (1991) Identification of *Thunnus* Tuna species by the polymerase chain reaction and direct sequence analysis of their mitochondrial cytochrome b genes. *Can. J. Fish. Aquat. Sci.*, **48**, 309–17.

Batuecas, B., Garesse, R., Cajella, M., Valverde, J.R. and Marco, R. (1988) Genome organisation of *Artemia* mitochondrial DNA. *Nucleic Acids Res.*, **16**, 6515–29.

Bermingham, E., Lamb, T. and Avise, J.C. (1986) Size polymorphism and heteroplasmy in the mitochondrial DNA of lower vertebrates. *J. Hered.*, **77**, 249–52.

Blot, M., Lendre, B. and Albert, P. (1990) Restriction fragment length polymorphism of mitochondrial DNA in sub-Antarctic mussels. *J. Exp. Mar. Biol. Ecol.*, **141**, 79–86.

Brasher, D.J., Ovenden, J.R., Booth, J.D. and White, R.W.G. (1992) Genetic subdivision of Australian and New-Zealand populations of *Jasus verreauxi* (Decapoda, Palinuridae) – preliminary evidence from the mitochondrial genome. *N. Z. J. Mar. Freshwat. Res.*, **26**, 53–8.

Brasher, D.J., Ovenden, J.R. and White, R.W.G. (1992) Mitochondrial DNA variation and phylogenetic relationships of *Jasus* spp (Decapoda, Palinuridae). *J. Zool.*, **227**, 1–16.

Brock, V. and Christiansen, G. (1989) Evolution of *Cardium* (*Cerastoderma*) *edule*, *Cardium lamarcki* and *Cardium glaucum* – studies of DNA variation. *Mar. Biol.*, **102**, 505–11.

Brown, B.L. and Paynter, K.T. (1991) Mitochondrial DNA analysis of native and selectively inbred Chesapeake Bay oysters, *Crassostrea virginica*. *Mar. Biol.*, **110**, 343–52.

Bucklin, D. and Kann, L. (1991) Mitochondrial DNA variation of copepods–markers of species identity. *Biol. Bull.*, **181**, 357.

Burke, T. (1989) DNA fingerprinting and other methods for the study of mating success. TREE, **4**, 139–44.

Burton, R.S. and Feldman, M.W. (1983) Physiological effects of an allozyme polymorphism: glutamate–pyruvate transaminase and response to hyperosmotic stress in the copepod *Tigriopus californicus*. *Biochem. Genet.*, **21**, 239–51.

Caetano-Anolles, G., Bassam, G.J. and Gresshof, P.M. (1991) High resolution DNA amplification fingerprinting using very short arbitrary oligonucleotide primers. *Biotechnology*, **9**, 553–6.

Cantatore, P., Roberti, M., Rainaldi, G., Gadaleta, M.N. and Saccone, C. (1989) The complete nucleotide sequence, gene organisation, and genetic code of the mitochondrial genome of *Paracentrotus lividus*. *J. Biol. Chem.*, **264**, 10965–75.

Chapman, R.W. and Powers, D.A. (1984) *A method for the rapid isolation of mitochondrial DNA of fishes*. Technical Report, Maryland Sea Grant Program, H.J. Patterson Hall, Room 1222, College Park, Maryland 20742, Publication number, Um-SG75-84-05.

Cohen, B.L., Balfe, P., Cohen, M. and Curry, G.B. (1991) Molecular evolution and morphological speciation in North Atlantic brachiopods (*Terebratulina* spp). *Can. J. Zool.*, **69**, 2903–11.

Crease, T.J. and Lynch, M. (1991) Ribosomal DNA variation in *Daphnia pulex*. *Mol. Biol. Evol.*, **8**, 62.

Crease, T.J., Lynch, M. and Spitze, K. (1990) Hierarchical analysis of population genetic variation in mitochondrial and nuclear genes of *Daphnia pulex*. *Mol. Biol. Evol.*, **7**, 444–58.

Crease, T.J., Stanton, D.J. and Hebert, P.D.N. (1989) Polyphyletic origins of asexuality in *Daphnia pulex*. 2. Mitochondrial DNA variation. *Evolution*, **43**, 1016–26.

Cunningham, C.W., Blackstone, N.W. and Buss, L.W. (1992) Evolution of king crabs from hermit crab ancestors. *Nature*, **355**, 539–42.

Cunningham, C.W., Buss, L.W. and Anderson, C. (1991) Molecular and geologic evidence of shared history between hermit crabs and the symbiotic genus *Hydractinia*. *Evolution*, **45**, 1301–16.

Degiorgi, C. (1988) Mitochondrial DNA polymorphism in the eggs of the sea-urchin *Arbacia lixula*. *Cell Biol. Int. Rep.*, **12**, 407–12.

Degiorgi, C., Lanave, C., Musci, M.D. and Saccone, C. (1991) Mitochondrial-DNA in the sea urchin *Arbacia lixula* – evolutionary inferences from nucleotide-sequence analysis. *Mol. Biol. Evol.*, **8**, 515–29.

Dover, G.A. (1986) Molecular drive in multigene families: how biological novelties arise, spread and are assimilated. *TIG*, **2**, 159–65.

Durica, D.S., Garza, D., Restrepto, M.A. and Hryniewicz, M.M. (1989) DNA-sequence analysis and structural relationships among the cytoskeletal actin genes of the sea-urchin *Strongylocentrotus purpuratus*. *J. Mol. Evol.*, **28**, 72–86.

Edwards, C.A. and Skibinski, D.O.F. (1987) Genetic variation of mitochondrial DNA in mussel (*Mytilus edulis* and *Mytilus galloprovincialis*) populations from South West England and South Wales. *Mar. Biol.*, **94**, 547–56.

Fisher, C. and Skibinski, D.O.F. (1990) Sex-biased mitochondrial DNA heteroplasmy in the marine mussel *Mytilus*. *Proc. R. Soc. Lond. Ser. B.*, **242**, 149–56.

Fu, Y., Hirayama, K. and Natsukari, Y. (1991) Genetic divergence between S-type and L-type strains of the rotifer *Brachionus plicatilis* Muller, O.F. *J. Exp. Mar. Biol. Ecol.*, **151**, 43–56.

Gardner, J.P.A. and Skibinski, D.O.F. (1991) Mitochondrial DNA and allozyme covariation in a hybrid mussel population. *J. Exp. Mar. Biol. Ecol.*, **149**, 45–54.

Gjetvaj, B., Cook, D.I. and Zouros, E. (1992) Repeated sequences and large-scale size variation of mitochondrial DNA – a common feature among scallops (Bivalvia, Pectinidae). *Mol. Biol. Evol.*, **9**, 106–24.

Gyllensten, U. and Erlich, H. (1988) Generation of single-stranded DNA by the polymerase chain reaction and its applications to direct sequencing of the HLA DQa locus. *Proc. Natl. Acad. Sci. USA*, **85**, 7652–56.

Gyllensten, U., Wharton, D., Josefsson, A. and Wilson, A.C. (1991) Paternal inheritance of mitochondrial DNA in mice. *Nature*, **352**, 255–7.

Hadrys, H., Balick, M. and Schierwater, B. (1992) Applications of random amplified polymorphic DNA (RAPD) in molecular ecology. *Mol. Ecol.*, **1**, 55–63.

Hall, H.G. (1986) DNA differences found between Africanised and European honeybees. *Proc. Natl. Acad. Sci.*, USA, **83**, 4847–77.

Hall, H.G. (1990) Parental analysis of introgressive hybridization between African and European honeybees using nuclear DNA RFLPs. *Genetics*, **125**, 611–21.

Hall, H.G. and Smith, D.R. (1991) Distinguishing African and European honeybee matrilines using amplified mitochondrial DNA. *Proc. Natl. Acad. Sci., USA*, **88**, 4548–52.

Hartl, D.L. and Clark, A.G. (1989) *Principles of Population Genetics*, 2nd edn, Sinauer Associates, Inc., Sunderland, Massachusetts.

Hauser, L., Carvalho, G.R., Hughes, R.N. and Carter, R.E. (1992) Clonal structure of the introduced freshwater snail *Potamopyrgus antipodarum* (Prosobranchia: Hydrobiidae), as revealed by DNA fingerprinting. *Proc. Roy. Soc. Lond.*, **B**, **249**, 19–25.

Hillis, D.M. and Moritz, C. (1990) *Molecular Systematics*, Sinauer Associates, Inc., Sunderland, Massachusetts.

Hoeh, W.R., Blakley, K.H. and Brown, W.M. (1991) Heteroplasmy suggests limited biparental inheritance of *Mytilus* mitochondrial DNA. *Science*, **251**, 1488–90.

Hoffmann, R.J., Boore, J.L. and Brown, W.M. (1992) A novel mitochondrial genome organization for the blue mussel, *Mytilus edulis*. *Genetics*, **131**, 397–412.

Hudson, T.J., Engelstein, M., Lee, M.K. *et al.* (1992) Isolation and chromosomal assignment of 100 highly informative human simple sequence repeat polymorphisms. *Genomics*, **13**, 622–9.

Jacobs, H.T., Asakawa, S., Araki, T. *et al.* (1989) Conserved transfer RNA gene cluster in starfish mitochondrial DNA. *Curr. Genet.*, **15**, 193–206.

Jeffreys, A.J., Macleod, A., Tamaki, K., Neil, D.L. and Monckton, D.G. (1991) Minisatellite repeat coding as a digital approach to DNA typing. *Nature*, **354**, 204–209.

Jeffreys, A.J., Wilson, V. and Thein, S.L. (1985) Hypervariable 'minisatellite' regions in human DNA. *Nature*, **314**, 67–73.

Karl, S.A. and Avise, J.F. (1992) Balancing selection at allozyme loci in oysters – implications from nuclear RFLPs. *Science*, **256**, 100–102.

Karl, S.A., Bowen, B.W. and Avise, J.C. (1992) Global population genetic structure and male mediated gene flow in the green turtle (*Chelonia mydas*) – RFLP analyses of anonymous nuclear loci. *Genetics*, **131**, 163–73.

King, C.E. (1989) Molecular genetics of rotifers: preliminary restriction mapping of the mitochondrial genome of *Brachionus plicatilis. Hydrobiol.,* **186/187**, 375–380.

Kirkpatrick, B.W. (1992) Detection of a 3-allele single-strand conformation polymorphism (SSCP) in the 4th intron of the bovine growth hormone gene. *Anim. Genet.,* **23**, 179–80.

Klein-Lankhorst, R.M., Vermunt, A., Weide, R., Liharska, T. and Zabel, P. (1991) Isolation of molecular markers for tomato (*Lycopersicon esculentum*) using random amplified polymorphic DNA (RAPD). *Theoret. App. Genet.,* **83**, 108–14.

Kocher, T.D., Thomas, W.K. and Meyer, A. (1989) Dynamics of mitochondrial DNA evolution in animals. *Proc. Natl. Acad. Sci. USA,* **86**, 6196–200.

Koehn, R.K. and Hilbish T.J. (1987) The adaptive importance of genetic variation. *Am. Sci.,* **75**, 134–41.

Kondo, R., Matsuura, E.T. and Chigusa, S.I. (1992) Further observation of paternal transmission of *Drosophila* mitochondrial DNA by PCR selective amplification method. *Genet. Res.,* **59**, 81–4.

Kreitman, M. and Hudson, R.R. (1991) Inferring the evolutionary histories of the *Adh* and *Adh-dup* loci in *Drosophila melanogaster* from patterns of polymorphisms and divergence. *Genetics,* **127**, 565–82.

Lansman, R.A, Shade, R.O., Shapira, J.F. and Avise, J.C. (1981) The use of restriction endonucleases to measure mitochondrial DNA sequence relatedness in natural populations III. Techniques and potential applications. *J. Mol. Evol.,* **17**, 214–16.

LaRoche, J., Snyder, M., Cook, D.I., Fuller, K. and Zouros, E. (1990) Molecular characterization of a repeat element causing large-scale size variation in the mitochondrial DNA of the sea scallop *Placopecten magellanicus. Mol. Biol. Evol.,* **7**, 45–64.

Lessa, E.P. (1992) Rapid surveying of DNA sequence variation in natural populations. *Mol. Biol. Evol.,* **9**, 323–30.

Lewontin, R.C. (1974) *The Genetic Basis of Evolutionary Change,* Columbia University Press, New York.

Losekoot, M., Fodde, R., Harteveld, C.L. *et al.* (1990) Denaturing gradient gel electrophoresis and direct sequencing of PCR amplified genomic DNA: a rapid and reliable diagnostic approach to beta thalassemia. *Br. J. Haematol.,* **76**, 269–74.

Love, J.M., Knight, A.M., McAleer, M.A. and Todd, J. (1990) Towards construction of a high resolution map of the mouse genome using PCR-analysed microsatellites. *Nucleic Acids Res.,* **18**, 4123–30.

Matsuura, E.T. (1991) Selective transmission of mitochondrial DNA in *Drosophila. Jap. J. Genet.,* **66**, 683–700.

McMillan, J. and Miller, D.J. (1989) Nucleotide sequences of highly repetitive DNA from Scleractinian corals. *Gene,* **83**, 185–6.

McMillan, J. and Miller, D.J. (1990) Highly repeated DNA sequences in the Scleractinian coral genus *Acropora* – evaluation of cloned repeats as taxonomic probes. *Mar. Biol.,* **104**, 483–7.

McMillan, J., Mahony, T., Veron, J.E.N. and Miller, D.J. (1991) Nucleotide sequencing of highly repetitive DNA from 7 species in the coral genus *Acropora* (Cnidaria, Scleractinia) implies a division contrary to morphological criteria. *Mar. Biol.,* **110**, 323–7.

Michelmore, R.W, Paran, I. and Kesseli, R.V. (1991) Identification of markers linked to disease-resistance genes by bulked segregant analysis. A rapid method to detect markers in specific genomic regions by using segregating populations. *Proc. Natl. Acad. Sci. USA,* **8**, 9228–32.

Milyutina, I.A. and Petrov, N.B. (1989) Divergence of unique DNA sequences in bivalve molluscs of subfamily Mytilinae (Bivalvia Mytilidae). *Mol. Biol.,* **23**, 1091–8.

Moore, S.S., Sargeant, L.L., King, T.J. *et al.* (1991) The conservation of dinucleotide microsatellites among mammalian genomes allows the use of heterologous PCR primer pairs in closely related species. *Genomics,* **10**, 654–60.

Mort, M.A. and Streit, B. (1992) Measuring molecular variation in zooplankton populations – DNA extraction from small Daphnia species. *Aquat. Sci.,* **54**, 77–84.

Myers, R.M., Maniatis, T. and Lerman, L.S. (1987) Detection and localisation of single base changes by denaturing gradient gel electrophoresis. *Methods Enzymol.,* **155**, 501–17.

Neigel, J.E., Felder, D.L., Chlan, C.A. and Laporte, R. (1991) Cloning and screening of DNA probes for genetic studies in stone crabs (Decapoda, Xanthidae, Menippe). *J. Crustacean Biol.,* **11**, 496–505.

Orita, M., Suzuki, Y., Sekiya, T. and Hayashi, K. (1989) Rapid and sensitive detection of point mutations and DNA polymorphisms using the polymerase chain reaction. *Genomics,* **5**, 874–9.

Ovenden, J.R., Brasher, D.J. and White, R.W.G. (1992) Mitochondrial DNA analyses of the red rock lobster *Jasus edwardsii* supports an apparent absence of population subdivision throughout Australasia. *Mar. Biol.,* **112**, 319–26.

Paabo, S. and Wilson, A.C. (1988) Polymerase chain reaction reveals cloning artefacts. *Nature,* **334**, 387–8.

Palumbi, S.R. and Kessing, B.D. (1991) Population biology of the trans-arctic exchange: mtDNA sequence similarity between pacific and atlantic sea urchins. *Evolution,* **45**, 1790–1805.

Palumbi, S.R. and Metz, E.C. (1991) Strong reproductive isolation between closely related tropical sea-urchins (Genus *Echinometra*). *Mol. Biol. Evol.*, **8**, 227–39.

Palumbi, S.R. and Wilson, A.C. (1990) Mitochondrial DNA diversity in the sea-urchins *Strongylocentrotus purpuratus* and *Strongylocentrotus droebachie*. *Evolution*, **44**, 403–15.

Palva, T.K. and Palva, E.T. (1985) Rapid isolation of animal mitochondrial DNA by alkaline extraction. *FEBS Lett.*, **192**, 267–70.

Powers, D.A. (1990) The adaptive significance of allelic isozyme variation in natural populations, in *Electrophoretic and Isoelectric Focusing Techniques in Fisheries Management* (ed. D.H. Whitmore), CRC Press, Boca Raton, Florida, pp. 324–39.

Quinn, T.W. and White, B.N. (1987) Identification of restriction fragment length polymorphisms in genomic DNA of the lesser snow goose (*Anser caerulescens caerulescens*). *Mol. Biol. Evol.*, **4**, 126–43.

Quinn, T.W., Quinn, J.S., Cooke, F. and White, B.N. (1987) DNA marker analysis detects multiple maternity and paternity in single broods of the lesser snow goose. *Nature*, **326**, 392–4.

Quiros, C.F., Hu, J., This, P., Chevre, A.M. and Delseny, M. (1991) Development and chromosomal localization of genome-specific markers by polymerase chain reaction in *Brassica*. *Theoret. App. Genet.*, **82**, 627–32.

Rand, D.M. and Harrison, R.G. (1989) Molecular population genetics of mtDNA size variation in crickets. *Genetics*, **121**, 551–69.

Reeb, C.A. and Avise, J.C. (1990) A genetic discontinuity in a continuously distributed species – mitochondrial DNA in the American oyster, *Crassostrea virginica*. *Genetics*, **124**, 397–406.

Riedy, M.F., Hamilton, W.J. and Aquadro, C.F. (1992) Excess of non-parental bands in offspring from known primate pedigrees assayed using RAPD PCR. *Nucleic Acids Res.*, **20**, 918.

Rollinson, D. and Kane, R.A. (1991) Restriction enzyme analysis of DNA from species of *Bulinus* (Basommatophora, Planorbidae) using a cloned ribosomal RNA gene probe. *J. Moll. Studies*, **57**, 93–8.

Saiki, R.K., Bugawan, T.L., Horn, G.T., Mullis, K.B. and Erlich, H.A. (1986) Analysis of enzymatically amplified b globin and HLA DQa DNA with allele specific oligonucleotide probes. *Nature*, **324**, 163–6.

Saiki, R.K., Gelfand, D.H., Stoffel, S. *et al.* (1988) Primer-directed enzymatic amplification of DNA with a thermostable DNA polymerase. *Science*, **239**, 487–91.

Sainz, J. and Cornudella, L. (1990) Preservation of a complex satellite DNA in 2 species of echinoderms. *Nucleic Acids Res.*, **18**, 885–90.

Sambrook, J., Fritsch, E.F. and Maniatis, T. (1989) *Molecular Cloning: A Laboratory Manual*, 2nd edn, 3 volumes. Cold Spring Harbor Laboratory, Cold Spring Harbor, New York.

Saunders, N.C., Kessler, L.G. and Avise, J.C. (1986) Genetic variation and geographic differentiation in mitochondrial DNA of the horseshoe crab, *Limulus polyphemus*. *Genetics*, **112**, 613–27.

Schlotterer, C., Amos, B. and Tautz, D. (1991) Conservation of polymorphic simple sequence loci in Cetacean species. *Nature*, **354**, 63–5.

Schlotterer, C. and Tautz, D. (1992) Slippage synthesis of simple sequence DNA. *Nucleic Acids Res.*, **20**, 211–15.

Sibley, C.G. and Alquist, J.E. (1981) The phylogeny and relationships of the ratite birds as indicated by DNA–DNA hybridization, in *Evolution Today* (eds C.G.E Scudder and J.L. Reveal), *Proceedings of the Second International Congress on Systematics and Evolutionary Biology*, Carnegie-Mellon University, Pittsburgh, PA, pp. 301–335.

Sibley, G.C. and Alquist, J.E. (1987) DNA hybridization evidence of hominoid phylogeny: results from an expanded data set. *J. Mol. Evol.*, **26**, 99–121.

Skibinski, D.O.F. (1985) Mitochondrial DNA variation in *Mytilus edulis* L. and the Padstow mussel. *J. Exp. Mar. Biol. Ecol.*, **92**, 251–8.

Skibinski, D.O.F. and Edwards, C.A. (1987) Mitochondrial DNA variation in marine mussels (*Mytilus*), in *Proceedings of the World Symposium on Selection, Hybridisation and Genetic Engineering in Aquaculture*, Vol. I (ed. K Tiews), Heeneman, Berlin, pp. 209–26.

Smith, M.F., Boom, J.D.G. and Raff, R.A. (1990) Single-copy DNA distance between 2 congeneric sea-urchin species exhibiting radically different modes of development. *Mol. Biol. Evol.*, **7**, 315–26.

Smith, M.F., Banfield, D.K., Doteval, K., Gorski, S. and Kowbel, D.J. (1989) Gene arrangement in sea star mitochondrial DNA demonstrates a major inversion event during echinoderm evolution. *Gene*, **76**, 181–5.

Smith, M.F., Banfield, D.K., Doteval, K., Gorski, S. and Kowbel, D.J. (1990) Nucleotide sequence of 9 protein-coding genes and 22 transfer-RNAs in the mitochondrial DNA of the sea star *Pisaster ochraceus*. *J. Mol. Evol.*, **31**, 195–204.

Snyder, M., Fraser, A.R., La Roche, J., Gartner-Kepkay, K.E. and Zouros, E. (1987) Atypical mitochondrial DNA from the deep-sea scallop *Placopecten magellanicus*. *Proc. Natl. Acad. Sci. USA*, **84**, 7595–9.

Stanton, D.J., Crease, T.J. and Hebert, P.D.N. (1991) Cloning and characterization of *Daphnia* mitochondrial DNA. *J. Mol. Evol.*, **33**, 152–5.

Steinbruck, G., Schlegel, M., Kramer, M., Kupfermann, H. and Willig, S. (1991) Identification and phylogenetic analysis of 4 *Tisbe* species (Copepoda, Harpacticoida) using DNA restriction site variation. *Z. Zool. Syst. Evol.*, **29**, 393–408.

Stine, O.C. (1989) *Cepaea nemoralis* from Lexington, Virginia: the isolation and characterization of their mitochondrial DNA, the implications for their origin and climatic selection. *Malacologia*, **30**, 305–15.

Tautz, D. (1989) Hypervariability of simple sequences as a general source for polymorphic DNA markers. *Nucleic Acids Res.*, **17**, 6463–71.

Tegelstrom, H. (1986) Mitochondrial DNA in natural populations: an improved routine for the screening of genetic variation based on sensitive silverstaining. *Electrophoresis*, **7**, 226–9.

Thomas, W.K., Paabo, S., Villablanca, F.X. and Wilson, A.C. (1990) Spatial and temporal continuity of kangaroo rat populations shown by sequencing mitochondrial DNA from museum specimens. *J. Mol. Evol.*, **31**, 101–12.

Turner, B.J., Elder, J.F. Jr and Laughlin, T.F. (1991) Repetitive DNA sequences and the divergence of the fish populations: some hopeful beginnings. *J. Fish Biol.*, **39**, 131–42.

Ward, R.D., Skibinski, D.O.F. and Woodwark, M. (1992) Protein heterozygosity, protein structure and taxonomic differentiation, in *Evolutionary Biology*, Vol. 26 (eds M.K. Hecht, B. Wallace and R.J. MacIntyre), Plenum Press, New York and London, pp. 73–159.

Weatherhead, P.J. and Montgomerie, R.D. (1991) Good news and bad news about DNA fingerprinting. *TREE*, **6**, 173–4.

Weber, J.L. and May, P.E. (1989) Abundant class of human DNA polymorphisms which can be typed using the polymerase chain reaction. *Am. J. Hum. Genet.*, **44**, 388–96.

Welsh, J. and McClelland, M. (1990) Fingerprinting genomes using PCR with arbitrary primers. *Nucleic Acids Res.*, **18**, 7213–18.

Williams, J.F.K., Kubelik, A.R., Livak, K.J., Rafalski, J.A. and Tingey, S.V. (1990) DNA polymorphisms amplified by arbitrary primers are useful as genetic markers. *Nucleic Acids Res.*, **18**, 6531–5.

Wirgin, I.I. and Maceda, L. (1991) Development and use of striped bass-specific RFLP probes. *J. Fish Biol.*, **39**, 159–67.

Wrischnik, L.A., Higuchi, R.G., Stoneking, M. *et al.* (1987) Length mutations in human mitochondrial DNA – direct sequencing of enzymatically amplified DNA. *Nucleic Acids Res.*, **15**, 529–42.

Zouros, E., Freeman, K.R., Ball, A.O. and Pogson, G.H. (1992) Direct evidence for extensive paternal mitochondrial DNA inheritance in the marine mussel *Mytilus*. *Nature*, **359**, 412–14.

4.2 INFERRED PHYLOGENETIC TREES OF ANTARCTIC BROOD-PROTECTING SCHIZASTERID ECHINOIDS FROM PARTIAL 28S RIBOSOMAL RNA SEQUENCES

Jean-Pierre Féral, Evelyne Derelle and Hervé Philippe

Abstract

Four genera of brood-protecting schizasterid echinoids that are endemic to Antarctic and sub-Antarctic waters were chosen for this study. Various regions of the 28S ribosomal RNA molecules were sequenced to locate a domain variable enough to be useful for studying short-range phylogenetic distances. The D2 variable domain provides an indication of the relative phylogenetic distances between species under study.

The genus Amphipneustes *appears to cluster with* Brachysternaster, *another schizasterid genus, and* Delopatagus, *a so-called asterostomatid. On the other hand, the genus* Abatus *appears not to be monophyletic. The 'molecular position' of* Tripylus abatoides *which represents the fourth schizasterid genus is not clear, as is the case when using morphological data. In the present state of knowledge, morphological and molecular phylogenies roughly agree.*

The 28S rRNA D2 domain seems to have an appropriate rate of evolution to resolve family relationships and for phylogenetic resolution within the family Schizasteridae to the species level.

4.2.1   Introduction

Sea urchins generally disseminate by means of a swimming planktotrophic echinopluteus larva. However, as with other marine invertebrate groups, over 50% of the south polar regular and irregular sea urchin species brood their young, which means that no larval dispersion occurs. Most of the irregular echinoids belong to the family Schizasteridae, which includes 16 genera and occurs in all oceans. Among them, four genera are endemic to the sub-Antarctic and Antarctic zones (36 °S to 75 °S from the Patagonian coast and sub-Antarctic islands to the Antarctica coast; Figure 4.2.1). All species of these genera brood embryos and juveniles in dorsal ambulacral pouches. Their geographic and bathymetric distributions are given in Schatt and Féral (1991). *Abatus* and *Tripylus* live in both sub-Antarctic and Antarctic waters while *Amphipneustes* and *Brachysternaster* live primarily in the Antarctic zone.

Studies have been made on the ecology, life and brooding cycles and physiological functions of a species endemic to the Kerguelen Islands, *Abatus cordatus* (Magniez, 1983;

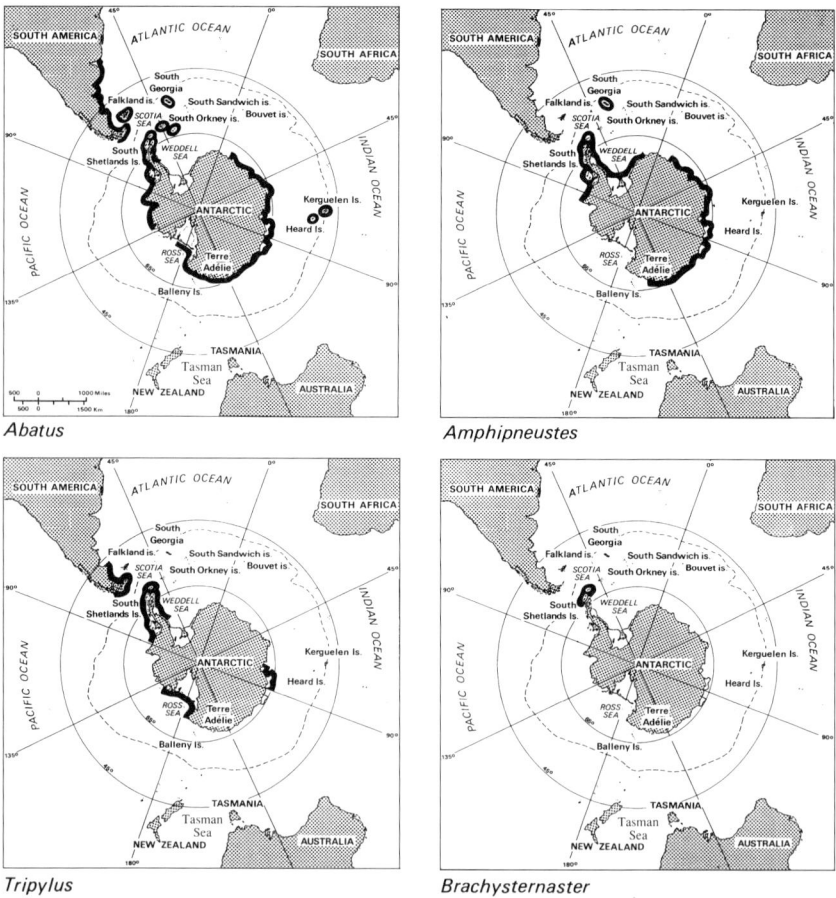

Abatus

Amphipneustes

Tripylus

Brachysternaster

Figure 4.2.1 Geographic distribution of the four genera of brood-protecting schizasterid echinoids, after data compiled by Schatt and Féral (1991).

Schatt, 1988; Magniez and Féral, 1988; Schatt and Féral, 1991) and of two other species, *A. nimrodi* and *A. shakletoni* (McClintock and Pearse, 1986; Pearse and McClintock, 1990). However, these studies cannot entirely explain the richness in brooding species within the Antarctic.

Current hypotheses invoke (1) the cold stenothermal environment, linked with an aseasonal reproduction (Orton, 1920) (but see Féral and Magniez (1985) and Schatt and Féral (1991) for opposite conclusions concerning sub-Antarctic echinoderm species living in the same environment) and (2) the latitude and depth, linked with the mode of development, namely pelagic and non-pelagic (Thorson, 1950). These hypotheses appear to be two of the principal paradigms in marine biology. However, we have noted that these hypotheses were inappropriate generalities and certainly lacked a temporal dimension (Féral and Derelle, 1991). Indeed, they are now clearly rejected, at least in the case of the larval biology of echinoderms (Pearse, McClintock and Bosh, 1991; Pearse, 1993) and of molluscs (Hain and Arnaud, 1992). The purpose of our research is to find out if the south polar endemism of brood-protecting schizasterids is the result of a common origin or if it is an opportunistic gathering in an environment presenting specific ecological characteristics. To resolve these hypotheses, on the one hand, their genetic structure should be understood and gene flow estimated to evaluate the mechanism and the importance of the radiation(s). We have some data on this point (Poulin & Féral unpublished data). On the other hand, their evolutionary history should be investigated. The lack of good fossil records (Wiedman, 1990) imposes the use of molecular methods. Comparative treatment of DNA or RNA sequence information is now considered to be an appropriate and effective method for reconstructing phylogenies (Hillis and Moritz, 1990).

Among nucleic acids, large sub-unit ribosomal RNAs appear particularly well suited as general indicators of phylogenetic relationships. They occur in all species and have largely conservative structure and function. In this section we present preliminary results on the reliability of comparison of 28S rRNA partial sequences at different taxonomic levels in echinoderms.

4.2.2 Materials and methods

Schizasterid sea urchins from the Kerguelen Islands (49° 21′ S to 70° 12′ E) (*Abatus cordatus* (Verrill 1876)) and from Terre-Adélie (66° 40′ S to 140° 01′ E) (*Abatus nimrodi* (Koehler 1911)), *Abatus ingens* Koehler 1926, *Abatus cavernosus* (Philippi 1845) and the echinid *Sterechinus neumayeri* (Meissner 1900) were collected from depths of up to 30 m by diving. They were dissected and frozen on the site. Schizasterids from the Weddell Sea (*Tripylus abatoides* (Clark 1925), *Amphipneustes lorioli* Koehler 1901, *Amphipneustes similis* Mortensen 1936, *Brachysternaster chesheri* Larrain 1985) as well as the so-called asterastomatid *Delopatagus brucei* Koehler 1908 were dredged and frozen during the 'EPOS 3' cruise (1989).

Echinocardium mortenseni Thiéry 1909 (Loveniidae), *Brissopsis lyrifera* (Forbes 1841) (Brissidae) and *Paracentrotus lividus* (Lamark 1816) (Parechinidae) were collected in Banyuls-sur-mer (Mediterranean Sea, 42° 29′ N to 03° 08′ E).

Preparation of the RNA was carried out as described in Qu, Michot and Bachellerie (1983) and Qu, Nicoloso and Bachellerie (1988). In brief, total RNA was extracted from ovaries or gut homogenized in guanidium thiocyanate medium at 60 °C and then in a phenol–chloroform–isoamyl alcohol mixture. The aqueous phase was extracted again with the phenol–chloroform mixture and twice in chloroform before ethanol precipitation.

RNA was sequenced by reverse transcriptase extension of the 5′-end-^{32}P-labelled DNA synthetic primer in the presence of chain terminator dideoxynucleotides (Sanger, Nicklen and Coulson, 1977). The oligonucleotide used, complementary to the evolutionary conserved 28S rRNA *D2* variable domain, was 5'CCTTGGTCCGTGTTTCAAGA3' (*Mus musculus* sequence) (Hassouna, Michot and Bachellerie, 1984). Sequences were performed at least five times on one or two specimens. In the case of *Abatus cordatus* the sequencing was performed on five individuals belonging to different populations. No polymorphism was detected in the domains *D2* and *D1*, and in the 5'-end of the 28S rRNA.

Distance matrix and parsimony analyses were carried out using the PAUP (Swofford, 1985), PHYLIP (Felsenstein, 1990) and MUST (Philippe, 1992) packages. Statistical tests of topological differences were undertaken using the bootstrap method (Felsenstein, 1988, 1990; Nei, 1991; Philippe, 1992).

The echinoid *D2* domain consists of about 430 nucleotides. Because of the partial sequencing of the RNA of certain species we present here preliminary results based on about 300 nucleotides.

4.2.3   Results and discussion

All groupings in the shortest tree inferred from a distance matrix calculated after the alignment of the 3'-end 140 first nucleotides by the neighbour-joining method (Saitou and Nei, 1987) make sense (Figure 4.2.2). Regular echinoids *P. lividus* (Parechinidae) and *S. neumayeri* (Echinidae) cluster together. Irregular echinoids (spatangoids) are separated according to the family to which they belong: Schizasteridae, Loveniidae and Brissidae. Even though short, this stretch of the 28S rRNA molecule contains a phylogenetic signal for taxa at the family level. However, branches are very short and it is difficult to assign significance. This is resolved in the multifurcated consensus tree (Figure 4.2.3) obtained with parsimony methods. Note that both tree construction methods yield basically similar topologies of unrooted trees (Figures 4.2.2. and 4.2.3).

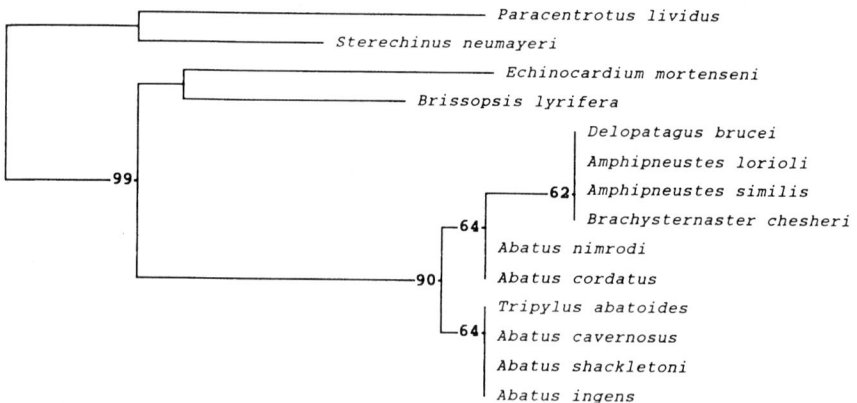

Figure 4.2.2 Tree inferred from a distance matrix by the neighbour-joining method (NJ program from MUST). Incompletely specified nucleotides were discarded, 123 on the 3'-end 140 first sites were kept, 19 sites were variable. For the Schizasteridae, the number at the forks gives the percentage of times the group of species to the right of that fork occurred among 1000 bootstrap replicates (NJBOOT program from MUST). The distance between two species is proportional to the sum of the projections on the *x* axis of the branch lengths, whereas distances on the *y* axis are arbitrary. This unrooted tree was rooted by *Paracentrotus lividus* and *Sterechinus neumayeri*.

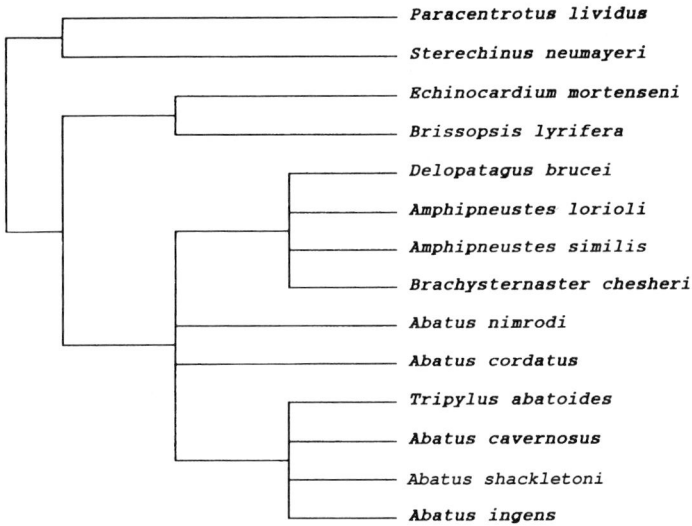

Figure 4.2.3 Strict consensus tree inferred from the same set of data as in Figure 4.2.4 after using PAUP. Ten sites were informative for parsimony (50 trees, CI = 0.923, total length = 13). Designated outgroup taxon = *Paracentrotus lividus*.

When 312 sites are used and *Paracentrotus lividus* is taken as the outgroup (Figure 4.2.4), a similar clustering is observed, at the family level, according to the neighbour-joining method (Figure 4.2.4(A)) or to parsimony methods (Figure 4.2.5). Concerning the schizasterids, two main groups of species stand out at the present state of the analysis: an '*Amphipneustes* group' and an '*Abatus* group'.

Amphipneustes similis and *A. lorioli* yield a cluster together with two other species presently belonging to two different genera. The first one, *Brachysternaster*, is a genus recently created by Larrain (1985), who proposed affinities of that genus to the genus *Amphipneustes* and the family Asterostomatidae, to which the second genus, *Delopatagus*, belongs. Both share characters with *Amphipneustes*. Moreover, Asterostomatidae is a polyphyletic taxon proposed by Fischer (1966) to accommodate a number of genera of obscure affinities. The point is that in spite of this similarity *Delopatagus* is an echinoid that seems not to be brood protecting and lives in the Antarctic abyssal zone (2000 m depth). This does not correspond with the bathymetric distribution of the four brood-protecting genera under study (Schatt and Féral, 1991).

The other remarkable point is the divergence within the genus *Abatus*, which seems to be a 'non-monophyletic' taxon (sequencing of all the species would be necessary to conclude paraphyly). *Tripylus abatoides* is clustered with the *Abatus* group. The species *abatoides* was referred by taxonomists to the genus *Tripylus*, but, even morphologically, we suggest that it should be referred to the genus *Abatus*, particularly because of the structure of the globiferous pedicellariae, whose valves terminate in two long hooks, unlike in other *Tripylus*, where the hooks are more numerous. It seems that Clark (1925), who described this species, gave it a name reflecting his own doubts concerning its generic position. This 'taxonomic' doubt is confirmed at the molecular level. It would not be surprising if a revision based on morphological data establishes that *T. abatoides* is related to the '*Abatus* group' or part of it. A similar doubt arose concerning the *Tripylus* fossil species *T. pseudoviviparus* (Lambert 1933), which should be also referred to *Abatus* (Wiedman, 1990).

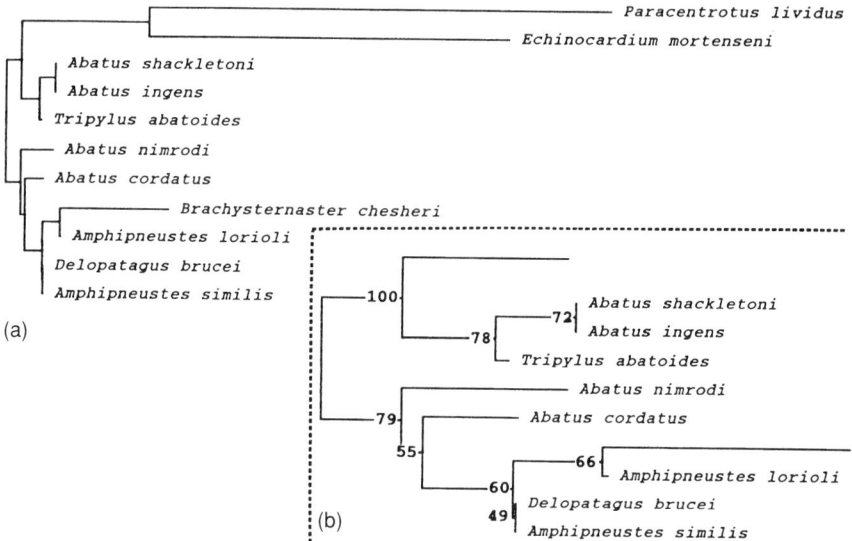

Figure 4.2.4 (a) Tree inferred from distance matrix by the neighbour-joining method after the alignment of the 3′-end 312 first sites. It is rooted by *Paracentrotus lividus* (278 sites of 312 are kept, 66 are variable). (b) Inset showing an enlargement of the Schizasteridae with the bootstrap values (for details see caption to Figure 4.2.2).

A bootstrapped parsimony analysis (1000 randomly sampled replicates) was used as another approach to characterize the phylogenetic signal (presence and strength) in the data (Figure 4.2.5(B)). If one interprets the significance of these results by collapsing all branches that failed to meet a 95% criterion, the bootstrap reveals a structure not very much different from that seen in the consensus tree (Figure 4.2.5(A)). The structures obtained with the molecular data are roughly congruent with the phylogenetic relationships inferred from other data sets (e.g. morphology) (Smith, 1992).

It should be noted that the species taken as the outgroup, *P. lividus*, is phylogenetically very distant from the schizasterids. A possible effect of an outgroup which is too distant is to root an ingroup on the longest branch (Wheeler, 1990). This may explain some of the surprising *Abatus* clustering (Figure 4.2.5).

The separation of *Abatus* into two groups is still weakly supported. The neighbour-joining tree (Figure 4.2.4(B)) shows a similar structure to the parsimony trees (Figure 4.2.5). The only topological difference (Figure 4.2.5(B)) concerns *Amphipneustes similis*, from which the next group arises (to the right of the fork). However, the rooting is not significant because this group of species occurred only 260 times among 1000 bootstrap replicates.

The number of species studied is not sufficient to be conclusive. So our results should be considered as preliminary and without a safe historical dimension in their present state. More extensive sequences as well as additional species are required, to fully understand the history of the south polar schizasterid brood-protecting species. All of them should be analysed and an appropriate outgroup has to be taken such as the non-brood-protecting schizasterids or hemiasterids.

Nevertheless, the *D2* divergent domain of the 28S rRNA seems suitable to study long-range and relatively short-range phylogenetic relationships among echinoids. The results of

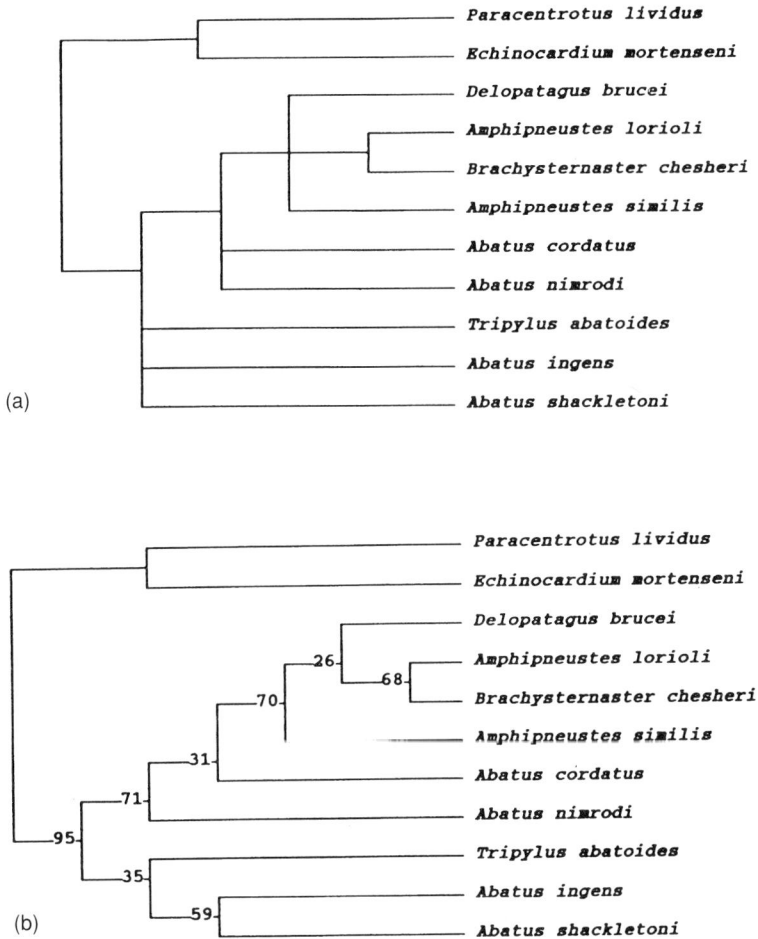

Figure 4.2.5 (a) Strict consensus tree inferred from the same set of data as in Figure 4.2.4 after using PAUP. Twelve sites were informative for parsimony (15 trees, CI = 0.737, total length = 15). Designated outgroup taxon = *Paracentrotus lividus*. (b) Bootstrapped parsimony tree inferred from the same set of data as in Figure 4.2.5 (a) after using DNABOOT from PHYLIP (see caption to Figure 4.2.4). Designated outgroup taxon = *Paracentrotus lividus*.

the present study indicate that even this relatively small fragment of rRNA contains significant phylogenetic signals for taxa whose divergence extends as far back as some ten million years (early Tertiary). This study will be continued using longer rRNA sequences, including the *D1* divergent domain and the 5'-end to add informative sites and to provide phylogenetic signals which will be useful at different taxonomic levels. As most of the enlargement of the 28S rRNA molecule during evolution is restricted to two of the divergent domains, *D2* and *D8* (Michot and Bachellerie, 1987), we will also sequence *D8* as well as noncoding parts of the DNA that may help either to confirm interspecific relationships or to study infraspecific ones.

It has recently been shown that *Delopatagus brucei* is a brood protecting species. This supports the idea that this genus belongs to the Schizasterids, the only brooding family in the spatangoids (De Ridder, David and Larrain, 1993).

Acknowledgement

This work was supported by grants from the TAAF 'BENTHOS/MAC' program and from the GDR ECOPROPHYCE (CNRS). We thank P. Arnaud, C. Ozouf-Costaz and J.-C. Hureau for collecting material during the EPOS cruise (leg 3) in the Weddell Sea, and the TAAF and the EPF for facilities in Kerguelen and Terre-Adélie. We are grateful to M.-J. Bodiou for drawing the maps. Thanks are also due to B. David, M. Jensen, P. Laboute, J. Mabit, P. Mespoulhé, R. Mooi, E. Poulin and P. Schatt for stimulating discussions or help in the field and to B. David and A. Larrain for the identification of some specimens.

References

Clark, H.L. (1925) *A Catalogue of the Recent Sea-urchins in the Collection of the British Museum (Natural History)*, Oxford University Press, London, pp. 1–250.

De Ridder, C., David, B. and Larrain, A. (1993) *Bull. Mus. Natl. Hist. Nat.*, 4e sér, **14**, 1992, A, (2), 405–41.

Felsenstein, J. (1988) Phylogenies from molecular sequences: inference and reliability. *Annu. Rev. Genet.*, **22**, 521–65.

Felsenstein, J. (1990) *PHYLIP manual, version 3.3*, University Herbarium, University of California, Berkeley, California 94720.

Féral, J.-P. and Derelle, E. (1991) Partial sequence of the 28S ribosomal RNA and the echinid taxonomy and phylogeny–application to the antarctic brooding schizasterids, in *Biology of Echinodermata*, Proceedings of the 7th International Echinoderm Conference, Atami, Japan, 9–14 Sept. 1990 (eds T. Yanagisawa, I. Yasumasu, C. Oguro, N. Suzuki and T. Motokawa), Balkema, Rotterdam, pp. 331–7.

Féral, J.-P. and Magniez, P. (1985) Level, content and energetic equivalent of the main biochemical constituents of the subantarctic molpadid holothurian *Eumolpadia violacea* (Echinodermata) at two seasons of the year. *Comp. Biochem. Physiol.*, **81A**, 415–22.

Fischer, A.G. (1966) Spatangoids, in *Treatise on Invertebrate Paleontology, Echinodermata*, 3 (U) (ed. R.C. Moore), Geological Society of America and University of Kansas Press, pp. 257–65.

Hain, S. and Arnaud, P.M. (1992) Notes on the reproduction of high-antarctic molluscs from the Weddell Sea. *Polar Biol.*, **12**, 303–12.

Hassouna, N., Michot, B. and Bachellerie, J.-P. (1984) The complete nucleotide sequence of mouse 28S rRNA gene. Implications for the process of size increase of the large subunit rRNA in higher eukaryotes. *Nucleic Acids Res.*, **12**, 3563–83.

Hillis, D.M. and Moritz C. (1990) An overview of applications of molecular systematics, in *Molecular Systematics* (eds D.M. Hillis and C. Moritz), Sinauer Associates Inc., Sunderland, Massachusetts, pp. 502–15.

Larrain, A.P. (1985) *Brachysternaster*, new genus, and *Brachysternaster chesheri*, new species of antarctic echinoid (Spatangoida, Schizasteridae). *Polar Biol.*, **4**, 121–4.

Magniez, P. (1983) Reproductive cycle of the brooding echinoid *Abatus cordatus* (Echinodermata) in Kerguelen (Antarctic Ocean): changes in the organ indices, biochemical composition and caloric contents of the gonads. *Mar. Biol.*, **74**, 55–64.

Magniez, P. and Féral, J.-P. (1988) The effect of somatic and gonadal size on the rate of oxygen consumption in the subantarctic echinoid *Abatus cordatus* (Echinodermata) from Kerguelen. *Comp. Biochem. Physiol.*, **90A**, 429–34.

McClintock, J.B. and Pearse, J.S. (1986) Organic and energetic content of eggs and juveniles of antarctic echinoids and asteroids with lecithotrophic development. *Comp. Biochem. Physiol.*, **85A**, 341–5.

Michot, B. and Bachellerie, J.-P. (1987) Comparisons of large subunit rRNAs reveal some eukaryote-specific element of secondary structure. *Biochimie*, **69**, 11–23.

Nei, M. (1991) Relative efficiencies of different tree-making methods for molecular data, in *Phylogenetic Analysis of DNA Sequences* (eds M.M. Miyamoto and J. Cracraft), Oxford University Press, New York.

Orton, J.H. (1920) Sea-temperature, breeding and distribution in marine animals. *J. Mar. Biol. Assoc., UK*, **12**, 339–66.

Pearse, J.S. (1993) Cold-water echinoderms break Orton's and Thorson's 'rules', in *Reproduction, Larval Biology and Recruitment in the Deep-sea Benthos* (eds K.J. Eckelbarger and C.M. Young), Columbia University Press.

Pearse, J.S. and McClintock, J.B. (1990) A comparison of reproduction by the brooding spatangoid echinoids *Abatus shackletoni* and *A. nimrodi* in McMurdo sound, Antarctica. *Invert. Reprod. Devel.*, **17**, 181–91.

Pearse, J.S., McClintock, J.B. and Bosh, I. (1991) Reproduction of antarctic benthic marine invertebrates: tempos, modes and timing. *Am. Zool.*, **31**, 65–80.

Philippe, H. (1992) *Manuel d'Utilisation du logiciel MUST, version 0.0*, Laboratoire de Biologie Cellulaire 4, Université Paris-Sud, 91405 Orsay Cedex, France.

Qu, L.H., Michot, B. and Bachellerie, J.-P. (1983) Improved methods for structure probing in large RNAs: a rapid 'heterologous' sequencing approach is coupled to the direct mapping of nuclease accessible sites. Application to the 5′ terminal domain. *Nucleic Acids Res.*, **11**, 5903–20.

Qu, L.H., Nicoloso, M. and Bachellerie, J.-P. (1988) Phylogenetic calibration of the 5′ terminal domain of large rRNA achieved by determining twenty eukaryotic sequences. *J. Mol. Evol.*, **28**, 113–24.

Saitou, N. and Nei, M. (1987) The neighbor-joining method: a new method for reconstructing phylogenetic trees. *Mol. Biol. Evol.*, **4**, 406–25.

Sanger, F., Nicklen, S. and Coulson, A.R. (1977) DNA sequencing with chain-terminating inhibitors. *Proc. Natl. Acad. Sci., USA*, **74**, 5463–7.

Schatt, P. (1988) Embryonic growth of the brooding sea urchin *Abatus cordatus*, in *Echinoderm Biology*, Proceedings of the 6th International Echinoderm Conference, Victoria BC, Canada, 23–28 Aug. 1987 (eds R.D. Burke, P.V. Mladenov, P. Lambert and R.L. Parsley), Balkema, Rotterdam, pp. 225–8.

Schatt, P. and Féral, J.-P. (1991) The brooding cycle of *Abatus cordatus* (Echinodermata: Spatangoida) at Kerguelen Islands. *Polar Biol.*, **11**, 283–92.

Smith, A.B. (1992) Echinoderm phylogeny: morphology and molecules approach accord. *TREE*, **7**, 224–9.

Swofford, D.L. (1985) *PAUP version 2.4.1*, Illinois Natural History Survey, Urbana-Champaign, Illinois 61801.

Thorson, G. (1950) Reproduction and larval ecology of marine bottom invertebrates. *Biol. Rev.*, **25**, 1–45.

Wheeler, W.C. (1990) Nucleic acid sequence phylogeny and random outgroups. *Cladistics*, **6**, 363–7.

Wiedman, L.A. (1990) The paleontology of selected antarctic and New Zealand Eocene fossils. PhD thesis, Kent State University.

4.3   CONTRASTING PATTERNS OF DNA POLYMORPHISM IN THE DEEP-SEA SCALLOP (*PLACOPECTEN MAGELLANICUS*) AND THE ATLANTIC COD (*GADUS MORHUA*) REVEALED BY RANDOM CDNA PROBES

Grant H. Pogson

Abstract

*Random clones isolated from cDNA libraries from the deep-sea scallop (*Placopecten magellanicus*) and Atlantic cod (*Gadus morhua*) were used as probes to screen for restriction site polymorphisms by Southern blot analysis. The proportion of cDNA probes screened that detected DNA polymorphism was similar in scallop and cod (65 % and 59 %, respectively). Three distinct classes of DNA polymorphism were observed: (1) restriction site variation; (2) VNTR polymorphisms; and (3) complex enzyme-dependent 'fingerprint' patterns. Although the percentage of probes that produced simple restriction site variation was comparable in two species, mean heterozygosity levels of this type of polymorphism were 49 % higher in cod compared to scallop. A considerably higher percentage of scallop cDNA probes produced VNTR polymorphisms (9 of 22, or 41 %) compared to cod (6 of 38, or 16 %). No differences were observed, however, in the mean heterozygosities of VNTR polymorphisms in the two species. Complex 'fingerprint' patterns, presumably caused by the probe belonging to a multigene family, were observed at more than twice the level in cod (22 %) relative to scallop (9 %). The observed heterozygosity levels of both restriction site and VNTR polymorphisms exhibited significant positive relationships with mean restriction fragment size. Possible explanations for the contrasting patterns of DNA polymorphism observed between the two species and the heterozygosity/fragment size correlation are discussed.*

4.3.1   Introduction

The molecular study of naturally occurring genetic variation was pioneered a quarter of a century ago by the application of protein electrophoresis to population studies by Lewontin and Hubby (1966) and Harris (1966). This technique provided geneticists with a simple, inexpensive method of scoring genetic polymorphisms based on the detection of differences in the charge properties of allelic isozymes, or allozymes. Surveys of allozyme variation have played a dominant role in shaping our views of the extent of genetic diversity in natural populations. However, at present it remains unclear to what extent selection, stochastic processes (i.e. mutation and drift) or constraints at the level of amino acid replacements have affected the levels and patterns of protein polymorphism in natural populations; see discussions in Kimura (1983), Nevo, Beiles and Ben-Shlomo (1984), Zera, Koehn and Hall (1985) and Gillespie (1991).

  Although the characterization of DNA polymorphism in natural populations is in its infancy, several recent studies incorporating measures of polymorphism at both the protein and nuclear DNA levels have noted discordant results concerning both the extent of this variation within single populations (e.g. Aquadro, Lado and Noon, 1988) and the distribution of this variation among populations (e.g. Karl and Avise, 1992). Comparison of the patterns and levels of polymorphism of nuclear DNA markers and allozymes allow for powerful indirect tests of neutrality, since the variation at both levels is expected to be affected to the same extent by stochastic processes like random drift, mutation and migration.

  The type of probe that appears best suited to comparing DNA and allozyme polymorphism within natural populations is that derived from transcribed structural genes (mRNA) that has been converted into cDNA (complementary DNA) by the enzyme reverse transcriptase. In this section, I describe the general patterns of DNA polymorphism detected by random cDNA probes in the deep-sea scallop, *Placopecten magellanicus*, and the Atlantic cod, *Gadus morhua*. Interesting differences are shown to occur between these species involving both the types of polymorphism uncovered by random cDNA probes and the observed heterozygosity levels associated with restriction site polymorphisms.

4.3.2   Materials and methods

(a)   Animals

Scallop spat were collected from a natural population in Pasmaquoddy Bay, New Brunswick, Canada as detailed in Zouros *et al.* (1992). Cod were captured live from Western Bank (44.3° lat., 64.3° long.) off the eastern coast of Nova Scotia, Canada in November 1991 as described in Nelson, Tang and Boutilier (1993).

(b)   Isolation of cDNA probes

cDNA libraries were commercially prepared from poly A^+ RNA extracted from adult scallop adductor muscle and adult cod liver in the phagemid vector lambda Bluemid (Clontech Inc.). Inserts were amplified from individually picked plaques by the polymerase chain reaction. Details of the amplification conditions will be published elsewhere (Pogson and Zouros, unpublished data).

(c)   DNA extractions

Cod genomic DNA was isolated from blood samples that had been stored in absolute ethanol by the salt-extraction procedure of Miller, Dykes and Polesky (1988). Scallop genomic DNA was phenol/chloroform extracted from approximately 0.2 g of adductor muscle that had been ground in liquid nitrogen.

(d) Screening for polymorphism

Samples of genomic DNA (7 μg) from seven unrelated individuals were digested with various restriction endonucleases according to the manufacturer (Pharmacia or Bethesda Research Laboratories) and separated on 0.8 % agarose gels for 20–24 h at 2 V/cm in 100 mM Tris-HCl, 100 mM borate, 2 mM EDTA, pH 8.3 buffer. The digested DNA was vacuum-blotted onto nylon membrane (Amersham Hybond N) and fixed by baking for 2 h at 80°C. Randomly selected cDNA fragments were labelled with digoxigenin-11-dUTP; details of this methodology will be published elsewhere (Pogson and Zouros, unpublished data).

The minimum criterion set for accepting DNA polymorphism was the observation of one heterozygote among seven unrelated individuals for genomic DNA samples digested with a minimum of four restriction endonucleases. The detection of one heterozygous individual among seven corresponds to setting an α level at 0.071, which is slightly more conservative than standard levels of either 0.05 or 0.01. Observed heterozygosities, mean restriction fragment sizes and numbers of alleles are estimates from this initial sample of seven individuals.

4.3.3 Results

(a) Patterns of DNA polymorphism

The proportion of cDNA probes screened that detected polymorphism was similar in the two species (65 % in scallop versus 59 % in cod). Three distinct classes of DNA polymorphism were observed. Restriction site polymorphisms (Figure 4.3.1(A)) were unique to a single enzyme and typically exhibited low levels of heterozygosity. The second class of polymorphism (Figure 4.3.1(B)) was characterized by highly polymorphic one- or two-banded genotypic patterns that were often replicated by more than one restriction enzyme but modified slightly in size. Being independent of restriction enzyme, this type of polymorphism is consistent with that expected by variable number of tandem repeat (VNTR) arrays located within introns or flanking regions of the structural gene probed. Although insertions and/or deletions can also produce length polymorphism that is independent of restriction enzyme, the high frequency of this type of variation conflicts with its expected deleterious nature (e.g. Golding, Aquadro and Langley, 1986), and the quasi-continuous distribution of fragment sizes is more consistent with that expected by VNTR arrays.

The third class of polymorphism involved highly complex enzyme-dependent 'fingerprint' patterns, often involving ten or more fragments per individual (Figure 4.3.1(C)). In cases where a limited number of restriction fragments are observed, this may be accounted for by the existence of internal restriction sites or introns breaking up the coding region probed. However, for most polymorphisms falling into this class the combined size of fragments detected exceeded the size of the probe by a factor of 30 or more. In these instances, the large number of fragments observed and the complexity of the polymorphism can be explained by the cDNA fragment either belonging to a multigene family or, perhaps, being a composite of two or more mRNAs.

The numbers of polymorphisms falling into these three classes are shown in Table 4.3.1. Of the probes detecting polymorphism, the percentage producing simple restriction site variation was similar in the two species (63 % in cod versus 50 % in scallops). However, VNTR polymorphisms were considerably more frequent in scallops (representing 41 % of the total versus only 16 % in cod) whereas for complex fingerprint patterns the opposite

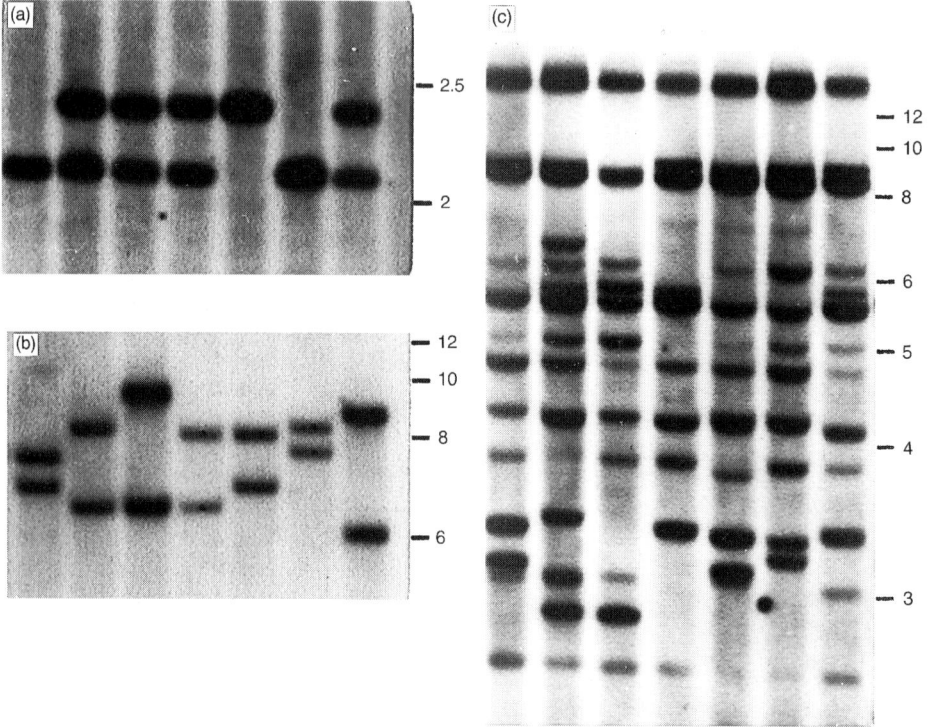

Figure 4.3.1 Examples of cDNA-based polymorphisms. (a) Restriction site polymorphism revealed by cod probe GM867 when hybridized to seven genomic DNA samples digested with *Pvu*II. (b) VNTR polymorphism exhibited by scallop probe PM322 when hybridized to seven genomic DNA samples digested with *Hae*III. (c) 'Fingerprint' polymorphism detected by cod probe GM311 when hybridized to seven genomic DNA samples digested with *Pst*I.

Table 4.3.1   Summary of the types of DNA polymorphism detected by cDNA probes in scallop and cod

Species	Total no. of probes screened	Class of polymorphism			
		Restriction site	VNTR	Fingerprint	Total
Scallop	34	11	9	2	22
Cod	64	24	6	8	38

pattern was found (22 % of cod polymorphisms falling into this category versus only 9 % in scallops). Because of the differences in proportions of VNTR and fingerprint patterns observed, a contingency test comparing the distribution of DNA polymorphism between the two species approached statistical significance ($\chi^2 = 5.12$, d.f. = 2, $P = 0.077$).

(b)   Levels of DNA polymorphism

It is not possible to directly estimate heterozygosity levels for probes detecting complex multibanded patterns since the allelic nature of this variation is unknown and, at present, untested. The genetic interpretation of the other types of DNA polymorphism is more straightforward. Heterozygotes for both restriction site and VNTR polymorphisms were exclusively two-banded and exhibited strict codominance. This indicates for both types of polymorphism that the variation cannot reside within the coding region represented by the cDNA probe itself, as otherwise heterozygotes with three restriction fragments would have been produced (with the sizes and intensities of the two smallest fragments summing to the largest seen). Where larger population samples have been obtained, genotypes for restriction site polymorphisms conform closely to Hardy–Weinberg expectations (Pogson and Zouros, unpublished data; Pogson *et al.*, unpublished data). Although the Mendelian nature of all cDNA-based polymorphisms has not been confirmed, in every instance where it has been examined in pair crosses of both species (at four loci in scallop and six in cod) alleles have been observed to segregate as proper genetic units (Pogson, unpublished data).

Table 4.3.2 details the restriction site polymorphisms detected by cDNA probes in scallop and cod. Differences between the two species in the number of polymorphisms associated with four- and six-cutter enzymes reflect differences in their use during the screening for polymorphism, not any inherent bias in their ability to produce polymorphism. In scallop, most polymorphisms were detected with four-cutter enzymes and these showed no difference in observed heterozygosity levels from the few six-cutter polymorphisms observed. In contrast, for cod the majority of polymorphisms were associated with six-cutter enzymes and the mean heterozygosity levels of these polymorphisms significantly exceeded those of the four-cutters (by paired *t*-test, $t = 2.88$, $P = 0.009$). Consequently, the overall observed heterozygosity level of restriction site polymorphisms in cod was significantly larger than in scallop ($t = 2.24$, $P = 0.033$). Physical sizes of cDNA probes did not differ between polymorphisms produced by the two enzyme types either within or between species. However, mean sizes of the restriction fragments detected by these probes differed, as expected from the probability of occurrence of recognition sites for four- and six-cutter enzymes. Unlike with restriction site polymorphisms, no differences were observed between species in the observed heterozygosities, cDNA probe sizes, or mean fragment sizes associated with VNTR polymorphisms (Table 4.3.3).

Table 4.3.2   Comparison of restriction site polymorphisms associated with four- and six-cutter enzymes in scallop and cod

Species	Enzyme type	N	Mean observed heterozygosity (SE)	Mean probe size (SE) (bp)	Mean fragment size detected (SE) (kb)
Scallop	4-cutter[a]	9	0.221 (0.040)	813.3 (54.2)	2.62 (0.41)
	6-cutter[b]	2	0.215 (0.072)	950.0	5.84 (0.57)
	Totals	11	0.220 (0.034)	838.2 (46.9)	3.21 (0.52)
Cod	4-cutter[c]	6	0.215 (0.032)	853.3 (51.6)	2.11 (0.28)
	6-cutter[d]	18	0.366 (0.042)	784.4 (25.4)	6.21 (0.71)
	Totals	24	0.328 (0.035)	801.7 (23.2)	5.19 (0.65)

[a]*Hae*III, *Hpa*II, *Rsa*I; [b]*Pst*I; [c]*Taq*I; [d]*Dra*I, *Pvu*II, *Pst*I, *Sac*I, *Bam*HI, *Bst*EII.

Table 4.3.3   Comparison of VNTR polymorphisms detected by cDNA probes in scallop and cod; mean fragment sizes have been averaged between the different enzymes producing the same genotypic patterns

Species	N	Mean observed heterozygosity (SE)	Mean probe size (SE) (bp)	Mean fragment size detected (SE) (kb)
Scallop	9	0.658 (0.062)	706.7 (40.1)	4.32 (0.81)
Cod	6	0.547 (0.073)	831.7 (101.5)	3.52 (0.80)

Combining data from both species, the relationship between observed heterozygosity and the mean size of the variable restriction fragments producing the polymorphism are presented in Figure 4.3.2. A significant positive correlation was observed between these variables for both restriction site ($r^2 = 0.257$, $F_{(1,33)} = 11.4$, $P = 0.002$) and VNTR polymorphisms ($r^2 = 0.650$, $F_{(1,13)} = 24.1$, $P < 0.001$). For restriction site polymorphisms, this relationship is primarily attributable to the differences between restriction enzymes noted in Table 4.3.2. However, similar patterns are present within enzyme types within species (e.g. for cod six-cutters $r^2 = 0.176$, $F_{(1,16)} = 3.42$, $P = 0.083$), suggesting that fragment size *per se* and not enzyme classification is responsible for this relationship. One factor that might have contributed to the heterozygosity–fragment size relationship was cDNA probe size. However, no relationship was apparent between the size of the cDNA probe and observed heterozygosity for restriction site ($r^2 = 0.005$, $F_{(1,33)} = 0.155$, $P = 0.697$) or VNTR polymorphisms ($r^2 = 0.001$, $F_{(1,13)} = 0.011$, $P = 0.919$). Similarly, cDNA probe size was not related to the mean size of restriction fragments detected ($r^2 = 0.002$, $F_{(1,33)} = 0.054$, $P = 0.818$ for restriction site polymorphisms; $r^2 = 0.111$, $F_{(1,13)} = 1.62$, $P = 0.225$ for VNTR polymorphisms).

4.3.4   Discussion

Randomly selected clones amplified from cDNA libraries proved to be a rich source of probes for detecting DNA polymorphism in both the deep-sea scallop, *Placopecten magellanicus*, and the Atlantic cod, *Gadus morhua*. The majority of cDNA probes detected at least one of three distinct classes of polymorphism when hybridized to genomic DNA samples from seven unrelated individuals that had been digested with at least four restriction endonucleases. In both species, the most common type of polymorphism observed was consistent with the simple gain or loss of restriction sites located either within, or in the vicinity of, single copy genes. The second class of DNA polymorphism can be explained by the presence of VNTR arrays closely associated with the structural gene probed, and the third by the cDNA probe belonging to a multigene family. Although a similar proportion of cDNA probes in the two species failed to discover any variation, it is possible that more extensive screening involving a larger number of restriction enzymes may have produced at least one polymorphism with virtually all probes tested. However, the rationale for the work was not to identify polymorphism with every probe but rather to produce a battery of probes that identify polymorphism with the same restriction enzyme. Once this objective is met, repeated hybridizations of the same blot with a number of different cDNA probes allows many independent DNA polymorphisms to be readily scored at the population level.

The different types of polymorphism detected by cDNA probes in scallop and cod are consistent with those reported in previous studies. Restriction site variation detected by cDNA probes in tomato (e.g. Bernatzky and Tanksley, 1986) and maize (e.g. Helentjaris, Weber and

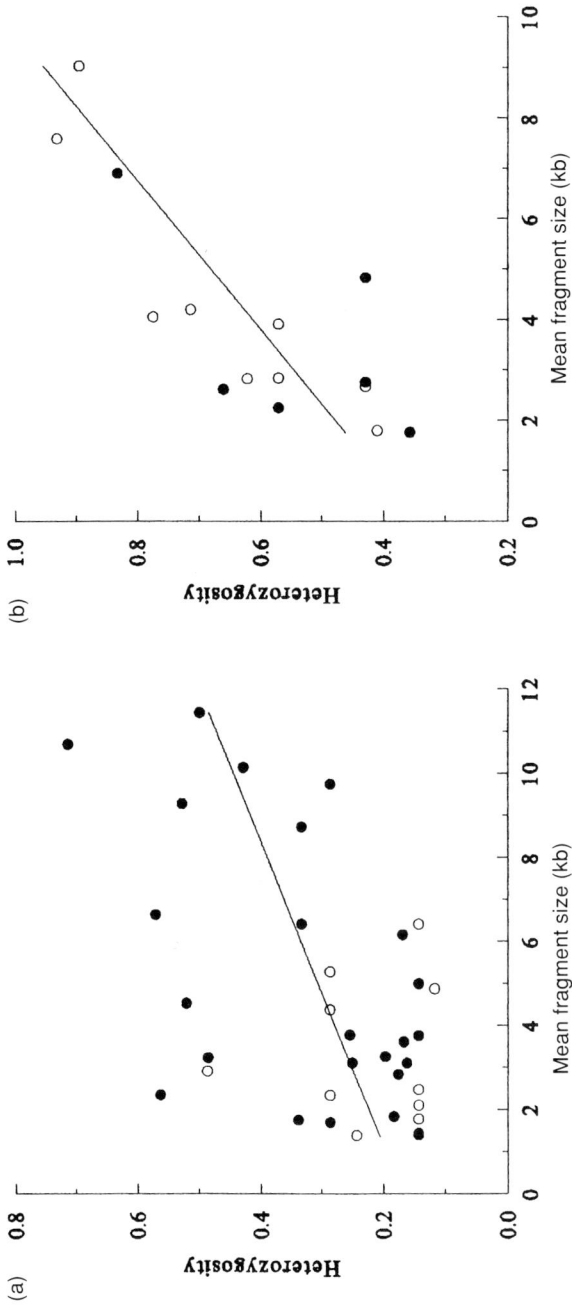

Figure 4.3.2 Relationships between observed heterozygosity and mean restriction fragment size for (a) restriction site polymorphisms and (b) VNTR polymorphisms. Open circles = scallop polymorphisms; closed circles = cod polymorphisms. See text for details.

Wright, 1988) has contributed significantly to the construction of high-resolution linkage maps in these species. Substantial restriction site polymorphism has also been found within the vicinity of structural genes in various *Drosophila* species. For example, in the nine chromosomal regions in *D. melanogaster* summarized by Kreitman (1991), approximately one in six restriction sites exhibits polymorphism. Tandemly repeated minisatellite arrays have also been discovered within the introns or flanking regions of a number of human genes (e.g. Bell, Selby and Rutter, 1982; Proudfoot, Gil and Maniatis, 1982; Weller *et al.*, 1984) and polymorphisms consistent with similar elements residing near structural genes have been described in *Drosophila simulans* (Begun and Aquadro, 1991), the meadow vole *Microtus pennsylvanicus* (Plante *et al.*, 1991), and the red-wing blackbird, *Agelaius phoeniceus* (Gibbs *et al.*, 1991). Two of the 13 human cDNA probes examined by Helentjaris and Gesteland (1983) produced complex multibanded patterns showing considerable restriction enzyme-dependent polymorphism which the authors suggested were the result of the probes belonging to a multigene family. These classes of polymorphism match those observed in scallop and cod and are clearly not mutually exclusive, for both single and multiple copy genes are capable of producing both restriction site and VNTR polymorphisms.

Although the types of cDNA-based polymorphism observed in the present study are consistent with past studies, the levels of variation uncovered by this type of probe are greater than described previously, particularly in humans. For example, all ten single copy human cDNA probes tested by Helentjaris and Gesteland (1983) failed to detect restriction site variation when DNA samples from nine unrelated individuals were digested with a minimum of seven different enzymes. The proportion of cDNA probes producing restriction site variation in scallop and cod was considerably greater than this (approximately one in three for both species), but generally lower than that detected by random clones of originating from genomic DNA in humans (e.g. Botstein *et al.*, 1980) as well as other species (e.g. Quinn and White, 1987; Hall, 1990). Another difference from the human data concerns the high percentage of cDNA clones that uncovered VNTR polymorphisms in scallop and cod (26 % and 9 % of all probes tested, respectively). In comparison, only a limited number of human genes have been found to harbour hypervariable regions (Wong *et al.*, 1987) and only about 2–4 % of unselected human genomic clones detect VNTR polymorphisms (Nakamura *et al.*, 1987). The considerable number of VNTR polymorphisms discovered by cDNA probes in the two aquatic species studied here suggests that this type of variation might be more commonly associated with structural genes than was previously thought from studies of the human genome.

The proportions of cDNA probes screened that detect VNTR and so-called 'fingerprint' polymorphisms differ considerably between the two species examined (Table 4.3.1). The frequent occurrence of VNTR polymorphisms in scallop could simply reflect the fact that such repeat arrays are relatively common within introns or flanking regions of their structural genes. However, this result may have been influenced by the inadvertent use of restriction enzymes that cut within repeating elements present in the cod genome. When this occurs, VNTR polymorphism will go unnoticed and the frequency of this type of variation will be underestimated. However, given the fact that VNTR polymorphisms were observed in both species with virtually all restriction enzymes tested, the extent of this bias is unlikely to be too large. The higher proportion of cDNAs in cod that apparently belong to multigene families may reflect the general trend towards the greater number and diversity of enzyme loci (in the form of tissue- and developmental stage-specific isozymes) and other proteins (such as the α and β-globins) that has occurred in the evolution of the vertebrates (*cf.* Ohno, 1970). Why such members of multigene families should produce higher levels of polymorphism than single copy genes is unclear, but may be related to diminished constraint or reflect the presence of pseudogenes.

Both restriction site and VNTR polymorphisms exhibit significant positive correlations between observed heterozygosity and mean restriction fragment size. One mechanism that may be responsible for this relationship is mutation pressure. Jeffreys and coworkers were the first to recognize that large minisatellite arrays were more highly polymorphic and exhibited higher rates of mutation than smaller arrays (Jeffreys, Wilson and Thein, 1985; Wong *et al.*, 1985). Irrespective of the mutational processes generating this variability, high rates of mutation should generate positive relationships between the size of a VNTR array, its observed heterozygosity, and the number of alleles segregating in a population. Positive relationships between allele size, heterozygosity and allele number are indeed evident at the 60 human VNTR polymorphisms (characterized for $n = 60$) summarized in Table 3 of Nakamura *et al.* (1987). Here, the number of alleles correlated strongly with both maximum allele size ($r^2 = 0.117$, $F_{(1,58)} = 7.70$, $P = 0.007$) and observed heterozygosity ($r^2 = 0.624$, $F_{(1,58)} = 96.4$, $P < 0.0001$). Since VNTR alleles detected by cDNA probes in the present study are independent of flanking restriction sites, and were sized by averaging across the different enzymes producing the same genotypic patterns, the length of the repeat array giving rise to the polymorphism should correlate positively with the mean fragment size observed. If this prediction holds, a similar trend is confirmed for the combined scallop and cod VNTR data as the number of alleles observed at each polymorphism is highly correlated with mean fragment size ($r^2 = 0.594$, $F_{(1,13)} = 19.1$, $P < 0.001$). These patterns suggest that the relationship between heterozygosity and mean fragment size for VNTR polymorphisms is driven by mutation.

Mutation can also account for the positive relationship observed between heterozygosity and fragment size for restriction site polymorphisms, since the number of segregating sites (and hence number of alleles) present along a particular fragment of DNA is expected to correlate positively with the size of the fragment itself. However, unlike with VNTR poly-morphisms the number of alleles at restriction site polymorphisms did not correlate with mean fragment size ($r^2 = 0.054$, $F_{(1,33)} = 1.88$, $P = 0.179$). This suggests that it is not the number of segregating restriction sites that is determining the level of polymorphism for this type of variation but rather the population frequency of the polymorphic site itself, and a significant factor determining this frequency is its apparent proximity to the region of the structural gene recognized by the cDNA probe. For restriction site polymorphisms, it seems reasonable to assume that the coding region recognized by a cDNA probe is randomly located along a particular restriction fragment, and that an equal probability exists for the polymorphism to reside at either of the two flanking restriction sites. If these two conditions are met, the larger the DNA fragment detected by a probe, the greater the probability that the variable site occurs, on average, further away from the transcribed region represented by the probe. Therefore, the positive relationship between heterozygosity and mean fragment size observed for this type of variation suggests that polymorphic restriction sites situated closer to the probed region (on smaller fragments) exhibit lower levels of presumably neutral polymorphism than those residing further away (on larger fragments).

The diminished level of polymorphism at restriction sites situated close to the transcribed region may reflect increased constraint and/or a reduced neutral mutation rate. This is certainly expected if the polymorphism falls within coding DNA, yet none of the restriction site polymorphisms observed in either species occurred within the region corresponding to the cDNA probe. This argues against the possibility that the variable sites detected happened to reside within transcribed regions that, by chance, did not overlap with that of the cDNA probe. Alternatively, this pattern of reduced neutral polymorphism is also consistent with 'selective sweeps' of advantageous mutations that have led to the local fixation of neutral polymorphisms by genetic hitchhiking (*cf.* Kaplan, Hudson and Langley, 1989). Recent studies comparing DNA sequence variation within and between various chromosomal regions in *Drosophila melanogaster* and its sibling species *D. simulans* have produced evidence favouring hitchhiking (Begun and Aquadro, 1991, 1992). It is not possible, however, to distinguish between explanations based on hitchhiking or differential levels of constraint without more detailed information on nucleotide polymorphism within and around these structural genes provided either by the mapping of multiple restriction sites or, preferably, by direct sequencing.

Acknowledgements

I would like to thank K. Mesa for technical assistance and two anonymous reviewers for providing helpful comments on the manuscript. Funding for the study was provided by a Killam Postdoctoral Fellowship, by the Marine Gene Probe Laboratory of Dalhousie University, and by the Ocean Production Enhancement Network (NSERC Canada).

References

Aquadro, C.F., Lado, K.M. and Noon, W.A. (1988) The *rosy* region of *Drosophila melanogaster* and *Drosophila simulans*. I. Contrasting levels of naturally occurring DNA restriction-map variation and divergence. *Genetics*, **119**, 875–88.

Begun, D.J. and Aquadro, C.F. (1991) Molecular population genetics of the distal portion of the X chromosome in *Drosophila*: evidence for genetic hitchhiking of the *yellow-achaete* region. *Genetics*, **129**, 1147–58.

Begun, D.J. and Aquadro, C.F. (1992) Levels of naturally occurring polymorphism correlate with recombination rates in *D. melanogaster. Nature*, **356**, 519–20.

Bell, G.I., Selby, M.J. and Rutter, W.J. (1982) The highly polymorphic region near the human insulin gene is composed of simple tandemly repeating sequences. *Nature*, **295**, 31–5.

Bernatzky, R. and Tanksley, S.D. (1986) Towards a saturated linkage map in tomato based on isozymes and random cDNA sequences. *Genetics*, **112**, 887–98.

Botstein, D., White, R.L., Skolnick, M. and Davis, R.W. (1980) Construction of a genetic linkage map in man using restriction fragment length polymorphisms. *Am. J. Human Genet.*, **32**, 314–31.

Gibbs, H.L., Boag, P.T., White, B.N., Weatherhead, P.J. and Tabak, L.M. (1991) Detection of a hypervariable DNA locus in birds by hybridization with a mouse MHC probe. *Mol. Biol. Evol.*, **8**, 433–46.

Gillespie, J.H. (1991) *The Causes of Molecular Evolution*, Oxford University Press, New York.

Golding, G.B., Aquadro, C.F. and Langley, C.H. (1986) Sequence evolution within populations under multiple types of mutation. *Proc. Natl. Acad. Sci., USA*, **83**, 427–31.

Hall, H.G. (1990) Parental analysis of introgressive hybridization between African and European honeybees using nuclear DNA RFLPs. *Genetics*, **125**, 611–21.

Harris, H. (1966) Enzyme polymorphism in man. *Proc. R. Soc. Lond. B*, **164**, 298–310.

Helentjaris, T. and Gesteland, R. (1983) Evaluation of random cDNA clones as probes for human restriction fragment polymorphisms. *J. Mol. Appl. Genet.*, **2**, 237–47.

Helentjaris, T., Weber, D. and Wright, S. (1988) Identification of the genomic locations of duplicate nucleotide sequences in maize by analysis of restriction fragment length polymorphisms. *Genetics*, **118**, 353–63.

Jeffreys, A.J., Wilson, V. and Thein, S.L. (1985) Hypervariable 'minisatellite' regions in human DNA. *Nature*, **314**, 67–73.

Kaplan, N.L., Hudson, R.R. and Langley, C.H. (1989) The 'hitchhiking' effect revisited. *Genetics*, **123**, 887–99.

Karl, S.A. and Avise, J.C. (1992) Balancing selection at allozyme loci in oysters: implications from nuclear RFLPs. *Science*, **256**, 100–102.

Kimura, M. (1983) *The Neutral Theory of Molecular Evolution*, Cambridge University Press, Cambridge.

Kreitman, M. (1991) Detecting selection at the level of DNA, in *Evolution at the Molecular Level* (eds R.K. Selander, A.G. Clark and T.S. Whittam), Sinuaer, Sunderland, MA, pp. 204–21.

Lewontin, R.C. and Hubby, J.L. (1966) A molecular approach to the study of genic heterozygosity in natural populations. II. Amount of variation and degree of heterozygosity in natural populations of *Drosophila pseudoobscura. Genetics*, **54**, 595–609.

Miller, SA., Dykes, D.D. and Polesky, H.F. (1988) A simple salting out procedure for extracting DNA from human nucleated cells. *Nucleic Acids Res.*, **16**, 1215.

Nakamura, Y., Leppert, M., O'Connell, P. *et al.* (1987) Variable number of tandem repeat (VNTR) markers for human gene mapping. *Science*, **235**, 1616–22.

Nelson, J.A., Tang, Y. and Boutilier, R.G. (1993) Differences in exercise physiology between two Atlantic cod (*Gadus morhua*) populations from different environments. *Physiol. Zool.* (in press)

Nevo, E., Beiles, A. and Ben-Shlomo, R. (1984) The evolutionary significance of genetic diversity: ecological, demographic and life history correlates, in *Evolutionary Dynamics of Genetic Diversity* (ed. G.S. Mani), Springer-Verlag, Berlin, pp. 13–213.

Ohno, S. (1970) *Evolution by Gene Duplication*, Springer-Verlag, Berlin.

Plante, Y., Boag, P.T., White, B.N. and Boonstra, R. (1991) Highly polymorphic genetic markers in meadow voles (*Microtus pennsylvanicus*) revealed by a murine major histocompatibility complex (MHC) probe. *Can. J. Zool.*, **69**, 213–20.

Proudfoot, N.J., Gil, A. and Maniatis, T. (1982) The structure of the human zeta-globin gene and a closely linked, nearly identical pseudogene. *Cell*, **31**, 553–63.

Quinn, T.W. and White, B.N. (1987) Identification of restriction-fragment-length polymorphisms in genomic DNA of the lesser snow goose (*Anser caerulescens caerulescens*). *Mol. Biol. Evol.*, **4**, 126–43.

Weller, P., Jeffreys, A.J., Wilson, V. and Blanchetot, A. (1984) Organization of the human myoglobin gene. *EMBO J.*, **3**, 439–46.

Wong, Z., Wilson, V., Jeffreys, A.J., and Thein, S.L. (1985) Cloning a selected fragment from a human DNA 'fingerprint': isolation of an extremely polymorphic minisatellite. *Nucleic Acids Res.*, **14**, 4605–16.

Wong, Z., Wilson, V., Patel, I., Povey, S. and Jeffreys, A.J. (1987) Characterization of a panel of highly variable minisatellites cloned from human DNA. *Ann. Human Genet.* **51**, 269–88.

Zera, A.J., Koehn, R.K. and Hall, J.G. (1985) Allozymes and biochemical adaptation, in *Comprehensive Insect Physiology, Biochemistry, and Pharmacology*, Vol. 10 (eds G.A. Kerkut and L.I. Gilbert), Pergamon, New York, pp. 633–74.

Zouros, E., Pogson, G.H., Cook, D.I. and Dadswell, M.J. (1992) Apparent selective neutrality of mitochondrial DNA size variation: a test in the deep-sea scallop *Placopecten magellanicus*. *Evolution* **46**, 1466–76.

5

DNA technology and phylogeny of fish

5.1 MOLECULAR PHYLOGENETIC STUDIES OF FISH

Axel Meyer

Abstract

Knowledge of the evolution of DNA, particularly mitochondrial DNA, of fish and the phylogenetic relationships among fish has increased rapidly since the invention of the polymerase chain reaction (PCR). This enzymatic cloning technique, together with direct sequencing, has simplified and dramatically accelerated the accumulation of DNA sequence information. Methods of data collection and data analysis for phylogenetic studies with particular emphasis on fish are outlined. Aspects of the biology of mitochondrial DNA that pertain to phylogenetic reconstruction are reviewed and advantages of DNA sequences over alternative DNA-based genetic markers are highlighted. Examples of phylogenetic work based on mitochondrial and nuclear DNA sequences are used to illustrate the methods, advantages and potential problems with techniques, choice of genes and analyses.

5.1.1 Introduction

There are several kinds of biochemical data that can be used to infer phylogenetic relationships among species. Allozyme, immunological and DNA–DNA hybridization data have been widely used but are now increasingly being replaced by several types of DNA-based data. Since the advent of the polymerase chain reaction (PCR) in 1985–1986 (Mullis *et al.*, 1986; Saiki *et al.*, 1985, 1988; Wrishnik, Higuchi and Stoneking, 1987), our knowledge about DNA of fish has increased dramatically. The impact of this technological innovation on the understanding of the evolution of DNA of fish and phylogenetic relationships among fish is the focus of this section. From my biased perspective, I will attempt to summarize the advantages and disadvantages of currently used DNA-based molecular data to deduce genealogical relationships among fish. Data collection and analysis will be briefly touched on and recent publications and reviews that provide more detailed information will be recommended.

Genetics and Evolution of Aquatic Organisms. Edited by A.R. Beaumont.
Published in 1994 by Chapman & Hall, 2–6 Boundary Row, London SE1 8HN.
ISBN 0-412-49370-5

5.1.2 Methods and kinds of DNA data

(a) Restriction endonuclease methods

With this method, DNA (usually mtDNA) is cut with restriction enzymes (endonucleases) and the resulting restriction fragment length polymorphisms (RFLP) are used as binary characters in a phylogenetic analysis (restriction-fragment data). Often, restriction maps are constructed from the RFLP patterns (after simultaneous digests with two enzymes) and then analysed in a phylogenetic analysis as binary characters, coding for absence or presence of particular restriction sites (restriction-site data). Usually enzymes with six-base recognition sites are used for evolutionary studies among more distantly related species and enzymes with four-base-pair (bp) recognition sites are used in studies that require more detailed information, e.g. in investigations among more closely related species and population-level questions within species. This method has enjoyed widespread application (reviews: Wilson, Thomas and Beckenbach, 1985; Avise, 1986; Avise *et al.*, 1987; Harrison, 1991; but see Wilson *et al.*, 1989) and continues to be applied for a wide range of taxonomic questions in fish.

Methodological problems with restriction-fragment data are that the assumption of independence (see below) is violated, and deletions and insertions are problematic and are potential sources of error when trying to establish homology between restriction fragments (e.g. Swofford and Olsen, 1990). For these reasons caution should be applied when interpreting RFLP patterns for phylogenetic analysis. Restriction-site data, a map derived from RFLP data, also have problems with phylogenetic reconstruction, due to the asymmetry with which restriction sites are gained and lost. The loss of a restriction site is much more likely than a gain. If a particular restriction site is 6 bp long, and a 6-bp stretch of DNA is different by only one nucleotide for the recognition sequence, only 1 out of 18 substitutions (substitutions of the same nucleotide at the same site remain undetected) is going to create this site, i.e. it is unlikely that sites will be gained. However, losing sites is going to be much more frequent, i.e. any of these 18 mutations at this restriction site is going to result in the loss of that site (Templeton, 1983a,b). A special case of parsimony (Dollo parsimony, see below) takes the asymmetry of gains and losses into consideration during the phylogenetic analysis of RFLP data.

An additional disadvantage of restriction data is that results of RFLP and even restriction sites are not immediately transferable between laboratories. That is because the same endonucleases are not always used and the same kinds of gels are not run by all researchers. In this respect, restriction data suffer from the same drawbacks as allozyme data; they tend to be laboratory-specific and often even project-specific results. Although endonuclease data contributed tremendously to our increased understanding of intra- and interspecific genetic variation and phylogenetic relationships among fish and remain a viable technique, I am biased in favour of actual DNA sequences (see also Wilson *et al.*, 1989). For population-level work, PCR (see below) and restriction analyses are sometimes combined: known DNA fragments (usually mtDNA) are amplified via PCR and then cut with restriction enzymes rather than sequenced. The advantage of this combined approach is that larger sample sizes can be screened than if every individual is sequenced.

(b) The polymerase chain reaction and direct sequencing

PCR is an enzymatic cloning technique that allows the amplification of any stretch of DNA (within size limits) that is flanked by synthetic oligonucleotide 'primers' (Saiki *et al.*, 1985, 1988). The primers are usually around 20 bp in length and define the 5′ and 3′ end of the double-stranded piece of DNA that is going to be amplified. The specificity of the amplification is accomplished through the need for an almost perfect fit of the primers to the template DNA. During each cycle of PCR, the number of copies of the DNA fragment delineated by the primers at either end is doubled. Usually 25–40 cycles are completed in a thermal cycler in about three hours. PCR is much faster and cheaper than conventional cloning techniques. First, a double-stranded PCR product is produced (Figure 5.1.1 (a)) that is then either sequenced (double-stranded sequencing, or, alternatively 'cycle-sequenced') or subcloned and then sequenced or cut with restriction enzymes (see above) or used as template DNA for a subsequent asymmetric amplification (Gyllensten and Erlich, 1988) or digested with an exonuclease to produce single-stranded DNA (Figure 5.1.1 (b)) for direct sequencing of single-stranded DNA. Sequencing gels of single-stranded DNA often allow one to read more base pairs than sequencing gels of double-stranded DNA. Single-stranded PCR-amplified DNA can be as clean as subcloned DNA and routinely more than 300–400 bp can be unambiguously determined from a single sequencing reaction. Figure 5.1.2 shows some partial cytochrome *b* sequences from three-spined sticklebacks that were produced through double- and single-stranded amplification and direct sequencing.

Although only a recent addition to the molecular toolbox, PCR and its application to evolutionary biology have already been reviewed (e.g. White, Arnheim and Erlich, 1989; Arnheim, White and Rainey, 1990; refs in Erlich, 1989; Innis *et al.*, 1990). A new journal entitled *PCR* first appeared in 1991. Details on PCR methodology, technical improvements and modifications and new applications of PCR technology can be found in journals like *PCR* and *BioTechniques*.

The determination of DNA sequences tends to be more time-consuming, costly and technically involved; however, DNA sequences of homologous mitochondrial and nuclear

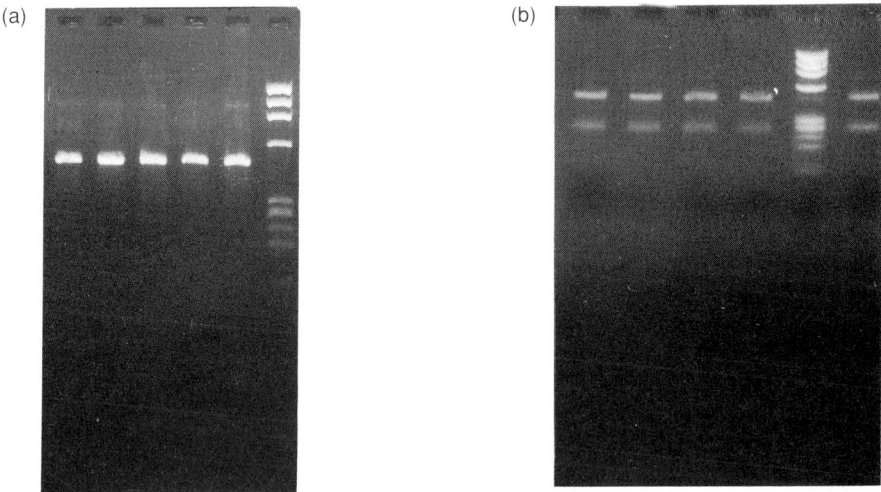

Figure 5.1.1 (a) Double-stranded PCR product of the 5' end of cytochrome *b* of five species of South American characins (five left lanes, amplified with L14725 and H15148, protocols from Kocher *et al.*, 1989). One major amplification product of the expected size, about 450 bp, is visible. The size standard is Phi-X-174 cut with *Hae*III of 1353, 1078, 872, 602, 310, 281, 271, 234, 194, 118 in lane 6. (b) Single-stranded PCR products produced by asymmetric PCR (Gyllensten and Erlich, 1988). L14725 was diluted 1 : 100. The same five fish as in (a) (size standard in lane 5). The targeted double-stranded product is seen above and the faster migrating single-stranded product is below. Data from G. Orti's unpublished study on the phylogenetic relationships among the major lineages of characin fish.

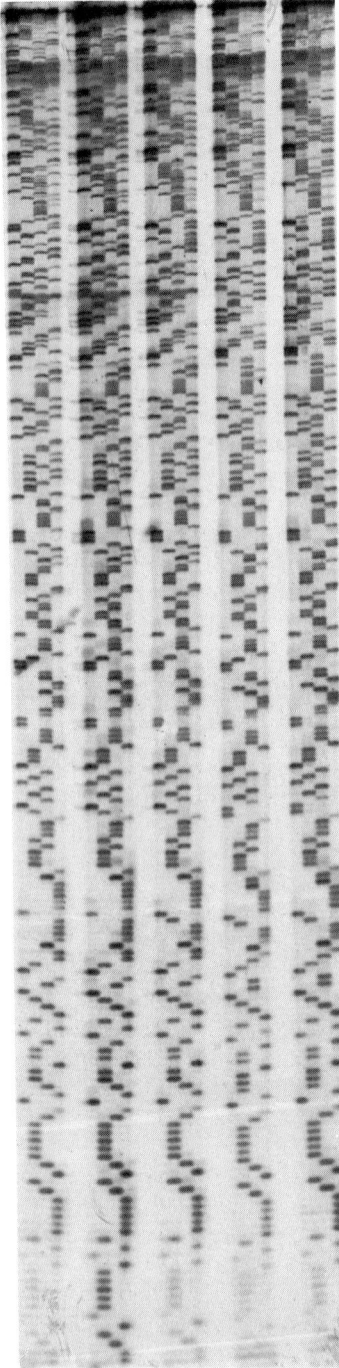

Figure 5.1.2 Example of the quality of direct sequencing of PCR-amplified DNA: 5′ end of the cytochrome *b* gene of five individuals of the three-spined stickleback (*Gasterosteus aculeatus*) from several circumpolar locations (from left to right: Quebec, New York, Japan, Rhode Island, Quebec) (Orti, Bell and Reimchen, 1994). Up to 350 bp can be read from this wedge gel. The loading sequence is G, A, T, C from left to right.

genes will allow direct comparisons and study of DNA from different species that has been obtained in different laboratories. DNA sequences can be stored in databanks (e.g. EMBL, GenBank) and are universally usable, powerful data. The increased costs of DNA sequences compared to RFLP data are far outweighed by their advantage as a universally retrievable and applicable type of data, since homologous data from independent laboratories can be used in direct comparisons for new studies.

(c)   RAPD-PCR

A recent PCR-based technique, random amplified polymorphic DNA (RAPD-PCR) (Welsh and McClelland, 1990; Williams *et al.*, 1990), has stirred up a lot of interest among researchers interested in population-level questions (e.g. stock identity, paternity, mating systems). This technique utilizes only one instead of two different PCR primers. This single primer is small, usually in the range of 10 bp, has a random sequence, and anneals on the genome at priming sites close to each other and in an inverted orientation. During the PCR amplification, at low annealing temperatures and with a relatively large number of cycles, random pieces of DNA are amplified and then visualized usually on ethidium-bromide-stained NuSieve-agarose gels (often 2–4% gels). Based on the number and position of bands on these gels, genetic relatedness between the samples is inferred. RAPD-PCR was greeted with enthusiasm, because it seems to provide very detailed genetic information quickly and relatively inexpensively.

Unfortunately, the large initial interest in RAPD-PCR has not been matched by a flood of successful studies. Several technical and analytical problems have dampened the initial excitement. Part of the appeal of PCR is its technical simplicity. However, in RAPD-PCR many factors (Mg^{2+}, primer, template DNA concentrations and exact thermal cycling conditions) need to be carefully controlled to ensure reproducibility (e.g. Hadrys, Balick and Schierwater, 1992). Analytically, the fact that DNA fragments of uncertain homology between individuals are amplified by RAPD-PCR makes the analysis less tractable. However, if these technical and analytical obstacles are solved in the future, this technique might very well become the panacea for population-level work. Any diagnostic RAPD marker can be used to generate specific probes for Southern analyses.

5.1.3   Methods of phylogenetic data analysis

The history of phylogenetic reconstruction has been turbulent and full of acrimony (review: Hull, 1988). There is a large body of literature on phylogenetic reconstruction and several excellent reviews provide an entry to this literature (Felsenstein, 1982, 1988; Swofford and Olsen, 1990; refs in Hillis and Moritz, 1990). All methods make underlying simplifying assumptions about how DNA sequences evolve. No consensus for a single method of phylogenetic reconstruction has been reached among researchers favouring rivalling methodologies. Philosophical as well as practical arguments are used by the proponents of particular methods to argue the superiority of one method over another. Simulation studies, intended to determine which method of phylogenetic reconstruction will provide the best estimate, tend to give preference to the favoured method of the researcher who conducted the simulation study.

Two types of molecular data can be used for phylogenetic analysis. Discrete characters are collected when DNA is sequenced or scored with restriction enzymes. These discrete data provide information about the DNA of a particular individual, often assumed to be characteristic for the species. Further assumptions are the independence and homology of nucleotide positions. If species rather than gene-trees are the purpose of the study, only orthologous rather than paralogous genes should be compared. Discrete data can be transformed into similarity or distance data, by pairwise comparisons of two sequences.

Distance methods of phylogenetic reconstruction are based on these pairwise distances (see below for different ways of calculating and correcting for 'multiple hits') of sequences and attempt to fit a tree to a distance matrix. The goodness of fit (e.g. least squares methods) of the observed distances to the expected distances (based on the tree) is measured and the topology that minimizes the discrepancy between expected and observed distances is chosen. Several distance methods exist (reviews: Felsenstein, 1988; Swofford and Olsen, 1990; Nei, 1991).

The parsimony method, 'the method of minimum net evolution', aims to find the evolutionary tree that requires the fewest changes of nucleotides to explain the evolution of the DNA sequences under consideration. Its philosophy is the hypothetico-deductive approach in which Ockham's principle is invoked, i.e. evolution is believed to proceed by the shortest, simplest pathway. Parsimony only considers so-called 'phylogenetically informative sites' in the calculation of the topology of the tree. Sites that do not require different numbers of changes on alternative trees of different topology (e.g. sites that are identical or sites that differ only in one of the species under consideration) are ignored. Both distance and parsimony methods only look at part of the information in the data.

Only the maximum-likelihood method attempts to use all the information contained in DNA sequences (Felsenstein, 1981) by using statistical criteria to distinguish between alternative trees. It uses a model of likelihoods of substitution changes and attempts to fit the data with a tree. The likelihood of the topology of a tree is the probability of the data, given the tree and the model. The maximum-likelihood method chooses a tree with a topology and branch length that has the highest likelihood. This method allows for unequal base composition and uneven transition–transversion rates, and does not require a molecular clock. The major practical drawback of the maximum-likelihood method is its inherent computational complexity due to exact probability models of sequence change, which in practice limits the number of sequences that can be analysed in a reasonable amount of time with currently available computer power.

Unfortunately, studies that test the power of various commonly used methods of phylogenetic reconstruction on known phylogenies are rare (Atchley and Fitch, 1991; Hillis *et al.*, 1992). They have failed to clearly identify a particular methodology as the best. However, the strengths and weaknesses of the alternative methods of phylogenetic reconstruction and the kinds of genes to be used and to be avoided have been made more clear by these studies (Felsenstein, 1988; Swofford and Olsen, 1990; Nei, 1991; Atchley and Fitch, 1991; Hillis *et al.*, 1992).

In practice, the data should be subjected to several methods of phylogeny reconstruction and differences in results will usually pinpoint areas of weakness in the phylogenetic tree. Congruent results of different phylogenetic methods will inspire confidence that a phylogenetic estimate has been found that is close to the true relationships. The robustness of a molecular phylogeny is also often judged by whether or not it is congruent with a 'well-established' phylogeny based on morphological data. Obviously, since the molecules are just a part of the whole organism they are expected to have experienced the same evolutionary history, and therefore report the same evolutionary information, as other parts (e.g. morphology) of the species. Generally, a molecular phylogeny is viewed as suspicious if it is in conflict with the traditional phylogeny (reviews: Patterson, 1987; Swofford, 1991).

Several methods testing the confidence in the phylogenetic estimate have been developed; the most commonly used one is the 'bootstrap' (Felsenstein, 1985). Several other methods that evaluate the statistical confidence of molecular phylogenies are available (review: Li and Gouy, 1991).

5.1.4 Biology of the mitochondrial genome

Because of the widespread popularity of mitochondrial DNA (mtDNA) for population and phylogenetic work, I will first review some of the basic biology of mtDNA. A more detailed review on the evolution of mtDNA of fish is available (Meyer, 1993). Fish are probably the least well-studied group of vertebrates in terms of their mtDNA, and to date only a handful of papers with actual mtDNA sequences from fish has been published.

The mitochondrial genome of vertebrates is a single, small, double-stranded, circular DNA molecule contained in mitochondria, and up to several thousand copies of the mitochondrial genomes are found per cell. Typically, the size of animal mitochondrial genomes is about 16 500– 500 bp (reviews: ± Brown, 1981, 1983, 1985; Clark-Walker, 1985; Moritz, Dowling and Brown, 1987).

The complete mitochondrial genome has been sequenced or the gene order determined in several invertebrates and vertebrates. The invertebrates include two sea urchins (Jacobs *et al.*, 1988; Cantatore *et al.*, 1989), sea stars (not complete) (Smith *et al.*, 1989, 1990; Jacobs *et al.*, 1989; Himeno *et al.*, 1987), the fruit fly, *Drosophila yakuba* (Clary and Wolstenholme, 1985), *Leishmania tarentolae* (not complete) (de la Cruz, Neckelmann and Simpson, 1984), *Artemia salina* (not complete) (Batuecas *et al.*, 1988), *Daphnia* (not complete) (Stanton, Crease and Herbert, 1991), the blue mussel, *Mytilus edulis* (not complete) (Hoffmann, Boore and Brown, 1992), the nematodes, *Ascaris suum* and *Caenorhabditis elegans* (Okimoto *et al.*, 1992), the platyhelminth *Fasciola* (not complete) (Garey and Wolstenholme, 1989) and the ciliate, *Paramecium* (Pritchard *et al.*, 1990). Among the vertebrates, complete sequences for human (Anderson *et al.*, 1981), cow (Anderson *et al.*, 1982), mouse (Bibb *et al.*, 1981), rat (Gadaleta *et al.*, 1989), fin whale (Arnason, Gullberg and Widegren, 1991), harbour seal (Arnason and Johnnson, 1992), chicken (Desjardins and Morais, 1990) and frog *Xenopus* (Roe *et al.*, 1985) have been published. Additionally, many other partial mitochondrial sequences, too numerous to be listed here, are known.

Two complete mitochondrial genomes for two cyprinid fish have been determined (Tzeng *et al.*, 1992) (*Crossostoma lacustre*: GenBank accession no. M91245; the complete sequence for the carp is also available in GenBank accession no. X61010). The next best known fish mtDNA genome is of the Atlantic cod (*Gadus morhua*) (Johansen, Guddal and Johansen, 1990; Johansen, personal communication) (Figure 5.1.3).These studies and restriction and partial sequence studies on carp (Araya *et al.*, 1984) and salmonids (Thomas, Withler and Beckenbach, 1986; Davidson, Birt and Green,1989) and unpublished data from chondrichthyes and agnathans from Wes Brown's laboratory (cf. Moritz, Dowling and Brown, 1987) indicate that the piscine mtDNA gene order complies with the general vertebrate condition. The mitochondrial gene order of animals is different in every phylum that has been studied. Even within phyla (e.g. echinoderms) differences in gene order exist (Smith *et al.*, 1989, 1990; Himeno *et al.*, 1987). Mitochondrial gene orders differ slightly among vertebrates (Yoneyama, 1987; Desjardins and Morais, 1990, 1991; Pääbo *et al.*, 1991; reviewed in Meyer, 1993). The piscine mitochondrial gene order does not differ from the vertebrate 'consensus' gene order (Figure 5.1.3)

The mitochondrial genome of vertebrates contains 13 genes coding for proteins, two genes coding for ribosomal RNAs (small 12S and large 16S rRNA), 22 genes coding for transfer RNAs (tRNAs) and one major noncoding region (control region) that contains the initiation sites for mtDNA replication and RNA transcription (Figure 5.1.3). The protein coding genes are seven subunits of NADH dehydrogenase (ND1, 2, 3, 4, 4L, 5, 6), cytochrome *b*, three subunits of cytochrome *c* oxidase (CO I, II, III) and two subunits of ATP synthetase (ATPase 6 and 8).

DNA technology and phylogeny of fish

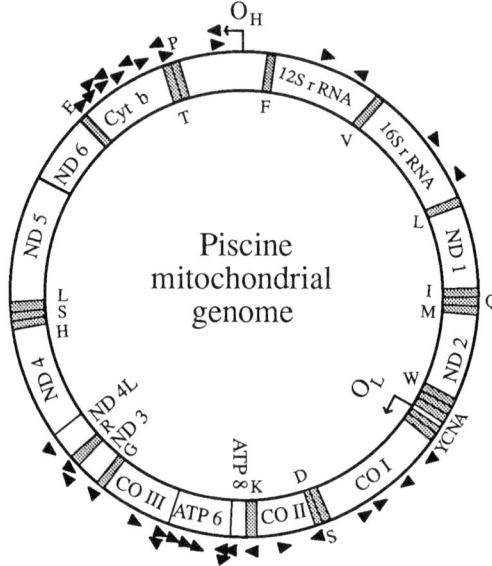

Figure 5.1.3 Piscine mitochondrial gene order. The origins of H- and L-strand replication are indicated in the figure. The origin of the H-strand is in the control region, and the origin of the L-strand replication is in the YCNAW tRNA gene cluster. Transfer RNA genes are shown in shaded boxes. The coding sequences (templates) of all proteins (except ND 6) and the majority of the tRNA genes are on the H-strand. The complete names for the abbreviated names of proteins are given in the text. The tRNA genes coded for by the L-strand are labelled on the outside of the circle, and the tRNA genes coded for by the H-strand are labelled on the inside. The arrows indicate the direction and approximate relative positions of 'universal' PCR primers, the sequence of which is listed in Table 5.1.1.

The two strands of the mitochondrial genome are designated light 'L' and heavy 'H'. These names reflect marked differences in their G + T content in vertebrates resulting in their different behaviour in caesium chloride density gradients. With few exceptions, all genes in the vertebrate mitochondrial genome are encoded by the H-strand; only ND 6 and eight tRNAs are encoded by the L-strand (Figure 5.1.3).

The largest mitochondrial genome (up to 39.3 kb) found so far in animals is from scallops, which also show large intraspecific size variation (Snyder *et al.*, 1987; Gjetvaj, Cook and Zouros, 1992). Often, large intraspecific variation is due to tandem duplications involving the control region (see below). In fish, intraspecific size differences can be as large as differences between species (e.g. Bentzen, Leggett and Brown, 1988; Bermingham, Lamb and Avise, 1986; Buroker *et al.*, 1990; reviewed in Brown, 1983, 1985; Moritz, Dowling and Brown, 1987).

Animal mtDNA is haploid and non-recombining (Hayashi, Tagashira and Yoshida, 1985; but Horak, Coon and Dawid, 1974; Olivo *et al.*, 1983; refs in Hurst, 1991) and appears to be almost exclusively maternally inherited. Paternal mitochondria appear to be actively degraded during fertilization (Vaughn, DeBonte and Wilson, 1980) or 'outreplicated' shortly thereafter (Meland *et al.*, 1991). Recent results (Koehler *et al.*, 1991) tend to support the 'bottleneck hypothesis', that maternal inheritance of mtDNA is mediated by the differential amplification of a small number of specific germ-line mtDNA molecules from the

Table 5.1.1 Primers for the polymerase chain reaction

Gene and primer name	Sequence	Source
Control region		
L15926	TCAAAGCTTACACAGTCTTGTAAACC	Kocher et al. (1989)
H16498	CCTGAAGTAGGAACCAGATG	Meyer et al. (1990)
L16518	CATCTGGTTCTTTCTTCAGGGCCAT	Meyer (unpublished)
12S rRNA[a]		
H1109	GTGGGGTATCTAATCCCAGTT	Meyer (unpublished)
L1091	AAAAAGCTTCAAACTGGGATTAGATACCCCACTAT	Kocher et al. (1989)
H1478	TGACTGCAGAGGGTGACGGGCGGTGTGT	Kocher et al. (1989)
16S rRNA[a]		
L2510	CGCCTGTTTATCAAAAACAT	Palumbi et al. (1991)
H3080	CCGGTCTGAACTCAGATCACGT	Palumbi et al. (1991)
CO I[a]		
L5950	ACAATCACAAAGAYATYGG	Normark, McCune and Harrison (1991)
L6586	OCTGCAGGAGGAGGAGAYCC	Palumbi et al. Cof, (1991)
H7086	CCTGAGAATARKGGGAATCAGTG	Palumbi, Coe, (1991)
H7196	AGAAAATGTTGWGGGAARAA	Normark, McCune and Harrison (1991)
CO II		
L7450	AAAGGAAGGAATCGAACCCCC	Normark, McCune and Harrison (1991)
H8055	GCTCATGAGTGGAGACGTCTT	Normark, McCune and Harrison (1991)
ATPase 8		
L8331	TAAGCRNYAGCCTTTTAAG	Meyer (unpublished)
ATPase 6		
H8517	GGGRACTTTGACTGGTACT	Meyer (unpublished)
L8531	CCCCYTGAAACTGACCATG	Meyer (unpublished)
L8580	AGCCCCACATACCTAGGTATCCC	Meyer (unpublished)
H8674	AARATTTGTTGHGTRAARCGRTT	Meyer (unpublished)
H8907	GGGGTTCCTTCAGGCAATAAATG	Meyer (unpublished)
H8969	GGGGNCGRATRAANAGRCT	Meyer (unpublished)
H9210	GTAKGCGTTGCTTGGTGTGCCAT	Meyer (unpublished)

Table 5.1.1 (contd)

Gene and primer name	Sequence	Source
CO III		
L9225	CACCAAGCACACGCATACCACAT	Meyer (unpublished)
H9407	AAAGTTCCTGTGGTGTGCGGGGG	Meyer (unpublished)
ND3		
L10028	AGTAYANGTRRCTTCCAA	Meyer (unpublished)
H10430	TTGAGCGAAATCAA	Meyer (unpublished)
ND4L		
L10420	AAYAARAYCNTTGATTTCGRCTCA	Meyer (unpublished)
H10720	TCYGTGCCRTGYGTYCGNGC	Meyer (unpublished)
Cytochrome *b*		
L14725	CGAAGCTTGATATGAAAAACCATCGTTG	Pääbo (1990)
L14841	AAAAAGCTTCCATCCAACATCTCAGCATGATGAAA	Kocher et al. (1989)
L15018	GCYAAYGGCGCATCCTTYTYTYT	Orti et al. (1994)
H15149	AAACTGCAGCCCCTCAGAATGATATTTGTCCTCA	Kocher et al. (1989)
L15162	GCAAGCTTCTACCATGAGGACAAATATC	Pääbo, et al.. 1991; Irwin et al. (1991)
L15225	TCCCATACATTGGGACAGACCT	Wilson lab
L15424	ATCCCATTCCACCCATACTACTC	Edwards, Arctander and Wilson (1991)
H15525	TTTGCAGGGGTAAAATTATCAGGAT	Orti et al. (1994)
H15547	AATAGGAAGTATCATTCGGGTTTGATG	Edwards, Arctander and Wilson (1991)
L15774	GTACATGAATTGGAGGACAACCAGT	Wilson lab, Irwin, Kocher and Wilson (1991)
H15915	AACTGCAGTCATCTCCGGTTTACAAGAC	Wilson lab, Irwin, Kocher and Wilson (1991)
H15243	TTRTCTACNGARAANCCNCCTCA`	Orti (unpublished)

Nuclear ribosomal genes:

18S

18A	TGGTTGATCCTGCCAGTAG	Wilson lab (unpublished)
18E	TACCATCGAAAGTTGATAGGGCAGA	Hamby *et al.* (1988)
18J	TCTAAGGGCATCACAGACCTGTTATTG	Hamby *et al.* (1988)
18L	CACCTACGGAAACCTTGTTACGACTT	Hamby *et al.* (1988)
		See also several other papers, e.g. Stock and Whitt (1992)

28S

	See several papers, e.g. Hillis and Dixon (1991), Hillis *et al.* (1992)

ITS 1

	GTTTCCGTAGCTGAACCTGC	Meyer (unpublished)
	AATTCTCCGAGCTAG	Meyer (unpublished)

ITS 2

	CTCTTAGCGGTGGAT	Meyer (unpublished)
	GTCTGATCTGAGGTC	Meyer (unpublished)

Only PCR primers are listed that have been successfully used with fish DNA. Primer names of mitochondrial PCR primers follow the Kocher *et al.* (1989) convention of naming (i.e. numbering) the primer by the most 3′ position of the primer in the human mtDNA sequence (Anderson *et al.*, 1981). L and H refer to the light and heavy strand respectively. If sequences have been published previously the reference is indicated; otherwise, primer sequences have not been published, and were designed and tested in my laboratory and should be referenced as such. Primer sequences are given 5′ to 3′. Y = C or T, R = A or G, K = G or T, H = not G, W = A or T. Several more primer sequences have been published in the *Simple Fool's Guide to PCR*, distributed by S.R. Palumbi's laboratory, University of Hawaii (Palumbi *et al.*, 1991). Primers for nuclear ribosomal PCR primers have been published in several sources, e.g. Hamby *et al.* (1988), Hillis and Dixon (1991) and Stock and Whitt (1992), and are likely to work in fish.

[a]Genes for which more PCR primer sequences have been published.

mitochondrial DNA genotype of the previous generation (Hauswirth and Laipis, 1982; Ashley, Laipis and Hauswirth, 1989; Koehler *et al.*, 1991).

Usually, only one type of mtDNA is found in an organism. However, reports of hetero-plasmy, the presence of more than one type of mtDNA in an individual, are accumulating rapidly (reviews: Moritz, Dowling and Brown, 1987; Moritz, 1991). Heteroplasmy has occasionally been found in most major groups of organisms, including several species of fish e.g. shad (Bentzen, Leggett and Brown, 1988), sturgeon (Buroker *et al.*, 1990; Brown, Beckenbach and Smith,1992), cod (Arnason and Rand, 1992) and bowfin (Bermingham, Lamb and Avise, 1986). In the reported cases of heteroplasmy in fish, the proportion of heteroplasmic individuals in the population was 5% in bowfins (Bermingham, Lamb and Avise, 1986), 12% in shad (Bentzen, Leggett and Brown, 1988), and up to 41% in sturgeon (Buroker *et al.*, 1990).

Simulation and modelling of the 'population genetic' behaviour of male mitochondrial genomes suggest that male mitochondria should only rarely be fixed in the population of oocyte mitochondria (Chapman *et al.*, 1982; Birky, Maruyama and Fuerst, 1983; Neigel and Avise, 1986). However, paternal leakage may occur at the low rate of 1 in a 1000 molecules. The universality and rate of parental leakage is still unknown (Gyllensten *et al.*, 1991). Horizontal transfer of mtDNA between species of animals has been observed – e.g. mice (Ferris *et al.*, 1983), and voles (Tegelstrom, 1987) – but appears to be a rare phenomenon.

The mitochondrial genetic code is more degenerate and thus less constrained than the 'universal' eukaryotic nuclear genetic code (review: Attardi, 1985). The piscine mito-chondrial code appears to be like the vertebrate mitochondrial code in that TGA and ATA align with tryptophan and methionine respectively (Kocher *et al.*, 1989; Johansen, Guddal and Johansen, 1990; Meyer and Wilson, 1990; Meyer *et al.*, 1990; Tzeng *et al.*, 1992; Meyer, 1993).

In contrast to the nuclear genome, the mitochondrial genome of animals (but not that of fungi and plants) is highly efficient (reviews: Attardi, 1985; Cantatore and Saccone, 1987; Gray, 1989), i.e. it rarely contains duplicate or noncoding sequences. Mitochondrial protein coding genes do not contain introns, and genes are usually separated by less than 10 bp. Two genes may abut directly without intergenic spacers, and in some cases these genes even overlap by several bases. Transfer RNA genes and protein coding genes may overlap considerably (Clary and Wolstenholme, 1985). Some genes that are coded for by different strands overlap as well. The reading frames of the ATPase 6 and the ATPase 8 genes overlap never by less than 6–7 bp in vertebrates. The two reading frames of the ATPase 6 and 8 genes differ by one nucleotide. A single transcript has been found for each of these reading frame pairs (Anderson *et al.*, 1981; Ojala, Montoya and Attardi, 1981); the overlapping reading frames in each pair of genes appear to be translated from a single mRNA.

In fish, several relatively large noncoding regions have been described. Johansen *et al.*, (1990) found a 74-bp insertion in the Atlantic cod (*Gadus morhua*) between the Thr-tRNA and the Pro-tRNA. In *Xenopus* this stretch of noncoding DNA is only 26 bp long (Roe *et al.*, 1985) and in sturgeon only 3 bp long (Gilbert *et al.*, 1988). In several dozen species of fish from several families there are no or less than 5 bp between the Thr-tRNA and Pro-tRNA (Meyer *et al.*, 1990; Sturmbauer and Meyer, 1992; Meyer, unpublished data). In the 15-spine stickleback (*Spinachia spinachia*) we (Meyer, Orti and Bell, unpublished) found a 109-bp insertion between the Glu-tRNA and the cytochrome *b* gene.

5.1.5 Rate of evolution

Several studies of mtDNA using restriction enzymes and actual sequences (Brown *et al.*, 1982) indicate that mtDNA generally evolves at elevated rates (5–10 times faster) compared to single copy nuclear genes (Brown, George and Wilson, 1979; Perler *et al.*, 1980). The faster mtDNA evolution is due to a higher frequency of point and length mutations (Brown *et al.*, 1982; Cann, Brown and Wilson, 1984; Wilson *et al.*, 1985). The divergence at silent sites of protein coding genes may be about 10% per 10^6 years (generalized from results from primates), ten times the rate found in nuclear protein coding genes. The overall rate of substitution for the complete mitochondrial genome of primates may be 0.5–1% per 10^6 years. The rate of silent substitutions (substitutions that do not result in amino acid changes), mainly transitions, is about 4–6 times that of replacement substitutions (Brown, George and Wilson, 1979; Brown *et al.*, 1982). However, among closely related species, if transitions are ignored and only transversions counted, the difference in rates between mitochondrial and nuclear genes would be much less pronounced. Several reasons for the higher rate of mtDNA evolution have been suggested; for references see Meyer (1993). The observed differences in rates of evolution between nuclear and mtDNA cannot be explained exclusively by the increased variation at third positions. Substitution rates at first and second positions and rRNA and tRNA genes are also increased. However, substitution rates at these sites may not be as different between mitochondrial and nuclear genes as rates at third positions.

(a) Tempo and mode of substitution

Of the three kinds of sequence changes, substitutions are more frequent than indels (additions or deletions), and rearrangements are the least common form of DNA change. Additions and deletions are most frequently observed in the control region and intergenic spacers. At a lower frequency, they occur in the tRNAs and rRNAs. Rarely have they been found in protein coding genes, but they do exist. A single-codon deletion was found among salmonids in the ATPase 6 gene (Thomas and Beckenbach, 1989). A three-codon addition and a one-codon addition in the ATPase 6 gene of percoid fish compared to salmonid fish (Thomas and Beckenbach, 1989) was also discovered (Meyer, Orti and Titus, unpublished data). As far as is known, slowly evolving protein coding genes like CO I, II, III and cytochrome *b* do not seem to contain any length variation among neopterygian fish (Normark, McCune and Harrison, 1991; Meyer, unpublished data), and no length variation has been detected among ray-finned or lobe-finned fish for cytochrome *b* and CO III (Meyer and Wilson, 1990; Meyer, unpublished data).

As sequencing techniques developed it became clear that the higher rate of mtDNA evolution is due mainly to transitional differences (changes from one purine to another or one pyrimidine to another) (Brown and Simpson, 1982; Brown *et al.*, 1982; Aquadro and Greenberg, 1983; Greenberg, Newbold and Sugino, 1983; Wolstenholme and Clary, 1985). Transitions often outnumber transversions (changes from a purine to a pyrimidine or vice versa) by a factor of 10 or 20 in within-species comparisons. The pronounced predominance of transitions over transversions was found in all positions of codons and all mitochondrial genes. This supports the idea that a mutational bias for transitions over transversions rather than selection and constraints may be largely responsible for the tempo and mode of evolution of the mitochondrial genome. This transition bias appears to decrease with increasing sequence divergence and therefore time since common ancestry.

Substitutions at third positions quickly accumulate and they become saturated with transitions. Mutations at first and second positions will continue to accumulate despite saturated third positions. For phylogenetic reconstruction purposes (see below) transitions at third positions will therefore not be reliable indicators of evolutionary relationships beyond about 30–40% sequence divergence at these positions (Figure 5.1.4).

The increase in observed number of transversions (e.g. Figure 5.1.5) is due to several factors (Brown *et al.*, 1982; Holmquist, 1983; DeSalle *et al.*, 1987; Jukes, 1987; Meyer,

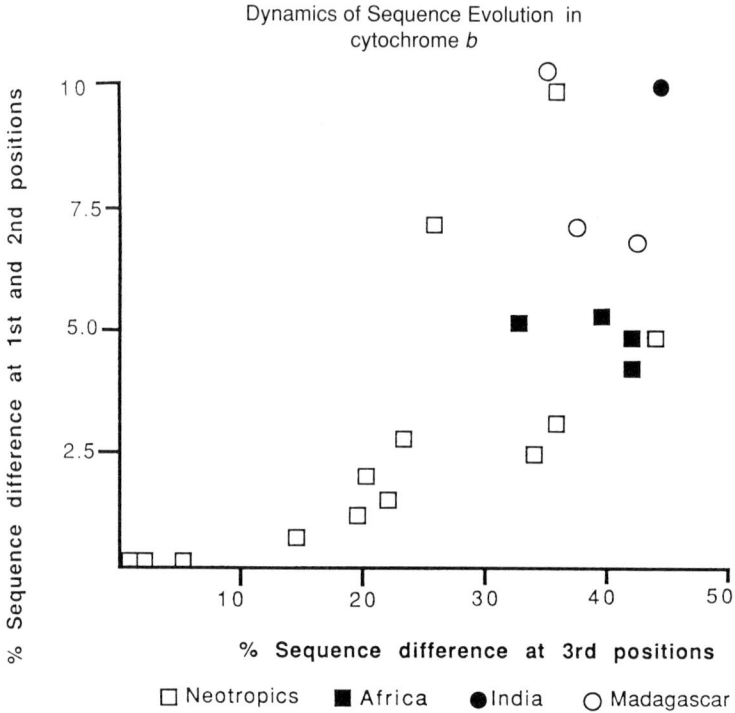

Figure 5.1.4 Mode of sequence evolution for the 5′ end (about 360 bp) of the cytochrome *b* gene for cichlid fish (Meyer and Orti, unpublished data). Cichlid fish are a family of fish found in South America, Africa, Madagascar and India. All species were compared to the neotropical species *Cichlasoma citrinellum*. Third position substitutions quickly accumulate and become saturated with substitutions at a maximum level of divergence of around 45%. Hence, for phylogenetic analyses, transitions in the third positions will not be reliable indicators of evolutionary relationships for the whole family, although they are informative among closely related groups of species. Mutations at first and second positions continue to accumulate up to a level of about 10% within the family Cichlidae.

1993). Asymmetry in the persistence of transversions and transitions exists (DeSalle *et al.*, 1987) and will produce an observed accumulation of scored transversions with time. Although transitions remain a more common type of mutation event between distantly related species, the percentage of all differences which are scored as transitions will decrease. The observed number of transversions will increase because transversions will become fixed at sites that had already fixed transitions and also at new sites that had not been substituted before.

Transversions, however, accumulate more slowly (Brown and Simpson, 1982; Brown *et al.*, 1982; Aquadro and Greenberg, 1983; Greenberg, Newbold and Sugino, 1983; Wolstenholme and Clary, 1985), eventually outnumbering transitions, i.e. the percentage transversions of the total number of substitutions observed between two species increases with increasing sequence divergence (Figure 5.1.5). The mode of change from transitions predominating to transversions becoming more prevalent differs between genes and may be related to functional constraints and base compositional biases (Holmquist, 1976, 1983; see below).

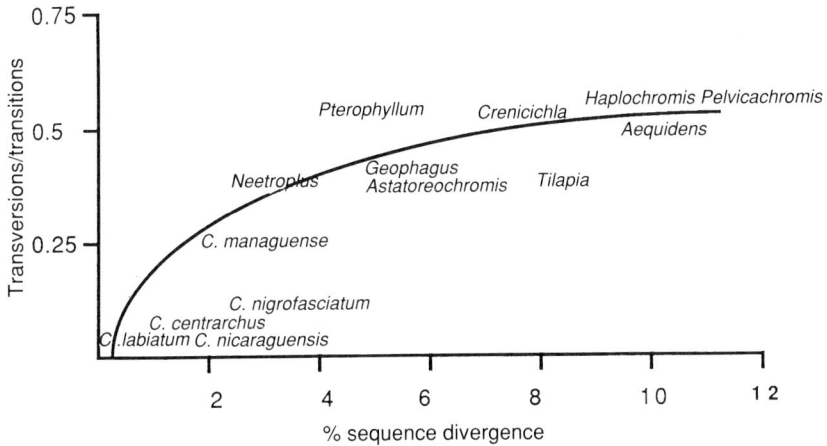

Figure 5.1.5 Dynamics of sequence evolution among cichlid fish for a conservative (about 300 bp) portion of the 12S rRNA gene (Meyer, unpublished data); for methods see Kocher *et al.* (1989) and Meyer *et al.* (1990). Graphed is the mode of sequence divergence and dynamics of the transversion/transition ratio from the neotropical species *Cichlasoma citrinellum*. Among closely related species of the genus *Cichlasoma*, transversions were only rarely observed, but the percentage of observed transversions quickly increases to an asymptotic level of about 50%.

These dynamics will have to be accounted for in phylogenetic analysis. Transversions will trace phylogenetic events more reliably because back-mutations will accumulate at a much lower rate. Often, they will be weighted higher in a parsimony analysis or in the calculation of the matrix of genetic divergences for a distance phylogenetic method (e.g. Mindell and Honeycutt, 1990; Meyer and Dolven, 1992; Sturmbauer and Meyer, 1992).

Transitions also occur more frequently among closely related species of fish, and transversions become apparent among more distantly related species (Kocher *et al.*, 1989; Meyer *et al.*, 1990; Sturmbauer and Meyer, 1992; Fajen and Breden, 1992). Among conspecific cichlid fish, transitions often outnumber transversions 10 to 1; among distantly related species this ratio changes to about one. Figure 5.1.5 shows this trend quite clearly in a conserved portion of the 12S ribosomal RNA gene for some cichlid fish. With increasing sequence divergence from a neotropical species, *Cichlasoma citrinellum*, the proportion of tranversion substitutions increases rapidly to an upper limit of about 50%. The level at which saturation is reached will depend on several factors, e.g. the base compositional bias (see below). The absolute time required to reach the plateau is dependent on the absolute rate with which transversions become fixed (see below). Among congeners, total sequence divergence tends not to exceed about 5%, and scored transversions are usually less than 25% of the total number of substitutions, whereas closely related species exhibit transition differences almost exclusively. Within the family Cichlidae the sequence divergence (uncorrected for multiple substitutions) in this portion of the 12S rRNA gene does not exceed about 15%. This is not the case for species of fish in the family Gasterosteidae (Orti *et al.*, 1994). Patterns like the one seen for the 12S rRNA gene in cichlid fish are typical (Figure 5.1.5). The observed ratio of transversions to the total number of substitutions between species increases relatively rapidly (beyond the level of the species?); it usually quickly reaches a plateau (among congeners?) with some fluctuation (Holmquist, 1983). Even quite distantly related species of fish do not exceed about 20% of total (uncorrected) sequence divergence for this conservative part of the 12S rRNA gene (Meyer and Wilson, 1990; Mindell and Honeycutt, 1990; Meyer and Dolven, 1992).

(b)   Protein coding genes

Substitution patterns in protein coding genes follow some relatively well-understood rules. This regularity in the way in which mutations accumulate makes protein coding genes attractive candidates for phylogenetic studies of fish. Mutations in third (and rarely in first) positions of codons that do not result in amino acid (AA) substitutions (silent or synonymous substitutions) accumulate much more rapidly than AA replacement substitutions (non-synonymous substitutions). The most frequently observed substitutions are transitions in third positions of codons; second most frequent are transversions in third positions and silent transitions in some first codon positions (Figure 5.1.6). In some codons (leucine), the first position is degenerate (both TTA/G and CTN codon families code for leucine). Furthermore, since transitions in first positions of codons will usually result in conservative AA substitutions and will tend to maintain a functional gene product, mutations will accumulate more quickly in first than in the most constrained second positions of codons (Figure 5.1.6). Knowledge of these rules allows one to weight (or exclude, as an extreme form of weighting) kinds of substitutions (transitions and transversions) and positions (first, second and third) differently based on the phylogenetic question that is being addressed. Obviously, among distantly related species transitions in third positions are going to be unreliable tracers of evolutionary descent and represent largely 'phylogenetic noise'. Transversions are rarer and have a higher chance of being reliable indicators of descent. The most conserved positions (second positions in codons) are going to be the most similar among closely related species but contain phylogenetic information among distantly related species.

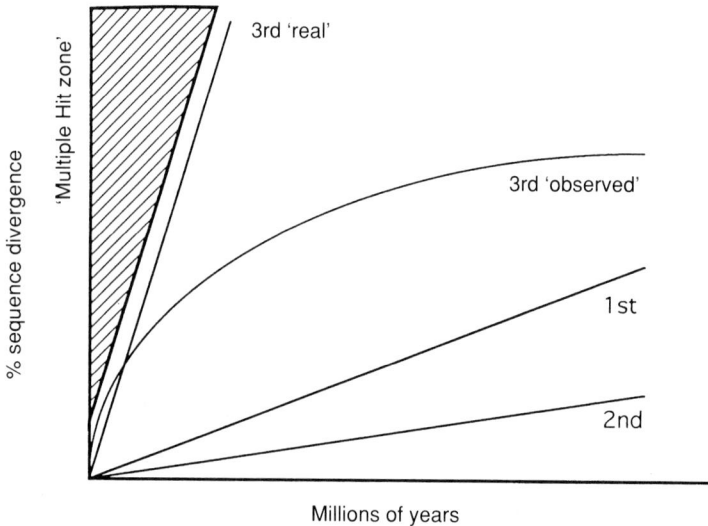

Figure 5.1.6 Hypothetical example of the dynamics of sequence evolution in a mitochondrial protein coding gene. The multiple hit zone is depicted to signify the observation that the likelihood of a multiple substitution at a particular nucleotide increases with sequence divergence. The probability of a multiple hit will not increase linearly but rather exponentially with increasing sequence divergence. Due to multiple substitutions the 'observed' sequence divergence at third positions (also, of course, at first and second positions, but more slowly) will 'decelerate' quickly compared to second and first positions in codons. The 'observed' rate of sequence divergence will increase in a relatively linear fashion for about the first 10–15% sequence divergence. The 'real' rate of evolution can be estimated from a tangent that is laid on the observed mode of divergence. The actual slopes will vary between particular mitochondrial protein coding genes.

Each mitochondrial protein coding gene has its own particular rate of evolution that will depend on factors such as functional constraints on the gene product and base compositional biases (Johansen *et al.*, 1990; Meyer, 1993, Table 1). Despite the fast rate of mtDNA evolution, some genes may be highly conserved; there may be a low ceiling for total divergence, which is partly due to nucleotide base compositional biases (e.g. DeSalle *et al.*, 1987) and strong functional constraints. The genes coding for the subunits of cytochrome oxidase and cytochrome *b* are the most conserved genes, and the most variable ones are some of the ND and the ATPase genes. These slowly evolving protein coding genes have been used to test relationships among distantly related neopterygian fish (Normark, McCune and Harrison, 1991), yet these genes turned out to have a rate of AA substitution that is too slow for this phylogenetic question. The early publication and availability of 'universal' PCR primers, particularly for 12S and cytochrome *b* (Kocher *et al.*, 1989), had the effect that cytochrome *b* is being sequenced in many different organisms for many different questions. Often, however, particularly for distantly related species, other genes might have been better choices (see below).

(c) Transfer RNA genes

All vertebrate mitochondrial genomes contain the usual set of 22 tRNAs. They are smaller than their cytoplasmic counterparts, usually 59–75 bp in length. Secondary structures of mitochondrial tRNAs are less conserved but still fold into the cloverleaf secondary structure (review: Sprinzl *et al.*, 1991). Mitochondrial tRNAs are surprisingly variable in their primary sequence (substitutions and, less frequently, small 1–2 bp insertions and deletions are observed) and secondary structure. Lengths of stems and loops of mitochondrial tRNAs are exceedingly variable compared with nuclear tRNA genes. Mitochondrial tRNA genes evolve at least 100 times faster than nuclear tRNA genes (Brown *et al.*, 1982). Despite the observed elevated rates of variation compared to the nuclear tRNAs, the tRNA genes are among the more slowly evolving genes of the mitochondrial genome, but variation exists in their relative rates. Because of their small size they may not be the best choice of genes for phylogenetic inquiries (although they have been used successfully); however, they are often good sites for the design of PCR primers (e.g. Kocher *et al.*, 1989), because of the conserved features of tRNA genes.

(d) Ribosomal RNA genes

Two ribosomal RNA (rRNA) genes are found in animal mitochondrial genomes: the small 12S (about 819–975 bp in vertebrates) rRNA gene and the large 16S (about 1571–1640 bp in vertebrates) rRNA gene. Nuclear rRNA (18S and 28S) and tRNA genes (see below) evolve about 100 times more slowly than their mitochondrial counterparts (Dawid, 1972) when the more variable expansion segments of the nuclear rRNA genes are excluded from the analysis (Mindell and Honeycutt, 1990).

The same basic rules of substitutions (see above) apply to rRNA genes (Brown *et al.*, 1982). Transitions are more frequent than transversions; this is most apparent among closely related species where the record of transitions has not been overlaid by more slowly accumulating and more persistent transversions (Figure 5.1.5, see above). Insertions and deletions are usually small, in the range of 1–5 bp. The 16S gene tends to contain more length variation than the 12S gene. Hence, DNA sequences of the 16S rRNA for distantly related species are more difficult to align than for the 12S rRNA gene. Length mutations are more frequent in rRNA than in protein coding genes. These length variations make alignment an issue for phylogenetic studies; different alignments, and therefore different hypotheses of homology, can result in drastically different phylogenetic inferences. Areas of questionably aligned sequence (criteria for what is well aligned are not always obvious; see Swofford and Olsen (1990) and Mindell (1991)) are best removed from the phylogenetic analysis. Alignment and delineation of which portion of the sequence should be excluded from the analysis and which should be used remain problematic issues.

Secondary structure models are available for both genes, and general agreement about the secondary structure exists (Glotz *et al.*, 1981; Dunon-Bluteau and Brun, 1986; Hixson and Brown, 1986). These secondary structures are conserved across large evolutionary distances. Among tetrapods, lungfish, the coelacanth and ray-finned fish, substitutions occur about four times more frequently in proposed loops than in stems; transversions are about nine times more frequent in loops than in stems (Meyer and Wilson, 1990; Meyer and Dolven, 1992; Meyer, unpublished data). Transversions appear to be indicators of phylogenetic relationships in a slowly evolving portion of the 12S rRNA gene for an evolutionary distance of more than 400 000 000 years (Meyer and Wilson, 1990; Mindell and Honeycutt, 1990; Meyer and Dolven, 1992; Meyer, 1993). Among closely related species of primates, stems are less likely to contain substitutions than loop regions (Hixson and Brown, 1986). Stems sometimes show 'compensatory mutations': a substitution in one strand of a stem region is compensated for by a change in the other strand of the stem, in order to maintain intrastrand base pairing and a stable secondary structure. The rate of sequence divergence appears higher among closely related species than among distantly related species due to saturation effects (Figure 5.1.5). At around $100–150 \times 10^6$ years, the rate of substitution appears to decrease (Mindell and Honeycutt, 1990; Meyer and Dolven, 1992).

The overall rRNA and tRNA substitution rates are about half those of the protein coding genes (Brown *et al.*, 1982; Hixson and Brown, 1986; Jacobs *et al.*, 1988; Meyer and Wilson, 1990; Hillis and Dixon, 1991; Mindell and Honeycutt, 1990), making them attractive genes for phylogenetic questions among distantly related species.

(e)   Control region

The control region is partially constrained in primary sequence or secondary structure to regulate replication and transcription (Clayton, 1991). The control region is characterized by the displacement loop (D-loop), a stretch of DNA that is complementary to the light L-strand; the D-loop strand displaces the H-strand. The initiation site of D-loop synthesis and the origin of H-strand replication are identical, but, the termination site(s) for the D-loop strand and the H-strand are separate.

Of all mitochondrial genes the control region has the highest substitution rate. The rate of evolution of the control region is two to five times higher than that of mitochondrial protein coding genes (Aquadro and Greenberg, 1983). Nucleotide substitutions occur five times more frequently than additions and deletions in the human control region; among closely related species of cichlid fish, substitutions outnumber deletions and additions (Meyer *et al.*, 1990; Sturmbauer and Meyer, 1992). The control region varies tremendously in length, often because of tandem duplications of 200–4100 bp (Brown, 1985), and is primarily responsible for the observed variation in the total length of the vertebrate mitochondrial genomes. It also contains the highest frequency of length mutations at the population level (Densmore, Wright and Brown, 1985; Harrison, Rand and Wheeler, 1985). In fish (Bermingham, Lamb and Avise, 1986; Bentzen, Leggett and Brown, 1988; Buroker *et al.*, 1990), tandem duplications and various numbers of repeats are usually found here when heteroplasmy is observed.

The characteristically high amount of sequence divergence in the control region is not evenly distributed; there are several regions that exhibit high levels of sequence conservation. Usually A–T-rich regions are found at both ends of the D-loop, with an evolutionarily conserved central domain. Although highly variable in sequence and struc- turally variable in vertebrates, the central domain is probably functionally similar, since conserved sequence blocks (CSBs) can usually be identified in all animal control regions (Brown *et al.*, 1986; Saccone, Attimonelli and Sbiza, 1987).

Because mutations accumulate fastest here, the control region is the molecule of choice for the study of population level phenomena and the study of phylogenetic relationships among closely related species and has been used for both of these purposes in fish (e.g. Meyer *et al.*, 1990; Sturmbauer and Meyer, 1992; Reinthal and Meyer, unpublished).

(f) Slower rate of evolution in ectotherms?

Much variation was expected in terms of amino acid differences in mitochondrial protein coding genes of fish, since up to 8.5% sequence divergence had been found within species of sunfish (Avise *et al.*, 1984a) and also some cichlid fish have high levels of intraspecific variation (Sturmbauer and Meyer, 1992). However, this is not always the case. There is little variation within or between species of salmonids (Thomas and Beckenbach, 1989) or cichlids (Kocher *et al.*, 1989; Meyer *et al.*, 1990; Meyer, Biermann and Pallson, unpublished data) in the more conserved protein coding genes.

Ectothermic animals may be up to five times slower than endothermic animals in the evolution of cytochrome *b* amino acid replacement substitutions (Kocher *et al.*, 1989), and perhaps even 10 times slower (Martin, Naylor and Palumbi, 1992). This may also be true for nucleotide substitutions. Given that synonymous changes in protein coding genes appear to accumulate so rapidly, it was surprising to find that amino acid substitutions in the most conservative mitochondrial genes (cytochrome oxidase and cytochrome *b* genes) accumu- late rather slowly in fish (Kocher *et al.*, 1989; Meyer and Wilson, 1990; Normark, McCune and Harrison, 1991). This slower rate allowed mtDNA to be used to investigate phylogenetic relationships among very distantly related species (Meyer and Wilson, 1990; Normark, McCune and Harrison, 1991; Meyer and Dolven 1992). The rate of amino acid substitution in the cytochrome *b* gene proved to be conservative enough to test relationships among groups of lobe-finned fish and ray-finned fish that have not shared a common ancestor for more than 400 000 000 years (Meyer and Wilson, 1990). The rate of amino acid substitution in CO and cytochrome b genes actually was too slow to clearly elucidate the evolutionary relationships among neopterygian fish (Normark, McCune and Harrison, 1991).

Although mtDNA evolution in ectotherms appears to be slower compared to birds and mammals (Kocher *et al.*, 1989; Martin, Naylor and Palumbi, 1992), their nuclear rate is still slower than the mtDNA rate. Comparisons of mtDNA (cytochrome *b*) and nuclear genes (*X-mrk*, *X-src*) between closely related species of poeciliid fish of the genus *Xiphophorus* showed that mtDNA evolves at least 2.5 times faster than these nuclear genes (Meyer, Morrisey and Schartl, unpublished data). This estimate for the rate of these nuclear genes includes noncoding regions (introns) and is therefore underestimating the relative mitochondrial rate. Based on earlier findings in primates (e.g. Brown, George and Wilson, 1979) it was expected that the mitochondrial rate generally should be at least five times the nuclear rate.

Whether mitochondrial and nuclear substitution rates are more similar in fish because of slower mtDNA molecular substitution rates or faster nuclear substitution rates is not clear and will require much more data than are currently available. If the mitochondrial rate is generally slower in fish and/or the rate of nuclear DNA evolution increased, we will require a knowledge of divergence dates. However, too few studies have calibration points from the fossil record to allow us to test these hypotheses.

5.1.6   Mitochondrial DNA and population-level questions

Until recently, mtDNA has been used mainly in population-level work and in studies of molecular relationships among closely related species (reviews: Wilson *et al.*, 1985; Avise, 1986; Avise *et al.*, 1987; Moritz, Dowling and Brown, 1987). The fast rate of evolution of mtDNA (Brown, George and Wilson, 1979) compared to nuclear DNA makes mtDNA useful for high-resolution analyses of recent evolutionary events. This fast rate of mtDNA evolution, coupled with maternal inheritance (but see Zouros *et al.* (1992) and above), has made mtDNA an extremely popular genetic system with which to study gene flow, hybrid zones, population structure and other population-level questions (reviews: Wilson *et al.*, 1985; Avise, 1986; Moritz, Dowling and Brown, 1987; Harrison, 1991). Mitochondrial DNA lends itself to the study of founder events and female-mediated gene flow, i.e. differences in dispersal between sexes will be apparent through comparisons of the geographic distribution of nuclear DNA and mtDNA (review: Harrison, 1991). Mitochondrial DNA also lends itself to the study of the origin of clonally reproducing species (e.g. Densmore, Wright and Brown, 1985; Quattro, Avise and Vrijenhoek, 1992; reviewed in Moritz, Dowling and Brown, 1987). Mitochondrial DNA, as well as allozyme markers, has been used to identify hybridization events (e.g. Avise and Saunders, 1984; Avise and Van Den Avyle, 1984; Herke *et al.*, 1990). The special mode of inheritance of mtDNA has consequences for its population genetic behaviour (Birky, Maruyama and Fuerst, 1983; Avise *et al.*, 1984b; Avise, Ball and Arnold, 1988; Neigel and Avise, 1986; Pamilo and Nei, 1988; see reviews in Avise, 1986, 1989; Avise *et al.*, 1987; Harrison, 1991).

Until recently most of the work on population structure and phylogenetic relationships of fish based on mtDNA data was based on the analysis of restriction fragment length polymorphisms (RFLPs) (e.g. Graves, Ferris and Dizon, 1984; Wilson *et al.*, 1985; Wilson, Thomas and Beckenbach, 1987; Bermingham, Lamb and Avise, 1986; Gyllensten and Wilson, 1987; Avise *et al.*, 1986; Ovenden, White and Sanger, 1988; Wirgin, Proenca and Grossfield, 1989; reviews in Avise, 1986; Avise *et al.*, 1987; Moritz, Dowling and Brown, 1987). High levels of polymorphism and geographic differentiation within species are sometimes found (see above). Extremely low levels of sequence variation within and between species have been found as well (e.g. Meyer *et al.*, 1990). Restriction analyses are usually unable to provide information about the patterns of substitutions that cause observed differences in RFLPs. Actual DNA sequences do provide this information. The estimates of sequence divergence based on mtDNA RFLPs are not always the same as DNA sequence analyses (e.g. Wilson *et al.*, 1985; Thomas and Beckenbach, 1989). In the case of intra-specific variation in salmonids, the divergence estimate from actual DNA sequences is about half that estimated from RFLP analyses. The opposite trend was observed for interspecific variation, in which sequence data tended to be larger than estimates based on RFLP data (Thomas and Beckenbach, 1989; Beckenbach, Thomas and Sohrabi, 1990).

Differences in the estimates of sequence divergence based on restriction enzymes and DNA sequences can be due to several factors; for example, the selection of restriction enzymes might have been biased to cut preferentially at particular nucleotides. If equal representation of all four nucleotides was expected, but strong base compositional bias existed, the estimates of sequence divergence based on restriction analysis would be inaccurate. Furthermore, restriction enzymes cut indiscriminately throughout the whole mtDNA molecule (including the rapidly evolving control region, as well as more slowly evolving rRNA and tRNA genes), whereas most often only portions of mitochondrial genome are sequenced, and estimates of sequence difference are influenced by the choice of gene(s) sequenced. Obviously, this has to be taken into consideration. In intraspecific comparisons the rapidly evolving control region might contribute disproportionately to the estimate of sequence divergence based on RFLP analysis of the complete mtDNA genome;

in interspecific studies, multiple hits (see below) in the control region might lead to an underestimate of sequence divergence compared to comparisons of actual DNA sequences.

Methods in population genetics and fisheries management change rapidly and mtDNA has found wide application (reviews: Hallermann and Beckmann, 1988; Ovenden, 1990; Ryman and Utter, 1987; Dizon *et al.*, 1992). RFLP analysis had been the predominant method in population studies (e.g. Wirgin, Proenca and Grossfield, 1989) but PCR has experienced increased application for these questions. Even conservative protein coding genes like those for cytochrome *b* tend to show intraspecific variation, mainly in third positions of codons that can be used to identify fish stocks (Bartlett and Davidson, 1991, 1992; Carr and Marshall, 1991; McVeigh, Bartlett and Davidson, 1991; Finnerty and Block, 1992; Orti *et al.*, 1994) or be used for phylogenetic studies among closely related species (Kocher *et al.*, 1989; Meyer *et al.*, 1990; Meyer, Kocher and Wilson, 1991; Sturmbauer and Meyer, 1992; Reinthal and Meyer, unpublished data).

When using mtDNA to study population differentiation and systematics of closely related species, i.e. recent divergence events, one needs to be aware of some potential difficulties. If the ancestral species contains more than one mtDNA haplotype, lineage sorting at speciation and subsequent random lineage extinctions may cause two species to contain some of the same mtDNA haplotypes; they may not exactly follow species boundaries (e.g. Avise, 1989; Avise *et al.*, 1984b; Avise, Ankney and Nelson, 1990). This is due to random sampling of mtDNA haplotypes at speciation and may introduce errors in phylogenetic reconstruction (see below). The danger of this happening is particularly large if the speciation event is recent and the ancestral species highly polymorphic. If the measured divergence between the mtDNA haplotypes of two young species is used to estimate the time since the origin of these species, this date may be overestimated, since the divergence of haplotypes might have predated the splitting of the species, i.e. the amount of sequence divergence was not zero at the time of speciation and needs to be corrected. This correction can be based on the currently observed level of intraspecific variation (cf. Harrison, 1991). However, current levels of intraspecific variation may not be an accurate reflection of the pre-speciation levels, and furthermore assumes that the rates of mtDNA evolution are the same in all lineages.

5.1.7   Mitochondrial DNA and phylogenetic questions

The study of fish mtDNA experienced a boost through the development of so-called 'universal' primers (see above) (Kocher *et al.*, 1989). These primers were designed, based on comparisons of published mtDNA sequences, to anneal to stretches of DNA that are conserved across a wide taxonomic range. Despite the generally fast substitution rate of mtDNA, conserved areas can be identified and primers designed that have applicability for a wide taxonomic range (e.g. a phylum or even beyond) (Kocher *et al.*, 1989). With these 'universal' primers mtDNA from most organisms can be amplified and the DNA sequence determined without prior sequence knowledge from the particular organism studied. The first 'universal' primers (Kocher *et al.*, 1989) and in particular the primers that amplify portions of the mitochondrial 12S ribosomal and cytochrome *b* genes currently enjoy widespread application in many evolutionary and phylogenetic studies from a wide range of taxonomic groups (reviews: Meyer, 1993; Esposti Degli *et al.*, 1993).

Universal PCR primers for mtDNA (Kocher *et al.*, 1989) led to an even further increase in the use of mtDNA for phylogenetic questions. However, the availability of universal PCR primers has led to their uncritical use for many groups of organisms for many different questions. Cytochrome *b* is a gene that has been targeted for many different inquiries, but it may not be the best gene for the study of evolutionary splits that are ancient enough so that third positions are saturated. The 5' end of cytochrome *b* tends to be very conserved in amino acid substitutions and it may not contain enough variation in first and second positions of codons once third positions are saturated. The 3' end of this gene appears to be somewhat less constrained in terms of amino acid substitutions (Irwin, Kocher and Wilson, 1991; reviewed in Esposti Degli, *et al.*, 1992, 1993).

Base compositional biases, saturation effects and the low variation in first and second positions of codons all contribute to problems in phylogenetic reconstruction. Similar difficulties in the use of cytochrome *b* were encountered in the reconstruction of the evolutionary relationships among groups of cichlid fish, believed to have diverged more than 80 000 000 years ago (Meyer and Orti, unpublished data). Although this gene has been very useful for studies involving populations (see above) and the phylogenetic study of very different groups of fish, it may not be the most appropriate gene for all purposes (review: Esposti Degli *et al.*, 1993). It would appear that for some phylogenetic problems faster evolving protein coding genes like ND2 and ATPase 6 and 8 or the ribosomal genes might provide more appropriate data. Primer sequences for many other genes have already been published or are available (Table 5.1.1, Figure 5.1.3).

More 'universal' mitochondrial, but also nuclear, DNA PCR primers are constantly being published. Table 5.1.1 presents a compilation of several published (e.g. Kocher *et al.*, 1989; Meyer *et al.*, 1990; Normark, McCune and Harrison, 1991; Palumbi *et al.*, 1991; Grachev *et al.*, 1992) and unpublished PCR primers for several mitochondrial and nuclear genes that are known to work for most species of fish. With these primers, mtDNA can be amplified and sequenced without prior sequence knowledge and they are expected to work for most groups of teleosts.

Mitochondrial DNA has found widespread use as a tool for phylogenetic analyses (reviews: Kornfield, 1984, 1991; Avise, 1986; Avise *et al.*, 1987; Moritz, Dowling and Brown, 1987; Wilson *et al.*, 1985). RFLP analyses were mainly conducted on closely related species since the homology among fragments was not always clear between distantly related species (more than about 10–15% sequence divergence). Through the advent of PCR it has become apparent that mtDNA sequences may also be a useful tool for the study of evolutionary relationships among more distantly related species (e.g. Meyer and Wilson, 1990; Normark, McCune and Harrison, 1991; Meyer and Dolven, 1992).

Multiple mutations at the same nucleotide position (multiple hits) become increasingly more likely and therefore accumulate with increasing sequence divergence (Figure 5.1.6). 'Multiple hits' are problematic for phylogenetic studies. They tend to obscure evolutionary relationships, and they also result in an observed upper limit of sequence divergence, a ceiling that is approached asymptotically (Figure 5.1.4 and 5.1.5). Estimates of sequence divergence (important for the calculation of distance matrices for distance methods, see above) between two species, particularly 'up' in the 'multiple hit zone', can severely underestimate 'real' divergence and need to be corrected. The more distantly related two species are, the more the simple estimate of sequence divergence (e.g. uncorrected percentage sequence divergence) will differ from an estimate that is derived from some correction that attempts to account for multiple substitutions.

Distance methods are based on a distance matrix, e.g. neighbour-joining (see above, Saitou and Nei, 1987). These distance matrices need to be corrected for multiple hits. Multiple hit corrections can be done in a number of ways; Jukes and Cantor's method (Jukes and Cantor, 1969) is the simplest but tends to underestimate the evolutionary divergence. More elaborate models are available and provide more reliable estimates of 'real' evolutionary divergence (Figure 5.1.6; Kimura, 1980, 1981; Takahata and Kimura, 1981; Gojobori, Ishii and Nei, 1982; Tajima and Nei, 1984). Multiple hits are also problematic for parsimony analysis and their detrimental effect on phylogenetic analyses needs to be counteracted by downweighting of transitions.

Since the dynamics and rules of substitution are somewhat clearer in protein coding genes than in tRNA and rRNA genes (e.g. the secondary structure and alignments of rRNA are not always clear), they are often a more straightforward choice for a gene used for the study of evolutionary relationships. However, the applicability and usefulness of mitochondrial as well as nuclear rRNA genes is not called into question (Meyer and Dolven, 1992; Meyer and Wilson, 1990; reviewed in Mindell and Honeycutt, 1990; Hillis and Dixon, 1991).

Other mitochondrial or nuclear genes may have substitution rates more appropriate for particular questions. On a per base pair basis the control region will provide more information about population-level questions and contain more (up to three times more) phylogenetic information than cytochrome *b* (e.g. Meyer *et al.*, 1990; Sturmbauer and Meyer, 1992). In a comparison between two closely related species of neotropical cichlid fish (*Cichlasoma citrinellum* and *Neetroplus nematodus*) the variation in a portion of the control region was comparable (18.6%) to the variation in third positions for cytochrome *b* (22.9%) and cytochrome oxidase subunit III (14.3%) (Meyer, unpublished data). The variation in first and second positions for these conservative genes was low (0.7% and 2.8% respectively). The Pro-tRNA was found to vary considerably (8.9%) and a conservative portion of the 12S rRNA gene contained few substitutions (2.4%) (Meyer, unpublished data).

The usefulness of mtDNA sequences is determined at one extreme by the stochasticity of the distribution of haplotype polymorphisms within species and at the other extreme by saturation effects due to base compositional biases, mutational bias and selective constraints on the gene product. Another complication is the potential for rate heterogeneity in the molecular clock even among closely related species, which might complicate phylogenetic analyses. Various phylogenetic methods are more or less sensitive to the effects of rate heterogeneity and the complicating effects they have on the reconstruction of evolutionary relationships (Nei, 1991).

5.1.8 Nuclear genes in molecular systematics of fish

(a)   Ribosomal genes

The large existing database and the ease of direct sequencing techniques of nuclear ribosomal genes have led to their widespread, successful use in molecular systematics (reviews: Hamby *et al.*, 1988; Mindell and Honeycutt, 1990; Hillis and Dixon, 1991). Both small (18S) and large (28S) ribosomal genes (Figure 5.1.7) have been used for phylogenetic work. Because of the extremely slow rate of evolution of these genes they have mainly been used for phylogenetic work on distantly related species, e.g. the origin of tetrapods (Hillis and Dixon, 1989; Hillis, Dixon and Ammer, 1989; Stock *et al.*, 1991; Meyer, unpublished data) and the monophyly of agnathan fish (Stock and Whitt, 1992; Meyer, unpublished data). Many PCR primers are available (refs in Table 5.1.1) that will ensure the continued use of

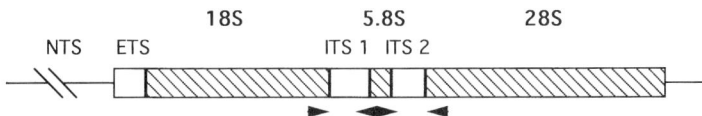

Figure 5.1.7 Arrangements of the elements of the nuclear ribosomal genes. Primer sequences of two primer pairs (Table 5.1.1) that amplify the ITS 1 and ITS 2 regions or both regions plus the 5.8S ribosomal RNA gene.

these genes for phylogenetic work. Some particular problems, e.g. alignment (see above), plague the use of these genes for phylogenetic work.

(b)   Growth hormone and prolactin

Growth hormone (GH), prolactin (PRL) and chorionic somatomammoptropin (CS) are members of a gene family of polypeptide hormones, believed to have evolved from a single ancestral gene. The widespread use of GH in transgenic fish work has led to the accumulation of several (about 19 so far) GH sequences from a variety of groups of fish (e.g. Schneider *et al.*, 1992; Watanabe *et al.*, 1992). The potential for phylogenetic information in GH is largely unexplored (but see Bernandi *et al.*, 1994).

(c)   Protein kinases

A large body of literature exists on several protein (particularly tyrosine) kinase genes in fish. This is largely due to the general interest in these genes as proto-oncogenes, and fish as model systems in cancer research. Several of these genes (e.g. *X-src*) provide useful phylogenetic information at the genus and family level (Meyer and Lydeard, 1993). PCR-based projects are ongoing (Meyer *et al.*, unpublished data) that highlight the universal applicability of nuclear protein coding genes for many phylogenetic questions in fish. These sequences can be used for phylogenetic questions among closely related species (introns contain enough variation) and distantly related species, for which the phylogenetic information contained in exons is utilized.

Acknowledgements

I thank G. Orti and S. Johansen for providing unpublished data, C. Davis for technical assistance and G. Orti for comments on the manuscript. Support from the National Science Foundation for my laboratory (BSR-9107838, BSR-9119867, BSR-9112367, INT-9117104) is gratefully acknowledged.

References

Anderson, S., Bankier, A.T., Barrell, B.G., *et al.*, (1981) Sequence and organization of the human mitochondrial genome. *Nature*, **290**, 457–65.

Anderson, S., de Bruijn, M.H.L., Coulson, A.R. *et al.*, (1982) Complete sequence of bovine mitochondrial DNA. *J. Mol. Biol.*, **156**, 683–717.

Aquadro, C.F. and Greenberg, B.D. (1983) Human mitochondrial DNA variation and evolution: analysis of nucleotide sequences from seven individuals. *Genetics*, **103**, 287–312.

Araya, A., Amthauer, R., Leon, G. *et al.* (1984) Cloning, physical mapping and genome organisation of mitochondrial DNA from *Cyprinus carpio* oocytes. *Mol. Gen. Genet.*, **196**, 43–52.

Arnason, E. and Rand, D.M. (1992) Heteroplasmy of short tandem repeats in mitochondrial DNA of Atlantic cod, *Gadus morhua. Genetics*, **132**, 211–20.

Arnason, U., Gullberg, A. and Widegren, B. (1991) The complete nucleotide sequence of the mitochondrial DNA of the fin whale, *Balaenoptera physalus. J. Mol. Evol.*, **33**, 556–68.

Arnason, U. and Johnsson, E. (1992) The complete mitochondrial DNA sequence of the harbor seal *Phaca vitulina. J. Mol. Evol.*, **34**, 493–505.

Arnheim, N., White, T. and Rainey, W.E. (1990) Application of PCR: organismal and population biology. *BioScience*, **40**, 174–82.

Ashley, M.V., Laipis, P.J. and Hauswirth, W.W. (1989) Rapid segregation of heteroplasmic bovine mitochondria. *Nucleic Acid. Res.*, **17**, 7325–31.

Atchley, W. and Fitch, W.M. (1991) Gene trees and the origins of inbred strains of mice. *Science*, **254**, 554–58.

Attardi, G. (1985) Animal mitochondrial DNA: an extreme example of genetic economy. *Int. Rev. Cyt.,* **93**, 93–145.

Avise, J.C. (1986) Mitochondrial DNA and the evolutionary genetics of higher animals. *Phil. Trans. R. Soc. Lond. B.,* **312**, 325–42.

Avise, J.C. (1989) Gene trees and organismal histories: a phylogenetic approach to population biology. *Evolution*, **43**, 1192–208.

Avise, J.C., Akney, C.D. and Nelson, W.S. (1990) Mitochondrial gene trees and the evolutionary relationship of mallard and black ducks. *Evolution*, **44**, 1109–19.

Avise, J.C. and Saunders, N.C. (1984) Hybridization and introgression among species of sunfish (*Lepomis*): analysis by mitochondrial DNA and allozyme markers. *Genetics*, **108**, 237–55.

Avise, J.C. and Van den Avyle, M.J. (1984) Genetic analysis of reproduction of hybrid White bass × Striped bass in the Savannah River. *Trans. Am. Fish. Soc.,* **113**, 563–70.

Avise, J.C, Bermingham, E., Kessler, L.G. and Saunders, N.C. (1984a) Characterisation of mitochondrial DNA variability in a hybrid swam between subspecies of bluegill sunfish (*Lepomis macrochir*). *Evolution*, **38**, 931–41.

Avise, J.C., Helfman, G.S., Avise, J.C., Neigel, J.E. and Arnold, J. (1984b) Demographic influences on mitochondrial DNA lineage survivorship in animal populations. *J. Mol. Evol.,* **20**, 99–105.

Avise, J.C., Helfman, G.S., Saunders, N.C. *et al.,* (1986) Mitochondrial DNA differentiation in North Atlantic eels: population genetic consequences of an unusual life history pattern. *Proc. Natl. Acad. Sci. USA,* **83**, 4350–54.

Avise, J.C., Arnold J., Ball R.M. *et al.* (1987) Intraspecific phylogeography: the mitochondrial DNA bridge between population genetics and systematics. *Annu. Rev. Ecol. Syst.,* **18**, 489–522.

Avise, J.C., Ball, R.M. and Arnold, J. (1988). Current versus historical population sizes in vertebrate species with high gene flow: a comparison based on mitochondrial DNA lineages and inbreeding theory for neutral mutations. *Mol. Biol. Evol.,* **5**, 331–44.

Bartlett, S.E. and Davidson, W.S. (1991) Identification of *Thunnus* tuna species by the polymerase chain reaction and direct sequence analysis of their mitochondrial cytochrome *b* genes. *Can. J. Aquat. Sci.,* **48**, 309–17.

Bartlett, S.E. and Davidson, W.S. (1992) FINS (forensically informative nucleotide sequencing): a procedure for identifying the animal origin of biological specimens. *Biotechniques*, **12**, 408–11.

Batuecas, B., Garesse, R., Calleja, M. *et al.* (1988) Genome organization of *Artemia* mitochondrial DNA. *Nucleic Acids, Res.,* **16**, 6515–29.

Beckenbach, A.T., Thomas, W.K. and Sohrabi, H. (1990) Intraspecific sequence variation in the mitochondrial genome of rainbow trout (*Oncorhynchus mykiss*). *Genome*, **33**, 13–15.

Bentzen, P., Leggett, W.C. and Brown, G.G. (1988) Length and restriction site heteroplasmy in the mitochondrial DNA of American shad (*Alosa sapidissima*). *Genetics*, **118**, 509–18.

Bermingham, E. and Avise, J.C. (1986) Molecular zoogeography of freshwater fish in the southeastern United States. *Genetics*, **113**, 939–65.

Bermingham, E., Lamb, T. and Avise, J.C. (1986) Size polymorphism and heteroplasmy in the mitochondrial DNA of lower vertebrates. *J. Hered.,* **77**, 249–52.

Bernardi, G., D'Onofrio, G., Caccio, S. and Bernardi, G. (1994) Molecular phylogeny of bony fishes based on the amino acid sequence of the growth hormone. *J. Mol. Evol.* **38** (in press).

Bibb, M.J., Van Etten, R.A., Wright, C.T. *et al.* (1981) Sequence and gene organization of mouse mitochondrial DNA. *Cell*, **26**, 167–180.

Birky, C.W. Jr, Maruyama, T. and Fuerst, P.A. (1983) An approach to population and evolutionary genetic theory for genes in mitochondria and chloroplasts and some results. *Genetics*, **103**, 513–27.

Brown, G.G. and Simpson, M.V. (1982) Novel features of animals' mtDNA evolution as shown by sequences of two rat cytochrome oxidase subunit II genes. *Proc. Natl. Acad. Sci. USA,* **79**, 3246–50.

Brown, G.G., Gadaleta, G., Pepe, G. *et al.* (1986) Structural conservation and variation in the D-loop containing region of vertebrate mitochondrial DNA. *J. Mol. Biol.,* **192**, 503–11.

Brown, J.R., Beckenbach, A.T. and Smith, M.J. (1992) Mitochondrial DNA length variation and heteroplasmy in populations of white sturgeon (*Acipenser tranmontanus*). *Genetics*, **132**, 221–28.

Brown, W.M. (1981) Mechanisms of evolution in animal mitochondrial DNA. *Ann. N.Y. Acad. Sci.,* **361**, 119–34.

Brown, W.M. (1983) Evolution of mitochondrial DNA, in *Evolution of Genes and Proteins* (eds M. Nei and R.K Koehn), Sinauer, Sunderland, pp. 62–88.

Brown, W.M. (1985) The mitochondrial genome of animals, in *Molecular Evolutionary Genetics* (ed. R.J. MacIntyre), Plenum Press, New York, pp. 95–130.

Brown, W.M., George, M. Jr and Wilson, A.C. (1979) Rapid evolution of mitochondrial DNA. *Proc. Natl. Acad. Sci. USA,* **76**, 1967–71.

Brown, W.M., Prager, E.M., Wang, A. *et al.* (1982) Mitochondrial DNA sequences of primates: tempo and mode of evolution. *J. Mol. Evol.*, **18**, 225–39.

Buroker, N.E., Brown, J.R., Gilbert, T.A. *et al.* (1990) Length heteroplasmy of sturgeon mitochondrial DNA: an illegitimate elongation model. *Genetics*, **124**, 157–63.

Cann, R.L., Brown, W.M. and Wilson, A.C. (1984) Polymorphic sites and the mechanism of evolution and human mitochondrial DNA. *Genetics*, **106**, 479–99.

Cantatore, P., and Saccone, C. (1987) Organization, structure, and evolution of mammalian mitochondrial genes. *Int. Rev. Cytol.*, **108**, 149–208.

Cantatore, P., Roberti, M., Rainaldi, G. *et al.* (1989) The complete nucleotide sequence, gene organization, and genetic code of the mitochondrial genome of *Paracentrotus lividus. J. Biol. Chem.*, **264**, 10965–75.

Carr, S.M. and Marshall, D. (1991) DNA sequence variation in the mitochondrial cytochrome *b* gene of Atlantic cod (*Gadus morhua*) detected by the polymerase chain reaction. *Can. J. Fish. Aquat. Sci.*, **48**, 48–52.

Chapman, R.W., Stephens, J.C, Lansman, R.A. *et al.* (1982) Models of mitochondrial DNA transmission genetics and evolution in higher eukaryotes. *Genet. Res.*, **40**, 41–57.

Clark-Walker, G.D. (1985) Basis of diversity in mitochondrial DNAs, in *The Evolution of Genome Size* (ed. T. Cavalier-Smith), Wiley, New York, pp. 277–97.

Clary, D.O. and Wolstenholme, D.R. (1985) The mitochondrial DNA molecule of *Drosophila yakuba*: nucleotide, gene organization and genetic code. *J. Mol. Evol.*, **22**, 252–71.

Clayton, D.A. (1991) Replication and transcription of vertebrate mitochondrial DNA. *Annu. Rev. Cell. Biol.*, **7**, 453–78.

Cruz de la, V.F., Neckelmann, C. and Simpson, L. (1984) Sequence of six genes and several open reading frames in the kinetoplast maxicircle DNA of *Leishmania tarentolae. J. Biol. Chem.*, **259**, 15136–47.

Davidson, W.S., Birt, T.P. and Green, J.M. (1989) Organization of the mitochondrial genome from Atlantic salmon (*Salmo salar*). *Genome*, **32**, 340–42.

Dawid, I.B. (1972) Evolution of mitochondrial DNA sequences in *Xenopus. Dev. Biol.*, **29**, 139–51.

Densmore, L.D., Wright, J.W. and Brown, W.M. (1985) Length variation and heteroplasmy are frequent in mitochondrial DNA from parthenogenetic and bisexual lizards (genus *Cnemidophorus*). *Genetics*, **110**, 689–707.

DeSalle, R., Freedman, T., Prager, E.M. *et al.* (1987) Tempo and mode of sequence evolution in mitochondrial DNA of Hawaiian *Drosophila. J. Mol. Evol.*, **26**, 157–64.

Desjardins, P. and Morais, R. (1990) Sequence and gene organization of the chicken mitochondrial genome. *J. Mol. Biol.*, **212**, 599–634.

Desjardins, P. and Morais, R. (1991) Nucleotide sequence and evolution of coding and noncoding regions of a quail mitochondrial genome. *J. Mol. Evol.*, **32**, 153–61.

Dizon, A.E., Lockyer, C., Perrin, W.F. *et al.* (1992) Rethinking the stock concept: a phylogenetic approach. *Conservation Biol.*, **6**, 24–36.

Dunon-Bluteau, D. and Brun, G. (1986) The secondary structure of the *Xenopus laevis* and human mitochondria small ribosomal subunit RNA are similar. *FEBS Lett.*, **198**, 333–38.

Edwards, S.V., Arctander, P. and Wilson, A.C., (1991) Mitochondrial resolution of a deep branch in the genealogical tree for perching birds. *Proc. R. Soc. Lond. B*, **243**, 99–107.

Erlich, H.A. (ed.) (1989) *PCR Technology: Principle and Applications for DNA Amplification*, Stockton Press, New York.

Esposti Degli, M., Ghelli, A., Crimi, M. *et al.* (1992) Cytochrome *b* of fish mitochondria is strongly resistant to funiculosin, a powerful inhibitor of respiration. *Arch. Biochem. Biophys.*, **295**, 198–204.

Esposti Degli, M., De Vries, S., Crimi, M. *et al.* (1993) Mitochondrial cytochrome *b*: evolution and structure of the protein. *Acta Biophys. Biochem.*, 243–7.

Fajen, A. and Breden, F. (1992) Mitochondrial DNA sequence variation among natural populations of the Trinidad Guppy, *Poecilia reticulata. Evolution*, **46**, 1457–65.

Felsenstein, J. (1981) Evolutionary trees from DNA sequences: a maximum likelihood approach. *J. Mol. Evol.*, **17**, 368–76.

Felsenstein, J. (1982) Numerical methods for inferring evolutionary trees. *Q. Rev. Biol.*, **57**, 379–404.

Felsenstein, J. (1985) Confidence limits on phylogenies: an approach using the bootstrap. *Evolution*, **39**, 783–91.

Felsenstein, J. (1988) Phylogenies from molecular sequences: inference and reliability. *Annu. Rev. Genet.*, **22**, 521–65.

Ferris, S.D., Sage, R.D., Huang, C.M. *et al.* (1983) Flow of mitochondrial DNA across a species boundary. *Proc. Natl. Acad. Sci. USA*, **80**, 2290–94.

Finnerty, J.R. and Block, B.A. (1992) Direct sequencing of mitochondrial DNA detects highly divergent haplotypes in blue marlin (*Makaira nigricans*). *Mol. Mar. Biol. Biotech.*, **1**, 206–14.

Gadaleta, G., Pepe, G., DeCandia, G. *et al.* (1989) The complete nucleotide sequence of the *Rattus norvegicus* mitochondrial genome: cryptic signals revealed by comparative analysis between vertebrates. *J. Mol. Evol.*, **28**, 497–516.

Garey, J.R. and Wolstenholme, D.R. (1989) Platyhelminth mitochondrial DNA: evidence for early evolutionary origin of a tRNA (Ser) AGN that contains a dihydrouridine arm replacement loop, and of serine-specifying AGA and AGG codons. *J. Mol. Evol.*, **28**, 374–87.

Gilbert, T.L., Brown, J.R., O'Hara, P.J. *et al.* (1988) Sequence of tRNA-Thr and tRNA-Pro from white sturgeon (*Acipenser transmontanus*) mitochondria. *Nucleic Acids Res.*, **16**, 11825.

Gjetvaj, B., Cook, D.I. and Zouros, E. (1992) Repeated sequences and large-scale size variation of mitochondrial DNA: a common feature among scallops (Bivalvia: Pectinidae). *Mol. Biol. Evol.*, **9**, 106–24.

Glotz, C., Zwieb, C., Brimacombe, R. *et al.* (1981) Secondary Structure of the large subunit ribosomal RNA from *Echerichia coli*, *Zea mays* chloroplast, and human and mouse mitochondrial ribosomes. *Nucleic Acids Res.*, **9**, 3287–306.

Gojobori, T., Ishii, K. and Nei, M. (1982) Estimation of average number of nucleotide substitutions when the rate of substitution varies with nucleotides. *J. Mol. Evol.*, **18**, 414–23.

Grachev, M.A., Slobodyanyuk, S. Ja., Kholodilov, N.G. *et al.* (1992) Comparative study of two protein coding regions of mitochondrial DNA from three endemic sculpins (Cottoidei) of Lake Baikal. *J. Mol. Evol.*, **34**, 85–90.

Graves, J.E., Ferris, S.D. and Dizon, A.E. (1984) Close genetic similarity of Atlantic and Pacific skipjack tuna (*Katsuwonis pelamis*) demonstrated with restriction endonuclease analysis of mitochondrial DNA. *Mar. Biol.*, **79**, 315–19.

Gray, M.W. (1989) Origin and evolution of mitochondrial DNA. *Annu. Rev. Cell. Biol.*, **5**, 25–50.

Greenberg, B.D., Newbold, J.E. and Sugino, A. (1983) Intraspecific nucleotide sequence variability surrounding the origin of replication in human mitochondrial DNA. *Gene*, **21**, 33–49.

Gyllensten, U.B. and Erlich, H.A. (1988) Generation of single-stranded DNA by the polymerase chain reaction and its application to direct sequencing of the HLA-DQa locus. *Proc. Natl. Acad. Sci. USA*, **85**, 7652–55.

Gyllensten, U. and Wilson, A.C. (1987) Mitochondrial DNA of salmonids: inter- and intraspecific variability detected with restriction enzymes, in *Population Genetics and Fishery Management* (eds N. Ryman and F. Utte), University of Washington Press, Seattle, pp. 301–17.

Gyllensten, U., Wharton, D., Josetsson, A. *et al.* (1991) Paternal inheritance of mitochondrial DNA in mice. *Nature*, **352**, 255–57.

Hadrys, H., Balick, M. and Schierwater, B. (1992) Applications of random amplified polymorphic DNA (RAPD) in molecular ecology. *Mol. Ecol.*, **1**, 55–63.

Hallerman, E.M and Beckmann, J.S. (1988) DNA-level polymorphism as a tool in fisheries science. *Can. J. Fish. Aquat. Sci.*, **45**, 1075–87.

Hamby, R.K., Sims, L., Issel, L. *et al.*, (1988) Direct ribosomal RNA sequencing: optimization of extraction and sequencing methods for work with higher plants. *Plant Mol. Biol. Reporter*, **6**, 175–92.

Harrison, R.G. (1991) Molecular changes at speciation. *Annu. Rev. Ecol. Syst.*, **22**, 281–308.

Harrison, R.G., Rand, D.M. and Wheeler, W.C. (1985) Mitochondrial DNA size variation within individual crickets. *Science*, **228**, 1446–48.

Hauswirth, W.W. and Laipis, P.J. (1982) Mitochondrial DNA polymorphism in a maternal lineage of Holstein cows. *Proc. Natl. Acad. Sci. USA*, **79**, 4686–90.

Hayashi, J-I., Tagashira, Y. and Yoshida, M.C. (1985) Absence of extensive recombination between inter- and intraspecies mitochondrial DNA in mammalian cells. *Exp. Cell. Res.*, **160**, 387–95.

Herke, S.W., Kornfield, I. Moran, P. *et al.* (1990) Molecular confirmation of hybridization between Northern Pike (*Esox lucius*) and Chain Pickerel (*E. niger*). *Copeia*, **1990**, 846–50.

Hillis, D.M. and Dixon, M.T. (1989) Vertebrate phylogeny: evidence from 28S ribosomal DNA sequences, in *Phylogenetic Analysis of DNA Sequences* (eds M.M. Miyamoto and J. Cracraft), Oxford University Press, New York, pp. 355–67.

Hillis, D.M. and Dixon, M.T. (1991) Ribosomal DNA: molecular evolution and phylogenetic inference. *Q. Rev. Biol.*, **66**, 411–53.

Hillis, D.M. and Dixon, M.T. and Ammerman, L.K. (1989) The relationships of the coelacanth *Latimeria chalumnae*: evidence from sequences of vertebrate 28S ribosomal genes, in *The Biology of Latimeria chalumnae and Evolution of Coelacanths* (eds J.A. Musick, M.N. Bruton and E.K. Balon), Kluwer Academic Publishers, Dordrecht, pp. 117–30.

Hillis, D.M. and Moritz, C. (eds) (1990) *Molecular Systematics*, Sinauer Associates Inc., Sunderland, MA.

Hillis, D.M., Bull, J.J., White, M.E., Badgett, M.R. and Molineux, I.J. (1992) Experimental phylogenetics: generation of a known phylogeny. *Science*, **255**, 589–592.

Himeno, H., Masaki, H., Kawai, T. *et al.* (1987) Unusual genetic codes and a novel gene structure for tRNA-AGY-Ser in starfish mitochondrial DNA. *Gene*, **56**, 219–30.

Hixson, J.E. and Brown, W.M. (1986) A comparison of the small ribosomal RNA genes from the mitochondrial DNA of the great apes and humans: sequence, structure, evolution and phylogenetic implications. *Mol. Biol. Evol.*, **3**, 1–18.

Hoffman, R., Boore, J.L. and Brown, W.M. (1992) A novel mitochondrial genome organization in the blue mussel, *Mytilus edulis. Genetics*, **131**, 397–412.

Holmquist, R. (1976) Solution to a gene divergence problem under arbitrary stable nucleotide transition probabilities. *J. Mol. Evol.*, **8**, 337–49.

Holmquist, R. (1983) Transitions and transversion in evolutionary descent: an approach to understanding. *J. Mol. Evol.*, **19**, 134–44.

Horak, I., Coon, H.G. and Dawid, I.B. (1974) Interspecific combination of mitochondrial DNA molecules in hybrid somatic cells. *Proc. Natl. Acad. Sci. USA.*, **71**, 1828–32.

Hull, D.L. (1988) *Science as a Process*, Chicago University Press, Chicago.

Hurst, L.D. (1991) Sex, slime and selfish genes. *Nature*, **354**, 23–4.

Innis, M.A., Gelfand, D.H., Sninsky, J.J. and White, T.J. (1990) *PCR Protocols: A Guide to Methods and Applications*, Academic Press, San Diego.

Irwin, D.M., Kocher, T.D. and Wilson A.C. (1991) Evolution of the cytochrome *b* of mammals. *J. Mol. Evol.*, **32**, 128–44.

Jacobs, H.T., Elliott, D.J., Math, V.B. *et al.* (1988) Nucleotide sequence and gene organization of sea urchin mitochondrial DNA. *J. Mol. Biol.*, **202**, 185–217.

Jacobs, H., Asakawa, S., Araki, T. *et al.* (1989) Conserved tRNA gene cluster in starfish mitochondrial DNA. *Curr. Genet.*, **15**, 193–206.

Johansen, S., Guddal, P.H. and Johansen, T. (1990) Organization of the mitochondrial genome of Atlantic cod, *Gardus morhua. Nucleic Acids, Res.*, **18**, 411–19.

Jukes, T.H (1987) Transitions, transversions,and the molecular evolutionary clock. *J. Mol. Evol.*, **26**, 87–98.

Jukes, T.H and Cantor, C.R. (1969) Evolution of protein molecules, in *Mammalian Protein Metabolism* (ed. H.N. Munro), Academic Press, New York, pp. 21–123.

Kimura, M. (1980) A simple method for estimating evolutionary rate of base substitutions through comparative studies of nucleotide sequences. *J. Mol. Evol.*, **16**, 111–20.

Kimura, M. (1981) Estimation of evolutionary distances between homologous nucleotide sequences. *Proc. Natl. Acad. Sci. USA*, **78**, 454–58.

Kocher, T.D., Thomas, W.K., Meyer, A. *et al.* (1989). Dynamics of mitochondrial DNA evolution in animals. *Proc. Natl. Acad. Sci. USA*, **86**, 6196–200.

Koehler, G.M., Lindberg, G.L., Brown, D.R. *et al.* (1991) Replacement of bovine mitochondrial DNA by sequence variant within one generation. *Genetics*, **129**, 247–55.

Kornfield, I. (1984) Descriptive genetics of cichlid fish, in *Evolutionary Genetics of Fishes* (ed. B.J. Turner), Plenum Press, New York, pp. 591–615.

Kornfield, I. (1991) Genetics, in *Cichlid Fish: Behaviour, Ecology and Evolution* (ed. M.H.A. Keenleyside), Chapman and Hall, New York, pp. 103–28.

Li, W.-H. and Gouy, M. (1991) Statistical methods for testing molecular phylogenies, in *Phylogenetic Analysis of DNA Sequences* (eds M.M. Miyamoto and J. Cracraft), Oxford University Press, Oxford, pp. 249–77.

Martin, A.P., Naylor, G.J.P. and Palumbi, S.R. (1992) Rates of mitochondrial DNA evolution in sharks are slow compared with mammals. *Nature*, **357**, 153–5.

McVeigh, H.P., Bartlett, S.E. and Davidson, W.S. (1991) Polymerase chain reaction/direct sequence analysis of the cytochrome *b* gene in *Salmo salar. Aquaculture*, **95**, 225–31.

Meland, S., Johansen, S., Johansen, T. *et al.* (1991) Rapid disappearance of one parental mitochondrial genotype after isogamous mating in the myxomycete *Physarum polycephalum. Curr. Genet.*, **19**, 55–60.

Meyer, A. (1993) Evolution of mitochondrial DNA in fish, in *Biochemistry and Molecular Biology of Fish*, Vol. 2 (eds P.W Hochachka and P. Mommsen), Elsevier Press, pp. 1–38.

Meyer, A. and Dolven, S.I. (1992) Molecules, fossils and the origin of tetrapods. *J. Mol. Evol.*, **35**, 102–13.

Meyer, A., Kocher, T.D. and Wilson, A.C. (1991) African fish–a reply. *Nature*, **351**, 467–68.

Meyer, A and Lydeard, C. (1994) The evolution of copulatory organs, internal fertilization, placentas, and viviparity in killifishes (Cyprinodontiformes) inferred from DNA phylogeny of the tyrosine kinase gene X-*SIC. Proc. Roy. Soc. Lond* B, (in press).

Meyer, A. and Wilson, A.C. (1990) Origin of tetrapods inferred from their mitochondrial DNA affiliation to lungfish. *J. Mol. Evol.*, **31**, 359–64.

Meyer, A., Kocher, T.D., Basasibwaki, P. *et al.* (1990) Monophyletic origin of Victoria cichlid fish suggested by mitochondrial DNA sequences. *Nature*, **347**, 550–53.

Mindell, D.P. (1991) Aligning DNA sequences: homology and phylogenetic weighting, in *Phylogenetic Analysis of DNA Sequences* (eds M.M. Miyamoto and J. Cracraft), Oxford University Press, New York, pp. 73–89.

Mindell, D.P. and Honeycutt, R.L (1990) Ribosomal RNA in vertebrates: evolution and phylogenetic applications. A*nnu. Rev. Ecol. Syst.*, **21**, 541–66.

Moritz, C. (1991) Evolutionary dynamics of mitochondrial DNA duplications in parthenogenetic geckos, *Heteronotia binoei. Genetics*, **129**, 221–30.

Moritz, C., Dowling, T.E. and Brown, W.M. (1987) Evolution of animal mitochondrial DNA: relevance for population biology and systematics. *Annu. Rev. Ecol. Syst.*, **18**, 269–92.

Mullis, K., Faloona, F., Scharf, S. *et al.* (1986). Specific enzymatic amplification of DNA in vitro: the polymerase chain reaction. *Cold Spring Harbor Symp. Quant. Biol.*, **51**, 263–73.

Nei, M. (1991) Relative efficiencies of different tree-making methods for molecular data, in *Phylogenetic Analysis of DNA Sequences* (eds M.M. Miyamoto and J. Cracraft), Oxford University Press, New York, pp. 90–128.

Neigel, J.E. and Avise, J.C. (1986) Phylogenetic relationships of mitochondrial DNA under various demographic models of speciation, in *Evolutionary Processes and Theory* (eds E. Nevo and S. Karlin), pp. 515–34.

Normark, B.B., McCune, A.R. and Harrison, R.G. (1991 Phylogenetic relationships of Neopterygian fish, inferred from mitochondrial DNA sequences. *Mol. Biol. Evol.*, **8**, 819–34.

Ojala, D., Montoya, J. and Attardi, G. (1981) tRNA punctuation model of RNA processing in human mitochondrial DNA. *Nature*, **290**, 470–74.

Okimoto, R., Macfarlane, J.L., Clary, D.O. *et al.* (1992) The mitochondrial genomes of two nematodes, *Caenorhabditis elegans* and *Ascaris suum. Genetics*, **130**, 471–98.

Olivo, P.D., Van de Walle, J.J., Laipis, J.P. *et al.* (1983) Nucleotide sequence evidence for rapid genotypic shifts in bovine mitochondrial DNA D-loop. *Nature*, **306**, 400–2.

Orti, G., Bell, M.A., Reimchen, T.E. *et al.* (1984) Global survey of mitochondrial DNA sequence variation in the threespine stickleback; evidence for recent colonizations. *Evolution* (in press).

Ovenden, J.R. (1990) Mitochondrial DNA and marine stock assessment: a review. *Aust. J. Mar. Freshwater Res.*, **41**, 835–53.

Ovenden, J.R., White, R.W.G. and Sanger, A.C. (1988) Evolutionary relationships of *Gadopsis* gspp. inferred from restriction enzyme analysis of their mitochondrial DNA. *J. Fish. Biol,*, **32**, 137–48.

Paabo, S. (1990). Amplifying ancient DNA, in *PCR Protocols: A Guide to Methods and Applications*, (eds M.A. Innes, D.H. Gelfand, J.J. Sninsky and T.J. White), Academic Press, San Diego, pp. 159–66.

Pääbo, S., Thomas, W.K., Whitfield, K.M. *et al.* (1991) Rearrangements of mitochondrial transfer RNA genes in marsupials. *J. Mol. Evol.*, **33**, 426–30.

Palumbi, S.R., Martin, A., Romano, S. *et al.* (1991). *Simple Fool's Guide to PCR*, Department of Zoology, University of Hawaii, Honolulu.

Pamilo, P. and Nei, M. (1988) Relationships between gene trees and species trees. *Mol. Biol. Evol.*, **5**, 568–83.

Patterson, C. (ed.) (1987) *Molecules and Morphology in Evolution: Conflict or Compromise?* Cambridge University Press, Cambridge.

Perler, F., Efstratiadis, A., Lomedico, P. *et al.* (1980) The evolution of genes: the chicken preproinsulin gene. *Cell*, **20**, 555–6.

Pritchard, A.E., Seilhamer, J.J., Mahalingam, R. *et al.* (1990) Nucleotide sequence of the mitochondrial genome of *Paramecium. Nucleic Acids Res.*, **18**, 173–80.

Quattro, J.M., Avise, J.C. and Vrijenhoek, R.C. (1992) Mode and origin and sources of genotypic diversity in triploid gynogenetic fish clones (Poeciliopsis: Poeciliidae). *Genetics*, **130**, 621–218.

Roe, B.A., Ma, D.-P., Wilson, R.K. *et al.* (1985) The complete nucleotide sequence of the *Xenopus laevis* mitochondrial genome. *J. Biol. Chem.*, **260**, 9759–74.

Ryman, N. and Utter, F. (ed.) (1987) *Population Genetics and Fishery Management*, University of Washington Press, Seattle.

Saccone, C., Attimonelli, M. and Sbiza, E. (1987) Structural elements highly preserved during the evolution of the D-loop-containing region in vertebrate mitochondrial DNA. *J. Mol. Evol.*, **26**, 205–11.

Saiki, R.K., Scharf, S., Faloona, F. *et al.* (1985) Enzymatic amplification of beta-globin genomic sequences and restriction site analysis for diagnosis of sickle cell anemia. *Science*, **230**, 1350–54.

Saiki, R.K., Gelfand, D.H., Stoffel, S. *et al.* (1988) Primer-directed enzymatic amplification of DNA with a thermostable DNA polymerase. *Science*, **239**, 487–91.

Saitou, N. and Nei, M. (1987) The neighbour-joining method: a new method for reconstructing phylogenetic trees. *Mol. Biol. Evol.*, **4**, 406–525.

Schneider, J.F., Myster, S.H., Hackett, P.B. *et al.* (1992) Molecular cloning and sequence analysis of the cDNA for another pike (*Esox lucinus*) growth hormone. *Mol. Mar. Biol. Biotechnol.*, **1**, 106–12.

Smith, M.F., Banfield, D.K., Doteval, K. *et al.* (1989) Gene arrangement in sea star mitochondrial DNA demonstrates a major inversion event during echinoderm evolution. *Gene*, **76**, 181–5.

Smith, M.F., Banfield, D.K., Doteval, K. *et al.* (1990) Nucleotide sequence of nine protein coding genes and 22 tRNAs in the mitochondrial DNA of the sea star *Piaster ochraceus. J. Mol. Evol.*, **31**, 195–204.

Synder, M., Fraser, A.R, LaRoche, J. *et al.* (1987) A typical mitochondrial DNA from the deep-sea scallop *Placopecten magellanicus. Proc. Natl. Acad. Sci. USA*, **84**, 7595–99.

Sprinzl, M., Dank, N., Nock, S. *et al.* (1991) Compilation of tRNA sequences and sequences of tRNA genes. *Nucleic Acid Res.*, **19**, 2127–71.

Stanton, D.J., Crease, T.J. and Herbert, P.D.N. (1991) Cloning and characterization of *Daphnia* mitochondrial DNA. *J. Mol. Evol.*, **33**, 152–5.

Stock, D.W. and Whitt, G.S. (1992) Evidence from 18S ribosomal RNA sequences that lampreys and hagfish form a natural group. *Science*, **257**, 787–9.

Stock, D.W., Moberg, K.D., Maxson, L.R. *et al.* (1991). A phylogenetic analysis of the 18S ribosomal RNA sequence of the coelacanth *Latimeria chalumnae*, in *the Biology of Latimeria chalumnae and Evolution of Coelacanths* (eds J.A. Musick, M.N. Bruton and E.K. Balon), Kluwer Academic Publishers, Dordrecht, pp. 99–117.

Sturmbauer, C. and Meyer, A. (1992) Genetic divergence, speciation and morphological stasis in a lineage of African cichlid fish. *Nature*, **358**, 578–81.

Swofford, D.L. (1991) When are phylogeny estimates from molecular and morphological data incongruent? in *Phylogenetic Analysis of DNA Sequences* (eds M.M. Miyamoto and J. Cracraft), Oxford University Press, New York, pp. 295–333.

Swofford, D.L. and Olsen, G.J. (1990) Phylogeny reconstruction, in *Molecular Systematics* (eds D.M. Hillis and C. Moritz), Sinauer, Sunderland, pp. 411–501.

Taberlet, P., Meyer, A. and Bouvet, J. (1992) Unusually large mitochondrial variation in populations of the blue tit, *Parus caeruleus. Mol. Ecol.*, **1**, 27–36.

Tajima, F. and Nei, M. (1984) Estimation of evolutionary distance between nucleotide sequences. *Mol. Biol. Evol.*, **1**, 269–85.

Takahata, N. and Kimura, M. (1981) A model of evolutionary base substitutions and its application with special reference to rapid change of pseudogenes. *Genetics*, **98**, 641–57.

Tegelstrom, J.T. (1987) Transfer of mitochondrial DNA from northern red-back vole (*Clethrionomys rutilus*) to the bank vole (*C. glareolus*). *J. Mol. Evol.*, **24**, 218–27.

Templeton, A.R. (1983a) Convergent evolution and non-parametric inferences from restriction fragment and DNA sequence data, in *Statistical Analysis of DNA Sequence Data* (ed. B. Weir), Marcel Dekker, New York, pp. 151–79.

Templeton, A.R. (1983b) Phylogenetic inference from restriction endonuclease site maps with particular reference to the humans and apes. *Evolution*, **37**, 221–44.

Thomas, W.K. and Beckenbach, A.T. (1989) Variation in salmonid mitochondrial DNA: evolution constraints and mechanisms of substitution. *J. Mol. Evol.*, **29**, 233–45.

Thomas, W.K., and Withler, R.E. and Beckenbach, A.T. (1986) Mitochondrial DNA analysis of Pacific salmonid evolution. *Can. J. Zool.*, **64**, 1058–64.

Tzeng, C.S., Hui, C.F., Shen, S.C. and Huang, P.C. (1992) The complete nucleotide sequence of the *Crossostoma lucustre* mitochondrial genome: conservation and variations among vertebrates. *Nucleic Acids Res.*, **20**, 4853–8.

Vaughn, K.C., DeBonte, L.R. and Wilson, K.G. (1980) Organelle alteration as a mechanism for maternal inheritance. *Science*, **208**, 196–7.

Watanabe, K., Igarashi, A., Noso, T. *et al.* (1992) Chemical identification of catfish growth hormone and prolactin. *Mol. Mar. Biol. Biotechnol.*, **1**, 239–49.

Welsh, J. and McClelland, M. (1990) Polymorphisms generated by arbitrarily primed PCR in the mouse: application with arbitrary primers. *Nucleic Acids Res.*, **18**, 7123–18.

White, T.J., Arnheim, N. and Erlich, H.A. (1989) The polymerase chain reaction. *TIG*, **5**, 185–9.

Williams, J.G.K., Kubelik, A.R., Livak, K.J. *et al.* (1990) DNA polymorphisms amplified by arbitrary primers are useful as genetic markers. *Nucleic Acids Res.*, **18**, 6531–35.

Wilson, A.C., Cann, R.L., Carr, S.M. *et al.* (1985) Mitochondrial DNA and two perspectives on evolutionary genetics. *Biol. J. Linn. Soc.*, **26**, 375–400.

Wilson, A.C., Zimmer, E.A., Prager, E.M. *et al.* (1989). Restriction mapping in the molecular systematics of mammals: a retrospective salute, in *The Hierarchy of Life* (eds B. Fernholm, K. Bremer and H. Jornvall), Elsevier, Amsterdam, pp. 407–19.

Wilson, G.M., Thomas, W.K. and Beckenbach, A.T. (1985) Intra- and interspecific mitochondrial DNA sequence divergence in *Salmo*: rainbow, steelhead, and cutthroat trouts. *Can. J. Zool.*, **63**, 2088–94.

Wilson, G.M., Thomas, W.K. and Beckenbach, A.T. (1987) Mitochondrial DNA analysis of Pacific Northwest populations of *Oncorhynchus tshawytscha. Can. J. Fish. Aquat. Sci.*, **44**, 1301–5.

Wirgin, I.I., Proenca, R. and Grossfield, J. (1989) Mitochondrial DNA diversity among populations of striped bass in the southeastern United States. *Can. J. Zool.*, **67**, 891–907.

Wolstenholme, D.R. and Clary, D.O. (1985) Sequence evolution of *Drosophila* mitochondrial DNA. *Genetics*, **109**, 725–44.

Wrishnik, L.A., Higuchi, R.G. and Stoneking, M. (1987) Length mutations in human mitochondrial DNA: direct sequencing of enzymatically amplified DNA. *Nucleic Acids Res.*, **15**, 529–42.

Yoneyama, Y. (1987) The nucleotide sequences of the heavy and light strand replication origins of the *Rana catesbeiana* mitochondrial genome. *Nippon Ika Daioaku Zasshi*, **54**, 429–40 (in Japanese).

Zouros, E., Freeman, K.R., Oberhauser Ball, A. *et al.* (1992) Direct evidence for extensive paternal mitochondrial DNA inheritance in the marine mussel *Mytilus. Nature*, **359**, 412–14.

5.2 COMPARISONS OF MITOCHONDRIAL DNA VARIATION IN FOUR ALOSID SPECIES AS REVEALED BY THE TOTAL GENOME, THE NADH DEHYDROGENASE I AND CYTOCHROME *b* REGIONS

Robert W. Chapman, John C. Patton and Brandon Eleby

Abstract

Comparison of mitochondrial DNA variation in four species of the genus Alosa *demonstrates that the cytochrome* b *and NADH dehydrogenase portions of the molecule are evolving at different rates. While this result was, to a degree, anticipated from sequence divergence data involving disparate taxa, the present study extends the results to closely related species. In addition, estimates of sequence divergence obtained from comparisons of mtDNA regions among species may serve as a guide for selecting regions that might reveal intraspecific variation.*

Comparison of the total mitochondrial genome using six-base digestion failed to distinguish Alosa alabamae *from* A. sapidissima *while four-base digestion of the* ND I *portion of the genome found two site changes the distinguished these forms. The analyses indicate the existence of an* A. pseudoharengus–A. aestivalis *clade and an* A. sapidissima– A. alabamae *clade. Within each of these clades, phylogenetic relationships are uncertain, as different segments of the mtDNA molecule suggest polyphyly among the species. A systematic revision of the genus is recommended to establish phylogenetic relationships and the species status of* A. alabamae.

5.2.1 Introduction

Surveys of mitochondrial DNA over the past decade have revealed extensive structuring of populations of many fish species. In some fish, however, mtDNA analyses have not successfully identified discrete populations or subdivisions. The 'failures' stem from: (1) limited variation in the species, where one or two haplotypes dominate most populations and a number of additional haplotypes are found only once or a few times (Kornfield and Bogdanowicz, 1987; Bentzen, Brown and Leggett, 1989); (2) length variation in the noncoding portion of the molecule which is difficult to assess phylogenetically, but may be used in frequency analyses (Bentzen, Brown and Leggett, 1989; Chapman, 1987, 1989, 1990; Mulligan and Chapman, 1989; Wirgin, Proenca and Grossfield, 1989); (3) variation concentrated within rather than among populations (Bowen and Avise, 1990; Gold and Richardson, 1991); or (4) a combination of these problems. In systematics, mtDNA has proven useful in reconstructing genealogical relationships (Dowling, Moritz and Palmer, 1990; Dowling *et al.*, 1992) although in many instances mtDNA restriction data are plagued by homoplasies that characterize other data as well. Overall, the rate of mtDNA evolution appears to be initially rapid, slows down beyond 15% sequence divergence and asymptotes around 30% divergence (Moritz, Dowling and Brown, 1987). This has resulted in the well-known curves relating time and sequence divergence (Brown *et al.*, 1982; Hixson and Brown, 1986; Meyer, 1993). Relationships among distantly related species may, therefore be difficult to assess with traditional RFLP methods for mtDNA analysis. It is generally accepted that this limitation stems from the boundaries that exist on permissible nucleotide changes (Moritz, Dowling and Brown, 1987), i.e. a gene may change only so much and still yield a functional product. It has been known for some time, however, that different genes in the mtDNA molecule evolve at different rates. Meyer (1993) has compared amino acid (AA) and nucleotide (N) sequences of the cod, *Gadus morhua*, to those of the frog, *Xenopus laevis*, and found a range similarity for 46% AA, 54% N in the *ND4L* gene to 93% AA, 79% N in the *COI* gene. This implies that rapidly evolving portions of the molecule may be more appropriate to studies of closely related species (or populations of a single species), and the slowly changing portions to studies of more distantly related species.

The ability of evolutionary biologists to select portions of the genome appropriate for study at a given taxonomic level was greatly enhanced by the advent of direct amplification and sequencing using the polymerase chain reaction (PCR). This tool is certainly one of the 'ultimate' weapons in the search for genetic variability and, by virtue of the data generated, is an excellent approach for systematists. One limitation, however, is the availability of oligonucleotide primers for the region of interest and this limitation is rapidly being eroded by the efforts of a number of laboratories. For population biologists needing large sample sizes, the approach is tedious due to the relatively slow process of sequencing DNA once it is amplified, and there is a lack of knowledge regarding the level of intraspecific variation that might be obtained from a selected region of mtDNA. While some portions of mtDNA appear to evolve more rapidly than others when disparate taxa are compared, this does not necessarily translate into intraspecific variation.

In this section, we address the relationships among four fish species (*Alosa sapidissima*, *A. alabamae*, *A. pseudoharengus* and *A. aestivalis*), using six-base restriction enzymes to assess mtDNA variation in the total genome and data obtained for four-base digestion of the two regions of the molecule (NADH dehydrogenase I (*ND I*) and cytochrome *b* (*cyt b*) genes). The effort is intended to compare rates of nucleotide substitution as well as intra- and interspecific variation. These species were selected because a large repository of *A. sapidissima* mtDNA was on hand and initial screening with six-base digests of the entire mtDNA molecules of *A. alabamae* failed to reveal any differences when compared to *A. sapidissima*. In addition to the issue of variation in different parts of the mtDNA, it was of interest to know (1) whether limited differences among closely related forms was a characteristic of alosids, (2) if *A. alabamae* might not warrant species status, or (3) if the fine-scale discrimination available by application of four-base restriction enzymes might find some differences. *Alosa aestivalis* and *A. pseudoharengus* were selected because of their close morphological similarity to each other and membership in the genus.

5.2.2   Methods and materials

Alosa sapidissima samples from the Connecticut and Susquehanna (Maryland) Rivers and *A. alabamae* from the Apalachicola River (Florida) were used in this study (Figure 5.2.1). These samples were part of a larger study on mtDNA variation in *A. sapidissima* (Chapman and Brown, unpublished data). Specimens of *A. pseudoharengus* and *A. aestivalis* were taken from Albemarle Sound, North Carolina.

Figure 5.2.1 Map of east coast of the US showing collection sites. *Alosa sapidissima* was collected from the Connecticut and Susquehanna Rivers, *A. pseudohargenus* and *A. aestivalis* from the Chowan River, and *A. alabamae* from the Appalachicola River.

Mitochondrial DNA was isolated using a slight modification of the procedures outlined by Chapman and Powers (1984). The critical modification was to homogenize ovarian tissue by two passes through syringe needles. The first pass was through a 20-gauge needle which fractured large eggs and eliminated strings of tissue. The second pass was made through a 22-gauge syringe needle which fractured the smaller and less mature eggs. The modification was necessary as other homogenization techniques resulted in limited or no mtDNA and substantial nuclear contamination. The problem was largely confined to *A. sapidissima* and *A. alabamae*. Once isolated, mtDNA was digested with a battery of restriction enzymes (Table 5.2.1) and the fragments separated on 0.8% agrose gels.

The *ND I* and *cyt b* portions of the molecule were amplified following standard protocols using oligonucleotide primers developed by one of us. The target of the *ND I* primer set was a 2.1-kb fragment extending from a position approximately 600 bases inside the 16S rRNA

Table 5.2.1   Size estimates in kilobases for the most common mtDNA fragment patterns produced by *Aat*I, *Apa*I, *Bgl*I, *Dra*I, *Eco*RI, *Kpn*I and *Pvu*II in *Alosa aestivalis*, *A. pseudoharengus*, *A. sapidissima* and *A. alabamae*

*Aat*I					*Apa*I				*Bgl*I				
N	*O*	*A*	*C*	*D*	*N*	*A*	*C*	*E*	*N*	*O*	*A*	*B*	*E*
8.8	8.8	11.4	11.4	11.4	16.5	10.0	18.0	8.0	5.6	5.6	8.7	11.5	8.7
4.8	3.6	3.3	4.8	3.3		8.0		5.5	4.2	4.2	3.4	3.4	4.3
1.2	1.2	1.5	1.1	1.8				4.5	3.3	3.3	2.8	1.8	3.4
0.9	1.2	1.1	0.7	1.5					1.8	1.6	1.8	1.3	1.8
(0.7)	0.9	0.7							1.6	1.2	1.3		1.3
	(0.7)									(0.6)			
16.4	16.4	18.0	18.0	18.0	16.5	16.5	18.0	18.0	16.5	15.6	18.0	18.0	18.0

*Dra*I				*Eco*RI				*Kpn*I				
N	*O*	*A*		*N*	*A*	*B*		*N*	*A*	*B*	*C*	*D*
9.4	9.4	8.8		9.6	9.6	7.2		16.5	6.4	12.3	7.9	18.0
3.4	2.5	4.2		7.2	7.2	5.4			6.1	6.1	6.1	
2.5	2.2	2.5			1.5	4.4			4.3		4.3	
1.4	1.5	1.5				1.5			1.5			
	1.2	1.2										
16.7	16.8	18.2	18.2	16.8	16.8	18.3		16.5	18.3	18.4	18.3	18.0

*Pvu*II		
N	*A*	*B*
7.9	8.3	10.2
4.9	6.6	8.3
2.3	3.6	
1.4		
16.5	18.5	18.5

Numbers in parentheses indicate one or more fragments inferred to exist, but not actually seen due to a lack of resolution in ethidium-bromide-stained gels. For example, the 0.6 *Bgl*I fragment (O pattern) may actually be two or more, but was scored as a single missing band based upon molecular weight estimates of the other bands. Fragment patterns A–D were found in *A. sapidissima* and *A. alabamae*, while patterns N and O were found in *A. pseudoharengus* and/or *A. aestivalis*.

gene through *ND 1* and terminating in the fmet tRNA. The target of the *cyt b* primers was a 1.3-kb portion beginning near the 5′ end of the reading frame and terminating in the proline tRNA. The amplified products were precipitated from the reaction mix with two volumes of ethanol, dried, and rehydrated in sterile water. The samples were then digested with a battery of four-base restriction enzymes and the fragments separated on 9% polyacrylamide gels. The gels were stained with ethidium bromide and photographed under UV illumination.

The data were analysed for nucleotide divergence within and among species using the REAP package kindly provided by McElroy *et al.* (1991). The data were encoded by making the assumption of most parsimonious intraconversion of fragment patterns. For example, in Table 5.2.1, it was assumed that *Aat*I patterns N and O differed by a single site change converting the 4.8-kb fragment of pattern N to the 3.6- and 1.2-kb fragment in pattern O or vice versa. For the six-base survey the process was greatly assisted by a restriction map (Bentzen, Leggett and Brown, 1988; Bentzen, personal communication). Restriction sites in the *cyt b* and most of the *ND I* region were easily assigned except for the *Hae*III digestion patterns. The REAP package also produced data matrices for analysis by the BOOT program contained in PHYLIP version 3.4 (Felsenstein, 1985). Two phylogenetic trees were rooted by creating an ancestral haplotype that (1) possessed only the restriction sites common to all haplotypes found in this study or (2) possessed restriction sites that were common to all species whether or not they were represented in all haplotypes of all species, i.e. a consensus ancestor.

5.2.3   Results

(a)   Total mtDNA analysis

Size estimates for mtDNA fragment patterns obtained with the seven restriction enzymes employed in this analysis are presented in Table 5.2.1. Mitochondrial DNAs from *A. pseudoharengus* and *A. aestivalis* are about 1.5 kb smaller than those of *A. sapidissima* and *A. alabamae* and are consistent with values obtained by Bentzen (personal communication). Composite restriction patterns (Table 5.2.2) for each enzyme revealed 14 haplotypes in *A. sapidissima*, two haplotypes in *A. alabamae*, and only one haplotype in *A. pseudoharengus* and *A. aestivalis*. The different numbers of haplotypes should not be viewed as indicating different levels of genetic variability in the latter species as the sample sizes and geographic coverage are not equivalent to those of *A. sapidissima*. In addition, the two haplotypes found in *A. alabamae* were identical to the two most common haplotypes found in *A. sapidissima* and this identity has held up in a more exhaustive survey of 13 six-base restriction digests of the entire mtDNA molecule (Chapman and Brown, unpublished data).

Estimates of nucleotide divergence within and among species (Table 5.2.3) show that *A. sapidissima* haplotypes are more diversified from each other than they are from *A. alabamae*. This is consistent with expectation, due to the fact that both of the haplotypes

Table 5.2.2   Composite mtDNA haplotypes of alosid species

Species	*Haplotype*	*N*
A. sapidissima 1	AAAAAAA	72
A. sapidissima 2	AAAAAAB	1
A. sapidissima 3	AAAAABA	8
A. sapidissima 4	AAAAABB	1
A. sapidissima 5	AAAAACA	2
A. sapidissima 6	AAAAADA	1
A. sapidissima 7	AAAABAA	1
A. sapidissima 8	AAEAAAA	2
A. sapidissima 9	AABAAAA	1
A. sapidissima 10	ACAAAAA	1
A. sapidissima 11	AEAAAAA	1
A. sapidissima 12	CAAAAAA	1
A. sapidissima 13	DAAAAAA	3
A. sapidissima 14	DAAAABA	1
A. alabamae 1	AAAAAAA	14
A. alabamae 2	AAAAABA	5
A. pseudoharengus	NNNNNNN	7
A. aestivalis	ONOONNN	20

Haplotypes found in the six-base survey and restriction profiles are given in Table 5.2.1. Composite haplotypes are constructed by listing enzymes in alphabetical order. *N* is the number of individuals observed with a given haplotype.

Table 5.2.3   Average mtDNA nucleotide divergence between alosid species

	A. sapidissima	A. alabamae	A. aestivalis	A. pseudoharengus
A. sapidissima	0.0025	0.0016	0.0220	0.0180
	(0.0011)	(0.0009)	(0.0014)	(0.0013)
A. alabamae		0.0010	0.0266	0.0183
		na	(0.0006)	(0.0004)
A. aestivalis			na	0.0046
				na

Numbers in parentheses indicate standard deviation of the estimates. na = comparison not possible due to one haplotype or only one pairwise comparison.

found in *A. alabamae* were represented in *A. sapidissima*. Comparisons of *A. pseudoharengus* and *A. aestivalis* showed lower sequence divergence estimates than comparisons of either of these species to *A. sapidissima* or *A. alabamae*. The estimates of divergence between *A. sapidissima* and *A. pseudoharengus* are somewhat smaller than the estimates of Bentzen (personal communication), but this may be due to the smaller number of enzyme digests used in this analysis.

This portion of the study generated 36 characters for phylogenetic analysis (Figure 5.2.2). Bootstrap treatment of the data matrix was able to statistically differentiate an

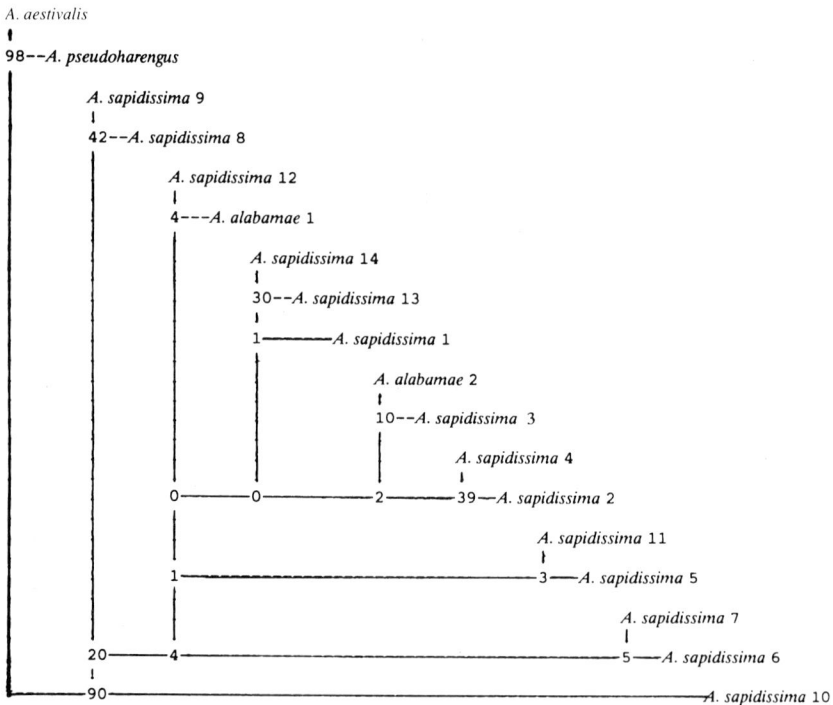

Figure 5.2.2 Results of phylogenetic analysis of six-base restriction digest data obtained from the entire mtDNA genome of alosids. Haplotypes are listed in Table 5.2.2. Numbers at the nodes indicate the number of times out of 100 that the haplotypes above and to the right were included in that branch. The tree was rooted by constructing an ancestral haplotype from the restriction sites found in all species, i.e. a consensus ancestor.

Table 5.2.4 Size estimates for fragment patterns produced by *Alu*I, *Cfo*I, *Hae*III, *Msp*I, *Rsa*I, and *Taq*I in the *ND I* portion of mtDNAs from *Alosa aestivalis*, *A. pseudoharengus*, *A. sapidissima* and *A. alabamae*

AluI		CfoI				TaqI			
A	N	A	B	C	D	A	B	C	N
948	778	951	677	1383	695	863	863	1342	1120
432	432	367	367	367	574	524	524	524	524
266	266	344	344	290	367	479	236	93	396
115	154	290	301	119	290	93	234	41	
97	115	119	290		119	41	93		
75	75		119				41		
67	67								
	51								
	45								
2000	1983	2071	2098	2159	2045	2000	1991	2000	2040

					HaeIII					
A	B	C	D	E	F	M	N	O	P	Q
386	408	386	408	386	386	386	386	386	543	386
317	386	317	386	317	317	335	335	270	386	270
240	240	226	240	240	240	205	226	240	226	226
226	226	215	226	226	226	203	205	226	205	205
203	203	203	203	203	203	167	203	205	203	203
161	161	161	161	161	179	161	167	203	167	167
97	97	152	88	97	97	97	161	167	161	161
96	88	97	72	96	96	96	97	161	97	97
88	72	88	65	88	88	72	96	97	96	96
72	57	72	57	72	72	58	72	96	58	72
57		(35)	30	37	57		57	72		58
			20	20				57		
1943	1938	1917	1936	1943	1961	1780	2005	2180	2142	1941

RsaI				MspI					
A	N	O	P	A	B	C	D	E	F
670	670	670	670	689	689	499	601	499	592
316	528	454	454	352	478	499	499	358	275
308	316	316	308	262	300	262	262	262	262
227[a]	308	308	227	203	186	203	203	186	186
151	227	227	200	186	129	186	186	171	171
		89	89	129	91	129	129	149	149
			50	91	71	91	71	129	129
			40	71	58	71	58	91	91
				58		58		71	71
								58	58
1899	2049	2064	2038	2041	2002	1998	2009	1974	1984

Variances of the regressions indicate that estimates have an error of 5%. Fragment patterns A–F were found in
A. sapidissima and/or *A. alabamae*, while patterns M–Q were found in *A. pseudoharengus* and/or *A. aestivalis*.
[a] The fragment is probably a doublet that could not be resolved on 9% polyacrylamide gels.

A. pseudoharengus–A. aestivalis clade from an *A. sapidissima–A. alabamae* clade (99 of 100 bootstrap replicates). No statistically significant differences were found in the nodes of the *A. sapidissima–A. alabamae* clade.

(b)   Four-base survey of the ND I region

Size estimates for the restriction fragments produced by the six enzymes used in this portion of the study are listed in Table 5.2.4. Inspection of this table shows that only minor differences might exist in the length of the region amplified across the four species. Some restriction profiles (e.g. *Hae*III pattern M) did not sum to a value close to the estimated weight of the fragment (approximately 2.1 kb) and many of the rest (e.g. *Hae*III A–F) summed to values slightly smaller than, but not statistically different from, 2.1 kb (Table 5.2.4). In the case of *Hae*III pattern M, we simply cannot account for the deficiency unless one or more of the observed bands are unresolved doublets (Figure 5.2.3). We suspect that the *Rsa*I A pattern contained such a doublet in the 308-bp fragment and, if true, this would indicate that the fragments must be very close, as the 316 and 308 bands were clearly distinct. For the other patterns we suspect, but cannot prove, that some of the shortages are due to small (< 30-bp) fragments that could not be resolved on 9% acrylamide gels.

Composite haplotypes for this portion of the survey are given in Table 5.2.5. Here we discerned eight distinct haplotypes for *A. sapidissima* compared to 14 in the six-base data, one for *A. alabamae* compared to two in the six-base data, and three and four haplotypes for *A. pseudoharengus* and *A. aestivalis*, respectively, versus one in the six-base data. Most significant in Table 5.2.5 is the complete differentiation of *A. sapidissima* and *A. alabamae*.

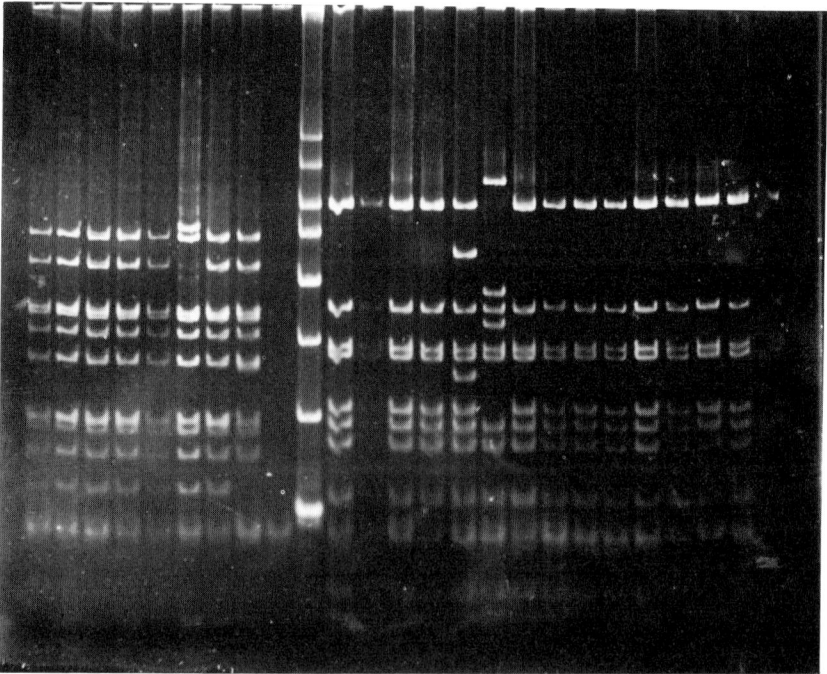

Figure 5.2.3 Restriction fragments obtained from the *ND I* region of the mtDNA molecules in *A. sapidissima*. A molecular weight standard is in lane 10. To the left of the standard are *Hae*III digests, and *Msp*I digests are to the right.

Table 5.2.5 Composite haplotypes for the specimens successfully amplified for the *ND I* region of the mtDNA molecule in alosids

Species	Haplotype	N
A. sapidissima	ACACAA	32
A. sapidissima	ACAFAA	1
A. sapidissima	ACBEAB	4
A. sapidissima	ACDCAB	1
A. sapidissima	ACECAA	1
A. sapidissima	ADFCAA	1
A. sapidissima	ACBCAB	1
A. sapidissima	ADCCAA	1
A. alabamae	ACADAC	12
A. pseudoharengus	NANANN	9
A. pseudoharengus	NBNANN	2
A. pseudoharengus	NAMANN	1
A. aestivalis	NAOANN	5
A. aestivalis	NAQANN	3
A. aestivalis	NAOBON	3
A. aestivalis	NAPAPN	1

Enzymes are listed in alphabetical order for the restriction patterns given in Table 5.2.4 and *N* is the number of individuals observed with a given haplotype.

Haplotypes for these species are differentiated by two site losses (*Taq*I pattern C and *Msp*I pattern D) in *A. alabamae* which were not observed in any of the *A. sapidissima*, despite the fact that these species share the most common six-base profiles. Clearly, this indicates a separate matriarchal origin for *A. alabamae* and homoplasy in the most common six-base profiles in *A. sapidissima*.

Estimated sequence divergences between the species are presented in Table 5.2.6 and in general reflect the results from the six-base survey, i.e. *A. sapidissima* and *A. alabamae* are more closely related to each other than either are to *A. pseudoharengus* or *A. aestivalis*. The within-species divergences are of the same order as those in the six-base survey, but the divergence between *A. sapidissima* and *A. pseudoharengus* is larger than the 4–5% values obtain by Bentzen (personal communication) in an extensive survey using six-base restriction enzymes.

This portion of the study generated 62 characters for input to PHYLIP and the resulting trees are presented in Figures 5.2.4 and 5.2.5. In general these trees match those of the six-base survey in that distinct clades are generated, one containing *A. sapidissima* and

Table 5.2.6 Average nucleotide divergence between alosid species based upon four-base digests of the *ND I* portion of the mtDNA molecule

	A. sapidissima	A. alabamae	A. pseudoharengus	A. aestivalis
A. sapidissima	0.0114 (0.0047)			
A. alabamae	0.0154 (0.0041)	na		
A. pseudoharengus	0.063 (0.0037)	0.0562 (0.0014)	0.0043 (0.0015)	
A. aestivalis	0.0733 (0.0078)	0.0692 (0.0081)	0.0208 (0.0074)	0.0232 (0.0137)

Numbers in parentheses indicate standard deviation of the estimates. na = comparison not possible due to one haplotype.

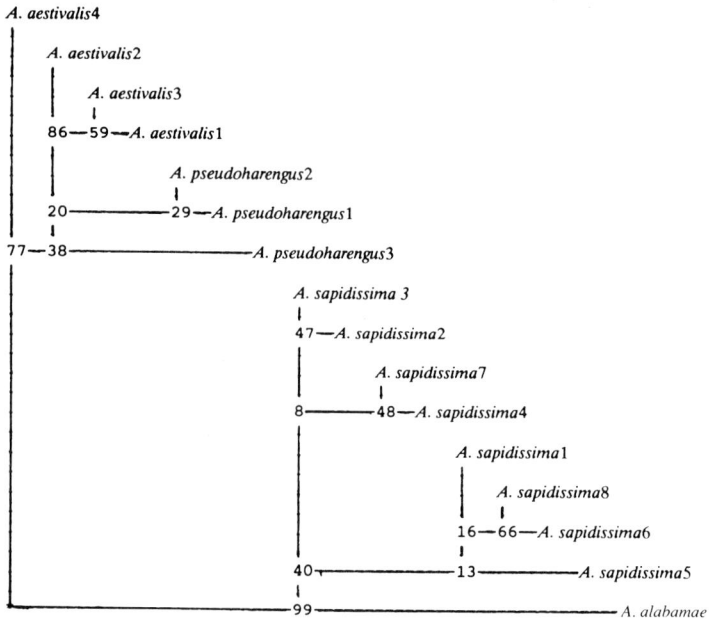

Figure 5.2.4 Results of phylogenetic analysis of four-base restriction digest data of the *ND I* portion of alosid mtDNA molecules. The tree was rooted by constructing an ancestral haplotype that possessed only the restriction sites found in all composite haplotypes. Haplotypes are listed in Table 5.2.4.

A. alabamae, the other containing *A. pseudoharengus* and *A. aestivalis*. Both of the trees (Figures 5.2.4 and 5.2.5) distinguish *A. alabamae* from *A. sapidissima* and an *A. alabamae*–*A. sapidissima* clade from an *A. pseudoharengus*–*A. aestivalis* clade. It should be noted that both trees suggest that *A. aestivalis* and *A. pseudoharengus* are polyphyletic.

(c)   Four-base survey of the cyt b *region*

The fragments produced by four-base digestion of the amplified *cyt b* region (approximately 1300 bp) are presented in Table 5.2.7. Most of the digests sum to more or less 1290 bp except for the *Hpa*II digests. We assume that this difference is due to small fragments that could not be resolved. The composite haplotypes and sample size for each species are also listed in Table 5.2.7 and show that fewer variant patterns were seen in this region compared to the *ND I* digests. In fact, only one intraspecific variant was noted in this region. Estimates of nucleotide divergence (Table 5.2.7) were smaller than those obtained from *ND I* portion and no difference was noted between *A. sapidissima* and *A. alabamae*. The divergence estimates do indicate that *A. pseudoharengus* and *A. aestivalis* are more closely related to each other than either is to *A. sapidissima* or *A. alabamae*. Divergence estimates between *A. aestivalis* and *A. pseudoharengus* were an order of magnitude smaller than those obtained from the *ND I* portion. Comparisons between *A. sapidissima* versus *A. pseudoharengus* or *A. aestivalis* were about one-third of those obtained from the *ND I* portion, but nearly the same as those obtained from the six-base data.

The restriction digests produced 23 characters for input to the PHYLIP program and the results of the bootstrap analysis are consistent with the trees generated from the previous data (Figure 5.2.5). Inspection of the composite haplotypes shows that *A. aestivalis* shares

(a)

```
A. aestivalis 4
|
44 - - A. pseudoharengus 3

        A. aestivalis 2

            A. aestivalis 3
            |
        93 — 69 — A. aestivalis 1
        |
        24 —————————— A. pseudoharengus 1
        |
    93 ————— 23 ————————————— A. pseudoharengus 2

                        A. sapidissima 7
                        |
                    49 — A. sapidissima 4

                            A. sapidissima 5

                                A. sapidissima 6
                                |
                            67 - - A. sapidissima 8
                            |
                    7 ————————— 14 — 17 ————————— A. sapidissima 1

                                        A. sapidissima 2
                                        |
                    40 ——————————————————————— 46 — A. sapisissima 3
                    |
                ——— 99 —————————————————————————————— A. alabamae
```

(b)

```
A. alabamae
|
99    . A. sapidissima

        A. aestivalis 1

            A. aestivalis 2
            |
        94 — 77 — A. pseudoharengus
```

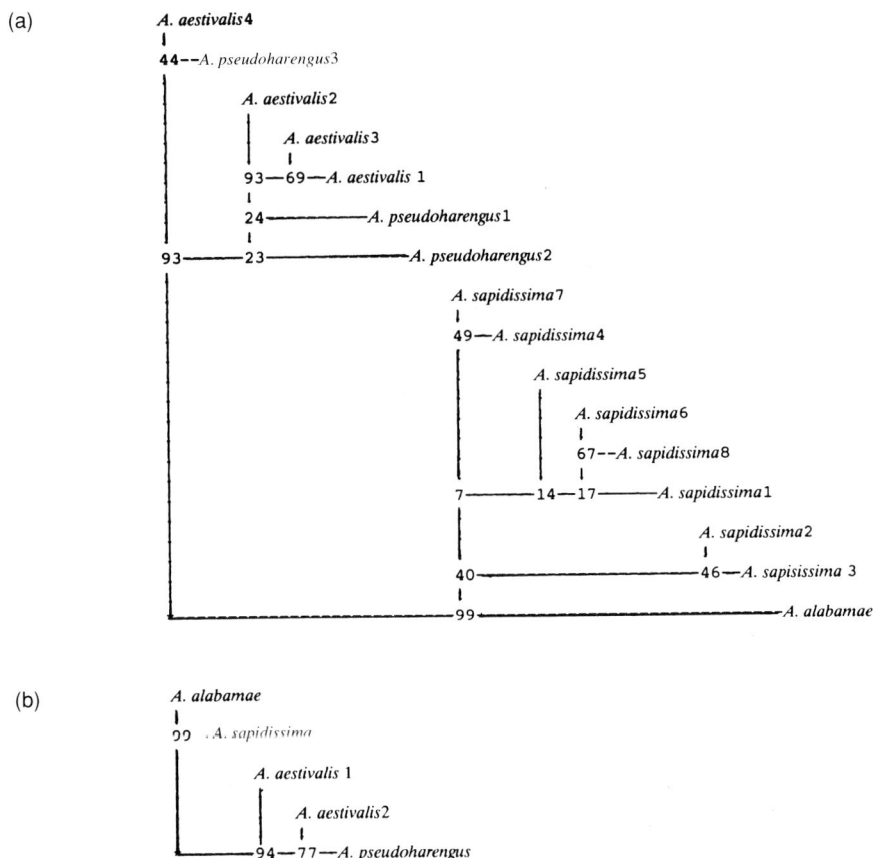

Figure 5.2.5 The upper tree (a) shows the results of phylogenetic analysis of four-base restriction digest of the *ND I* portion of alosid mtDNA molecules. The trees were rooted by constructing an ancestral haplotype that possessed restriction sites found in each species, i.e. a consensus haplotype. The lower tree (b) is based upon four-base digests of the *cyt b* portion of the genome.

the *Hpa*II digestion profile with *A. sapidissima* and *A. alabamae* and this may indicate a retained ancestral condition. A single specimen of *A. pseudoharengus* shared the *Hae*III A restriction profile with *A. aestivalis*, while the remainder possess a unique site loss, e.g. pattern B.

5.2.4 Discussion

The results of this study indicate that the *ND I* portion of the mtDNAs in the genus *Alosa* are evolving at a more rapid rate than *cyt b* and reveal greater genetic variation and genetic distances within and among the species than a limited six-base survey of the entire genome. This observation is, in part, consistent with the results of Meyer (1993), where comparison of nucleotide sequences between *Xenopus laevis* and *Gadus morhua* found slightly more divergence in the *ND I* genes (70% sequence similarity) than in the *cyt b* gene (72% sequence similarity). However, the sequence divergence estimates in *Alosa* indicate that the *ND I* region is evolving at a more rapid pace compared to the *cyt b* than might be predicted

Table 5.2.7   Size estimates (top), composite haplotypes (middle) and sequence divergence estimates (bottom) for fragment patterns produced by *Alu*I, *Cfo*I, *Hae*III, *Hpa*II, *Rsa*I and *Taq*I in the *cyt b* portion of mtDNAs from *Alosa aestivalis*, *A. pseudoharengus*, *A. sapidissima*, and *A. alabamae*

AluI		CfoI		HaeIII		
A	B	A	B	A	B	C
645	1118	668	501	664	664	405
280	112	501	435	329	329	329
169		118	244	151	282	266
115			118	131		151
82						131
1291	1220	1287	1298	1275	1275	1282

HpaII		RsaI		TaqI	
A	B	A	B	A	B
580	1037	909	909	1288	967
457	86	361	272		322
86	55		93		
55					
1178	1178	1260	1274	1288	1289

Species	Composite haplotypes	Sample size
A. pseudoharengus1	AABAAA	11
A. pseudoharengus2	AAAAAA	1
A. aestivalis	AAABAA	12
A. sapidissima	BBCBBB	42
A. alabamae	BBCBBB	12

	A. pseudoharengus1	A. pseudoharengus2	A. aestivalis	A. sapidissima
A. pseudoharengus2	0.0022			
A. aestivalis	0.0022	0.0048		
A. sapidissima	0.0201	0.0273	0.01777	
A. alabamae	0.0201	0.0273	0.01777	0.0000

Variances of the regressions indicate that estimates have an error of 5%.

from Meyer (1993). The regions amplified in the present study covered larger segments of the mtDNA molecule than just the 432 *ND I* and 255 *cyt b* nucleotides used by Meyer (1993). This may indicate that the regions flanking the 255 *cyt b* nucleotides used in Meyer's (1993) comparison may be more highly conserved than existing sequence data indicate. Conversely, the regions flanking the 432 *ND I* nucleotides may be less conserved. Alternatively, the *ND I* regions analysed in this study may actually evolve at a more rapid pace than the data of Meyer (1993) indicate, but boundary conditions exist upon the range of obtainable states. It has been known for some time (Chapman, Avise and Asmussen, 1979) that when boundaries are imposed on the character states, phylads will eventually saturate the space and erode our ability to make clear inferences on evolutionary histories. In other words, the similarity between sequence divergences *ND I* and *cyt b* inferred from Meyer (1993) may result from comparison of disparate taxa and reflect the boundary conditions imposed upon nucleotide changes in these portions of the genome. The much larger difference in rates of evolution in *ND I* and *cyt b* reported here may reflect the actual rate of nucleotide change, as it is unlikely that these more closely related species have managed to saturate the available character space. These conclusions are consistent with the curves relating time and sequence divergence (Brown *et al.*, 1982; Hixson and Brown, 1986; Meyer, 1993). A final possibility for the apparent rate differences between the *ND I* and *cyt b* regions may relate to the different sizes of the regions amplified, 2.1 versus 1.3 kb. The larger *ND I* region produced more restriction fragments, and thus it is possible to detect more base changes. However, the methods used to estimate sequence divergence do not depend upon the length of the DNA analysed, but only upon the number of changes recorded versus the number of fragments observed. The estimates should therefore be relatively free from bias due to the length of DNA.

Perhaps more important, the present data suggest that comparisons involving closely related lineages can be used as a guide for selecting portions of the mtDNA molecule that may be appropriate for studies at varying taxonomic levels. In other words, substantial nucleotide divergence between closely related taxa within a given region of mtDNA may indicate that the region contains substantial within-species variability. The *ND I* region revealed fixed differences among *Å. alabamae* and *A. sapidissima*, whereas no differences were found between these species in the six-base survey or the *cyt b* region. In addition, we found greater levels of within-species variability in the *ND I* region compared to the *cyt b* portion (in all species) and a comparable number of haplotypes to that found in the six-base survey of *A. sapidissima*, given differences in the sample sizes. We stress that the increased variability is not simply a function of four-base versus six-base digestions, which has been demonstrated in *A. sapidissima* (Nolan, Grossfield and Wirgin, 1991), but that some regions of the molecule reveal more intraspecific variation.

Analysis of the *ND I* region suggests that *A. aestivalis* and *A. pseudoharengus* are polyphyletic species, while the six-base data clearly indicate monophyletic relationships. If the polyphyletic relationship is correct, it could indicate that *A. pseudoharengus* and *A. aestivalis* arose from a common ancestor and random lineage sorting has yet to produce reciprocal monophyly (Avise, Neigel and Arnold, 1984; Neigel and Avise, 1986; Ball, Neigel and Avise, 1990). The *cyt b* data would support this hypothesis, i.e. that the *Hae*III restriction site producing the 151- and 131-bp fragments is ancestral to all four species and this site has been lost in most, but not all, *A. pseudoharengus*. Taken at face value, the six-base data are difficult to rationalize with a hypothesis of polyphyly for these species. Some of the difficulty may be associated with the *Hae*III digest of the *ND I* region (pattern M), where we could not account for nearly 300 bases. It is also possible that homoplasies in the *ND I* region begin to appear among closely related species. In effect, *Hae*III digests of the *ND I* region confuse rather than resolve interspecific relationships. Alternatively, it could be argued that the sample size and number of restriction enzymes employed in the six-base survey failed to find evidence of intraspecific variation and polyphyletic relationships were therefore precluded. As the present study was not an *a priori* test of specific hypotheses, no strong conclusions should be drawn from the present data regarding relationships among these taxa. We do recommend that additional studies examine the possibility with alternative regions of mtDNA and more extensive examination (perhaps sequencing) of the *ND I* and *cyt b* regions. A more in-depth treatment of this species with additional specimens and six-base enzymes may be a useful approach, but we suspect that amplification and more exhaustive examination of selected portions of the mtDNA molecule may be more informative.

As stated in section 5.2.1, one purpose of this study was to resolve some questions regarding the systematic status of *A. alabamae* relative to *A. sapidissima* by comparing the level of genetic divergence to that found in a well-recognized, but morphologically similar, species pair in the same genus. The issue is certainly complicated by the uncertain relations of *A. aestivalis* to *A. pseudoharengus*, but the absence of fixed differences in the six-base survey and *cyt b* portions of the mtDNA molecule compared to *A. aestivalis* and *A. pseudoharengus* would suggest that species recognition of *A. alabamae* is problematical and a revision of the genus *Alosa* is in order. If forced to express an opinion, we doubt that *A. alabamae* merits separate species status and would refer it to a distinct subgroup of *A. sapidissima*.

An unanticipated aspect of the present data concerns the use of six-base mtDNA surveys to assess stock structure of commercially important species and the contribution of this stock to mixed fisheries. Only two haplotypes were found in the six-base survey of *A. alabamae* and both of these are common in Atlantic Coast populations of *A. sapidissima*. A more extensive 13-enzyme database found a third haplotype in *A. alabamae* (somewhat rare in *A. sapidissima*), but this individual also possessed the two site losses in the *ND I* region. This strongly indicates that significant homoplasies exist in the evolution of alosid mtDNAs when variation is assessed by six-base digestion of the entire molecule. An alternative hypothesis would be that the six-base data reflect ancestral relationships of *A. alabamae*, and the *Msp*I and *Taq*I site losses are homoplasies. This interpretation is less likely, as it requires two parallel site losses in at least three lineages. In addition, it is well known that nucleotide substitutions are more likely in the third base of a codon (Aquadro and Greenberg, 1983; Meyer, 1993) and that six-base restriction enzymes cover two third base nucleotides, while four-base enzymes cover on the average 1.33. Thus, it is less probable that specific four-base sites will undergo parallel losses in distinct lineages relative to six-base sites. In addition, the composite haplotypes in Table 5.2.2 show that homoplasies characterize the six-base data (cf. haplotypes 1,3 versus 13, 14) and this inference is supported by the data of Bentzen, Brown and Leggett (1989). To the extent that assignment of stock structure is dependent upon unambiguous determination of shared matriarchal ancestry, the present data would not recommend the use of six-base data as a tool for management of *A. sapidissima* populations. While it is possible to make qualitative assessments of statistical differences among populations (Chapman, 1987, 1989, 1990; Bentzen, Brown and Leggett, 1989) when genealogical relationships are uncertain, quantitative determinations of the degree of differences would be unwarranted. Assessment of contributions of spawning stocks to mixed populations could be completely misleading for this reason as well.

Acknowledgements

This study was supported in part by Maryland Sea Grant R/F-66 and the Maryland Chesapeake Bay Research Program. Special thanks are due to Ms Sara Winslow for providing samples and Dr Trip Lamb for a review. Technical assistance was provided by Dr B.L. Brown during the early phases of this study.

References

Aquadro, C.F. and Greenberg, B.D. (1983) Human mitochondrial DNA variation and evolution: analysis of nucleotide sequences from seven individuals. *Genetics*, **103**, 287–312.

Avise, J.C., Neigel J.E. and Arnold J. (1984) Demographic influences on mitochondrial DNA lineage survivorship in animal populations *J. Mol. Evol.*, **20**, 99–105.

Ball, R.M. Jr, Neigel, J.E. and Avise, J.C. (1990) Gene genealogies within the organismal pedigrees of random mating populations. *Evolution*, **44**, 360–70.

Bentzen, P., Brown, G.G. and Leggett, W.C. (1989) Mitochondrial DNA polymorphisms, population structure, and life history variation in American Shad (*Alosa sapidissima*). *Can. J. Fish. Aquat. Sci.*, **46**, 1446–54.

Bentzen, P., Leggett, W.C. and Brown, G.G. (1988) Length and restriction site heteroplasmy in the mitochondrial DNA of American shad (*Alosa sapidissima*). *Genetics*, **118**, 509–18.

Bowen, B.W. and Avise, J.C. (1990) Genetic structure of Atlantic and Gulf of Mexico populations of sea bass, menhaden, and sturgeon: influences of zoogeographic factors and life history patterns. *Mar. Biol.*, **107**, 371–81.

Brown, W.M., Prager, E.M., Wang, A. and Wilson, A.C (1982) Mitochondrial DNA sequences in primates: tempo and mode of evolution. *J. Mol. Evol.*, **18**, 225–39.

Chapman, R.W. (1987) Changes in the population structure of male striped bass, *Morone saxatilis*, spawning in three areas of Chesapeake Bay from 1984 to 1986. *Fish. Bull.*, **85**, 167–71.

Chapman, R.W. (1989) Spatial and temporal variation of mitochondrial DNA haplotype frequencies in the striped bass 1982 year class. *Copeia*, **1989**, 344–8.

Chapman, R.W. (1990) Mitochondrial DNA analysis of striped bass populations in Chesapeake Bay. *Copeia*, **1990**, 355–66.

Chapman, R.W. and Powers, D.A. (1984) A technique for the rapid isolation of mitochondrial DNA from fishes. *Maryland Sea Grant Program Tech. Rept.*

Chapman, R.W., Avise, J.C. and Asmussen, M.A. (1979) Character space restrictions and boundary conditions in the evolution of quantitative multistate characters. *J. Theor. Biol.*, **80**, 51–64.

Dowling, T.E., Hoeh, W.R., Smith, G.R., and Brown, W.M. (1992) Evolutionary relationships of shiners in the genus *Luxilus* (Cyprinidae), as determined by analysis of mitochondrial DNA. *Copeia*, **1992**, 306–22.

Dowling, T.E., Moritz, C. and Palmer, J.D. (1990) Nucleic Acids II. Restriction site analysis, in *Molecular Systematics* (eds D.M. Hillis and C. Moritz), Sinauer Associates Inc. Sunderland, MA, USA, pp. 250–318.

Felsenstein, J. (1985) Confidence limits on phylogenies: an approach using the bootstrap. *Evolution*, **39**, 783–91.

Gold, J.R. and Richardson, L.R., (1991) Genetic studies of marine fishes. IV. An analysis of population structure in the red drum (*Sciaenops ocellatus*) using mitochondrial DNA. *Fish. Res.*, **12**, 213–41.

Hixson, J.E. and Brown, W.M (1986) A comparison of the small RNA genes from the mitochondrial DNA of the great apes and humans: sequence, structure, evolution and phylogenetic implications. *Mol. Biol. Evol.*, **3**, 1–18.

Kornfield, I. and Bogdanowicz, S.R. (1987) Differentiation of mitochondrial DNA in Atlantic herring, *Clupea harengus. Fish. Bull.*, **85**, 561–8.

McElroy, D., Moran, P., Bermingham, E. and Kornfield, I. (1991) *The Restriction Enzyme Analysis Package*, Version 4, University of Maine, Orono, Maine.

Meyer, A. (1993) Evolution of mitochondrial DNA in fishes, in *Biochemistry and Molecular Biology of Fishes*, (eds P.W. Hochachka and T.P. Mommsen), Elsevier Press, pp. 1–38.

Moritz, C., Dowling, T.E. and Brown, W.M. (1987) Evolution of animal mitochondrial DNA: relevance for population biology and systematics. *Annu. Rev. Ecol. Syst.*, **18**, 26–92.

Mulligan, T.J. and Chapman, R.W. (1989) Mitochondrial DNA analysis of Chesapeake Bay white perch, *Morone americana. Copeia*, **1989**, 679–88.

Neigel, J.E. and Avise, J.C. (1986) Phylogenetic relationships of mitochondrial DNA under various demographic models of speciation, in *Evolutionary Processes and Theory* (eds E. Nevo and S. Karlin), Academic Press, New York, pp. 515–34.

Nolan, K., Grossfield, J. and Wirgin, I. (1991) Discrimination among Atlantic coast populations of American shad (*Alosa sapidissima*) using mitochondrial DNA. *Can. J. Fish. Aquat. Sci.*, **48**, 1724–34.

Wirgin, I.I., Proenca, R. and Grossfield, J. (1989) Mitochondrial DNA diversity among populations of striped bass in the southeastern United States. *Can. J. Zool.*, **67**, 891–907.

5.3 CLONING OF HIGHLY VARIABLE MINISATELLITE DNA SINGLE-LOCUS PROBES FOR BROWN TROUT (*SALMO TRUTTA* L.) FROM A PHAGEMID LIBRARY

Paulo A. Prodöhl, John B. Taggart and Andrew Ferguson

Abstract

*Highly polymorphic minisatellite DNA regions are a potential source of informative genetic markers for familial and individual identification and for population studies. A relatively simple cloning strategy for the isolation of minisatellite DNA locus-specific probes for brown trout (*Salmo trutta L.) is described. This involved the cloning of a size-selected fraction of brown trout MboI DNA fragments (1.5–4.4 kb) into a pBluescript II phagemid vector. The resultant library was simultaneously screened with Jeffreys' 33.6 and 33.15 human* *minisatellite probes. Five minisatellite DNA locus-specific probes, which should be informative at the population level, were quickly isolated from a small fraction of this library. This technique is probably applicable to many other organisms, especially when only relatively few minisatellite DNA single-locus probes are required.*

5.3.1 Introduction

Electrophoretic studies of polymorphic enzyme loci have proved valuable for obtaining information concerning the genetic structure of brown trout (*Salmo trutta* L.) populations (e.g. Ryman, 1983; Krueger and May, 1988; Ferguson, 1989; Ferguson and Taggart, 1991). This approach, however, has often been limited by the relatively low levels of detectable isozyme polymorphism observed within many populations. Furthermore, with generally only 2–3 alleles segregating, such loci provide few potential genetic markers for familial or individual identification. These would be useful for more detailed ecological and behavioural studies of this species.

Recently, DNA fingerprinting techniques (Jeffreys, Wilson and Thein, 1985 a,b; Vassart *et al.*, 1987) have attracted considerable attention as an alternative source of highly polymorphic genetic markers for a wide range of organisms. This and other DNA-based approaches, e.g. mtDNA, microsatellites and RAPDS, are reviewed in section 4.1 (Skibinski, 1993). The occurrence and detection of minisatellite DNA loci in the salmonid genome is now well documented (Lloyd, Fields, and Thorgaard, 1989; Fields, Johnson and Thorgaard, 1989; Gyllensten *et al.*, 1989; Taggart and Ferguson, 1990a; Bentzen, Harris and Wright, 1991). However, due to their high complexity, such multilocus profiles are only of limited value for salmonid population studies (Prodöhl, Taggart and Ferguson, 1992). Individual 'alleles' cannot be easily attributed to specific loci. Consequently, locus-specific gene and genotype frequencies, the basis of the conventional population analyses, are usually unknown (Lynch, 1991).

Single-locus minisatellite probes (SLPs) provide a more practical approach for the detection and large-scale screening of these highly variable regions of DNA. Unlike 'universal' multilocus DNA fingerprinting probes, however, SLPs need to be specifically isolated from the study organism (Wong *et al.*, 1987). The probes comprise both ubiquitous minisatellite sequences and unique species/genus-specific flanking regions of DNA. Following high-stringency post-hybridization washes of Southern blots, only the latter locus-specific sequences remain annealed to homologous target DNA, enabling detection of single-locus variation. To date only a few SLPs have been cloned and characterized for salmonids, specifically the Atlantic salmon, *Salmo salar* L. (Taggart and Ferguson, 1990b; Bentzen, Harris and Wright, 1991). While some of these have sufficient sequence similarity to detect single minisatellite loci in the congeneric brown trout, additional SLPs for this species were sought.

Our previous cloning strategy for Atlantic salmon minisatellite DNA loci utilized bacteriophage lambda (EMBL3) as the vector (Taggart and Ferguson, 1990b). This was very successful but involved relatively tedious and complex protocols. For isolation of brown trout minisatellite DNA probes a much simpler protocol, cloning into a phagemid vector, was successfully used. The methodology followed and provisional results obtained are described below. The same procedures are likely to be applicable to many other species.

5.3.2 Materials and methods

(a) Library construction

Routine DNA extraction (3 h proteinase K digestion; 1 × phenol, 3 × phenol/chloroform, 1 × chloroform extractions) was carried out on 400 mg of liver tissue, pooled from 27 brown trout. An aliquot of the DNA (1.2 mg) was fully digested with the restriction enzyme *Mbo*I (10 units/μg), and then size fractionated by sucrose density gradient centrifugation (Hadfield, 1987). The resultant fractions were ethanol precipitated and resuspended in TE buffer (10 mM Tris, 1 mM EDTA; pH 8.0). Fifty nanograms of $\approx$ 1.5–4.4 kilobase-pair (kb) fraction were ligated to 100 ng of dephosphorylated, *Bam*HI-digested, pBluescript II KS (–) phagemid vector (Stratagene Ltd) in 10 μl ligation buffer. One-fifth of this ligation mix was used to transform 50 μl 'supercompetent' XL1-Blue host cells (Stratagene Ltd, 'Epicurion Coli'), according to the supplier's instructions. Ligation efficiency was shown to be > 70% (estimated by blue/white colour selection), yielding a partial genomic library of $\approx$ 8 $\times$ 10^3 recombinants.

(b) Library screening

Transformed cells were grown overnight at low density on LB agar plates containing ampicillin (200 μg/ml) and tetracycline (12.5 μg/ml) at 37 °C. Colonies were blotted onto Hybond-N membranes (Amersham International plc) according to the manufacturer's instructions. The LB plates were reincubated to allow colonies to recover, then stored at 4 °C. The membranes were simultaneously hybridized with the human minisatellite probes 33.6 and 33.15 (Jeffreys, Wilson and Thein, 1985a). Details of ^{32}P probe labelling are given elsewhere (Wells, 1988). Hybridization was performed in 3 $\times$ SSPE (0.54 M NaCl, 0.03 M NaH$_2$PO$_4$, 3 mM EDTA Na$_2$ salt; pH 7.7), 0.5% dried milk, 1% sodium dodecyl sulphate (SDS) and 6% polyethylene glycol (PEG) 8000 at 65 °C overnight (Dalgleish, 1987). High molecular weight *E. coli* XL1-Blue genomic DNA (0.5 μg/ml) and pBluescript II phagemid (1 μg/ml) DNA were included as competitor. Post-hybridization washes were carried out, first, in 2 $\times$ SSC (0.3 M NaCl, 0.03 M Na$_3$-citrate; pH 7.0), 0.1% SDS for 20 min at room temperature followed by two washes in 1 $\times$ SSC, 0.1% SDS at 65 °C for 30 min each. After overnight autoradiography, positive recombinants were identified and isolated from the original LB plates. (No secondary screening of potential positives was undertaken.) Each clone was grown in 5 ml LB medium containing ampicillin (200 μg/ml). A 1-ml aliquot of each was removed for storage (in 15% glycerol) at –70 °C. Recombinant phagemids were recovered from remaining cultures using a small-scale plasmid DNA preparation protocol (Sambrook, Fritsch and Maniatis, 1989).

(c) Probe testing

Each positive recombinant was digested with *Sau*3AI to release the cloned insert DNA. These were separated from small vector DNA fragments by electrophoresis through 0.65% low-melting-point agarose and excised from the gel. Approximately 10 ng insert DNA was ^{32}P-labelled by random oligonucleotide priming (Dalgleish, 1987). These potential probes were hybridized overnight to a panel of six *Pal*I-digested brown trout genomic DNA samples. DNA extraction, gel electrophoresis and Southern blotting details are published elsewhere (Taggart and Ferguson, 1990a; Taggart *et al.*, 1992). Hybridization conditions employed were 1.5 $\times$ SSPE, 0.5% dried milk, 1% SDS, 6% PEG at 65 °C, with 10 μg/ml sonicated brown trout genomic DNA as competitor. Following hybridization, membranes were washed to high stringency (0.5–0.1 $\times$ SSC, 0.1% SDS, 65 °C, 2 $\times$ 30 min) and autoradiographed for 48–96 h at –70 °C in the presence of intensifying screens. A further 50 individuals, representing three discrete wild populations, together with a single family (two parents and ten progeny), were then screened for each positively identified probe.

5.3.3 Results

Approximately 1600 clones (i.e. ≈ 1120 recombinants) were screened with the human multi-locus probes, of which 36 gave a strong positive signal. Twenty-four of these, selected at random, were expanded and the insert DNA used to probe test membranes. Five distinct single-locus probes were apparent, i.e. hybridizing to one or two DNA fragments per individual. A typical example is shown in Figure 5.3.1. Subsequent familial analyses have confirmed simple Mendelian inheritance of these fragments, e.g. Figure 5.3.2. A summary of the 'allele' fragment data for 50 brown trout is given in Table 5.3.1. Although these are only very provisional data, it is interesting to note that most discrete 'allele' classes observed were common to the three populations. No high-frequency population diagnostic bands were apparent. All probes detected highly polymorphic minisatellite loci with observed heterozygosity values of at least 50%.

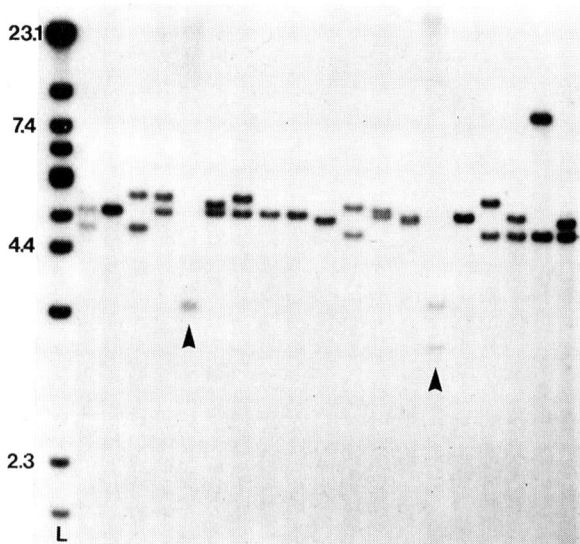

Figure 5.3.1 Typical autoradiograph pattern for 17 brown trout revealed by the locus-specific probe Str-A5. L denotes the lambda DNA marker lane. Two micrograms of *Pal*I-digested genomic DNA was separated through a 0.7% agarose gel. (Two misidentified juvenile Atlantic salmon individuals are arrowed. Str-A5 detects an apparent single minisatellite locus in this species too.)

Table 5.3.1 Summary of basic data available for the five locus-specific probes; 'allele' statistics are based on 50 brown trout sampled from three different populations

Probe name	Probe size (kb)	'Allele' size range (kb)	Total no. 'alleles'	No. 'alleles' per population	Mean observed heterozygosity
Str-A1	1.9	2.0–3.0	6	5–6	59%
Str-A3	4.4	2.3–7.4	11	2–11	79%
Str-A5	4.3	3.5–7.5	8	5–8	62%
Str-A9	1.8	2.0–21.0	7	6–7	49%
Str-A22	1.9	3.0–5.6	9	6–9	69%

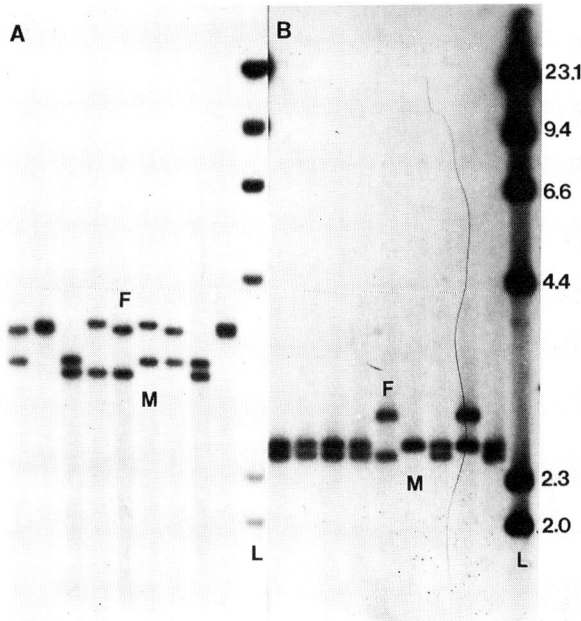

Figure 5.3.2 Autoradiographs from inheritance studies of two locus-specific probes. A = Str-A22; B = Str-A5. In both cases a single brown trout family comprising father (**F**), mother (**M**) and seven progeny is shown. **L** denotes the lambda DNA marker lane. Electrophoretic conditions are as given in caption to Figure 5.3.1.

A further nine probes revealed slightly more complex patterns (2–6 bands per individual). Multibanded patterns (i.e. generally > 10 bands per individual) were detected by another nine probes, many of them giving clearer resolution (and fewer bands) than either the 33.6 or 33.15 human minisatellite probes (see Figure 5.3.3 for example). Only one probe failed to detect any bands at all. The cloning procedure and initial characterization of probes was completed within 3 months.

5.3.4 Discussion

Armour *et al.* (1990) have described a systematic SLP isolation protocol for humans, based on the use of charomid vectors and sequential multilocus probe screening. This approach, comprehensively described by Bruford *et al.* (1992), has since been used for SLP isolation from a number of bird species (e.g. Hanotte *et al.*, 1991; Bruford and Burke, 1991). However, such an efficient approach is only necessary when very large numbers of probes are required, e.g. in linkage studies. For most studies, a maximum of five SLPs is a realistic number for screening comparatively large sample sizes. As such, the relatively simple phagemid cloning strategy employed in this study has proved to be more than adequate. With only 14% of the partial DNA library having been examined, and with no two probes, of the 24 analysed, identifying the same locus, it is likely that more SLPs could be isolated from this library. While not a drawback in this study, it should be noted that pBluescript II can accommodate only relatively short inserts (max. ≈ 8 kb for efficient cloning). For efficient screening of a large proportion of the genome, bacteriophage lambda or charomid vectors would be more appropriate.

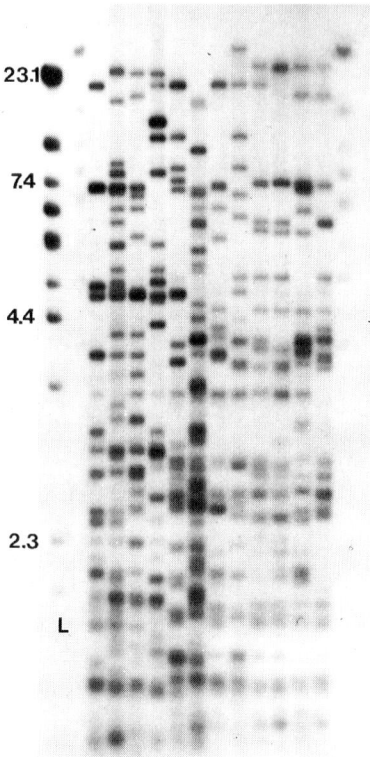

Figure 5.3.3 Example of multilocus patterns revealed by the minisatellite probe Str-20 for 12 brown trout. Electrophoretic conditions are as given in caption to Figure 5.3.1.

All SLP recombinants identified have been re-amplified from glycerol stocks and have been found to retain full-length inserts. Thus the vector-host system used is suitable for the propagation and maintenance of at least some common minisatellite classes recognized by either 33.6 or 33.15 human probes. The system offers a number of other practical advantages. No further subcloning of fragments is necessary. With a high insert to vector ratio ($\approx$ 50%; Bluescript II is 2.96 kb in length), small-scale phagemid preparations (50 ml) can yield sufficient DNA for hundreds of probings. Direct RNA transcription of inserts is feasible (e.g. for probe labelling) and the phagemid can be recovered in either single- or double-stranded form for simplified DNA sequencing protocols.

Multilocus DNA fingerprints from different organisms tend to show a markedly lower level of band sharing among larger minisatellite fragments, indicative of a positive correlation between minisatellite length and mutation rate (Jeffreys *et al.*, 1988). Previous SLP isolation studies have invariably involved the cloning of relatively large (> 4 kb) DNA fragments. This size selection both increases the likelihood of detecting highly variable loci, and enriches the library for repeat sequences. Our initial attempts at isolating SLPs from Atlantic salmon followed this approach, with fragments of size range 9–23 kb being cloned. This produced SLPs with 3–7 alleles segregating per population (Taggart and Ferguson, 1990b). It was assumed that by cloning a smaller size fraction of brown trout DNA, SLPs with fewer segregating alleles would be obtained. However, to date, the SLPs isolated appear to be at least as variable as our existing Atlantic salmon probes, both in number of alleles and observed heterozygosity levels. Furthermore, the 1.5–4.4 kb fraction of brown trout DNA still appears to be considerably enriched for minisatellite sequences. Approximately 3% of the library has tested positive for minisatellite DNA.

Whereas the number of segregating 'alleles' is similar, there is a significant difference in the size of alleles detected by probes isolated from the two libraries. Not surprisingly, the Atlantic salmon SLPs detect larger minisatellites, two out of four having alleles in the size range 13–30 kb (Taggart and Ferguson, 1990b). In contrast, four of the five brown trout SLPs detect alleles between 2 and 8 kb in length. From a population screening viewpoint, the latter probes are of more practical value since all such minisatellite loci can be optimally resolved from a Southern blot of a single 0.7% agarose gel run.

From the limited analyses undertaken to date, it would appear that some of the 'complex' SLP patterns observed are due to infrequent *Pal*I restriction sites within the minisatellite 'loci' detected by these probes, since in these cases simple single-locus patterns are obtained with *Mbo*I digested genomic DNA. The initial choice of *Pal*I, as isoschizomer of *Hae*III, for genomic DNA digestion (Taggart and Ferguson, 1990b) was based primarily on financial grounds. Presented with numerous positive recombinants, it was considered prudent to select those that could be used in combination with a relatively inexpensive restriction enzyme. The probes have not been systematically screened against DNA digested with other restriction enzymes. The simultaneous detection of homologous loci in this tetraploid derivative lineage may be another factor accounting for 'complex' SPL patterns. Some of the multilocus probes isolated in this study may also prove to be of value as individual, familial or even populational markers in salmonid studies. The simpler DNA fingerprints obtained, compared to those produced using Jeffreys' human multilocus probes (Taggart and Ferguson, 1990a; Prödohl, Taggart and Ferguson, 1992), are more suitable for practical analyses.

The cloning strategy described in this section is probably applicable to many other organisms. It may be particularly useful when only relatively few SLPs are required. Using a commercially available 'ready to use' vector–host system is both convenient and time-saving. Furthermore, we have since found that insert DNA isolation is unnecessary. Probes produced from whole phagemid constructs are adequate to resolve minisatellite DNA profiles. No expensive specialized equipment is required. Size fractionation of DNA, carried out by density gradient centrifugation in this study, could be simply replaced by electro-elution from agarose gel (Sambrook, Fritsch and Maniatis, 1989). Similarly, radioactive labelling could be replaced by alternative non-isotopic methods.

Acknowledgements

The authors wish to thank Professor A.J. Jeffreys FRS for the provision of the human minisatellite probes, and the Lister Institute for permission to use the probes. We are also grateful to R. Hynes and S. Martin for useful discussion on preliminary drafts of the manuscript and to C. Maggs for photographic assistance. The award of a Postgraduate Studentship to P.A.P. by the Brazilian Federal Agency of Postgraduate Education (CAPES – No. 487/89–5), and an advanced research fellowship to J.B.T. from the Natural Environment Research Council, are gratefully acknowledged.

References

Armour, J.A.L., Povey, S., Jeremiah, S. and Jeffreys, A.J. (1990) Systematic cloning of human minisatellites from ordered array charomid libraries. *Genomics*, **8**, 501–12.

Bentzen, P., Harris, A.S. and Wright, J.M. (1991) Cloning of hypervariable minisatellite and simple sequence microsatellite repeats for DNA fingerprinting of important aquacultural species of salmonids and tilapias, in *DNA Fingerprinting: Approaches and Applications* (eds T. Burke, G. Dolf, A.J. Jeffreys and R. Wolff), Birkhäuser Verlag, Basel, pp. 243–62.

Bruford, M.W. and Burke, T. (1991) Hypervariable DNA markers and their applications in the chicken, in *DNA Fingerprinting: Approaches and Applications* (eds T. Burke, G. Dolf, A.J. Jeffreys and R. Wolff), Birkhäuser Verlag, Basel, pp. 230–42.

Bruford, M.W., Hanotte, O., Brookfield, J.F.Y. and Burke, T. (1992) Single-locus and multilocus DNA fingerprinting, in *Molecular Genetic Analysis of Populations: A Practical Approach* (ed. A.R. Hoelzel), Oxford University Press, New York, pp. 225–69.

Dalgleish, R. (1987) Southern blotting, in *Gene Cloning and Analysis: A Laboratory Guide* (ed. G.J. Boulnois), Blackwell Scientific Publications, Oxford, pp. 45–60.

Ferguson, A. (1989) Genetic differences among brown trout, *Salmo trutta*, stocks and their importance for the conservation and management of species. *Freshwater Biol.*, **21**, 35–46.

Ferguson, A. and Taggart, J.B (1991) Genetic differences among the sympatric brown trout (*Salmo trutta*) populations of Lough Melvin, Ireland. *Biol. J. Linn. Soc.*, **43**, 221–37.

Fields, R.D., Johnson, K.R. and Thorgaard, G.H. (1989) DNA fingerprints in rainbow trout detected by hybridization with DNA of bacteriophage M13. *Trans. Am. Fish. Soc.*, **118**, 78–81.

Gyllensten, U.B., Jakobsson, S., Temrin, H. and Wilson, A.C. (1989) Nucleotide sequence and genomic organization of bird minisatellites. *Nucleic Acids Res.*, **17**, 2203–14.

Hadfield, C. (1987) Genomic library cloning – with plasmid, lambda and cosmid vectors, in *Gene Cloning and Analysis: A Laboratory Guide* (ed. G.J. Boulnois) Blackwell Scientific Publications, Oxford, pp. 61–105.

Hanotte, O., Burke, T., Armour, J.A. and Jeffreys, A.J. (1991) Hypervariable minisatellite DNA sequences in the Indian peafowl *Pavo cristatus*. *Genomics*, **9**, 587–97.

Jeffreys, A.J., Wilson, V. and Thein, S.L. (1985a) Hypervariable 'minisatellite' regions in human DNA. *Nature*, **314**, 67–73.

Jeffreys, A.J., Wilson, V. and Thein, S.L. (1985b) Individual-specific 'fingerprints' of human DNA. *Nature*, **316**, 76–9.

Jeffreys, A.J., Royle, N., Wilson, V. and Wong, Z. (1988) Spontaneous mutation rates to new length alleles at tandem repetitive hypervariable loci in human DNA. *Nature*, **332**, 278–81.

Krueger, C.C. and May, B. (1988) Stock identification of naturalised brown trout (*Salmo trutta*) in Lake Superior tributaries: differentiation based on allozyme data. *Trans. Am. Fish. Soc.*, **116**, 785–94.

Lloyd, M.A., Fields, M.J. and Thorgaard, G.H. (1989) BKm minisatellite sequences are not sex associated but reveal DNA fingerprint polymorphisms in rainbow trout. *Genome*, **32**, 865–8.

Lynch, M. (1991) Analysis of population genetic structure by DNA fingerprinting, in *DNA Fingerprinting: Approaches and Applications* (eds T. Burke, G. Dolf, A.J. Jeffreys and R. Wolff), Birkhäuser Verlag, Basel, pp. 113–26.

Prodöhl, P.A., Taggart, J.B. and Ferguson, A. (1992) Genetic variability within and among sympatric brown trout (*Salmo trutta*) populations: multilocus DNA fingerprint analysis. *Hereditas*, **117**, 45–50.

Ryman, N. (1983) Patterns of distribution of biochemical genetic variation in salmonids: differences between species. *Aquaculture*, **33**, 1–21.

Sambrook, J., Fritsch, E.F. and Maniatis, T. (1989) *Molecular Cloning: A Laboratory Manual*, Cold Spring Harbor Laboratory Press, New York.

Taggart, J.B. and Ferguson, A. (1990a) Minisatellite DNA fingerprints of salmonid fishes. *Anim. Genet.*, **21**, 377–89.

Taggart, J.B. and Ferguson, A. (1990b) Hypervariable minisatellite DNA single locus probes for the Atlantic salmon, *Salmo salar* L. *J. Fish Biol.*, **37**, 991–93.

Taggart, J.B., Hynes, R.A., Prodöhl, P.A. and Ferguson, A. (1992) A simplified protocol for routine total DNA isolation from salmonid fishes. *J. Fish Biol.*, **40**, 963–65.

Vassart, G., Georges, M., Monsieur, R. et al. (1987) A sequence in M13 phage detects hypervariable minisatellites in human and animal DNA. *Science*, **235**, 683–84.

Wells, R.A. (1988) DNA fingerprinting, in *Genome Analysis: A Practical Approach* (ed. K.E. Davies), IRL Press, Oxford, pp. 153–70.

Wong, Z., Wilson, V., Patel, I., Povey, S. and Jeffreys, A.J. (1987) Characterization of a panel of highly variable minisatellites cloned from human DNA. *Ann. Human Genet.*, **51**, 269–88.

5.4 CHARACTERIZATION AND DISTRIBUTION OF GENOMIC REPEAT SEQUENCES FROM ARCTIC CHARR (*SALVELINUS ALPINUS*)

Sheila E. Hartley and William S. Davidson

Abstract

*The genome of Arctic charr (*Salvelinus alpinus*) contains several families of repetitive elements ranging in size from 70 base pairs (bp) to 1.7 kilobase pairs (kb) detected when DNA is digested with AluI, BglII, BstEII, DraI, EcoRI, EcoRV, HindIII, MboI and RsaI. AluI, MboI and RsaI generate repeats approximately 100 bp in length while the DraI repeat is of the order of 200 bp long, and these have been analysed by cloning, sequencing and hybridization studies. The AluI and RsaI repeats are part of the same 127-bp tandemly repetitious family. AluI repeats are present in all members of the* Salvelinus *genus and in two members of* Oncorhynchus *but are absent from representatives of* Salmo *and* Coregonus. *The DraI repeat is 172 bp long, is tandemly repetitious and is present in Dolly Varden (*Salvelinus malma*) and bull charr (*S. confluentus*) as well as Arctic charr but not brook charr (*S. fontinalis*), lake charr (*S. namaycush*) or Japanese charr (*S. leucomaenis*). The 78-bp repeat of the* MboI *family is based on a doublet of a 33-mer. This repetitive family has been found to be absent from a population of Arctic charr in Scotland. MboI repeats are present in lake charr and Japanese charr but not brook charr, Dolly Varden or bull charr. The implications of these findings for phylogenetics of* Salvelinus *are discussed.*

5.4.1 Introduction

The salmonid fish have a widespread distribution in the northern hemisphere and because of their commercial importance have been the subjects of intense study over the years by biologists, physiologists, endocrinologists and geneticists. Genetic studies are complicated by the ancient tetraploid origin of these fish and by reorganizations that have taken place within the genome during diploidization. Relatively little is known of the molecular organization of salmonid genomes although it is estimated that up to 60% of the genome may be composed of repetitive sequences (Gharret, Simon and McIntyre, 1977; Hanham and Smith, 1980).

The genomes of eukaryotic organisms contain three classes of repeated DNA sequences: multigene families such as the histone and ribosomal RNA genes, interspersed repetitive SINE (short interspersed elements) and LINE (long interspersed elements) families and tandemly repeated satellite DNA sequences (Singer, 1982). All three of these classes of repeated DNA sequences are currently being studied to further understanding of phylogenetic relationships within the salmonids. Phillips and her collaborators (Phillips, Pleyte and Ihssen, 1989; Phillips *et al.*, 1989; Phillips and Pleyte, 1991) are concentrating on the ribosomal RNA genes and their cytological location, the nucleolar organizer regions (NORs). Okada and his coworkers are elucidating the organization of SINE families which are derived from tRNA genes (Matsumoto, Murakami and Okada, 1986; Kido *et al.*, 1991; Koishi and Okada, 1991). We are studying the organization of highly repeated or satellite DNA sequences which are amongst the fastest evolving sequences in the nuclear genome (Brutlag, 1980).

Although such sequences have not been detected in salmonid genomes by isopycnic gradient centrifugation (Hudson *et al.*, 1980; Bernardi and Bernardi, 1990), their presence is inferred because all salmonids examined to date possess centromeric and telomeric constitutive heterochromatin bands, the cytological sites of satellite DNA sequences (Pardue and Gall, 1970), which can be identified by C, Q or restriction enzyme banding (Hartley, 1987; Lloyd and Thorgaard, 1988; Pleyte, Phillips and Hartley, 1989; Hartley, 1991a,b;

Sanchez *et al.*, 1991). An alternative technique for detecting satellite DNA sequences is restriction enzyme digestion of nuclear DNA followed by electrophoresis through agarose or acrylamide gels and ethidium bromide staining, when satellite sequences show up as distinct bands against a background smear (Cooke, 1976; Manuelidis, 1976). Using the technique we have discovered several repeated DNA sequences in Arctic charr, *Salvelinus alpinus*, and Atlantic salmon, *Salmo salar* (Goodier and Davidson, 1993).

In this section we review the characterization and distribution of three families of repeated sequences from Arctic charr and discuss their phylogenetic implications, particularly for the subgenus *Salvelinus*, where relationships are largely unresolved.

5.4.2   Materials and methods

(a)   DNA isolation, digestion and electrophoresis

High molecular weight DNA was isolated from livers of Arctic charr by standard procedures involving proteinase K digestion, phenol/chloroform extraction and ethanol precipitation (Sambrook, Fritsch and Maniatis, 1989). Restriction enzyme digestion (Gibco-BRL or Pharmacia) followed the manufacturer's instructions and fragments were separated by electrophoresis through 1.5% or 3% agarose gels in Tris-borate buffer (Sambrook, Fritsch and Maniatis, 1989).

(b)   Cloning and sequencing

Bands of interest were excised from preparative 3% low melting point agarose gels and ligated into an appropriate site of pTZ18R. For *Alu*I, *Dra*I and *Rsa*I bands this was the *Sma*I site, and for the *Mbo*I band the *Bam*HI site was used. *E. coli* DH5α cells were transformed with the ligation reaction and plasmid DNA was extracted from recombinant clones using the alkaline lysis method of Sambrook, Fritsch and Maniatis, (1989). Sequencing was carried out by the dideoxy chain termination procedure (Sanger, Micklen and Coulson, 1977) using universal and reverse primers and a 'Sequenase' kit from the US Biochemical Corporation.

(c)   Southern blotting and hybridization

Restriction enzyme digests separated on agarose gels were transferred to Hybond N nylon membrane (Amersham) following denaturation in 1.5 M NaCl/0.5 M NaOH and using 1.5 M NaCl/0.25 M NaOH as the transfer buffer. Filters were prehybridized in 50% formamide, 6 × SSC, 5 × Denhardt's, 50 mM sodium phosphate pH 6.5, 0.1% SDS at 40 °C for 4 h and hybridized with random primer radiolabelled probe (Feinberg and Vogelstein, 1983) in the same solution overnight at 40 °C. Following hybridization, filters were washed three times for 10 min each in 2 × SSC/0.1% SDS at 40 °C and three times for 10 min each in 1 × SSC/0.1% SDS at 50 °C. Autoradiography was carried out at –70 °C with intensifying screens.

5.4.3   Results

(a)   Repeated sequences in Arctic charr

The restriction enzymes which reveal repeated sequences in Arctic charr are *Alu*I, *Bgl*II, *Bst*EII, *Dra*I, *Eco*RI, *Eco*RV, *Hin*dIII, *Mbo*I and *Rsa*I. These are seen as distinct bands in ethidium bromide stained agarose gels and range in size from approximately 70 bp to 1.7 kb

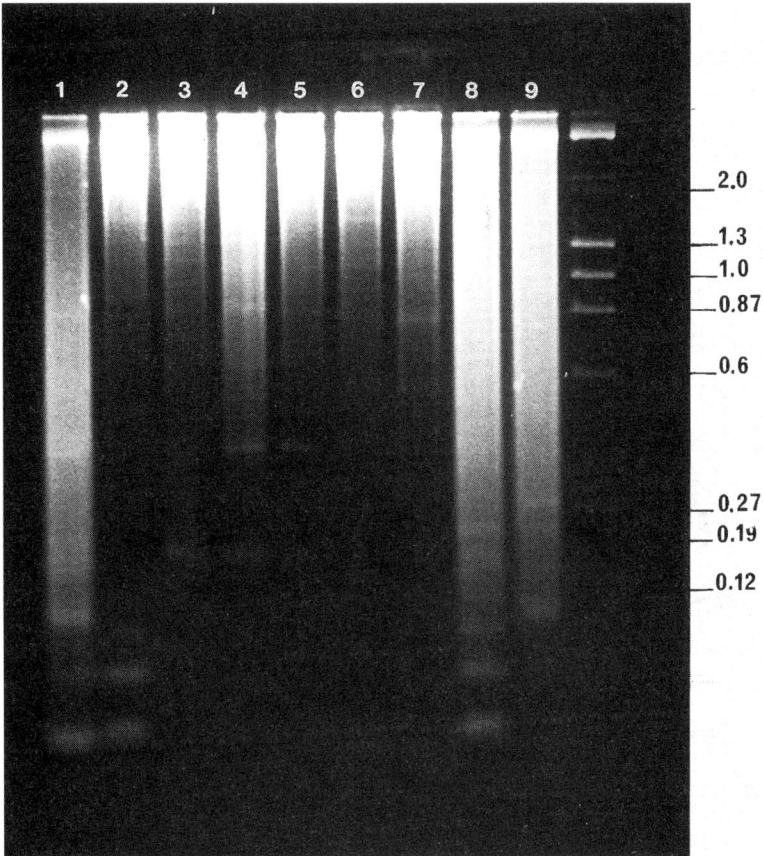

Figure 5.4.1 Ethidium-bromide-stained 3% agarose gel of Arctic charr DNA digested with (1) *Alu*I, (2) *Bgl*II, (3) *Bst*EII, (4) *Dra*I, (5) *Eco*RI, (6) *Eco*RV, (7) *Hin*dIII, (8) *Mbo*I, (9) *Rsa*I. Size markers (kb) are from λ DNA digested with *Hin*dIII and ΦX174 DNA digested with *Hae*III.

(Figure 5.4.1). Bands from *Alu*I, *Dra*I, *Mbo*I and *Rsa*I digests have been cloned and sequenced. Sequencing and hybridization studies show that these repeats are from three distinct, unrelated families: an *Alu*I/*Rsa*I family, a *Dra*I/*Bst*EII family and a *Mbo*I/*Bgl*II family. Partial digestion of DNA with these restriction enzymes shows that all three families are tandemly repetitive.

(b)   The Alu*I/*Rsa*I family*

The repeated sequences detected with *Alu*I and *Rsa*I form part of the same 127-bp tandemly repetitive family (Figure 5.4.2). Hybridization of one of the clones generated from the *Alu*I repeat (pChA2), which contains a dimer of the 127-bp sequence, to Southern blots of Arctic charr DNA digested with all the enzymes that reveal repeated sequences confirms this relationship (Figure 5.4.3). When DNA from the other species of *Salvelinus*, brook charr

```
1        10        20        30        40        50        60
                                                   RsaI
CTCTAAATCGTGTTTAATGCACTCTGTTAGTGAAATTTTTTGCAGTACTGCTTAAACTAG

        70        80        90       100       110       120
AGGATAGGAACCAATTTCAGCGAGTTTCAAGCATGAAATTCAAAAAAAAACACTTTTCCTT

       127
AAACAAG
```

Figure 5.4.2 Nucleotide sequence of the 127-bp repeat of the *Alu*I/*Rsa*I family. This sequence has been deposited with GenBank under accession number L00985.

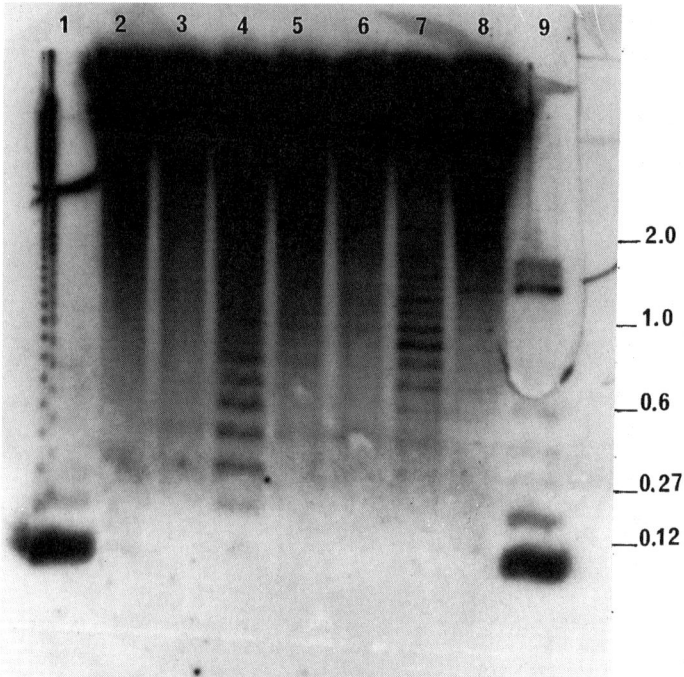

Figure 5.4.3 Hybridization of clone pChA2 to Arctic charr DNA digested with (1) *Alu*I, (2) *Bgl*II, (3) *Bst*EII, (4) *Dra*I, (5) *Eco*RI, (6) *Eco*RV, (7) *Hind*III, (8) *Mbo*I, (9) *Rsa*I. Size markers (kb) are from λ DNA digested with *Hind*III and ΦX174 DNA digested with *Hae*III.

(*S. fontinalis*), Dolly Varden (*S. malma*), lake charr (*S. namaycush*), bull charr (*S. confluentus*) and Japanese charr (*S. leucomaenis*), and representatives of other salmonid genera was digested with *Alu*I, repeated sequences were observed in all the *Salvelinus* species and in two members of the *Oncorhynchus*, chinook salmon (*O. tshawytscha*) and coho salmon (*O. kisutch*), but not in representatives of the *Salmo* and *Coregonus* genera (Figure 5.4.4).

(c)   The Dra*I/Bst*EII *family*

The nucleotide sequence of the 172-bp *Dra*I repeat is shown in Figure 5.4.5. When the clone designated pChDr3 is hybridized to Southern blots of Arctic charr DNA digested with the

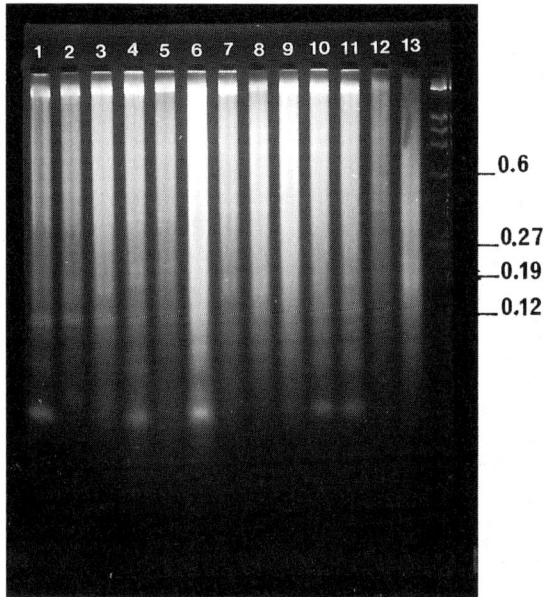

Figure 5.4.4 Ethidium-bromide-stained 3% gel of *Alu*I-digested DNA from (1) *Salvelinus alpinus*, (2) *S. fontinalis*, (3) *S. malma*, (4) *S. namaycush*, (5) *S. confluentus*, (6) *S. leucomaenis*, (7) *Salmo salar*, (8) *Salmo trutta*, (9) *Oncorhynchus mykiss*, (10) *O. tshawytscha*, (11) *O. kisutch*, (12) *O. gorbuscha*, (13) *Coregonus lavaretus*. Size markers (kb) are from λ DNA digested with *Hin*dIII and ΦX174 DNA digested with *Hae*III.

```
1         10        20        30        40        50        60
                                      AluI
AAATGGGCTACACACCTCAATGCCCGCTATGGGAGCTACATACTGCGGGCACAAGTCAGT

          70        80        90        100       110       120

AGGGGCCGGCCTGAGATGTCTGGACCAATTTACCAGCCCGTGCACCTGGTGTTTTTTTCC

          130       140       150       160       170
RsaI                    BstEII
GTACGCCTGGTATTTTTTTTAGGTTACCAGGCCCACAATTTTATAATGGTTT
```

Figure 5.4.5 Nucleotide sequence of the 172-bp *Dra*I repeat.
This sequence has been deposited with GenBank under accession number L00991.

nine restriction enzymes, hybridization is with the *Bst*EII and *Dra*I repeats, confirming that these repeats form part of the same family (Figure 5.4.6). *Dra*I digestion of DNA from the other *Salvelinus* species and other salmonids reveals repeated sequences in all the *Salvelinus* species (Figure 5.4.7 (a)) and Atlantic salmon, *Salmo salar* (not shown), but only the Dolly Varden (*S. malma*) and bull charr (*S. confluentus*) repeats are homologous to the Arctic charr repeat (Figure 5.4.7 (b)).

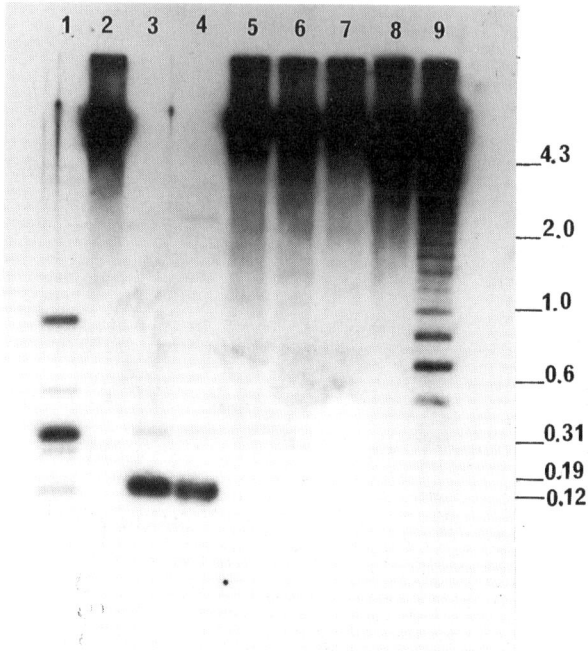

Figure 5.4.6 Hybridization of clone pChDr3 to Arctic charr DNA digested with (1) *Alu*I, (2) *Bgl*II, (3) *Bst*EII, (4) *Dra*I, (5) *Eco*RI, (6) *Eco*RV, (7) *Hin*dIII, (8) *Mbo*I, (9) *Rsa*I. Size markers (kb) are from λ DNA digested with *Hin*dIII and ΦX174 DNA digested with *Hae*III.

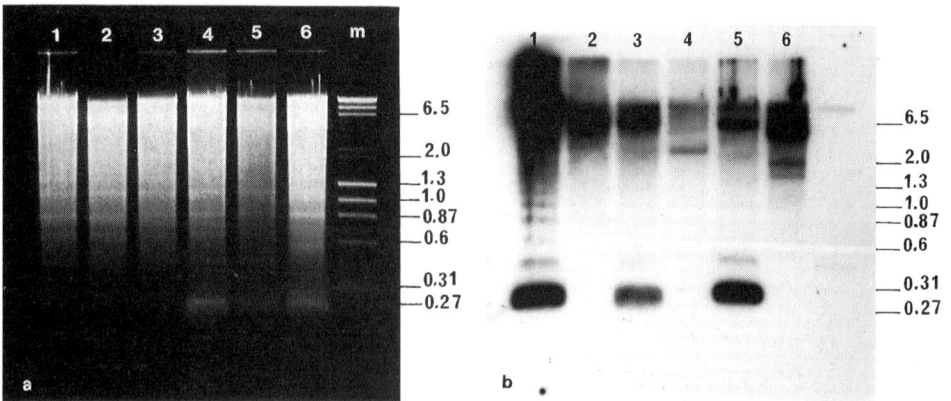

Figure 5.4.7 (a) Ethidium-bromide-stained 1.5% agarose gel of *Dra*I-digested DNA from (1) *Salvelinus alpinus*, (2) *S. fontinalis*, (3) *S. malma*, (4) *S. namaycush*, (5) *S. confluentus*, (6) *S. leucomaenis*. The marker lane (m) is λ DNA digested with *Hin*dIII and ΦX174 DNA digested with *Hae*III. Sizes are in kb. (b) Hybridization of clone pChDr3 to Southern blot of (a).

1 60
*Mbo*I *Alu*I
GATCTACTGTCTCTATTATATTATGACAGACCAGCTAGATATACTGTCTCTATTATATTA

*Alu*I *Mbo*I
TGACAGACCAGCAGCTAGATC

Figure 5.4.8 Nucleotide sequence of the 78-bp *Mbo*I repeat. The internally repeated 33-mer is underlined. This sequence has been deposited with GenBank under accession number L01078.

(d) The Mbo*I*/Bgl*II family*

The consensus sequence of the tandemly repetitious *Mbo*I family is a 78-bp repetitive element composed of a doublet of a 33-mer (Figure 5.4.8). The reference clone pChMb17 hybridizes to the *Mbo*I and *Bgl*II repeats, confirming the relationship of these repeats which was inferred from the recognition sequences of the two enzymes. The internal four bases of the *Bgl*II recognition sequences are the same (GATC) as the recognition sequence for *Mbo*I. This repeat has been found to be absent from a population of Arctic charr in Wester Ross,

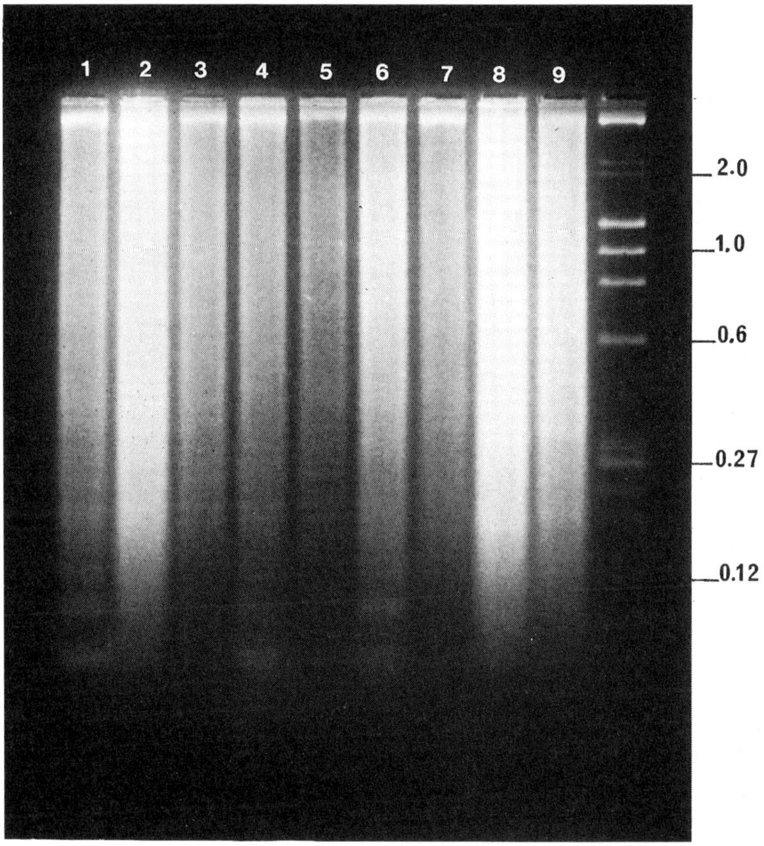

Figure 5.4.9 Ethidium-bromide-stained 3% agarose gel of *Mbo*I-digested DNA from (1) *Salvelinus alpinus*, (2) *S. fontinalis*, (3) *S. malma*, (4) *S. namaycush*, (5) *S. confluentus*, (6) *S. leucomaenis*, (7) *Salmo salar*, (8) *Salmo trutta*, (9) *Oncorhynchus mykiss*. Size markers (kb) are from λ DNA digested with *Hin*dIII and ΦX174 DNA digested with *Hae*III.

Scotland but is present in other Scottish populations (S.E. Hartley and A.A. Bell, un-published data). In addition to Arctic charr, *Mbo*I detects repeats in the DNA of lake charr (*S. namaycush*) and Japanese charr (*S. leucomaenis*) but not brook charr, Dolly Varden or bull charr (Figure 5.4.9).

5.4.4 Discussion

The phylogenetic relationships within the *Salvelinus* genus are complex. On morphological, behavioural and ecological grounds, Behnke (1980, 1984) recognizes three subgenera, *Baione*, *Cristivomer* and *Salvelinus*. The brook charr, *S. fontinalis*, is the only member of *Baione* while *Cristivomer* is represented by the lake charr, *S. namaycush*. All remaining species are assigned to the subgenus *Salvelinus* and two species complexes are recognized by Behnke (1980): the Dolly Varden complex comprising Dolly Varden (*S. malma*), bull charr (*S. confluentus)* and Japanese charr (*S. leucomaenis*) and the Arctic charr complex comprising all the subspecies of *S. alpinus*. The Arctic charr complex is further divided into three major groups with origins in Europe, Siberia and the Chukotsk Sea–Bering Sea region of the Arctic Ocean (Behnke, 1984). However, on both morphological and cytogenetic grounds, Cavender (1980, 1984) groups Dolly Varden (*S. malma*) more closely with Arctic charr (*S. alpinus*) than the bull charr (*S. confluentus*) and patterns of NOR differences also group Dolly Varden and Arctic charr together (Phillips, Pleyte and Ihssen, 1989). Other karyotypic data (Cavender, 1984) suggest that Japanese charr (*S. leucomaenis*) is closer to brook charr (*S. fontinalis*) and lake charr (*S. namaycush*) than to Dolly Varden, Arctic charr and bull charr which all have derived karyotypes. Mitochondrial DNA analysis readily resolved the three subgenera and distinguished two groups in the *Salvelinus* subgenus – one containing two subspecies of Arctic charr (*S. a. alpinus* from Britain and *S. a. oquassa* from Maine, USA) and Dolly Varden, and the other containing bull charr and a third subspecies of Arctic charr (*S. a. stagnalis* from NWT, Canada) (Grewe, Billington and Hebert, 1990). Japanese charr was not included in the study so its position based on mtDNA analysis is not known. While the relationships of brook charr and lake charr with respect to one another and the other *Salvelinus* species are not in dispute, the relationships between the members of the *Salvelinus* subgenus are largely unresolved.

As satellite DNA sequences are amongst the fastest evolving sequences in the nuclear genome (Brutlag, 1980), their characterization and distribution present an opportunity to explore phylogenetic relationships between closely related species and to study the evolution of repeated sequences. Within *Salvelinus*, *Alu*I and *Dra*I repeats are present in all species. This suggests an ancestral origin for these repeats and their maintenance in all species al-though the finding that homologous sequences to the Arctic charr *Dra*I repeat are found only in Dolly Varden and bull charr suggests significant divergence within the repeat during the evolution of the *Salvelinus* genus. This distribution splits Japanese charr (*S. leucomaenis*) from the other members of the *Salvelinus* subgenus and supports a close relationship among Dolly Varden, Arctic charr and bull charr. This is in accord with other genetic evidence from cytogenetic studies (Cavender, 1984; Phillips, Pleyte and Ihssen, 1989a), restriction fragment length polymorphisms of mtDNA (Grewe, Billington and Hebert, 1990) and mitochondrial cytochrome *b* gene sequencing (H.P. McVeigh and W.S. Davidson, unpublished data).

The distribution of *Mbo*I repeats in Arctic charr, lake charr and Japanese charr appears contradictory. Either the repeat was originally present in all species and has been lost through time from brook charr, Dolly Varden and bull charr, or it has arisen recently and independently in the three species where it is found. Its absence from a putative ancestral population of Arctic charr in northwest Scotland (S.E. Hartley and A.A. Bell, unpublished data) provides evidence for the latter possibility. The Arctic charr from which the repeats have been isolated are all members of the *S. a. alpinus* subspecies, and thus the *Mbo*I repeat is a possible candidate for resolving subspecific relationships within the Arctic charr complex. It is also probable that further characterization of other repetitive elements in *Salvelinus* will reveal additional phylogenetic relationships.

Acknowledgements

We are grateful to Dr R.B. Phillips, University of Wisconsin-Milwaukee, USA, who kindly provided DNA samples from lake charr, bull charr, Dolly Varden and Japanese charr and to R.B. Greer and A.F. Walker, SOAFD Freshwater Fisheries Laboratory, Pitlochry, Scotland, who provided Arctic charr. This work was supported by grants from the Natural Environment Research Council of Great Britain (SEH), the Natural Sciences and Engineering Research Council of Canada (WSD) and a NATO Collaborative Research Grant.

References

Behnke, R.B. (1980) A systematic review of the genus *Salvelinus*, in *Charrs: Salmonid Fishes of the Genus* Salvelinus (ed. E.K. Balon), Dr W. Junk, The Hague, pp. 441–81.

Behnke, R.B. (1984) Organizing the diversity of the Arctic charr complex, in *Biology of the Arctic Charr* (eds L. Johnson and B.L. Burns), University of Manitoba, Winnipeg, pp. 3–21.

Bernardi, G. and Bernardi, G. (1990) Compositional patterns in the nuclear genome of cold-blooded vertebrates. *J. Mol. Evol.*, **31**, 265–81.

Brutlag, D.L. (1980) Molecular arrangement and evolution of heterochromatin. *Annu. Rev. Genet.*, **14**, 121–44.

Cavender, T.M (1980) Systematics of *Salvelinus* from the North Pacific basin, in *Charrs: Salmonid Fishes of the Genus* Salvelinus (ed. E.K. Balon), Dr W. Junk, The Hague, pp. 295–322.

Cavender, T.M (1984) Cytotaxonomy of North American *Salvelinus*, in *Biology of the Arctic Charr* (eds L. Johnson and B.L. Burns), University of Manitoba, Winnipeg, pp. 431–45.

Cooke, H.J. (1976) Repeated sequence specific to human males. *Nature*, **262**, 182–86.

Feinberg, C. and Vogelstein, B. (1983) A technique for radiolabelling DNA restriction endonuclease fragments to high specific activity. *Analyt. Biochem.*, **132**, 6–13.

Gharret, A.J., Simon, R.C. and McIntyre, J.D. (1977) Reassociation and hybridization properties of DNAs from several species of fish. *Comp. Biochem. Physiol.*, **56B**, 81–5.

Goodier, J.L. and Davidson, W.S. (1993) A repetitive element in the genome of Atlantic salmon, *Salmo salar*. *Gene* (in press).

Grewe, P.M., Billington, N. and Hebert, P.D.N. (1990) Phylogenetic relationships among members of *Salvelinus* inferred from mitochondrial DNA divergence. *Can. J. Fish. Aquat. Sci.*, **47**, 984–91.

Hanham, A.F. and Smith, M.J. (1980) Sequence homology in the single-copy DNA of salmon. *Comp. Biochem. Physiol.*, **65B**, 333–8.

Hartley, S.E. (1987) The chromosomes of salmonid fishes. *Biol. Rev.*, **62**, 197–213.

Hartley, S.E. (1991a) Restriction enzyme banding in Atlantic salmon (*Salmo salar*) and brown trout (*Salmo trutta*). *Genet. Res.*, **57**, 273–78.

Hartley, S.E. (1991b) C, Q and restriction enzyme banding of the chromosomes in brook trout (*Salvelinus fontinalis*) and Arctic charr (*Salvelinus alpinus*). *Hereditas*, **114**, 253–61.

Hudson, A.P., Cuny, G., Cortadas, J. *et al.* (1980) An analysis of fish genomes by density gradient centrifugation. *Eur. J. Biochem.*, **112**, 203–10.

Kido, Y., Aono, M., Yamaki, T. *et al.* (1991) Shaping and reshaping of salmonid genomes by amplification of tRNA-derived retroposons during evolution. *Proc. Natl. Acad. Sci. USA*, **88**, 2326–30.

Koishi, R. and Okada, N. (1991) Distribution of salmonid *Hpa*I family in the salmonid species demonstrated by in vitro runoff transcription assay of total genomic DNA: a procedure to estimate repetitive frequency and sequence divergence of a certain repetitive family with a few known sequences. *J. Mol. Evol.*, **32**, 43–52.

Lloyd, M.A. and Thorgaard, G.H. (1988) Restriction endonuclease banding of rainbow trout chromosomes. *Chromosoma*, **96**, 171–77.

Manuelidis, L. (1976) Repeating restriction fragments of human DNA. *Nucleic Acids Res.*, **3**, 3063–78.

Matsumoto, K., Murakami, K. and Okada, N. (1986) Gene for lysine tRNA$_1$ may be a progenitor of the highly repetitive and transcribable sequences present in the salmon genome. *Proc. Natl. Acad. Sci. USA*, **83**, 3156–60.

Pardue, M.L. and Gall, J.G. (1970) Chromosomal localization of mouse satellite DNA. *Science*, **168**, 1356–58.

Phillips, R.B. and Pleyte, K.A. (1990) Nuclear DNA and salmonid phylogenetics. *J. Fish Biol.*, **39A**, 259–75.

Phillips, R.B., Pleyte, K.A. and Ihssen, P.E. (1989) Patterns of chromosomal nucleolar organizer region (NOR) variation in fishes of the genus *Salvelinus. Copeia*, **1**, 47–53.

Phillips, R.B., Pleyte, K.A., Van Ert, L.M. and Hartley S.E. (1989) Evolution of nuclear organizer regions and ribosomal RNA genes in *Salvelinus. Physiol. Ecol. Japan, Special Volume*, **1**, 429–47.

Pleyte, K., Phillips, R.B. and Hartley, S.E. (1989) Q-band chromosomal polymorphisms in arctic charr (*Salvelinus alpinus*). *Genome*, **32**, 129–33.

Sambrook, J., Fritsch, E.F. and Maniatis, T. (1989) *Molecular Cloning. A Laboratory Manual*, 2nd edn, Cold Spring Harbor Laboratory, New York.

Sanchez, L., Martinez, P., Bouza, C. and Vinas, A. (1991) Chromosomal heterochromatin differentiation in *Salmo trutta* with restriction enzymes. *Heredity*, **66**, 241–49.

Sanger, F., Micklen, S. and Coulson, A.R. (1977) DNA sequencing with chain termination inhibitors. *Proc. Natl. Acad. Sci. USA*, **74**, 5463–67.

Singer, M.F. (1982) Highly repeated sequences in mammalian genomes. *Int. Rev. Cytol.*, **76**, 67–112.

5.5 DEMOGRAPHIC AND LIFE-HISTORY CHARACTERISTICS INFLUENCE THE CYTONUCLEAR GENETIC COMPOSITION OF MOSQUITOFISH HYBRID POPULATIONS

Kim T. Scribner and John C. Avise

Abstract

Experimental laboratory crosses and population experiments reveal significant differences in individual life-history traits and population demography between two related species of mosquitofish, Gambusia affinis *and* G. holbrooki. *With respect to life-history traits, progeny from* G. holbrooki *parents exhibit larger size at birth and earlier age at sexual maturity than do progeny from* G. affinis *parents. With respect to demography, populations of* G. holbrooki *exhibit higher recruitment and carrying capacity and lower overwinter mortality than do populations of* G. affinis. *These differences help to explain the dramatic changes in cytonuclear genotype frequency observed in replicated experimental hybrid populations of* Gambusia *monitored over 52 weeks. These experimental results are interpreted in the context of introgression patterns previously studied indirectly from distributions of cytonuclear genotypes in a natural mosquitofish hybrid zone.*

5.5.1 Introduction

Geographic surveys of mitochondrial (mt) DNA within several freshwater fish species in the southeastern USA have demonstrated the importance of both historical biogeography and contemporary gene flow in shaping intraspecific population genetic structure (Bermingham and Avise, 1986; Avise *et al.*, 1987; Avise, 1992). The geographic concordance of major phylogenetic discontinuities across taxa emphasizes the overriding importance of historical (Pleistocene) events: most of the major genetic separations for freshwater species distinguish eastern from western populations, with the dividing line in the region of Mobile Bay or the Apalachicola drainage. Along this boundary zone, hybridization frequently takes place. Within either region, finer-scale genetic subdivision is also evident, and probably reflects contemporary restrictions on gene flow and/or the effects of environmental heterogeneity and local selection.

Genotype-specific variation in life-history traits may affect rates of recruitment and mortality and thus may be of critical importance in understanding short-term changes in population genetic architecture. Life-history traits presumably coevolve to optimize fitness at each age through differential allocation of resources to maintenance, growth and reproduction (Partridge and Harvey, 1988). Differences in life-histories between species or between alternative genotypes may affect competitive interactions and demographics, and hence the genetic composition of populations. Such effects might be magnified in secondary hybrid zones.

To address the role of species-specific demographies in shaping population genetic structure, we have conducted research on natural and experimental populations of the mosquitofish *Gambusia holbrooki* and *G. affinis*. These species inhabit eastern and western drainages, respectively, in the southeastern USA, but also overlap across a broad central area in which hybridization and introgression apparently take place (Wooten, Scribner and Smith, 1988; Wooten and Lydeard, 1990; Scribner and Avise, 1993). The species are known to differ in a number of life-history traits, e.g. offspring size (Reznick, 1981), juvenile growth rates and size and age at sexual maturity (Scribner, 1993), and these potentially could influence genetic outcomes in the areas of secondary contact. Our general objectives have been to: (1) describe the genetic composition of natural populations of *Gambusia* in the southeastern USA with particular focus on the area of hybridization; (2) utilize controlled laboratory crosses and experimental pool populations to characterize life-history traits, population demographies and temporal genetic changes in *Gambusia*; and (3) use the experimental results as an aid to interpreting the genetic patterns observed in nature.

5.5.2 Materials and methods

(a) Genetic patterns in nature

Spatial distributions of nuclear and mitochondrial genotypes were based on collections of fish from 45 locations in seven states (SC, GA, FL, AL, MS, LA and TX). Collections of approximately 100 individuals per location were made using dip nets during the spring and summer of 1989. Five to ten individuals per locale were returned live to the laboratory for mtDNA analysis, while the remaining specimens were frozen in liquid nitrogen for allozyme assays. Mitochondrial DNA was isolated from fresh tissues using standard procedures involving CsCl density gradient centrifugation (Lansman *et al.*, 1981). Restriction digests were conducted overnight. Digestion fragments were end-labelled with appropriate ^{35}S-radionucleotides, separated on 1.0% agarose gels, and revealed by autoradiography (Brown, 1980; Maniatis, Fritsh and Sambrook, 1982). All differences in fragment patterns could be interpreted as gains or losses of particular restriction sites. In addition, the nuclear locus encoding aspartate aminotransferase (M-AAT-A; EC 2.6.1.1) was scored in 30 individuals from each location (Scribner *et al.* (1992); see McClenaghan, Smith and Smith (1985) and Wooten and Lydeard (1990) for techniques).

(b) Genotypic variation in life-history traits

We estimated life-history characters (gestation length and brood size) for females using experimental crosses. Allopatric stocks of adult female *G. affinis* and *G. holbrooki* were obtained from LA and SC, respectively, and virgin progeny from these females were used for the breeding experiments. The six crosses conducted were as follows (the first letter refers to the female): parental *G. affinis* × *G. affinis* (AA) and *G. holbrooki* × *G. holbrooki* (HH), reciprocal F_1s (AH and HA), and backcrosses in two directions (HA × HH and AH × AA). In each case, individual females were inseminated artificially, using sperm from a different male (Scribner, 1992, 1993), techniques are described in Stearns (1983). Female life-history characteristics were determined from 20 females from each cross. All progeny from parental and F_1 crosses were raised in individual plastic containers and were fed under *ad lib* conditions (Scribner, 1993). One female offspring from each female of each cross was chosen at random, and length was recorded at day 1, at day 15, and at the date of sexual maturity. Measurements were recorded from photographs using the computer program MORPHOSYS (Meecham and Duncan, 1989).

(c)   Population dynamics in experimental populations

To document temporal changes in population numbers, recruitment and mortality rates, and population cytonuclear genotypic composition over time, replicate experimental populations ($N = 6$) were established in each of three treatments (pure *G. affinis*, pure *G. holbrooki* and mixed *G. affinis* + *G. holbrooki*). Replicate aquatic communities were established in plastic wading pools approximately 2.4 m in diameter and 30 cm in depth. Pools were self-supporting 'mesocosms' established with organic debris, and inoculated monthly with zoo-plankton to maintain a seasonally diverse planktonic prey source. Each replicate pool contained predator and prey sources, producers, structural complexity and refuge areas for juveniles. Apart from population sampling, all pools were left unaltered, thus allowing for natural diurnal and seasonal fluctuations in environmental conditions. Population size was estimated during each of four 6-week periods (weeks 6, 12 and 18 during 1990 and week 52 during 1991). Recruitment and overwinter mortality rates were determined from per capita changes in population size. Two additional replicates of the mixed treatment were sampled completely during each sampling period. Mitochondrial DNA haplotypes were determined for each individual using the species-diagnostic enzyme *Hin*dIII. Each individuals was also scored for five species-specific allozyme loci: adenosine deaminase (ADA), EC 3.5.3.3; aspartate aminotransferase (M-AAT-A), EC 2.6.1.1; malate dehydrogenase (S-MDH-A), EC 1.1.1.37; peptidase-B (PEP-B), EC 3.4.11; and aconitase-1 (ACON-1), EC 4.2.1.3. Estimates of composite gametic disequilibrium were calculated for each replicate population as described by Weir (1979) and Weir and Cockerham (1989). Estimates of gametic cytonuclear disequilibrium were calculated as described by Asmussen, Arnold and Avise (1987).

5.5.3   Results

(a)   Genetic patterns in nature

Five mtDNA haplotypes at *Spe*I were observed among *Gambusia* from the 45 sampling locations (Figure 5.5.1): one haplotype (*C*) in *G. affinis*, and four (*B, D, E* and *F*) in *G. holbrooki*. Genotypes were geographically clustered (Figure 5.5.1), and at each locale a single haplotype was observed (in admittedly small samples, usually $N = 4$). At the nuclear locus for *M-AAT-A**, allopatric populations of *G. affinis* and *G. holbrooki* were also fixed or nearly fixed for different allozyme alleles (the '100' and '90' alleles versus '112' and '124', respectively). However, in population samples (usually $N = 30$) from the central overlap region (Figure 5.5.2), *M-AAT-A** alleles characteristic of both species were frequently shared, and in patterns suggestive of hybridization. Specifically, populations 7–15, 17, 21 and 23 sampled in eastern Alabama and western Georgia appeared to be hybrid populations. Furthermore, differences in the geographic distribution of mtDNA and allozyme alleles point

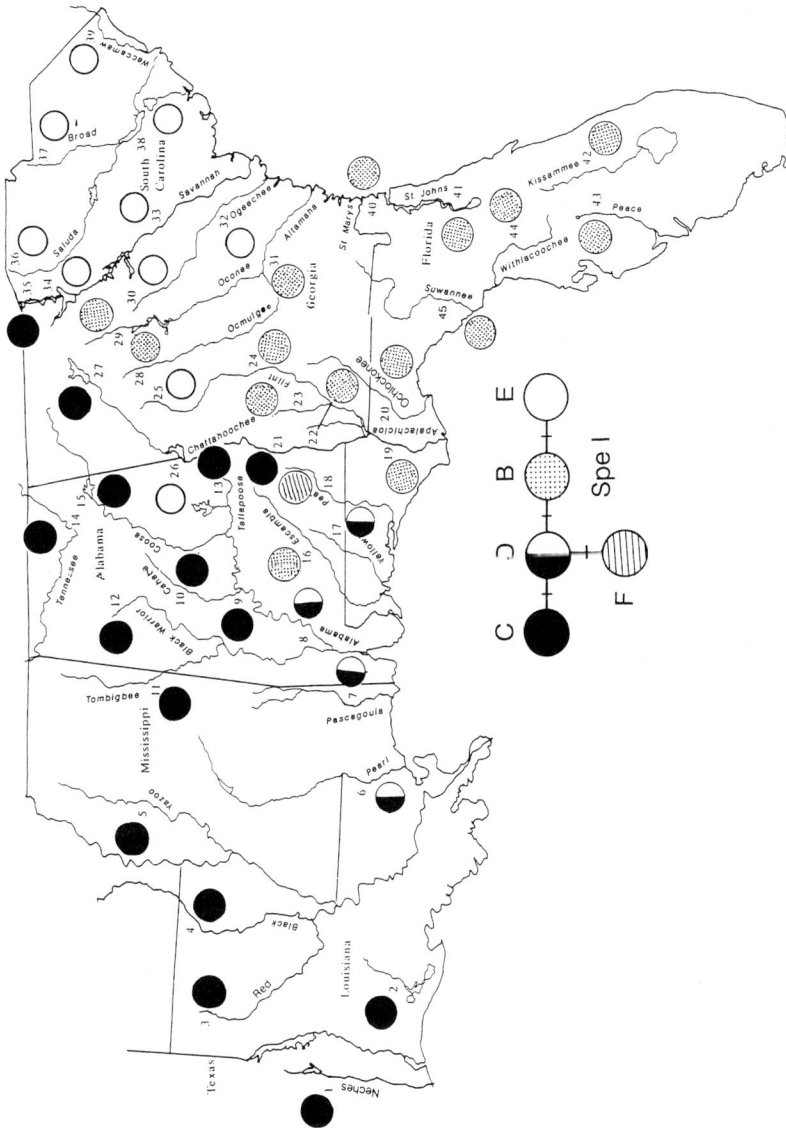

Figure 5.5.1 Geographic distributions of mtDNA haplotypes revealed by *Spe*I in mosquitofish in the southeastern USA. Numbers represent sampling locations.

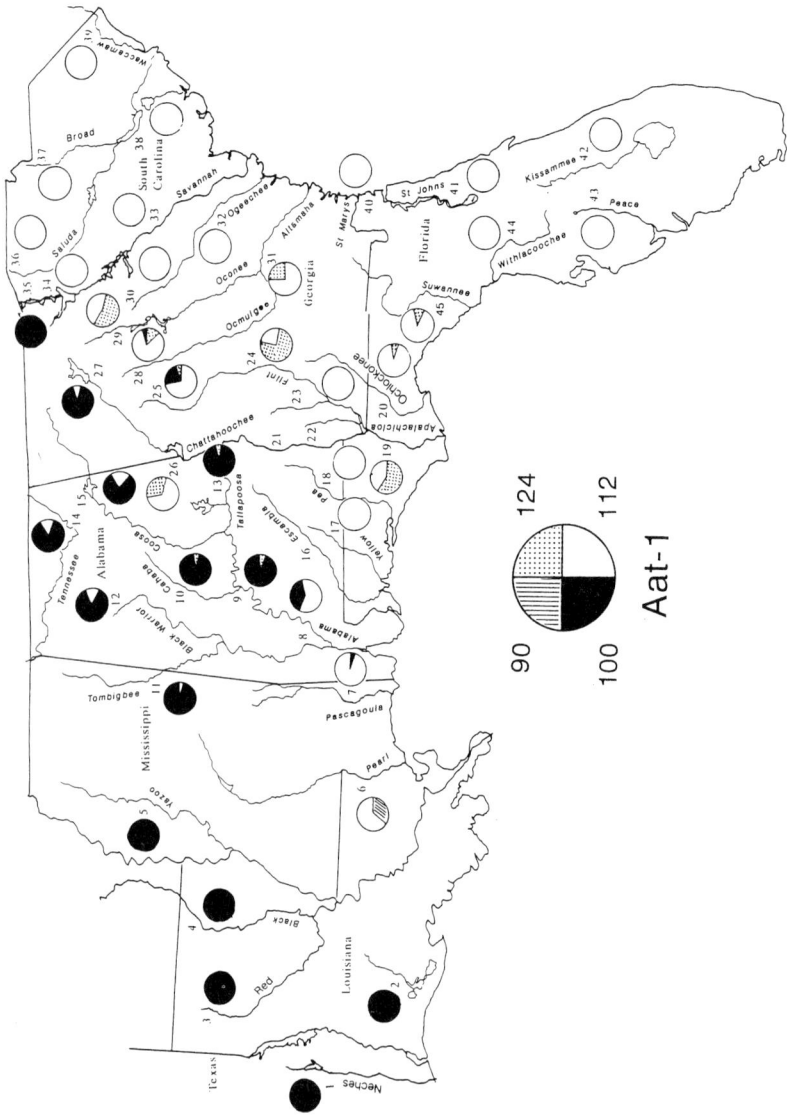

Figure 5.5.2 Geographic distributions of nuclear alleles at the locus encoding M-AAT-A in mosquitofish in the southeastern USA. Numbers represent sampling locations.

to the likelihood of differential cytoplasmic and nuclear gene flow through the hybrid region (Scribner and Avise, 1993).

(b)   Life-history traits in experimental crosses

Laboratory crosses revealed significant differences in gestation length and brood size among the adult females from parental, F_1 and backcross matings (Table 5.5.1). In particular, gestation lengths were slightly longer for females produced from HH and F_1 (HA) matings than for females from AH and AA matings (and backcrosses). Brood size was dramatically lower (and many progeny were born dead) in progeny from the HA × HH backcross.

Juvenile life-history traits also differed significantly among some experimental crosses (Figure 5.3.3 (a)). Female progeny from AA crosses were smaller at birth and at 15 days, grew at significantly slower rates, and matured at significantly smaller sizes and later dates than did female progeny from HH or F_1 crosses. Genotype-specific differences in fertility and viability may thus confer an advantage to *G. holbrooki* in mixed populations.

(c)   Population demography in experimental populations

Changes in population numbers in experimental pools of *G. affinis*, *G. holbrooki* and mixed samples over four sampling periods (Figure 5.5.3 (b)) can be divided roughly into three stages: stage A, with rapid growth (approximately to weeks 12–18); stage B, where carrying capacity is presumably reached; and stage C, involving overwinter mortality. No clear differences among treatments were observed in the intrinsic rates of population increase (stage A). However, carrying capacity was significantly lower for *G. affinis* populations (stage B), and when coupled with overwinter mortality led to significantly lower population sizes in *G. affinis* in the spring of the second year.

(d)   Temporal genetic changes in experimental populations

Changes in the cytonuclear composition of mixed populations were observed over the same time period (Table 5.5.2). From week 6 onwards, extensive hybridization is indicated by the high frequency of fish carrying AH nuclear genotypes. However, most hybrids contained mtDNA from *G. holbrooki*, suggesting a strong bias in the contribution by *G. holbrooki* females. Overall, the frequencies of *G. holbrooki* mtDNA and nuclear alleles increased dramatically over 52 weeks (Table 5.5.2), a result qualitatively consistent with findings from the life-history and population demography experiments in which *G. affinis* females appear to be at a competitive disadvantage.

Measures of gametic disequilibrium (D_{AB}) and cytonuclear disequilibrium (D) declined over the period of the study from the initial stocking conditions ($D_{AB} = D = 0.25$) to values of $D_{AB} = 0.11$ and $D = 0.04$ (Table 5.5.2). The rates of decay in both disequilibrium measures observed in replicate experimental populations were significantly different from expectations under random or assortative mating (Scribner and Avise, 1993), a result apparently reflecting the differential recruitment of offspring from parental and hybrid matings.

5.5.4   Discussion

The laboratory experiments on *Gambusia affinis* and *G. holbrooki* were conducted to examine life-history traits and demographic factors that might have a bearing on microevolutionary genetic processes operating within natural mosquitofish hybrid zones. Several potential agents of selection have been identified.

Table 5.5.1 Estimates of gestation length (days) and brood size for adult female mosquitofish resulting from various experimental crosses

| | Cross | | | | | |
| | Parental[a] | | F1[a] | | Backcross[a] | |
Trait	holbrooki (HH)	affinis (AA)	holbrooki × affinis (HA)	affinis × holbrooki (AH)	(HA) × (HH)	(AH) × (AA)
Sample size[b]	20	20	20	20	6	14
Gestation length	36.8 ± 1.4	34.8 ± 1.4	39.9 ± 1.5	32.8 ± 1.2	33.7 ± 4.3	34.9 ± 6.3
Number of offspring	24.4 ± 2.1	26.2 ± 2.9	24.6 ± 2.8	26.2 ± 3.0	6.2 ± 2.5[c]	29.9 ± 10.5

[a] The first letter in each cross corresponds to the female. Data and analytical procedures outlined in Scribner (1992, 1993).
[b] Number of broods from each cross used in the analysis.
[c] Mean based solely on live-born offspring.

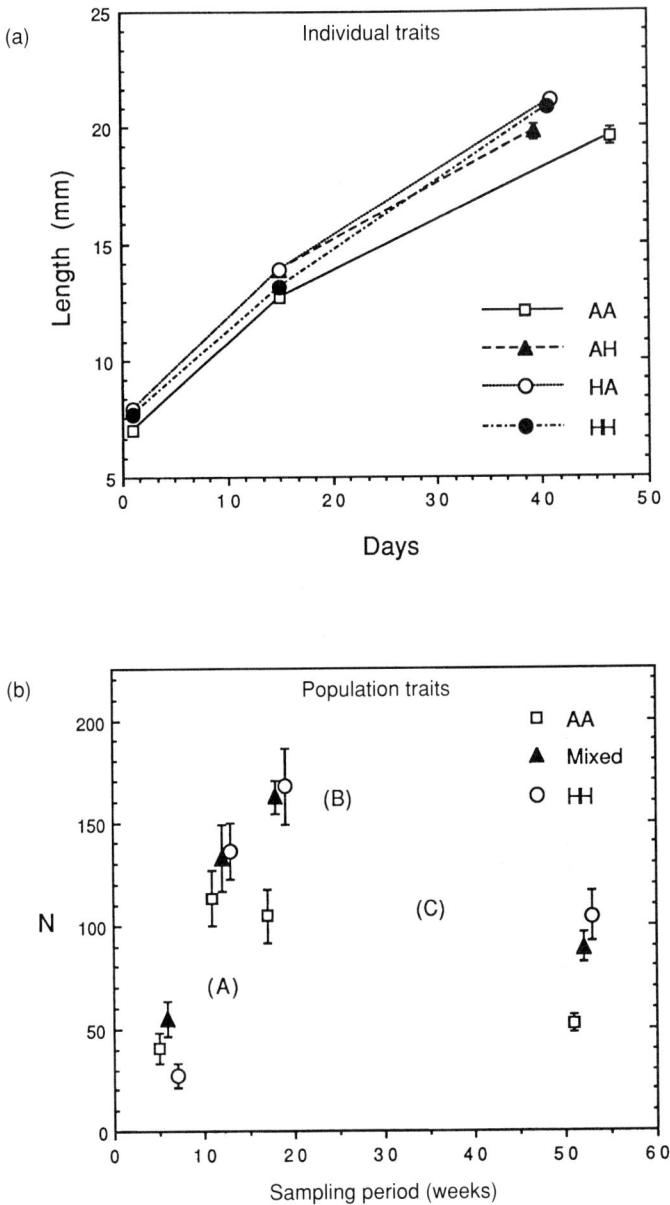

Figure 5.5.3 Individual life-history traits and population size trajectories in experimental populations of *Gambusia*. (a) Body lengths at various times since birth for progeny of crosses involving pure *G. affinis* (AA), pure *G. holbrooki* (HH) and the F₁ hybrids (AH and HA). (b) Sizes of experimental populations of pure *G. affinis* (AA) and pure *G. holbrooki* (HH), and mixture of the two species following their founding by $N = 12$ individuals each.

Table 5.5.2 Genetic characteristics of *Gambusia* from mixed experimental populations during sampling periods in 1990 and 1991

	Period (weeks)				
Variable[a]	0	6	12	18	52
Sample size	24[b]	152	193	221	135
% *holbrooki* alleles	0.500	0.654	0.651	0.648	0.813
% *holbrooki* mtDNA	0.500	0.788	0.824	0.872	0.850
% A/AA[c]	0.500	0.092	0.145	0.113	0.002
% A/AH[c]	0.000	0.026	0.052	0.018	0.126
% H/AH[c]	0.000	0.500	0.358	0.480	0.163
% H/HH[c]	0.500	0.382	0.445	0.389	0.689
D_{AB}[d]	0.250	0.180	0.174	0.156	0.106
D[e]	0.250	0.059	0.099	0.073	0.044

[a] Values for each variable represent means over two replicate experimental populations sampled during each period.
[b] Each experimental population was established with three virgin male and three virgin female adult *G. affinis* and *G. holbrooki*.
[c] Frequencies of each of four cytonuclear genotypes. The first single letter corresponds to the mtDNA cytotype (A = *affinis*; H = *holbrooki*) while the second two-letter code refers to the multilocus nuclear genotype (AA = parental *affinis*, HH = parental *holbrooki*, AH = F_1 or higher filial generation hybrid (Scribner and Avise, 1993).
[d] Weir's (1979) composite measure of gametic disequilibrium (Scribner and Avise, 1993)).
[e] Asmussen, Arnold and Avise (1987) measure of gametic cytonuclear disequilibrium (Scribner and Avise, 1993).

With respect to juvenile and adult female life-history (Table 5.5.1; Figure 5.5.3 (a)), large differences exist between progeny from parental, F_1 and backcross matings. In particular, gestation lengths were longer for females from HH and HA crosses than for females from AA or AH crosses, and these longer gestation lengths appear to translate into larger sizes at birth and at sexual maturity. Such size and growth rate differences may place *G. holbrooki* maternal lineages at an advantage in environments characterized by high juvenile mortality rates, high density and resource scarcity, conditions that probably characterize many natural populations as well. For example, the small size and slow growth of *G. affinis* individuals probably place them within the threshold of cannibalism and interspecific predation for longer periods of time.

With respect to population demography, large differences exist between pure *G. affinis*, pure *G. holbrooki* and mixed populations in recruitment, carrying capacity and overwinter mortality (Figure 5.5.3 (b)). The reduced carrying capacity and lower recruitment rates of *G. affinis* were probably the result of high juvenile mortality rates, because no evidence of lower fecundity was uncovered (Table 5.5.1; Scribner, 1993).

These differences between *G. holbrooki* and *G. affinis* in life-history and demography probably contribute to the dramatic temporal genetic changes observed in mixed experimental populations over 52 weeks (Table 5.5.2). Extensive hybridization occurs shortly after experimental contact between the two species, and F_1 and certain backcross progeny are fully fertile, although backcross (HA × HH) progeny do exhibit reduced viability (Table 5.5.1). Yet cytonuclear frequencies shift quickly. Though populations were established with small numbers of individuals ($N = 12$), the rapidity of genetic change and the consistency in the direction across population replicates argue that natural selection (rather than genetic drift) was the primary force responsible. The near fixation by *G. holbrooki* genotypes within 52 weeks of population founding probably results in large measure from the life-history and demographic differences between the species discussed above.

Nonetheless, the repeatable and directional changes in mtDNA composition in the experimental populations appear inconsistent with some other observed differences between

G. affinis and *G. holbrooki* in behavioral and life-history traits. First, female mate-choice experiments revealed that female *G. holbrooki* mate assortatively with regard to male genotype (*G. affinis* versus *G. holbrooki*), whereas female *G. affinis* mate randomly (Scribner, unpublished data). This suggest that, should introgression occur, hybrids should possess predominantly *G. affinis* mtDNA, a prediction clearly at odds with the experimental genetic results (Table 5.5.2). Second, reduced fertility *G. holbrooki* backcross females (Table 5.5.1) should inhibit introgression of *G. holbrooki* mtDNA, but this appears not to have happened, at least in the experimental populations. Perhaps the intensity of viability selection in the form of high juvenile mortality rates for *G. affinis* overrides the effects of reduced *G. holbrooki* backcross fecundity.

To date, molecular characterizations of aquatic species in the southeastern USA have shown that major genetic discontinuities exist in a general east to west orientation. These findings argue strongly for the importance of historical events in structuring populations of many otherwise independent species. Similar phylogeographic structuring is observed for *Gambusia*. However, population substructure within *G. affinis* and *G. holbrooki* (Figure 5.5.1) and clinal latitudinal changes in allele frequency within several drainages (Figure 5.5.2), provide evidence for recent or contemporary genetic contact in a central region of secondary hybridization and introgression. Differences in the geographic dispersion of mtDNA and nuclear markers is suggestive of differential introgression of female versus biparentally inherited genes, resulting either from sex-biased gene flow or asymmetry in the direction of selection. Presence of *G. affinis* nuclear alleles in predominantly *G. holbrooki* populations may represent the evolutionary footprints of a historically more widespread *G. affinis* distribution, and movements in hybrid zone position. Hybrid zone theory proposes that hybrid zones tend to diffuse from regions of high to low population density and from high to low fitness (Barton and Hewitt, 1985). The experimental results summarized here suggest that species-specific differences in life-histories and density-dependent population regulation can be important factors influencing the genetic architecture of a hybrid zone.

Acknowledgements

This work was supported by the Savannah River Ecology Laboratory under contract number DE-AC09-765R00819 between the US Department of Energy and the University of Georgia's Institute of Ecology and through NSF grants to J.C. Avise. We would like to thank G. Helfman for allowing us to use his laboratory facilities to rear fish. Gary Meffe and P. Leberg contributed many valuable suggestions during the initial phases of the experimental laboratory breeding and population studies. This research was conducted while K. T. Scribner was a graduate student in the Department of Zoology at the University of Georgia.

References

Asmussen, M., Arnold, J. and Avise, J.C. (1987) Definition and properties of disequilibrium statistics for associations between nuclear and cytoplasmic genotypes. *Genetics*, **115**, 755–68.

Avise, J.C. (1992) Molecular population structure and the biogeographic history of a regional fauna: a case history with lessons for conservation. *Oikos*, **62**, 62–76.

Avise, J.C., Arnold, J., Ball, R.M *et al.* (1987) Intraspecific phylogeography: the mitochondrial DNA bridge between population genetics and systematics. *Annu. Rev. Ecol. Syst.*, **18**, 489–522.

Barton, N.H. and Hewitt, G.M. (1985) Analysis of hybrid zones. *Annu. Rev. Ecol. Syst.*, **16**, 113–48.

Bermingham, E. and Avise, J.C. (1986) Molecular zoogeography of freshwater fishes in the southeastern United States. *Genetics*, **113**, 939–65.

Brown, W.M. (1980) Polymorphism in mitochondrial DNA of humans as revealed by restriction endonuclease analysis. *Proc. Natl. Acad. Sci. USA*, **77**, 3605–9.

Lansman, R.A., Shade, R.O., Shapiro, J.F. and Avise, J.C. (1981) The use of restriction endonucleases to measure mitochondrial DNA sequence relatedness in natural populations. III. Techniques and potential applications. *J. Mol. Evol.*, **17**, 214–26.

Maniatis, T., Fritsh, E.F. and Sambrook, J. (1982) *Molecular Cloning*, Cold Spring Harbor Laboratory, Cold Spring Harbor, NY.

McClenaghan, L.R. Jr, Smith, M.H. and Smith, M.W. (1985). Biochemical genetics of mosquitofish. IV. Changes in allele frequency through space and time. *Evolution*, **39**, 451–60.

Meecham, R.H. and Duncan, T. (1989) *Morphosys: Morphometrics Analysis Package, V. 1.26*. Exeter Publishing Co. Ltd, Setauket, NY.

Partridge, L. and Harvey, P.H. (1988) The ecological context of life history evolution. *Science*, **241**, 1449–55.

Reznick, D. (1981) 'Grandfather effects'. The genetics of interpopulation differences in offspring size in the mosquitofish. *Evolution*, **35**, 941–53.

Scribner, K.T. (1992) Molecular and demographic characterization of hybridization in natural and experimental populations of *Gambusia*. PhD dissertation, Univ. Georgia.

Scribner, K.T. (1993) Hybrid zone dynamics are influenced by genotype-specific variation in life history traits: experimental evidence from hybridizing *Gambusia* species. *Evolution*, **47**, 632–46.

Scribner, K.T. and Avise, J.C. (1993) Cytonuclear genetic architecture in mosquitofish populations and the possible roles of introgressive hybridization *Mol. Ecol.*, **3**, 139–49.

Scribner, K.T., Wooten, M.C., Smith, M.H., Kennedy, P. and Rhodes, O.E. (1992) Variation in life history and genetic traits of Hawaiian mosquitofish populations. *J. Evol. Biol.*, **5**, 267–88.

Stearns, S.C. (1983) A natural history experimental of life-history evolution: field data on the introduction of mosquitofish (*Gambusia affinis*) that shared ancestors in 1905. *Evolution*, **37**, 618–27.

Weir, B.S. (1979) Inferences about linkage disequilibrium. *Biometrics*, **35**, 235–54.

Weir, B.S. and Cockerham, C.C. (1989) Complete characterization of disequilibrium at two loci, in *Mathematical Evolutionary Theory* (ed. M.W. Feldman), Princeton University Press, Princeton, NJ, pp. 86–110.

Wooten, M.C. and Lydeard, C. (1990) Allozyme variation in a natural contact zone between *Gambusia affinis* and *Gambusia holbrooki*. *Biochem. Syst. Ecol.*, **18**, 169–73.

Wooten, M.C., Scribner, K.T. and Smith, M.H. (1988) Genetic variability and systematics of *Gambusia* in the southeastern United States. *Copeia*, **1988**, 283–9.

6

Genetics of aquatic clonal organisms

6.1   EVOLUTIONARY GENETICS OF AQUATIC CLONAL INVERTEBRATES:
CONCEPTS, PROBLEMS AND PROSPECTS

Gary R. Carvalho

Abstract

Aquatic animals that reproduce without the intervention of syngamy and recombination at some stage in their life-cycle (amictic), so-called 'clonal animals', are richly represented among animal phyla, including the Porifera, Cnidaria, Platyhelminthes, Rotifera, Annelida, Mollusca, Arthropoda, Bryozoa, Hemichordata and Vertebrata, though evolutionary genetic studies have so far concentrated on two major groups, the cladocerans and anthozoans. Despite the wide diversity in form, aquatic clonal animals may be classified by a few fundamental reproductive and morphological criteria: (1) whether development proceeds from gametes ('gametic' clonal animals, i.e. parthenogenetic) or unsegregated somatic tissues ('agametic' clonal animals, e.g. fission, budding, laceration); (2) whether they are solitary (e.g. cladocerans) or colonial (e.g. most bryozoans); and (3) whether sexual reproduction (mixis) occurs at some stage in the otherwise clonal life-cycle. The hierarchical components of population structure in clonal animals ('genets', 'ramets', 'modules', 'colonies') are defined, and their evolutionary significance in relation to demography and selection briefly considered. Several characteristics of cloning which exert evolutionary pressures distinct from those operating in unitary animals are discussed, including clone longevity, prolific reproductive rates, inheritance of intact genomes, and the dispersion of clones in time and space; such features render clonal animals especially suitable for investigating genotypic variance in ecologically measurable traits. Methods for discriminating clones, with emphasis on recently developed DNA markers, are reviewed, together with a standard approach for determining clonal diversity. The value of clonal animals in evolutionary studies is illustrated by considering three fundamental problems: the evolution of breeding systems, the relationship between reproduction and population structure, and the maintenance of genetic variation. Several noteworthy contributions have modified our views on the role of temporal variation in maintaining genetic variation, the effects of dispersal on population differentiation, the evolution of niche breadth, and the geographic distribution of amictic and bisexual forms. Prospects for future studies are

Genetics and Evolution of Aquatic Organisms. Edited by A.R. Beaumont.
Published in 1994 by Chapman & Hall, 2–6 Boundary Row, London SE1 8HN.
ISBN 0-412-49370-5

presented, with emphasis on models using empirical data, the in situ *manipulation and long-term monitoring of clonal dynamics, and the use of advanced molecular techniques.*

6.1.1 Introduction

(a) Evolutionary genetics and genotypic variability

Although evolutionary theory relies on a sound knowledge of how genotypes interact with their environment, the integration of population genetics with ecology has been slow. Examination of genotype fitness in ecologically realistic circumstances is crucial when predicting the relative ability of genotypes to gain representation in future generations. Although such data were initially scattered, findings which document the relationship between genotypic variation and ecological traits have begun to emerge (review: Travis and Mueller, 1989). Indeed, the degree to which genotypes differ in their response to environmental factors determines not only the nature and extent of adaptive responses, but also the rate of evolutionary change (Fisher, 1930; Figure 6.1.1). It follows, then, that among the primary aims of evolutionary genetics is the estimation of genotypic variation in fitness, together with an analysis of the factors determining its extent and dynamics in natural populations (Lomnicki, 1988).

This section considers how and why aquatic animals that incorporate an amictic phase, i.e. without the involvement of syngamy and recombination (e.g. fragmentation, fission, budding, apomictic parthenogenesis) in their life-cycle, i.e. 'clonal animals' (Hughes, 1989a), have provided such valuable material for examining the ecological significance of

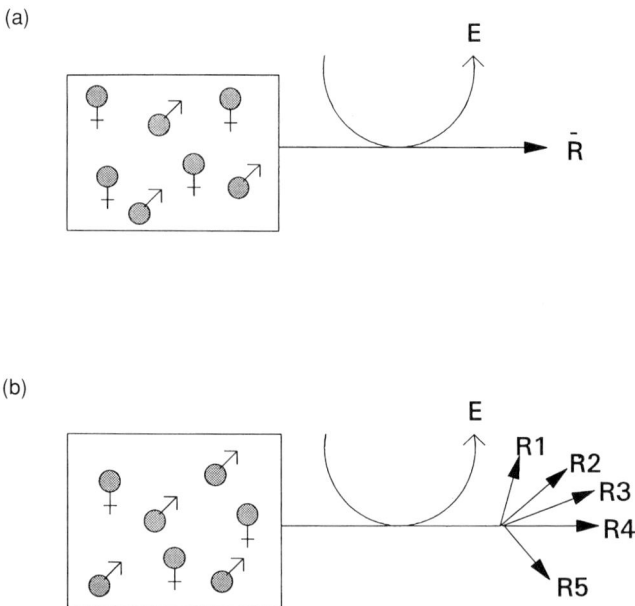

Figure 6.1.1 Contrasting models of genotypic variance in fitness showing genotypic response (R) to an environmental factor (E). In the classical population 'type' approach (a), the genotypic variance in response is averaged to produce a population mean. The intrapopulation approach (b) incorporates the genotypic range in response, producing several phenotypes (R1–R5). Each genotype will show phenotypic plasticity, which can be measured accurately using clonal lineages.

genotypic variation in fitness. At first sight it may seem that a group so diverse as animals ranging from solitary *Daphnia* and sea anemones to colonial ascidians and corals would defy any attempt to identify evolutionary trends. Naturally, there are fundamental differences between such divergent taxa, but the challenge here will be to identify and explain those features relevant to evolutionary genetics that arise from their clonal lifestyle. An exhaustive review of aquatic clonal animals is not provided; several accounts dealing with specific groups already exist (Jackson, 1985; Hebert, 1987a; Shick, 1991; Hughes, Ayre and Connell, 1992). My approach is to develop certain themes through examining some recent studies.

I focus on six main questions.

1. *What are the criteria for providing a functional classification of aquatic clonal animals?* Although aquatic clonal animals encompass a wide range of phyla, a few basic criteria can be used to organize their biological diversity.

2. *Why study clonal animals?* What is the justification for studying clonal animals as a functional group (Hughes, 1989a), both in terms of their distribution and abundance, and also in terms of their fundamental distinction from unitary organisms?

3. *What characteristics are peculiar to animals that clone?* Several life-history characters can be identified amongst clonal species that exert common evolutionary pressures.

4. *How can we detect and measure the genetic diversity of clones?* Clones and their products must be separable in natural populations using reliable methodology, and some standard approach for quantifying clonal diversity should be employed.

5. *How have aquatic clonal animals contributed to our understanding of certain fundamental evolutionary problems?* The value of clonal material for addressing evolutionary problems will be explored, including the evolution of breeding systems, the relationship between reproduction and population structure, and the maintenance of genetic variation.

6. *Where do we go from here?* What are some of the evolutionary problems of clonal animals that remain, and how can changes in experimental design and approach, as well as recent improvements in clonal discrimination, help?

(b)   The study of clonal animals: scope and definitions

The study and identification of clonal organisms as a fundamentally different and identifiable evolutionary group owes much to Harper (1977, 1981), though the study of animals that clone began much earlier with the discovery of parthenogenesis in aphids (Bonnet, 1745). Harper (1977) pointed out that most ecological and evolutionary theory was based on unitary, or aclonal, organisms, i.e. those species with sexually derived individuals, that exhibit a unitary morphology and whose life-cycle proceeds from birth to adult, with one or more bouts of mixis (genetic rearrangement through meiosis or syngamy, usually both), terminating in senescence and death of the individual (Figure 6.1.2). In many plants, however, and in representatives of two-thirds of the metazoan phyla, including sponges, anthozoans, cladocerans, molluscs, aphids and unisexual fish (Bell, 1982), the genotype develops into many iterated copies which may either expand indefinitely through amictic replication, or undergo sexual reproduction (mixis, Figure 6.1.2). It is critical at the outset to appreciate that most clonal animals at some stage in their life-cycle undergo mixis (Bell, 1982), and that the combination of amictic and sexual phases within a life-cycle necessitates the construction of modified models of resource allocation and life-history evolution (Caswell, 1985; Hughes, 1987, 1989a). A wide range of distinctive phenomena arising from the clonal habit, including modular growth (Coates and Jackson, 1985; Hughes, 1992), the rarity of senescence (Harper, 1977), the absence of a segregated germplasm in some species

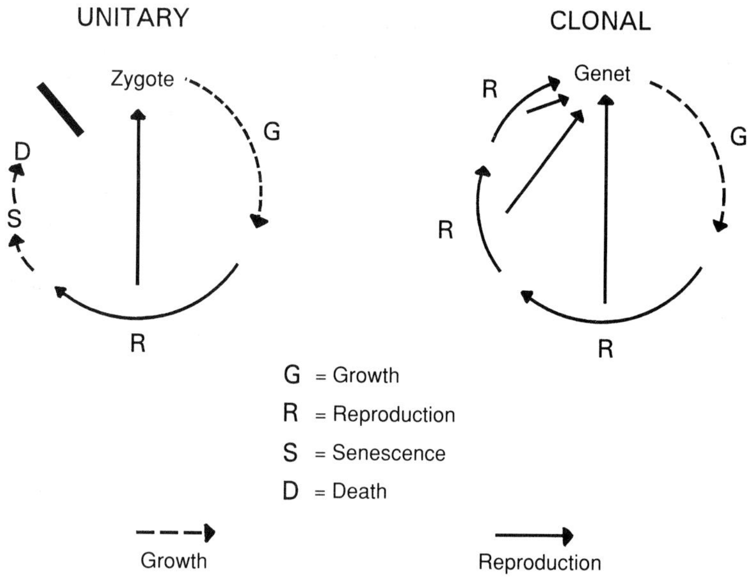

Figure 6.1.2 Schematic representation of unitary and clonal life-cycles. Unitary animals are characterized by senescence and death of individuals, whereas clonal organisms have potentially infinite reproductive potential through amictic iteration of the genome.

(Hughes and Cancino, 1985; Hughes, 1989a) and the effects of branching structure on capture of resources (Harper and Bell, 1979; Jackson, 1979; Harper, 1985), all set clonal organisms apart from those that are unitary. So fundamental and distinct are the consequences of clonality that Harper (1981) disclaimed any separate status for the population biology of plants and animals, but suggested a new division between clonal and unitary organisms. A number of recent reviews have applied and developed the ideas of Harper to clonal animals (Larwood and Rosen, 1979; Jackson, 1985; Hebert, 1987b; Hughes, 1989a; Shick, 1991; Hughes, 1992).

Unfortunately, such are the diversity of approaches and range of phyla with clonal lifestyles that the terminology used to delimit their various levels of biological organization has become confused (Vuorisalo and Tuomi, 1986; Pearse, Pearse and Newberry, 1989). It is critical that the various terms are used unambiguously because different units of clonal organization may be exposed to contrasting evolutionary pressures, and so contribute differentially to genotype fitness (Tuomi and Vuorisalo, 1989).

The term 'asexual' should be used with caution, since some workers (e.g. Bell, 1982) view parthenogenesis as an aberrant form of sexual reproduction because it involves development from gametes (eggs). In this section, the term 'amictic' is preferred, which defines any form of reproduction lacking syngamy and recombination, and which thus results in the production of genetically identical lineages. Furthermore, the genetic consequences of parthenogenesis, which is more usually regarded as a form of 'asexual' reproduction (Hughes, 1989a), depend on the precise mode of egg development, and may not always lead to production of identical offspring. Parthenogenesis involves development from eggs without fertilization from another individual, and can be either apomictic, where both meiosis and syngamy are absent, or automictic, where fusion between gametes or gametic pronuclei from the same individual can generate variants (Bell, 1982). Coverage of parthenogenesis in this section is exclusively concerned with apomictic development.

A clone is an assemblage of genetically identical individuals that can function and survive on their own (Hughes, 1989a); 'cloning' is the process of multiplication, resulting in the production of genetically identical offspring. Thus, members of a clone are genetically identical, although new clones may eventually arise through automixis (Bell, 1982; Browne and Hoopes, 1990), mixis (Hebert, 1974a) or the accumulation of mutations (Lynch, 1985). Each iterated genotype arising through clonal replication is termed a genet (Harper, 1977) and is considered a unit of selection. Thus a multiclonal population of solitary *Daphnia*, sea anemones or hydras will comprise a population hierarchy at two levels: first, the number of distinct genets (= clones), and second, the number of individuals belonging to each genet (= 'ramets') (Kays and Harper, 1974; Figure 6.1.3 (b)(i)). Genet fitness depends markedly on

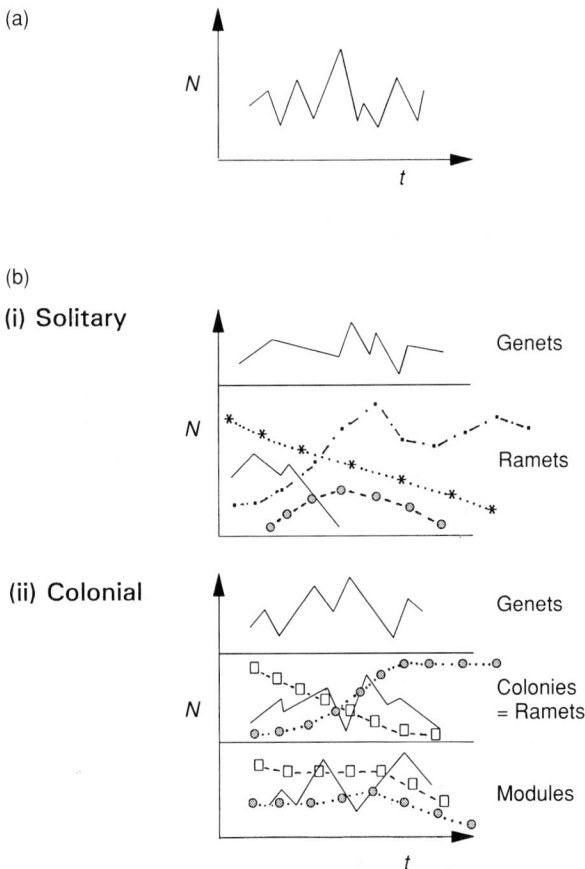

Figure 6.1.3 Population hierarchy in aquatic clonal animals. Populations of unitary animals (a) comprise a single level of organization where the number of individuals is usually equal to the number of genotypes. Clonal animals (b) have either two or three levels of population hierarchy, depending on whether ramets are detached (solitary) or remain connected (colonial) after replication. Fluctuations in the abundance of the hypothetical modules, ramets and genets shown here demonstrate that changes can be independent of each other, and that demographic studies should incorporate data at each level.

the survivorship and fecundity of individual ramets, which in turn determine their proportional representation in future generations, via either mixis or obligate cloning.

Colonial animals (Jackson, 1985) such as sponges, soft corals, bryozoans and ascidians produce units that remain attached after replication to form colonies, i.e. assemblages that are organically and functionally integrated. The units of colonial architecture, or 'modules' (Chapman and Stebbing, 1980), are repeated, multicellular units of development, which jointly make up larger physically coherent units capable of resource capture and reproduction, and which themselves are not entirely subdivisible into iterated units (Tuomi and Vuorisalo, 1989). Animal modules (Hughes, 1989a) include the aquiferous modules of sponges, polyps of cnidarians and zooids of bryozoans, and thus exclude those solitary clonal animals that separate after replication (cf. Harper, 1977).

Hughes (1989a) suggests reserving the term 'modular animals' for only those species that produce colonies, thus emphasizing the fundamental evolutionary division between physically separated and connected iterated units. Thus, colonial animals such as corals have three levels of organization (Figure 6.1.3(b)(ii)): firstly, the modules, or polyps, which are morphological individuals with their own mouth, gastric cavity or gonads; second, clumps of connected polyps which make up the colonies, often fragmented in space (each colony is a ramet) and which typically have an integrated physiology specialized for feeding, reproduction or defence; and third, the genetic individuals, or genets, which comprise polyps and ramets that are derived from the same zygote.

Hierarchical organization in colonial animals results in replication at three distinct levels: polyp replication, colony replication and genet replication. Such complexity in form has led to confusion regarding the concepts of growth and reproduction in clonal organisms (Harper, 1977, 1985; Jackson, 1985; Pearse, Pearse and Newberry, 1989). The argument centres on whether an increase in the number of ramets, without any corresponding increase in the number of genets (Figure 6.1.3), should be viewed as reproduction or merely growth. Effects on fitness are most clearly appreciated if some distinction is made between solitary and colonial strategies, such that three types of proliferation, namely 'somatic growth', 'modular iteration' and 'reproduction', are defined. Somatic growth is any increase in the size or mass of an individual, i.e. a physically recognizable organism, whether it be a polyp or individual *Daphnia*. Modular iteration is a term reserved for colonial animals only, and involves an increase in the number of physically connected polyps, i.e. colony size. Reproduction, which may be amictic or sexual, involves an increase in the number of ramets, be it the separation of clumps or polyps in a coral through fragmentation, or the production of a parthenogenetic brood in an ostracod. Somatic growth relates mainly to an increase in the mass of an individual unit, whereas modular iteration and reproduction enhances the fecundity of a genet.

(c)   Aquatic clonal animals: a functional classification

A wide range of clonal animals inhabit freshwater, estuarine and marine environments (Table 6.1.1), though the genetic information available for each group differs markedly (Hughes, 1989a). By far the two most thoroughly studied from an evolutionary perspective are cladocerans in fresh waters (Hebert, 1987a,b; Mort, 1991) and sea anemones in the marine environment (Shick, 1991), with a more scattered analysis of rotifers (King, 1972, 1980; Snell and Hawkinson, 1983; Zhao and King, 1989), ostracods (Havel and Hebert, 1989; Chaplin and Ayre, 1989; Havel, Hebert and Delorme, 1990a,b), colonial ascidians (Sabbadin, 1979; Grosberg, 1987; 1988a,b), bryozoans (Jackson, 1985; Hughes, 1989b, 1992) and corals (Hughes, Ayre and Connell, 1992). The major reasons for a concentration of research effort on cladocerans and anemones is due both to their display of diverse

Table 6.1.1 Major aquatic animal groups that clone (modified after Hughes, 1989a)

Phylum	Habitat	Clonal reproduction	Clonal category (A, G, S, C)
Porifera	M/FW	Gemmulation, budding	A, C
Placozoa	M	Fission, budding	A, S
Cnidaria			
Hydrozoa	M/FW	Budding, fragmentation	A, S, C
Scyphozoa	M	Fission, laceration, budding	A, C
Anthozoa[a]	M	Fragmentation, laceration, fission, budding, parthenogenesis	A, G, S, C
Ctenophora	M	Laceration	A, S
Platyhelminthes			
Turbellaria	M/FW	Fission, laceration, parthenogenesis	A, G, S
Trematoda	E	Parthenogenesis/polyembryony?	A, G?, S
Cestoda	E	Budding (larvae)	A, S
Nemertinea	M/FW	Fission	A, S
Gastrotricha	M/FW	Parthenogenesis	G, S
Rotifera[a]	M/FW	Parthenogenesis	G, S
Nematoda	M/FW	Parthenogenesis	G, S
Annelida			
Polychaeta	M/FW	Fission, parthenogenesis	A, G, S
Oligochaeta	M/FW	Fission, parthenogenesis, polyembryony	A, G, S
Mollusca			
Gastropoda[a]	M/FW	Parthenogenesis	G, S
Bivalvia	M	Parthenogenesis (autogynogen)[b]	G, S
Arthropoda			
Arachnida	FW	Parthenogenesis	G, S
Crustacea[a]	M/FW	Parthenogenesis	G, S
Insecta	FW	Parthenogenesis	G, S
Pogonophora	M	Fission	A, S
Sipuncula	M	Fission	A, S
Tardigrada	FW	Parthenogenesis	G, S
Phoronida	M	Fission, budding	A, C
Bryozoa[a]	M/FW	Fragmentation, encapsulation	A, S (rare), C
Entoprocta	M	Budding	A, S
Echinodermata			
Asteroidea	M	Fission, autotomy parthenogenesis	A, G, S
Ophiuroidea	M	Fission	A, S
Holothuroidea	M	Fission	A, S
Hemichordata			
Enteropneusta	M	Fission	A, S
Pterobranchia	M	Fragmentation	A, C
Chordata			
Urochordata[a]	M	Budding, fragmentation	A, C, S
Vertebrata			
Pisces[a]	M/FW	Parthenogenesis	G, S
Amphibia[a]	FW	Parthenogenesis	G, S

M = marine; FW = fresh water; E = endoparasitic; A = agametic; G = gametic; S = solitary; C = colonial.
[a] Major groups with population genetic data.
[b] See section 1.1.

cloning mechanisms (Hughes, 1989a), and the comprehensive and detailed ecological data that are available on natural populations (reviews: Hebert, 1978; Peters and de Bernardi, 1987; Shick, 1991). Most empirical tests have been conducted on cladocerans because of their short generation times, and ease of experimental manipulation in the field and

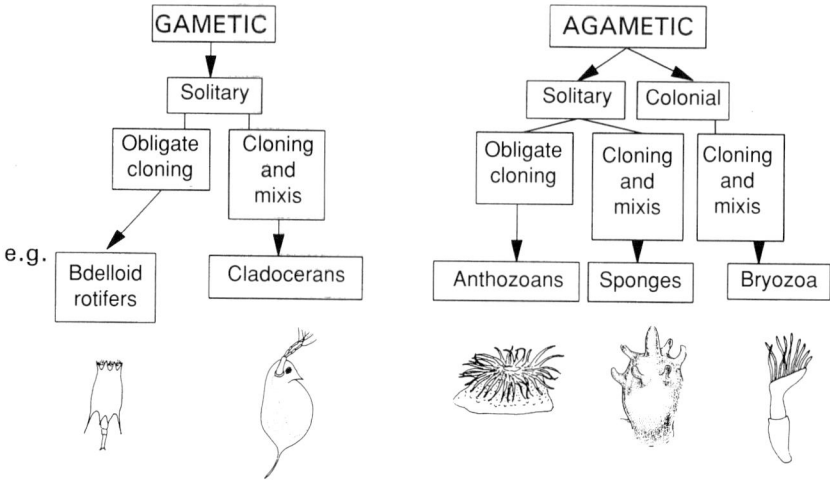

Figure 6.1.4 Functional classification of aquatic clonal animals. Species may be classified according to whether or not development proceeds from gametes, whether units remain attached or are separated after replication, and whether or not sexual reproduction occurs in the life-cycle. The animal groups identified may contain species with a range of breeding systems.

laboratory (Mort, 1991). Cladocerans are also particularly easy to collect in the wild. The array of cloning mechanisms and life-histories among aquatic clonal animals are summarized, together with detailed examples, by Hughes (1989a).

Aquatic clonal animals may be classified according to certain common morphological and reproductive criteria (Figure 6.1.4). First, reproduction may involve either gametes, so-called 'gametic' cloners (apomictic parthenogenesis), or propagules derived from the soma, the 'agametic' cloners (fragmentation, laceration, fission, budding, polyembryony) with no involvement of a germ-line (Hughes, 1989a). The major influence of a segregated germ-line in gametic cloners is to ensure continuity of gametic mutations between generations, provided that the mutation-bearing individuals reproduce. It is indeed the separation of clonal animals into gametic and agametic forms that Hughes (1989a) adopts as the primary functional division.

Second, a fundamental division occurs between those species with solitary ramets (e.g. *Daphnia*) and those which are interconnected, or colonial (e.g. some ascidians). Typically, detached solitary ramets are mobile and may actively swim or drift over considerable distances, whereas colonial species are usually sessile, even though their reproductive propagules may disperse (Hughes and Cancino, 1985). The division between solitary and colonial forms will affect profoundly the degree of dispersal and gene flow, as well as the nature of ecological interactions among ramets, and the risk of genet mortality. Hughes and Cancino (1985) discuss the life-history correlates of mobile and sedentary clonal habits.

Finally, clonal life-histories may be entirely amictic, as in obligate cloners, or exhibit bouts of mictic reproduction, either with a frequency that is genetically fixed or, more commonly, on a facultative basis (Hughes, 1989a). The periodic input of mixis will affect not only levels of clonal diversity within populations, but also evolutionary rates and the extent of ecological differentiation among clones.

6.1.2 Why study clonal animals?

(a) Value of cloning in evolutionary studies

The extensive use of clonal animals in evolutionary studies (Hughes, 1989a) exemplifies their merit as valuable investigative tools of evolution. The advantages to a geneticist of using a clonal organism stem from its ability to produce genetic replicas. In a population of conventional unitary organisms where every individual is genetically unique, except monozygotic twins, heterogeneity within a genotypic class greatly complicates estimation of genotype fitness due to complex genotype–environment interactions (Browne *et al.*, 1983; Fletcher, 1984). The ability to monitor the dynamics and performance of a specific genotype in a range of natural (e.g. Grosberg, 1988a) and experimental (c.g. Carvalho, 1987) selective regimes facilitates the estimation of genotype–environment interactions, allowing measures of phenotypic plasticity within genets (Stearns, 1989). Since it is impossible to measure the survivorship and reproductive value of an individual unitary organism, a cohort of individuals must be examined, so imparting an associated variance in such factors as survival, brood size and growth rate. Ramets of a single genet, on the other hand, allow a precise partitioning of environmental and genetic variance in fitness (Lynch, 1984a; Carvalho, 1987; Grosberg, 1987; Hughes, 1992), as well as determining within-genotype values for probabilistic traits (e.g. l_x, m_x and r) (Grosberg, 1988a). As such, clonal propagation provides the only opportunity for measuring the reproductive value of specific genotypes (Hughes and Cancino, 1985).

Phases of mixis occurring in many clonal animals afford the opportunity for investigating the genetic basis of clonal differences in life-histories (Lynch, 1984a). Moreover, clonal performance can be compared in different environments (Pace, Porter and Feig, 1984; Shick and Dowse, 1985; Weider and Lampert, 1985; Weider, 1987; Rossi and Menozzi, 1990; Hughes, 1992), so simplifying empirical tests on the maintenance of genetic variation.

Detecting the role of selection in wild populations of unitary organisms is confounded by the fact that for polymorphism to exist at all, directional selection is often so weak that gene frequency changes in samples of tractable size are slight (Endler, 1986). In those clonal species that combine mixis with amictic reproduction, the effects of selection are magnified relative to recombination (Young, 1979; Brookfield, 1984), so increasing the chances of detecting selective effects. During sexual reproduction, the additive component of the variance in fitness can be exploited, but during amictic replication, the dominance and interaction components can be utilized as well, ensuring the cumulative effect of any fitness advantage. This, taken together with the usually greater longevity of a genet (Jackson, 1985) and rapid rate of proliferation (Allan, 1976), can drive rapid, and thus detectable, changes in genotype frequency. Carvalho (1988), for example, studying a permanent population of *Daphnia magna*, documented an increase in the frequency of an allozyme-marked genotype from 0 to 0.56 in just 4 months.

Although clonal animals undoubtedly provide valuable material for studies on selection in the wild, they also have drawbacks when it comes to examining how the bulk of genetic variation is maintained. While it is undisputed that a portion is maintained by selective forces, assessing the relative incidence of neutral and selective variants in the genome is difficult even in bisexual populations, let alone clonal animals. In the latter, the unit of selection is the entire genome, and it is this which is hazarded intact during amictic propagation to future generations. The complete genome undoubtedly includes both neutral and selectively significant variants, thereby making it extremely difficult to identify locus-specific effects. Despite this limitation of using clonal material, observations on fluctuations in clonal genotypes over time or space can still provide valuable indicators of the biological and environmental constraints that operate to maintain genetic diversity in natural populations.

Where mixis and cloning co-occur within the same life-cycle, as in many rotifers (King, 1980), cladocerans (Hebert, 1987b), anemones (Shick 1991; Section 6.5) and marine colonial invertebrates (Jackson, 1985), it is possible to explore the evolutionary consequences of different reproductive modes. Furthermore, within a taxon, species may exhibit both obligate and cyclic cloning (Hebert, 1987a), thus facilitating studies on the evolutionary origins of different breeding systems. Where a species inhabits a diverse array of habitats, as typified by *Daphnia* (Hebert, 1978), a rare chance for investigating the relationship between reproduction, environment and population structure is also offered.

Unlike unitary organisms undergoing regular mixis where genotypes are disrupted and reassembled each generation, clonal populations with extended bouts of amictic replication may accumulate ecological divergence more readily due to prolonged selection at the level of the genome (Harper, 1977; Templeton, 1979). Since clonal selection may enhance evolution-ary adaptation to local environmental conditions (Bell, 1982), the environmental and genetic conditions favouring ecological divergence may be readily explored. Clones may also be used as model systems to examine the evolutionary mechanisms underlying species assemblage patterns (Wilson and Hebert, 1992).

(b) Evolution and the clonal habit

What are the characteristic evolutionary features of clonal animals compared to unitary?

Genet longevity

Longevity of a genet is typically much greater than for a unitary individual due to the theoretically infinite production of ramets, at least until mixis intervenes. Delayed senescence derives from the continued production of genetically identical young from either germinal cells (gametic cloners) or from mixtures of differentiated tissues and undifferentiated generative cells (agametic cloners). Clonal persistence exposes identical genotypes to a wide variety of temporal and spatial patchiness, which may occasionally facilitate local adaptation, depending upon input from sex and the intensity of selection pressures (King, 1980; Carvalho, 1984; Ayre, 1985; Rossi and Menozzi, 1990, 1992). Fitness differentials will be magnified as indicated by marked linkage disequilibria (Hebert 1974b; Black and Johnson, 1979; Ayre 1983; Shaw, Beardmore and Ryland, 1987; Ayre and Willis, 1988); a slight reduction in fitness in each clonal generation will, if continued indefinitely, prove lethal. Genet immortality will usually, however, be prevented by the onset of sexual reproduction (Banta, 1939), environmental harshness (Shick and Lamb, 1977) or accumulation of deleterious mutations (Muller, 1964).

Extended longevity, and an associated large genet size, especially where mixis is rare (e.g. some corals), may provide variants for within-genet (i.e. zygote to zygote) or 'somatic' selection (Buss, 1987; Hughes, Ayre and Connell, 1992). The evolutionary role of somatic mutations does, however, depend markedly on the mode of inheritance, which differs fundamentally between gametic and agametic cloners (Hughes, 1989a).

The scale of longevity differs markedly between gametic and agametic cloners; a parthenogenetic cladoceran or rotifer will exhibit the same lifespan as its unitary counterpart, whereas many sponges and corals commonly live for one to several centuries (Hughes and Jackson, 1980). In contrast to parthenogenetic ramets, each with determinate growth and senescence, fissiparous polyps, for example, may continue to divide indefinitely.

Reproductive rates

Reproductive rates in gametic cloners are typically high (Allan, 1976) due to the two-fold advantage of parthenogenesis (Maynard Smith, 1978). Comparisons are more complex when

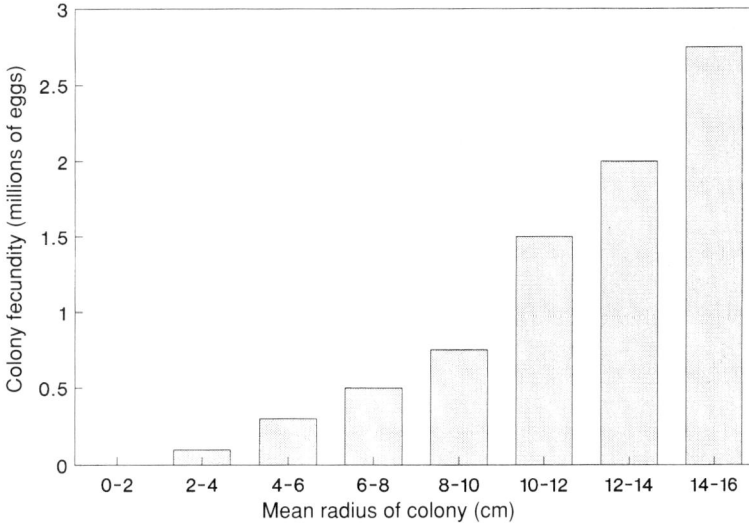

Figure 6.1.5 Skewed fecundity distribution in the braincoral, *Goniastrea aspera*. Note the large output by a few larger colonies which contribute disproportionately to the total output of the population. (Redrawn from Babcock, 1984.)

reproduction and form of offspring differ widely, e.g. between egg laying (mictic) and fission (amictic) in triclads (Calow, Beveridge and Sibly, 1979). A high population growth rate may not only produce rapid changes in genotype frequency, thus increasing the chances of gene fixation, but may also increase genet fitness through enhanced fecundity (Hughes and Cancino, 1985). Further, the frequency distribution of genets may become log-normal, with a few individuals contributing disproportionately to the gene pool (Harper and Bell, 1979; Babcock, 1984; Figure 6.1.5). Babcock (1984) calculated that 25% of the annual egg production of the coral *Goniastrea aspera* was generated by colonies constituting only 3% of the population. Such skewed distributions may increase rates of evolutionary change under selection pressure (Levin, 1977).

Effective population size
Clonal life-histories where cloning and mixis are combined may result in an effective population size, N_e (an idealized measure of the number of genetic individuals that contribute to succeeding generations and the evenness of their contribution) being several orders of magnitude smaller than the census population size (the number of ramets counted) due to such factors as skewed sex ratios and genet fecundities (Hebert, 1978; Stoddart, 1984), overlapping generations (Hughes and Connell, 1987), self-fertilization and inbreeding (Grosberg, 1987), and periodic catastrophes (Hughes, Ayre and Connell, 1992). The net effect of a small N_e will be to increase the chances of founder effect and genetic drift, producing locally divergent subpopulations, though such effects will depend on the relative contributions of sexual and amictic recruitment (Hebert, 1974a,b; Ayre, 1984). Significant effects of inbreeding may arise more commonly in colonial forms, where dispersal is often apparently low (Jackson, 1985).

It is worthwhile noting that genetic drift in clonal animals could produce genetic structures similar to those attributed to selection (linkage disequilibria, heterozygote excesses, patchy clonal distributions), and that detection of such phenomena should not always be attributed to deterministic forces. In many natural populations of clonal animals (e.g. cladocerans), however, unless there has been some recent or persistent population bottleneck, drift is unlikely to be a key force in determining genetic structure due to large population sizes, as often indicated by correlations of fecundity with clone type (Hebert, 1974a,b; Carvalho, 1988; Hughes, 1992).

Heterozygosity and fitness

An association between overall heterozygosity and genet fitness (Hebert, Ferrari and Crease, 1982) will be perpetuated clonally, sometimes producing heterozygote excesses, especially where mixis is rare (Hebert, 1974b; Section 6.5).

Clonal reproduction and adaptation

A phase of cloning avoids disruption of coadapted gene complexes, which, when combined with periodic mixis, may facilitate response to directional selection (Templeton, 1979), and thus favour local adaptation. The extended longevity of some genets, together with their likely accumulation of heterozygosity, may further increase response to selection at mixis through the release of hidden genetic variance (Lynch and Gabriel, 1983).

Exclusively clonal taxa may suffer from the reduced opportunities to form coadapted complexes (Vrijenhoek, 1984), especially since the relationship between rates of environmental change and evolution may determine the incidence of localized adaptation (Bradley, 1981). Nevertheless, abundant evidence exists for the adaptive nature of phenotypic variation in many obligate cloners (Hughes, 1989a).

Risks of mortality

Separation of genets into many spatially or temporally dispersed ramets greatly decreases the risk of genet mortality, so extending their lifespan. A predator or catastrophe is unlikely to decimate a genet, unless there is density-independent mass mortality (Hughes, Ayre and Connell, 1992). Risks of genet mortality will, however, differ between solitary and colonial cloners (Hughes, 1989a), since aggregation of modules will render genets more vulnerable to localized mortality. Indeed, Jackson (1985) concludes that the risk of genet mortality is the primary selective force shaping the geometry and integration of ramets in aquatic colonial animals.

In summary, evolutionary change in clonal animals is dominated by the consequences arising from rapid rates of amictic proliferation and the extended longevity of genets. The associated skewed fecundity distributions and reduced genet mortality may promote localized adaptation, especially where extended bouts of clonal reproduction precede mixis. Furthermore, the organization of genetic variation into iterated units that persist across a range of heterogeneous environments provides ample opportunity for empirical investigation of genotypic variance in ecologically measurable traits.

6.1.3 Detection and measurement of clonal diversity

(a) How to distinguish clones

Most clones within a species are inseparable on the basis of morphology, even though there may be polymorphism within clones, especially in aquatic colonial animals (Jackson, 1985). A major challenge has been to develop a widely applicable method for separating the products of distinct genets. Various techniques are now available, including the use of molecular (e.g. Loaring and Hebert, 1981; Carvalho and Crisp, 1987; Ayre and Willis, 1988; Rossi and Menozzi, 1990; Carvalho *et al.*, 1991), behavioural (e.g. Brace, 1981; Ayre, 1983) and immunologically based (e.g. Ivker, 1972; Niegel and Avise, 1983; Heyward and Stoddart, 1985; Grosberg, 1988b) characters.

Allozyme electrophoresis is the technique most commonly employed, where screening of several polymorphic loci within an individual enables the grouping of animals into composite genotypes referred to as 'electrophoretic clones' (Carvalho and Crisp, 1987). Each electrophoretic clone is thus characterized by a unique array of alleles at the loci studied, though the actual number of genets may be greater due to undetected differentiation at other loci. Electrophoretically marked multilocus genotypes have identified ecologically meaningful assemblages whose frequency changes in natural populations have been related to specific selective factors (review: Hughes, 1989a).

Recently, a new source of highly variable DNA markers has been discovered scattered throughout the genome of many invertebrate and vertebrate groups (Jeffreys, Wilson and Thein, 1985; Burke *et al.*, 1991). The technique of multilocus DNA fingerprinting exploits diversity at minisatellite regions which comprise short (4–40 base pair (bp)) tandem repeat sequences differing in the number of repetitions and giving rise to extensive polymorphism at many loci. The resultant complex multibanded profiles are usually individual-specific and thus highly sensitive for genotypic discrimination. In clonal organisms, genetic fingerprinting has already revealed hidden genetic heterogeneity within electrophoretic clones (Turner, Elder and Davis, 1990; Carvalho *et al.*, 1991; Black *et al.*, 1992), demonstrating its high resolving power for separating genets. Brookfield (1992) examined the ability of DNA fingerprinting to discriminate clones based on the expected band-sharing values, and concluded that it was effective in clonal discrimination provided the frequency of sexual reproduction is considerably greater than the minisatellite mutation rate.

(b)   Measures of clonal diversity

Two broad types of measure are available to describe clonal population structure: clonal richness (N), the number of detectable genets in a sample, or clonal diversity, which is a combination of two independent characteristics, namely clonal richness (N) and clonal evenness, the proportionate representation of clones amongst the individuals of the population (p_i). Although richness is related to diversity, the latter provides a more meaningful indicator of clonal dynamics since it incorporates variance in genet size.

A range of diversity measures has been used to study clones, including Simpson's index, D (Parker, 1979), the Shannon–Weiner function, H (Carvalho and Crisp, 1987), and the genotype diversity ratio, GDR (Ward Tollit and Bickerton, 1991), but the most widely employed statistic is the genotypic diversity measure, G_o (Stoddart, 1983; Stoddart and Taylor, 1988), which may be quantified as:

$$G_o = 1 / \sum_{i=1}^{k} g_i^2$$

where: g_i is the relative frequency of the ith multilocus genotype, and k is the number of genotypes. G_o will vary from a minimum of 1, where there is only a single genotype, to a maximum of k, when genotypes are evenly distributed. Since multilocus genotypes are inherited intact, some possible combinations will not be expressed, causing uneven representation in the population and deviations from Hardy–Weinberg expectations and linkage equilibria. It may therefore be useful to calculate a Hardy–Weinberg expected value, G_e^* (Stoddart, 1983) which effectively pools all genotypes that have an expectation of occurring in less than one individual per sample, and produces a sum of their contributions as a number of unique genotypes.

Deviations of the ratio G_o/G_e^* from unity can be used as a qualitative index of the combined effects of departures from Hardy–Weinberg equilibria at single loci, and multiple-locus linkage disequilibria (Black and Johnson, 1979; Ayre, 1984). For example, a genetic-ally variable population with high levels of amictic recruitment should display a low ratio of G_o/G_e^* (Ayre and Willis, 1988; Chaplin and Ayre, 1989).

Ward, Tollit and Bickerton (1991) suggested an alternative approach, which, although similar in principle to the above, removes the need to pool rare genotypes by using Monte Carlo techniques. The GDR is obtained by dividing the observed number of electrophoretic clones by the expected mean number at Hardy–Weinberg equilibrium in a sample of defined size and allele frequencies. GDR should be close to 1 in a sexually reproducing population, but will be considerably less than 1 in an asexual population.

6.1.4 Aquatic clonal animals and evolutionary problems

Evolutionary studies on aquatic clonal animals encompass many themes, including the evolution of breeding systems (e.g. Hebert, 1987a), the genetic consequences of alternative reproductive modes (e.g. Shick, 1991), the evolution of life-histories (e.g. Caswell, 1985), the ecological significance of polyploidy (e.g. Zhang and Lefcort, 1991), the maintenance of genetic variation (e.g. Grosberg, 1988a), and the evolutionary dynamics of interspecific hybridization (e.g. Wolf, 1987; Weider and Wolf, 1991). Here, I will focus only on three topics which serve to illustrate the evolutionary value of clonal material and associated evolutionary trends.

(a) The evolution of breeding systems

Empirical tests for the evolution of sex
The diversity of breeding systems in aquatic clonal animals, taken together with the evolutionary characteristics of the clonal habit, have ensured that such animals have played a major role in discussions of the relative merits of amictic and sexual reproduction (Williams, 1975; Hebert, 1978; Maynard Smith, 1978; Bell, 1982; Stearns, 1987; Hughes, 1989a). Despite the pervasive claim that clonal organisms, especially cyclical parthenogens, may hold the key to resolving the 'queen of evolutionary problems' (Bell, 1982), only rarely have studies specifically examined the short-term advantage of sex (Bierzychudek, 1987); most have shown the consequences of different reproductive modes, which nevertheless provide valuable comparative data (Hughes, 1989a). Here, I outline two recent studies that illustrate approaches that may be employed using clonal animals, and conclude the section by considering the polyphyletic origin of obligate cloning in *Daphnia*.

One major group of theories for the evolution of sex emphasizes the advantages to the individual of producing genetically diverse young (Bell, 1987). Such advantage may accrue from having a few highly fit recruits, or from possessing a broader adaptive range of individ-uals exhibiting high genotypic variance. Two contrasting models of the environment, the 'Tangled Bank', which highlights spatial changes (Bell, 1982, 1987), and the 'Red Queen' (Van Valen, 1973; Hamilton, 1980), which stresses temporal changes, have been proposed to predict the short-term advantage of sex. According to the Tangled Bank, genetically diverse progeny have higher fitness than amictic counterparts in a spatially heterogeneous environ-ment because they compete less intensely due to their broader range of niches. Alternatively, the Red Queen hypothesis places emphasis on temporal variation in the effects of predators, pathogens and parasites which are continually evolving to exploit their hosts more effectively. Novel combinations of genes are required on the part of the 'victims' to maintain their fitness, as well as in the antagonists.

Examination of both hypotheses requires studies of genotype–environment (G–E) interactions in fitness traits (Bell, 1987), such that variance in genotype fitness may be compared in spatially or temporally variable environments. Such a study has been carried out on the sessile colonial bryozoan *Celleporella hyalina* (Hughes, 1992). Clones were propagated to provide cohorts of identical colonies reared in the field under three different water flow treatments (spatial variation) and two seasonal periods (temporal variation), and clonal fitness was estimated by counting the number of zooids at the end of each season. The aim was to discover whether variance in genotypic fitness was affected more by variation among sites than over time. Differences of up to four-fold were observed in zooid abundance, together with significant G–E interactions. Seasonal effects had a greater impact on fitness than flow regime, lending some support for the importance of genotypic fitness variation over time. Hughes' (1992) study illustrates the experimental design needed to test assumptions underlying the maintenance of sex, especially the need to define the adaptive range of genets, and the relative intensity of competition within and between them.

There are difficulties in trying to test empirically the predictions of the Red Queen hypothesis, since coevolutionary interactions are difficult to monitor (Bierzychudek, 1987). One approach is to examine a clonal species that reproduces both mictically and amictically, and to investigate associations between reproductive mode and the incidence of such factors as pathogenic or parasitic infection. Such a case is provided by the dioecious freshwater snail *Potamopyrgus antipodarum* (Lively, 1992). In New Zealand, parthenogenetic (amictic) and bisexual (mictic) populations are infected differentially by digenetic trematodes. Lively (1987) documented an association between trematode infestation and reproductive mode, such that parthenogenetic populations had significantly lower levels of infection than those that were bisexual, suggesting that parthenogenetic females have replaced bisexuals where parasites are rare. In relation to the Red Queen, the proposal would be that antagonistic coevolution favours the production of diverse offspring via mixis, through the generation of a time-lagged selection against abundant phenotypes, so favouring cross-fertilizing females.

Lively (1992) determined whether the previously disclosed association between parasite infection and sex was absolute, or whether it derived from some other correlate such as 'reproductive assurance', which predicts that cloning will be favoured in sparse populations where mates are difficult to find. Results revealed a negative association between mixis and snail density (the opposite predicted for reproductive assurance), and a significantly positive correlation between sex and infection, indicating that sex had persisted in populations where parasites are common. Although the association between low parasitism and parthenogenesis could have arisen from an unknown further factor, present data do support frequency selection by parasites against common genotypes, as predicted by the Red Queen hypothesis.

Alternatively, such relationships between parasitism and *Potamopyrgus* could be interpreted as evidence against the Red Queen hypothesis. If the adaptive significance of mixis is to provide greater diversity to protect against parasitism, then it could be predicted that parasitism should be lower in sexual populations, not higher, though it is difficult to conceive of a situation where parasite and host would reach a stable equilibrium. Moreover, such factors as differences in the distributions of parasite and host reproductive modes, parasite epidemiology and the influence of other selective influences may preclude such a scenario.

In addition to Lively's (1992) findings, the study identifies further possibilities for examining the coevolutionary effects of parasites. For example, monitoring the association between parasites and common host genotypes in nature (Lively, Craddock and Vrijenhoek, 1990) may reveal the temporal nature of genetic and demographic responses between antagonist and host.

Obligate parthenogenesis in Daphnia

One of the fundamental assumptions underlying the use of clonal animals in studies on breeding systems is that the distribution of reproductive modes is related, in some way, to selective factors imposed by the environment (Stearns, 1987). Some enlightening studies using *Daphnia* (Hebert, 1981; Hebert, 1987a; Innes and Hebert, 1988) have, however, shown that factors such as bisexual–clonal hybridization and polyploidy may complicate the use of geographic parthenogenesis (Glesener and Tilman, 1978; Maynard Smith, 1978) in arguments of sexual advantage.

Most species of *Daphnia* reproduce by cyclic parthenogenesis (apomictic), though some species are obligately parthenogenetic, producing resting eggs amictically, either exclusively within a species (*D. middendorffiana*), or with local variation in reproductive mode among populations (*D. pulex, D. cephalata*) (Hebert, 1987a). There is now evidence (Innes and Hebert, 1988) that males from obligately parthenogenetic clones can transmit a meiosis suppressor gene(s) to sexual females through interbreeding, thus changing an originally bisexual clone into an obligately parthenogenetic one. High clonal diversity detected in such permanently amictic populations of *D. pulex* (Weider and Hebert, 1987) thus arises through multiple matings between males from obligately parthenogenetic clones and cyclically parthenogenetic females. Indeed, Lynch, Spitze and Crease (1989) emphasize that polyphyletic origin and incomplete reproductive isolation are not unique to *Daphnia*, but are properties of the majority of parthenogens (Lynch, 1984b), and must be incorporated into theories on the phylogenetic and geographic distribution of sex.

The question may arise, then, why have not most populations of *Daphnia* adopted obligate parthenogenesis, especially in view of its predicted two-fold advantage (Maynard Smith, 1978)? The complete transition from bisexuality to obligate parthenogenesis is not easily accomplished, since the production of both amictic eggs and the removal of males requires gene substitutions at several loci. Studies on *Daphnia* breeding systems, in conjunction with evidence that hybridization may drive the transition of parthenogens from bisexuals (Moore, 1984), support the view that the incidence of parthenogenesis may be related more to internal genetic factors than natural selection, though this may depend markedly on the species in question. Even the geographic association of parthenogenesis with certain environments (Glesener and Tilman, 1978) may be complicated by the observation that many parthenogens are polyploid (Bell, 1982), and that high latitudes may favour polyploidy. Indeed, recent experimental evidence on the parthenogenetic brine shrimp *Artemia parthenogenetica* (Zhang and Lefcort, 1991) demonstrated a greater tolerance to temperature extremes in polyploid populations compared with diploid, correlating well with the distribution of apomictic and mictic forms. The association between polyploidy and body-size variation in rotifers (Walsh and Zhang, 1992) further indicates a possible source of fitness variation that may not be directly related to reproductive mode.

In summary, not only does the existence of different breeding systems in clonal animals provide a basis for examining theories on the evolution of sex, but the occurrence of different breeding systems within a species provides a rare opportunity to test empirically the underlying mechanisms of their origin.

(b)　Reproduction and population structure

Population genetic aspects of reproduction largely relate to whether or not recombination occurs, with a consideration of consequences on population structure (distribution and levels of genetic variability) and genotypic variance in fitness. In addition to mutation, clonal ancestry (Parker, 1979) can affect population clonal structure, depending on whether clones have evolved from a related sexual species (polyphyletic) or are mutationally derived from a single ancestral lineage (monophyletic). In polyphyletic clones of multiple origin, clonal diversity and ecological differentiation are predicted to be greater than in monophyletically derived clones, though the latter can exhibit greater diversity if they hybridize with sexual relatives (Hebert, 1987a).

A plethora of studies has been undertaken on genetic population structure in aquatic clonal animals (Hughes, 1989a), and here I identify certain trends that have emerged.

In many cyclically parthenogenetic cladocerans and rotifers, the environment exerts a strong influence on the relative contributions of amictic and mictic reproduction. Cloning usually proceeds when conditions for genet expansion are favourable, and mictic development is induced when conditions deteriorate (Hughes, 1989a), resulting in the production of resting eggs. Using allozymes, the effects of sexual reproduction in restoring genotype frequencies to Hardy–Weinberg equilibrium have been well shown in populations of freshwater crustaceans (e.g. Hebert, 1974a; Young, 1979; Mort and Wolf, 1985; Chaplin and Ayre, 1989; Ward, Tollit and Bickerton, 1991), rotifers (King, 1980), and sea anemones (e.g. Sole-Cava, Thorpe and Kaye, 1985; Shaw, Beardmore and Ryland, 1987). Extended amictic phases typically produce marked deviations from Hardy–Weinberg equilibria, often but not exclusively as a consequence of heterozygote excess (Bucklin and Hedgecock, 1982) and associated shifts in clonal frequencies and significant linkage disequilibria among allozymic loci (Black and Johnson, 1979; Ayre, 1983; Shaw, Beardmore and Ryland, 1987). In a permanent population of *Daphnia magna* (Carvalho and Crisp, 1987), where sex was rare, extensive fluctuations in electrophoretic clone frequencies occurred, some of which were seasonally predictable. Such patterns contrast with lake populations of *Daphnia*, where regular annual bouts of mixis are associated with temporally stable genotype frequencies (Mort and Wolf, 1985, 1986); but see Jacobs (1990) and section 6.4.

Although the relationship between genetic equilibria and sexual recruitment is well established, the effects of periodic sex and cloning on genotypic diversity are less clear. In some circumstances, notably cyclic parthenogens, diversity may be high, even where sex is rare, with populations apparently consisting of up to tens of thousands of clones (Carvalho and Crisp, 1987). In contrast, extended amictic recruitment in anemones has led both to domination by just a few clones (Francis, 1976; Ayre, 1984) and high clonal diversity (Hoffmann, 1986; Shaw, 1991). At one extreme is the obligate fissiparous anemone *Haliplanella lineata*, where populations are usually dominated by a single clone, with distinct clones in different localities (Shick and Lamb, 1977). Population growth is usually rapid, with expansive areas becoming covered in just a few years, though locally, populations may crash. Among species where sex is periodic, genotypic diversity may remain high even where cloning predominates. In the anemone *Sargartia elegans*, for example, high clonal diversity was detected even though recruitment was mainly from pedal laceration (Shaw, 1991). Local factors may also modify levels of genotypic diversity. In *Actinia tenebrosa*, populations on stable rocky shores exhibit low genotypic diversity owing to predominance of a few clones (Ayre, 1984), whereas on unstable boulder habitats, periodic storms may decimate adults, and the correspondingly greater larval recruitment is associated with higher genotypic diversity.

It appears, therefore, that the consequences of alternative reproductive modes on population structure will be strongly mediated by disturbance, temporal fluctuations in selective factors, the longevity of individual ramets, and intensity of competitive interactions (Sebens and Thorne, 1985; Weider, 1992). Nevertheless, where sex is frequent, clonal diversity within populations tends to be high; the converse is not always true.

In many colonial animals, especially in soft corals, some sea anemones and sponges, population dynamics are usually dominated by the births and deaths of amictic ramets (McFadden, 1991), with only occasional sexual recruitment (Ayre and Willis, 1988). Matrix projection models (Caswell, 1986) have been developed from empirical data on colony growth, mortality, fission, sexual reproduction and recruitment for the soft coral *Alcyonium* (McFadden, 1991) to assess the relative importance of amictic and sexual reproduction. The analysis revealed that sex contributed less than 1% of fitness, while more than 40% of fitness was accounted for by transitions between the upper size classes, either by fission or modular iteration. Eliminating sexual reproduction from the life-cycle had a negligible effect on fitness; eliminating fission greatly reduced fitness and suggested rapid extinction for all populations.

Why has sex been retained at all in such situations? It seems likely that selection for dispersal may be related to the maintenance of sex, since sexually derived propagules are also the agents of dispersal. Gene flow increases localized genetic heterogeneity by mixing ramets from distinct genets, even where overall genotypic diversity is low, so reducing the chances of population extinction due to genetic uniformity (Shick and Lamb, 1977). Such a view adds credence to Williams' (1975) 'strawberry-coral' model, which predicts that where organisms reproduce both amictically and sexually, the former preserves intact highly adapted genotypes, whereas the latter, despite its costs, may enhance fitness in the long term by producing diverse progeny that are dispersed. Observations on the population structure of sea anemones also provides empirical support, though direct evidence for localized adaptation is scant (Shick, 1991).

The effect of gene flow, and its interaction with other microevolutionary forces, is fundamental to the evolutionary diversification and mating structure of natural populations. The discontinuous nature of ponds and lakes would be expected to promote genetic differentiation among populations, especially if opportunities for passive dispersal are limited. Indeed, available data on cladocerans (Hebert, 1974c; Hebert and Moran, 1980; Korpelainen, 1984; Crease, Lynch and Spitze, 1990) have shown genetic divergence even among adjacent populations, due both to gene frequency differences, and occasional allelic substitution (Hebert, 1974c; Crease, Lynch and Spitze, 1990). Such heterogeneity may arise from restricted dispersal of diapausing eggs, founder effects, intense localized selection within habitats, or interclonal competition among resident and immigrant clones. The importance of restricted dispersal and founder effects is indicated by the observation that most differentiation occurs within, rather than between, regions (Crease, Lynch and Spitze, 1990). Such findings are concordant with multiple colonization events, with founding of populations from a small number of propagules. Under such circumstances, skewed fecundity distributions, the perpetuation of highly fit genotypes, and prolific population growth will all serve to exaggerate evolutionary divergence.

In the marine environment, where habitat continuity affords greater opportunities for dispersal, gene flow via larvae is expected to overcome diversification and inbreeding. While allelic homogeneity is common in many unitary marine invertebrates, the few sessile benthic cloners so far studied are characterized by microgeographic variation in gene frequencies (Burton, 1983; Shick, 1991). Some colonial animals such as coral species on the Great Barrier Reef evidently undergo extensive dispersal of planula larvae, but there is increasing evidence (Jackson, 1985, 1986; Jackson and Coates, 1986) that restricted, rather than wide-scale, larval dispersal is the norm for modular benthic invertebrates.

A number of morphological, behavioural and reproductive adaptations operate to induce settlement in close proximity to the parent (Hughes, 1989a), leading to philopatry (Shields, 1982). Compared to many unitary marine animals which spawn gametes that develop into feeding pelagic larvae, most colonial taxa brood embryos that undergo a brief, non-feeding pelagic or benthic larval existence before settlement (Jackson, 1985). High genetic relatedness of individuals in local colonial populations may ensue, due both to amictic replication of genotypes and restricted larval dispersal. Under such circumstances, inbreeding is likely, leading to microgeographic differentiation. Indeed, evidence of both inbreeding and out-breeding depression in different populations of the ascidian *Botryllus schlosseri* (Sabbadin, 1979; Grosberg, 1987) demonstrate the crucial role that dispersal and settlement patterns can play in determining genet fitness.

Some particularly informative work on gene flow in the colonial ascidian *Botryllus schlosseri* (Grosberg, 1987, 1991) has, however, shown that the relationship between dispersal and population structure in colonial benthic invertebrates may not always be clear-cut. Using an *in situ* mark–recapture experiment, Grosberg (1987) showed that the majority of sibling larvae settled within 1 m of their respective parents. The pattern of restricted dispersal was supported by results from histocompatibility assays showing a reduction in intergenotypic fusion in experimental grafts with increasing distances from parental colonies. Laboratory studies of mating success as a function of proximity among colonies further suggested microgeographic structuring of natural populations of *B. schlosseri*. Subsequent analysis using allozymes of the same population (Grosberg, 1991), however, detected little evidence of fine spatial structuring of gene frequencies. In the same study, direct measures of sperm-mediated gene flow were made from paternity analyses in the field by tracking the distribution of three rare allozymic markers. The frequencies of matings declined with increasing separation, as predicted from the diffusion concentration gradient of released sperm, but more fertilizations than expected occurred at distances exceeding 50 cm.

Grosberg (1991) explained the enhanced gene flow in terms of the absolute number (rather than concentration) of gametes and larvae dispersing from a point source which, depending on diffusion and geometric assumptions, may actually increase with distance. Indeed, despite the observed decline in sperm concentration and larvae with distance, genetic homogeneity indicated sufficiently high levels of gene flow, at least during the 5 months of the experiment (= five generations), to prevent differentiation. Genetic incompatibility could further influence patterns of fertilization success, especially late in the season when inbreeding intensifies, and adjacent colonies are more likely to share incompatibility alleles. Gametic incompatibility may oppose the effects of high concentrations of sibling sperm, leading to higher levels of gene flow later in the season, with maximum male fertilization success at some intermediate distance from the source colony (Grosberg, 1991).

The above study cautions against making predictions based on dispersal capacity alone, and emphasizes the effects that occasional gene flow, self-incompatibility and local factors may have in determining population differentiation. There is an urgent need to examine genetic population structure of a wider range of sessile colonial animals, together with direct and indirect measures of gene flow. The effects of restricted dispersal may be less obvious than genetic differentiation, involving the structuring of breeding systems and evolution of social behaviour (Uyenoyama, 1984).

(c) The maintenance of genetic variation

Many clonal populations, especially those that are mictic, exhibit levels of genetic diversity comparable to those of unitary organisms (King, 1980; Hebert, 1987a; Shick, 1991; Hughes, Ayre and Connell, 1992), quantifiable as heterozygosity, allozyme polymorphism or clonal diversity. Extensive electrophoretic studies have shown that the classical notion (Asher and Nace, 1971; Lokki, 1976) that amictic organisms have highly depressed levels of variation is the exception rather than the rule, and depends on clonal ancestry (Parker, 1979), hybridiza-tion with sexual relatives (Lynch, Spitze and Crease, 1989), or the frequency of mixis in the life-cycle (Hughes, 1989a). Although the organization of genetic variation in cloners is profoundly affected by such things as linkage disequilibria, enhanced heterozygosity, epistatic interactions, and the periodic release of additive genetic variance after mixis, some generalizations can be made concerning its maintenance. Indeed, due to the evolutionary characteristics of cloning (section 6.1.2(b)), many recent studies implicating selection in the wild have involved aquatic clonal animals (Lynch, 1983; Pace, Porter and Feig, 1984; Ayre, 1985; Hoffman, 1986, 1987; Carvalho, 1987, 1988; Carvalho and Crisp, 1987; Weider and Hebert, 1987; Grosberg, 1988a; Zhao and King, 1989; Browne and Hoopes, 1990; Hebert and Emery, 1990; Jacobs, 1990; Rossi and Menozzi, 1990, 1992; Wilson and Hebert, 1992). Here, I consider two aspects of genetic variation that centre on the evolutionary consequences of ecological differentiation among genets; first, how temporally variable selective pressures may maintain genotypic diversity, and second, the related problem of how clones coexist.

Clonal succession and replacement
The impact of temporal variation in selective factors depends upon the balance between the relative number of generations over which a selective regime persists, and the selective differentials among the regimes (Hedrick, 1986). If the variance in genotype fitness is marked (i.e. greater than 10–20%) between regimes, variation over time can maintain polymorphism at equilibrium. Through tracking identical clonal genotypes over time, and comparing their performance either in the laboratory (e.g. Carvalho, 1987; Rossi, Rozzi and Menozzi, 1991) or *in situ* through manipulation experiments (e.g. Wilson and Hebert, 1992), it is possible to estimate genotype-specific fitness, and identify putative selective factors.

Clonal frequency changes have been related to environmental factors such as temperature (Carvalho, 1987; Rossi, Rozzi and Menozzi, 1991), oxygen (Weider and Lampert, 1985), photoperiod (Rossi and Menozzi, 1992), predator abundance (Pace, Porter and Feig, 1984; Parejko and Dodson, 1991), food concentrations (Mort, 1989), and light intensity (DeMeester, 1991). In a permanent population of *Daphnia magna* (Carvalho and Crisp, 1987), three dominant electrophoretic clones exhibited seasonal succession over 2 years ('seasonal clones'), predominating respectively in summer, autumn and winter. Subsequent laboratory studies (Carvalho, 1987) disclosed marked differences in thermal response among seasonal clones, with differences in clone temperature preferenda of 5–10 °C between summer and winter clones. Seasonal clones were adapted to ambient temperature, as shown by their respective survivorship and fecundity at different temperatures. Such fitness differences were supported from field observations on temporal changes in fecundity (Carvalho, 1988), where up to 10-fold differences in the relative fecundity of seasonal clones were detected. Similar seasonal changes in clonal frequencies have been recorded in the obligately parthenogenetic ostracod *Heterocypris incongruens* (Rossi and Menozzi, 1990; Rossi, Rozzi and Menozzi, 1991), where seasonal electrophoretic clones again exhibited adaptation to ambient temperature.

Although the biology of *D. magna* and *H. incongruens* differ in ways other than their breeding systems, the above studies demonstrate that in both cyclic and obligate cloners, specialized clones can evolve. Marked genotypic variance in obligate cloners is not difficult to envisage, since genets are isolated and genetic differences can readily accumulate. In cyclic parthenogens such as *D. magna*, the evolution and persistence of clonal divergence is more difficult to understand; even if sex is infrequent, as in the population studied (Carvalho and Crisp, 1987), recombination and segregation would probably disrupt coadapted gene complexes, especially where clones exhibit widely divergent reaction norms (Stearns, 1989). It seems likely that persistence of such ecological differentiation in cyclic parthenogens may derive from survival of parthenogenetic females at low frequency between seasons, tight linkage between coadapted loci, or temporal separation of sexual reproduction among clones. Indeed, there is evidence (Young, 1975; Carvalho and Hughes, 1983; Yampolsky, 1992) that genetic variation exists in the timing of both sexual females and males, thus providing the potential for mating among more closely related clones. Linkage relationships among coadapted complexes would, however, be a crucial factor in determining the ecological consequences of the diversity generated following mixis. Such constraints on mating systems may indeed determine the incidence of temporal adaptive change. Considerably more attention is needed on the dynamics of mixis and recruitment from sexually derived resting eggs.

Two widely divergent morphs detected and monitored in a population of a colonial sea squirt *Botryllus schlosseri* (Grosberg, 1988a) demonstrate the extent of fitness differences that can evolve in a clonal animal. One morph produces semelparous colonies (single reproductive event: $r_{max} = 0.99$) that die immediately after reproduction, have early age and rapid growth to first reproduction, and exhibit high reproductive effort. The other morph produces at least three colonies (iteroparous; several reproductive events) before dying ($r_{max} = 0.72$), postpones mixis until a larger size, has a lower growth rate, and invests roughly 75% less in reproductive effort than semelparous colonies (Figure 6.1.6). Semelparous colonies dominate during the midsummer, after which iteroparous recruits become more abundant. The genetic control of such marked genotypic variance in fitness expressed as life-history polymorphism indicates that the repeated, single shift in morph frequencies is underlain by temporal variation in selection, which in this case is mediated by interspecific competition with another ascidian, *Botrylloides leachi*. The value of using clonal animals in such studies is to determine readily whether seasonal variation in life-history traits represents genetic variation due to varying selection, or phenotypic plasticity.

The role of spatially varying selective pressure has also been demonstrated in clonal animals either using reciprocal transplant experiments (e.g. Carvalho, 1984; Shick, Hoffman and Lamb, 1979; Ayre, 1985), or by correlating phenotype distribution with putative selective factors (Weider and Hebert, 1987; Hebert and Emery, 1990; Parejko and Dodson, 1991; Wilson and Hebert, 1992). Although it is generally held (e.g. Hedrick, 1986) that spatial variation plays a more important role in maintaining genetic variation than temporal

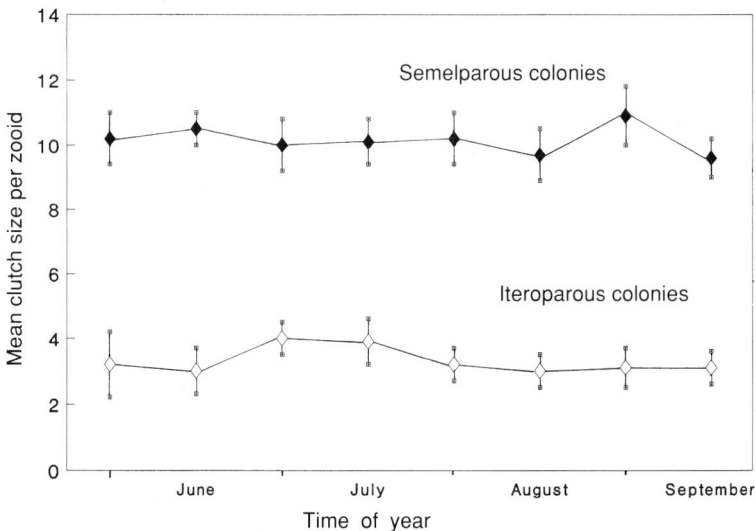

Figure 6.1.6 Differences in the seasonal patterns of reproductive effort in iteroparous and semelparous colonies of the ascidian *Botryllus schlosseri*. (Redrawn from Grosberg, 1988a.) Error bars show standard errors.

changes, studies on aquatic clonal animals have illustrated repeatedly the importance of the latter. It is, however, difficult to know whether the apparent importance of temporal variation is due to such things as the greater tractability of clonal material (section 6.1.2(a)), the significant difficulties of tracking mobile genets in space, or whether the clonal habit (e.g. high reproductive rates and marked ecological differentiation among genotypes) may itself render clonal animals more likely to respond to selective changes over time.

Clonal coexistence

Variable selection pressures in time and space can explain how specific shifts in clonal frequencies occur, but they do not account for how clones coexist (Hebert and Crease, 1980; Lynch, 1983; Carvalho and Crisp, 1987), or the mechanisms determining clonal richness and diversity. Hebert and Crease (1980) pointed out that since clones of a species probably utilize similar resources, niche overlap and intense competition should exclude all but a few ecologically divergent clones. Such divergence among genets may indeed reduce competition among obligate parthenogens, where each clone is genetically isolated and only a few genetically distant clones are found in each habitat (Shick and Lamb, 1977; Hebert and Crease, 1983; Hauser *et al.*, 1992), but in populations containing hundreds of clones, niche separation is more difficult to envisage, especially during phases of prolonged amictic replication (Hebert, 1974b; Carvalho and Crisp, 1987). Several factors may account for clonal coexistence, including the generation of new genets from mixis, temporal and spatial variation in selection, and extreme similarity among some clones (Young, 1979; Hebert, 1982; Sebens and Thorne, 1985).

Periodic recruitment to clonal populations from resting egg banks (e.g. cladocerans and rotifers (Carvalho and Wolf, 1989; Wolf and Carvalho, 1989)), or regular sexual recruitment from larvae (e.g. bryozoans), may replenish clonal diversity, as well as modify competitive interactions and dominance hierarchies among clones (Sebens and Thorne, 1985). The effects of sexual recruitment will depend on its frequency in relation to cloning, the longevity of individual ramets, and the degree of disturbance. In those cloners where sex occurs annually, a regular input of new genets may prevent domination by a few clones, and promote diversity (Section 6.5), though strong selection pressure can lead to rapid clonal replacement (Pace, Porter and Feig, 1984; Carvalho, 1988; Grosberg, 1988a).

The frequency of sexual recruitment in benthic colonies is relatively straightforward to measure, where recruitment from larvae gives rise to identifiable cohorts. In rotifers and cladocerans, where resting eggs remain viable in sediments for many years (Herzig, 1985; Moritz, 1987; Carvalho and Wolf, 1989), the effects of mixis are more difficult to quantify. Continuous or periodic hatching of sexual resting eggs may replenish clonal diversity, as indicated by *in situ* hatching experiments (Herzig, 1985; Wolf and Carvalho, 1989), and so greatly inflate estimates at any given time.

Extended ramet longevity such as occurs in the polyps of some sea anemones and corals will reduce the contribution that mixis makes to clonal diversity, mainly because population demography will depend more on changes in the number of ramets (McFadden, 1991). In the anemone *Metridium senile* (Hoffman, 1987), clonal genotypes remained remarkably stable over a 6-year period, implying clonal coexistence, at least during this interval. Extensive spread of long-lived polyps, agonistic behaviour from non-clonemates and predation from nudibranchs reduced the scope for sexual recruitment. Indeed, novel genets were detected only outside the study area when the epifauna was artificially cleared from a substrate, supporting the notion that sexual recruitment contributes to clonal diversity, especially in disturbed areas (Sebens and Thorne, 1985). On a much longer timescale, periodic catastrophes that disturb coral communities (Hughes, Ayre and Connell, 1992) may be evolutionarily significant in determining inputs of new genets by colonizing sexual recruits.

The major influence of sex, then, is to determine the number of genets that enter a population, which sets an upper limit to clonal richness. Subsequent demographic changes in ramets and their degree of ecological differentiation will have a greater bearing on the dynamics of clonal diversity.

Direct empirical tests of factors affecting clonal coexistence and diversity are rare, and among aquatic clonal animals are again limited to *Daphnia*. The effect of food availability on clonal coexistence has been examined in *Daphnia galeata* within closed natural systems using *in situ* enclosure experiments (Mort, 1989). Experiments undertaken in different years, each for 4 months, revealed temporal stability of clone frequencies, with no marked effect of the supplemental food, except in an overall increase in *Daphnia* density as expected. Despite the short-term nature of the study, the data demonstrate clonal coexistence, probably due to insufficient ecological divergence among clones, as supported by previous observations on natural populations of lake *Daphnia* (Mort and Wolf, 1985, 1986). Lake *Daphnia* may comprise general-purpose genotypes selected for average responses to seasonal environments, as suggested by Lynch (1983), though there are exceptions (Jacobs, 1990; Section 6.4). Where taxa exhibit fitness differences but share a similar ecology, as in many clonal assemblages (Lynch, 1983; Weider and Lampert, 1985; Carvalho and Crisp, 1987; Grosberg, 1988a; Rossi and Menozzi, 1990), clonal displacement should be rapid.

A model of clone displacement by competitive exclusion depends on the continued presence of a fitness advantage of a superior clone, but temporal changes that are known to be selectively significant may prevent any clone from remaining long enough in optimal conditions to displace all others. Sebens and Thorne (1985) considered the effect of local (mortality of ramets from physical or biological causes) and large-scale (physical clearing of large areas of habitat) disturbance on coexistence of clones and clonal diversity. They predicted from computer simulations that clonal coexistence and diversity would be greatest at intermediate levels of disturbance.

Some recent evidence in support of the intermediate disturbance hypothesis was provided by laboratory microcosm experiments on clonal coexistence and diversity in *Daphnia pulex* (Weider, 1992). Cultures were disturbed by varying the frequency of dilution volumes, i.e. by removing a given volume of medium, and replacing it with fresh. The frequency of disturbance had a profound effect on clonal diversity, with the highest diversity maintained at low to intermediate disturbance frequencies, through pre-emption of competitive exclusion between genotypes. Although one or two clones dominated within each experiment, some rarer clones persisted throughout. According to Parker's (1979) views on clonal origins, ecological differentiation among polyphyletically derived clones may facilitate coexistence. The clones which persisted longer in Weider's (1992) experiment were indeed polyphyletic, having arisen independently from presumed cyclically parthenogenetic ancestors, and were thus genetically differentiated (Crease, Stanton and Hebert, 1989).

Wilson and Hebert (1992) examined factors affecting the distribution and abundance of polyphyletically derived *D. pulex* clones using natural observations and manipulation experiments. Clonal coexistence in single habitats was prevented by marked fitness differences, though within the study area clonal diversity was promoted by environmental heterogeneity favouring a range of specialized clones. The persistence of three dominant clones was dependent upon variation in pond salinities, predation regimes and competitive abilities, and represented obligate lineages that were well adapted to prevalent habitat conditions.

In summary, the extent of ecological divergence among genets, determined in part by ancestry, breeding system and intensity of selection pressures, will determine the rate at which clonal displacement occurs. Clonal coexistence may depend not only on the differential response of clones to heterogeneity in time and space, but also on the frequency of disturbance. Moderate levels of disturbance may promote diversity due to prevention of clonal dominance, though empirical tests of this hypothesis are rare.

Studies on clonal coexistence and diversity also illustrate a further point regarding the evolution of specialized and generalized clonal genotypes. The relationship between clonal origins and subsequent clonal divergence is more complex than originally thought (Parker, 1979), since marked clonal divergence can occur in both mono- and polyphyletically derived clones (Lynch, Spitze and Crease, 1989). The assumption that obligate amixis leads to the evolution of general-purpose genotypes (Baker, 1965) should be received with caution, since the nature of local selection pressures appears to play a predominant role. Cyclic parthenogens, for example, that undergo frequent mixis may exhibit both general-purpose (Mort and Wolf, 1985, 1986; Mort, 1989) and specialized genotypes (DeMeester, 1991), evidently even in the same population (Carvalho and Crisp, 1987). Conversely, obligate parthenogens may comprise clones that are broadly adapted (Hauser *et al.*, 1992) as well as those that are adapted to a narrow range of conditions (Wilson and Hebert, 1992). It is evident, therefore, that further studies are needed to quantify clonal differentiation and breadth of adaptation in relation to habitat type and breeding system, especially in natural populations.

6.1.5   Aquatic clonal animals and genetics: future prospects

Although numerous studies have been undertaken on the evolutionary genetics of aquatic clonal animals, many outstanding problems remain. Throughout this section I make recommendations, but here summarize some topics and approaches that may contribute usefully not only to our understanding of clonal animals, but also generally to evolutionary theory

(a)   Taxonomic coverage

It is evident that coverage of clonal genetics among aquatic groups is limited to just a few (Table 6.1.1). It is important to extend studies, especially to colonial agametic cloners where, for example, among corals there are virtually no population genetic data on the extent of population differentiation or gene flow (Hughes, Ayre and Connell, 1992). Population genetic studies on a wider range of agametic cloners would also provide information on the evolutionary consequences of the fundamental division between agametic and gametic cloners in terms of genetic variability and extent of ecological differentiation (Buss, 1987; Hughes, 1989a).

(b)   Techniques

Developments in DNA methodology, particularly the advent of multilocus DNA fingerprinting (Jeffreys, Wilson and Thein, 1985), have opened up new possibilities for discriminating genets with high resolution. The higher number of visualized bands and greater polymorphism compared with allozymes provides a powerful discriminatory tool. Previously undetected genetic heterogeneity within electrophoretic clones may not only obscure patterns of clonal distribution in time or space, but also prevent the tracking of a specific clonal lineage in the field. Although DNA fingerprinting has so far been applied only rarely to clonal animals (Turner, Elder and Davis, 1990; Carvalho *et al.*, 1991; Black *et al.*, 1992; Hauser *et al.*, 1992), it has demonstrated a high sensitivity and stability in clonal recognition. Hauser *et al.* (1992), for example, identified three major clonal genotypes in the apomictic hydrobiid snail *Potamopyrgus antipodarum*, from widely separated localities within the UK, using an RNA derivative of Jeffreys' (Jeffreys, Wilson and Thein, 1985)

33.15 minisatellite core sequence. Such conformity within clonal profiles could provide a valuable approach for estimating dispersal of genets where rare electrophoretic markers are not available, and for making assessments of genetic relatedness (Lynch, 1988). The Mendelian inheritance of multilocus fingerprints (Burke *et al.*, 1991) provides a means of exploring such topics as paternity analysis and kin selection, which may be particularly relevant to the spatial organization and allocation of resources in colonial species (Uyenoyama, 1984). It remains to be seen, however, whether the technique can separate clones in natural populations that exhibit high clonal diversity (Brookfield, 1992).

In addition to effective discrimination of genets, applications of DNA methodology may offer an approach to examining individuals or those stages of the life-cycle (e.g. larvae) normally too small for electrophoretic analysis. Tracking larval release and settlement may be particularly important where high post-settlement mortality occurs. The use of the polymerase chain reaction (PCR) and the recently developed random amplification of polymorphic DNA (RAPD) (Black *et al.*, 1992; Hadrys, Balick and Schierwater, 1992) is particularly promising, since profiles can be obtained from nanogram quantities of DNA, through PCR amplification of small inverted repeats using commercially available short oligonucleotide primers of random sequence.

The widescale adoption of such technology by population biologists is, however, hindered by the degree of molecular expertise required, and its time-consuming nature, though recent advances in the use of non-isotopic probes (Kricka, 1992) and use of the RAPD may significantly reduce sample processing time. It is worth stating in this current age of 'DNA mania' that allozymes still continue to offer a valuable source of genetic markers, especially in determining population structure at the macroscale (see Section 4.1). DNA markers should be employed only where specificity of clonal identity or meagre amounts of material make it necessary.

(c)   Fundamental processes

Demography of genets
One of the major challenges in studying the population genetics of clonal animals is to record reproduction and mortality among ramets in the field, thus quantifying the fitness of individual genets. Although the process is difficult among mobile solitary genets due to dispersal, examination of sessile colonies may be more tractable. The use of appropriate models based on empirical data (McFadden, 1991) or long-term monitoring of genet fitness (Hughes, 1992) could yield important data on the consequences of different reproductive schedules, as well as genotype–environment interactions. Advances in methods for clonal discrimination may for the first time allow measurements of the reproductive output of specific genets under natural conditions as well as providing more unequivocal estimates of dispersal (Carvalho, 1990).

Evolutionary significance of somatic mutations
The evolutionary significance of somatic mutations has received considerable attention in plant populations, where vegetative propagation can perpetuate somatic variants even if they are not incorporated into the germ-line. Extended longevity and large genet size could in theory allow ramets to evolve through diplontic or somatic selection (Buss, 1987), where differential cell growth may result from the effects of mutations, such that the variant is more or less vigorous than the wild type. In corals, for example, variation due to somatic mutation may be heritable, because mobile undifferentiated cells that mutate could eventually become part of the germ-line. Pelagic tunicates are unusual in that they budd from a stolon which incorporates gonadal precursors (J.S. Ryland, personal communication), thus

making it possible to have some continuity in the transmission of mutants. The significance of differential transmission of mutations, and the role of somatic mutations has thus far received no empirical attention in clonal animals. Somatic mutation may be important if, as theory suggests (Buss, 1987), it endows genets with an additional opportunity for selection (zygote to zygote), thus enhancing evolutionary rates. Studies to examine the incidence of somatic mutations, together with their patterns of inheritance throughout a clonal lineage, could prove valuable. A recent study on somatic mutations in the nuclear ribosomal DNA of white clover (Capossela *et al.*, 1992) illustrates an appropriate molecular approach for such studies.

(d) Experimental design

Many of the experiments carried out on aquatic clonal animals have thus far been laboratory-based, with few field manipulation studies. There is, for example, an abundance of information on genotype–environment interactions in the laboratory (e.g. Loaring and Hebert, 1981; Carvalho, 1987), but there are relatively few such experiments undertaken in the field (Hughes, 1992). The time is now ripe, with a plethora of testable ideas (e.g. intermediate disturbance hypothesis, restricted dispersal and microgeographic differentiation in colonial species, Tangled Bank versus Red Queen hypotheses), and enhanced powers for distinguishing genets and their constituent ramets in the field, to link laboratory investigations with natural observations and measurements.

Manipulation experiments involving reciprocal transplants of genets (Ayre, 1985) or examining interclonal competition (Wilson and Hebert, 1992) could reveal much about the effects of local adaptation and stochastic factors in determining clonal abundance and distribution. To minimize disturbance to natural habitats, *in situ* enclosures could be more fully exploited (Mort, 1989), with the added potential of manipulating clonal diversity or environmental factors. Where sexual recruitment is rare or unpredictable (e.g. resting egg seedbanks), long-term monitoring of populations is required to document genet dynamics effectively, such that the effects of periodic disturbance on clonal diversity, and associations between clonal dynamics and putative selective factors, can be estimated. To test hypotheses, it is, however, important that appropriate laboratory experiments are integrated with field observations, much in the same way that has been suggested for demonstrating selection at enzyme coding loci (Clarke, 1975).

Exploring the effects of pathogenic or parasitic infestations on clonal frequencies would provide a tractable approach to examining the effects of temporal variation on the advantages of sexual reproduction. Species that have easily identifiable genets (e.g. using molecular markers), that have mictic and clonal populations and that have each been affected to some extent by an antagonistic organism would provide model systems. Manipulation studies where clonal frequencies and infection are varied experimentally may test hypotheses relating to genotype frequency and sexual advantage (Bell, 1982).

6.1.6   Concluding remarks

The wide distribution of aquatic clonal animals throughout the Metazoa, taken together with the value of clonal material in genetic analysis, will ensure their continued use to examine fundamental evolutionary issues. The ability to organize such diversity in form through only a few morphological and reproductive criteria not only facilitates meaningful comparisons across species, but also, with accumulating data, allows generalizations about such things as the maintenance of genetic variation, the ecological and genetic implications of contrasting reproductive strategies, and the determinants of ecological differentiation. Already, aquatic clonal animals have modified some of our views on evolutionary phenomena, such as the role that temporal variation can play in maintaining genetic variation, the complexity of factors that interact to determine population differentiation, the evolution of niche breadth, and the geographic distribution of sexual and amictic forms. The use of novel theoretical approaches (non-equilibrium population genetics and demographic models), and the *in situ* manipulation and long-term monitoring of genet dynamics, with judicious use of recent highly sensitive molecular techniques, will undoubtedly elucidate both various aspects of clonal population genetics and outstanding evolutionary problems.

It is worth concluding by emphasizing the need to undertake genetic studies within the ecological setting of the species concerned. Data on such factors as substrate preferences, growth and mortality rates, spatial structure, species interactions, limiting resources and larval biology are all basic to evolution, and will determine genotype–environment interactions, and thus the subsequent fate of alternative genets. Indeed, exploiting the evolutionary characteristics of the clonal habit may provide one of the best opportunities of securing effective integration of ecological and genetic theory.

Acknowledgements

I wish to thank R.N. Hughes, J. Pijanowska, L.R. Weider and an anonymous reviewer for valuable comments on the manuscript, and also S.B. Piertney for help with the figures.

References

Allan, D.J. (1976) Life history patterns in zooplankton. *Am. Nat.*, **100**, 165–80.

Asher, J.H. and Nace, G.W. (1971) The genetic structure and evolutionary fate of parthenogenetic amphibian populations as determined by Markovian analysis. *Am. Zool.*, **11**, 381–98.

Ayre, D.J. (1983) The effects of asexual reproduction and inter-genotypic aggression on the genotypic structuring of the sea anemone, *Actinia tenebrosa. Oecologia*, **57**, 158–65.

Ayre, D.J. (1984) The effects of sexual and asexual reproduction on geographic variation in the sea anemone *Actinia tenebrosa. Oecologia*, **69**, 222–29.

Ayre, D.J. (1985) Localised adaptation of clones of the sea anemone *Actinia tenebrosa. Evolution*, **39**, 1250–60.

Ayre, D.J. and Willis, B.L. (1988) Population structure in the coral *Pavona cactus*: clonal genotypes show little phenotypic plasticity. *Mar. Biol.*, **99**, 495–505.

Babcock, R.C. (1984) Reproduction and distribution of two species of *Goniastrea* (Scleractinia) from the Great Barrier Reef Province. *Coral Reefs*, **2**, 187–95.

Baker, H.G. (1965) Characteristics and modes of origin of weeds, in *The Genetics of Colonising Species* (eds H.G. Baker and G.L. Stebbins), Academic Press, New York, pp. 147–68.

Banta, A.M. (1939) *Studies on the Physiology, Genetics and Evolution of some Cladocera,* Carnegie Institution, Washington.

Bell, G. (1982) *The Masterpiece of Nature: The Evolution and Genetics of Sexuality,* Croom Helm, London and Canberra.

Bell, G. (1987) Two theories of sex and variation, in *The Evolution of Sex and its Consequences* (ed. S.C. Stearns), Birkhauser Verlag, Basel and Boston, pp. 117–34.

Bierzychudek, P. (1987) Experimental tests, in *The Evolution of Sex and its Consequences* (ed. S.C. Stearns), Birkhauser Verlag, Basel and Boston, pp. 163–74.

Black, R. and Johnson, M.S. (1979) Asexual viviparity and population genetics of *Actinia tenebrosa. Mar. Biol.*, **53**, 27–31.

Black, W.C. IV, DuTeau, N.M., Puterka, G.J., Nechols, J.R. and Petteroini, J.M. (1992) Use of the random amplified polymorphic DNA polymerase chain reaction (RAPD-PCR) to detect DNA polymorphisms in aphids. *Bull. Entomol. Soc.*, **82**, 151–60.

Bonnet, C. (1745) *Traite d'insectologie, ou Observations sur els Pucersons*, Chez Durand, Paris.

Brace, R.C. (1981) Intraspecific aggression in the colour morphs of the anemone *Phymactis clematis* from Chile. *Mar. Biol.*, **64**, 85–93.

Bradley, B.P. (1981) Models for physiological and genetic adaptation to variable environments, in *Evolution and Genetics of Life Histories* (eds. H. Dingle and J. Hegman), 33–50. Springer Verlag, New York.

Brookfield, J.F.Y. (1984) Measurement of the intraspecific variation in population growth rate under controlled conditions in the clonal parthenogen *Daphnia magna. Genetica*, **63**, 161–74.

Brookfield, J.F.Y. (1992) DNA fingerprinting in clonal organisms. *Mol. Ecol.*, **1**, 21–6.

Browne, R.A. and Hoopes, C.W. (1990) Genotype diversity and selection in asexual brine shrimp (*Artemia*). *Evolution*, **44**, 1035–51.

Browne, R.A., Sallee, S.E., Grosch, D.S., Segreti, W.O. and Purser, S.M. (1983) Partitioning genetic and environmental components of lifespan in *Artemia. Ecology*, **65**, 949–60.

Bucklin, A. and Hedgecock, D. (1982) Biochemical genetic evidence for a third species of *Metridium* (Coelenterata, Actiniaria). *Mar. Biol.*, **66**, 1–7.

Burke, T., Dolf, G., Jeffrey, A.J. and Wolff, R. (1991) *DNA Fingerprinting: Approaches and Applications*, Birkhauser, Berlin.

Burton, R.S. (1983) Protein polymorphism and genetic differentiation of marine invertebrate populations. *Mar. Biol. Lett.*, **4**, 193–206.

Buss, L.W. (1987) *The Evolution of Individuality*, Princeton University Press, New Jersey.

Calow, P., Beveridge, M. and Sibly, R. (1979) Heads and tails: adaptational aspects of asexual reproduction in freshwater triclads. *Am. Zool.*, **19**, 715–27.

Capossela, A., Silander, J.A., Jansen, R.K., Bergen, B. and Talbot, D.R. (1992) Nuclear ribosomal DNA variation among ramets and genets of white clover. *Evolution*, **46**, 1240–7.

Carvalho, G.R. (1984) Haemoglobin synthesis in *Daphnia magna* Straus (Crustacea: Cladocera): ecological differentiation between neighbouring populations. *Freshwater Biol.*, **14**, 501–6.

Carvalho, G.R. (1987) The clonal ecology of *Daphnia magna* (Crustacea: Cladocera). II. Thermal differentiation among seasonal clones. *J. Anim. Ecol.*, **56**, 469–78.

Carvalho, G.R. (1988) Differences in the frequency and fecundity of *PGI*-marked genotypes in a natural population of *Daphnia magna* (Crustacea: Cladocera). *Funct. Ecol.*, **2**, 453–62.

Carvalho, G.R. (1990) Molecular population genetics of marine animals: some future lines of research, in *The Genetics of Marine and Estuarine Organisms* (ed. H. Hummel), Delta Institute, Yerseke, pp. 27–31.

Carvalho, G.R. and Crisp, D.J. (1987) The clonal ecology of *Daphnia magna* (Crustacea: Cladocera). I. Temporal changes in the clonal structure of a natural population. *J. Anim. Ecol.*, **56**, 453–68.

Carvalho, G.R. and Hughes, R.N. (1983) The effect of food availability, female culture-density and photoperiod on ephippia production in *Daphnia magna* Straus (Crustacea: Cladocera). *Freshwater Biol.*, **13**, 37–46.

Carvalho, G.R. and Wolf, H.G. (1989) Resting eggs of lake-*Daphnia*. I. Distribution, abundance and hatching of eggs collected from various depths in lake sediments. *Freshwater Biol.*, **22**, 459–70.

Carvalho, G.R., Maclean, N., Wratten, S.D., Carter, R.E. and Thurston, J.P. (1991) Differentiation of aphid clones using DNA fingerprints from individual aphids. *Proc. R. Soc. Lond., Series B*, **243**, 109–14.

Caswell, H. (1985) The evolutionary demography of clonal reproduction, in *Population Biology and Evolution of Clonal Organisms* (eds J.B.C. Jackson, L.W. Buss and R.E. Cook), Yale University Press, New Haven and London, pp. 187–224.

Caswell, H. (1986) Life cycle models for plants. *Lect. Math. Life Sci.*, **18**, 171–233.

Chaplin, J.A. and Ayre, D.J. (1989) Genetic evidence of variation in the contributions of sexual and asexual reproduction to populations of the freshwater ostracod, *Candononcypris novaezelandiae. Freshwater Biol.*, **22**, 275–84.

Chapman, G. and Stebbing, A.R.D. (1980) The modular habit–a recurring strategy, in *Developmental and Cellular Biology of Coelenterates* (eds P. Tardent and R. Tardent), Elsevier/North Holland Biomedical Press, Amsterdam, pp. 157–62.

Clarke, B. (1975) The contribution of ecological genetics to evolutionary theory: detecting the direct effects of natural selection on particular polymorphic loci. *Genetics*, **79**, 101–13.

Coates, A.G. and Jackson, J.B.C. (1985) Morphological themes in the evolution of clonal and aclonal marine invertebrates, in *Population Biology and Evolution of Clonal Organisms* (eds J.B.C. Jackson, L.W. Buss and R.E. Cook), Yale University Press, New Haven and London, pp. 67–106.

Crease, T.J., Lynch, M. and Spitze, K. (1990) Hierarchical analysis of population genetic variation in mitochondrial and nuclear genes of *Daphnia pulex. Mol. Biol. Evol.*, **7**, 444–58.

Crease, T.C., Stanton, D.J. and Hebert, P.D.N. (1989) Polyphyletic origins of asexuality in *Daphnia pulex*. II. Mitochondrial-DNA variation. *Evolution*, **43**, 1016–26.

DeMeester, L. (1991) An analysis of the phototactic behaviour of *Daphnia magna* clones and their sexual descendants. *Hydrobiologia*, **225**, 217–27.

Endler, J.A. (1986) *Natural Selection in the Wild*, Princeton University Press, New Jersey.

Fisher, R.A. (1930) *The Genetical Theory of Natural Selection*, Clarendon Press, Oxford.

Fletcher, W.J. (1984) Variability in the reproductive effort of the limpet, *Cellana tramoserica*. *Oecologia*, **61**, 259–64.

Francis, L. (1976) Social organisation within clones of the sea anemone *Anthopleura elegantissima*. *Biol. Bull.*, **150**, 361–76.

Glesener, R.R. and Tilman, D. (1978) Sexuality and the components of environmental uncertainty: clues from geographical parthenogenesis in terrestrial animals. *Am. Nat.*, **112**, 659–73.

Grosberg, R.K. (1987) Limited dispersal and proximity-dependent mating success in the colonial ascidian *Botryllus schlosseri*. *Evolution*, **41**, 372–84.

Grosberg, R.K. (1988a) Life-history variation within a population of the colonial ascidian *Botryllus schlosseri*. I. The genetic and environmental control of seasonal variation. *Evolution*, **42**, 900–20.

Grosberg, R.K. (1988b) The evolution of allorecognition specificity in clonal invertebrates. *Q. Rev. Biol.*, **63**, 377–412.

Grosberg, R.K. (1991) Sperm-mediated gene flow and the genetic structure of a population of the colonial ascidian *Botryllus schlosseri*. *Evolution*, **45**, 130–42.

Hadrys, H., Balick, M. and Schierwater, B. (1992) Applications of random amplified polymorphic DNA (RAPD) in molecular ecology. *Mol. Ecol.*, **1**, 55–63.

Hamilton, W.D. (1980) Sex versus non-sex versus parasite. *Oikos*, **35**, 282–90.

Harper, J.L. (1977) *Population Biology of Plants*, Academic Press, London.

Harper, J.L. (1981) The concept of population in modular organisms, in *Theoretical Ecology: Principles and Application*, 2nd edn (ed. R.M. May), Blackwell Scientific Publications, Oxford, pp. 53–7.

Harper, J.L. (1985) Modules, branches and the capture of resources, in *Population Biology and Evolution of Clonal Organisms* (eds J.B.C. Jackson, L.W. Buss and R.E. Cook), Yale University Press, New Haven and London, pp. 1–34.

Harper, J.L. and Bell, A.D. (1979) The population dynamics of growth form in organisms with modular construction, in *Population Dynamics* (eds R.M. Anderson, E.D. Turner and L.R. Taylor), Blackwell Scientific Publications, Oxford, pp. 29–52.

Hauser, L., Carvalho, G.R., Hughes, R.N. and Carter, R.E. (1992) Clonal structure of the introduced freshwater snail *Potamopyrgus antipodarum* (Prosobranchia: Hydrobiidae), as revealed by DNA fingerprinting. *Proc. R. Soc. Lond., Series B*, **249**, 19–25.

Havel, J.E. and Hebert, P.D.N. (1989) Apomictic parthenogenesis and genotypic diversity in *Cypridopsis vidua* (Ostracoda: Cyprididae). *Heredity*, **62**, 383–92.

Havel, J.E., Hebert, P.D.N. and Delorme, L.D. (1990a) Genetics of sexual Ostracoda from a low arctic site. *J. Evol. Biol.*, **3**, 65–84.

Havel, J.E., Hebert, P.D.N. and Delorme, L.D. (1990b) Genotypic diversity of asexual Ostracoda from a low Arctic site. *J. Evol. Biol.*, **3**, 391–410.

Hebert, P.D.N. (1974a) Enzyme variability in natural populations of *Daphnia magna*. III. Genotypic frequencies in intermittent populations. *Genetics*, **77**, 335–41.

Hebert, P.D.N. (1974b) Enzyme variability in natural populations of *Daphnia magna*. II. Genotypic frequencies in permanent populations. *Genetics*, **77**, 323–34.

Hebert, P.D.N. (1974c) Enzyme variability in natural populations of *Daphnia magna*. I. Population structure in East Anglia. *Evolution*, **28**, 546–56.

Hebert, P.D.N. (1978) The population biology of *Daphnia*. *Biol. Rev.*, **53**, 387–426.

Hebert, P.D.N. (1980) The genetics of Cladocera, in *Evolution and Ecology of Zooplankton Communities* (ed. W.C. Kerfoot), University Press of New England, Hanover, pp. 329–36.

Hebert, P.D.N. (1981) Obligate asexuality in *Daphnia*. *Am. Nat.*, **117**, 784–9.

Hebert, P.D.N. (1982) Competition in zooplankton communities. *Ann. Zool. Fennici*, **19**, 349–56.

Hebert, P.D.N. (1987a) Genotypic characteristics of cyclical parthenogens and their obligately asexual derivatives, in *The Evolution of Sex and its Consequences* (ed. S.C. Stearns), Birkhauser, Basel, pp. 175–95.

Hebert, P.D.N. (1987b) Genotypic characteristics of the Cladocera. *Hydrobiologia*, **145**, 183–93.

Hebert, P.D.N. and Crease, T.J. (1980) Clonal coexistence in *Daphnia pulex* Leydig: another planktonic paradox. *Science*, **207**, 1373–5.

Hebert, P.D.N. and Crease, T.J. (1983) Clonal diversity in populations of *Daphnia pulex* reproducing by obligate parthenogenesis. *Heredity*, **51**, 353–69.

Hebert, P.D.N. and Emery, C.E. (1990) The adaptive significance of cuticular pigmentation in *Daphnia*. *Funct. Ecol.*, 4, 703–10.

Hebert, P.D.N., Ferrari, D.C. and Crease, T.J. (1982) Heterosis in *Daphnia*: a re-assessment. *Am. Nat.*, **119**, 427–34.

Hebert, P.D.N. and Moran, C. (1980) Enzyme variability in natural populations of *Daphnia carinata* King. *Heredity*, **45**, 313–21.

Hedrick, P.W. (1986) Genetic polymorphism in heterogeneous environments: a decade later. *Annu. Rev. Ecol. Syst.*, **17**, 535–66.

Herzig, A. (1985) Resting eggs – a significant stage in the life cycle of the crustaceans *Leptodora kindti* and *Bythotrephes longimanus*. *Verhandlungen Int. Vereinigung Theoret. Angewandte Limnol.*, **22**, 3088–98.

Heyward, A.J. and Stoddart, J.A. (1985) Genetic structure of two species of *Montipora* on a patch reef: conflicting results from electrophoresis and histocompatibility. *Mar. Biol.*, **75**, 117–22.

Hoffman, R.J. (1986) Variation in contributions of asexual reproduction to the genetic structure of populations of the sea anemone *Metridium senile. Evolution*, **40**, 357–65.

Hoffman, R.J. (1987) Short-term stability of genetic structure of populations of the sea anemone *Metridium senile. Mar. Biol.*, **93**, 499–507.

Hughes, D.J. (1989b) Variation in reproductive strategy among clones of the bryozoan, *Celleporella hyalina. Ecol. Monogr.* **59**, 387–403.

Hughes, D.J. (1992) Genotype–environment interactions and relative clonal fitness in a marine bryozoan. *J. Anim. Ecol.*, **61**, 291–306.

Hughes, R.N. (1987) The functional ecology of clonal animals. *Funct. Ecol.*, **1**, 63–9.

Hughes, R.N. (1989a) *A Functional Biology of Clonal Animals*, Chapman & Hall, London and New York.

Hughes, R.N. and Cancino, J.M. (1985) An ecological overview of cloning in Metazoa, in *Population Biology and Evolution of Clonal Organisms* (eds J.B.C. Jackson, L.W. Buss and R.E. Cook), Yale University Press, New Haven and London, pp. 153–86.

Hughes, T.P., Ayre, D. and Connell, J.H. (1992) The evolutionary ecology of corals. *Trends Ecol. Evol.*, **9**, 292–8.

Hughes, T.P. and Connell, J.H. (1987) Population dynamics based on size or age? A reef coral analysis. *Am. Nat.*, **129**, 818–29.

Hughes, T.P. and Jackson, J.B.C. (1980) Do corals lie about their age? Some demographic consequences of partial mortality, fission and fusion. *Science*, **209**, 713–15.

Innes, D.J. and Hebert, P.D.N. (1988) The origin and genetic basis of obligate parthenogenesis in *Daphnia pulex. Evolution*, **42**, 1024–35.

Ivker, F.B. (1972) A hierarchy of histo-compatibility in *Hydractinia echinata. Biol. Bull.*, **143**, 162–74.

Jackson, J.B.C. (1979) Morphological strategies of sessile animals, in *Biology and Systematics of Colonial Organisms* (eds G. Larwood and B.R. Rosen), Academic Press, London and New York, pp. 499–555.

Jackson, J.B.C. (1985) Distribution and ecology of clonal and aclonal benthic invertebrates, in *Population Biology and Evolution of Clonal Organisms* (eds J.B.C. Jackson, L.W. Buss and R.E. Cook), Yale University Press, New Haven and London, pp. 297–356.

Jackson, J.B.C. (1986) Modes of dispersal of benthic invertebrates: consequences for species distributions and genetic structure of local populations. *Bull. Mar. Sci.*, **39**, 588–606.

Jackson, J.B.C. and Coates, A.G. (1986) Life cycles and evolution of clonal (modular) animals. *Phil. Trans. R. Soc. Lond., Series B*, **313**, 7–22.

Jacobs, J. (1990) Microevolution in predominantly clonal populations of pelagic *Daphnia* (Crustacea: Phyllopoda): selection, exchange and sex. *J. Evol. Biol.*, **3**, 257–282.

Jeffreys, A.J., Wilson, V. and Thein, S.W. (1985) Hypervariable minisatellite regions in human DNA. *Nature, Lond.*, **314**, 67–73.

Kays, S. and Harper, J.L. (1974) The regulation of plant and tiller density in a grass sward. *J. Ecol.*, **63**, 97–105.

King, C.E. (1972) Adaptation of rotifers to seasonal variation. *Ecology*, **53**, 408–18.

King, C.E. (1980) The genetic structure of zooplankton populations, in *Evolution and Ecology of Zooplankton Communities*, University Press of New England, Hanover, pp. 315–28.

Korpelainen, H. (1984) Genic differentiation of *Daphnia magna* populations. *Hereditas*, **101**, 209–16.

Kricka, L.J. (1992) *Nonisotopic DNA Probe Techniques*, Academic Press, London.

Larwood, G. and Rosen, B.R. (1979) *Biology and Systematics of Colonial Organisms*. The Systematics Association Special Volume no. 11, Academic Press, London, New York and San Francisco.

Levin, D.A. (1977) The genetic implication of different modes of reproduction in plants in relation to their environment, in *A Synthesis of Demographic and Experimental Approaches to the Functioning of Plants*, International Symposium, Wageningen, Holland.

Lively, C.M. (1987) Evidence from a New Zealand snail for the maintenance of sex by parasitism. *Nature Lond.*, **328**, 519–21.

Lively, C.M. (1992) Parthenogenesis in a freshwater snail: reproductive assurance versus parasitic release. *Evolution*, **46**, 907–13.

Lively, C.M., Craddock, C. and Vrijenhoek, R.C. (1990) Red Queen hypothesis supported by parasitism in sexual and clonal fish. *Nature*, **344**, 864–66.

Loaring, J.M. and Hebert, P.D.N. (1981) Ecological differences among clones of *Daphnia pulex*. *Oecologia*, **51**, 162–8.

Lokki, J. (1976) Genetic polymorphism and evolution in parthenogenetic animals. VII. The amount of heterozygosity in diploid populations. *Hereditas*, **83**, 57–64.

Lomnicki, M. (1988) *Population Ecology of Individuals*, Princeton University Press, New Jersey.

Lynch M. (1983) Ecological genetics of *Daphnia pulex*. *Evolution*, **37**, 358–74.

Lynch, M. (1984a) The limits to life history evolution in *Daphnia*. *Evolution*, **38**, 465–82.

Lynch, M. (1984b) Destablising hybridization, general-purpose genotypes and geographical parthenogenesis. *Q. Rev. Biol.*, **59**, 257–90.

Lynch, M. (1985) Spontaneous mutations for life-history characters in an obligate parthenogen. *Evolution*, **39**, 804–18.

Lynch, M. (1988) Estimation of relatedness by DNA fingerprinting. *Mol. Biol. Evol.*, **5**, 584–99.

Lynch, M. and Gabriel, W. (1983) Phenotypic evolution and parthenogenesis. *Am. Nat.*, **122**, 745–64.

Lynch, M., Spitze, K. and Crease, T.J. (1989) The distribution of life history variation in the *Daphnia pulex* complex. *Evolution*, **43**, 1724–36.

Maynard Smith, J. (1978) *The Evolution of Sex*, Cambridge University Press, Cambridge.

McFadden, C.S. (1991) A comparative demographic analysis of clonal reproduction in a temperate soft coral. *Ecology*, **72**, 1849–66.

Moore, W.S. (1984) Evolutionary ecology of unisexual fishes, in *Evolutionary Genetics of Fish* (ed. B.J. Turner), Plenum Press, New York, pp 329–98.

Moritz, C. (1987) A note on the hatching and viability of *Ceriodaphnia* ephippia collected from lake sediment. *Hydrobiologia*, **145**, 309–14.

Mort, M.A. (1989) Ecological genetics of *Daphnia*: response of coexisting genotypes to resource manipulation. *Arch. Hydrobiol.*, **117**, 141–61.

Mort, M.A. (1991) Bridging the gap between ecology and genetics: the case of freshwater zooplankton. *Trends Ecol. Evol.*, **6**, 41–4.

Mort, M.A. and Wolf, H.G. (1985) Enzyme variability in large lake *Daphnia* populations. *Heredity*, **55**, 27–36.

Mort, M.A. and Wolf, H.G. (1986) The genetic structure of large lake *Daphnia* populations. *Evolution*, **40**, 756–66.

Muller, H.J. (1964) The relation of recombination to mutational advance. *Mutat. Res.*, **1**, 2–9.

Niegel, J.E. and Avise, J.C. (1983) Clonal diversity and population structure in a reef building coral, *Acropora cervicornis*: self-recognition analysis and demographic interpretation. *Evolution*, **37**, 437–53.

Pace, M.L., Porter, K. and Feig, Y.S. (1984) Life history variation within a parthenogenetic population of *Daphnia parvula* (Crustacea: Cladocera). *Oecologia*, **63**, 43–51.

Parejko, K. and Dodson, S.I. (1991) The evolutionary ecology of an anti-predator reaction norm: *Daphnia pulex* and *Chaoborus americanus*. *Evolution*, **45**, 1665–74.

Parker, E.D. (1979) Ecological implications of clonal diversity in parthenogenetic morphospecies. *Am. Zool.*, **19**, 753–62.

Pearse, J.S., Pearse, V.B. and Newberry, A.T. (1989) Telling sex from growth: dissolving Maynard Smith's paradox. *Bull. Mar. Sci.*, **45**, 433–46.

Peters, R.H. and de Bernardi, R. (1987) *Daphnia*, Isituto Italiano di Idrobiologia, Verbania Pallanza.

Rossi, V. and Menozzi, P. (1990) The clonal ecology of *Heterocypris incongruens* (Ostracoda). *Oikos*, **57**, 388–98.

Rossi, V. and Menozzi, P. (1992) The clonal ecology of *Heterocypris incongruens* (Ostracoda): life-history traits and photoperiod. *Funct. Ecol.*, **7**, 177–82.

Rossi, V., Rozzi, M.C. and Menozzi, P. (1991) Life strategy differences among electrophoretic clones of *Heterocypris incongruens* (Crustacea: Ostracoda). *Verhandlungen Int. Vereinigung Theoret. Angewandte Limnol.*, **24**, 2816–19.

Sabbadin, A. (1979) Colonial structure and genetic patterns in ascidians, in *Biology and Systematics of Colonial Organisms* (eds G. Larwood and B.R. Rosen), Academic Press, London and New York, pp. 433–44.

Sebens, K.P. and Thorne, B.L. (1985) Coexistence of clones, clonal diversity, and the effects of disturbance, in *Population Biology and Evolution of Clonal Organisms* (eds J.B.C. Jackson, L.W. Buss and R.E. Cook), Yale University Press, New Haven and London, pp. 357–98.

Shaw, P.W. (1991) Effects of asexual reproduction on population structure of *Sagartia elegans* (Anthozoa: Actinaria). *Hydrobiologia*, **216/217**, 519–25.

Shaw, P.W., Beardmore, J.A. and Ryland, J.S. (1987) *Sagartia troglodytes* (Anthozoa: Actinaria) consists of two species. *Mar. Ecol. Prog. Series*, **41**, 21–8.

Shick, J.M. (1991) *A Functional Biology of Sea Anemones*, Chapman & Hall, London.

Shick, J.M. and Dowse, H.B. (1985) Genetic basis of physiological variation in natural populations of sea anemones: intra- and inter-clonal analyses of variance, in *Proceedings of the Nineteenth European Marine Biology Symposium* (ed. P.E. Gibbs), Cambridge University Press, Cambridge, pp. 465–79.

Shick, J.M., Hoffman, R.J. and Lamb, A.N. (1979) Asexual reproduction, population structure, and genotype–environment interactions in sea anemones. *Am. Zool.*, **19**, 699–713.

Shick, J.M. and Lamb, A.N. (1977) Asexual reproduction and genetic population structure in the colonizing sea anemone *Haliplanella luciae*. *Biol. Bull.*, **153**, 604–17.

Shields, W.M. (1982) *Philopatry, Inbreeding and Evolution of Sex*, State University of New York Press, Albany, New York.

Snell, T.W. and Hawkinson, C.A. (1983) Behavioural reproductive isolation among populations of the rotifer *Brachionus plicatilis*. *Evolution*, **37**, 1294–305.

Sole-Cava, A.M., Thorpe, J.P. and Kaye, J.G. (1985) Reproductive isolation with little genetic divergence between *Urticina* (= *Tealia*) *felina* and *U. eques* (Anthozoa: Actinaria). *Mar. Biol.*, **85**, 279–84.

Stearns, S.C. (1987) *The Evolution of Sex and its Consequences*, Birkhauser, Berlin.

Stearns, S.C. (1989) The evolutionary significance of phenotypic plasticity. *Bioscience*, **39**, 436–45.

Stoddart, J.A. (1983) A genotypic diversity measure. *J. Hered.*, **74**, 489–90.

Stoddart, J.A. (1984) Genetical structure within populations of the coral *Pocillopora damicornis*. *Mar. Biol.*, **81**, 19–30.

Stoddart, J.A. and Taylor, J.F. (1988) Genotypic diversity: estimation and prediction in samples. *Genetics*, **118**, 705–11.

Templeton, A.R. (1979) The unit of selection in *Drosophila mercatorum* II. Genetic revolutions and the origin of coadapted genomes. *Genetics*, **92**, 1265–82.

Travis, J. and Mueller, L.D. (1989) Blending ecology and genetics: progress toward a unified population biology, in *Perspectives in Ecological Theory* (eds J. Roughgarden, R.M. May and S.A. Levin, Princeton University Press, New Jersey, pp. 101–24.

Tuomi, J. and Vuorisalo, T. (1989) Hierarchical selection in modular organisms. *Trends Ecol. Evol.*, **4**, 209–13.

Turner, B.J., Elder, J.F. and Davis, W.P. (1990) Genetic variation in clonal vertebrates detected by simple-sequence DNA fingerprinting. *Proc. Natl. Acad. Sci., USA*, **87**, 5653–7.

Uyenoyama, M. (1984) Inbreeding and the evolution of altruism under kin selection. *Evolution*, **38**, 778–95.

Van Valen, L. (1973) A new evolutionary law. *Evol. Theory*, **1**, 1–30.

Vrijenhoek, R.C. (1984) Ecological differentiation among clones: the frozen niche variation model, in *Population Biology and Evolution* (eds K. Wöhrmann and V. Loeschke), Springer-Verlag, Berlin, Heidelberg, New York and Tokyo, pp. 217–31.

Vuorisalo, T. and Tuomi, J. (1986) Unitary and modular organisms: criteria for ecological division. *Oikos*, **47**, 382–5.

Walsh, E.J. and Zhang, L. (1992) Polyploidy and body size variation in a natural population of the rotifer, *Euchlanis dilatata. J. Evol. Biol.*, **5**, 345–53.

Ward, R.D., Tollit, R.M. and Bickerton, M.A. (1991) Genetic variation and reproductive mode in populations of *Simocephalus vetulus* (Branchiopoda: Cladocera) from the East Midlands of England. *Freshwater Biol.*, **25**, 41–50.

Weider, L.J. (1987) Life history variation among low-Arctic clones of obligately parthenogenetic *Daphnia pulex*: a diploid–polyploid complex. *Oecologia*, **73**, 251–56.

Weider, L.J. (1992) Disturbance, competition and the maintenance of clonal diversity in *Daphnia pulex*. *J. Evol. Biol.*, **5**, 505–21.

Weider, L.J. and Hebert, P.D.N. (1987) Ecological and physiological differentiation of low Arctic clones of *Daphnia pulex. Ecology*, **68**, 188–98.

Weider, L.J. and Lampert, W. (1985) Differential response of *Daphnia* genotypes to oxygen stress: respiration rates, hemoglobin content and low oxygen tolerance. *Oecologia*, **65**, 487–91.

Weider, L.J. and Wolf, H.G. (1991) Life-history variation in a hybrid species complex of *Daphnia*. *Oecologia*, **87**, 506–13.

Williams, G.C. (1975) *Sex and Evolution*, Princeton University Press, Princeton, New Jersey.

Wilson, C.C. and Hebert, P.D.N. (1992) The maintenance of taxon diversity in an asexual assemblage: an experimental analysis. *Ecology*, **73**, 1462–72.

Wolf, H.G. (1987) Interspecific hybridization between *Daphnia hyalina*, *D. galeata* and *D. cucullata* and seasonal abundances of these species and their hybrids. *Hydrobiologia*, **145**, 213–17.

Wolf, H.G. and Carvalho, G.R. (1989) Resting eggs of lake *Daphnia*. II. *In situ* observations on the hatching of eggs and their contribution to population and community structure. *Freshwater Biol.*, **22**, 471–8.

Wolf, H.G. and Mort, M.A. (1986) Inter-specific hybridization underlies phenotypic variability in *Daphnia* populations. *Oecologia*, **68**, 507–11.

Yampolsky, L.Y. (1992) Genetic variation in the sexual reproduction rate within a population of a cyclic parthenogen, *Daphnia magna*. *Evolution*, **46**, 833–7.

Young, P.J.W. (1975) Enzyme polymorphisms and reproduction in *Daphnia magna*, PhD thesis, University of Cambridge.

Young, P.J.W. (1979) Enzyme polymorphism and cyclical parthenogenesis in *Daphnia magna*. II. Evidence of heterosis. *Genetics*, **92**, 971–82.

Zhang, L. and Lefcort, H. (1991) The effects of ploidy level on the thermal distributions of brineshrimp *Artemia parthenogenetica* and its ecological implications. *Heredity*, **66**, 445–52.

Zhao, Y. and King, C.E. (1989) Ecological genetics of the rotifer, *Brachionus plicatilis* in Soda Lake Nevada, USA. *Hydrobiologia*, **185**, 175–81.

6.2 HABITAT PARTITIONING IN *DAPHNIA*: COEXISTENCE OF *DAPHNIA MAGNA* CLONES DIFFERING IN PHOTOTACTIC BEHAVIOUR

Luc De Meester

Abstract

A competition experiment was carried out in a simplified laboratory environment, using three Daphnia magna *clones differing in phototactic behaviour. Genotype frequencies were determined after 94 days. Interactions were asymmetric, with the positively phototactic clone being dominant over the others. The negatively phototactic clone was strongly suppressed by both the intermediately and the positively phototactic clone, but survived at low density in all di-cultures. The vertical distribution in the culture jars proved to be strongly genotype-dependent, in pure as well as in mixed cultures.*

The prolonged coexistence of genotypes with different vertical distribution under simplified and small-scale laboratory conditions strongly suggests that genotypes with different vertical distribution and/or migration behaviour may often coexist in natural populations, their coexistence probably in part being mediated by habitat selection itself.

6.2.1 Introduction

The variability in vertical migration patterns of zooplankton is striking, as is made abundantly clear by reviews on the topic (e.g. Bayly, 1986; Haney, 1988). For several species, differences in migration behaviour have been reported between (Gliwicz, 1986; Ohman, 1990; Neill, 1992) as well as within populations, the latter mostly related to, often seasonal, changes in predator regimes (Stich and Lampert, 1981; Bollens and Frost, 1989; Neill, 1990; Ohman, 1990; Ringelberg, 1991). Such variability may be the result of phenotypic plasticity (Johnsen and Jakobsen, 1987; Neill, 1990; Ringelberg, 1991) as well as of selection for genotypes with different vertical migration patterns (Weider, 1984; Gliwicz, 1986; Dumont and De Meester, 1990; Neill, 1992). The latter implies significant intrapopulation genetic variability for the trait. It has been shown that the variability in phototactic behaviour in *D. magna* has a significant genetic component (De Meester and Dumont, 1988; De Meester, 1989, 1991a), with evidence for both within- and between-population genetic differences for this trait (De Meester, 1991a,b).

It has been hypothesized that habitat selection through diurnal vertical migration may be a mechanism to lower competitive interactions (Healey, 1967; Roe, 1974; Lane, 1975; Arndt and Heerkloss, 1989). Hebert (1982) and DeMott (1989) have pointed out, however, that migrating and non-migrating zooplankton essentially compete for the same food, such that vertical migration may not be a mechanism by which exploitative competition can be lowered.

The results of competition experiments with cladocerans have been reviewed by Bengtsson (1987). Studies dealing with competition between species are more numerous than studies of competition between genotypes of a given species. Loaring and Hebert (1981) observed coexistence for 80 days for only one of all possible pairwise combinations of four clones of *Daphnia pulex*, and only under particular environmental circumstances. Perrin, Baird and Calow (1992) carried out competition experiments with two *D. magna* clones differing in intrinsic rate of increase but not in carrying capacity. Their results suggest that competitive success in high-density cultures is related to the resistance to starvation. Competitive exclusion was, however, rarely observed. Weider (1992) observed that intermediate levels of disturbance (lowering of the density through dilution) promoted coexistence of *D. pulex* clones in laboratory microcosms. In his study of competition between different *Asplanchna brightwelli* clones, Snell (1979) reports clonal replacement in 5 to 16 generations.

In an effort to determine whether genotypes with different vertical distribution can coexist, we carried out a competition experiment with three *D. magna* clones that differ in phototactic behaviour. Coexistence in simple culture jars would indicate opportunities for coexistence in natural populations.

6.2.2   Materials and methods

We studied pure and mixed cultures of the positively phototactic clone C_1, the intermediately phototactic clone P_181, and the negatively phototactic clone P_112 (De Meester, 1991a). Clone C_1 was isolated from a small fishless city pond in Ghent (Eastern Flanders), whereas clones P_112 and P_181 are full sibs from intraclonal mixis within clone P_1, which was isolated from a pond (Driehoekvijver) in Heusden (Eastern Flanders).

Cultures were started on 27 March 1991, by placing six female neonates less than 24 h old in glass jars containing 0.5 l medium. Neonates were taken from cultures that were grown in 1 l jars at 20 °C ($\pm$ 1 °C), low density (< 25 adults/l) and abundant food ($3–5 \times 10^5$ *Scenedesmus acutus* cells per ml).

In addition to the pure cultures with each of the three clones established as controls, mixed cultures were started with all two-genotype combinations ($C_1 + P_112$, $C_1 + P_181$ and $P_181 + P_112$; these di-cultures are typed [+–], [+×] and [–×] respectively from here on) and with the three clones together ($C_1 + P_181 + P_112$; [+×–]). For each of the two-genotype cultures, three neonates of each of the two clones were used; two neonates of each clone were inoculated for the three-genotype culture. All pure and mixed cultures were established in triplicate. Cultures were kept at 20 °C ($\pm$ 1 °C) and at long day photoperiod (14 h light, 10 h dark; fluorescence tubes, 103×10^{-3} W/m²). The medium at the start of the competition experiments consisted of dechlorinated tapwater enriched with 1×10^5 *Scenedesmus acutus* cells per ml. Three times a week, half of the medium was replaced with fresh dechlorinated tapwater, and food density was restored to $3–3.5 \times 10^5$ *Scenedesmus* cells per ml. To reduce minor differences in culture conditions that may accumulate in the course of the experiment, the medium of all cultures was mixed three times a week. This was done as follows: after the animals had been transferred (on a 60 μm mesh screen) to the fresh medium (250 ml, containing $6–7 \times 10^5$ *Scenedesmus* cells per ml), the 'old' medium of all the cultures was mixed, filtered over 60 μm, and redistributed. Medium and initial food concentrations were therefore identical for all cultures. Density was not controlled. The number of adult parthenogenetic and sexual females, males and juveniles was determined at day 5, 16, 36, 65, 78 and 94. For parthenogenetic females, the numbers of eggs were determined; barren females > 2 mm were considered to be adults. They were considered to be parthenogenetic, unless the development of an ephippial case was clearly visible. Deposited ephippia were counted and removed three times a week.

To get an idea of the vertical distribution of the animals in the culture jars (14 cm high), the animals were counted in two batches of 250 ml medium for each culture, separated by decantation. Percentages of animals in the upper and lower half could thus be determined. This method is admittedly crude, but the overall tendencies can be compared with more precise results from experiments in a laboratory set-up (De Meester, 1991a, 1993b) and outdoor containers (De Meester, 1993a).

At day 94, the experiment was terminated, and genotype frequencies in the mixed cultures were determined by enzyme electrophoresis. The three clones could be recognized by their banding patterns at the malate dehydrogenase (*Mdh*; cellulose acetate electrophoresis, Helena Laboratories) (Hebert and Beaton, 1989) and esterase (*Est*; native polyacrylamide gel electrophoresis, using Phastsystem, Pharmacia) loci. Due to technical difficulties, separation at the *Est* locus, necessary to distinguish clones P_112 and P_181, could only be done for two of the three replicates. For each culture, 13 adult females were randomly picked out from the upper 250 ml, and 26 from the lower 250 ml. At day 94, therefore, the relative abundance of the adults of the different clones in the mixed cultures, as well as their vertical distribution, could be reconstructed.

Life-history characteristics of the three clones were determined in a life table experiment. Six neonates less than 24 h old were individually raised in 1 l jars, and kept on a 14 h light/10 h dark photoperiod at 20 °C ($\pm$ 1 °C). Four-fifths of the medium was replaced with fresh dechlorinated tapwater daily, at a final food concentration of 5×10^5 *Scenedesmus* cells per ml. In this experiment, too, all medium was mixed and redistributed over the cultures. The experiment was run until the death of all animals. The intrinsic rate of natural increase (r_{max}) was determined by solving iteratively Euler's equation, and variability was estimated using the jackknife technique (Meyer *et al.*, 1986).

All statistical analyses were done using the SYSTAT program (Wilkinson, 1990).

6.2.3 Results

(a) Competition experiments

After about 2 weeks, the number of animals in the cultures can be considered more or less in equilibrium (Figure 6.2.1). The number of adult females is significantly different between the pure cultures of the different genotypes, with the intermediately phototactic genotype having the highest density, and the negatively phototactic the lowest (Table 6.2.1; a repeated measurements ANOVA, incorporating all observations on days 16, 36, 65, 78 and 94, also yields a significant effect, $p < 0.001$, of culture type on total density as well as the number of adult females per culture). All mixed cultures have a similar density of about 260 adult females/l, which is lower than the average density of the pure cultures of the intermediately phototactic clone (about 280 adult females/l). However, this difference is not significant at the $p = 0.05$ level. The equilibrium density of the [+−] mixed culture tends to be higher than that of both pure cultures (Figure 6.2.1). The difference is, however, only significant for the comparison with the negatively phototactic clone.

N (>2mm)

(a)

Experimental time (days)

+ — +−

N (>2mm)

(b)

Experimental time (days)

+ x +x +x−

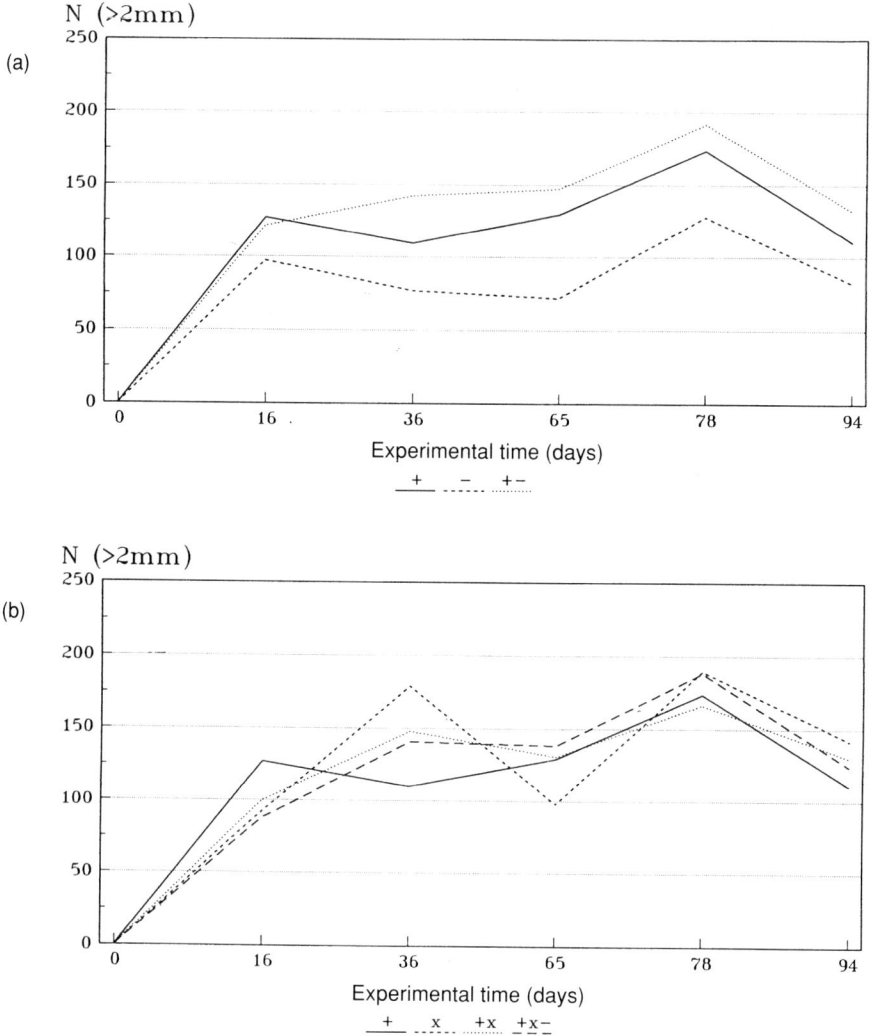

Figure 6.2.1 The changes in average number of adult females in pure cultures compared to that of mixed cultures; (a) pure cultures of the negatively (−) and the positively phototactic (+) clone, and their di-cultures; (b) the intermediately phototactic (×) and the positively phototactic (+) clone in pure cultures, di-cultures and tri-cultures.

The genetic composition of the mixed cultures at day 94 reveals a superiority of the positively phototactic clone to both other clones (Table 6.2.1). In none of the mixed cultures is the density of clone C_1 significantly lower than in the pure cultures, though there is a tendency for lower numbers in the presence of the intermediately phototactic clone P_181. As a matter of fact, the average density of clone C_1 in the di-cultures with the negatively phototactic clone P_112 is slightly higher than that in the pure cultures. Density of the clone P_181 is suppressed in the presence of both clones C_1 and P_112, though it is by far more strongly suppressed in the presence of the positively phototactic clone than in the presence of the negatively phototactic one. Finally, clone P_112 is very strongly suppressed in the

Table 6.2.1 Equilibrium density, population structure, and average number of adult females of the three genotypes in pure and mixed cultures, at the end of the experiment (day 94)

	+	×	−	+−	+×	×−	+×−	*Significance*
$N_{> 2\ mm}$	110.7	142.3	82.0	132.0	130.3	132.3	124.3	
	(9.3)	(4.9)	(10.0)	(4.6)	(14.6)	(8.0)	(5.0)	
	A	B	C	A, B	A, B	A, B	A, B	***
N_{all}	114.3	154.7	111.0	136.3	137.3	146.0	132.0	
	(9.5)	(7.2)	(13.9)	(4.9)	(14.4)	(7.0)	(1.0)	
	A	B	A	A, B	A, B	B	A, B	***
%♂	0.9	7.7	27.8	1.8	3.0	7.8	3.3	
	(0.1)	(1.9)	(4.9)	(0.9)	(1.2)	(2.4)	(2.3)	
	A	B	C	A	A, B	A, B	A, B	***
$N(C_1)$	110.7			127.5	84.4		70.5	
	(9.3)			(7.9)	(13.4)		(31.7)	
	A			A, B	A		A, C	*
$N(P_181)$		142.3			45.9	125.8	71.9	
		(4.9)			(2.2)	(0.9)	(6.0)	
		A			B	C	D	***
$N(P_112)$		82.0		4.5		10.8		
		(10.0)		(3.4)		(4.0)	ND	
		A		B		B		***

Standard deviations are given between parentheses. $N_{>2\ mm}$ = number of females > 2 mm (adults) per culture; N_{all} = total number of animals per culture, including males and juveniles; % ♂ = percentage of males among adults; $N(×)$ = number of adult females of genotype ×; Significance = significance level of differences in a one-way ANOVA; $^*p < 0.05$; $^{**}p < 0.01$; $^{***}p < 0.001$; A, B, C = results of Tukey pairwise comparisons, different characters indicating significant differences at the $p = 0.05$ level; ND = none detected in sample.

presence of both other clones. However, this clone is still present at low frequency (less than 10% of the adult females) in all di-cultures at day 94. We were unable to detect the presence of P_112 adults in any of the tri-cultures.

(b) Sexual reproduction

The percentage of males among the adults at the end of the experiment is significantly different between the pure cultures of the different genotypes (Table 6.2.1). Clone C_1 has a low frequency (less than 1%) of males, whereas the pure cultures of clone P_112 have, on average, more than 25% males. The occurrence of males is intermediate in clone P_181. In the mixed cultures the proportion of males is rather low, reflecting the dominance of the clones C_1 and P_181 in these cultures (Table 6.2.1).

Clone C_1 produced significantly less ephippia during the 94-day period (about 125/culture) than the P_112 and P_181 sister clones (about 200/culture; Figure 6.2.2). The number of ephippia produced in the [+−] and [+×] di-cultures and the tri-cultures (range: 91–147) again reflects the numerical dominance of clone C_1 in the mixed cultures. Figure 6.2.3 depicts the average number of ephippia produced in the different pure and mixed cultures at a weekly interval. In most cultures the first ephippial cases are found in the fourth week (in one of the P_112 cultures one ephippial case was removed during the third week). This is shortly after the equilibrium density is reached, and about 2 weeks after a peak in juvenile numbers. The total number of animals at day 16 ranged from 289 (a P_112 pure culture) to

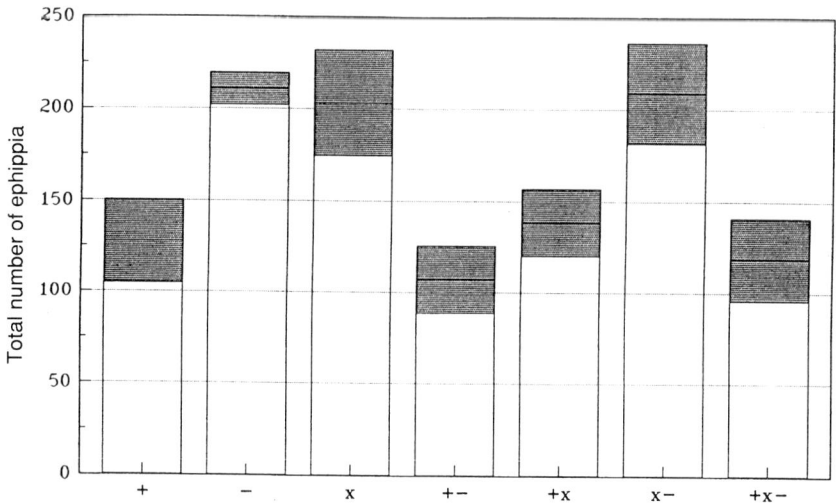

Figure 6.2.2 The total number of ephippia collected during the experimental period (94 days) plotted as the average per culture for the different pure and mixed cultures. The hatched bar area indicates the zone of ±1 standard deviation.

578 (a $P_1 81$ pure culture), which is more than twice the density at day 36 (or at day 94). Overall, the different cultures follow rather similar patterns of ephippial egg production, but the absolute numbers are different. Pure cultures of clone $P_1 12$ produce more ephippia than pure cultures of clone C_1 during the whole experimental period. Initially (week 4), the mixed culture of these clones yields an intermediate number of ephippia, but from then on its pattern is very similar to that of the pure cultures of clone C_1 (Figure 6.2.3(a)). Similarly, in the [×–] mixed cultures, clone $P_1 81$ takes over after about 5 or 6 weeks (Figure 6.2.3(b)).

(c)　Vertical distribution

The vertical distribution in the culture jars was clone-dependent: whereas more than 40% of the adult females of the positively phototactic clone C_1 were found in the upper 250 ml at day 94, not a single adult female of the negatively phototactic clone $P_1 12$ was found there

Table 6.2.2　Vertical distribution (%U: percentage in upper 250 ml, separated through decantation) of adult females of the positively phototactic clone C_1 (+), the intermediately phototactic clone $P_1 81$ (×), and the negatively phototactic clone $P_1 12$ (–) in pure and mixed cultures at day 94

	Pure	*Pure (♂)*	*+–*	*+×*	*×–*	*+×–*	*Significance*
% U(C_1)	43.2	66.7	38.7	58.8		62.1	*
	(9.0)	(70.7)	(1.6)	(12.1)		(8.0)	
% U ($P_1 81$)	15.6	40.6		10.2	24.3	25.2	NS
	(8.8)	(18.6)		(17.7)	(3.7)	(10.5)	
% U ($P_1 12$)	0.0	6.0	0.0		0.0		
	(0.0)	(3.3)	(0.0)		(0.0)	ND	

Standard deviations are given between parentheses. Pure (♂) = the percentage of adult males found in the upper 250 ml in the pure cultures of clone C_1 (three males only), clone $P_1 81$ (12 males) and clone $P_1 12$ (31 males); Significance = significance level of differences in adult female distribution, by one-way ANOVA; *$p < 0.05$; ND = none detected in sample.

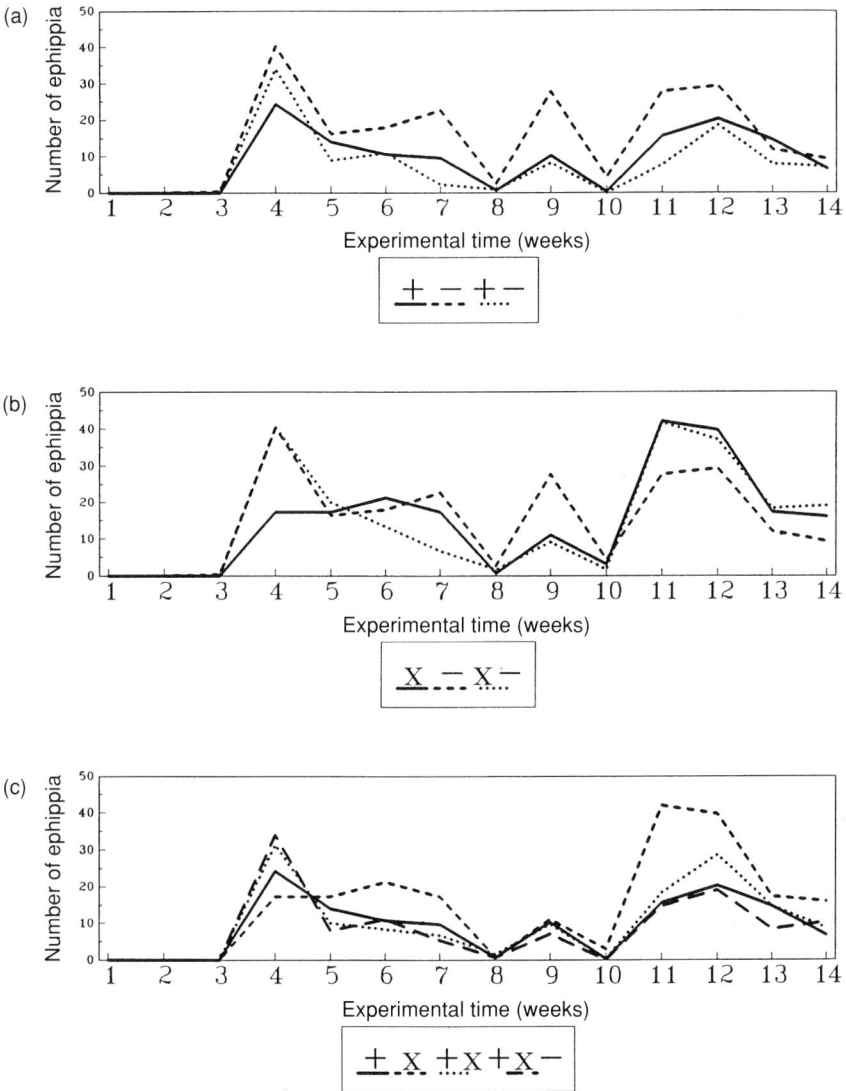

Figure 6.2.3 The average number of ephippia collected in the different pure and mixed cultures, plotted at a 1-week interval. (a) Comparison of the pure cultures of the positively phototactic clone C_1 and the negatively phototactic clone $P_1 12$, with their mixed culture [+–]; (b) comparison of the pure cultures of the intermediately phototactic clone $P_1 81$ and the negatively phototactic clone $P_1 12$, with their mixed culture [×–]; (c) comparison of the pure cultures of the positively phototactic clone C_1 and the intermediately phototactic clone $P_1 81$, with their mixed culture [+×] and the tri-culture [+×–].

(Figure 6.2.4). A one-way ANOVA on the observations of day 94 indicates an overall effect of culture type on the vertical distribution of adult females of clone C_1 (Table 6.2.2). The tendency of adult females of clone C_1 to be more predominant in the upper 250 ml in the presence of the intermediately phototactic clone is significant ($p = 0.022$) for the comparison between the tri-culture and the [+–] culture, and nearly so ($p = 0.064$) when the percentage of animals in the lower 250 ml in the [+×] culture is compared to that of the [+–] culture (no

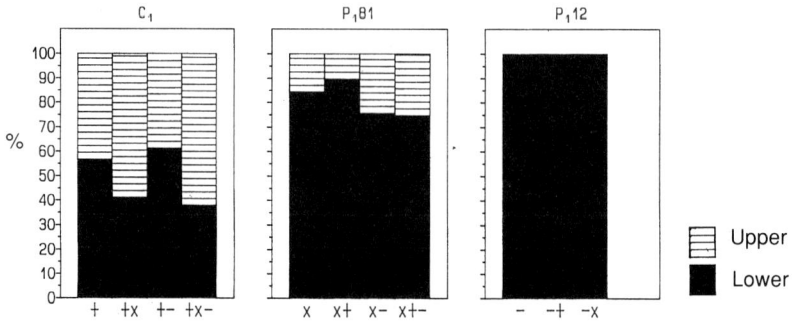

Figure 6.2.4 The vertical distribution of the positively phototactic clone C_1, the intermediately phototactic clone $P_1 81$ and the negatively phototactic clone $P_1 12$ in the various pure and mixed cultures, as determined on day 94. The blackened zone indicates the percentage of adult females found in the lower 250 ml.

other pairwise comparisons are significantly different). There is no significant effect of the presence of other clones on the vertical distribution of adult females of clone $P_1 81$ (Table 6.2.2).

(d) Life-history characteristics

There are significant differences between the three genotypes studied in several life-history characteristics (Table 6.2.3). The negatively phototactic clone is on average 24 h later than the other clones in releasing its first brood. The total reproductive output is largest by far in the intermediately phototactic clone $P_1 81$. The intrinsic rate of natural increase, however, is not significantly different between the positively phototactic clone C_1 and the intermediately phototactic clone $P_1 81$ (Table 6.2.3). This is probably in part due to the somewhat longer adult instar duration for clone $P_1 81$ (from release of the first to the release of the sixth brood, an average of 2.82 days/instar) than for clone C_1 (2.60 days/instar; the corresponding value

Table 6.2.3 Life-history characteristics of the positively phototactic clone C_1 (+), the intermediately phototactic clone $P_1 81$ (×), and the negatively phototactic clone $P_1 12$ (−)

	+	×	−	*Significance*
r_{max}	0.419	0.433	0.364	
	(0.018)	(0.013)	(0.047)	
	A	A	B	**
R_0	362.3	869.5	170.0	
	(292.9)	(196.3)	(119.0)	
	A	B	A	***
Time	8.7	8.7	9.8	
	(0.5)	(0.5)	(0.5)	
	A	A	B	**

Standard deviations are given between parentheses. r_{max} = intrinsic rate of natural increase, as determined by iteratively solving Euler's equation; R_0 = reproductive effort, the total number of female offspring produced; time = number of days until release of the first brood; Significance = significance level of differences in a one-way ANOVA (Kruskal-Wallis analysis of variance in the case of r_{max}); *$p < 0.05$; **$p < 0.01$; ***$p < 0.001$; A, B, C = visualized results of Tukey pairwise comparisons, different characters indicating significant differences at the $p = 0.05$ level (Mann-Whitney U-tests in the case of r_{max}).

for clone $P_1 12$ is 3.44 days/instar). The value for r_{max} is much lower in the negatively phototactic clone $P_1 12$ than in the other clones.

6.2.4 Discussion

The intermediately phototactic clone $P_1 81$ has a higher equilibrium density in the pure cultures than either of the extreme phenotypes. We note that the carrying capacity realized in our experimental jars is about twice that reported by Perrin, Baird and Calow (1992), despite the higher food level (daily restoration of food concentration to 2.5×10^6 *Chlorella* cells/ml) provided by these authors. It is possible that this discrepancy is related to the higher turnover rate of the medium in our study (Perrin, Baird and Calow (1992) refreshened the medium completely once a week). The relatively low equilibrium density of the negatively photo-tactic clone in comparison to that of her intermediately phototactic sister clone was to be expected, as negatives only use part of the available space. Indeed, by far the largest part of the population of clone $P_1 12$ remains at or near the bottom, similar to the behaviour of this clone in experimental set-ups (De Meester, 1991a,b) and in outdoor containers (De Meester, 1993a). Our present results indeed confirm the existence of a genotype-dependent habitat selection in *Daphnia*, in concordance with differences in phototactic behaviour. The number of males is too low to draw any firm conclusions, but, in concordance with results from experiments on phototactic behaviour (De Meester, 1992), we note a tendency for males of the intermediately phototactic clone to be more often found in the upper 250 ml than the females of that clone (Table 6.2.2).

The *Daphnia* clones used in the present study are not only significantly different in their behaviour related to vertical distribution; they also differ significantly in life-history characteristics. Furthermore, our results indicate that the different clones have a different tendency to produce males and ephippial eggs: the $P_1 12$ and $P_1 81$ clones produced on average about twice as many ephippia as clone C_1, and clone $P_1 12$ seems to be more prone to produce males than clone $P_1 81$. Such genotype-dependent sensitivity to stimuli that induce sexual reproduction, here related to density and food depletion, has been reported by several authors (e.g. Ferrari and Hebert, 1982; Carvalho and Hughes, 1983). Perrin, Baird and Calow (1992) also observed a difference in the amount of males and ephippia produced in the two *D. magna* clones they studied. However, in contrast to their results, we do not observe that the clone which tends most quickly to switch to sexual reproduction also has the highest intrinsic rate of increase when raised individually and at high food concentration.

Competition is strong in all cultures, as can be deduced from the low number of eggs pro-duced: at day 94, the total number of parthenogenetic eggs averages 3.3 in the $P_1 12$ pure cultures and 21.7 in the C_1 pure cultures, with all other cultures having values in between. The outcome of the competition experiments seems to be quite deterministic, as was also noted by Loaring and Hebert (1981). For instance, the negatively phototactic clone coexists, at low frequency, with the other clones in all of the 12 di-cultures, but is eliminated from all replicate tri-cultures. Clone C_1 is numerically dominant in all di-cultures, but dominance is for all replicates more expressed in the [+−] cultures than in the [+×] cultures.

Competitive interactions between the three genotypes studied are highly asymmetric, with the positively phototactic clone dominant over both the intermediately and the negatively phototactic clone, and the intermediately phototactic clone being dominant over the neg-atively phototactic one. Moreover, the interaction between the positively and negatively phototactic clones shows evidence for niche complementarity: the number of animals in the mixed culture is higher than that of either pure culture (though the difference is not significant with respect to the C_1 cultures), and the number of C_1 adults is not suppressed in the mixed culture compared to the pure culture. If anything, the density of C_1 animals tends to be higher in the presence of clone $P_1 12$ compared to the pure cultures. In fact, the interaction between positives and other genotypes in the present study is more appropriately typed as amensalism (0-interaction; Arthur and Mitchell, 1989) than as competition (— interaction). The interaction between the intermediately and the negatively phototactic clone is also asymmetrical. In this case, however, there is no indication of niche complementarity, and the density of both clones is significantly suppressed in the mixed cultures.

Though the results of the competition trials are deterministic, they cannot be adequately predicted from the performance of the different genotypes in pure cultures. The superiority of C_1 and $P_1 81$ is not expected from equilibrium densities in pure cultures. Furthermore, results of preliminary starvation experiments indicate that neonates of clone $P_1 81$ are able to survive on average for almost 2 days longer than neonates of the clone C_1 in the absence of food (average survival time was 4.37 days for clone C_1 neonates, 6.33 days for clone $P_1 81$ neonates, and 5.33 days for clone $P_1 12$ neonates; data based on four replicate batches of 10 animals). These results indicate that the outcome of the present competition experiments is probably not due to a superior starvation resistance of clone C_1 (Rothhaupt, 1990; Perrin, Baird and Calow, 1992). We suggest that the high competitive ability of clone C_1 in our experiments is the result of an efficient cropping of the food added three times a week. This is not necessarily directly related to its vertical distribution, because the food was homogeneously distributed in the culture jars, as it was added when the medium was mixed and redistributed. However, the relationship may be indirect, in that emphasis on efficient cropping of available resources is a trait that is coadapted to positive phototaxis. If there is a trade-off between cropping efficiency and starvation resistance, emphasis in the intermediately and negatively phototactic clones would tend to be more on starvation resistance. Co-selection of these traits seems logical in view of the vertical differences in quantity and quality of algal food under natural conditions.

While the inferiority of clone $P_1 12$ in the competition experiments was to be expected from its performance in pure cultures, its coexistence for almost 100 days in all di-cultures is less obvious. We suggest that this prolonged coexistence is the result of frequency-dependent selection, mediated through habitat partitioning. Though it is difficult to prove, our results indicate that the negatively phototactic clone is the superior competitor in the marginal habitat in which it occurs. This hypothesis is mainly inspired by the observation of a genotype-dependent vertical distribution mediated by genetic differences in phototactic behaviour (De Meester 1993a), together with the evidence for niche complementarity in the [+−] interaction.

Our observations on the changes of ephippial egg production with time in the different cultures provides indirect evidence of a rather abrupt clonal replacement, in 2 or 3 weeks from the fourth week onward. This also implies that the subdominant clone is able to coexist in the mixed cultures, at low frequency, for several (6–8 at least) weeks after it has been largely replaced by the dominant clone.

Though we admit that our study does not represent a real test of the hypothesis that differences in phototactic behaviour are the mechanism through which coexistence in *Daphnia* clones can be mediated, our results are suggestive. We indeed observe habitat segregation and coexistence of *Daphnia* genotypes with different phototactic behaviour in 0.5 l culture jars which are only 14 cm in height. The experimental period (94 days) equals about 10 generations (time required for juvenile development until release of the first brood), and is comparable to the growing season of many intermittent *Daphnia* populations. It is to be expected that opportunities for coexistence are much higher in natural habitats. It is also evident that the coexistence of the negatively phototactic clone with the other clones in the present study is not in the first place mediated by ecological similarity (Hebert, 1982), but is the result of habitat heterogeneity and genotype–environment interactions with respect to fitness (Vrijenhoek, 1979; Bell, 1991). How the superiority of the negatively phototactic clone in the near-bottom water layer is mediated is at the moment unclear. We suggest that these genotypes may show coadapted characteristics to a near-benthic lifestyle, such as a higher haemoglobin content (Landon and Stasiak, 1983). Our results indicate that the negatively phototactic genotype studied might persist in evolutionary time despite its low intrinsic capacity for increase. This persistence would be mediated through habitat selection and frequency-dependent selection, as well as by a high tendency to shift to sexual reproduction.

Vertical migration probably has not evolved as a mechanism to reduce competitive interactions between species and/or genotypes (Lampert, 1989). However, differences in vertical distribution and/or migration may facilitate coexistence (DeMott, 1989). Moreover, competitive interactions may have resulted in a modification of distribution or migration patterns. For instance, with respect to the vertical segregation of three *Daphnia* species (*D. cucullata*, *D. cristata* and *D. hyalina*) reported by Pijanowska and Dawidowicz (1987), predation pressure by fish can explain why the smaller species occur in shallower waters than the larger ones, whereas the displacement of the smaller species by the larger ones in the deeper waters probably reflects competitive abilities.

Our results indicate that coexistence of genotypes with different vertical daytime distribution and/or vertical migration in natural populations may be rather the rule than the exception. This suggests a capacity for rapid evolution of the trait by clonal selection in natural *Daphnia* populations. Our competition experiments consisted of systems with no predators. The coexistence at low frequency of negatively phototactic genotypes in populations that are dominated by positives and/or intermediates may enable a rapid genetic response to a change in (visual) predation pressure and other environmental changes (Dumont and De Meester, 1990).

Acknowledgements

Many thanks to Piet Spaak and Rob Hoekstra (Nieuwersluis, The Netherlands) for making me acquainted with the cellulose acetate electrophoresis technique during a short but nice stay at their laboratory. I thank two anonymous reviewers for constructive comments. Special thanks also to A. Beaumont for organizing the conference. The author is Senior Research Assistant with the National Fund for Scientific Research (Belgium). Financial travel support by the National Fund for Scientific Research is gratefully acknowledged.

References

Arndt, H. and Heerkloss, R. (1989) Diurnal variation in feeding and assimilation rates of planktonic rotifers and its possible ecological significance. *Int. Rev. Ges. Hydrobiol.*, **74**, 261–72.

Arthur, W. and Mitchell, P. (1989) A revised scheme for the classification of population interactions. *Oikos*, **56**, 141–3.

Bayly, I.A.E. (1986) Aspects of diel vertical migration in zooplankton, and its enigma variations, in *Limnology in Australia* (eds P. De Deckker and W.D. Williams), Junk, Dordrecht, pp. 349–68.

Bell, G. (1991) The ecology and genetics of fitness in *Chlamydomonas* III. Genotype-by-environment interaction within strains. *Evolution*, **45**, 668–79.

Bengtsson, J. (1987) Competitive dominance among Cladocera: are single-factor explanations enough? *Hydrobiologia*, **145**, 245–57.

Bollens, S.M. and Frost, B.W. (1989) Predator-induced diel vertical migration in a planktonic copepod. *J. Plankton Res.*, **11**, 1047–65.

Carvalho, G.R. and Hughes, R.N. (1983) The effect of food availability, female culture-density and photoperiod on ephippia production in *Daphnia magna* Straus (Crustacea: Cladocera). *Freshwat. Biol.*, **13**, 37–46.

De Meester, L. (1989) An estimation of the heritability of phototaxis in *Daphnia magna* Straus. *Oecologia*, **78**, 142–4.

De Meester, L. (1991a) An analysis of the phototactic behaviour of *Daphnia magna* clones and their sexual descendants. *Hydrobiologia*, **225**, 217–27.

De Meester, L. (1991b) Evidence for intra-population genetic variability for phototactic behaviour in *Daphnia magna* Straus, 1820. *Biol. Jb. Dodonaea*, **58**, 84–93.

De Meester, L. (1992) The phototactic behaviour of male and female *Daphnia magna*. *Anim. Behav.*, **43**, 696–8.

De Meester, L. (1993a) The vertical distribution of *Daphnia magna* genotypes selected for different phototactic behaviour: outdoor experiments. *Arch. Hydrobiol*, **39**, 137–55.

De Meester, L. (1993b) On the genetical ecology of phototactic behaviour in *Daphnia magna* (Crustacea: Cladocera). *Acad. Analecta* (in press).

De Meester, L. and Dumont, H.J. (1988) The genetics of phototaxis in *Daphnia magna*: existence of three phenotypes for vertical migration among parthenogenetic females. *Hydrobiologia*, **162**, 47–55.

DeMott, W.R. (1989) The role of competition in zooplankton succession, in *Plankton Ecology, Succession in Plankton Communities* (ed. U. Sommer), Springer Verlag, London, pp. 195–252.

Dumont, H.J. and De Meester, L. (1990) Are contrasting patterns of vertical migration in zooplankton the result of differential natural selection? *Rev. Brasil. Biol.*, **50**, 867–74.

Ferrari, D.C. and Hebert, P.D.N. (1982) The induction of sexual reproduction in *Daphnia magna*: genetic differences between arctic and temperate populations. *Can. J. Zool.*, **60**, 2143–8.

Gliwicz, M.Z. (1986) Predation and the evolution of vertical migration in zooplankton. *Nature*, **320**, 746–8.

Haney, J.F. (1988) Diel patterns of zooplankton behaviour. *Bull. Mar. Sci.*, **43**, 583–603.

Healey, M.C. (1967) The seasonal and diel changes in distribution of *Diaptomus leptopus* in a small eutrophic lake. *Limnol. Oceanogr.*, **12**, 34–9.

Hebert, P.D.N. (1982) Competition in zooplankton communities. *Ann. Zool. Fenn.* **19**, 349–56.

Hebert, P.D.N. and Beaton, M.J. (1989) *Methodologies for Allozyme Analysis using Cellulose Acetate Electrophoresis*, Helena Laboratories, Beaumont, Texas.

Johnsen, G.J. and Jakobsen, P.J. (1987). The effect of food limitation on vertical migration in *Daphnia longispina*. *Limnol. Oceanogr.*, **32**, 873–80.

Lampert, W. (1989) The adaptive significance of diel vertical migration of zooplankton. *Funct. Ecol.*, **3**, 21–7.

Landon, M.S. and Stasiak, R.H. (1983) *Daphnia* haemoglobin concentration as a function of depth and oxygen availability in Arco Lake, Minnesota, *Limnol. Oceanogr.*, **28**, 731–7.

Lane, P.A. (1975) The dynamics of aquatic systems: a comparative study of the structure of four zooplankton communities. *Ecol. Monogr.*, **45**, 307–36.

Loaring, J.M. and Hebert, P.D.N. (1981) Ecological differences among clones of *Daphnia pulex* Leydig. *Oecologia*, **51**, 162–8.

Meyer, J.S., Ingersoll, C.G., McDonald, L.L. and Boyce, M.S. (1986). Estimating uncertainty in population growth rates: jackknife vs. bootstrap techniques. *Ecology*, **67**, 1156–66.

Neill, W.E. (1990) Induced vertical migration in copepods as a defence against invertebrate predation. *Nature*, **345**, 524–26.

Neill, W.E. (1992) Population variation in the ontogeny of predator-induced vertical migration of copepods. *Nature*, **356**, 54–7.

Ohman, M.D. (1990) The demographic benefits of diel vertical migration by zooplankton. *Ecol. Monogr.*, **60**, 257–81.

Perrin, N., Baird, D.J. and Calow, P. (1992) Resource allocation, population dynamics and fitness: some experiments with *Daphnia magna* Straus. *Arch. Hydrobiol.*, **123**, 431–49.

Pijanowska, J. and Dawidowicz, P. (1987) The lack of vertical migration in *Daphnia*: the effect of homogeneously distributed food. *Hydrobiologia*, **148**, 175–81.

Ringelberg, J. (1991) Enhancement of the phototactic reaction in *Daphnia hyalina* by a chemical mediated by juvenile perch (*Perca fluviatilis*). *J. Plankton Res.*, **13**, 17–25.

Roe, H.S.J. (1974) Observations on the diurnal vertical migrations of an oceanic animal community. *Mar. Biol.*, **28**, 99–113.

Rothhaupt, K.O. (1990) Resource competition of herbivorous zooplankton: a review of approaches and perspectives. *Arch. Hydrobiol.*, **118**, 1–29.

Snell, T.W. (1979) Intraspecific competition and population structure in rotifers. *Ecology*, **60**, 494–502.

Stich, H.B. and Lampert, W. (1981) Predator evasion as an explanation of diurnal vertical migration by zooplankton. *Nature*, **293**, 396–8.

Vrijenhoek, R.C. (1979) Factors affecting clonal diversity and coexistence. *Am. Zool.*, **19**, 787–97.

Weider, L.J. (1984) Spatial heterogeneity of *Daphnia* genotypes: vertical migration and habitat partitioning. *Limnol. Oceanogr.*, **29**, 225–35.
Weider, L.J. (1992) Disturbance, competition and the maintenance of clonal diversity in *Daphnia pulex*. *J. Evol. Biol.*, **5**, 505–22.
Wilkinson, L. (1990) *SYSTAT: The System for Statistics*, Systat Inc., Evanston, IL.

6.3   VARIATION IN REPRODUCTION AND SEX ALLOCATION AMONG CLONES OF *DAPHNIA PULEX*

David J. Innes and Doreen R. Singleton

Abstract

To evaluate reproductive variation in Daphnia pulex, *10 clones were examined for variation in fecundity and the proportion of males produced under four densities (1, 2, 4 and 8 females in 60 ml) at 20 °C and constant photoperiod (16L : 8D). Cultures were fed aquarium-cultured algae (15 000 cells/ml per individual) and changed every 2 days. Brood release was approximately synchronous. Number and sex of the total progeny were counted for the first five broods. There was significant variation in fecundity among the clones, and fecundity decreased with increased density. No males were produced at the lowest density and the proportion of males increased with increasing density. The proportion of males varied significantly among clones. Three clones produced only 1–3% males while the remaining seven averaged 9–23% males at the highest densities. Male production was not restricted to earlier or later broods. The presence of among-clone variation for male production suggests the potential for evolutionary shifts in the pattern of sex allocation in* Daphnia.

6.3.1   Introduction

Daphnia characteristically reproduce by cyclical parthenogenesis in which parthenogenetic reproduction alternates with sexual reproduction. Sex in *Daphnia* appears to be determined by the environment (environmental sex determination, 'ESD') in contrast to the more commonly encountered genotypic sex determination (GSD) (Bull, 1983; Korpelainen, 1990). As long as environmental conditions are favourable, *Daphnia* reproduce parthenogenetically. The switch to sexual reproduction appears to be stimulated by factors associated with a deterioration in the environment (Hebert, 1978). During the sexual phase of the life-cycle, females produce parthenogenetic eggs that develop into males, and diapausing eggs that require fertilization in order to develop. A diapausing eggs is the only stage able to survive unfavourable conditions such as the pond freezing or drying up. Research by Agar (1914), Berg (1931, 1934) and Banta (1939) first demonstrated that the onset of sexual reproduction in *Daphnia* and other Cladocera was strongly influenced by environmental conditions. Their research refuted earlier suggestions that the transition from parthenogenetic to sexual reproduction was under internal control and that sexual reproduction would only occur after a fixed number of parthenogenetic generations independent of external conditions (Berg, 1934). These early studies, as well as more recent research, suggest that male and diapause egg production can be stimulated by changes in photoperiod, temperature, food supply or conditions connected with increased population density (Berg, 1931, 1934; Banta, 1939; Leary, 1967; Stross, 1969a,b; Stimpfl, 1971; Carvalho and Hughes, 1983; Korpelainen, 1989; Hobaek and Larsson, 1990).

Laboratory studies on sex expression in *Daphnia* have focused primarily on determining the level of various conditions (i.e. temperature, photoperiod, density) that elicit a switch from parthenogenetic to sexual reproduction (Leary, 1967; Stross, 1969a,b; Stimpfl, 1971; Carvalho and Hughes, 1983; Korpelainen, 1989; Hobaek and Larsson, 1990; Kleiven, Larsson and Hobaek, 1992). It is clear from these studies that conditions stimulating sexual reproduction can not only vary among species of *Daphnia* but can also vary among different populations of the same species. For example, the occurrence of sexual reproduction in Arctic populations tends to be more responsive to variation in photoperiod than density compared to temperate populations (Edmondson, 1955; Stross, 1969a,b; Frey, 1982; Ferrari and Hebert, 1982). Thus, although sex expression in *Daphnia* is strongly influenced by the environment this does not preclude genetic variation for the degree of response to environmental conditions. Genetic variation for sex expression is also supported by the observation of variation in sex expression among clones within populations (Carvalho and Hughes, 1983; Korpelainen, 1986, 1989; Yampolsky, 1992). The presence of genetic variation for sex expression suggests that studies on the influence of environmental factors on sex expression based on a single clone should be interpreted with caution (Stimpfl, 1971; Stross, 1969a; Hobaek and Larsson, 1990; Kleiven, Larsson and Hobaek, 1992).

Although *Daphnia pulex* in southern Ontario consists primarily of populations reproducing by obligate parthenogenesis in which there is no sexual reproduction, cyclically parthenogenetic populations are common in a few isolated areas (Herbert, Ward and Weider, 1988; Innes, 1991). Recent studies have uncovered variation in the ability to produce males among clones isolated from these populations (Innes and Dunbrack, 1993). On average about 37% of the clones from eight populations failed to produce males under crowded conditions in the laboratory. However, no information was collected on variation in the allocation to male and female progeny among the male-producing clones. Because of the lack of information on clonal variation in sexual reproduction for species of *Daphnia*, the present study was initiated to examine fecundity and sex allocation variation among clones of *D. pulex.*

6.3.2   Materials and methods

Clones of *Daphnia pulex* were established from samples collected from four cyclically parthenogenetic populations (LP-7, LP-8A, LP-8B, LP-9A) near Long Point, Ontario (Innes, 1991). Variation in fecundity and male production was observed in 10 male-producing clones (clones 1, 2 and 3 from LP-7; clones 4 and 5 from LP-8A; clone 6 and 7 from LP-8B; clones 8, 9 and 10 fro LP-9A) that had been growing in the laboratory for more than 1 year. These clones were chosen to represent a sample of male-producing clones from the Long Point area rather than to estimate variation in male production in clones from different populations. Prior to experimentation, single juvenile females from stock cultures were raised under experimental conditions. Neonates from the third or fourth brood of each clone were placed in a plastic beaker with 60 ml of zooplankton medium (Lynch, Weider and Lampert,1986) at four densities (1, 2, 4 and 8 individuals per 60 ml) with three replicates for each density treatment. The medium in each culture was replaced every second day with aquarium-cultured algae (primarily *Ankistrodemus* sp. and *Scenedesmus* sp.) added at a constant amount of 15 000 cells/ml per individual. Thus, at 2-day intervals food density was restored to 15 000 cells/ml in the culture with one individual, 30 000 cells/ml for two individuals, 60 000 cells/ml for four individuals and 120 000 cells/ml for eight individuals. Food was never completely depleted within any of the experimental containers. All cultures were placed in an incubator at 20 ± 1 °C and constant photoperiod (16L : 8D). This photoperiod has a slightly longer day than the photoperiod experienced by the natural populations at the time of sexual reproduction (May–June).

Brood release occurred at approximately 2-day intervals. Although male production and fecundity could only be determined for females maintained individually, brood release among multiple females within a breaker was approximately synchronous. Therefore, the offspring could be counted, sexed and assigned to each of the first five broods. Fecundity was expressed on a per female basis (total number of offspring produced in the first five broods divided by the number of females within a beaker). Variation in the fecundity of each female was examined in a two-way ANOVA (clone X density) following a square root transformation of the data (Sokal and Rohlf, 1981). Sex allocation for each clone at each density was expressed as the proportion of males among the total progeny produced in the first five broods for each replicate. A two-way ANOVA examined the effect of clone and density on variation in the proportion of males following an arcsin–square root transformation (Sokal and Rohlf, 1981). The pattern of sex allocation variation among broods was also examined for six clones showing the highest male production at densities of four and eight females per 60 ml.

6.3.3 Results

(a) Fecundity variation

There was large variation in fecundity among the ten clones. For example, the total progeny produced in the first five broods varied from an average of 54 (clone 5) to 104 (clone 10) for the lowest-density treatment (Table 6.3.1). Fecundity was depressed with increased density despite a constant amount of food added per individual (Table 6.3.1). The ANOVA showed that clone ($F_{[9, 78]} = 10.2$), density ($F_{[3, 27]} = 34.6$) and clone $\times$ density interaction ($F_{[27, 78]} = 2.8$) were significant ($p < 0.001$) sources of variation for fecundity.

(b) Sex ratio variation

Clone ($F_{[9, 78]} = 11.9$), density ($F_{[3, 27]} = 18.9$) and clone $\times$ density interaction ($F_{[27, 78]} = 3.0$) were significant (all $p < 0.001$) sources of variation for the proportion of males observed among the total progeny produced in the first five broods. No males were produced by any clone at the lowest density (Table 6.3.2). The proportion of males among the progeny of the first five broods generally increased with increasing density but the pattern of increase, as

Table 6.3.1 Fecundity (number of offspring per female) for 10 clones of *Daphnia pulex* raised at four densities

| Clone | Density (No. of females per 60 ml) | | | |
	1	2	4	8
1	80.7 (9.71)	81.2 (14.18)	63.0 (8.21)	42.8 (9.08)
2	87.3 (9.71)	69.3 (7.01)	48.4 (2.77)	40.8 (6.23)
3	97.3 (1.16)	86.8 (2.26)	77.7 (3.77)	58.8 (3.62)
4	100.7 (9.07)	77.8 (9.17)	61.9 (7.66)	55.6 (4.81)
5	54.0 (7.07)	58.7 (3.62)	57.3 (6.30)	40.5 (5.99)
6	72.0 (4.24)	74.5 (4.50)	60.5 (5.86)	49.2 (1.40)
7	79.0 (11.27)	70.5 (26.63)	72.9 (3.50)	59.3 (4.12)
8	79.7 (2.08)	76.0 (6.06)	59.2 (5.18)	63.5 (4.65)
9	84.0 (3.46)	78.0 (12.76)	61.4 (2.71)	54.5 (5.69)
10	104.3 (9.87)	77.7 (8.51)	64.9 (8.96)	42.3 (5.69)

Fecundity is based on the total progeny produced in the first five broods (mean of three replicates). Standard deviation in parentheses.

Table 6.3.2   Percentage of males (mean of three replicates) observed for 10 clones of *Daphnia pulex* raised at four densities

	Density (No. of females per 60 ml)			
Clone	1	2	4	8
1	0.0 (0.00)	3.0 (5.20)	19.7 (10.60)	22.7 (2.52)
2	0.0 (0.00)	0.0 (0.00)	0.0 (0.00)	2.3 (2.52)
3	0.0 (0.00)	0.0 (0.00)	0.0 (0.00)	2.7 (2.08)
4	0.0 (0.00)	5.0 (4.58)	11.7 (5.51)	15.0 (12.53)
5	0.0 (0.00)	0.0 (0.00)	2.0 (3.46)	8.7 (2.89)
6	0.0 (0.00)	1.3 (2.31)	23.3 (18.82)	17.3 (10.50)
7	0.0 (0.00)	0.0 (0.00)	1.0 (1.73)	0.0 (0.00)
8	0.0 (0.00)	10.0 (11.79)	20.0 (5.57)	22.7 (1.53)
9	0.0 (0.00)	0.0 (0.00)	19.3 (4.16)	11.3 (9.61)
10	0.0 (0.00)	3.0 (4.36)	10.7 (9.24)	22.0 (8.54)

Percentage of males based on the total progeny produced in the first five broods. Standard deviation in parentheses.

indicated by the significant clone × density interaction, varied dramatically among the 10 clones (Table 6.3.2). Clone 5 showed the greatest male production only at the highest density. Five clones (1, 4, 6, 8, 9) showed the greatest male production at a density of 4 in 60 ml with only a slight increase or decrease in the proportion of males at the highest density. Three clones (2, 3, 7) produced very few males even at the highest density. Clone 10 showed an approximately linear increase in the proportion of males with increasing density.

Male production in clones 1, 4, 6, 8, 9 and 10 showed little evidence of any pattern among the first five broods for either of the two highest densities (Figure 6.3.1). Males were produced in all broods with no distinction between earlier broods or later broods. There was no evidence for an alternation between predominantly male and female broods with respect to brood sequence. However, such an alternation of brood sex by individual females may be obscured by examining the sex of the combined progeny produced by four or eight females in the same container.

6.3.4   Discussion

Similar to previous studies (Loaring and Hebert, 1981; Brookfield, 1984; Lynch, 1984; Korpelainen, 1986, 1989; Weider and Hebert, 1987), the results presented here demonstrated significant among-clone variation in fecundity for *Daphnia pulex*. Although clonal variation in fecundity in three of the densities was based on groups of females rather than individual females, the synchronous brood release suggests that the observed variation is an accurate reflection of among-clone fecundity variation. However, the decrease in fecundity with increased density, despite a constant amount of food added per individual, is probably due to the static culture system producing food limitations at the highest densities. The presence of a significant interaction between clone and density indicates that no one clone had a superior or inferior fecundity at all densities. However, clone 5 generally showed the lowest fecundity and clone 3 had one of the highest fecundities at all densities.

The present study confirmed that *Daphnia pulex* responds to increasing density by increasing the proportion of males produced in parthenogenetic broods (Berg, 1931, 1934; Banta, 1939; Leary, 1967; Stimpfl, 1971; Larsson, 1991; Hobaek and Larsson, 1990; Kleiven, Larsson and Hobaek, 1992; Yampolsky, 1992). It was not possible to separate the effects of density from the effects of variation in food concentration in this static culture

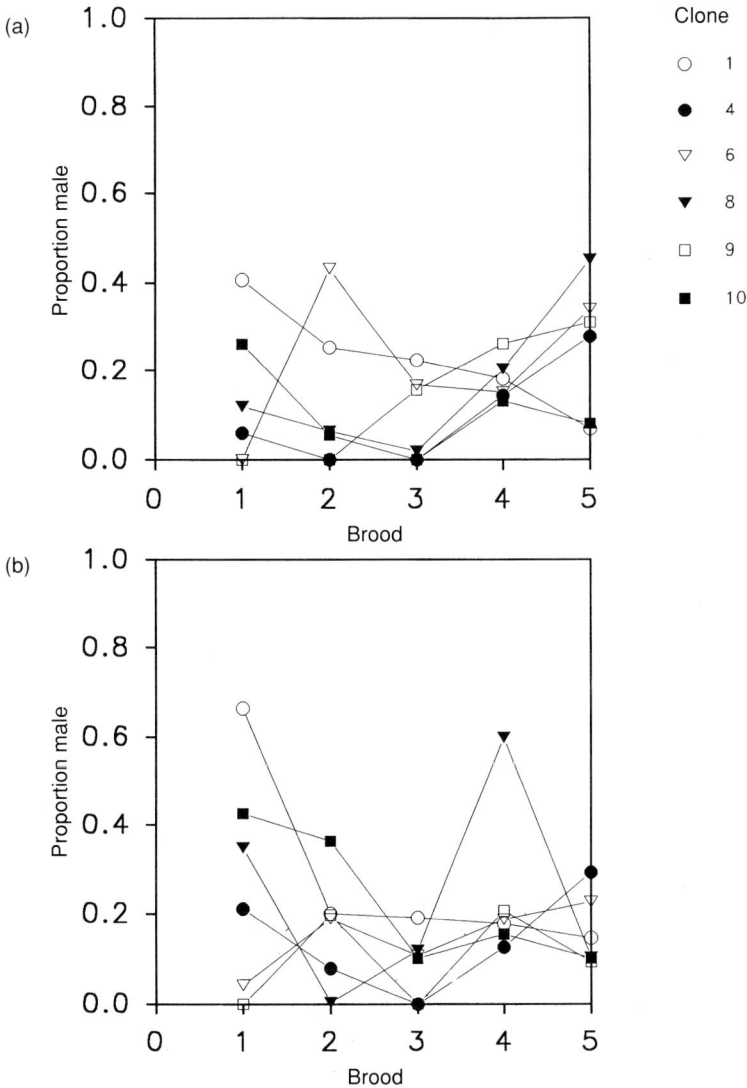

Figure 6.3.1 Proportion of males produced in the first five broods for six clones (1, 4, 6, 8, 9, 10) of *Daphnia pulex* raised at densities of (a) 4 individuals per 60 ml, and (b) 8 individuals per 60 ml.

system. The increase in males at the higher density may actually be a response to a greater fluctuation in food level in the higher-density treatment than in the low-density treatment (D'Abramo, 1980). However, recent experiments with a flow-through system suggested no effect of fluctuating food conditions on male production in *Daphnia magna* (Hobaek and Larsson, 1990). Male production was found to be stimulated directly through chemically mediated crowding (Hobaek and Larsson, 1990; Kleiven, Larsson and Hobaek, 1992). A combination of chemically mediated crowding, limited food and short-day photoperiod greatly stimulated the production of males in *D. magna* (Kleiven, Larsson and Hobaek, 1992).

Regardless of the actual mechanism responsible for increased male production under crowding, the response was highly variable among clones, with some clones producing very few males even at the highest density. Therefore, although density generally stimulates male production in *D. pulex* from the populations sampled in southern Ontario, some clones showed only a very weak response. Interclonal variation in male production in these populations appears to show almost continuous variation from clones with no male production (Innes and Dunbrack, 1993) to clones producing abundant males under high-density conditions.

Among-clone variation in sex allocation has also been reported for *D. pulex* by Larsson (1991). Clones were exposed to a short-day photoperiod (8L : 16D) in water that came from a crowded culture. One clone produced 17 broods of males and only one brood of females. A few clones produced no male broods. Yampolsky (1992) tested 30 clones of *D. magna* raised under high and low density. The results revealed a significant interaction between clone and density for the proportion of males produced by each clone. Some clones produced only females while others produced predominantly males independent of the density treatment. Korpelainen (1986, 1989) also examined variation in male production among five clones of *D. magna*. The greatest male production was under conditions of low temperature and 16L : 8D photoperiod (compared with 10L:14D and 22L:2D). All clones produced males but there was significant variation in lifetime male production among the five clones (8–45%). Although not tested, there appeared to be an interaction between clone and photoperiod, and clone and temperature (Korpelainen, 1986, Table 4).

These studies suggest that populations of *Daphnia* exhibit genetic variation for the relative proportion of males and females produced under a particular set of environmental conditions. The presence of such variation has been interpreted as being consistent with the general observation of moderate levels of genetic variation within *Daphnia* populations as measured by enzyme electrophoresis (Korpelainen, 1986). Similar to the explanations for the maintenance of enzyme variation, the maintenance of genetic variation for sex expression may be an adaptation to a heterogeneous environment (Korpelainen, 1986). However, there have been a few recent attempts to explain variation in sex expression in *Daphnia* with respect to sex allocation theory (Barker and Hebert, 1986; Hobaek and Larsson, 1990; Larsson, 1991; Kleiven, Larsson and Hobaek, 1992; Yampolsky, 1992).

The optimum investment in males and females by a particular clone will be a function of the investment pattern of other clones in the population. For example, a female-biased sex ratio in a population will favour clones with a greater investment in males. Larsson (1991), Yampolsky (1992) and Innes and Dunbrack (1993) suggested that observed variation in male production indicates that a population may contain a mixture of clones specialized as either male producers or female producers. The primary evidence for this is the observation of clones that failed to produce any males when exposed to conditions that normally stimulate male production. Conversely, evidence for an increased investment in males by some clones is more difficult to establish due to the sensitivity of male production to environmental conditions. There is limited evidence for the occurrence of clones with very high investment in males (Barker and Hebert, 1986; Hobaek and Larsson, 1990; Larsson, 1991). Yampolsky (1992) observed some clones that produced '... almost only males ...' but provided no data on the proportion of males produced. Kleiven, Larsson and Hobaek (1992) have recently observed individual females that produced up to six consecutive all-male broods under conditions of crowded water, short day and low food. The greatest male production observed for *D. pulex* in the present study was only about 23% using density as the primary stimulus. However, it may be difficult to relate the proportion of males produced by a clone under laboratory conditions to male production in the natural population, and a requirement for such studies clearly exists.

The presence of variation for sex allocation, demonstrated among the clones of *D. pulex* tested here, suggests that breeding system variation within and among population of *Daphnia* could be more extensive than previously thought. Sex allocation theory should be useful for explaining patterns of parthenogenetic and sexual reproduction observed in *Daphnia* species. Sex allocation theory (Charnov, 1982) can be applied to cyclical partheno-genesis since sexual reproduction is required in temporary habitats and equal investment in male and female progeny is still expected for the purposes of sexual reproduction (Innes and Dunbrack, 1993).

Acknowledgements

We would like to thank A. Snellen and L. Fitzgerald for assistance with the experiments. L. Hermanutz provided comments on the manuscript. This research was supported by funds from NSERC and Memorial University of Newfoundland.

References

Agar, W.E. (1914) Parthenogenetic and sexual reproduction in *Simocephalus vetulus. J. Genet.,* **3**, 179–94.

Banta, A.M. (1939) Studies on the physiology, genetics and evolution of some Cladocera. Paper 39, Department of Genetics, Carnegie Institution of Washington, Publication number 513, Washington, D.C.

Barker, D.M. and Hebert, P.D.N. (1986) Secondary sex ratio of the cyclic parthenogen *Daphnia magna* (Crustacea: Cladocera) in the Canadian Arctic. *Can. J. Zool.,* **64**, 1137–43.

Berg, K. (1931) Studies on the genus *Daphnia* O.F. Muller with special reference to the mode of reproduction. *Vidensk. Medd. Dan. Naturhist. Foren.,* **92**, 1–222.

Berg, K.(1934) Cyclic reproduction, sex determination and depression in the Cladocera. *Biol. Rev.,* **9**, 139–74

Brookfield, J.F.Y. (1984) Measurement of the intraspecific variation in population growth rate under controlled conditions in the clonal parthenogen *Daphnia magna. Genetica,* **63**, 161–74.

Bull, J.J. (1983) *Environmental Sex Determination Mechanisms,* Benjamin/Cummings, Melano Park, California.

Carvalho, G.R. and Hughes, R.N. (1983) The effect of food availability, female culture-density and photoperiod on ephippial production in *Daphnia magna* Straus (Crustacea: Cladocera). *Freshwat. Biol.,* **13**, 37–46.

Charnov, E.L. (1982) *The Theory of Sex Allocation,* Princeton University Press, Princeton.

D'Abramo, L.R. (1980) Ingestion rate decrease as the stimulus for sexuality in populations of *Moina macrocopa. Limnol. Oceanogr.,* **33**, 143–9.

Edmondson, W.T. (1955) The seasonal life history of Daphnia in an Arctic lake. *Ecology,* **36**, 439–55.

Ferrari, D.C. and Hebert, P.D.N. (1982) The induction of sexual reproduction in *Daphnia magna*: genetic differences between arctic and temperate populations. *Can. J. Zool.,* **60**, 2143–8.

Frey, D.G. (1982) Contrasting strategies of gamogenesis in northern and southern populations of Cladocera. *Ecology,* **63**, 223–41.

Hebert, P.D.N. (1978) The population biology of *Daphnia. Biol. Rev.,* **53**, 387–426.

Hebert, P.D.N., Ward, R.D. and Weider, L.J. (1988) Clonal-diversity patterns and breeding-system variation in *Daphnia pulex,* an asexual–sexual complex. *Evolution,* **42**, 147–59.

Hobaek, A. and Larsson, P. (1990) Sex determination in *Daphnia magna. Ecology,* **71**, 2255–68.

Innes, D.J. (1991) Geographic patterns of genetic differentiation among sexual populations of *Daphnia pulex. Can. J. Zool.,* **69**, 995–1003.

Innes, D.J. and Dunbrack, R.L. (1993) Sex allocation variation in *Daphnia pulex. J. Evol. Biol.* (in press).

Kleiven, O.T., Larsson, P. and Hobaek, A. (1992) Sexual reproduction in *Daphnia magna* requires three stimuli. *Oikos,* **64**, 197–206.

Korpelainen, H. (1986) The effects of temperature and photoperiod on life history parameters of *Daphnia magna* (Crustacea: Cladocera). *Freshwat. Biol.,* **16**, 615–20.

Korpelainen, H. (1989) The effects of periodically changing temperature and photoperiod conditions on reproduction and sex ratio of *Daphnia magna* (Crustacea: Cladocera). *Zool. Beitr.,* **32**, 247–60.

Korpelainen, H. (1990) Sex ratios and conditions required for environmental sex determination in animals. *Biol. Rev.*, **65**, 147–84.

Larsson, P. (1991) Intraspecific variability in response to stimuli for male and ephippia formation in *Daphnia pulex. Hydrobiologia*, **225**, 281–90.

Leary, D. (1967) Induction of males and ephippial eggs in *Daphnia pulex*. MSc thesis, University of Maryland, College Park.

Loaring, J.M. and Hebert, P.D.N. (1981) Ecological differences among clones of *Daphnia pulex* Leydig. *Oecologia*, **51**, 162–8.

Lynch, M. (1984) The genetic structure of a cyclical parthenogen. *Evolution*, **38**, 186–203.

Lynch, M., Weider, L.J. and Lampert, W. (1986) Measurement of carbon balance in *Daphnia. Limnol. Oceanogr.*, **31**, 17–33.

Sokal, R.R. and Rohlf, F.J. (1981) *Biometry*, 2nd edn. W.H. Freeman, San Francisco.

Stimpfl, K.J. (1971) The effects of four environmental variables on the induction of gamogenesis in *Daphnia pulex* Leydig. PhD thesis, Indiana University, Bloomington.

Stross, R.G. (1969a) Photoperiod control of diapause in *Daphnia*. II. Induction of winter diapause in the Arctic. *Biol. Bull.*, **136**, 264–73.

Stross, R.G. (1969b) Photoperiod control of diapause in *Daphnia*. III. Two-stimulus control of long-day, short-day induction. *Biol. Bull.*, **137**, 359–74.

Weider, L.J. and Hebert, P.D.N. (1987) Ecological and physiological differentiation among low-arctic clones of *Daphnia pulex. Ecology*, **68**, 188–98.

Yampolsky, L. Yu. (1992) Genetic variation in the sexual reproduction rate within a population of a cyclic parthenogen, *Daphnia magna. Evolution*, **46**, 833–7.

6.4 INFLUENCE OF DIFFERENTIAL NATALITY AND MORTALITY ON TEMPORAL FLUCTUATIONS OF *DAPHNIA* GENOTYPES IN NATURAL POPULATIONS

Jakob Müller and Alfred Seitz

Abstract

Recent papers have revealed the existence of high genotypic diversity in most pelagic Daphnia *populations. This study addresses the problem of clonal coexistence in a temporally fluctuating environment. Over the 2 years studied, pronounced temporal fluctuations in the clonal composition were found for two* Daphnia *populations, i.e.* Daphnia galeata *and the hybrid* Daphnia galeata × cucullata. *The intensity of clonal shifts exhibited a seasonal trend with maxima in spring and early summer. A multiple correlation analysis revealed that clonal shifts were directed more by differential mortality than by differential natality. Differential recruitment of sexually produced resting eggs might also have influenced clonal shifts.*

6.4.1 Introduction

Daphnia populations, which inhabit the pelagic zone of large lakes, show high genotypic diversity (Mort and Wolf, 1986; Giessler, 1987). Because these daphnids are cyclically parthenogenetic, their populations are essentially an assemblage of many coexisting clones, which intermix during occasional events of sexual reproduction. Therefore, the methods which should be applied to study clonal coexistence should be a mixture taken from the fields of interspecies and intrapopulational studies. Models have been presented explaining the coexistence of similar zooplankton species (Jacobs, 1977a; Seitz, 1991). These models try to explain coexistence by differential adaptation to natality and mortality in a temporally fluctuating and spatially structured environment. It is convenient to reduce environmental influence solely to natality and mortality, because causal connections between environmental factors and life-history traits are highly complex. For example, body size, which is a crucial parameter for both birth and death rates (Seitz, 1980), is influenced among others, by food availability, temperature and therefore vertical migration behaviour, predation pressure and genotype. In addition, most of these factors are interrelated.

In a previous paper (Müller and Seitz, 1993), coexistence of *Daphnia* genotypes has been partly explained by spatial habitat partitioning, in respect to water depth. In the present study, we examined temporal fluctuations of *Daphnia* genotypes in terms of differential natality and mortality.

6.4.2 Materials and methods

Our study site is lake Neuhofener Altrhein, an old oxbow lake from the river Rhine, which was separated from it about 380 years ago. Its surface area is 0.65 km^2 and its depth structure is complex with deep furrows down to 13 m and shallow ridges up to 3 m. This eutrophic lake is very rich in species, including all forms of the *Daphnia galeata/hyalina/cucullata* complex. In this work, the most frequent *Daphnia types* – *D. galeata* and the hybrid *D. galeata × cucullata* – were analysed. Daphnids were identified morphologically according to Flössner (1972) and Flössner and Kraus (1986) and genetically according to Wolf and Mort (1986). Potential predators in the lake consist of fish (mean density: 0.07 ± 0.01 individuals /m^3; predominantly planctivorous cyprinids; see Wagner (1992), the midge larvae *Chaoborus flavicans*, and, less importantly, *Leptodora sp.* and *Cyclops spp.*

We took samples at the deepest site by vertical hauls with a plankton net (100 μm mesh size) for the temporal survey (May 1989 until June 1991) and in vertical series at 1-m intervals for the spatial survey in May 1990, October 1990 and May 1991. The latter was done by pumping water from every metre interval through a plankton net. Separate samples for a counting survey (preserved in formalin-sucrose) and a genetic survey were taken. The genetic samples were brought alive to the laboratory, sorted and stored in liquid nitrogen. Adult individual daphnids were assayed by enzyme electrophoresis at six polymorphic allozyme loci: aldehyde oxidase (*Ao*) EC 1.2.3.1, phosphoglucose isomerase (*Pgi*) EC 5.3.1.9, mannose phosphate isomerase (*Mpi*) EC 5.3.1.8, malate dehydrogenase (*Mdh*) EC 1.1.1.37, phosphoglucomutase (*Pgm*) EC 2.7.5.1 and glutamate oxaloacetate transaminase (*Got*) EC 2.6.1.2. The *Got* locus discriminates between the *Daphnia* forms (Müller, Menne and Seitz, 1991). Electrophoresis was performed on cellulose acetate plates following the method of Hebert and Beaton (1989). We decided to create multilocus genotypes as a combination of only three loci (*Pgi*, *Mpi* and *Pgm*) to get sufficiently large sample sizes for statistical tests. These multilocus genotypes were used as markers for clonal groups. From the genetic survey we determined relative frequencies of the multilocus genotypes and from the counting survey we calculated the absolute abundances of the *Daphnia* forms.

Before electrophoresis, the number of subitaneous eggs in the brood pouch of each female was counted. Instantaneous birth rates (*b*) per individual per day were calculated by a program of S. Bast according to the formula of Paloheimo (1974). For the calculation of *b*, the duration of egg development in days (*D*) must be known, which is simply a function of temperature with constants taken from Vijverberg (1980). To improve the estimation of *D*, diurnal vertical migration of egg-bearing females was taken into consideration. During 1989 and 1990, the migration pattern observed in May 1990 and October 1990 was applied and during 1991, the May 1991 migration pattern was used. Growth rates (*r*) per individual per day were calculated from density values N_1 and N_2 at successive sampling times t_1 and t_2:

$$r(t_2t_1) = (\ln N_2 - \ln N_1)/(t_2 - t_1)$$

The mortality rate (m) for a given time interval was simply inferred:

$$m(t_2t_1) = b(t_1) - r(t_2t_1)$$

Relative birth and death rates were calculated as the common logarithm of the ratio between the clone-specific rate and the mean of all clonal rates (Seitz 1980).

As a measure of clonal fitness superiority or inferiority for a given time interval, relative frequency changes were calculated:

$$\Delta Q(t_2t_1) = [Q(t_2) - Q(t_1)]/(t_2 - t_1)$$

with Q being the relative frequency of a given genotype.

Heterozygosities (H) and deviations from Hardy–Weinberg equilibrium were calculated for all loci with the G-Stat program of Siegismund (unpublished). The negative common logarithm of the probability of the G-test for agreement with Hardy–Weinberg expectations ($-\log P(G)$) was taken as a measure for Hardy–Weinberg deviation according to Hebert *et al.* (1988). Again, $\Delta -\log P(G)$ and ΔH are changes of these parameters for a given time interval (see ΔQ).

6.4.3   Results

Based on three loci, a large number of coexisting clonal groups were readily distinguished. In the total survey, there were 23 different multilocus genotypes for the parental species *D. galeata* and 48 for the hybrid *D. galeata* × *cucullata*. For *D. galeata*, only four genotypes were included in the following analysis, because they were abundant throughout the year. For the hybrid, there were five abundant genotypes. These genotypes comprised 47–92% of the total clonal composition at different times for *D. galeata* and 54–93% for the hybrid respectively.

Temporal fluctuations in the relative frequencies of the *D. galeata* genotypes are given (Figure 6.4.1(a)). Dramatic shifts, but no clear seasonal oscillations, were observed. For instance, genotype I dominated during the first half of the study period, whereas it was less abundant for the rest of the study. The other three genotypes exhibited only weak seasonal trends: for example, genotype II was most frequent in June and genotype III during summer and winter. All distributions of $\Delta Q \pm$ SD, averaged over time for all four genotypes separately, included zero. This indicates that, although significant selection periodically operated on genotypes, the genotypes were quasi-neutral averaged over the 2-year period.

To find out when the intensity of clonal shifts was most pronounced, a mean value of the absolute amounts of the genotype-specific ΔQs for each time interval (mean ΔQ) was calculated. The temporal dynamics of this mean ΔQ showed a significant seasonal trend (Figure 6.4.1(b)) Each mean ΔQ is depicted at the beginning of each time interval and is therefore valid for the following interval. Peaks in clonal shifts occurred between June and July 1989, between May and June 1990 and between March and April 1991. Similar results were found for the hybrid, with no clear seasonal changes in genotype frequencies but a seasonal trend in mean ΔQ, at least for the second half of the study period (Figure 6.4.2). Peaks were found between April and June both in 1990 and 1991. Again, all distributions of $\Delta Q \pm$ SD, averaged over time for each genotype, included zero, thus indicating long-term quasi-neutral genotypes.

To reveal whether the clonal fitness differences were due to differential natality or mortality, a multiple and partial correlation analysis between relative frequency changes per month (ΔQ) and relative birth rates (b_{rel}) and relative death rates (m_{rel}) was performed. This

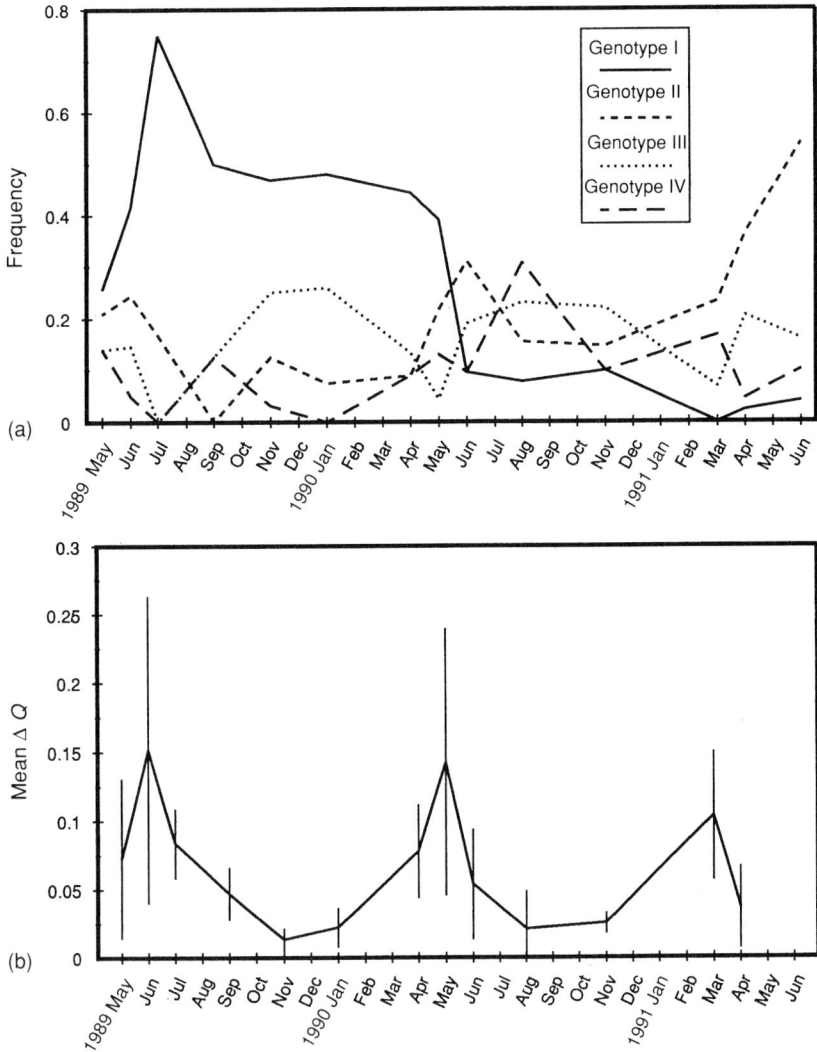

Figure 6.4.1 *D. galeata* (a) Temporal fluctuations in relative frequencies of the four most abundant genotypes; (b) temporal dynamics of the intensity of clonal shifts (mean $\Delta Q \pm SD$).

was done separately for each genotype and all genotypes of one *Daphnia* type together (Table 6.4.1). Only the partial correlation coefficients are pointed out. Five significant correlations were found for the relative death rates (genotypes III and IV of *D. galeata* and genotypes II, III and V of the hybrid) and only one for the relative birth rates (genotype III of *D. galeata*). If all genotypes of a *Daphnia* type were lumped together, both correlations for the relative death rates and only one for the relative birth rates were significant. Similar ratios could be seen if tested with the more restricted sequential Bonferroni method (Rice, 1989) (Table 6.4.1). These results suggest that most of the genotype frequency shifts can be explained by differential death rates. But for some genotypes differential birth rates were

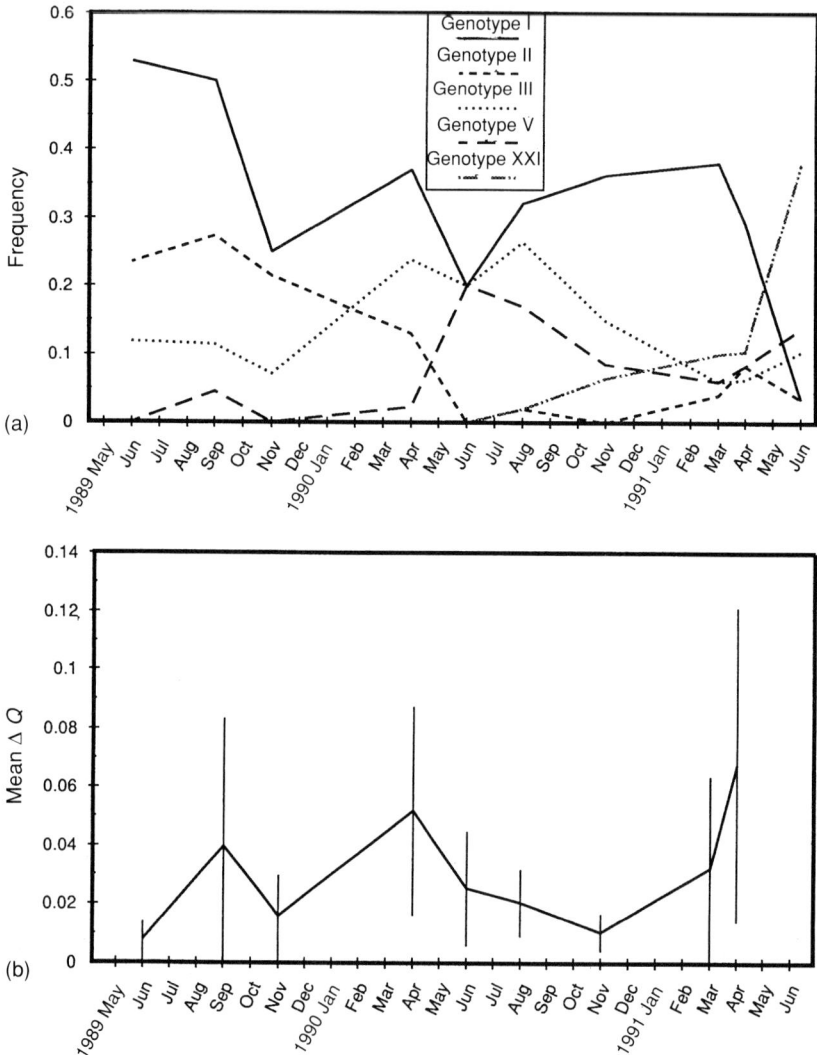

Figure 6.4.2 *D. galeata* × *cucullata*. (a) Temporal fluctuations in relative frequencies of the five most abundant genotypes; (b) temporal dynamics of the intensity of clonal shifts (mean $\Delta Q \pm SD$).

also determinant, as for genotype III of *D. galeata*, which showed the most distinct seasonal trend.

While calculating the rates, some negative death rates were observed. This could be due to the fact that the calculation of birth rates and therefore the deduced death rates also were based only on subitaneous egg numbers and ephippial recruitment was not taken into account. Deviations from Hardy–Weinberg equilibrium were observed in all loci for both *Daphnia* forms, but were stronger for the hybrid (Müller, Menne and Seitz, 1991). In order to examine ephippial recruitment more closely, a correlation analysis between the intensity of clonal shifts (mean ΔQ) and the change in Hardy–Weinberg deviation (Δ–log $P(G)$) respectively the change in heterozygosity (ΔH) was performed (Figure 6.4.3). For both

Table 6.4.1   *D. galeata* and *D. galeata* × *cucullata* hybrids: correlation analysis between changes in relative genotype frequencies per month (ΔQ) as dependent variable, and genotype-specific relative birth rates (b_{rel}), and relative death rates (m_{rel}) as independent variables

Species	Genotypes	N	R	r of b_{rel}	r of m_{rel}
D. galeata	All	46	0.458[*+]	0.285[*]	–0.439[*+]
	I	12	0.397	0.315	–0.393
	II	12	0.375	–0.010	–0.289
	III	12	0.961[*+]	0.940[*+]	–0.956[*+]
	IV	10	0.743	0.243	–0.641[*]
D. galeata ×	All	39	0.472[*+]	0.175	–0.449[*+]
D. cucullata	I	10	0.571	0.257	–0.550
(hybrid)	II	7	0.907[*]	0.415	–0.907[*+]
	III	10	0.735	0.369	–0.616[*]
	V	8	0.879[*]	0.340	–0.829[*]
	XXI	4	0.924	–0.915	0.920

R = multiple correlation coefficient; r = partial correlation coefficient; N = number of cases included; [*]$p < 0.05$; [+] $p < 0.05$ sequential Bonferroni method (Rice, 1989).

Daphnia forms, a negative correlation with the change in Hardy–Weinberg deviation was found, but was significant only for the hybrid ($r = -0.32$ for *D. galeata*; $r = -0.81$, $p < 0.05$ for the hybrid). In other words, deviation from Hardy–Weinberg equilibrium decreased when clonal shifts were high whereas it increased during periods of low to moderate clonal shifts. Correlation with change in heterozygosity was significantly positive for *D. galeata* ($r = 0.55$; $p < 0.05$) but significantly negative for the hybrid ($r = -0.87$; $p < 0.05$).

6.4.4   Discussion

In the studied lake, pronounced temporal shifts in clonal composition were observed and the intensity of these shifts showed a seasonal regularity with maxima in spring or early summer. Pronounced temporal genotypic changes in *Daphnia* populations have also been documented in other lakes (Hebert and Ward, 1976; Weider, 1985; Carvalho and Crisp, 1987; Jacobs, 1990), but only a few studies have documented a seasonal trend. Weider (1985) found distinct genotype successions in spring and early summer and Carvalho and Crisp (1987) described summer-, autumn- and winter-adapted genotypes. One reason why most studies did not detect any seasonal trend might be that only one- or two-locus genotypes were analysed and hence more clones were pooled within a marked genotype than within genotypes marked by three or more loci. When fewer loci are investigated perhaps the seasonal trends of single clones might be concealed. On the other hand, the occurrence of seasonal trends might be the result of lake structure, like, for example, extensive shallow waters to enable ephippia hatching or pronounced seasonal trends in environmental parameters, such as fish predation, oxygen content, food concentration and temperature. All these features were evident in the studied lake (Wagner, 1992) and it requires further analyses to find out which one has the most influence.

We think that genetic drift can be ruled out as a cause of the genotypic fluctuations because even at the lowest population densities, the total population size was estimated to be more than 3×10^7 individuals. Furthermore, it has been argued that horizontal spatial heterogeneity of clones could account for some of the temporal fluctuations. Even if not estimated here, we think this influence is unlikely to be very strong, because it has been shown that most of the spatial variation is due to the vertical component, i.e. depth (Weider, 1985) and we sampled always over the same depth interval. Therefore, some selective forces seem to

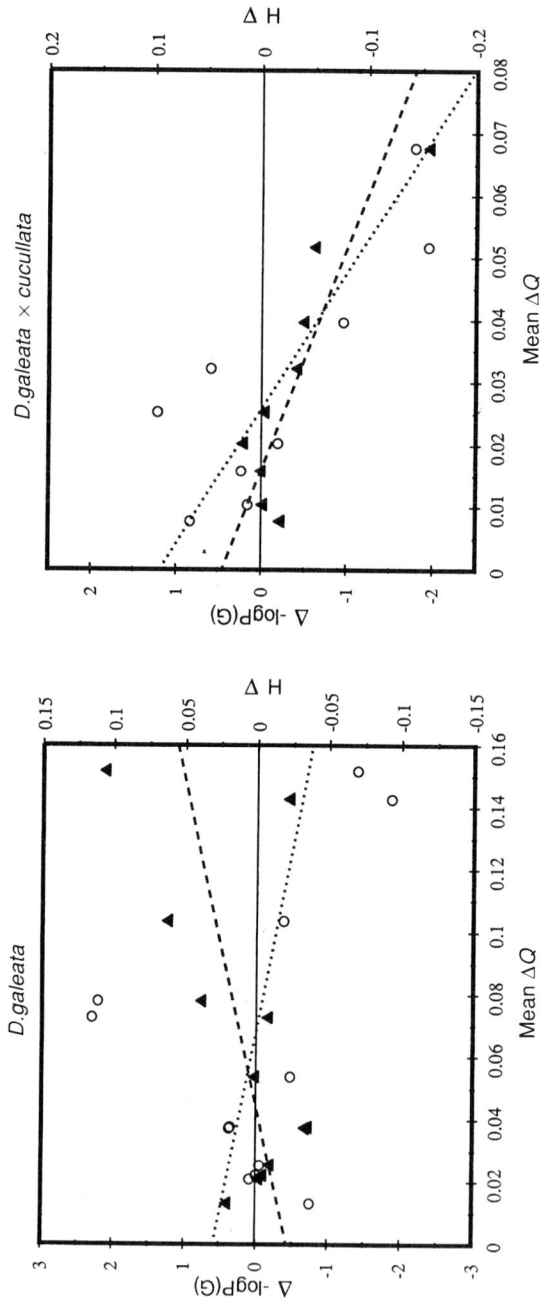

Figure 6.4.3 Correlation between intensity of clonal shifts (mean ΔQ) and change in Hardy–Weinberg deviation ($\Delta -\log P(G)$, dotted line with open circles) or change in heterozygosity (ΔH, dashed line with filled triangles); for correlation coefficients see text.

act upon the clones and ΔQ can be taken as a measure of clonal fitness differences. Our results support the idea of quasi-neutral genotypes (Lynch, 1987). No long-term directional selection on genotypes could be seen. This might be due to seasonal fluctuations (e.g. genotype III of *D. galeata*) or to the chance development of linkage disequilibria between the marker genes and fitness-relevant genes during sexual reproduction. However, strong short-term clonal fitness differentials were observed and it was shown that they were not randomly distributed over time (Lynch, 1987). On the contrary, they showed a distinct seasonal trend. Fitness differences can be split into differences in natality and differences in mortality. In the present study the main determinant for clonal shifts seems to be differential mortality, though the influence of natality cannot be ruled out. In a study of coexistence of similar daphnids, Jacobs (1977b) also found death rate to be the main determinant for population growth rate. Differential mortality can be achieved by, for example, differential escape capabilities from predators or differential tolerances to hunger or parasites. The idea that genotypic variation in the susceptibility to predation is higher than genotypic variation in natality is also supported by Hebert (1984). In addition, coexistence of similar *Daphnia* species by selective mortality, in particular size-selective fish predations, has been previously described by Seitz (1980).

In addition to differential natality and mortality there is a third factor which could account for clonal shifts: differential recruitment of ephippia, which is not considered in the calculation of instantaneous birth rates. Ephippia – resting eggs – are normally (at least for the parental species *D. galeata*) produced by sexual reproduction. That means that during high ephippial recruitment, the deviation from Hardy–Weinberg equilibrium should decrease. Therefore, data presented in Figure 6.4.3 suggest ephippial recruitment for both *Daphnia* forms during periods of high clonal shifts. We do not have direct estimates of ephippial recruitment like Wolf and Carvalho (1989), but we have estimates of ephippial production in the studied lake (S. Bast, personal communication). In the *D. galeata* population, males and ephippia-bearing females coexisted in 10 of the 26 studied months. During that time, they comprised on average 18.4% of the total population, with a mean sex ratio of 11 females to 1 male. The corresponding data for the hybrid were: coexistence of males and ephippia-bearing females in 6 out of 26 months; during that time 15.6% of the total; mean sex ratio 20 females to 1 male. These data suggest that both *Daphnia* forms – more so in *D. galeata* – have the potential for sexual ephippia production. In addition, for the hybrid, heterozygosity decreased during periods of high clonal shifts as well, suggesting that clonal recruitment originates from sexually produced eggs, as an F_2 generation or as the result of backcrossing. These latter events might occasionally occur (J. Müller and A. Seitz, unpublished data), but there is much evidence that they are very rare. For instance, the hybrid *D. galeata* × *cucullata* forms a morphologically distinct group with few smooth transitions to the parental species. Therefore, it is more realistic to propose that during moderate clonal shifts, heterozygosity accumulated via selection against homozygotes, while during high clonal shifts, heterozygosity decreased again because of recruitment from ephippia, which were newly produced by the parental species as the F_1 generation.

Acknowledgements

Special thanks are due to Susanne Rieder for technical assistance, to Siegmar Bast for help in calculating birth and death rates and to two anonymous reviewers. This work is supported by the German Federal Ministry for Science and Technology (Project number: 0339200 D).

References

Carvalho, G.R. and Crisp, D.J. (1987) The clonal ecology of *Daphnia magna* (Crustacea: Cladocera). I. Temporal changes in the clonal structure of a natural population. *J. Anim. Ecol.*, **56**, 453–68.

Flössner, D. (1972) Krebstiere, Crustacea: Kiemen- und Blattfüßer, Branchiopoda, Fischläuse, Branchiura, in *Die Tierwelt Deutschlands*, 60 Teil (eds M. Dahl and F. Peus) Fischer Verlag, Jena, pp. 104–202.

Flössner, D. and Krauss K. (1986) On the taxonomy of the *Daphnia hyalina–galeata* complex (Crustacea: Cladocera). *Hydrobiologia*, **137**, 97–115.

Giessler, S. (1987) Mikroevolution und Populationsgenetik im *Daphnia galeata/hyalina/cucullata*-Komplex, eine Freilandanalyse. PhD Thesis, L-M-University of Munich, Germany.

Hebert, P.D.N. (1984) Demographic implications of genetic variation in zooplankton populations, in *Population Biology and Evolution* (eds K. Wöhrmann and V. Loeschcke), Springer Verlag, Berlin, pp. 195–207.

Herbert, P.D.N. and Beaton, M.J. (1989) *Methodologies for Allozymes Analysis using Cellulose Acetate Electrophoresis*, Helena Laboratories, Beaumont, Texas.

Hebert, P.D.N. and Ward, R.D. (1976) Enzyme variability in natural populations of *Daphnia magna*. IV. Ecological differentiation and frequency changes of genotypes at Audley End. *Heredity*, **36**, 331–41.

Hebert, P.D.N., Ward, R.D. and Weider, L.J. (1988) Clonal-diversity patterns and breeding-system variation in *Daphnia pulex*, an asexual–sexual complex. *Evolution*, **42**, 147–59.

Jacobs, J. (1977a) Coexistence of similar zooplankton species by differential adaptation to reproduction and escape in an environment with fluctuating food and enemy densities. I. A model. *Oecologia*, **29**, 233–47.

Jacobs, J. (1977b) Coexistence of similar zooplankton species by differential adaptation to reproduction and escape in an environment with fluctuating food and energy densities. II. Field data analysis of *Daphnia*. *Oecologia*, **30**, 313–29.

Jacobs, J. (1990) Microevolution in predominantly clonal populations of pelagic *Daphnia* (Crustacea: Phyllopoda): selection, exchange and sex. *J. Evol. Biol.*, **3**, 257–82.

Lynch, M. (1987) The consequences of fluctuating selection for isozyme polymorphisms in *Daphnia*. *Genetics*, **115**, 657–69.

Mort, M.A. and Wolf, H.G. (1986) The genetic structure of large-lake *Daphnia* populations. *Evolution*, **40**, 756–66.

Müller, J., Menne, B. and Seitz, A. (1991) Die räumlich-zeitliche Verteilung der genetischen Struktur von Zooplanktonpopulationen in Seen unterschiedlichen Trophiegrades. *Verh. Ges. Ökologie*, **20/2**, 601–7.

Müller, J. and Seitz, A. (1993) Habitat partitioning and differential vertical migration of some *Daphnia* genotypes in a lake. *Arch. Hydrobiol. Beih.*, **39**, 167–74.

Paloheimo, J.E. (1974) Calculation of instantaneous birth rate. *Limnol. Oceanogr.*, **19**, 692–4.

Rice, W.R. (1989) Analyzing tables of statistical tests. *Evolution*, **43**, 223–5.

Seitz, A. (1980) The coexistence of three species of *Daphnia* in the Klostersee. II. The stabilizing effect of selective mortality and conclusions for the stability of the system. *Oecologia*, **47**, 333–339.

Seitz, A. (1991) The coexistence of three species of Daphnia in the Klostersee III. The simulation model CODA, in *Analyse dynamischer Systeme in Medizin, Biologie and Ökologie* (eds D.P.F Moeller and O. Richter), Informatik Fachberichte 275, Springer Verlag, Berlin, pp. 185–91.

Vijverberg, J. (1980) Effect of temperature in laboratory studies on development and growth of Cladocera and Copepoda from Tjeukemeer, The Netherlands. *Freshwat. Biol.*, **10**, 317–40.

Wagner, A.R. (1992) Untersuchungen zur Verteilungsdynamik pelagischer Fische in fünf Seen unterschiedlicher trophischer Struktur, unter besonderer Berücksichtigung der Populationsbiologie von Flußarsch (*Perca fluviatilis* L., Pisces, Percidae) und Rotauge (*Rutilis rutilis* L., Pisces, Cyprinidae). PhD thesis, J-G-University of Mainz, Germany.

Weider, L.J. (1985) Spatial and temporal genetic heterogeneity in a natural *Daphnia* population. *J. Plank. Res.*, **7**, 101–23.

Wolf, H.G. and Carvalho, G.R. (1989) Resting eggs of lake-Daphnia. II. In situ observations on the hatching of eggs and their contribution to population and community structure. *Freshwat. Biol.* **22**, 471–8.

Wolf, H.G. and Mort, M.A. (1986) Inter-specific hybridization underlies phenotypic variability in *Daphnia* populations. *Oecologia*, **68**, 507–11.

6.5   POPULATION GENETIC PARAMETERS WITHIN A SEA ANEMONE FAMILY
(SAGARTIIDAE) ENCOMPASSING CLONAL, SEMICLONAL AND ACLONAL
MODES OF REPRODUCTION

Paul Shaw, John S. Ryland and John A. Beardmore

Abstract

Anthozoans display a diverse array of reproductive modes, offering opportunities to examine the effects of reproductive methods on intra- and interpopulation genetic structuring.

Four species from the family Sagartiidae (Actiniaria), with reproductive modes ranging from obligately asexual to obligately sexual, were assayed allozymically to determine whether several population genetic parameters fitted to expectations based on their reproductive mode.

Most parameters fitted well to expectations. The aclonal species displayed maximal genotypic diversity, good fits to Hardy–Weinberg expectations at all loci, and no significant differentiation in allele frequencies between geographic populations. In contrast, the clonal species displayed extensive replication of a small number of genotypes, deviations from Hardy–Weinberg expectations at all variable loci, and extensive differentiation between populations. The two semiclonal species display characteristics intermediate between the two extremes, depending on the nature of the asexual method used (pedal laceration or parthenogenesis).

Levels of genetic variability were uniformly high across all four species ($P_{0.95}$ = 61–66%; H_o = 0.27–0.28), except for much higher heterozygosity in the obligate parthenogenetic species (H_o = 0.47) and higher mean number of alleles in the obligately sexual species (3.1 against 2.0–2.1).

6.5.1   Introduction

Sea anemones (Anthozoa: Actiniaria) display a diverse array of reproductive methods both within and between species (Chia, 1976; Shick, 1991). Many species utilize both asexual and sexual methods (semiclonal), whilst some are obligately aclonal or clonal. Finally, several species have been shown to employ different reproductive combinations in different habitats or parts of their geographic range (Rossi, 1975; Schafer, 1981). Such diversity of methods offers opportunities to examine the effects of reproductive mode on intra- and interpopulation genetic structuring, both within and across species.

The present study examines genotypic and genetic diversity across four species (*Sagartia troglodytes* Price, *Sagartia ornata* Holdsworth, *Sagartia elegans* Dalyell, and *Cereus pedunculatus* Pennant) from a single family, the Sagartiidae. These species encompass between them the complete range of reproductive styles from aclonal to obligately clonal. The four study species are similar in size and morphology (Manuel, 1981), are all widespread within the geographic area sampled (Southwestern British Isles – Figure 6.5.1), and show substantial overlap in habitat choice – all four can be found together on rockfaces near the low-water mark in some localities (personal observation). Such ecological similarities help reduce the problems of comparison across species.

S. troglodytes is dioecious, 96% of individuals sampled in this study being sexually mature, with a distinct annual cycle of gametogenesis and synchronized spawning (Shaw,

Figure 6.5.1 Location of sampling sites within the southwestern UK for the 28 populations of the four species of sagartiid sea anemones examined. ○ = *Sagartia troglodytes*; ● = *Sagartia elegans*; □ = *Sagartia ornata*; ▲ = *Cereus pedunculatus*.

1989). The species has a short-lived (7–17 days), probably lecithotrophic, pelagic larva (Nyholm, 1943; Riemann-Zurneck, 1969). No asexual method of reproduction has been reported for this species (Stephenson, 1929; Shaw, 1989).

S. ornata was until recently regarded as a variety of *S. troglodytes* but has been shown to be a separate species (Shaw, Beardmore and Ryland, 1987). This species is thought to be a polyploid obligately asexual 'morphospecies', with all-female populations and brooding of parthenogenetically produced offspring (Shaw, Beardmore and Ryland, 1987; Shaw, 1988, 1989).

S. elegans is dioecious with an annual cycle of gametogenesis, but with fewer sexually mature adults (63%) and less well synchronized spawning than *S. troglodytes* (Shaw, 1989). Asexual propagation by means of pedal laceration is common (Stephenson, 1929), producing localized clones (Shaw, 1991).

C. pedunculatus in different parts of its range is variously hermaphrodite or dioecious, viviparous or oviparous, having a pelagic planktotrophic or lecithotrophic planula larva or brooded actinula larvae (Stephenson, 1929; Rossi, 1971, 1975; Chia, 1976; Schafer, 1981). In the UK, populations comprise a mix of hermaphrodites and females, have poorly syn-chronized annual gamete development and spawning, and brood asexually (probably par-thenogenetically) and, rarely, sexually produced offspring (Heap, 1979; Shaw, 1988; 1989).

6.5.2 Methods

Genetic variability within the four species was assayed at 14 gene loci using standard techniques of allozyme electrophoresis (details in Shaw, Beardmore and Ryland, 1987; Shaw, 1988). Samples were collected from a total of 28 populations at 17 locations in the southwestern British Isles (Figure 6.5.1), all four species being represented at localities within both southern England and South Wales. Individuals were collected across as wide an area as possible within sites. Repeated sampling within small groups of phenotypically identical individuals (possible clonal groups) was avoided in order to reduce the number of animals sacrificed to detect enough different genotypes for genetic diversity analysis. Whilst such sampling may overestimate the absolute level of genotypic diversity in a site, it should not grossly distort the overall picture of degree of cloning within spatially extensive populations. After electrophoretic typing, identical 14-locus allozyme genotypes were identified and two measures of the relative genotypic diversity within samples calculated: the number of different genotypes divided by sample size, d; and the 'evenness' of abundance of clones, E (Fager, 1972), which takes into account the number of genotypes occurring and their frequency, scaled to account for sample size, and ranging from nearly zero when one genotype comprises most of the sample, to unity when all genotypes occur at equal frequency (Parker, 1979). It was assumed that identical multilocus phenotypes represented true clonemates based on calculations of the probabilities of such phenotypes occurring two or more times within a certain sample size using expansion of the binomial distribution (Sokal and Rohlf, 1981). The highest probability found was 0.00367, which was considered small enough in the circumstances to accept replicates as clonemates (Shaw, 1988). It is difficult to decide whether or not to include clonal replicates in calculations of genetic variability measures from samples taken from populations known to have the capacity for asexual propagation, as they may represent the 'true gene pool' of the population. In the present study it was felt prudent to take the more conservative path of removing them, although inclusion of replicates in fact rarely has a significant effect on genetic variability estimates in these species (Shaw, 1988).

After removal of clonal replicates, samples were used to calculate for each population: proportion of loci polymorphic at the 0.95 criterion for the common allele; mean number of alleles per locus; mean observed (H_O) and Nei's (1978) unbiased expected (H_E) heterozygosities. Departure of single-locus genotype frequencies from Hardy–Weinberg expectations for outcrossing was tested using χ^2 with Levene's (1949) correction, checked for the effect of rare genotypes using Fisher's exact test on pooled frequencies.

Heterogeneity between geographic populations was determined using χ^2 contingency tests on allele frequencies, checked for the effects of small expected frequencies using Nass's (1959) test, and estimation of genetic similarity using Nei's (1972) genetic distance D. Allele frequencies used in these calculations are taken from Shaw, Beardmore and Ryland (1987) and Shaw (1988). The BIOSYS-1 package (Swofford and Selander, 1981) was used in the calculation of some of the above statistics.

6.5.3 Results

Within-species mean values for all of the genotypic and genetic variability estimates calculated for the four species studied, together with number of populations studied and mean sample sizes, are presented in Table 6.5.1.

All individuals of *S. troglodytes* sampled possessed different multilocus genotypes, giving maximal values for d and E, whereas the other three species all showed some degree of replication of genotypes. Genotypic diversity was lowest in *S. ornata* and varied widely between samples, from a monoclonal condition ($d = 0.01$) to a diverse mix ($d = 0.89$).

Table 6.5.1 Within-species means of genotypic and genetic diversity measures in four species of sagartiid sea anemones estimated from 14 allozyme loci (SE or range in brackets)

	S. troglodytes	*S. ornata*	*S. elegans*	*C. pedunculatus*
Populations	6	6	8	8
Sample size	55.2	48.7	73.5	59.4
	(36.4)	(28.0)	(34.0)	(18.7)
d	1.00	0.43	0.61	0.54
	–	(0.01–0.89)	(0.41–0.73)	(0.06–0.69)
E	1.00	0.61	0.58	0.47
	–	(0.30–1.00)	(0.18–0.77)	(0.07–0.73)
$P_{0.95}$	0.61	0.62	0.64	0.66
	(0.06)	(0.10)	(0.06)	(0.08)
No. alleles	3.1^{**}	2.0	2.1	2.1
	(0.4)	(0.5)	(0.1)	(0.3)
H_O	0.27	0.47^{**}	0.27	0.28
	(0.01)	(0.13)	(0.03)	(0.06)
H_E	0.28	0.32	0.28	0.29
	(0.01)	(0.06)	(0.02)	(0.02)
D	0.018	0.159	0.016	0.048

d = proportion of different genotypes per sample; E = 'evenness' of abundance of clones (Fager, 1972); $P_{0.95}$ = proportion of loci polymorphic at the 0.95 criterion for the common allele; No. alleles = mean number of alleles per locus over 14 loci; H_O = observed heterozygosity; H_E = expected heterozygosity (unbiased estimate – Nei (1978)); D = interpopulation genetic distance (Nei, 1972); **significantly ($p < 0.01$) different from other species.

C. pedunculatus has a similarly wide range in *d*, whereas *S. elegans* exhibited much less variation between samples. *S. ornata* samples on average show more even genotype prolif- eration, whereas *S. elegans* and *C. pedunculatus* both include samples showing numerical dominance by several clones ($E = 0.18$ and 0.07 respectively), this being more pronounced in both degree and occurrence in *C. pedunculatus*.

Levels of genetic variability are very similar across the four species. Proportions of loci polymorphic were not significantly different between species. *S. troglodytes* exhibited sig- nificantly more alleles per locus than the other three species. Mean observed heterozygosity per locus (H_O) is significantly higher in *S. ornata*. Between-sample variation in H_O is low in *S. troglodytes* (SE of mean = 0.01), but progressively larger in *S. elegans* (SE = 0.03) and *C. pedunculatus* (SE = 0.06), and high in *S. ornata* (SE = 0.13).

Allele frequencies, departures of single-locus genotype frequencies from outcrossing expectations and heterogeneity in allele frequencies between samples are given in Shaw (1988), and are summarized below.

Genotype frequencies were not significantly different from expectations for random mating at any locus in any sample of *S. troglodytes*. As would be expected for an obligately asexual species, almost all loci in all samples of *S. ornata* exhibited extreme deviations from outcrossing predictions, giving a mean of 6.2 significant deviations (out of 8.7 polymorphic loci per sample). Most samples of both *S. elegans* and *C. pedunculatus* displayed significant ($P < 0.05$) departures from outcrossing predictions at one or more loci, although on average these only amounted to around one departure in nine polymorphic loci per sample (means of 0.8 and 1.3 respectively).

No significant heterogeneity in allele frequencies was detected at any locus among the six samples of *S. troglodytes*. In contrast, all 10 loci polymorphic in *S. ornata* displayed extreme heterogeneity between samples ($P \ll 0.001$). In *S. elegans* 9 out of 12 variable loci displayed significant geographic heterogeneity ($\chi^2 = 18.8$–196.0, d.f. = 7–35, $P = 0.009$ to $\ll 0.001$). However, of these only one was due to consistent differences in the frequency of the common alleles across all eight populations. In *C. pedunculatus* 9 out of 10 variable loci were significantly heterogeneous $\chi^2 = 25$–79, d.f. = 5–15, $P < 0.05$ to $\ll 0.001$), but unlike with the *S. elegans* data, seven of these resulted from distinct frequency differences in common alleles both within and between South Wales and Southern England. The differences in the level of allele frequency heterogeneity between the four species is illustrated by the mean estimates of interpopulation genetic distance, D (Nei, 1972). Mean D in *S. troglodytes* and *S. elegans* was very similar (0.018 and 0.016 respectively), whereas in *C. pedunculatus* D was three times larger (0.048), and in *S. ornata* mean D was very high for an intraspecific value (0.159) (Table 6.5.1).

6.5.4 Discussion

On the whole, the four sagartiid sea anemone species examined in this study displayed population genetic parameters that can be explained by what is known about their reproductive strategies. *S. troglodytes* is known to have well-synchronized gametogenesis and spawning, with no asexual method of propagation (Nyholm, 1949; Riemann-Zurneck, 1969; Shaw, 1989). In accordance with this mode of reproduction, local populations display maximal genotypic diversity and evidence of successful outcrossing (no deviations from Hardy–Weinberg expectations for random mating), and effective gene flow between populations (no heterogeneity in allele frequencies). In contrast, *S. ornata* is thought to be an obligate parthenogen brooder (Shaw, Beardmore and Ryland, 1987), and displays substantial replication of individual genotypes, with no evidence of sexual outcrossing within populations, or gene flow between populations ('fixed heterozygosity' – see Shaw, Beardmore and Ryland (1987) – and significant allele frequency heterogeneity between samples). Between these two extremes, *S. elegans* and *C. pedunculatus* utilize both sexual and asexual methods (semiclonal). Both species display cloning of individual genotypes and, despite apparently employing the same style of sexual reproduction as *S. troglodytes* (Shaw, 1989), display departure from outcrossing predictions within populations and breakdown of gene flow between populations (allele frequency heterogeneity). Variation in genetic structure within and differentiation between populations appears to be more pronounced in *C. pedunculatus* than in *S. elegans*, as indicated by the greater range in values of d and H_O and the greater significance of allele frequency heterogeneity and higher interpopulation mean genetic distance ($D = 0.048$) in the former. Reduced rates of outcrossing within populations of these two species compared to *S. troglodytes* may reflect the effects of cloning on population evolution (inbreeding within or between a few very abundant genotypes) or less successful recruitment from sexually produced larvae compared to asexually produced propagules. In the case of *C. pedunculatus*, reduced sexual recruitment might be explained by the breakdown of effective sexual reproduction (optimal synchronization of spawning, gametogenesis and larval development) due to unfavourable conditions near to the margins of the species' geographic and/or environmental range (Rossi, 1975; Heap, 1979; Shaw, 1988, 1989). In *S. elegans* reduced sexual reproductive efficiency might result, as suggested by Sebens (1981) and Bucklin (1987) for other actinians, by pedal laceration producing a high proportion of individuals below the minimum size for gonad production – 37% of individuals immature compared to 4% in *S. troglodytes* (Shaw, 1989).

S. ornata, *S. elegans* and *C. pedunculatus* all display substantial replication of individual genotypes in random population samples. Genotypic diversity is lowest in *S. ornata* and highest in *S. elegans*. It is presumed that this pattern results from the different capability for input into populations of new genotypes from sexual reproduction – possibly absent in *S. ornata*, and less in *C. pedunculatus* than in *S. elegans* as indicated by the greater number and significance of deviations from outcrossing predictions. In a theoretical consideration of the factors affecting genotypic diversity in sedentary clonal organisms, Sebens and Thorne (1985) concluded that input of new genotypes from sexual reproduction was probably the most important determinant of genotypic diversity in semiclonal species. Surprisingly high E values in *S. ornata* may stem from the high degree of genetic (and therefore presumably phenotypic) similarity between clones, a situation previously suggested for clonal coexistence in obligate parthenogens (Lynch, 1984; Hebert, 1987), associated with the mode of origin of such species (Parker, 1979). Differences between *S. elegans, C. pedunculatus* and *S. ornata* may result from asexuality in *S. elegans* being part of an integrated mixed reproductive strategy displayed throughout its range, whereas in the other two species it represents an alternative to sexual reproduction brought about by adverse conditions in geographically marginal habitats ('geographic parthenogenesis' in *C. pedunculatus*) or necessity arising from hybrid/polyploid origins (*S. ornata*).

Levels of genetic variability in the species in the present study were high compared to most previously studied organisms (Burton, 1983; Nevo, Beiles and Ben-Shlomo 1984), but consistent with other anthozoan species (Sole-Cava and Thorpe, 1991). Sole-Cava and Thorpe (1991) could not conclude whether the observed high levels of genetic variability in anthozoans, also seen in sponges, were a result of the evolutionary age of these groups or their potentially very large effective population sizes. The present study has shown that local populations of several species are not effectively reproducing sexually and show significant substructuring within local geographic areas. So if genetic variability in these species is maintained purely by stochastic processes (large N_E and gene flow) then these processes must be operating effectively only over long timescales. This proposition is acceptable given the potential longevity of individual sea anemones (and especially of genets in the cloning species) and stability of local populations, as described by Hoffmann (1987).

Levels of polymorphism ($P_{0.95}$) are not significantly different between the four species. No real pattern has emerged in previous studies on actinians except that P is extremely variable (Sole-Cava and Thorpe, 1991) – a suggestion of higher P in aclonal over clonal species is opposite to the trend seen in this study (aclonal *S. troglodytes* has the lowest value). However, in terms of number of alleles per locus there is a significant difference in polymorphism, the sexual *S. troglodytes* having the higher value. This result fits well the theoretical predictions that sexual species should maintain an advantage over obligate or facultative asexuals in terms of genetic variability from the view of both selectionist (spatial/temporal flexibility – Bell's (1982) 'tangled bank' and Maynard Smith's (1978) 'Red Queen' models) and neutralist ideas (larger N_E in freely panmictic species; Kimura (1983).

Observed heterozygosity is not significantly different between *S. troglodytes, S. elegans* and *C. pedunculatus*, but is significantly higher in *S. ornata*. This is largely due to the occurrence of 'fixed heterozygosity' in *S. ornata*, where all individuals possess the same heterozygous genotype at particular loci (Shaw, Beardmore and Ryland, 1987; Shaw, 1988). Elevated H_O in *S. ornata* is not surprising in view of the apparently obligate parthenogenetic polyploid nature, and possibly hybrid origins of this 'morphospecies' (Shaw, Beardmore and Ryland, 1987). Accumulation of mutations is expected in parthenogens (White, 1977), and high levels of 'fixed heterozygosity' have been observed in other hybrid species (Roose and Gottlieb, 1976; Lynch, 1984). Of course, direct comparison of H_O between *S. ornata* and the other species may not be entirely valid as the unit of segregation in *S. ornata* is the composite genome. Nei (1975) suggested that H_E (referred to as 'gene diversity') was a better estimate of genetic variation in such cases. Interestingly, H_E is higher in *S. ornata* than in the other species, but not significantly so ($p > 0.10$).

The conformity in mean levels of heterozygosity among the species in this study, comprising aclonal, semiclonal and clonal forms, differs from suggestions in previous studies of actinians. Bucklin and Hedgecock (1982) and Sole-Cava (1986) found higher H_O in aclonal over semiclonal *Metridium, Urticina* and *Actinia* species, whereas Smith and Potts (1987) found the opposite in *Anthopleura* species. Lower H_O in semiclonal species could be produced through inbreeding effects by crossing within localized clones or swamping of the gamete pool by one dominant genotype. Additionally, lower H_O could result from reduced N_E occurring through asexual amplification of genotypes and subdivision of populations into more or less isolated units by localized differentiaton (Kimura, 1983). Alternatively, higher H_o might occur in semiclonal species through the effects of increased population stability acting as a buffer against the decay of variation. Additionally, the ability for local colonization by 'selected' clones could promote the effects of increased 'grain' in the species environment (Levins, 1968), a condition promoting increased genetic variability according to Selander and Kaufman (1973) and Nevo (1983). Whether such processes are important, or even occurring at all, is impossible to conclude given the variation in genetic variability estimates produced to date – Selandar and Kaufman (1973) noted that sampling error among loci could be a significant source of variation given the variable contribution of individual loci to estimates of overall heterozygosity (Ayala and Powell, 1972; Woodwark, Skibinski and Ward, 1992).

In conclusion, the main finding of this study is that four closely related and morphologically and ecologically similar species utilizing four different combinations of reproductive methods, and showing local population structuring consistent with expectations from breeding systems, maintain approximately the same overall level of genetic variation within the species as a whole, although the way in which this variation is distributed within and between populations differs significantly from species to species.

Acknowledgements

This work was carried out during tenure of a NERC studentship to P. Shaw.

References

Ayala, F.J. and Powell, J.R. (1972) Enzyme variability in the *Drosophila willistoni* group. VI. Levels of polymorphism and the physiological function of enzymes. *Biochem. Genet.*, **7**, 331–45.

Bell, G. (1982) *The Masterpiece of Nature*, Croom Helm, London.

Bucklin, A. (1987) Adaptive advantages of patterns of growth and asexual reproduction of the sea anemone *Metridium senile* (L.) in intertidal and submerged populations. *J. Exp. Mar. Biol. Ecol.*, **110**, 225–43.

Bucklin, A. and Hedgecock, D. (1982) Biochemical genetic evidence for a third species of *Metridium* (Coelenterata: Actiniaria). *Mar. Biol.*, **66**, 1–7.

Burton, R.S. (1983) Protein polymorphism and genetic differentiation of marine invertebrate populations. *Mar. Biol, Lett.*, **4**, 193–206.

Chia, F.S. (1976) Sea anemone reproduction: patterns and adaptive radiations, in *Coelenterate Ecology and Behaviour* (ed. G.O. Mackie), Plenum Press, New York, pp. 261–70.

Fager, E.W. (1972) Diversity: a sampling study. *Am. Nat.*, **106**, 293–310.

Heap, A.S. (1979) Population ecology, reproductive biology and some physiological effects of temperature in the sea anemone *Cereus pedunculatus* Pennant 1777. PhD Thesis, University of Reading.

Hebert, P.D.N. (1987) Genotypic characteristics of the Cladocera. *Hydrobiologia*, **145**, 183–93.

Hoffmann, R.J. (1987) Short-term stability of genetic structure in populations of the sea anemone *Metridium senile. Mar. Biol.*, **93**, 499–507.

Kimura, M. (1983) *The Neutral Theory of Molecular Evolution*, Cambridge University Press, London.

Levene, H. (1949) On a matching problem arising in genetics. *Ann. Math. Stat.*, **20**, 91–4.

Levins, R. (1968) *Evolution in Changing Environments*, Princeton University Press, Princeton, New Jersey.

Lynch, M. (1984) The genetic structure of a cyclical parthenogen. *Evolution*, **38**, 186–203.

Manuel, R.L. (1981) *British Anthozoa*, Linnean Society, London.

Maynard Smith, J. (1978) *The Evolution of Sex*, Cambridge University Press, Cambridge.

Nass, C.A.G. (1959) The chi-squared test for small expectations in contingency tables, with special reference to accidents and absenteeism. *Biometrika*, **46**, 365 -85.

Nei, M. (1972) Genetic distance between populations. *Am. Nat.*, **106**, 283–92.

Nei, M. (1975) *Molecular Population Genetics and Evolution*, North-Holland, Amsterdam.

Nei, M. (1978) Estimation of average heterozygosity and genetic distance from a small number of individuals. *Genetics*, **89**, 583–90.

Nevo, E. (1983) The adaptive significance of protein variation, in *Protein Polymorphism: Adaptive and Taxonomic Significance* (ed. G.S. Oxford and D. Rollinson), Academic Press, London, pp. 239–82.

Nevo, E., Beiles, A. and Ben-Shlomo, R. (1984) The evolutionary significance of genetic diversity, ecological, demographic and life history correlates. *Lect. Notes Biomath.*, **53**, 13–213.

Nyholm, K.-G. (1943) Zur Entwicklung und Entwicklungsbiologie der Cerianthare und Aktinien. *Zool. Bidr. Upps.*, **22**, 87–248.

Nyholm, K.-G. (1949) On the development and dispersal of Athenaria Actinia with special reference to *Halcampa duodecimcirrata*. *Zool. Bidr. Upps.*, **27**, 465–506.

Parker, E.D. (1979) Ecological implications of clonal diversity in parthenogenetic morphospecies. *Am. Zool.*, **19**, 753–62.

Riemann-Zurneck, K. (1969) *Sagartia troglodytes* (Anthozoa). Biologie and Morphologie einer Schlickbewohnenden Aktinie. *Veroff. Inst. Meeresforsch. Bremerhaven*, **12**, 169–230.

Roose, M.L. and Gottlieb, L.D. (1976) Genetic and biochemical consequences of polyploidy in *Tragopogon*. *Evolution*, **30**, 818–30.

Rossi, L. (1971) Thelytochous parthenogenesis in *Cereus pedunculatus*. *Experientia*, **27**, 349–51.

Rossi, L. (1975) Sexual races in *Cereus pedunculatus* (Boad). *Pubbli. Staz. Zool. Napoli*, **39**, 462–470.

Schafer, W. (1981) Fortpflanzung und Sexualitat von *Cereus pedunculatus* und *Actinia equina* (Anthozoa, Actiniaria). *Helgolander Meeresunters*, **34**, 451–61.

Sebens, K.P. (1981) Reproductive ecology of the intertidal sea anemones *Anthopleura xanthogrammica* (Brandt) and *Anthopleura elegantissima* (Brandt): body size, habitat and sexual reproduction. *J. Exp. Mar. Biol. Ecol.*, **54**, 225–250.

Sebens, K.P. and Thorne, B.L. (1985) Coexistence of clones, clonal diversity and the effects of disturbance, in *Population Biology and Evolution of Clonal Organisms* (eds J.B.C. Jackson, L.W. Buss and R.E. Cook), Yale University Press, London, pp. 357–98.

Selander, R.K. and Kaufman, D.W. (1973) Genic variability and strategies of adaptation in animals. *Proc. Nat. Acad. Sci. USA*, **70**, 1875–7.

Shaw, P.W. (1988) Ecological genetics of anthozoans. PhD Thesis, University of Wales, Swansea.

Shaw, P.W. (1989) Seasonal patterns and possible long-term effectiveness of sexual reproduction in three species of Sagartiid sea anemones, in *Reproduction, Genetics and Distributions of Marine Organisms* (eds J.S. Ryland and P.A. Tyler), Olsen & Olsen, Fredensborg, Denmark, pp. 189–99.

Shaw, P.W. (1991) Effects of asexual reproduction on population structure of *Sagartia elegans* (Anthozoa: Actiniaria). *Hydrobiologia*, **216/217**, 519–25.

Shaw, P.W., Beardmore, J.A. and Ryland, J.S. (1987) *Sagartia troglodytes* (Anthozoa: Actiniaria) consists of two species. *Mar. Ecol. Prog. Ser.*, **41**, 21–8.

Shick, J.M. (1991) *A Functional Biology of Sea Anemones*, Chapman & Hall, London.

Smith, B.L. and Potts, D.C. (1987) Clonal and solitary anemones (*Anthopleura*) of western North America: population genetics and systematics. *Mar. Biol.*, **94**, 537–46.

Sokal, R.R. and Rohlf, F.J. (1981) *Biometry. The Principles and Practice of Statistics in Biological Research*, 2nd edn, W.H. Freeman & Co, New York.

Sole-Cava, A.M. (1986) Studies of biochemical genetics and taxonomy of coelenterates and sponges. PhD Thesis, University of Liverpool.

Sole-Cava, A.M. and Thorpe, J.P. (1991) High levels of genetic variation in natural populations of marine lower invertebrates. *Biol. J. Linn. Soc.*, **44**, 65–80.

Stephenson, T.A. (1929) On methods of reproduction as specific characters. *J. Mar. Biol. Ass. UK*, **16**, 131–72.

Swofford, D.L. and Selander, R.B. 1981 BIOSYS-1: a FORTRAN program for the comprehensive analysis of electrophoretic data in population genetics and systematics. *J. Hered*, **72**, 281–3.

White, M.J.D. (1977) *Animal Cytology and Evolution*, 3rd edn, Cambridge University Press, Cambridge.

Woodwark, M., Skibinski, D.O.F. and Ward, R.D. (1992) A study of interlocus allozyme heterozygosity correlations: implications for neutral theory. *Heredity*, **69**, 190–8.

6.6 DELETERIOUS MUTATIONS CAN ACCOUNT FOR THE MAINTENANCE OF THE HAPLO-DIPLOID CYCLE

Sophie Richerd, Véronique Perrot, Denis Couvet, Myriam Valero and Alexey S. Kondrashov

Abstract

The problem of the evolution of the alternation of haploid and diploid phases in life-cycles is of special interest in the algae, because they display a wide range and variability of life-cycles, including haplo-diploid cycles with maintenance of two free-living phases. Partial avoidance of the cost of sex provides to the haplo-diploid cycle a two-fold advantage over haploid and diploid cycles. A haplo-diploid cycle is thus maintained when the ratio of haploid to diploid individual fitness is within a range of $\frac{1}{2}$ to 2. The present model uses a genetic determination of the fitness, based on selection against many deleterious mutations, to demonstrate that the advantage of the haplo-diploid cycle holds for a large range of dominance and mutation parameters. Haplo-diploidy is favoured at low genomic mutation rate (total mutation rate per generation U < 0.4), or intermediate expression of deleterious mutations (dominance degree h = 0.5). When mutation rate increases, diploidy is favoured when deleterious mutations are more recessive (h < 0.4), and haploidy when deleterious mutations are less recessive (h > 0.55).

6.6.1 Introduction

The algae are of special interest for the study of the evolution of life-cycles, i.e. the alternation of haploid and diploid phases during the sexual reproductive cycle caused by the regular alternation of meiosis and syngamy. The algae display a wide range of cycles, along a continuum between the two extreme cycles, where only one of the phases, either haploid or diploid, is free-living and individualized. They include, at a high frequency, cycles where both phases are free-living generations, referred to hereafter as haplo-diploid cycles. Occurrence of wide morphological differences between the two generations has led some authors to describe them as separate species (heteromorphic cycle). In other species the haploid and diploid individuals are indistinguishable, for morphology, ecology and physiology (isomorphic cycle) (Lobban and Wynne, 1981).

It is difficult to find any clear evolutionary trend in the distribution of the different cycles among algal taxa. Things are not helped by the common use of the life-cycle as a taxonomic character. The first molecular phylogenies show that life-cycle evolution is not irreversible (Ragan, personal communication). Moreover, cycle variability is found at low taxonomic levels: within genera or species. Intraspecies variability has a genetic basis in some species (Van der Meer and Todd, 1977), or is caused by environmental factors (Abdel-Rahman and Magne, 1991). In the former case this would allow for cycle evolution in populations.

Knowing that the cycle can still evolve in some cases, and thus the maintenance of a particular cycle in a given species is not necessarily due only to phylogenetic constraints, we can ask the following questions:

1. What determines the evolution towards one or another life-cycle?
2. More specifically, how can a haplo-diploid cycle involving the obligate alternation of individuals of two different ploidy levels be maintained?

Several hypotheses have been suggested to answer the first question, proposing genetic or ecological advantages either to haploidy or to diploidy (review: Valero *et al.*, 1992). But the problem remains unsolved for the haplo diploid cycle: how can it maintain both haploid and diploid individuals if there is an advantage either to diploidy or to haploidy?

In a previous paper (Richerd, Couvet and Valero, 1993) we pointed out that the maintenance of haplo-diploid cycles could be related to the cost of sex. As long as one assumes that the lifespan of an individual is the same whatever its cycle, the haplo-diploid cycle involves half the meiosis and fertilization events of either the haploid or diploid cycles, for the same number of individualized phases (generations). The haplo-diploid cycle thus benefits from the same advantage as a cycle involving the alternation of one sexual and one asexual generation: namely a halving of the cost of sex.

This two-fold advantage was demonstrated with a model using fixed fitnesses for haploid and diploid individuals. The haplo-diploid cycle is maintained against haploid and diploid cycles, not only for equal haploid and diploid fitnesses, but when the fitness ratio haploid to diploid W_h/W_d lies between $\frac{1}{2}$ and 2, i.e. as long as the advantage of haploid or diploid individuals is less than two-fold.

However, it is important to test the relevance of this range $\frac{1}{2}$ to 2, both experimentally and theoretically, when fitness is not fixed but genetically determined. We present here a model where fitness of haploid and diploid individuals is determined by the number of deleterious mutations in their genomes. We follow Crow (1970) and Kondrashov (1988) in assuming that detrimental mutations represent a major evolutionary force.

The aim of this section is to test if the range of haploid to diploid fitness ratio from $\frac{1}{2}$ to 2, providing an advantage to the haplo-diploid cycle, is possible and easy to obtain, when genetic parameters like the kind of selection, dominance degree and mutation rate vary.

6.6.2   The model

The numerical model employed here is based on the method of Kondrashov (1982) to follow the distribution of an infinite number of unlinked loci undergoing mutation and selection in a haplo-diploid cycle. The fitness of a haploid or a diploid individual is a function of the number of deleterious mutations it carries in its genome.

The cycle begins with selection in the haploid phase. The next step is the mutation process in haploids: new mutations appear randomly at non-mutated loci, following a Poisson distribution of parameter U. U is the mutation rate per haploid genome per generation. Then syngamy occurs: haploid gametes (produced by mitosis) fuse randomly, adding their respective number of mutations in the zygote. Syngamy is followed by selection in the diploid phase, and the mutation process in diploids, following a Poisson law of parameter $2U$, because diploids have twice the genome size of haploids. Finally there is meiosis, involving independent segregation of all loci and random transmission of the non-mutated or of the mutated copy for each locus (binomial distribution of parameters $\frac{1}{2}$ and total number of loci). The cycle then repeats.

The underlying assumptions of the model are as follows. The population size is infinite. The number of loci per individual is infinite (practically, numerical iterations involve up to 100 mutable loci). Mutations are partially dominant, mutation rate per locus is small (since the total number of mutable loci is large), mating is at random, and there is free recombination among loci: hence the occurrence of homozygotes is neglected in diploids.

The fitness of an individual (w) is a function of the number of mutations in its genome (i). Different fitness functions are used, corresponding to different actions of selection on the mutated loci (Crow, 1970):

1. Mutations are independently selected, i.e. selection at one locus does not depend on selection at the other loci in the genome. The fitness function is called multiplicative or exponential; a is the selection coefficient:

$$w(i) = (1-s)^i = e^{i \ln(1-s)} = e^{-ai} \qquad \text{(Haldane, 1937)}$$

2. There is a positive epistatic action between loci: the effect of the $n + 1$th mutation on fitness is greater than the effect of the nth mutation. The fitness function is synergic, a being the exponential coefficient, and b the quadratic coefficient:

$$w(i) = 1 - ai - bi^2 \qquad \text{(Kimura, 1961)}$$

or, to avoid negative values of w if i is large:

$$w(i) = e^{-ai-bi^2} \qquad \text{(Charlesworth, 1990)}$$

3. An extreme case of synergic selection, simpler to handle, is truncation selection with a threshold: individuals with more than a given number T of mutations are lethal, while the others are equally fit (Crow and Kimura, 1979; Kondrashov, 1982, 1988):

$$w(i) = 1 \text{ if } i \leqslant T$$

$$w(i) = 0 \text{ if } i > T$$

In diploids, with heterozygote selection, each mutation has a smaller effect on fitness than with hemizygote selection in haploids. The dominance degree of the deleterious mutations, h, corresponds to the proportion of mutations expressed in diploids. A diploid with i/h mutations has the same fitness as a haploid with i mutations (h is between 0 and 1, so that i/h is greater than i). A haploid is equivalent to a homozygous diploid, and, for $h = 0$, haploid and diploid fitnesses are equal (Kondrashov and Crow, 1991).

We iterate numerically the equations in the Appendix, beginning with a diploid population with no mutations, until mutation–selection equilibrium is reached. Mean haploid fitness and mean diploid fitness are calculated, to obtain their ratio W_h/W_d. This is done for various values of selection (exponential selection $a = 0.1$, synergic selection $a = 0, 0.1$, and $b = 0.1$, 0.01; truncation selection $T = 10$), genomic mutation rate U (0.25, 0.5, 0.75, 1, 1.25, 1.5, 1.75, 2), and dominance degree h (0.01, 0.25, 0.5, 0.75, 1).

6.6.3 Results

The mean fitness ratio haploid to diploid W_h/W_d varies as a function of both mutation rate per generation U and dominance degree of the mutations h for exponential selection (Figure 6.6.1), for synergistic selection (Figure 6.6.2) and for truncation selection (Figure 6.6.3). W_h/W_d increases with h, i.e. when intensity of selection in diploids increases. W_h/W_d is a negative function of U when h is small, and a positive function when h is large.

Figure 6.6.4 shows the limits where W_h/W_d is equal to $\frac{1}{2}$, 1, or 2. There is a large range for both dominance and mutation rate, allowing W_h/W_d to range between $\frac{1}{2}$ and 2. Mean haploid and diploid fitnesses are equal when h is about $\frac{1}{2}$ whatever U and the selection. The limits $W_h = 2W_d$ or $W_d = 2W_h$ vary with h and U, and to a smaller degree with selection.

The haplo-diploid cycle is thus favoured:

1. whatever the dominance if mutation rate is low, i.e. less than about 0.4 per generation per haploid genome;
2. whatever the mutation rate when there is intermediate dominance for deleterious mutations (h about $\frac{1}{2}$);
3. for a range of dominance, getting narrower as mutation rate increases.

The haploid cycle is favoured when $U > 0.6$ and $h = 1$, and for h decreasing as U increases, until h is higher than about 0.55 for $U = 2$.

Exponential selection

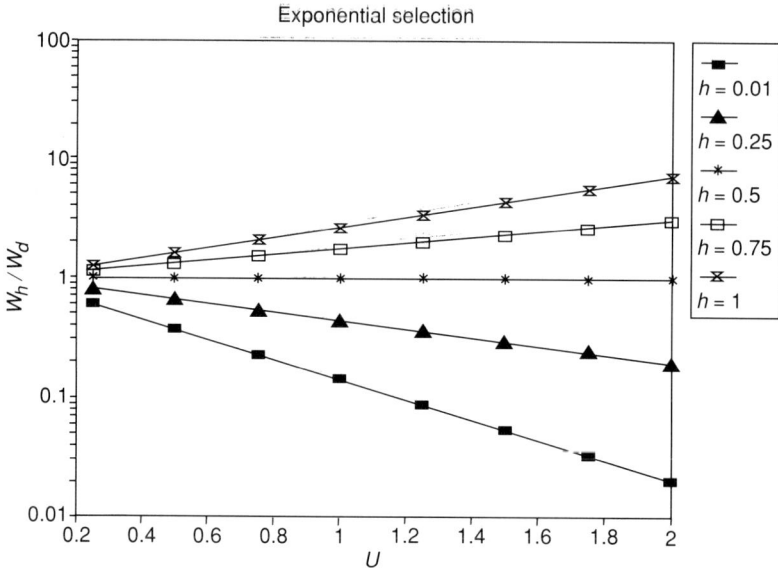

Figure 6.6.1 Mean haploid to diploid fitness ratio W_h/W_d, according to the mutation rate per haploid genome per generation, U, and for different degrees of dominance of deleterious mutations, h, under exponential selection. Selection coefficient $a = 0.1$, $w_d(i) = \exp(-ai)$, and $w_h(i) = \exp(-hai)$.

Synergic selection 2

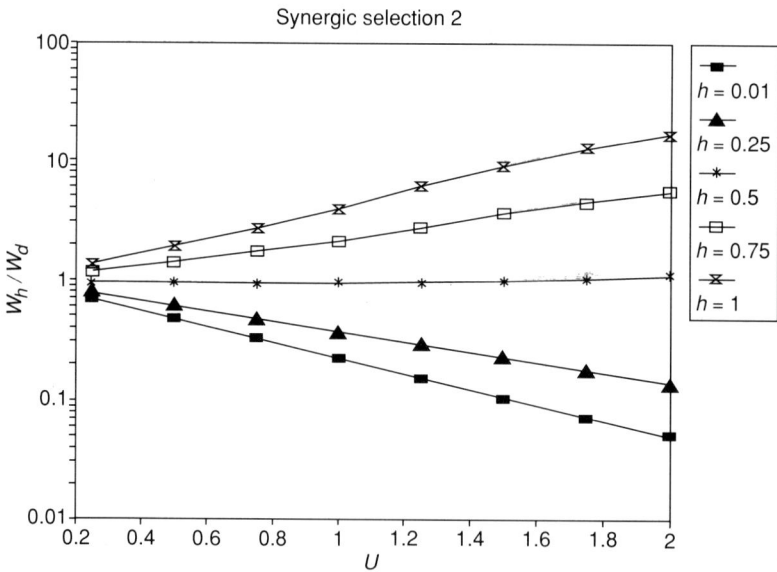

Figure 6.6.2 Mean haploid to diploid fitness ratio W_h/W_d, according to the mutation rate per haploid genome per generation, U, and for different degrees of dominance of deleterious mutations, h, under synergic selection. Selection coefficient $a = 0.1$, $b = 0.01$, $w_d(i) = \exp(-ai-bi^2)$, and $w_h(i) = \exp(-hai-hbi^2)$.

Truncation selection

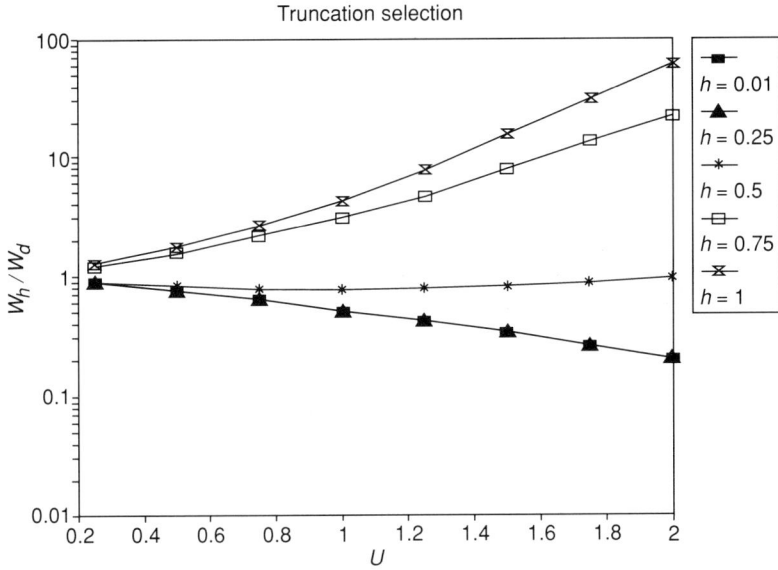

Figure 6.6.3 Mean haploid to diploid fitness ratio W_h/W_d, according to the mutation rate per haploid genome per generation, U, and for different degrees of dominance of deleterious mutations, h, under truncation selection. Threshold $T = 10$, $w_d(i) = 1$ if $i <= T$, else $w_d(i) = 0$, $w_h(i) = 1$ if $i < = hT$, else $w_h(i) = 0$.

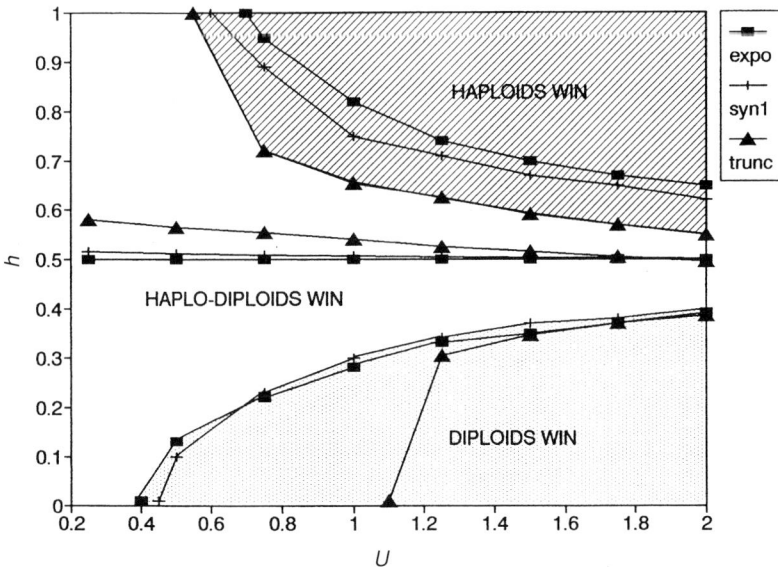

Figure 6.6.4 Domains of advantage of the three different life-cycles, according to dominance degree h and the mutation rate U, for different types of selection. The lower lines correspond to $W_h/W_d = 1/2$ and delimit the domain of advantage of the diploid cycle; the mid-lines correspond to $W_h/W_d = 1$; the upper lines correspond to $W_h/W_d = 2$ and delimit the domain of advantage of the haploid cycle. expo = exponential selection, $a = 0.1$; syn = synergic selection, $a = 0.1$ and $b = 0.01$; trunc = truncation selection, $T = 10$.

The diploid cycle is favoured when $U > 0.4$ (synergic and exponential selection) or $U > 1$ (truncation selection), and for h increasing as U increases until h is less than about 0.4 for $U = 2$.

Results are globally consistent among the three kinds of selection and the different intensities of selection. Increasing the strength of epistatic interactions between loci (from exponential, synergic to truncation selection) has two main effects: there is more space for the haploid cycle, because the limit is shifted relative to both h and U; and there is less space for the diploid cycle, because the minimal mutation rate increases.

At selection–mutation equilibrium, the distribution of mutations in diploids has approximately twice the mean and variance compared to that in haploids. The mean fitness of the population ($W_h.W_d$) decreases when h and U increase (data not shown).

6.6.4 Discussion

This model shows that a genetic determination of fitness, based on selection against many deleterious mutations, furnishes a large range of values of selection and mutation parameters, that can account for the two-fold advantage of the haplo-diploid cycle (Richerd, Couvet and Valero, 1993).

Haplo-diploidy is favoured for a low genomic mutation rate per generation. Although the relation between genome size and gene number is far from clear ('paradox of C value', Cavalier-Smith (1985)), a low genomic mutation rate could be more easy to obtain when the genome is small and contains few genes. Available data (review: Cavalier-Smith, 1985) are insufficient to test for a difference of genome size between haplo-diploid organisms and others. Recently, studies by Drake (1991a) on mutation rate in haploid microbes (phage, bacteria, yeast and fungus), have shown that, although genome size varies 6500-fold, the total genomic mutation rate is surprisingly constant for all these organisms, close to 0.003 per replication. However, the number of replications per sexual cycle is likely to vary, so that we cannot infer from this the mutation rate per phase.

Diploidy is advantageous for low values of dominance ($h < 0.4$) and high mutation rate ($U > 0.4$ for exponential and synergic selection, but $U > 1.1$ for truncation selection). The value for dominance usually reported in *Drosophila* is about 0.02 for lethals, and 0.2 for midly deleterious genes. Total genomic mutation rate per generation in metazoans is close to or higher than 1 (Kondrashov, 1988; Drake, 1991b). These estimates fit the conditions of advantage of the diploid cycle in our model, for exponential or synergic selection, but correspond to the limit between haplo-diploids and diploids for truncation selection.

Considering exponential, synergic or truncation selection in multilocus selection models leads to very different evolutionary patterns for various features such as mutational load (Haldane, 1937; Kimura, 1961; Crow and Kimura, 1979), inbreeding depression (Charlesworth, 1990), evolution of sexual reproduction (Kondrashov, 1988) or evolution of diploidy from haploidy (Kondrashov and Crow, 1991). In this model the action and intensity of selection against deleterious mutations have relatively weak effects on the mean fitness ratio of haploids to diploids. However, truncation selection significantly increases the threshold mutation rate necessary for the diploid cycle to be advantageous. This is another demonstration that the way in which deleterious mutations may be eliminated by selection has an influence on the evolution of diploid cycles (Kondrashov and Crow, 1991).

The degree of dominance, h, of mutant alleles is the only difference between haploids and diploids in our model. We assume that genes expressed in haploid and diploid phases are the same, and that hemizygous and homozygous selection is similar. Thus it applies mainly to isomorphic haplo-diploid cycles. In the cases where a fraction of the genes are expressed only in one of the phases (e.g. heteromorphic cycles of algae, or pollen selection in higher plants), or if mutation expression differs between hemizygous and homozygous loci, due to gene redundancy, then the coefficient h should include both this difference and the true dominance degree (Charlesworth and Charlesworth, 1992). Further evidence comparing homozygous and hemizygous gene functioning is needed to estimate the importance of the dominance degree of deleterious mutations in life-cycle evolution.

In conclusion, similar haploid and diploid fitnesses are easy to obtain in a haplo-diploid cycle whatever the selection type. On the one hand, selection against mutations is always weaker in diploids than in haploids, because of complementation. On the other hand, as haploids and disploids produce each other, they have a similar number of mutations per haploid genome and a similar mutation rate, which is $3U$ for the total cycle. Diploids have on average twice as many mutations in their genomes, because their genome is twice as big. The equality of haploid and diploid fitnesses is thus obtained for dominance of about $\frac{1}{2}$. This limit of $h = \frac{1}{2}$ is also found by Perrot, Richerd and Valero (1991) when haploids and diploids interbreed, in a single-locus selection model. Moreover, a low mutation rate ($U < 0.4$) produces only a small difference in fitness between haploids and diploids, which cannot exceed a two-fold factor. Likewise, Kondrashov and Crow (1991) demonstrated that under truncation selection a minimal mutation rate of one per generation is necessary to ensure a significant difference of genetic load between diploids and haploids.

The intensity of interloci interactions, the number of mutations per generation, and the dominance degree of deleterious mutations are important factors likely to influence life-cycle evolution, and more empirical work is required to validate our theory.

Acknowledgements

This work was partly carried out during a stay of S.R. and V.P. in the Genetic Laboratory, University of Wisconsin, Madison, USA. W.R. Engels and J.F. Crow kindly let us use their fast MacIntoshes. S.R. benefited from CNRS travel and accommodation funding, and MRT grant 89329.

References

Abdel-Rahman, M.H. and Magne, F. (1991) Variation du carposporophyte chez les Acrochaetiales (Rhodophyta). *Cryptog. Algol.*, **11**, 23–30.
Cavalier-Smith, T. (1985) Eukaryotic gene numbers, non-coding DNA and genome size, in *The Evolution of Genome Size* (ed. T. Cavalier-Smith), John Wiley & Sons Ltd, Chichester, pp. 69–105.
Charlesworth, B. (1990) Mutation-selection balance and the evolutionary advantage of sex and recombination. *Genet. Res. Camb.*, **55**, 199–221.
Charlesworth, D. and Charlesworth, B. (1992) The effects of selection in the gametophyte stage on mutational load. *Evolution*, **46**, 703–20.
Crow, J.F. (1970) Genetic loads and the cost of natural selection, in *Mathematical Topics in Population Genetics* (ed. K. Kojima), Springer Verlag, Berlin, pp. 129–77.
Crow, J.F. and Kimura, M. (1965) Evolution in sexual and asexual populations. *Am. Nat.* **99**, 439–50.
Crow, J.F. and Kimura, M. (1979) Efficiency of truncation selection. *Proc. Nat. Acad. Sci. USA*, **76**, 396–99.
Drake, J.W. (1991a) A constant rate of spontaneous mutation in DNA-based microbes. *Proc. Nat. Acad. Sci. USA*, **88**, 7160–4.
Drake, J.W. (1991b) Spontaneous mutation. *Ann. Rev. Genet.*, **25**, 125–46.
Haldane, J.B.S. (1937) The effect of variation of fitness. *Am. Nat.*, **71**, 337–49.
Kimura, M. (1961) Some calculations on mutation load. *Jap. J. Genet.*, **365**, 179–90.
Kondrashov, A.K. (1982) Selection against harmful mutations in large sexual and asexual populations. *Genet. Res. Camb.*, **40**, 325–32.
Kondrashov, A.S. (1988) Deleterious mutations and the evolution of sexual reproduction. *Nature*, **336**, 435–40.
Kondrashov, A.S. and Crow, J.F. (1991) Haploidy or diploidy: which is better? *Nature*, **351**, 314–15.

Lobban, C.S. and Wynne, M.J. (eds) (1981) *The Biology of the Seaweeds*, Blackwell Scientific Publications, Oxford.

Perrot, V., Richerd, S. and Valero M. (1991) Transition from haploidy to diploidy. *Nature*, **351**, 315–17.

Richerd, S., Couvet, D. and Valero, M. (1993) Evolution of the alternation of haploid to diploid phases in life cycles. II. Maintenance of the haplo-diplontic cycle. *J. Evol. Biol.*, **6**, 263–80.

Valero, M., Richerd, S., Perrot, V. and Destombe, C. (1992) Evolution of the alternation of haploid and diploid phases in life cycles. Review and new perspectives. *TREE*, **7**, 25–9.

Van der Meer, J.P. and Todd, E.R. (1977) Genetics of Gracilaria tikvahiae (Rhodophyceae, Gigartinales). IV. Mitotic recombination and its relationship to mixed phases in life history. *Can. J. Bot.* **55**, 2810–17.

Appendix 6.6.A   Equations of the multilocus model for a haplo-diploid cycle

We note:

$p_d(i)$: frequency of diploids with i heterozygous mutations $\left(\sum_{i=0}^{n} p_d(i) = 1 \right)$

$p_h(i)$: frequency of haploids with i hemizygous mutations $\left(\sum_{i=0}^{n} p_h(i) = 1 \right)$

$w_d(i)$: fitness of a diploid individual with i heterozygous mutations
$w_h(i)$: fitness of a haploid with i hemizygous mutations

The haplo-diploid cycle involves: selection in the diploid stage – mutation in diploid cells – meiosis – selection in the haploid stage – mutation in haploid cells – syngamy.

We begin with only diploids, and at the first generation all individuals are mutation-free:

$$p_d(0) = 1$$

After selection in diploid stage:

$$p_d'(i) = p_d(i).w_d(i)/W_d \qquad W_d = \sum_{i=0}^{n} p_d(i).w_d(i)$$

After mutation in diploids:

$$p_d''(i) = \sum_{j=0}^{i} p_d'(j)\, M_d(j-i)$$

$M_d(j)$ is the mutation operator in diploids, or the probability of obtaining j new mutations per cell in one generation. It follows a Poisson law, with a mean of $2U$ in diploids. U is thus the mean mutation rate per haploid genome.

After meiosis:

$$p_h(i) = \sum_{j=0}^{i} p_d''(j).R(j,i)$$

$R(j,i)$ is the meiosis operator, or the probability of obtaining by meiosis i mutations among the j mutations of the parent. It follows a binomial law, with a probability of obtaining the mutated copy per locus.

$$R(j,i) = \frac{j!}{i!(j-1)!}.\left(\frac{1}{2}\right)^j$$

After selection in the haploid stage:

$$p'_h(i) = p_h(i).w_h(i)/W_h \qquad W_h = \sum_{i=0}^{n} p_h(i).w_h(i)$$

After mutation in haploids:

$$p''_h(i) = \sum_{j=0}^{i} p'_h(j) \, M_h(j-i)$$

After synagmy:

$$p_d(i) = \sum_{j=0}^{i} \sum_{k=0}^{i} p''_h(j) \, p''_h(k) \, S(i,j,k)$$

$S(i,j,k)$ is the syngamy operator. It is equal to 1 if $j + k = i$, else to 0.

7

Chromosomal genetics

7.1 ADVANCES IN CYTOGENETICS OF AQUATIC ORGANISMS

Catherine Thiriot-Quiévreux

Abstract

The aim of this section is to show how our knowledge of cytogenetics of aquatic organisms has advanced by investigations on: (1) chromosome number, (2) chromosome morphology and karyotype, and (3) chromosome banding. Chromosomal data of selected groups of aquatic invertebrates or fish will be used to illustrate evolutionary inferences. The evolution of chromosome number is described in Platyhelminthes (Turbellaria), Annelida (Polychaeta, Oligochaeta, Hirudinea), Crustacea (Branchiopoda, Copepoda, Peracarida, Eucarida), Mollusca (Polyplacophora, Gastropoda, Bivalvia, Cephalopoda) and Chordata (Larvacea, Ascidiacea, Thaliacea). The evolution in chromosome morphology and karyotypes is described in Turbellaria (Tricladia, Neorhabdocoela), Gastropoda (Archaeogastropoda, Thecosomata), Bivalvia (Veneroida) and fish (Characiformes). Chromosome banding is exemplified in Turbellaria Tricladia (Dugesia), Bivalvia Ostreidae and fish (NOR-banding, e.g. Cyprinidae; C-banding, e.g. Blennidae; restriction endonuclease banding, e.g. Salmonidae).

The evolutionary trends in chromosome number and morphology, and the use of banding techniques to enhance our understanding of the speciation process, are discussed.

7.1.1 Introduction

The study of eukaryote aquatic organisms has contributed significantly to our understanding of vertebrate evolution. The chromosomes of higher organisms have undergone diverse modifications during evolution. Several chromosomal features are indeed of phylogenetic and evolutionary significance. The first investigations of chromosomes were made by means of histological sections in the late nineteenth century. Then, squash techniques were performed which permitted a more accurate count of chromosome number. Later, the air-drying technique combined with colchicine and hypotonic treatment (to accumulate cells at metaphase and to ease chromosome spreading) greatly increased our knowledge not only of

Genetics and Evolution of Aquatic Organisms. Edited by A.R. Beaumont.
Published in 1994 by Chapman & Hall, 2–6 Boundary Row, London SE1 8HN.
ISBN 0-412-49370-5

chromosome number, but also of chromosome morphology. Karyotyping, i.e. pairing of the chromosomes according to their size and morphology, then became feasible. Recent development of new differential staining techniques has allowed the demonstration of characteristic banding of chromosomes. These techniques have led to a better understanding of the fine characterization of individual chromosomes and to a better standardization of karyotypes.

Aquatic invertebrates have been investigated mostly with squash or air-drying techniques from tissues, while in fish, since 1960 there has been a boom in cytogenetic studies due mainly to the application of human cell culture techniques. Because of this, there is more information on banding techniques in fish than in invertebrate species.

The aim of this section is to show how our knowledge of the cytogenetics of aquatic organisms has advanced by investigations on chromosome number, chromosome morphology and karyotype, and banding techniques. In each section, chromosomal data of selected groups of invertebrates or fish will be used to illustrate evolutionary inferences.

7.1.2 Evolution in chromosome number

There has been a biased recording of chromosome numbers in aquatic organisms depending on the group studied. Looking at the aquatic invertebrates, some phyla are poorly investigated, e.g. the sponges or the cnidarians. With regard to minor pseudocoelomate or eucoelomate animals, published data are even more rare. In fish, although there is a considerable literature, the chromosome numbers are known for only about 1000 species out of the 20 000 species actually recorded (Ojima, 1983). In other vertebrates, aquatic forms are scattered through different phyla and hence are not considered in this section.

Therefore, five groups are selected here to illustrate studies in chromosome numbers: the platyhelminths and the lower chordates, because of their phylogenetic position in the classification, either primitive or close to the vertebrates respectively, and the annelids, the crustaceans and the molluscs, which are the most studied invertebrate phyla. Table 7.1.1 summarizes the chromosome numbers recorded in those groups. Polyploid species, sex chromosomes and supernumerary chromosomes are excluded, as are data which give approximative chromosome counts or are considered very doubtful by some authors.

(a) Platyhelminthes

The combined features of the platyhelminths may represent a set of derived morphological traits marking major advancements in the evolution of metazoa. This phylum includes free-living aquatic flatworms from the class Turbellaria and endoparasitic worms such as the classes Trematoda and Cestoda. Only the free-living forms (marine, brackish and fresh water) will be considered here. Chromosome numbers of five orders of Turbellaria are examined here.

In the Acoela, believed to be the most ancient order within turbellarians, three species have been investigated showing diploid chromosome numbers of 12, 14 and 16 (Birstein, 1990). These numbers are close to the number 12 recorded in *Trichoplax*, which belongs to the most primitive forms of invertebrates, the Placozoa (Birstein, 1989).

In the Macrostomida, the haploid chromosome number ranges from 2 to 8 for 14 species studied, with a majority of species in the family Macrostomidae showing $n = 3$ (Benazzi and Benazzi Lentati, 1976).

In the Proseriata, which exhibit a cosmopolitan distribution with representatives in various marine habitats and brackish waters, 70 species have been studied with haploid chromosome numbers from 2 to 12. The family Monocelidae constitutes a monophyletic taxon (Martens,

Table 7.1.1 Chromosome numbers in different phyla of invertebrates (see text for references and details)

Taxon	No. species studied	Range of haploid chromosome number	Modal number
Platyhelminthes			
Turbellaria			
Acoela	3	6, 7, 8	None
Macrostomida	14	2–8	3
Proseriata	70	2–12	3, 6
Tricladida	67	3–22	8
Polycladida	15	6–10	10
Neorhabdocoela	57	2–8	3
Annelida			
Polychaeta	104	3–18	9, 10
Oligochaeta	21	9–25	24
Hirudinea	17	7–16	8
Crustacea			
Branchiopoda			
Anostraca (Artemia)			21
Cladocera (Daphniidae)	12	7–12	10
Maxillopoda			
Copepoda			
Calanoida	34	3–17	16
Harpacticoida	25	5–12	12
Cyclopoida	35	2–12	7
Malacostraca			
Peracarida			
Isopoda	35	6–36	28
Amphipoda	36	9–32	26
Mysidacea	16	5–56	None
Eucarida			
Euphausiacea	2	16, 17	
Decapoda	53	32–188	None
Mollusca			
Polyplacophora	15	6–13	12
Gastropoda			
Prosobranchia	134		
Archaeogastropoda	80	8–21	9, 12, 18
Mesogastropoda	29	13–35	17
Neogastropoda	25	28–36	35
Opisthobranchia	105	7–18	13, 17
Bivalvia	164	6–23	10, 14, 19
Cephalopoda	10	28–52	46
Lower chordates			
Hemichordata			
Enteropneusta	1		23
Chordata			
Urochordata			
Larvacea	5	4–8	8
Ascidiacea	75		
Enterogona	41	6–26	10
Pleurogona	34	8–24	16
Thaliacea	6	11–13	12
Cephalochordata	2		19

1984) with a haploid modal chromosome number of $n = 3$ (Curini-Galletti, Puccinelli and Martens, 1989). The non-monocelidid Proseriata include chromosome sets with $n = 6$ (Martens, Curini-Galletti and Van Oostveldt, 1989). Thus, the trend of chromosome number evolution must have been $n = 3$ to $n = 6$.

In the Tricladida, these freshwater and marine worms show haploid chromosome numbers from $n = 3$ to $n = 22$ in 67 species studied (Benazzi and Benazzi Lentati, 1976; Galleni and Puccinelli, 1979, 1984; Puccinelli and Deri, 1991). The modal number is 8, which corresponds to the planarian *Dugesia* where several species have been investigated. But in another dendrocelid family, the majority of representatives exhibit $n = 14$ to 16, while in Maricola it is $n = 6$ to $n = 8$ (Galleni and Puccinelli, 1979). There is no clear trend of evolution in chromosome number within this order because of the many cases of polyploidy, not considered here.

The Polycladida are of great interest because from the embryological point of view they seem to be the most conservative group within turbellarians (Galleni and Gremigni, 1988). In this order, 15 species show a haploid number of $n = 6$ to $n = 10$ (Galleni and Puccinelli, 1981). There seems to be some uniformity of chromosome number within families, with $n = 9$ in the Cotylea Pseudoceridae and $n = 10$ in the Acotylea Stylochidae and in five out of seven species of Leptoplanidae. The primitive suborder Cotylea shows the lowest chromosome number (Galleni and Puccinelli, 1981), so in this case, the evolutionary trend is an increase of chromosome number.

In the Neorhabdocoela, considered to be the most phylogenetically advanced order within turbellarians (Ax, 1963), 57 species have been investigated with a haploid chromosome number of $n = 2$ to $n = 8$ (Birstein, 1991). The modal number is $n = 3$, mostly due to the subfamily Gyratrinae, which has been particularly well studied. According to Birstein (1991), as freshwater neorhabdocoels have originated from marine forms with $n = 8$, and as there are species with $n = 8$ in all its suborders, it seems probable that the ancestral forms could have had $n = 8$. This relatively high chromosome number is found not only in marine neorhabdocoels but also in marine polyclads. The evolution of different phyletic Neorhabdocoela lines obviously involved a decrease in chromosome number (Birstein, 1991).

Thus, within the different orders of turbellarians, increase, as well as decrease, of chromosome number is observed.

(b) Annelida

Aquatic annelids include the marine polychaetes, the freshwater oligochaetes and the leeches.

In the polychaetes (Christensen, 1980; Vitturi, Maiorca and Carollo, 1984; Grassle, Gelfman and Mils, 1987; Curini-Galletti, Lardicci and Regoli, 1991; Robotti *et al.*, 1991; Sato and Ikeda, 1992), 104 species studied show haploid chromosome numbers of $n = 3$ to $n = 18$ with modal numbers of $n = 9$ and $n = 10$, which correspond to some Capitellidae and Serpulidae respectively. The Sedentaria have in general higher chromosome numbers than the two other groups, the errant polychaetes and the archiannelids. The low chromosome numbers $n = 2$ to $n = 6$ are found in species belonging to the Dorvilleidae, Tomopteridae, Syllidae, Protodrilidae and Nerillidae, presumably the most primitive families. This indicates that an increase in the number of chromosomes has occurred in phylogeny (Christensen, 1980).

In the oligochaetes (Christensen, 1980), haploid chromosome numbers range from $n = 9$ to $n = 25$ for 21 aquatic species studied. According to Christensen (1980), both increases and decreases in chromosome number have occurred within oligochaete species of non-polyploid origin.

In the hirudineans (Davies and Singhal, 1987), 17 species studied show haploid chromosome numbers of $n = 7$ to $n = 18$, with a modal number of $n = 8$, which has been suggested to be the primitive chromosome number.

Thus, within the different classes of annelids, the range of variation in chromosome number is more or less identical. Haploid chromosome numbers lower than $n = 8$ are mainly found within the polychaetes, and the general tendency is towards lower chromosome numbers in the polychaetes compared with the oligochaetes.

(c) Crustacea

Crustacean species are found at all depths in every marine, brackish and freshwater environment; there are only a few successful terrestrial species. They exhibit a remarkable diversity in form and habitat. Ten orders will be considered here either because numerous data on chromosome number have been reported, or because even with few data they are of interest in the classification.

In the branchiopod Anostraca, the genus *Artemia* has a worldwide distribution and comprises bisexual and parthenogenetic forms, with many sibling or polymorphic races. The chromosome number of $n = 21$ is usual for the diploid forms (Abatzopoulos, Kastritsis and Triantaphyllidis, 1986).

In the branchiopod Cladocera, the freshwater family Daphniidae show haploid chromosome numbers of $n = 7$ to $n = 12$ with a modal number of $n = 10$ (Trentini, 1980).

In the copepod Calanoida, 34 species studied show $n = 3$ to $n = 17$ with a modal number of 16 (Colombera and Lazzaretto-Colombera, 1978). The Harpacticoida have $n = 5$ to $n = 12$ for 25 species studied (Colombera and Lazzaretto-Colombera, 1978; Lazzaretto, 1983; Lazzaretto and Libertini, 1986), with a modal number of 12. The Cyclopoida have $n = 2$ to $n = 12$ for 35 species studied (Colombera and Lazzaretto-Colombera, 1978), with a modal number of 7. Within the Calanoida Copepoda, the Cyclopoida, the most primitive order, possess the highest chromosome number. Evolution within the Copepoda therefore seems to be characterized by a preferential reduction of chromosome number (Colombera and Lazzaretto-Colombera, 1978).

In the malacostracean peracarid Isopoda, the haploid chromosome numbers range from $n = 6$ to $n = 36$ for 35 species investigated (Niiyama, 1959; Salemaa, 1986). A modal number of $n = 28$ is found in eight species (belonging mostly to the suborders Oniscoida and Valvatifera). In the Amphipoda, 36 species have been investigated with $n = 9$ to $n = 32$ (Niiyama, 1959; Laval and Lécher, 1975; Salemaa, 1986). The modal number of $n = 26$ corresponds to the Gammaridae. In the Mysidacea, 16 species show a haploid chromosome number of $n = 5$ to $n = 56$ (Mauchline, 1980; Salemaa, 1986). There is no modal number because of considerable chromosome number variation in *Archaeomysis*, *Neomysis* and *Mysis*. Furthermore among the three orders of peracarids, a high level of genetic diversity has been observed, as well as the occurrence of supernumerary chromosomes. According to Salemaa (1986), low allozyme variability has been related to temporal environmental instability, as in the Baltic and other brackish waters. But, karyologically, these eurytopic species seem to exhibit an extraordinary capacity for evolutionary responses, mirrored by their chromosomal changes (Salemaa, 1986).

In the malacostracean eucarid Euphausiacea, two species only have been investigated, *Euphausia pacifica* with $n = 16$ (Yabu and Kawamura, 1981) and *Euphausia superba* with $n = 16$ (Yabu and Kawamura, 1984) or $n = 17$ (Ngan *et al.*, 1989). This variation in chromosome number may be due to differences in methodology, but may also be due to genetic differences between populations.

In the Decapoda, a checklist of chromosomal data (Nakamura *et al.*, 1988) gives a range of haploid chromosome numbers from $n = 32$ to $n = 188$ for the 53 species studied. The number of $n = 188$ is the highest chromosome number recorded in aquatic organisms and has been observed in the shrimp *Astacus trowbridgei* (Niiyama, 1959). There is no obvious modal chromosome number in this order; $n = 50$ to 58 only represent 14% of the species recorded (See Table 7.1.1).

Looking at the crustacean chromosome numbers, it is clear that lower taxa show lower chromosome numbers (branchiopods and maxillipods versus malacostraceans). But this general evolutionary trend following an increase of chromosome number is in contradiction with the converse trend observed, for instance, within copepods, where the chromosome number decreases from primitive to advanced groups.

(d) Mollusca

Molluscs include some of the best-known invertebrates, and many are harvested commercially. This is reflected in the greater amount of karyological data on this phylum compared to other aquatic invertebrates. An increasing number of studies have been performed in recent years. For instance, looking at the class Bivalvia, 23 species had been investigated up to 1969 (review: Patterson, 1969), 125 species up to 1985 (review: Nakamura, 1985) and 164 species up to 1992.

Four classes will be considered here, chromosomal information on the other minor molluscan classes, Aplacophora, Monoplacophora and Scaphopoda, being scarce or absent. The terrestrial forms such as land snails and slugs are also excluded.

In the Polyplacophora, the haploid chromosome numbers range from $n = 6$ to $n = 13$. Nine out of the 15 species studied showed $n = 12$, which can be considered as the modal number (Nakamura, 1986).

In the gastropod prosobranchs, 80 species of Archaeogastropoda have been studied (Vitturi and Catalano, 1984; 1989; Nakamura, 1986; Cervella *et al.*, 1988; Komatsu, 1988). Haploid chromosome numbers range from $n = 8$ to $n = 21$ and show a trimodal distribution, with $n = 9$ being represented mostly by the Patellacea, $n = 12$ by the Neritidae, and $n = 18$ being the most frequent number within this order. According to Nakamura (1986), there is no clear trend in the evolution of chromosome number; both increases and decreases occur. In the Mesogastropoda, haploid chromosome numbers range from $n = 13$ to $n = 35$ in 29 species studied (Stern, 1975; Kasinathan and Natarajan, 1975; Vitturi, Rasotto and Farinella-Ferruza, 1982; Vitturi and Catalano, 1984; Komatsu, 1985; Vitturi, Catalano and Macaluso, 1986). Most of the species show a relatively uniform chromosome number of $n = 13$ to $n = 18$. The highest numbers ($n = 31$ and $n = 35$) are recorded in one species of Capulidae and one species of Cymatiidae, families which are more advanced in the mesogastropod classification. The modal number of $n = 17$ corresponds to that observed for Littorinidae. In the Neogastropoda, the haploid chromosome numbers range from $n = 28$ to $n = 36$ in the 25 species studied since the review by Patterson (1969). The modal number is $n = 35$ in the Muricidae and Buccinidae families (Vitturi *et al.*, 1987; Vitturi and Catalano, 1990). In total, 134 prosobranch species have been investigated, with an evolutionary trend for haploid chromosome numbers to increase from the Archaeogastropoda to the Neogastropoda.

In the gastropod Opisthobranchia, haploid chromosome numbers range from $n = 7$ to $n = 18$ in 105 species studied (Natarajan, 1970; Vitturi *et al.*, 1982; Curini-Galletti, 1985, 1988; Thiriot-Quiévreux, 1988; 1990), with two modal numbers, $n = 13$ and $n = 17$. A conservation in chromosome numbers is evident within the Opisthobranchia in spite of the evolution of major morphological diversities within the various groups (Schmekel, 1985). There are,

Heterodonta

Lasaeidae	1	–	–	–	–	–	–	–	–	–	–	–	–	–	1	–	–	–	–	–	–
Cyamiidae	2	–	–	–	–	–	–	–	–	–	–	–	–	–	1	1	–	–	–	–	–
Gaimardiidae	1	–	–	–	–	–	–	–	–	–	–	–	–	–	–	1	–	–	–	–	–
Carditidae	1	–	–	–	–	–	–	–	1	–	–	–	–	–	–	–	–	–	–	–	–
Cardiidae	3	–	–	–	–	–	1	–	–	–	–	–	–	–	–	2	–	–	–	–	–
Mactridae	7	–	–	–	–	–	–	–	–	–	–	–	–	–	2	5	–	–	–	–	–
Tellinidae	2	–	–	–	–	–	–	–	–	–	–	–	–	–	–	2	–	–	–	–	–
Semelidae	1	–	–	–	–	–	–	–	–	–	–	–	–	–	–	1	–	–	–	–	–
Scrobiculariidae	1	–	–	–	–	–	–	–	–	–	–	–	–	–	–	1	–	–	–	–	–
Donacidae	2	–	–	–	–	–	–	–	–	–	–	–	–	–	1	2	–	–	–	–	–
Corbiculidae	2	–	–	–	–	–	–	–	–	–	–	–	–	–	–	1	–	–	–	–	–
Pisidiidae	–	–	–	–	–	–	–	–	–	–	–	–	–	–	1	–	–	–	–	–	–
Veneridae	14	–	–	–	–	–	–	–	–	–	–	–	–	–	–	14	–	–	–	–	–
Petricolidae	1	–	–	–	–	–	–	–	–	–	–	–	–	–	–	1	–	–	–	–	–
Myidae	1	–	–	–	–	–	–	1	–	–	–	–	–	–	–	–	–	–	–	–	–
Pholadidae	6	–	–	–	–	–	–	2	–	–	–	–	–	–	–	4	–	–	–	–	–
Teredinidae	1	–	–	–	–	–	–	–	–	–	–	–	–	–	–	1	–	–	–	–	–

Anomalodesmata

Laternulidae	1	–	–	–	–	–	–	–	–	–	–	–	–	–	–	–	1	1	–	–	–
Total	164	1	1	–	1	2	24	3	1	7	21	7	8	3	7	77	1	1	–	–	1

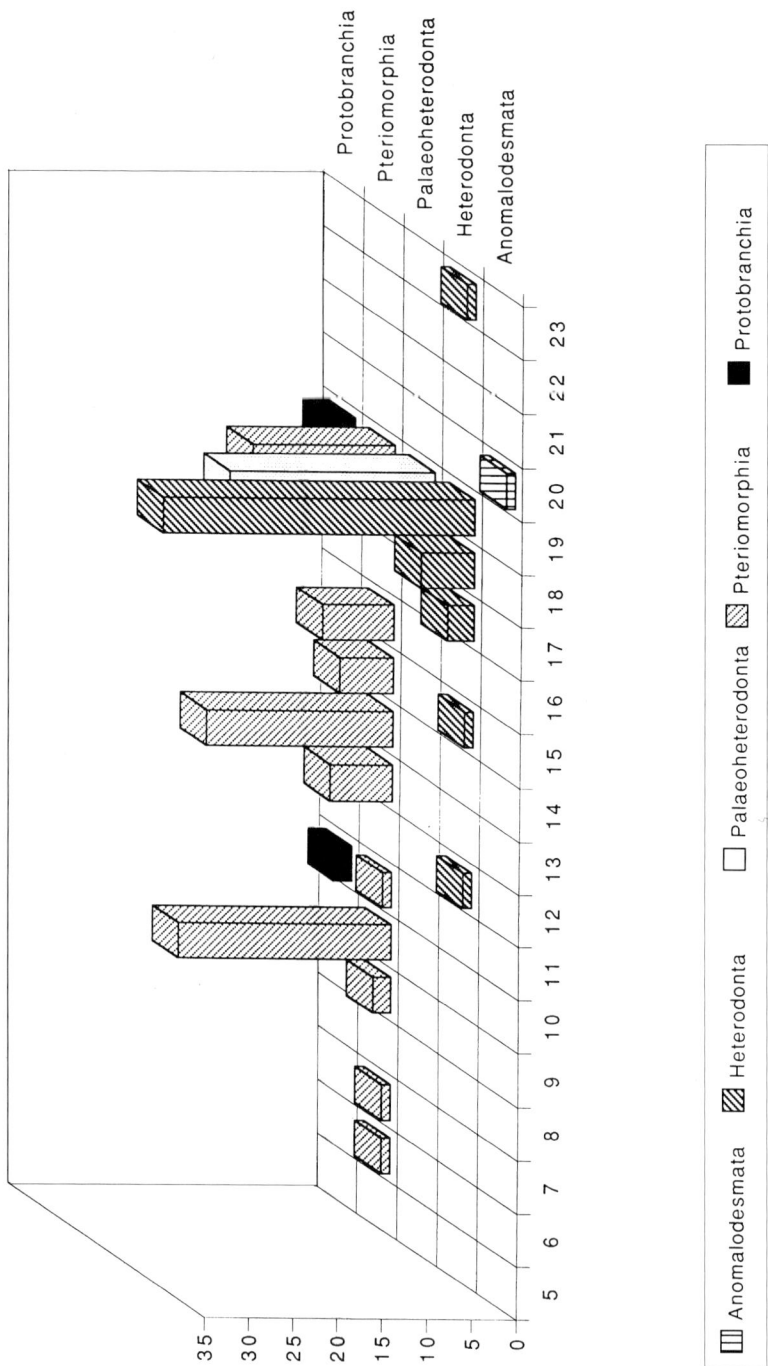

Figure 7.1.1 Haploid chromosome numbers in the different subgroups of bivalves. Horizontal axis refers to the haploid chromosome number while the vertical axis is the number of species.

7.1.3 Evolution in chromosome morphology and karyotypes

Advances in techniques, e.g. the use of colchicine, hypotonic treatment and air-drying cell suspensions, have allowed more detailed analyses of chromosome morphology. A karyotype is characterized by its different morphological types of chromosomes, i.e. metacentric, submetacentric, subtelocentric or telocentric (sometimes called acrocentric) according to the position of the centromere. Measurements of chromosomes which give an accurate location of the position of the centromere have contributed to an important gain of knowledge, entitled beta-level karyology (White, 1978), as opposed to alpha-level karyology in which only the number of chromosomes is reported. These new techniques have permitted interspecific comparison of karyotypes and thus the identification of chromosomal rearrangements, which are an important or significant factor in evolution (White, 1973).

For instance, in turbellarians, the comparison of karyotypes in five genera of marine triclads and polyclads allowed identification of three different types of chromosomal rearrangements: (1) those involving centromere position, i.e. para- or pericentric inversions, (2) rearrangement of whole chromosome arms, and (3) Robertsonian mechanisms, i.e. fusion of chromosomes (Galleni and Puccinelli, 1986). In neorhabdocoels, in the course of evolution of different phyletic lines the decrease in chromosome number evidently occurred mostly by Robertsonian fusions (Birstein, 1991).

In the mollusc Archaeogastropoda, Nakamura (1986) reported karyometric analyses of 23 species of the family Neritidae, which inhabit marine, brackish and fresh waters. In all marine species, the karyotypes are quite similar with a few chromosomes of subtelocentric to telocentric types. Exceptions occur in brackish and freshwater species, where there are high numbers of telocentric chromosomes. Nakamura (1986) suggests that the above karyological instability of the Neritidae may be related to their expanding habitat from the sea to brackish and fresh water.

In the opisthobranch Thecosomata, the evolution of chromosome number (from $n = 10$ to $n = 17$) is paralleled by an evolutionary trend in the increase of subtelocentric and telocentric chromosomes (Table 7.1.3), probably due to dissociations of chromosomes. This phenomenon may be related to their adaptation to a specialized pelagic environment (Thiriot-Quiévreux, 1988, 1990).

Table 7.1.3 Karyological data in pteropod Thecosomata (modified from Thiriot-Quiévreux, 1990)

Thecosomata	*Haploid no.*	*No. of chromosome pairs*			
		m	*sm*	*st*	*t*
Euthecosomata					
Limacina inflata	10	10			
Creseis acicula	10	5	5		
Creseis virgula	10	7	2	1	
Clio pyramidata	11	5	3	1	1
Cavolinia inflexa	12	6	1	5	
Hyalocylix striata	14	7	2	5	
Pseudothecosomata					
Peraclis reticulata	14	6	4	2	
Cymbulia peroni	17	6	3	7	1

m = metacentric; sm = submetacentric; st = subtelocentric; t = telocentric.

In bivalves, five species of the subfamily Tapetinae have been investigated by karyometric analyses (Borsa and Thiriot-Quiévreux, 1990; Insua and Thiriot-Quiévreux, 1992). Clearly, there are interspecific differences in the proportions of morphological types of chromosomes and each species shows a specific karyological formula (Figure 7.1.2). Comparison of chromosome measurements of the five species by means of principal component analysis and hierarchical classification reveals that *Venerupis pullastra* and *Ruditapes decussatus* are most closely related followed by *Venerupis aurea*, while *Ruditapes philippinarum* and *Venerupis rhomboides* constitute two separate stems with a considerably dissimilarity. This highlights the role that karyological information can play in inferring species relationships.

Cell culture techniques, initially developed for the investigation of human chromosomes, have now been applied to fish. The example of the study on Characiformes by Arefjev (1990a,b) has been chosen to illustrate how karyotypic divergence may be related to phylogenetic considerations. From a comparison of chromosome formulae in 97 species of Characiformes, Arefjev suggested a presumptive phylogenetic tree of this order (Figure 7.1.3). Over evolutionary time there is an increase as well as a decrease in chromosome number, both of which are connected with pericentric inversions and, in more advanced families, with sex chromosome differentiation, variability of supernumerary chromosomes and subchromosomal differentiation.

7.1.4 Chromosome banding

Differential banding patterns of chromosomes, usually observed at specific regions, were initially developed for the analysis of human chromosome segments. These bands are seen as low- and high-intensity regions under the fluorescence microscope or as differentially stained areas under the light microscope. These methods were then extended to different animals and later to plant chromosomes, and have recently been employed in locating marker segments of chromosomes, particularly those with repetitive DNA sequences (Sharma and Sharma, 1980). Thus, high-resolution banding techniques make it possible to establish standard banding karyotypes and to detect more precise chromosomal rearrangements.

In aquatic invertebrates, reproducible C-banding, which reveals the extent and location of blocks of constitutive heterochromatin, and nucleolus organizer region (NOR) character-ization have begun to be successfully applied in some taxa.

For instance, in the triclad turbellarian *Dugesia*, the C-banding pattern observed in *Dugesia lugubris* is quite different from that of *Dugesia polychroa*. The differences in karyotypes between these closely related species cannot be explained by simple chromo-somal rearrangements, but reflect different mechanisms of genome evolution (Galleni *et al.*, 1986, 1989).

In molluscs, silver-NOR patterns were observed in two species of prosobranch (Vitturi and Catalano, 1989, 1990). In the bivalve *Ostrea denselamellosa* (Insua and Thiriot-Quiévreux, 1991), karyometric analysis, together with C- and NOR-banding, has allowed the identification of all the different chromosome pairs, and has also provided chromosome markers for use in future genetic manipulation (Figure 7.1.4). Moreover, patterns of NOR variation have been studied in Ostreidae, which showed interspecific heteromorphism in the number of Ag-stained NORs per genome, their chromosomal location and their position within karyotypes (Thiriot-Quiévreux and Insua, 1992).

The development of these techniques has been particularly successful in fish following the improvement of methods for direct chromosome preparations (Kligerman and Bloom, 1977) and, especially, the progress in fish cell culture (Amemiya, Bickham and Gold, 1984). For instance, interspecific variants in chromosomal NORs may be used to construct phylogenetic

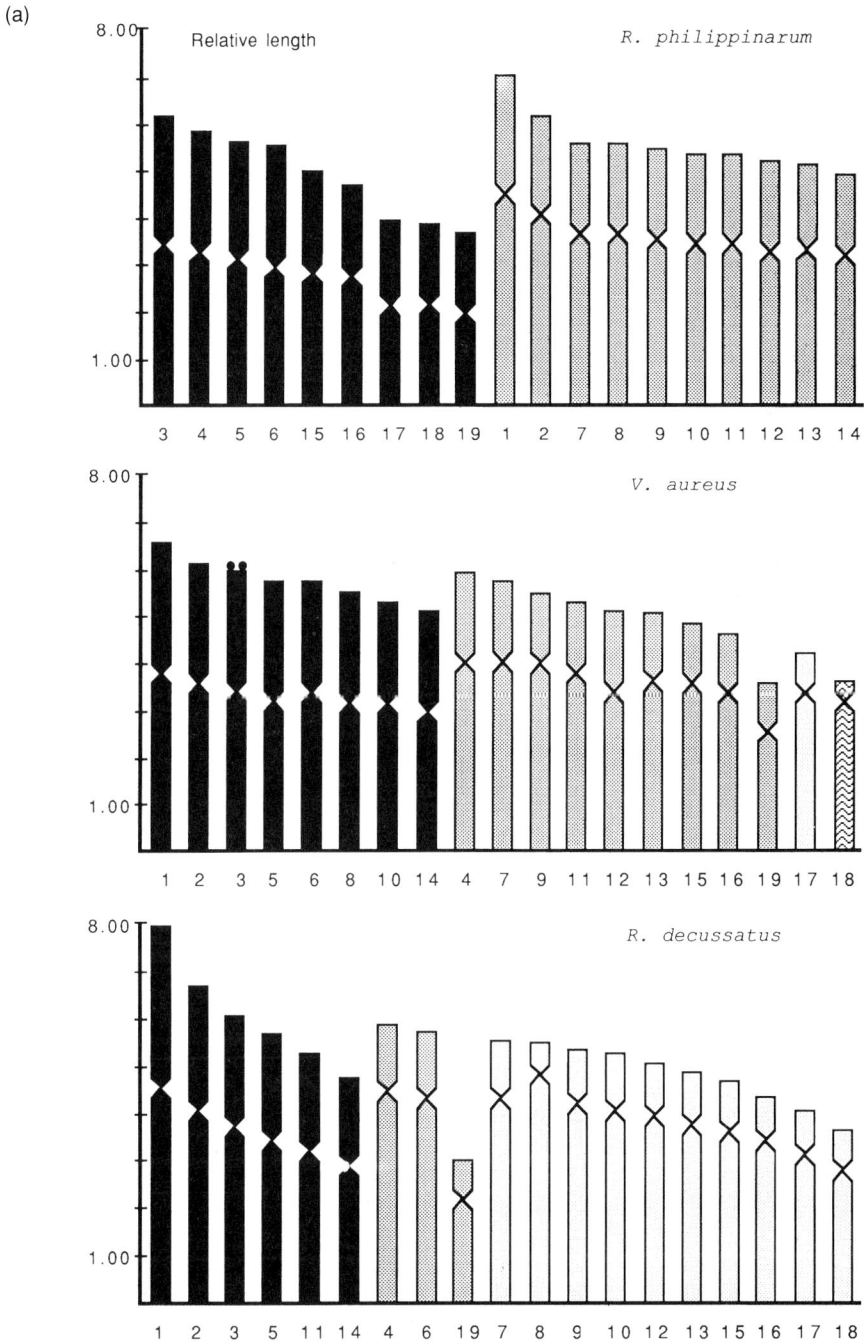

Figure 7.1.2 Idiograms for five species of Tapetinae (Bivalvia, Veneridae). (Redrawn from Borsa and Thiriot-Quiévreux, 1990, and Insua and Thiriot-Quiévreux, 1992.)

(b)

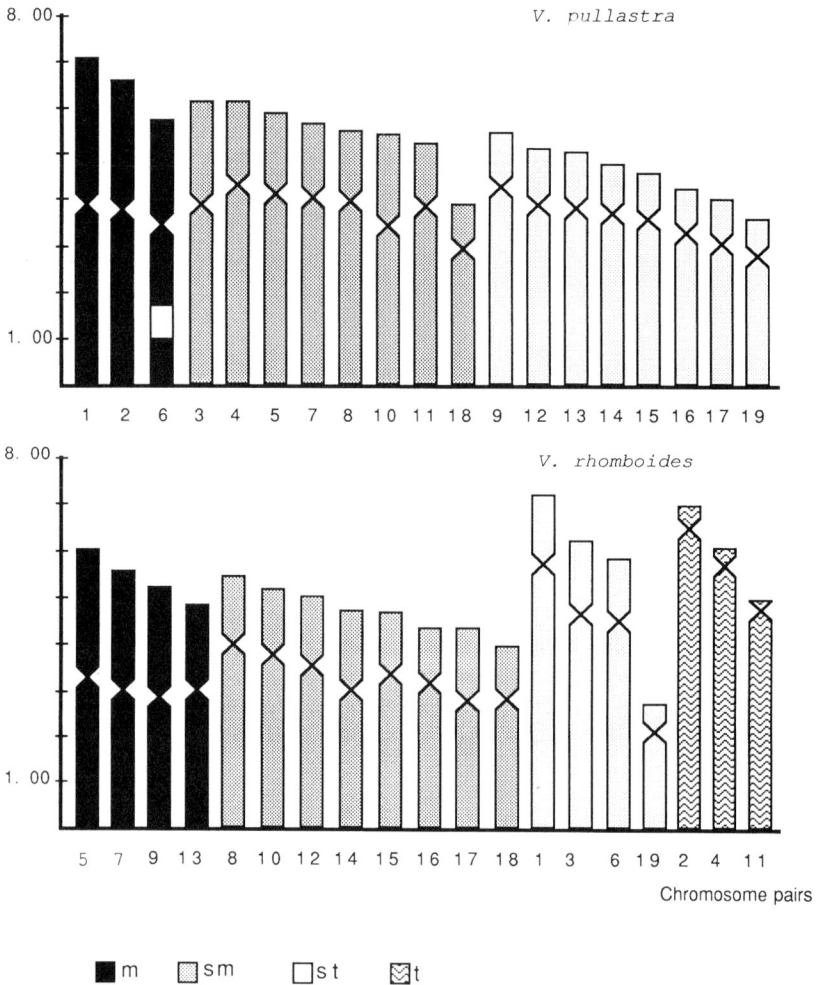

Figure 7.1.2 *Continued*

hypotheses of species relationships. Amemiya and Gold (1990) have documented the patterns of chromosomal NORs in 50 species of North American cyprinid fish. Twelve different NOR phenotypes were observed. These authors suggested that a single NOR located terminally on the short arm of a small acrocentric chromosome represents the plesiomorphic character, and all other NOR conditions are apomorphic. Phylogenetic inferences are drawn from these chromosomal NOR data and provide a framework upon which future cytosystematic studies in North American cyprinids can be based.

In blenniid fish (Garcia, Alvarez and Thode, 1987), the C-banding technique has been very useful in detecting interspecific differences. The heterochromatin patterns give rise to two different karyoevolutionary pathways, which agree with the accepted classification of the blenniids.

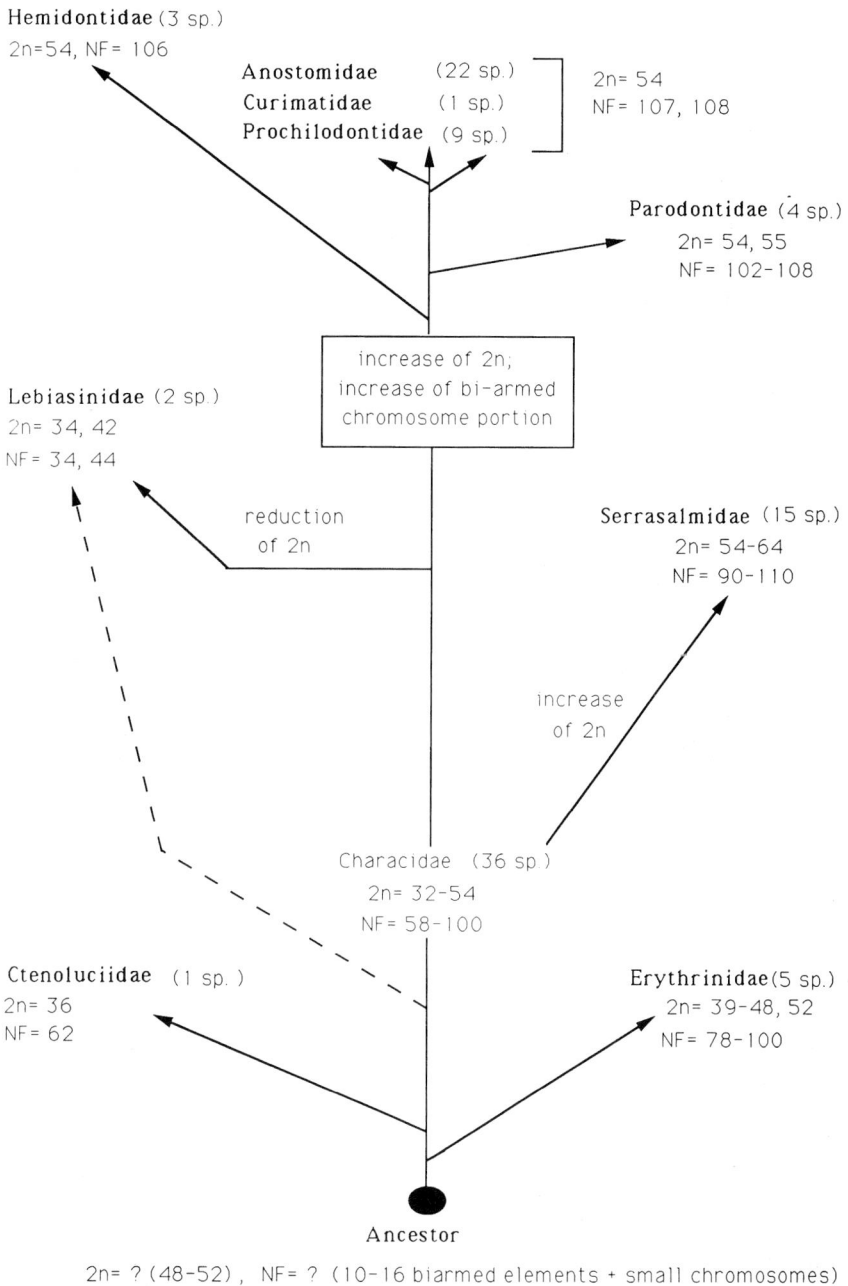

Figure 7.1.3 Presumptive phylogenetic tree of Characiformes fish based on karyological data. NF = fundamental number.(Redrawn and modified from Arefjev, 1990a,b)

Relative length

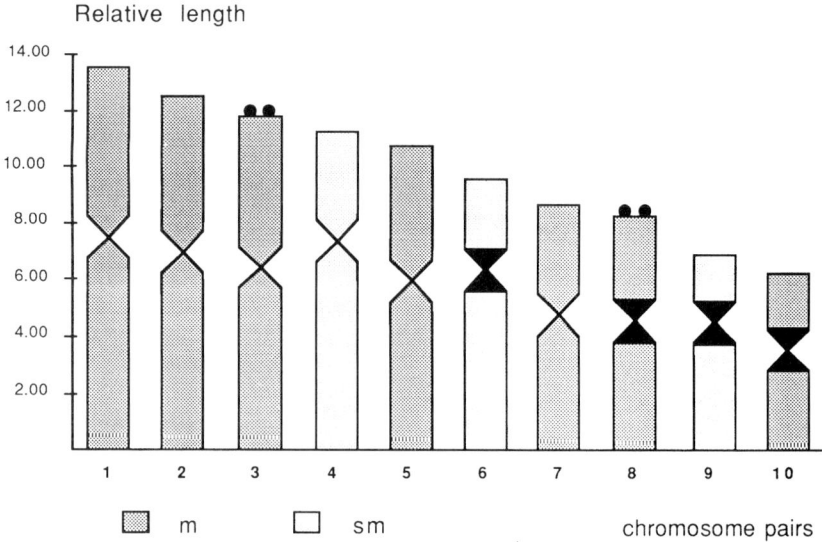

Figure 7.1.4 Ideogram of *Ostrea denselamellosa* (Bivalvia, Ostreidae) including marked chromosomes with C-banding (black insert) and NOR characterization (black dots). (Redrawn from Insua and Thiriot-Quiévreux, 1991.)

Most of the published banding studies in fish are restricted either to C-bands or to NOR characterization. The only species for which a clear G-banding pattern has been published is the European eel (Wiberg, 1983), and Q- and R-banding have been observed in three species of fish (Medrano *et al.*, 1988). Moreover, the use of *in vivo* 5-bromodeoxyuridine (BrdU) made it possible to induce highly reproducible replication bands (called RBA type) in two species of Cypriniformes (Hellmer, Voiculescu and Schempp, 1991).

Restriction endonucleases also produce reproducible bands by cutting the DNA at specific base pairs with the subsequent removal of the fragments resulting in diminished staining by Giemsa. Such methods involving restriction endonuclease digestion have now become an important tool in molecular cytogenetics. In salmonids, which constitute one of the most cytogenetically studied fish groups, restriction banding has been performed, e.g. in rainbow trout (Lloyd and Thorgaard, 1988), in brown trout (Sanchez *et al.*, 1990; Hartley, 1991) and in the coho salmon (Lozano, Ruiz Rejón and Ruiz Rejón, 1991). These authors emphasize that restriction endonuclease digestion induces more specific banding patterns and thus results in a better definition of fish chromosomes than that obtained with conventional banding methods.

7.1.5 Discussion

The haploid chromosome numbers of species of the different aquatic invertebrates recorded here lie between $n = 2$ (in some turbellarian Proseriata and Neorhabdocoela, and in some copepod Cyclopoida) to $n = 188$ (in one species of Decapoda). The mean of modal numbers here recorded is 14.3 (see Table 7.1.1), which agrees with the haploid chromosome numbers from 6 to 20 recorded for most animal species (White, 1973).

The modal (or type) number is sometimes inferred as ancestral (White, 1973), but in many cases the modal number of a group is simply the chromosome number characteristic of that subgroup which has been predominantly investigated at the present time. Nevertheless, below the rank of class or subclass, the concept of a type number may be correct, as for instance $n = 10$ in the bivalve Ostreidae.

The turbellarians, which are the most primitive group discussed in this section show the lowest chromosome numbers, followed by the lower chordates, the molluscs and the crustaceans. In considering the processes involved in the evolution of chromosome numbers, different evolutionary trends are evident within each of these groups. There is either an increase in chromosome number, e.g. from the prosobranch Archaeogastropoda to the Neogastropoda, or a decrease in chromosome number, e.g. from the copepod Calanoida to the Cyclopoida; while both an increase and decrease are evident within the Bivalvia. An evolutionary mechanism involving an increase of chromosome number in more specialized or evolved species has been supported by Patterson (1969) and Patterson and Burch (1978), while the alternative decrease hypothesis has been supported by Butot and Kiauta (1969), Colombera and Lazzaretto-Colombera (1978) and Vitturi, Rasotto and Farinella-Ferruza (1982). Nakamura (1986), in review on the chromosomes of Archaeogastropoda, does not infer any evolutionary tendency for change in chromosome number, since both increases and decreases can occur. Overall the pattern of chromosomal evolution seems to be rather different from one group to another and it is better to assume that no general rule can be inferred.

It is worth noting that marine species frequently exhibit greater numbers of chromosomes than those of freshwater species, as illustrated by the Neorhabdocoela and the Copepoda. However, this does not occur among bivalves. The freshwater Palaeoheterodonta all have an chromosome number of $n = 19$, similar to some of the more advanced marine groups, e.g. the Heterodonta. Here again, there is no general trend.

The advances in morphological and karyotypic knowledge allow us to infer evolutionary trends within limited taxa, such as families or even orders. But the processes in adaptation and speciation seem to be accompanied by different mechanisms according to the group studied. Fusions or dissociations of chromosomes may occur simultaneously, as in the fish Characiformes, or separately as in the turbellarian Neorhabdocoela. According to White (1978), it is fairly certain that in many species of animals various types of chromosomal rearrangements are directly adaptive to certain types of habitats and ecological niches. This has been illustrated in Neritidae and in the pelagic Thecosomata.

The introduction of banding techniques has provided increased information, the degree of which depends on the systematic group under study. This has been done exceptionally with C-banding and NOR characterization in aquatic invertebrates,and more commonly in fish, but the determination of G-banding (or R-banding) is still unsuccessfully in these organisms, in contrast to studies on primates (Dutrillaux, Couturier and Viegas-Péquignot, 1981). Studies involving restriction endonucleases are promising but still very rare. The achievement of standard banding karyotypes is necessary to understand the role of chromosomal rearrangements which may have occurred during the evolutionary speciation process. As reported in this section, karyological knowledge on aquatic organisms is increasing and promising, allowing some limited evolutionary inferences. Perhaps general trends in chromosomal evolution, which cannot presently be discerned in the scatter of available information, may emerge from further studies, particularly karyotype-banding studies, of the phylogenetically diverse aquatic invertebrates.

Acknowledgements

Thanks are due to E. Gosling for English corrections and helpful comments and to V. Thiriot for the illustrations.

References

Abatzopoulos, Th.J., Kastritsis, C.D. and Triantaphyllidis, C.D. (1986) A study of karyotypes and heterochromatic associations in *Artemia*, with special reference to two N. Greek populations. *Genetica*, **71**, 3–10.

Allen, J.A. (1985) The recent Bivalvia: their form and evolution, in *The Mollusca*, Vol. 10, *Evolution* (eds K.M. Wilbur, E.R. Trueman and M.R. Clarke), Academic Press, Orlando, pp. 337–403.

Amemiya, C.T., Bickham, J.W. and Gold, J.R. (1984) A cell culture procedure for chromosome preparation in cyprinid fish. *Copeia*, **1984**, 232–35.

Amemiya, C.T. and Gold, J.R. (1990) Cytogenetic studies in North American minnows (Cyprinidae). XVII. Chromosomal NOR phenotypes of 12 species, with comments on cytosystematic relationships among 50 species. *Hereditas*, **112**, 231–47.

Anonymous (1977) Appendix 2. Classification of recent cephalopods. *Symp. Zool. Soc. Lond.*, **38**, 575–79.

Arefjev, V.A. (1990a) Karyotypic diversify of characid families (Pisces, Characidae). *Caryologia*, **43**, 291–304.

Arefjev, V.A. (1990b) Problems of karyotypic variability in the family Characidae (Pisces, Characiformes) with the description of somatic karyotypes for six species of tetras. *Caryologia*, **43**, 305–19.

Ax, P. (1963) Relationships and phylogeny of the Turbellaria, in *The Lower Metazoa* (ed. E.C. Daugherty), University of California Press, Berkeley, pp. 191–224.

Benazzi, M. and Benazzi Lentati, G. (1976) Platyheminthes, in *Animal Cytogenetics 1* (ed. B. John), Gebrüder Borntraeger, Berlin and Stuttgart.

Birstein, V.J. (1989) On the karyology of *Trichoplax*. sp. (Placozoa). *Biol. Zentralbl. Leipzig*, **108**, 63–7.

Birstein, V.J. (1990) First contribution to karyology of two acoels (Turbellaria) and a dinophilid (Annelida). *Biol. Zentralbl. Leipzig*, **109**, 169–74.

Birstein, V.J. (1991) On the karyotypes of the Neorhabdocoela species and karyological evolution of Turbellaria. *Genetica*, **83**, 107–20.

Borsa, P. and Thiriot-Quiévreux, C. (1990) Karyological and allozymic characterization of *Ruditapes philippinarum*, *R. aureus* and *R. decussatus* (Bivalvia, Veneridae). *Aquaculture*, **90**, 209–27.

Butot, J.B. and Kiauta, B. (1969) Cytotaxonomic observation in the stylommatophoran family Helicidae, with considerations on the affinities within the family. *Malacologia*, **9**, 261–2.

Cervella, P., Ramella, L., Robotti, C.A. and Sella G. (1988) Chromosome analysis of three species of *Patella* (Archaeogastroda). *Genetica*, **77**, 97–103.

Christensen, B. (1980) Annelida, in *Animal Cytogenetics 2* (ed. B. John), Gebrüder Borntraeger, Berlin and Stuttgart.

Colombera, D. (1982) New development in vertebrate cytotaxonomy. VI. Cytotaxonomy and evolution of lower chordates. *Genetica*, **58**, 97–102.

Colombera, D., Borgato, A. and Brunetti, R. (1987) The male chromosomes of *Clavelina sabbadini*. *Caryologia*, **40**, 355–57.

Colombera, D. and Lazzaretto-Colombera, I. (1978) Chromosome evolution in some marine invertebrates, in *Marine Organisms: Genetics, Ecology and Evolution* (eds J.A. Beardmore and B. Battaglia), Plenum Press, New York, pp. 487–527.

Colombera, D., Nakauchi, M. and Tagliaferri, F. (1987) A preliminary report on the karyology of nine pleurogonic ascidians from the Pacific coast of Japan. *Rep. Usa. Mar. Biol. Inst., Kochi Univ.*, **9**, 39–44.

Curini-Galletti, M.C. (1985) Chromosome morphology of *Philinoglossa praelongata* (Gastropoda Cephalaspidea). *J. Mol. Stud.*, **51**, 220–22.

Curini-Galletti, M.C. (1988) Analyse du caryotype de *Runcina coronata* (Gastropoda Cephalaspidea). *Cah. Biol. Mar.*, **29**, 313–18.

Curini-Galletti, M.C., Lardicci, C. and Regoli, F. (1991) A contribution to the karyology of Syllidae (Polychaeta). *Ophelia*, **5**, 599–606.

Curini-Galletti, M.C., Puccinelli, I. and Martens, P.M. (1989) Karyotype analysis of ten species of Monocelididae (Proseriata Platyhelminthes) with remarks on the karyological evolution of the subfamily. *Genetica*, **78**, 169–78.

Davies, R.W. and Singhal, R.N. (1987) The chromosome numbers of five North American and European leech species. *Can. J. Zool.*, **65**, 681–4.

Dutrillaux, B., Couturier, J. and Viegas-Péquignot, E. (1981) Chromosomal evolution in primates. *Chromosomes Today*, **7**, 176–91.

Galleni, L. and Gremigni, V. (1988) Platyhelminthes – Turbellaria, in *Reproductive Biology of Invertebrates. IV. Fertilization, Development and Parental Care* (eds R.G. Adiyodi and K.G. Adiyodi), Oxford and I.B.H. Publishers, Oxford, pp. 63–9.

Galleni, L. and Puccinelli, I. (1979) The karyology of the genus *Procerodes* (Tricladida: Maricola) in British waters. *J. Mar. Biol. Assoc. UK*, **59**, 961–68.

Galleni, L. and Puccinelli, I. (1981) Karyological observations on Polyclads. *Hydrobiologia*, **84**, 31–44.

Galleni, L. and Puccinelli, I. (1984) Karyology of five species of Turbellaria from Øresund, Denmark. *Ophelia*, **23**, 141–8.

Galleni, L. and Puccinelli, I. (1986) Chromosomal evolution in marine triclads and polyclads (Turbellaria). *Hydrobiologia*, **132**, 239–42.

Galleni, L., Canovai, R., Gualandi, M. *et al.* (1986) Fine characterization of turbellarian chromosomes. I. Giemsa and quinacrine banding in *Dugesia polychroa* (O. Schmidt). *Genetica*, **71**, 47–50.

Galleni, L., Canovai, R., Esposito, A. and Stanyon, R. (1989) Characterization of turbellarian chromosomes. II. C-banding in *Dugesia lugubris* (Tricladida: Paludicola). *Trans. Am. Micros. Soc.*, **108**, 304–8.

Gao, Y.M. and Natsukari, Y. (1990) Karyological studies on seven cephalopods. *Venus*, **49**, 126–45.

Garcia, E., Alvarez, C. and Thode, G. (1987) Chromosome relationships in the genus *Blennius* (Blenniidae Perciformes) C-banding patterns suggest two karyoevolutional pathways. *Genetica*, **72**, 27–36.

Grassle, J.P., Gelfman, C.E. and Mils, S.W. (1987) Karyotypes of *Capitella* sibling species and of several species in the related genera *Capitellides* and *Capitomastus* (Polychaeta). *Bull. Biol. Soc. Wash.*, **7**, 77–88.

Hartley, S.E. (1991) Restriction enzyme banding in Atlantic salmon (*Salmo salar*) and brown trout (*Salmo trutta*). *Genet. Res., Camb.*, **57**, 273–8.

Hellmer, A., Voiculescu, I. and Schempp, W. (1991) Replication banding studies in two cyprinid fish. *Chromosoma*, **100**, 524–31.

Insua, A. and Thiriot-Quiévreux, C. (1991) The characterization of *Ostrea denselamellosa* (Mollusca, Bivalvia) chromosomes: karyotype, constitutive heterochromatin and nucleolus organizer regions. *Aquaculture*, **97**, 317–25.

Insua, A. and Thiriot-Quiévreux, C. (1992) Karyotypes of *Cerastoderma edule*, *Venerupis pullastra* and *Venerupis rhomboides* (Bivalvia, Veneroida). *Aquat. Living Resour.*, **5**, 1–8.

Kasinathan, R. and Natarajan, R. (1975) Chromosome numbers of five species of Cyclophoridae (Prosobranchia: Mesogastropoda). *Malacol. Rev.*, **8**, 109–10.

Kligerman, A.D. and Bloom, S.E. (1977) Rapid chromosome preparations from solid tissues of fish. *J. Fish. Res. Bd Can.*, **34**, 266–9.

Komatsu, S. (1985) Karyotypes of two species in two families of Prosobranchia. *Venus*, **44**, 49–54.

Komatsu, S. (1988) Karyotype of *Omphalius rusticus* (Gmelin) (Gastropoda: Trochidae). *Venus*, **47**, 57–61.

Laval, P. and Lécher, P. (1975) Caryotypes, chromosomes surnuméraires, parthénogénèse rudimentaire et polyploïdie chez deux espèces du genre *Phronima* (Crustacés Amphipodes). *Can. J. Genet. Cytol.*, **17**, 405–12.

Lazzaretto, I. (1983) Karyology and chromosome evolution in the genus *Tisbe* (Copepoda). *Crustaceana*, **45**, 85–95.

Lazzaretto, I. and Libertini, A. (1986) Karyological comparison among different Mediterranean populations of the genus *Tigriopus* (Copepoda Harpacticoida). *Boll. Zool.*, **53**, 197–202.

Lloyd, M.A. and Thorgaard, G.H. (1988) Restriction endonuclease banding of rainbow trout chromosomes. *Chromosoma*, **96**, 171–7.

Lozano, R., Ruiz Rejón, C. and Ruiz Rejón, M. (1991) An analysis of coho salmon chromatin by means of C-banding, AG- and fluorochrome staining, and *in situ* digestion with restriction endonucleases. *Heredity*, **66**, 403–9.

Mauchline, J. (1980) The biology of mysids and euphausiids. *Adv. Mar Biol.*, **18**, 1–677.

Martens, E.E. (1984) Ultrastructure of the spines in the copulatory of some Monocelididae (Turbellaria, Proseriata). *Zoomorphology*, **104**, 261–5.

Martens, E.E., Curini-Galletti, M.C. and Van Oostveldt, P. (1989) Polyploidy in Proseriata (Platyhelminthes) and its phylogenetic implications. *Evolution*, **43**, 900–7.

Medrano, L., Bernardi, G., Couturier, J., Dutrillaux, B. and Bernardi, G. (1988) Chromosome banding and genome compartmentalization in fish. *Chromosoma*, **96**, 178–83.

Nakamura, H.K. (1985) A review of molluscan cytogenetic information based on the CISMOCH – Computerized Index System for Molluscan chromosomes. Bivalvia, Polyplacophora and Cephalopoda. *Venus*, **44**, 193–225.

Nakamura, H.K. (1986) Chromosomes of Archaeogastropoda (Mollusca: Prosobranchia) with some remarks on their cytotaxonomy and phylogeny. *Publ. Seto Mar. Biol. Lab.*, **31**, 191–207.

Nakamura, H.K., Machii, A., Wada, K.T., Awaji, M. and Townsley, S.J. (1988) A checklist of decapod chromosomes (Crustacea). *Bull. Natl. Res. Inst. Aquaculture, Nansei*, **13**, 1–9.

Natarajan, R. (1970) Cytological studies of Indian mollusks: chromosomes of some opisthobranchs from Porto Novo, South India. *Malacol. Rev.*, **3**, 19–23.

Niiyama, H. (1959) A comparative study of the chromosomes in Decapods, Isopods and Amphipods, with some remarks on cytotaxonomy and sex-determination in the Crustacea. *Mem. Fac. Fish., Okkaido Univ.*, **7**, 1–60.

Ngan, P.V., Gomes, V., Suzuki, H. and Passos, M.J.A.C.R. (1989) Preliminary study on chromosomes of Antarctic krill, *Euphausia superba* Dana. *Polar Biol.*, **10**, 149–50.

Ojima, Y. (1983) Fish cytogenetics, in *Chromosomes in Evolution of Eukaryotic Groups*, Vol. I (eds A.K. Sharma and A. Sharma), CRC Press, Boca Raton, pp. 111–46.

Patterson, C.M. (1969) Chromosomes of molluscs, in *Proceedings of the 2nd Symposium on Mollusca*, Ernakulam, Cochin, India. Marine Biological Association, Bangalore, India, pp. 635–86.

Patterson, C.M. and Burch, J.B. (1978) Chromosomes of pulmonate molluscs, in *Pulmonate*, Vol. II (eds V. Fretter and J. Peake), Academic Press, New York, pp. 171–217.

Puccinelli, I, and Deri, P. (1991) Comparative karyological analysis of some American planarians belonging to the genus *Dugesia* (subgenus *Girardia*) (Platyhelminthes, Tricladida). *Caryologia*, **44**, 225–32.

Robotti, C.A., Ramella, L., Cervella, P. and Sella, G. (1991) Chromosome analysis of nine species of *Ophryotrocha* (Polychaeta: Dorvilleidae). *Ophelia*, **5**, 625–32.

Salemaa, H. (1986) Karyology of the Northern Baltic peracaridan. *Sarsia*, **71**, 17–25.

Sánchez, L., Martinez, P., Viñas, A and Bouza, C. (1990) Analysis of the structure and variability of nucleolar organizer regions of *Salmo trutta* by C-, Ag- and restriction endonuclease banding. *Cytogenet. Cell Genet.*, **54**, 6–9.

Sato, M. and Ikeda, M. (1992) Chromosomal complements of two forms of *Neanthes japonica* (Polychaeta, Nereididae) with evidence of male-heterogametic sex chromosomes. *Mar. Biol.*, **112**, 299–307.

Schmekel, L.(1985) Aspects of evolution within the opisthobranchs, in *The Mollusca*, Vol. 10. *Evolution* (eds K.M. Wilbur, E.R. Trueman and M.R. Clarke), Academic Press, Orlando, pp. 221–68.

Sharma, A.K. and Sharma, A. (1980) *Chromosomes Techniques*, 3rd edn, Butterworths, London.

Stern, E.M (1975) Chromosome numbers of five species of Cyclophoridae (Prosobranchia, Mesogastropoda) from South India. *Malacol. Rev.*, **8**, 107–8.

Thiriot-Quiévreux, C. (1988) Chromosome studies in pelagic opisthobranch molluscs. *Can. J. Zool.*, **66**, 1460–73.

Thiriot-Quiévreux, C. (1990) Karyotype analysis in several pelagic gastropods. *Am. Malacol. Bull.*, **8**, 37–44.

Thiriot-Quiévreux, C. and Insua, A. (1992) Nucleolar organiser region variation in the chromosomes of three oyster species. *J. Exp. Mar. Biol. Ecol.*, **157**, 33–40.

Thiriot-Quiévreux, C., Soyer, J. and Bouvy, M. (1987) Etude des chromosomes du Bivalve *Malletia sabrina* Hedley, 1916. *Vie Milieu*, **37**, 175–80.

Thiriot-Quiévreux, C., Albert, P. and Soyer, J. (1991) Karyotypes of five subantarctic bivalve species. *J. Moll. Stud.*, **57**, 59–70.

Trentini, M. (1980) Chromosome numbers of nine species of Daphniidae (Crustacea, Cladocera). *Genetica*, **54**, 221–3.

Vitturi, R. and Catalano, E. (1984) Spermatocyte chromosomes in 7 species of the sub-class Prosobranchia (Mollusca, Gastropoda). *Biol. Zentralbl., Leipzig*, **103**, 69–76.

Vitturi, R. and Catalano, E. (1989) Spermatocyte chromosomes and nucleolus organizer regions (NORs) in *Tricolia speciosa* (Mühfield, 1824) (Prosobranchia, Archaeogastropoda). *Malacologia*, **31**, 211–6.

Vitturi, R. and Catalano, E. (1990) Spermatocyte chromosome banding studies in *Buccinulum corneum* (Prosobranchia: Neogostropoda): variation in silver-NOR banding pattern. *Mar. Biol.*, **104**, 259–63.

Vitturi, R., Catalano, E. and Macaluso, M. (1986) Chromosome studies in three species of the gastropod superfamily Littorinoidea. *Malacol. Rev.*, **19**, 53–60.

Vitturi, R., Catalano, E., Macaluso, M. and Maiorca, A. (1987) Spermatocyte chromosomes in six species of Neogastropoda (Mollusca, Prosobranchia). *Biol. Zentralbl., Leipzig*, **106**, 81–8.

Vitturi, R., Maiorca, A. and Carollo, T. (1984) The chromosomes of *Hydroides elegans* (Haswell, 1883) (Annelida, Polychaeta). *Caryologia*, **37** 105–13.

Vitturi, R., Rasotto, M.B. and Farinella-Ferruza, N. (1982) The chromosomes of 16 molluscan species. *Boll. Zool.*, **49**, 61–71.

Vitturi, R., Rasotto, M.B., Parrinello, N. and Catalano, E. (1982) Spermatocyte chromosomes in some species of the family Aplysiidae (Gastropoda, Opisthobranchia). *Caryologia*, **35**, 327–33.

White, M.J.D. (1973) *Animal Cytology and Evolution*, 3rd edn, Cambridge University Press, Cambridge.

White, M.J.D. (1978) *Modes of Speciation*, W.H. Freeman and Company, San Francisco.

Wiberg, U.H. (1983) Sex determination in the European eel (*Anguilla anguilla* L.). *Cytogenet. Cell Genet.*, **15**, 306–31.

Yabu, H. and Kawamura, A. (1981) Chromosomes of *Euphausia pacifica* Hansen (Crustacea, Euphausiidae). *Chromosome Information Service*, **31**, 8–9.

Yabu, H. and Kawamura, A. (1984) Chromosomes of *Euphausia superba* Dana. *Bull. Plankton Soc. Jpn*, **31**, 61–3.

7.2 THE NATURE OF ROBERTSONIAN CHROMOSOMAL POLYMORPHISM IN
NUCELLA LAPILLUS: A RE-EXAMINATION
David R. Dixon, Philip L. Pascoe, Peter E. Gibbs and Juan J. Pasantes

Abstract

Over the past half century several different classes of large-scale chromosomal rearrange-ments have been reported for many different animal groups. In the dog-whelk Nucella lapillus *(Mollusca: Gastropoda) this variation takes the form of Robertsonian polymorphism (centric fusion or fission), the diploid chromosome number varying between 26 and 36 depending on the number of median/subterminal chromosomes in the karyotype.*

Historical data relating to the polymorphism exhibited by N. lapillus are reviewed; new data regarding numerical and structural chromosomal variation are also presented. A high level of intra-individual variation in chromosome number has been discovered in some populations: the link between chromosome number and adaptation to specific environmental conditions, proposed by early investigators, thus seems in doubt at least for some populations.

7.2.1 Introduction

Nucella (= *Thais*) *lapillus* (L.), the common dog-whelk, is widely distributed on rocky shores on both sides of the North Atlantic, in the east from northern Russia to Portugal, and in the west from Southern Newfoundland to New York. Its ancestors are likely to have been Pacific in origin, these migrating into the North Atlantic during the Miocene or Pliocene periods when milder climatic conditions prevailed (Briggs, 1970). The species has a direct larval development; this takes place over 3–4 months within a durable capsule attached to rock, the juveniles emerging as miniature adults. Dispersal of the species is thus limited by its lack of a planktonic larval phase and also by the fact that the movement of individual adults during their lifetime (> 6 years) is restricted to a few metres (Hughes, 1972). Several lines of evidence suggest that genetic interchange between *N. lapillus* populations is restricted. These include high levels of phenotypic, viz. shell morphology (e.g. Crothers, 1985), genotypic (allozymic and ribosomal DNA polymorphism) (e.g. Day, 1990; Patton and Dixon, unpublished data) and karyotypic (e.g. Staiger, 1950a, 1954) variation.

Staiger (1950a, 1954) was the first to show that *N. lapillus* populations in the region around Roscoff on the Channel coast of France contained a numerical (Robertsonian) chromosomal polymorphism (he used the term 'dimorphism') affecting five pairs of metacentric chromosomes in the $2n = 26$ form, which were represented by 10 pairs of acrocentrics in the $2n = 36$ form. Staiger (1950a, 1954) examined individuals from 48 sites in the Roscoff region of Brittany, representing a range of exposure conditions, and concluded that the degree of chromosomal heterogeneity correlated well with environmental factors to the extent that high chromosome numbers were characteristic of, 'although not exclusively found on', shores sheltered from wave action, whereas a low chromosome number was commonly found at exposed sites typified by strong wave action; animals with intermediate chromosome numbers occupied shores of intermediate exposure. What adaptive significance there is in such chromosome number variation and how this translates into differences in fitness between the different chromosomal genotypes under differing conditions of exposure remains a matter of conjecture. Recently, it has been suggested that crossover suppression around the centromeres in heterozygotes will assist these regions of the chromosomes to evolve independently in the two types of homozygotes ($2n = 26$ and $2n = 36$) to become adapted to different ecological niches within the same environment (racial separation), with the heterozygote possibly becoming better adapted to intermediate habitats (heterozygote advantage) (Searle, 1988a,b).

Bantock and Cockayne (1975) and Bantock and Page (1976) carried out a chromosome survey of dog-whelk populations on the Channel coast of southern England and reported finding a similar relationship to that found by Staiger (1954) between chromosome numbers and ecological conditions. Subsequently, Hoxmark (1970) examined Norwegian populations in an attempt to repeat Staiger's work but was only able to find $2n = 26$ populations on all shore types. Mayr (1969) stated that only the $2n = 26$ type occurs on the Atlantic coast of North America, an observation not supported by any data. Ahmed (1974) reported finding no evidence of any similar polymorphism in other species of *Nucella* on the Pacific coast of North America. One of us (J.J.P.) has examined *N. lapillus* from five localities (three exposed; two sheltered) around La Coruña, northwestern Spain, and could find only the $2n = 26$ form (Table 7.2.1). It would appear, therefore, that the $2n = > 26$ form is restricted to one relatively small, central part of a large geographic range, within which *N. lapillus* occupies a broader niche compared with the other (Pacific) species, presumably due to the lack of competing gastropod species on Atlantic shores (Kitching, 1985). If one accepts the argument put forward by Staiger and others that chromosomal polymorphism confers some adaptive advantage to dog-whelk populations along the English and French Channel coasts, then there remains a need to explain how *N. lapillus* succeeds throughout the greater part of its geographic range, and under a full range of exposure conditions, with a karyotype consisting of only $2n = 26$ chromosomes.

This section describes results of a study of karyotype variation in dog-whelk populations of intermediate and low chromosome number in southwestern England and elsewhere which cast new light on the nature of Robertsonian polymorphism in *N. lapillus* and its possible evolutionary significance.

7.2.2   Materials and methods

Samples of adult *N. lapillus* and/or egg capsules were collected from sites representing a range of exposure conditions in southern England, northwestern Spain and western Norway (Svartholmane, off the Biological Station, Bergen, i.e. a site of intermediate exposure which featured previously in Hoxmark (1970)). The animals were housed temporarily in a recirculating seawater aquarium where they were fed small *Mytilus edulis*. Egg capsules laid in the aquarium were kept for 4–6 weeks prior to analysis.

(a)   Preparation of mitotic metaphase spreads

Chromosome spreads were prepared from adult testes or early-veliger-stage embryos (4–6 weeks old) taken from intact egg capsules. Testes samples were dissected away from the

Table 7.2.1   Chromosome number counts for adult *N. lapillus* from five sites around La Coruña, northwestern Spain (Galician coast), based on testis preparations

Locality	Chromosome number			
	24	*25*	*26*	*n*
Reira (exposed)	1	1	61	10
Caion (exposed)	1	4	105	10
Sabon (exposed)	1	2	123	11
Muinos (sheltered)	3	3	83	11
Santa Christina (sheltered)	2	1	81	15

Numbers refer to the total number of metaphases counted from *n* number of individuals.

underlying digestive gland and cut into small (1 mm square) pieces; embryos were teased from the early shell and remaining yolk. Both tissues were given two combined colchicine and hypotonic treatments: 0.08% colchicine (Sigma) in 50% seawater, 30–45 min followed by 0.04% colchicine in 25% seawater, 30–45 min. After a further hypotonic treatment, 0.075 M KCl, 2×5 min, the tissues were fixed in fresh, cold, Carnoy's solution (3 : 1, ethanol/glacial acetic acid), three or four changes for a minimum of 1 h.

Slide making was based on the procedure for solid tissues described by Kligerman and Bloom (1977). Following dissociation of the Carnoy's fixed tissue in 60% acetic acid, concentric rings of spread nuclei were produced by pippeting drops of the cell suspension onto the surface of a warm slide at 40 °C. Slides were stained for 10 min in freshly prepared and filtered 10% Giemsa (Gurr's Improved R66) in phosphate buffer (Gurr's pH 6.8 buffer tablets). After rinsing and blueing in tapwater containing a few drops of ammonium hydroxide, the slides were blotted dry. Photographs were taken using a Zeiss photomicroscope ($63 \times$ or $100 \times$ oil immersion lens) on Kodak Panatomic-X/Technical Pan film, employing a green filter to enhance contrast. In excess of 1000 cells were counted during the course of this study.

(b) Karyotyping

Karyotyping followed the standard procedure based on arm length parameters. Chromosomes were first grouped by arm ratio, then arranged according to length, and finally matched by eye, rearranging to improve pairing where necessary. Measurements of chromosome arm length were made with the aid of a Kontron IBAS Image Analyser. Total length (nearest 0.01 µm), relative length ($100 \times$ chromosome length/total haploid length) and arm ratio (length of short arm, p/length of long arm, q) were computed based on the average length values for each pair of chromosomes, excluding centromeric distances. Centromere position was described according to the nomenclature of Levan, Fredga and Sandberg (1964). Chromosomes were classified according to centromere position as follows: 'median' (m), AR 1–0.59; 'submedian' (sm), AR 0.59–0.33; 'subterminal' (st), AR 0.33–0.14; and 'terminal' (t), AR 0.14–0. In the text, use is made also of the more widely employed, though less precise, terms metacentric (m), submetacentric (sm) and acrocentric (st-t), to aid comparison with previous work.

7.2.3 Results

(a) The 2n = 26 karyotype

The $2n = 26$ karyotype of *N. lapillus* consists of five groups of chromosomes (Figure 7.2.1), based on relative chromosome size and centromere position: group A (m, $n = 4$ pairs); group B (sm, $n = 2$ pairs); group C (m, $n = 4$ pairs); group D (st, $n = 1$ pair); and group E (m and sm, $n = 2$ pairs). Specific chromosome pairs, with the exception of subterminal pair ll, could not be unequivocally identified due to only small gradations in size within and between groups, and an absence of any consistent secondary features (e.g. secondary constrictions). This confirms the basic karyotype proposed for *N. lapillus* by Page (1988). Table 7.2.2 shows the results of measurements of total length, relative length and arm ratio for each of the 13 pairs of chromosomes in the karyotype of *N. lapillus* from Whitsand Bay, southeastern Cornwall, UK. There was no difference between this population with respect to the composition of the $2n = 26$ karyotype and other $2n = 26$ populations from eastern and southern England (Polzeath, Bude and Norfolk), southwestern Brittany (Pointe de la Torche), northwestern Spain (Galicia) and Norway (Svartholmane, near Bergen). These additional data are available from the authors.

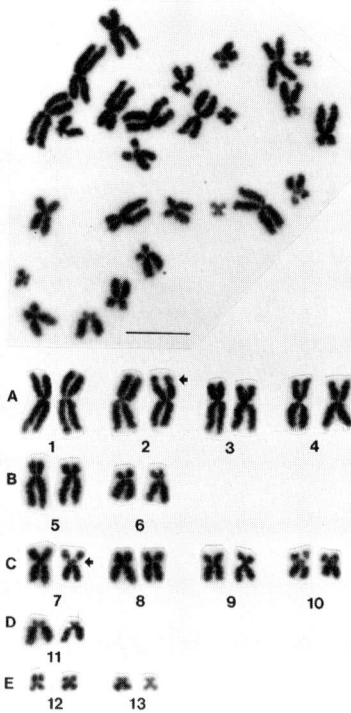

Figure 7.2.1 The $2n = 26$ karyotype of *N. lapillus* (here collected from West Runton, Norfolk) comprises five groups based on chromosome size and arm length parameters. In order of decreasing size, group A, median (AR = 1–0.59), 4 pairs; group B, submedian (AR = 0.59–0.33), 2 pairs; group C, median, 4 pairs; group D, subterminal (AR = 0.33–0.14), 1 pair; and group E, median/submedian, 2 pairs. Individual identification is possible only for pair 11. Note prominent secondary constrictions on single chromosomes in pairs 2 and 7 (arrowed). Scale bar = 10 μm.

Table 7.2.2   Total chromosome lengths, relative lengths and arm ratios (where x represents the mean) for *N. lapillus* (testes, $n = 13$) from Whitsand Bay, southeastern Cornwall

Chromosome pair no.	Total length (μm) ($x \pm 2\,SE$)	Relative length (%) ($x \pm 2\,SE$)	Arm ratio ($x \pm 2\,SE$)	Type
1	10.23 + 1.37	13.9 + 0.4	0.81 + 0.03	m
2	9.28 + 1.23	12.5 + 0.2	0.88 + 0.02	m
3	8.30 + 1.05	11.2 + 0.4	0.64 + 0.02	m
4	7.79 + 1.02	10.5 + 0.2	0.71 + 0.04	m
5	6.70 + 0.81	9.1 + 0.4	0.37 + 0.02	sm
6	5.20 + 0.64	7.0 + 0.2	0.35 + 0.04	sm
7	5.37 + 0.63	7.3 + 0.2	0.79 + 0.06	m
8	4.57 + 0.54	6.2 + 0.2	0.60 + 0.06	m
9	4.51 + 0.51	6.1 + 0.1	0.63 + 0.04	m
10	3.99 + 0.40	5.4 + 0.2	0.75 + 0.04	m
11	3.81 + 0.34	5.2 + 0.2	0.24 + 0.02	st
12	2.39 + 0.22	3.3 + 0.2	0.75 + 0.05	m
13	1.62 + 0.15	2.2 + 0.2	0.61 + 0.06	m/sm

Chromosome type: m = median; sm = submedian; st = subterminal.

(b)   The 2n = > 26 karyotype

Figure 7.2.2 shows $2n = 30$ and $2n = 33$ karyotypes from Thurlestone, southern Devon, a site of intermediate chromosomal heterozygosity. (Note: both spreads are from the same animal.) The fundamental number (nombre fondamental, NF; Matthey, 1945), the number of major chromosome arms in the $2n$ karyotype, was the same as that recorded for the $2n = 26$ form, i.e. 48. This supports the view that the observed numerical variation was a result of a

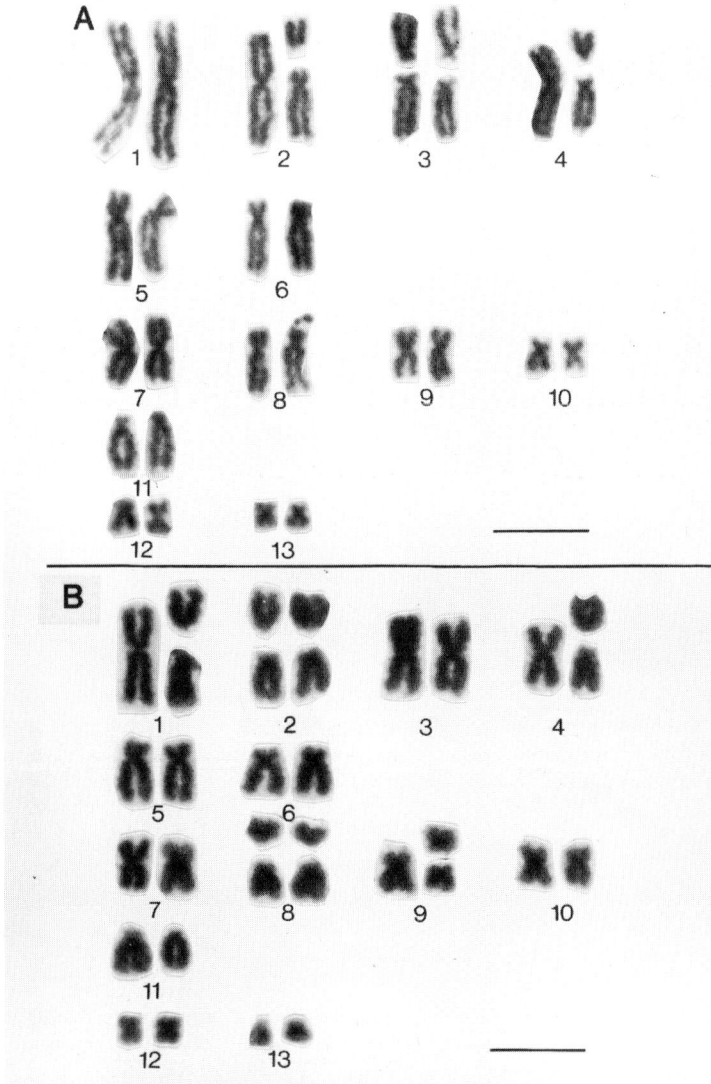

Figure 7.2.2 Contrasting karyotypes from a single *N. lapillus* (T4, the same individual featured in Figure 7.2.3 (b)) (A) $2n = 30$ cell, putatively polymorphic for chromosome pairs 2, 3 and 4; (B) $2n = 33$ cell, putatively polymorphic for pairs 1, 2, 4, 8 and 9. Scale bars = 10 μm.

Robertsonian rearrangement (i.e. centric fusion or fission). (Note: in this study, short arms such as those of pairs ll and 13 (Figure 7.2.1) were omitted from the NF calculation; this phenomenon will be addressed in more detail in a later paper (Pascoe and Dixon, 1993).) A significant proportion of $2n = > 26$ karyotypes analysed during this study were checked to ensure that the NF did not deviate from 48 to avoid including any that contained 'extra' chromosomes (i.e. hyperdiploid spreads) resulting from cell mixing during slide manufacture. Our investigation revealed that of the five pairs of metacentric chromosomes contributing to the polymorphism, three (or possibly four) were from group A and the remaining two pairs were from group C. This contrasts with the findings of Page (1988), who reported that two pairs from each of these groups and one pair from group B were involved in the polymorphism: this undoubtedly reflects population differences. The precise identification of chromosomes contributing to the Robertsonian polymorphism awaits the successful application of chromosome banding or related techniques. Work of this nature is currently in progress in our laboratory.

A most interesting feature of the $2n = > 26$ karyotype of *N. lapillus* was the presence of 'rabbit-ear' acrocentrics (terminology after Levan, Hsu and Stich, 1962), i.e. products of centric fission which display a short arm in addition to the predicted long arm. In fact, the majority of acrocentric chromosomes examined during this study were of the 'rabbit-ear' type (e.g. Figure 7.2.2, chromosomes 3 and 4).

Figure 7.2.3(a) shows pooled chromosome counts for 20 adult *N. lapillus* (257 cells; 1–37 counts performed per individual) from Thurlestone, South Devon. This pattern was typical of all those 'high' chromosome number populations we examined. No pure $2n = 36$ populations or, for that matter, animals, were discovered during our survey of sites on the English Channel coast (Pascoe, Gibbs and Dixon, unpublished data). This absence can be attributed to the effects of tributyltin antifouling paint which has seriously decimated dog-whelk populations around the UK and elsewhere during the past decade (Gibbs and Bryan, 1986), particularly those at sheltered sites (boat harbours) where the high chromosome number form appears to have been dominant, e.g. Salcombe (Bantock and Cockayne, 1975). Figure 7.2.3(b) shows the range of chromosome number counts from one adult animal from Thurlestone. When checked for fundamental number, about 25% of spreads were found to be the result of cell mixing and were consequently omitted from the analysis.

Figures 7.2.3(c) and 7.2.3(d) show chromosome counts for dog-whelk embryos from egg capsules collected at Thurlestone. These showed the same overall pattern of between-cell variation as was recorded for adults, although the range of numerical variation per individual embryo was somewhat reduced. A Mann-Whitney U-test revealed a significant difference ($P < 0.05$) between the range of intra-individual chromosome number variation in dog-whelk embryos and adults from Thurlestone.

7.2.4 Discussion

This investigation has revealed significant levels of intra and inter-individual variation in chromosome number in dog-whelks collected from several geographically separated sites on the south coast of England. Robertsonian transformations refer to chromosome structural changes due to centric fusion or centric fission of non-homologous acrocentric or telocentric chromosomes (Robertson, 1916). Such a mechanism involves changes in the number of centromeres without a major change in NF, i.e. the total number of chromosome arms. Centric fusions are considered to be one of the main ways by which chromosome numbers have changed in animal karyotype evolution (White, 1973; John, 1976), and are generally considered to be the more frequent (White, 1965) since there are fewer morphological obstacles to their formation, i.e. no need for new centromeres and telomeres.

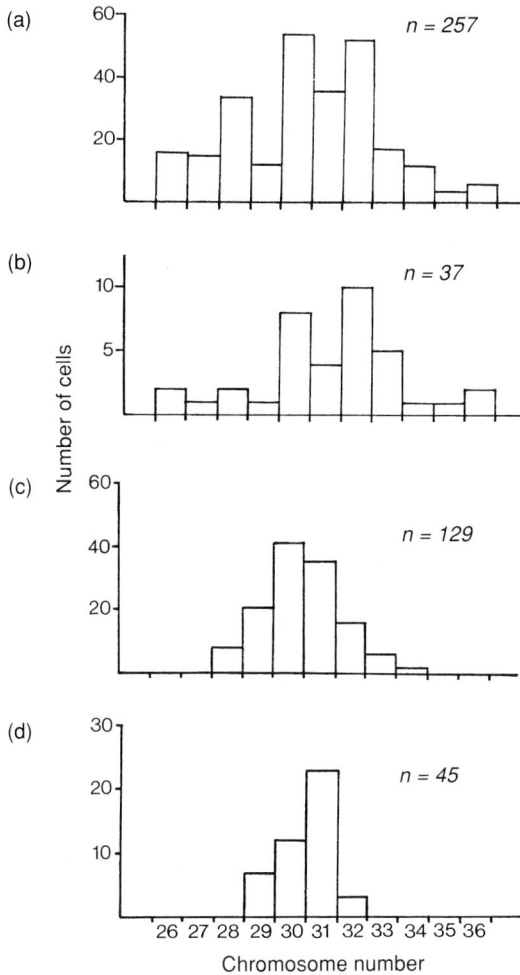

Figure 7.2.3 Chromosome number–frequency histograms for *N. lapillus* from Thurlestone, Southwestern Devon, a 2*n* = > 26 population. (a) Pooled counts based on 20 adults; (b) counts from a single adult showing a high level of intra-individual variation; (c) pooled embryo counts for the same population; and (d) counts from a single embryo showing a reduced level of numerical variation.

The conservation of fundamental number ignores the formation of so-called 'rabbit-ear' (i.e. short-armed) acrocentrics which we report here for *N. lapillus* and which have already been described for several other species exhibiting Robertsonian polymorphism, e.g. the coccinellid beetle *Chilocorus stigma* (White, 1973) and the pocket mouse *Perognathus goldmani* (Patton, 1969). Interestingly, Staiger (1954) described the chromosomes of the 2*n* = 36 form of *N. lapillus* on the Brittany coast as consisting of a mixture of metacentric and 'rod-shaped' (not strictly acrocentric, for the most part) chromosomes, which he depicted as having short arms (e.g. Staiger, 1954, Figure 18). This fact appears to have been overlooked by more recent investigators. White (1973) could see no reason why there should be any difference between centric fusions involving chromosomes with terminal centromeres (acrocentrics) and those, such as in the cases described above, including *N. lapillus*, where a short length of chromatin is lost. In other organisms (Patton, 1969), the short arms of 'rabbit-ear' acrocentrics have been shown to consist of heterochromatin, i.e. genetically inert (or dispensable) portions of chromosome arms which seemingly disappear upon fusion. This heterochromatin may serve to stabilize naked centromeres (Cooper, 1959). The presence of short arms on a large proportion of those acrocentrics involved in the polymorphism provides unequivocal evidence that the significant level of intra-individual variation in chromosome number reported here cannot be simply accounted for by chromosome breakage during slide making.

It remains to be shown whether there is any structural difference between the centromeres of monomorphic (i.e. $2n = 26$) and polymorphic (i.e. $2n = > 26$) populations of *N. lapillus*. The centromere is the primary constriction at which metaphase chromatids (chromosome arms) are held together. Using special staining techniques it is possible to visualize four dense structures, two in each (metacentric) chromatid (Bostock and Sumner, 1978), the precise nature of which is unknown. The presence of four in metacentric and two in telocentric chromosomes suggests the ready conversion of one to the other by splitting or fusion (Bostock and Sumner, 1978). The very low level of numerical variation recorded for the $2n = 26$ form (Figure 7.2.4) suggests that it may have reached a level of centromeric fixation not found in $2n = > 26$ populations. Whether this reflects the presence of a gene responsible for centromeric fragility/stability remains to be discovered. An inherent stability of the $2n = 26$ karyotype is strongly indicated by the results of a dog-whelk transplant experiment in which adults from the exposed site at Bude were moved to a sheltered site in Plymouth Sound (Batten Bay to the north of Renney Rocks, a high chromosome number site) where the native population had ceased breeding because of TBT-induced sterilization (Gibbs and Bryan, 1986). Significantly, examples of the F_1 generation (4-year-old males) produced by the Bude transplants resembled their parents in being $2n = 26$ ($n = 7$ individuals, 54 spreads, Gibbs, unpublished data). This indicates that the parental karyotype was conserved despite the altered environmental conditions, and supports the idea of centromeric fixation in the $2n = 26$ form.

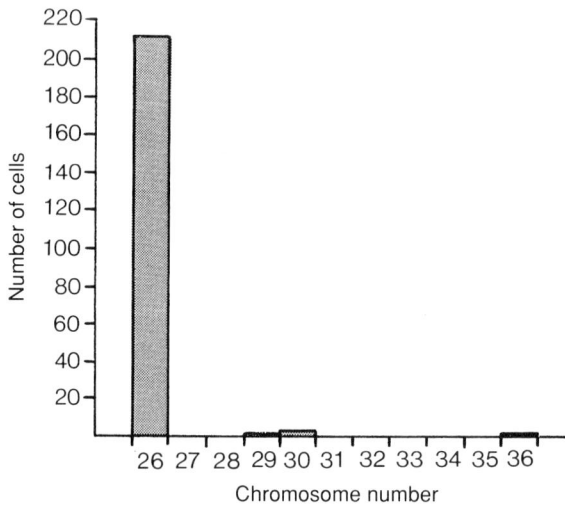

Figure 7.2.4 Chromosome number–frequency histogram for adult *N. lapillus* ($n = 11$ animals: 217 cells) from Whitsand Bay, southeastern Cornwall, a typical $2n = 26$ population.

The high level of intra-individual variation in *N. lapillus* casts considerable doubt on any supposed link between chromosome number and environmental conditions, viz. degree of exposure and/or shelter from wave action, as originally suggested by Staiger (1954) and Bantock and Cockayne (1975). White (1973) found it surprising that the adaptive properties of five different, non-homologous chromosomes were all similar in that they helped adapt the individual to 'exposed' environments. Hoxmark (1970) proposed the idea of two chromosomal races – form 13 (haploid, *n*), with a northern distribution, and form 18, with a southern distribution. He thought that the coast of Brittany and the Channel form an overlapping zone between the two forms with free mating between them. This racial zonation hypothesis can be discounted because the more southerly Spanish material described here (Table 7.2.1) proved to be $2n = 26$ on all shore types. We have been unable to locate any data addressing the phenomenon of intra-individual variation in Robertsonian populations.

White (1973) distinguished between cytologically **polymorphic** species in which many or all demes (breeding groups) contain several different karyotypes, and **polytypic** species which consist of two or more chromosomal races differing in karyotype. Two models have been suggested by Patton (1969) to explain the types of Robertsonian variation observed in a wide variety of animal species: (1) geographically separated chromosome races showing limited degrees of hybridization where the racial zones overlap at the margins (e.g. Patton, 1969; Ford and Hamerton, 1970; Nevo and Shaw, 1972; Baker, 1981; Greenbaum, 1981; Porter and Sites, 1985; Searle, 1988b); and (2) dominance of one race over another with limited expression of one pure form and a greater amount of hybridization through introgression. Physiographic and ecological barriers dominate in the type 1 model, but have broken down in the type 2 situation, resulting in evolutionary 'footprints' (Patton, 1969). The pattern of chromosomal variation described here for *N. lapillus* appears to fit more closely with the type 2 model and not with the type 1 model favoured by some previous investigators (e.g. Hoxmark, 1970).

The hypothesis we propose to explain the chromosome number variation in *N. lapillus*, in the light of the new facts presented here, is that this represents an example of chromosomal evolution in action brought about by centric fusions. Related species in the Pacific, where the ancestor of *N. lapillus* has its origins (Kitching, 1985), have diploid chromosome numbers of 60 or 70 (Nishikawa, 1962; Ahmed, 1974), which suggests that the reduction in chromosome number in *N. lapillus* commenced after its ancestor invaded the North Atlantic from the Pacific, via the Bering Strait and Arctic Ocean, in the late Miocene/Pliocene (Briggs, 1970). It is interesting to record that Staiger (1950b) drew attention to unusually low chromosome number in *N. lapillus* compared to other stenoglossan prosobranchs.

Centric fusions are recognized to be the main way by which chromosome numbers have been changed during animal karyotype evolution (White, 1965, 1973; John and Lewis, 1968; Mayr, 1969). The fact that the $2n = 26$ form thrives on all shore types over the large part of its range, which extends on both sides of the Atlantic from northern Norway to Portugal (Hoxmark, 1970; Mayr, 1969), indicates that variation from this is not a necessary prerequisite for survival under any specific set of environmental conditions as determined by the level of wave action or other physical factors. The functional significance of chromosome number variation in *N. lapillus* is currently being investigated at the structural and molecular levels of chromosomal organization.

Acknowledgements

We wish to express our grateful thanks to Prof. Annetrudi Kress (University of Basel) for her painstaking translations of Staiger's papers. We would also like to thank Dr John Green (Plymouth Marine Laboratory) for providing the Norwegian material, and Dr Eve Southward (Marine Biological Association) who kindly commented on an earlier draft of the manuscript.

References

Ahmed, M. (1974) Chromosome variation in three species of the marine gastropod *Nucella. Cytologia*, **39**, 597–607.

Baker, R.J. (1981) Chromosome flow between chromosomally characterized taxa of a volant mammal, *Uroderma bilobatum* (Chiroptera: Phyllostomatidae). *Evolution*, **35**, 296–305.

Bantock, C.R. and Cockayne, W.C. (1975) Chromosomal polymorphism in *Nucella lapillus. Heredity*, **34**, 231–45.

Bantock, C.R. and Page, C.M. (1976) Chromosomal polymorphisms in the dog whelk (*Nucella lapillus*) (L.), in *Current Chromosome Research* (eds K. Jones and P.E. Brandham), Elsevier/North Holland, Amsterdam, pp. 159–66.

Bostock, C.J. and Sumner, A.T. (1978) *The Eukaryotic Chromosome*, North-Holland Publishing Company, Amsterdam.

Briggs, J.C. (1970) A faunal history of the North Atlantic Ocean. *Syst. Zool.*, **19**, 19–34.

Cooper, K.W. (1959) Cytogenetic analysis of major heterochromatic elements (especially Xh and Y) in *Drosophila melanogaster* and the theory of 'heterochromatin'. *Chromosoma*, **10**, 535–88.

Crothers, J.H. (1985) Dog-whelks: an introduction to the biology of *Nucella lapillus* (L.), *Field Stud.*, **6**, 291–360.

Day, A.J. (1990) Microgeographic variation in allozyme frequencies in relation to the degree of exposure to wave action in the dogwhelk *Nucella lapillus* (L.) (Prosobranchia: Muricacea). *Biol. J. Linn. Soc.*, **40**, 245–61.

Ford, C.E. and Hamerton, J.L. (1970) Chromosome polymorphism in the common shrew, *Sorex araneus. Symp. Zool. Soc. Lond.*, **26**, 223–36.

Gibbs, P.E. and Bryan, G.W. (1986) Reproductive failure in populations of the dog-whelk, *Nucella lapillus*, caused by imposex induced by tributyltin from antifouling paints. *J. Mar. Biol. Assoc. UK*, **66**, 767–77.

Greenbaum, I.F. (1981) Genetic interactions between hybridising cytotypes of the tent-making bat (*Uroderma bilobatum*). *Evolution*, **35**, 306–21.

Hoxmark, R.C. (1970) The chromosome dimorphism of *Nucella lapillus* (Prosobranchia) in relation to wave action. *Nytt Magasin Zool.*, **18**, 229–38.

Hughes, R.N. (1972) Annual production of 2 Nova Scotian populations of *Nucella lapillus* (L.). *Oecologia*, **8**, 356–70.

John, B. (1976) *Population Cytogenetics*, Edward Arnold Ltd, London.

John, B. and Lewis, K.R. (1968) The chromosome complement, *Protoplasmatologia*, **6A**, 1–207.

Kitching, J.A. (1985) The ecological significance and control of shell variability in dogwhelks from temperate rocky shores, in *The Ecology of Rocky Coasts* (eds P.G. Moore and R. Seed), Hodder and Stoughton, London, pp. 234–48.

Kligerman, A.D. and Bloom, S.E. (1977) Rapid chromosome preparations for solid tissues of fishes. *J. Fish. Res. Bd Can.*, **34**, 266–9.

Levan, A., Fredga, K. and Sandberg, A.A. (1964) Nomenclature for centromeric position on chromosomes. *Hereditas*, **52**, 201–20.

Levan, A., Hsu, T.C.S. and Stich, H.F. (1962) The ideogram of the mouse (*Mus musculus*). *Hereditas*, **48**, 677–88.

Mayr, E. (1969) *Animal Species and Evolution*, Harvard University Press, Massachusetts.

Matthey, R. (1945) L' evolution de la formule chromosomiale chez les vertebres. *Experientia*, **1**, 50–6, 78–86.

Nevo, E. and Shaw, C. (1972) Genetic variation in a subterranean mammal. *Biochem. Genet.*, **7**, 235–41.

Nishikawa, S. (1962) A comparative study of the chromosomes in marine gastropods with some remarks on cytotaxonomy and phylogeny. *J. Shimonoseki Coll. Fish.* **11**, 149–86.

Page, C. (1988) The chromosome complement of *Nucella lapillus* (L.), Mollusca: Gastropoda: Prosobranchia. *Caryologia*, **41**, 79–91.

Pascoe, P.L. and Dixon, D.R. (1993) Structural chromosomal polymorphism in the dog-whelk *Nucella lapillus* (Mollusca: Neogastropoda). *Mar. Biol.* (in press)

Patton, J.L. (1969) Chromosome evolution in the pocket mouse, *Perognathus goldmani* Osgood. *Evolution*, **23**, 645–62.

Porter, C.A. and Sites, J.W. (1985) Normal disjunction in Robertsonian heterozygotes from a highly polymorphic lizard population. *Cytogenet. Cell Genet.*, **39**, 250–7.

Robertson, W. (1916) Chromosome studies. I. Taxonomic relationships shown in the chromosomes of Tettigidae and Acrididae. V-shaped chromosomes and their significance in Acrididae, Locustidae, and Gryllidae. *J. Morphol.*, **27**, 179–331.

Searle, J.B. (1988a) Selection and Robertsonian variation in nature: the case of the common shrew, in. *The Cytogenetics of Mammalian Autosomal Rearrangements.* Chapter 15, Alan R. Liss, New York, pp. 507–31.

Searle, J.B. (1988b) Karyotypic variation and evolution in the common shrew, *Sorex araneus. Kew Chromosome Conference III*, HMSO, 97–107.

Staiger, H. (1950a) Chromosomenzahl-varianten bei *Purpura lapillus. Experientia*, **6**, 140–5.

Staiger, H. (1950b) Chromosomenzahlen stenoglosser Prosobranchier. *Experientia*, **6**, 54–9.

Staiger, H. (1954). Der chromosomendimorphism beim Prosobranchier *Purpura lapillus* in Beziehung zur Okologie der Art. *Chromosoma*, **6**, 419–78.

White, M.J.D. (1965) Principles of karyotype evolution in animals, in *Genetics Today*, Vol. 2 (ed. S.J. Geerts), Pergamon Press, Edinburgh/London, pp. 391–7.

White, M.J.D. (1973) *Animal Cytology and Evolution*, Cambridge University Press, London.

7.3 KARYOTYPE DESCRIPTION AND THE POSITION OF THE NUCLEOLAR ORGANIZER REGION (NOR) IN THE CHILEAN OYSTER *TIOSTREA CHILENSIS* (PHILIPPI) CHANLEY AND DINAMANI

Bruno Ladrón de Guevara, Federico Winkler and Claudio Palma

Abstract

The chromosomes of the brooding Chilean oyster Tiostrea chilensis *were studied using traditional karyological analysis and silver staining techniques. This species has a diploid number of* 2n = 20. *Pairs 1, 2, 3, 5, 6, 8 and 10 are metacentric and pairs 4, 7 and 9 are submetacentric. The nucleolar organizer regions (NOR) were detected in terminal position in pairs 2 and 9. Variability in the number of NORs was observed between and within individuals, suggesting heteromorphism in the activity of NOR sites. There are differences in chromosome morphology and NOR position between this species and two other brooding oyster species for which data have been published.*

7.3.1 Introduction

Tiostrea chilensis is a larviparous oyster. It has been separated from its original genus, *Ostrea*, and placed in the new genus *Tiostrea*, which is distributed in Southern Chile and New Zealand, based on characteristics of its larval development (Chanley and Dinamani, 1980). However, in a recent taxonomic review of the oysters, Harry (1985) returned this species to the genus *Ostrea*, proposing that it is a junior synonym of *Ostrea (Eostrea) puelchana* which has a circumglobal distribution.

Usually, systematic studies in oysters have been carried out based on shell morphology and anatomical, ecological and reproductive features. Comparative karyology, however, has not been extensively used for this purpose, mainly because the diploid number and the gross morphology of the karyotypes is quite constant within the group ($2n = 20$) apart from *Dendrostrea folium* ($2n = 18$) (Thiriot-Quiévreux, 1984a; Ieyama, 1990; Insua and Thiriot-Quiévreux, 1991). In this context, karyological comparisons, especially in this group, should consider a detailed description of chromosome morphology and other cytogenetic characteristics such as the active nucleolar organizer region (NOR) position within the karyotype, which can be informative about the group's phylogenetic relationships.

In the present work a detailed description of chromosomal morphology and the position of the active NOR of *T. chilensis* is presented as a contribution to our knowledge of the karyology and phylogeny of oysters.

7.3.2 Materials and methods

Juvenile *T. chilensis* (10–20 mm) were conditioned in warm water (19 °C) and fed *ad libitum* with *Chaetoceros calcitrans* for at least 15 days. After this, animals were held in 0.05% colchicine in seawater for 16 h. A total of 17 oysters were dissected, and the gill tissue was hypotonized in 10% diluted seawater for 30 min and then fixed in Carnoy (methanol: acetic acid 3:1) at 4 °C for at least 30 min.

Slides were prepared using the squash technique. For analysis of chromosome morphology, staining was carried out using Giemsa 4% in phosphate buffer pH 6.8 for 25 min. Active NORs were stained using a silver nitrate precipitation technique adapted by Palma (1990) from Goodpasture and Bloom (1975), Rufas *et al.* (1982) and Mehra, Brokrus and Butler (1985). Giemsa- and NOR-stained slides were dehydrated and permanently mounted with Entellan.

Suitable metaphase plates were photographed from which chromosome measurements were taken using a caliper, and relative arm lengths of each chromosome pair were then calculated over the total haploid complement length. Homologous chromosomes were paired, and finally, the complete karyotype was arranged according to the Denver Conference (1966). The chromosomes were classified using the centromeric index (Levan, Fredga and Sandberg, 1964), and, to aid future comparisons, means and confidence intervals of relative arm lengths of each chromosome pair were plotted in a karyo-idiogram (Spotorno, Fernández-Donoso and Pincheira, 1979).

7.3.3 Results

A diploid complement of $2n = 20$ chromosomes was found in 50 metaphases from 12 individuals of *Tiostrea chilensis* (Figure 7.3.1). There are no visible secondary constrictions or other conspicuous features in the karyotype. Table 7.3.1 shows the mean relative arm length and centromeric index of each chromosome pair, with confidence intervals and standard deviations. Pairs 1, 2, 3, 5, 6, 8 and 10 are metacentric, and pairs 4, 7 and 9 submetacentric. The relative arm lengths and the confidence intervals of chromosomes are plotted in Figure 7.3.2.

Two NOR carrier chromosomes were found out of the 42 cells studies from four individuals. Both NORs are present in the telomeric region, one in the metacentric pair 2, and the other in the long arm of the submetacentric pair 9 (Figure 7.3.3). In this species both the number of active NORs and the intensity of silver staining were variable but the most frequent situation observed was of one NOR on each of the two chromosomes 2 and 9 (Table 7.3.2).

7.3.4 Discussion

T. chilensis has the same diploid number as other species of the family Ostreidae ($2n = 20$) (Thiriot-Quiévreux, 1984a; Insua and Thiriot-Quiévreux, 1991), except *Dendrostrea folium* (Lophinae), which has a complement of $2n = 18$ (Ieyama, 1990). Nevertheless, the chromosome morphology of *T. chilensis* is different. Based on the chromosome classification given by the centromeric index (Levan, Fredga and Sandberg, 1964), the karyotypes of *Ostrea edulis* (Thiriot-Quiévreux, 1984b) and *T. chilensis* differ with respect of the chromosome pairs 4, 6, 7, 8 and 10. Between *Ostrea denselamellosa* (Insua and Thiriot-Quiévreux, 1991) and *T. chilensis* there are only morphological differences in pair 6. Comparisons based upon the classification of Levan, Fredga and Sandberg (1964), however, may give biased information about the chromosome differences between species, because it separates continuous arm length variations into discrete classes (metacentric, telocentric,

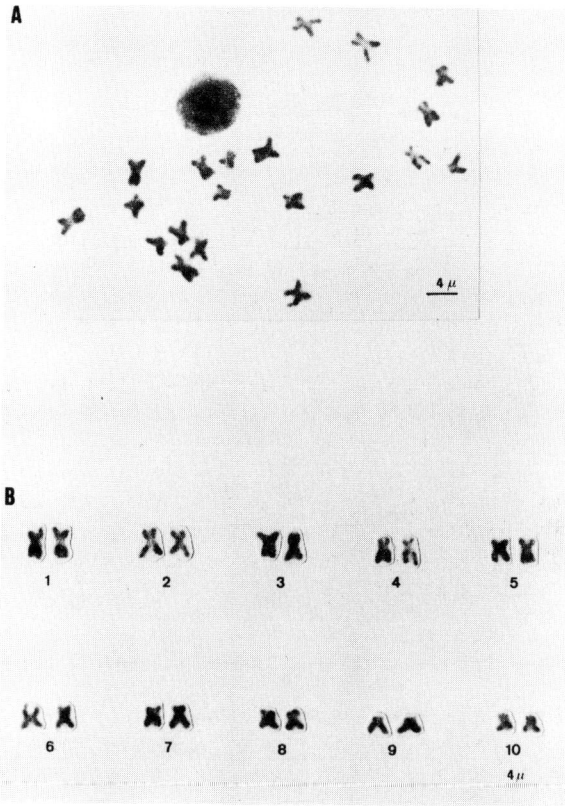

Figure 7.3.1 Chromosomes of *T. chilensis*. A = metaphase plate, Giemsa staining; B = karyotype of *T. chilensis*.

Table 7.3.1 *T. chilensis* chromosomes, mean relative length, 95% confidence interval (CI), and standard deviation (SD) of the long arm (LA) and short arm (SA)

Chrom. pair	$SA \pm CI$	SD	$LA \pm CI$	SD	C	SD	Morph. class
1	5.94 ± 0.37	0.79	7.52 ± 0.47	1.00	44.18	3.97	m
2	5.31 ± 0.28	0.60	6.79 ± 0.31	0.67	43.88	4.35	m
3	5.05 ± 0.29	0.62	6.50 ± 0.29	0.62	43.68	4.21	m
4	3.87 ± 0.35	0.75	6.89 ± 0.35	0.75	35.88	6.23	sm
5	4.59 ± 0.26	0.55	5.85 ± 0.27	0.57	43.94	4.52	m
6	4.04 ± 0.26	0.56	5.47 ± 0.26	0.57	42.44	5.26	m
7	3.28 ± 0.27	0.58	5.92 ± 0.31	0.66	35.66	5.66	sm
8	3.71 ± 0.31	0.67	4.79 ± 0.25	0.53	43.51	5.32	m
9	2.51 ± 0.29	0.63	5.10 ± 0.36	0.77	32.75	7.67	sm
10	2.79 ± 0.29	0.61	4.07 ± 0.30	0.64	40.58	6.63	m

Morphological classification according to the centromeric index (*C*); 26 metaphase plates analysed; 52 chromosomes measured for each mean value. m = metacentric; sm = submetacentric.

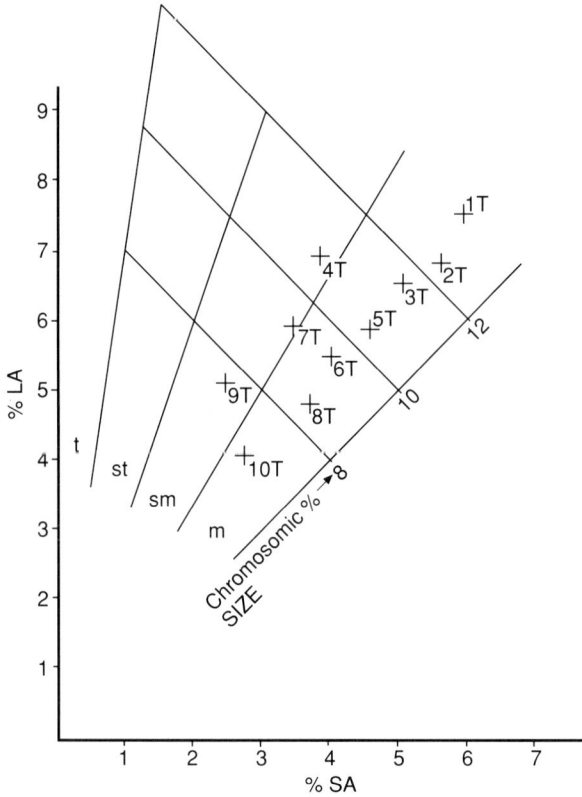

Figure 7.3.2 Mean karyo-idiogram of *T. chilensis*. Each point in the graph represents a mean value of the length of the short and the long arm of 52 chromosomes (26 homologous pairs), Bars represent 95% confidence intervals of the means. t = telocentric; st = subtelocentric; sm = submetacentric; m = metacentric; LA = long arm; SA = short arm.

etc.) and does not consider total length. Certain chromosomes may have large differences in their centromeric index, but belong to the same class and are therefore considered to have the same morphology. In contrast, other chromosomes may have very similar centromeric indexes, but be placed in adjoining classes and are therefore categorized as having a different morphology. This procedure restricts the phylogenetic value of information available from such analysis. Morphological data presented as mean lengths of short and long arms with standard deviations enable more objective comparison of chromosome morphology with data from other species using simple statistical methods (i.e. '*t*' test).

   The positions of active NORs within the karyotypes of related species have a comparative importance due to the possibility of homologation of carrier chromosomes and, by deduction, of the rest of the karyotype. *T. chilensis* shows a variable number and intensity of NORs between cells within individuals and between individuals. This phenomenon can be related to (1) some kind of regulation mechanism that causes preferential rRNA transcription in one of the two homologues within a pair, or in one of the two NOR carrier pairs; (2) mechanisms of dosage compensation (Goldman *et al.*, 1983); or (3) polymorphism in the number of ribosomic cistrons segregating in the population (McGregor, Vlad and Barnett 1977; Miller and Brown, 1969; Sato, Ohta and Kuroki, 1980).

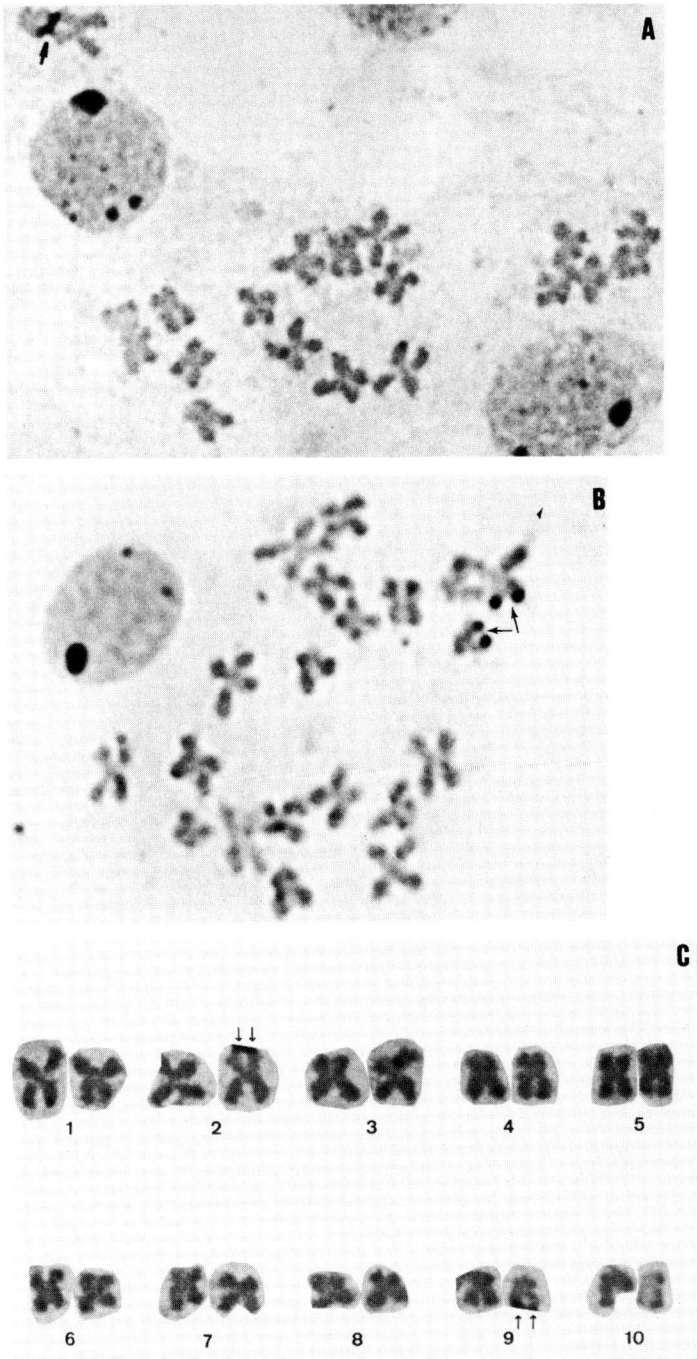

Figure 7.3.3 (A,B) Metaphase plates silver stained to determine active NORs. (A) The top arrow (top left) indicates the two active NORs in two different chromosomes, associated by recently synthesized nucleolar components. (B) Each arrow indicates one of the two active NORs. (C) Karyotype silver stained. NOR carrier chromosomes identified as chromosomes 2 and 9 (arrows). Only one homologous chromosome of each pair stained for synthetic activity in the previous interphase.

Table 7.3.2 *T. chilensis* chromosomes: frequency and distribution of silver-staining NORs according to their position and number on homologous chromosome pairs 2 and 9

Pair 2	Pair 9	No. of cells found	Percentage
1	0	2	4.76
1	Doubtful (0)	11	26.19
Doubtful (0)	1	2	4.76
1	1	22	52.38
2	2	5	11.90

The occasional doubtful characterization of NORs is due to very weak staining and possible confusion with other staining particles inside the metaphase plate.

The situation of variable number and size of NORs within oyster species has been demonstrated in the two other larviparous oyster species studied (Thiriot-Quiévreux and Insua, 1992). Nevertheless, despite the heteromorphism, the position of NORs within karyotypes is generally species specific. However, strong variation is sometimes observed in the location of the NOR on the chromosome among species. According to our results, in *T. chilensis* the NORs are both terminally located in the metacentric pair 2 and in the long arm of the submetacentric pair 9. The NOR locations in other oyster species so far studied are: *O. denselamellosa*, terminal in metacentric pairs 3 and 8; *O. edulis*, terminal in the long arms of the submetacentric pairs 9 and 10; and *Crassostrea gigas*, terminal in the metacentric pair 10 (Insua and Thiriot-Quiévreux, 1991; Thiriot-Quiévreux and Insua, 1992). *O. edulis* shares the NOR positions in chromosome pair 10 with *C. gigas* and in pair 9 with *T. chilensis*.

Small differences in chromosome size and morphology, and the inherent measurement error of the technique, can lead to the location of the same chromosome pair in different positions in the karyotype. Chromosome pairs 2 of *T. chilensis* and 3 of *O. denselamellosa* are very similar in size and could be homologous chromosomes. Pairs 8 and 9 show different size and shape among these species, suggesting that they are not homologous. On the other hand, pair 9 has the same morphology in *T. chilensis* and *O. edulis* (Thiriot-Quiévreux, 1984b), and the NOR is located in an equivalent position.

The comparison of chromosome morphology between brooding oysters suggests that *T. chilensis* and *O. denselamellosa* are more closely related to each other than either are to *O. edulis*. NOR position data, by contrast, put *T. chilensis* closer to *O. edulis* than *O. denselamellosa*. More extensive studies are necessary to define the evolutionary relationships between oyster species based on karyological data.

Acknowledgements

The authors are grateful to Andy Beaumont, Dr Catherine Thiriot-Quiévreux, Dr Lafayette Eaton and Dr Elisabeth von Brand for reviewing the manuscript. This research has been supported partly by the first author's own funds and partly by DGI UCN No. 021.

References

Chanley, P. and Dinamani, P. (1980) Comparative description of some oyster larvae from New Zealand and Chile, and a description of a new genus of oyster, *Tiostrea*. *N.Z. J. Mar. Freshwat. Res.*, **14**, 103–20.

Denver Conference (1966) in *Chicago Conference: Standardization in Human Cytogenetics, Birth Defects*. Original Article Series, **2**, 12–6.

Goldman, M.A., Lo Verde, P.T., Chrisman, C.L. *et al.* (1983) Nucleolar organizer regions in *Biomphalaria* and *Bulinus* snail. *Experientia*, **39**, 911–13.

Goodpasture, C. and Bloom, S.E. (1975) Vizualization of nucleolar organizer regions in mammalian chromosomes using silver staining. *Chromosoma*, **52**, 37–50.

Harry, H. (1985) Synopsis of the supraspecific classification of living Oysters (Bivalvia, Grypheidae and Streidae). *Veliger*, **28**, 121–58.

Ieyama, H. (1990) Chromosomes of the oysters, *Hyotissa imbricata* and *Dendrostrea folium* (Bivalvia, Pteriomorphia). *Venus*, **49**, 63–8.

Insua, A. and Thiriot-Quiévreux, C. (1991) The characterization of *Ostrea denselamellosa* (Mollusca, Bivalvia) chromosomes: karyotype, constitutive heterochromatin and nucleolus organizer regions. *Aquaculture*, **97**, 317–25.

Levan, A., Fredga, A. and Sandberg, A. (1964) Nomenclature for centromeric position of chromosomes. *Hereditas*, **52**, 201–20.

McGregor, H., Vlad, M. and Barnett, L. (1977) An investigation of some problems concerning nucleolus organizers in salamanders. *Chromosoma*, **59**, 283–9.

Mehra, R.C., Brokrus, S. and Butler, M.G. (1985) A simple two-step procedure for silver staining nucleolus organizer regions in plant chromosomes. *Can. J. Genet. Cytol.*, **27**, 255–7.

Miller, L. and Brown, D. (1969) Variation in the activity of nucleolar organizers and their ribosomal gene content. *Chromosoma*, **28**, 430–44.

Palma, C. (1990) *Orígen y algunas consecuencias fenotípicas a nivel celular de una poliploidía natural en especies chilenas de Phicella (Amarillidaceae)*. MSc thesis, Universidad de Chile, Santiago, Chile.

Rufas, J.S., Iturra, P., De Sousa, W. and Sponda, P. (1982) Simple silver staining procedure for the location of nucleolus and nucleolar organizer under light and electron microscopy. *Arch. Biol.*, **93**, 267–74.

Sato, S., Ohta, S. and Kuroki, Y. (1980) Heteromorphic appearance of acrocentric nucleolus organizer in *Nothoscordum fragans*. *Cytologia*, **45**, 87–96.

Spotorno, A.E., Fernández-Donoso, R. and Pincheira, J. (1979) Similitud cromosómica: un nuevo método cuantitativo de descripción. *Arch. Biol. Med. Exp.*, **12**, 223.

Thiriot-Quiévreux, C. (1984a) Les caryotypes de quelques Ostreidae et Mytilidae. *Malacologia*, **25**, 465–76.

Thiriot-Quiévreux, C. (1984b) Analise comparée des caryotypes d'Ostreidae (Bivalvia). *Cah. Biol. Mar.*, **XXV**, 407–18

Thiriot-Quiévreux, C. and Insua, A. (1992) Nucleolar organiser region variation in the chromosomes of three oyster species. *J. Exp. Mar. Biol. Ecol.*, **157**, 33–40.

7.4 CHROMOSOMAL RELATIONSHIPS OF THE ENDEMIC AMPHIPODA (CRUSTACEA) IN THE ANCIENT LAKES OHRID AND BAIKAL

Heikki Salemaa and Ravil Kamaltynov

Abstract

Lake Ohrid in Macedonia and Siberian Baikal are deep freshwater basins with exceptional evolutionary age and rich endemic invertebrate fauna. Nine species of Gammarus *occur in the Ohrid valley, seven of them belonging to the endemic* G. ochridensis *species group. The chromosome number of* G. roeseli *and* G. balcanicus, *species with wider European distribution, is a typical gammaridean haploid one of* n = 26. *A different chromosomal constitution was found in the* G. ochridensis *complex. A new species,* G. salemaai *G. Kar., has the lowest chromosome number,* n = 12, *ever observed in Gammaridae. Other exceptional numbers were found in* G. macedonicus *(n = 21) and in* G. lychnidensis *(n = 34). Haploid number* n = 25 *was counted for the four species of the* G. ochridensis *group. The observed haploid series* n = 12, 25 *and* 34, *suggesting polyploidy, is highly different to the modal number (*n = 26) *established for members of the genus* Gammarus *so far investigated.*

The amphipod species living in the Siberian lake Baikal display extreme morphological and ecological diversification when compared to other freshwater gammarids. However, after investigating the chromosome numbers in 32 species and two subspecies in 18 endemic *genera (*Micruropus, Poekilogammarus, Hyallellopsis, Pallasea, Acanthogammarus, Philolimnogammarus, Eulimnogammarus, Spinacanthus, Parapallasea, Plesiogammarus, Garjajewia, Boeckaxelia, Ommatogammarus, Echiuropus, Heterogammarus, Brandtia, Gmelinoides *and* Macrohectopus*), corresponding to 13% of the Baikalian species and 37% of the genera, practically only one basic number was found. This is* n = 26, *typical of the freshwater genus* Gammarus. *The only exceptional karyotype in lake Baikal was found in the genus* Echiuropus, *with the haploid number* n = 32.

Our results suggest different roles of karyological mechanisms in the speciation process of endemic species groups inhabiting lakes Ohrid and Baikal.

7.4.1 Introduction

Crustaceans display extraordinary complexity in their structural and developmental organization compared to other Arthropoda. On the karyological level, remarkable variability is also characteristic, their haploid chromosome numbers varying between $n = 3$ in the copepod genera *Centropages* and *Temora* (Goswami and Goswami, 1974) to $n = 188$ in the crayfish *Astacus trowbridgei* (Niiyama, 1962) – the widest range found in the animal kingdom (White, 1973). Correlation between karyological features and higher taxonomical categories is evident in Crustacea; high and variable haploid numbers in Decapoda (Niiyama, 1959) do not overlap with the numbers observed in Peracarida (Isopoda and Amphipoda), Maxillopoda (Copepoda and Cirripedia) or Branchiopoda (Cladocera) (Hedgecock, Tracey and Nelson, 1982; Section 7.1). Modal numbers are distinct in most intensively examined species groups. Exceptionally high chromosome numbers found in lower crustaceans have also been found in the polyploidic parthenogenetic genus *Artemia* (Barigozzi, 1974) and in Mysidacea (Holmquist, 1959; Salemaa, 1986), which seem to be the connecting link between peracarids and higher decapods based on the karyological evidence.

Chromosome studies in Amphipoda have focused on the Gammaridae, with only a few species being studied in other families. The early studies published before 1980 by Poisson and Le Calvez (1948), Le Calvez (1949), Le Calvez and Certain (1951) and Orian and Callan (1957) strongly indicated low rates of karyological evolution, the haploid set ($n = 25–32$) varying around the modal number $n = 26$. Also, supernumerary chromosomes were found in many species (Orian and Callan, 1957). Salemaa and Heino (1990) found the haploid $n = 26$ in Finnish populations of *Pallasea quadrispinosa*, the glacial immigrant species of Baikalian origin (cf. Segerstråle, 1957). Studies in families Melitidae (Le Calvez, 1949), Talitridae (Poission and Le Calvez, 1948; Le Calvez and Certain, 1951), Photidae and Calliopiidae (Salemaa, 1986) and in suborders Hyperiidea (Laval and Lecher, 1975) and Corophiidea (Salemaa, 1986) have revealed much wider variation in chromosome numbers ($n = 9–30$) in marine amphipods than in the gammarid family which originated from fresh water. Another example of striking karyological difference between closely related amphipod species (cf. Bousfield, 1989) was found by Salemaa (1984) in the genus *Pontoporeia* (= *Monoporeia*, Bousfield, 1989) in the Baltic Sea. The sub-Arctic marine species *P. femorata* has a haploid chromosome number of $n = 14$ and the glacial immigrant, *P. affinis*, $n = 26$. Salemaa (1984) proposed that *P. affinis* evolved from *P. femorata* in the Ice Age, suggesting occasional polyploidy as an evolutionary mechanism in some amphipod crustaceans. Väinölä and Varvio (1989) have rejected this hypothesis and indicated in their allozyme studies that the divergence between these two species took place several million years ago.

The endemic species group of Gammaridae are typical inhabitants of the Eurasian ancient lakes (Stancovic, 1960; Kozhov, 1963). Endemic species swarms in old lakes have been considered as examples of autochtonous intralacustrine speciation (cf. Mayr, 1965) and provide exceptional cases for karyological studies. The aim of this study is to review results on the karyological features of gammaridean fauna in these extraordinary mountain lakes. Karyological features of the gammarids and endemic *Asellus* species from lake Ohrid have been reviewed by Salemaa (1985). Karyological observations from the Baikalian amphipod fauna have not been published before.

Macedonian Lake Ohrid in the middle of the Balkan Peninsula, the oldest of the European freshwater basins, harbours the *G. ochridensis* complex of seven endemic *Gammarus* species (Karaman, 1977, 1985). Another two species, *G. roeseli* and *G. balcanicus*, with wider geographic distribution, also occur in the Ohrid valley. The species have characteristic habitat preferences, and are specialized to live in profundal communities in depths more than 285 m up to sublittoral *Chara* and *Dreissena* communities and karstic sources in the vicinity of the shore.

Siberian Baikal is the deepest lake in the world (more than 1640 m), containing 20% of all fresh water on the earth. The number of endemic animal species exceeds those in all other isolated lake basins. The diversity of the gammaridean crustaceans is extraordinary: Lake Baikal contains 46 genera with more than 259 species and 80 additional subspecies of Amphipoda (Bazikalova, 1945; Barnard and Barnard, 1983; Kamaltynov, 1992). It means that about 20% of all amphipod species occurring in fresh waters inhabit lake Baikal only. Significant ecological and morphological diversification is typical of Baikalian gammarid fauna, with a variety of species of different sizes, colours, spination and habitats from the abyssal waters to the shallow littoral and even open pelagic areas. The origin of Baikalian fauna has been a matter of debate for a long time (cf. Brooks, 1950; Kozhov, 1963). We agree with the view that the high biodiversity of the benthos in the deep ancient lakes is maintained by their geo-ecological stability, but emphasize, on the other hand, the role of habitat heterogeneity, especially in the littoral zone. Whether the amphipod genera evolved in continental fresh waters or whether they originated from marine progenitors was not known until recently. Barnard and Barnard (1983) concluded that the family Gammaridae evolved in fresh waters.

7.4.2 Materials and methods

Investigations on the chromosome numbers of gammaridean amphipods were carried out in the Ohrid area in the early summer of 1984 (Salemaa, 1985) and at lake Baikal in the autumn of 1988. Amphipods were collected from their characteristic habitats using hand nets in the littoral or by dredging from deeper water. Individuals used in the investigations were mature males. Standard karyological techniques were applied for crustacean material (Salemaa, 1979, 1984, 1986) according to to the following schedule.

Testicular tissues were dissected in the natural lake water, immersed for 30 s in distilled water for hypotonic shock and fixed in acetoethanol (1 : 3). Tubular testicles were stained for 15 min to 3 h in propionic-orcein (2% orcein dissolved in 45% propionic acid). Material was squashed gently in a drop of 45% acetic acid under the coverslip and mounted in glycerin. This simple technique made it possible to make reliable chromosome counts from mitotic divisions and meiotic configurations even in the field. Several metaphase plates from each species were examined.

7.4.3 Results

The studied species, their characteristic habitats and the haploid chromosome numbers are given in Tables 7.4.1 and 7.4.2 and examples of metaphase plates are illustrated in Figures 7.4.1 and 7.4.2.

Table 7.4.1 Chromosomal relationships of the endemic Amphipoda in lake Ohrid

Species and chromosome numbers	Habitat preferences

Gammarus ochridensis Schäferna,
  n = 25
	Lake Ohrid, littoral, under stones,
	  less than 0.5 m

Gammarus parechiniformis G. Karaman,
  n = 25
	Karstic sources at St Naum,
	  littoral vegetation

Gammarus stankokaramani G. Karaman, *n* = 25	Lake Ohrid, benthic 30–50 m

Gammarus solidus G. Karaman,
  n = 25
	Lake Ohrid, deep benthic,
	  below 70 m

Gammarus lychnidensis (Schell),
  n = 34
	Lake Ohrid, *Dreissena* zone,
	  20–30 m

Gammarus macedonicus G. Karaman,
  n = 21
	Lake Ohrid, *Dreissena* and
	  Chara zone, 10–30 m

Gammarus salemaai G. Karaman,
  n = 12
	Lake Ohrid, littoral and shallow
	  sublittoral, under stones

Other gammarids:
Gammarus roeseli Gervais
  f. *triacantha, n* = 26
  f. *meridionalis, n* = 26
	
	Karstic sources at St Naum and
	  Studencista, littoral vegetation
	Lake Ohrid, littoral and *Chara* zone

Gammarus balcanicus Schäferna,
  n = 26
	Karstic sources and streams,
	  under stones and among vegetation

Habitat preferences according to Salemaa (1985).

Spermatogenesis and associated meiotic cell divisions are essentially similar in gammarids from Ohrid and Baikal lakes when compared to earlier investigations of Amphipoda. In mitotic divisions of spermatogonia, mostly metacentric and submetacentric elements were observed. The mitotic metaphase chromosomes are 2–10 μm in length, the equatorial diameter of the metaphase plates varying between 30 and 50 μm. Acrocentric pairs with satellites and associated nucleolar elements were found in many species as well as acentric supernumerary bodies outside the metaphase plates. The final haploid counts were easy to make from meiotic divisions (Figures 7.4.1 and 7.4.2). Depending on the stage of meiotic I division, the shape and size of the bivalents varied from bipartite dumbbells to crosses and rings with two chiasmata. Metaphase bivalents in gammarids are tight, 1–3 μm in size. Because of the large number of chromosomes and their small size it was not possible to use special staining techniques or identify individual chromosomes.

The chromosome numbers (Table 7.4.1 and Figure 7.4.1) in the endemic *ochridensis* species group were different from those of *G. roeseli* and *G. balcanicus*, species with a wider European distribution, which have the typical gammaridean haploid number of *n* = 26. An exceptional chromosomal constitution was found in other *Gammarus* species living in the Ohrid valley. A new species, *G. salemaai* G. Kar, was found with the lowest chromosome number, *n* = 12, ever observed in Gammaridae. Other exceptional numbers were found in *G. macedonicus* (*n* = 21) and in *G. lychnidensis* (*n* = 34). Haploid number *n* = 25 was counted for *G. ochridensis*, *G. solidus*, *G. parechiniformis* and *G. stankokaramani*, the four species of the *G. ochridensis* species group. The observed haploid series *n* = 12, 25 and 34 in the endemic congeners is in contradiction to the modal number of *n* = 26 previously established for the genus *Gammarus*, which strongly suggests polyploidy in this isolated group.

Table 7.4.2 Chromosomal relationships of the endemic Amphipoda in lake Baikal

Species and chromosome numbers	Habitat preferences
Micruropus wohli platycercus Dybowsky, $n = 26$	All Baikal, most common 0–5 m, in canyons 10–154 m, sand bottoms
Micruropus vortex vortex Dybowsky, $n = 26$	All Baikal, rivers Angara and Jenisei, 0.5–88 m, most common 10 m, stone and sand bottoms
Echiuropus macronychus macronychus Sowinsky, $n = 32$	All Baikal, 1–100 m, sand bottoms
E. morawitzi Dybowsky, $n = 26$	All Baikal, 1.5–108 m, most common 10 m, sand and silt bottoms
E. rhodophthalamus microphthalamus Dybowsky, $n = 26$	All Baikal, 1–1350 m, most common 10–100 m, silt bottoms
E. seidlitzi Dybowsky, $n = 26$	All Baikal, 7–1130 m, most common 50 m, silt bottoms
Gmelinoides fasciatus Stebbing, $n = 26$	All Baikal, rivers Angara and Jenisei, 0–5 m, in canyons to 192 m, nectobenthic, sand and stone bottoms
Brandtia lata latior Dybowsky, $n = 26$	All Baikal, 0.5–50 m, stone bottoms
Hyalellopsis taczanowskii Dybowsky, $n = 26$	Southern Baikal, 10–150 m, sand bottom
Boeckaxelia carpenteri carpenteri Dybowsky, $n = 26$	All Baikal, 1–336 m, most common 20–100 m, silt bottoms
B. carpenteri elegans Dybowsky, $n = 26$	Northern Baikal, Maloe More, 18–106 m, silt and sand bottoms
Acanthogammarus (A.) victorii victorii Dybowsky, $n = 26$	Southern Baikal, 1.5–156 m, most common 100 m sand and stone bottoms
A. (A.) v. maculosus Dorogostaisky, $n = 26$	Northern Baikal, Maloe More, 5–150 m, most common 10–60 m, sand and stone bottoms
A. (A.) grewingki Dybowsky, $n = 26$	Baikal, sublittoral
A. (Brachyuropus) flawus flawus Garjajeff, $n = 26$	Maloe More, 5–61 m, sand bottom
Spinacanthus parasiticus Dybowsky, $n = 26$	All Baikal, 1.5–60 m, ectoparasites on Porifera
Garjajewia cabanisi cabanisi Dybowsky, $n = 26$	All Baikal, 35–1250 m, most common 100 m, silt bottoms
Plesiogammarus gerstaeckeri Dybowsky, $n = 26$	All Baikal, 3.5–1610 m, most common 40 m, silt bottoms
Pallasea (Pallasea) cancelloides Gerstfeldt, $n = 26$	All Baikal, river Angara, 0.3–178 m, most common 1–10 m, sand bottoms
Pallasea (P.) kessleri Dybowsky, $n = 26$	All Baikal, 1–61 m, most common 20 m, sand and stone bottoms
Pallasea (P.) grubei Dybowsky, $n = 26$	All Baikal, 1–175 m, most common 40 m, sand bottoms
Pallasea (Pentagonurus) brandti Dybowsky, $n = 26$	All Baikal, 1.5–442 m, most common 10–60 m, sand and stone bottoms
Poekilogammarus (Onychogammarus) jedorensis Bazikalova, $n = 26$	Maloe More, 40–55 m, sand bottoms
Macrohectopus branickii Dybowsky, $n = 26$	All Baikal, pelagic, 0–1620 m
Palicarinus puzylli puzylli Dybowsky, $n = 26$	All Baikal, 2.5–310 m, most common 50–200 m, sand and silt bottoms
P. p. carinulata Dorogostaisky, $n = 26$	Middle and northern Baikal, 3.5–250 m, most common 30–60 m, sand and silt bottoms
Eulimnogammarus verrucosus Gerstfeldt, $n = 26$	All Baikal, 0–12 m, most common 0.2–1.5 m, under stones
E. lividus Dybowsky, $n = 26$	All Baikal, 0–10 m, under stones
E. maacki Gerstfeldt, $n = 26$	All Baikal, river Angara, 0–15 m, under stones
Philolimnogammarus marituji Bazikalova, $n = 26$	Southern Baikal, 0.2–3 m, under stones
Ph. vittatus Dybowsky, $n = 26$	All Baikal, most common 0.5–3 m, under stones

Table 7.4.2   (*contd*)

Species and chromosome numbers	Habitat preferences
Ph. cyaneus Dybowsky, $n = 26$	All Baikal, river Angara, 0–20 m, most common 0–0.2 m, under stones
Eurybiogammarus ussolzewi Dybowsky, $n = 26$	All Baikal, 36–697 m, silt bottoms
Ommatogammarus flavus Dybowsky, $n = 26$	All Baikal, 2.5–1313 m, most common 100–600 m, nectobenthic, silt bottoms

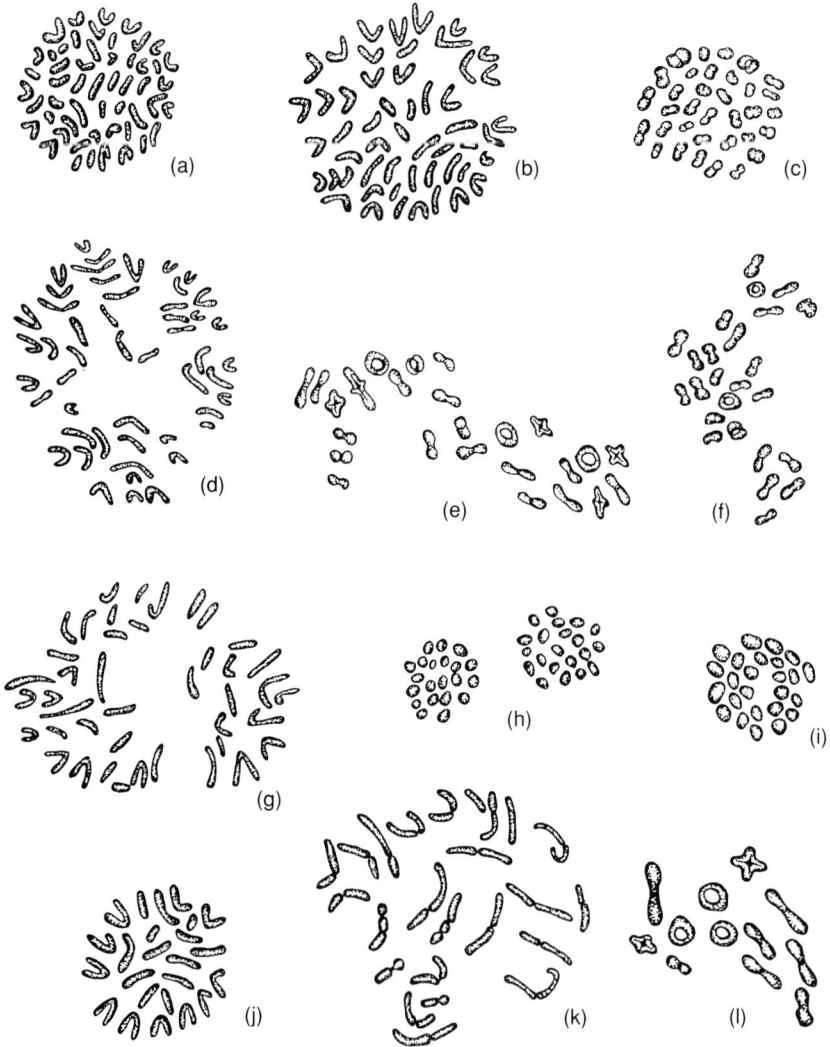

Figure 7.4.1 Chromosomes of endemic gammarids from Lake Ohrid. *Gammarus balcanicus*, $2n = 52$ (a); *G. roeseli f. triacantha*, $2n = 52$ (b); *G. lychnidensis*, $n = 34$ (c); *G. ochridensis*, $2n = 50$ (d) and $n = 25$ (e); *G. stankokarami*, $n = 25$ (f); *G. macedonicus*, $2n = 42$ (g) and $n = 21$ (h); *G. solidus*, $n = 25$ (i); and *G. salemaai*, $2n = 24$ (j,k) and $n = 12$ (l). The diameters of the I meiotic metaphase average 15–20 μm and goniomitosis 20–30 μm.

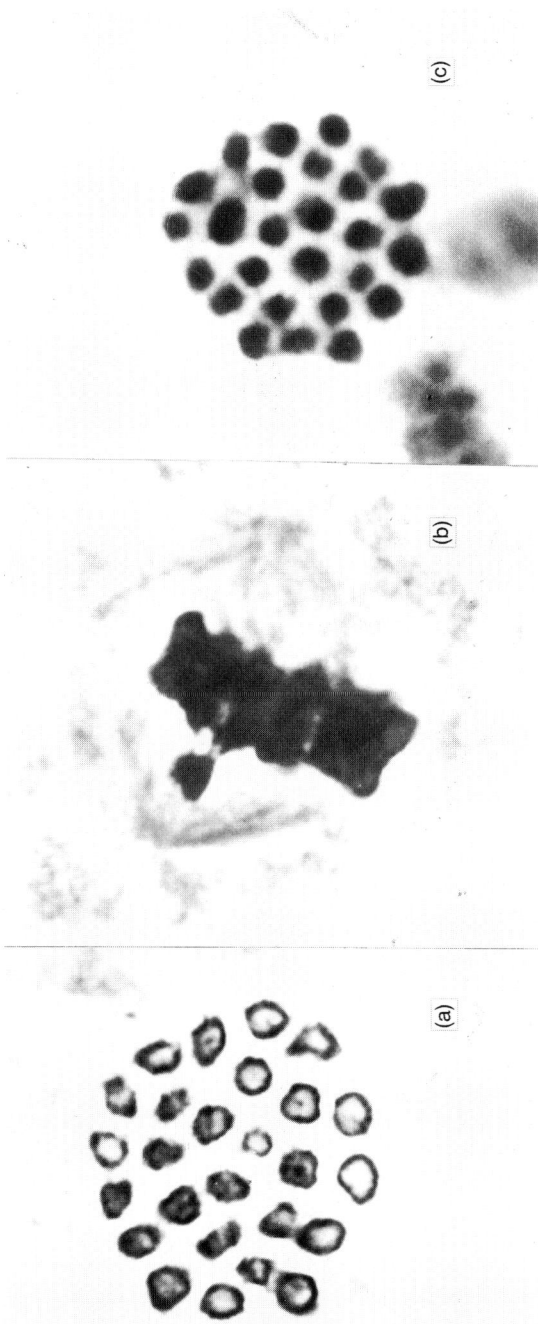

Figure 7.4.2 Chromosomes of Baikalian amphipods *Eulimnogammarus verrucosus*, $n = 26$. (a), *Eurybiogammarus ussolzewi*; (b) I meiotic metaphase (a side view with B-chromosome); and (c) *Macrohectopus branicki*, $n = 26$. The diameters of I meiotic metaphase average 15–20 µm.

The amphipod species living in lake Baikal display extreme morphological and ecological diversification when compared to all other freshwater gammarids. Consequently, the karyological inertia found in the Baikalian endemics is highly surprising. After investigating the chromosome numbers in 32 species and two subspecies (Table 7.4.2) in 18 endemic genera (*Micruropus, Poekilogammarus, Hyallellopsis, Pallasea, Acanthogammarus, Philolimnogammarus, Eulimnogammarus, Spinacanthus, Parapallasea, Plesiogammarus, Garjajewia, Boeckaxelia, Ommatogammarus, Echiuropus, Heterogammarus, Brandtia, Gmelinoides* and *Macrohectopus*) corresponding to 13% of the Baikalian species and 37% of the genera, practically only one basic number was found (Table 7.4.2, Figure 7.4.2). This is $n = 26$, typical of the freshwater genus *Gammarus*. The only exceptional karyotype found in Baikal is that in the genus *Echiuropus* with the haploid number $n = 32$.

7.4.4 Discussion

Before our investigations in lake Baikal, information on the karyology in Amphipoda was limited to less than 60 species (1% of total number of the species) in 19 genera. Consequently, the knowledge of chromosomal relationships in these crustaceans, especially for marine families, is far from complete.

Our results confirm the existence of stable symmetrical karyotypes in freshwater gammarids: the morphologically and ecologically diversified Baikalian amphipod genera revealed the same basic number established in the most intensively investigated genus, *Gammarus*. Karyological stability in Baikalian gammarid genera ($n = 26$) is exceptionally complete when compared with other examined species groups. Lop (1989) studied karyological features of Iberian populations of seven species in the *Echinogammarus berilloni* group and found varying haploid numbers: $n = 25 – 28$ for three species and $n = 27$ for four species. Similarly, four species of the *ochridensis* complex have a slightly different number of $n = 25$, established also in Talitridae (Poisson and Le Calvez, 1948; Le Calvez and Certain, 1951). In addition, supernumerary chromosomes, which were frequently found also in Baikalian species, cause slight variation in the modal number. Observations of the marine amphipod genera *Anisogammarus* ($n = 27$) (Niiyama, 1950), *Calliopius* ($n = 9$), *Gammaracanthus* ($n = 19$), *Pontoporeia* ($n = 14–26$), *Leptocheirus* ($n = 11$) and *Corophium* ($n = 14$) (Salemaa, 1984, 1986; Salemaa and Heino, 1990) suggest that karyological differentiation took place at an early stage of Amphipod evolution. On the other hand, this also explains the uniformity of Baikalian karyotypes provided that the gammarid fauna in lake Baikal share the same ancient freshwater origin. Our results support this possibility.

The role of karyological evolution in the genetic diversification of endemic Gammaridae is evidently different in lake Ohrid and lake Baikal. Although both lakes are old, deep and stable they are different in terms of the evolution of benthic animals. Heterogeneous environments with littoral springs and accessibility to underground karstic waters increase the possibility for isolated refuges during unfavourable epochs in the Ohrid valley. It is also possible that the karyologically diversified gammarids in Ohrid originate from several ancient immigration waves, *G. salemaai* with lowest haploid number ($n = 12$) representing the oldest palaeoendemic type surviving in the area. At the moment the most tempting explanation for the exceptional series of chromosome numbers $n = 12, 21, 25$ and 34 is polyploidy. The only deviation from the basic number in lake Baikal, i.e. $n = 32$ in *Echiuropus macronychus*, is also too great to be explained by means of B-chromosomes, aneuploidy or Robertsonian variation.

Acknowledgements

The authors are grateful to Professor J. Serafimova, the Hydrobiological Station, Ohrid, and Professor M. Grachev, Institute of Limnology, Siberian Branch of the Russian Academy of Sciences, Irkutsk, for facilities to carry out this investigation.

References

Barigozzi, C. (1974) *Artemia*: a survey of its significance in genetic problems. *Evol. Biol.*, **7**, 221–52.

Barnard, J.L. and Barnard, C.M. (1983) *Freshwater Amphiphoda of the World: I Evolutionary Patterns, II Handbook and Bibliography*, Hayfield Associates, Mt Vernon, Virginia.

Bazikalova, A. (1945) Les amphipodes du Baikal. *Tr. Baik. Limnol. Stn. Akad. Nauk. SSSR*, **11**, 3–440.

Bazikalova, A. (1962) Sistematika, ekologiia i rasprostranenie rodov *Micruropus* Stebbibg i *Pseudomicruropus* nov. gen. (Amphipoda, Gammaridea). Sistematika i Ekologiia Rakoobraznykh Baikala. *Tr. Limnol. Instituta*, **2** (22) 3–140.

Bazikalova, A. (1971) Donnaia fauna. Limnologiia prideltovykh prostranstv Baikala. *Tr. Limnol. Instituta*, **12** (32), 95–114.

Bazikalova, A. (1975) K sistematike baikalskikh amfipod (rody *Carinogammarus* Stebbing, *Eucarinogammarus* Sowinsky, *Echiuropus* (Sow.) i *Asprogammarus* gen.n.). Novoe o faune Baikala. *Tr. Limnol. Instituta*, **18** (38), 31–81.

Bekman, M. (1984) Glubokovodnaia fauns amfipod. *Sistematika i evolutsilla bespozvonochnykh Baikala*, Nauka C, pp. 114–23, Novosibirsk.

Bousfield, E.L. (1989) Revised morphological relationships within the amphipod genera *Pontoporeia* and *Gammaracanthus* and the 'glacial relict' significance of their postglacial distributions. *Can. J. Fish. Aquat. Sci.*, **46**, 1714–25.

Brooks, J.L. (1950) Speciation in ancient lakes. *Q. Rev. Biol.*, **25**, 30–176.

Goswami, U. and Goswami, S.C. (1974) Cytotaxonomical studies on some calanoid copepods. *The Nucleus*, **17**, 109–13.

Hedgecock, D., Tracey, M.L. and Nelson, K. (1982) Genetics, in *The Biology of Crustacea. 2. Embryology, Morphology and Genetics* (ed. L.G. Abele), pp. 283–403, Academic Press, New York.

Holmquist, C. (1959) *Problems on Marine-glacial Relicts on Account of Investigations on the Genus* Mysis, Berlingska Boktryckeriet, Lund, Sweden.

Kamaltynov, R. (1992) On the present state of systematics of the Lake Baikal amphipods (Crustacea, Amphipoda), *Zool. Zhu.*, **71**, 150–3.

Kaplina G.S. (1970) Zoobenthos Iuchnogo Baikala v raione Utulik-Murina. Izv. *Biologo-geografıcheskogo nauchno issledovatelskogo Insituta pri Irkutskom gosuniversitete*, **23**(1), 42–64.

Karaman, G.S. (1977) Contribution to the knowledge of the Amphipoda 77. *Gammarus ochridensis* Schäf. species complex of Ohrid lake. *The Montenegrin Academy of Sciences and Arts. Glasnik of the Section of Natural Sciences*, **2**, 49–89.

Karaman, G.S. (1985) Contribution to the knowledge of the Amphipoda 151. *Gammarus salemaai*, new species from lake Ohrid (Fam. Gammaridae). *Fragm. Balcan.*, **12**, 155–68.

Kozhov, M. (1963) Lake Baikal and its life. *Monogr. Biol.*, **XI**.

Laval, Ph. and Lécher, P. (1975) Caryotypes, chromosomes surnumeraires, parthenogenese rudimentaire et polyploidie chez deux espèces du genre *Phronima* (Crustacés, Amphipodes). *Can. J. Genet. Cytol.*, **17**, 405–12.

Le Calvez, J. (1949) Quelques nouveaux caryotypes de Crustacés Amphipodes. *C.R. Acad. Sci., (Paris)*, **228**, 427–8.

Le Calvez, J. and Certain, P. (1951) *Gammarus chevreuxi* Sext. et la caryologie des gammariens. *Arch. Zool. Exp. Gen.*, **88**, 131–41.

Lop, A.F. (1989) Karyotypes of the *Echinogammarus berilloni*-group (Crustacea, Amphipoda) from Spain. *Genetica*, **79**, 37–43.

Mayr, E. (1965) *Animal Species and Evolution*, The Belknap Press of Harvard University Press, Cambridge.

Niiyama, H. (1950) The X-Y mechanism of the sex chromosome in the male of *Anisogammarus anadalei. Annot. Zool. Jap.*, **23**, 58–62.

Niiyama, H. (1959) A comparative study of the chromosomes in decapods, isopods and amphipods, with some remarks on cytotaxonomy and sex-determination in the Crustacea. *Mem. Fac. Fish. Hokkaido Univ.*, **7**, 1–60.

Niiyama, H. (1962) On the unprecedentedly large number of chromosomes of the crayfish *Astacus trowbridgei* Stimpson. *Annot. Zool. Jap.*, **35**, 229–33.

Orian, A.J.E. and Callan, G. (1957) Chromosome numbers in gammarids. *J. Mar. Biol. Assoc. UK*, **36**, 129–42.

Poisson, R. and Le Calvez, J. (1948) La garniture chromosomique de quelques Crustaces Amphipodes. *C.R. Acad. Sci. (Paris)*, **227**, 228–30.

Salemaa, H. (1979) The chromosomes of *Asellus aquaticus* (L.) – a technique for isopod karyology. *Crustaceana*, **36**, 316–18.

Salemaa, H. (1984) Polyploidy in the evolution of the glacial relict *Pontoporeia* spp. (Amphipoda, Crustacea). *Hereditas*, **100**, 56–60.

Salemaa, H. (1985) Karyological studies in *Gammarus* and *Asellus* species from lake Ohrid. *Station hydrobiologique-Ohrid, Edition jubilaire livre* I, 245–54.

Salemaa, H. (1986) Karyology of the northern Baltic peracaridan Crustacea. *Sarsia*, **71**, 17–25.

Salemaa, H. and Heino, T. (1990) Chromosome numbers of Fennoscandian glacial relict Crustacea. *Ann. Zool. Fenn.*, **27**, 207–10.

Segerstråle, S.G. (1957) On immigration of the glacial relicts of Northern Europe, with remarks on their prehistory. *Soc. Sci. Fenn. Commentat. Biol.*, **16**, 1–117.

Stankovic, S. (1960) The Balkan lake Ohrid and its living world. *Monogr. Biol.*, IX.

White, M.J.D. (1973) *Animal Cytology and Evolution*, Cambridge University Press, Cambridge.

Väinölä, R. and Varvio, S.-L. (1989) Molecular divergence and evolutionary relationships in *Pontoporeia* (Crustacea: Amphipoda). *Can. J. Fish. Aquat. Sci.*, **46**, 1705–13.

7.5 SYMPATRIC SPECIATION IN THE FAMILY CHIRONOMIDAE (DIPTERA)

Paraskeva Michailova

Abstract

The path of sympatric speciation in some groups and sibling species in the family Chironomidae is studied. Based on an analysis of inversions in different chromosome arms, a schematic representation of chromosome evolution in the 'plumosus' group (genus Chironomus*) is given. It was found that the floating inversions in initial standard groupings became fixed in a derived karyotype.*

The role of heterochromatin in sympatric speciation is considered. A change in the amount and localization of heterochromatin has occurred. In addition, hybridization tests revealed that pre and postmating isolating mechanisms were operating.

It is concluded that the species with overlapping ranges are characterized by more complex chromosome changes.

7.5.1 Introduction

Speciation is ultimately an adaptive process that involves establishment of intrinsic barriers to gene flow between closely related populations by development of reproductive isolating mechanisms. A study of speciation is, to a considerable extent, a study of the genetics and evolution of premating and postmating reproductive isolating mechanisms (Mayr, 1970; Bush, 1975; Ayala and Kiger, 1984). In the development of those mechanisms, gene flow is restricted or completely suppressed, thus resulting in the accumulation of gene differences and allowing evolutionary divergence of the two resulting genomes. It is most important to understand the way in which reproductive isolating mechanisms are established. Speciation may occur either by spatial separation of populations (allopatric speciation) or by the emergence of reproductive isolating mechanisms within a population (sympatric speciation). In sympatric speciation a premating reproductive isolating mechanism may arise before a population shifts to a new niche (Bush, 1975). Speciation is therefore related to the biological features of a group of specimens in the population – for instance in their ability to adapt to a certain area or niche in a biotope. It is sometimes possible to trace the way in which a certain species has formed by revealing the reproductive relationships between individuals in a population.

For the family Chironomidae there are no data available on reproductive isolating mechanisms within populations or on the ways in which they might develop. The Chironomidae demonstrate an unusually wide range of ecological adaptations and many species are found in different kinds of water bodies. For this reason, studies on the Chironomidae could play an important role in establishing some general principles of sympatric speciation. The purpose of this section is to explore the reproductive relationships in model species of the family Chironomidae and to investigate to what extent and under what circumstances karyotype changes may initiate sympatric speciation.

7.5.2 Material and methods

The results of analysis were based on examination of two model groups: first, the sibling species of the *plumosus* group of the genus *Chrionomus* which are symbiotopic, and second, the sibling species of genus *Glyptotendipes* which are allobiotopic. The methodology includes external morphological analysis, cytology and hybridization.

(a) External morphological analysis

The species were determined after a detailed analysis of developmental stages (larva, pupa and imago). The pupa and imago were obtained in the laboratory from live larva collected from the wild. For preparations the method described in the literature by Schlee (1968) was employed.

(b) Cytological methods

Routine aceto-orcein staining
Both live larvae and those fixed in glacial acetic acid were used in karyological studies. Polytene chromosomes of salivary glands dissected from live and fixed material were examined using aceto-orceinic staining (Keyl and Keyl, 1959). Preparations of salivary gland polytene chromosomes and larval preparations for the analysis of external morphology were both made from the same individual larvae. The chromosome map of *C. plumosus* prepared by Maximova (1979) was used for comparing banding patterns of species incorporated in the *plumosus* group. Chromosome arms were designated: Ist-AB; IInd-CD; IIIrd-EF; and IVth-G. For sibling species of the genus *Glyptotendipes* the standard chromosome map of *G. barbipes* (Michailova, 1987a) was used.

Differential staining
Particularly important for the appearance of heterochromatin are the procedures of differential staining (C- and Q-banding). For this purpose I used a modified variant of differential staining (Michailova, 1987b). These methods help to obtain information on the biochemical and molecular organization of polytene chromosomes at the cytological level.

(c) Hybridization

Hybridization tests were carried out on species in the *plumosus* group (Fischer, 1969) and on sibling species of *Glyptotendipes* following the modified method of Michailova (1985).

Thus to investigate sympatric speciation in the family Chironomidae a mixture of methodologies has been applied involving a detailed analysis of the external morphology of separate metamorphic stages, cytogenetic analysis using aceto-orcein and differential staining methods, and tests of hybridization.

7.5.3 Results and discussion

Chironomus plumosus is an example in which two aspects of evolution are observed: phyletic evolution and evolution resulting in the formation of sibling species united in a *plumosus* group. Phyletic evolution of *C. plumosus* is related to gradual changes in the karyotype which ensure the adaptive potential of the species and the microevolutionary differentiation of populations. In a number of European populations (Switzerland, Finland, Hungary, Bulgaria, Russia) the species is not considered to be polytypic (Maximova, 1979) but to be highly polymorphous, characterized by the presence of a polymorphous system of inversions which involve the transition of a standard homozygous type through heterozygotization into another homozygous type (Figure 7.5.1; Michailova and Petrova, 1991). Sympatric speciation from the homozygous inversions in the polymorphous system of *C. plumosus* has produced the eight sibling species (*C. agilis, C. balatonicus, C. bonus, C. borokensis, C. entis, C. muratensis, C. nudiventris* and *C. vancouveri*) incorporated in the *plumosus* group. In these species the band pattern of some arms has been formed by fixed homozygous inversions, while in other arms it has resulted from several intermediate homozygous inversions. In both cases these inversions participate in the polymorphous system of *C. plumosus*. The Ist chromosome (AB) of *C. balatonicus* (Figure 7.5.2 (b)) is differentiated from those of *C. plumosus* (Figure 7.5.2(a)) by complex homozygous inversions (Figure 7.5.2). The homozygous inversion in arm A of *C. balatonicus* (Figure 7.5.2 (b)) has been fixed in that state having passed through an intermediate homozygous inversion (AII) involved in the polymorphous system of *C. plumosus* (Figure 7.5.2). The IInd chromosome (CD) (Figure 7.5.2(b)) has gone through transformation, with the homozygous sequence (D_4) being part of the polymorphous system of *C. plumosus* (Figure 7.5.2(a)) and fixed in that state in *C. balatonicus*. In the IIIrd chromosome (EF), a homozygous inversion is observed in the F arm, thus distinguishing the two species (Figure 7.5.2). The same process to that described above can be observed in the formation of *C. muratensis* and *C. nudiventris* but in *C. nudiventris* ($2n = 6$) a translocation of the centromere–telomere fusion type also takes part. All other sibling species in the *plumosus* group (Kerkis *et al.*, 1988; Shobanov and Demin, 1988) have been formed through homozygotization of their karyotypes.

The above examples show that in the formation of sympatric species of the *plumosus* group the fluctuating homozygous and heterozygous inversions of *C. plumosus* turn into fixed homozygous ones characteristic of the derived species. Therefore sympatric species are formed from species of marked polymorphism. In this case polymorphism is a starting point which favours the speciation of symbiotopic species in the *plumosus* group.

The same method of speciation can be observed in allobiotopic species. The sibling species *Glyptotendipes salinus* and *G. barbipes* are examples of species which exhibit allobiotopic distribution within a sympatric range. These species have many karyotypic features in common, e.g. the number of chromosomes, the appearance of the centromere region like a large heterochromatin block, the equal band patterns, and the location of the centromere region. Both species most probably originated from a strongly polymorphous species. With the gradual occupation of different habitats – *G. salinus* in brackish water, *G. barbipes* in freshwater ponds – transformations have occurred in the initial species, resulting in the divergence of both species. The two species are differentiated mainly by homozygous inversions (Michailova, 1987a).

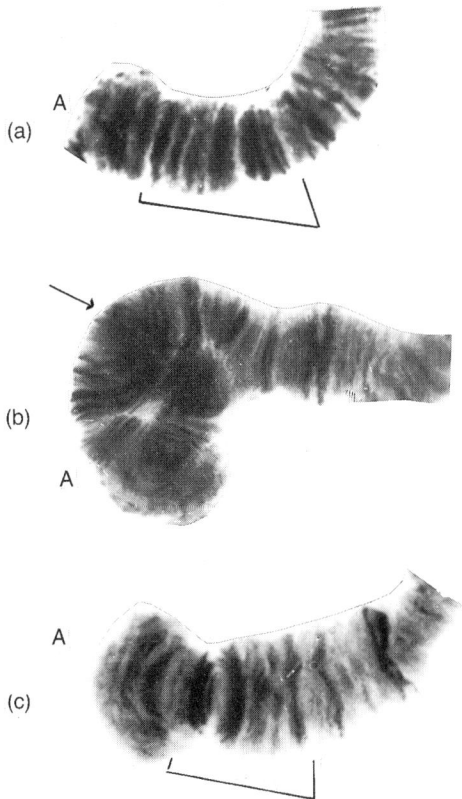

Figure 7.5.1 1st chromosome, arm A of *C. plumosus.* (a) A homozygous inversion; (b) a heterozygous inversion; (c) an alternative homozygous inversion.

How does the process of formation of these species occur? Initially, groups of individuals, each of them with a definite genotypic structure, developed in the population. Depending on the habitat, the standard homokaryotype, heterokaryotype or inverted homokaryotype is predominant. Thus chromosome changes appear to be a protective mechanism for the gene pool of the population. For instance, in heterozygotes of adaptive value, crossing-over is

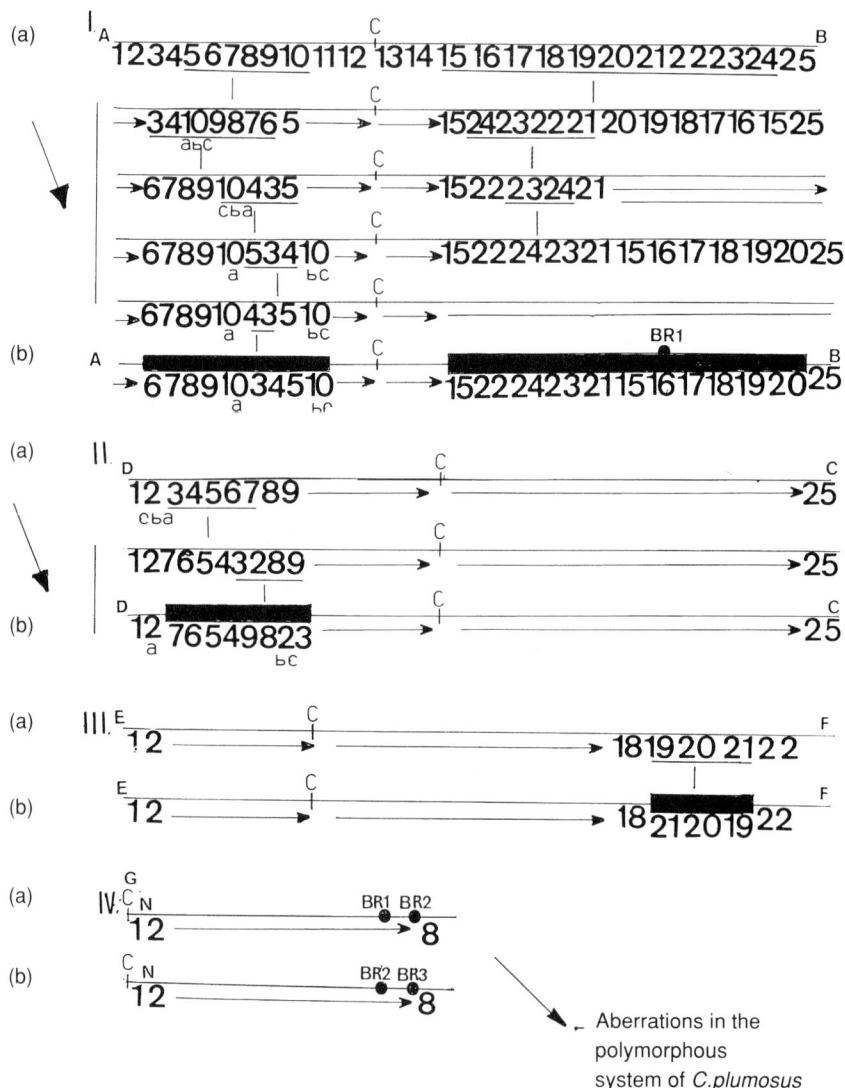

Figure 7.5.2 A scheme of formation of derived species – *C. balatonicus*. (a) Band sequences of 1st (**AB**), IInd (**CD**), IIIrd (**EF**) and IVth (**G**) chromosomes of *C. plumosus*. (b) Band sequences of 1st (**AB**), IInd (**CD**), IIIrd (**EF**) and IVth (**G**) chromosomes of *C. balatonicus*. C, centromere region; N, nucleolus; BR, Balbiani ring.

suppressed, thus allowing the preservation of coadapted gene complexes in the population. Later on, as the result of positive assortative mating, one or the other of the groups of a particular genetic structure is stabilized. In this way heterozygotes pass into a homozygous state. Under certain conditions new to the species these homozygous inversions of adaptive priority are then selected, and will spread and become fixed in the population. The occurrence of a new homozygous karyotype is associated with the restriction of panmixia with the initial alternative homozygous form. This brings about the enhancement of reproductive barriers between alternative homozygous groups in a population and to their independent evolution, notwithstanding the fact that they live in the same range. Thus, at a certain time polymorphism turns into monomorphism; for instance, in the *plumosus* group some homozygous inversions taking part in the polymorphous system of *C. plumosus* become constant characteristics of the karyotypes of sibling species.

This pattern differentiates the family Chironomidae from other families in the order Diptera where species with monomorphous chromosomes are considered to generate speciation (Stegnii, 1984).

Apart from interspecific differences in chromosomal rearrangements, distinctive chromosome heterochromatin is also a species-specific cytological character. There are two primary processes occurring in the formation of these species: (1) transformation of euchromatin; and (2) amplification of heterochromatin. When comparing C-bands of chromosomes among different species in the *plumosus* group, striking heterochromatin differences can be found. In *C. plumosus*, heterochromatin is only found in the centromere regions (Figure 7.5.3(a,b)). Telomeric and interstitial C-bands were consistently found in the chromosomes of *C. balatonicus* (Michailova, 1987b), (Figure 7.5.3 (c)). Many bands which in *C. plumosus* are in the euchromatic state are heterochromatic in *C. balatonicus*. The same process of transformation is observed in the sibling species *G. barbipes* and *G. salinus* (Michailova, 1987c). Thus, a transformation process is an evolutionarily derived state. It is possible that these staining differences reflect either some heterochromatin protein modification between species or differential amounts of repetitive DNAs or both.

The significance of heterochromatin amplification is clearly seen in species of *Glyptotendipes*. Significant differences were found in the amount of centromere DNA for the Ist, IInd and IIIrd chromosomes in comparing *G. salinus* with *G. barbipes* as well as in comparing the homologues in a hybrid (Figure 7.5.4(a,b)). These species were distinguished not only by the amount of centromeric heterochromatin but also by the quality of the heterochromatin (Michailova and Nikolov, 1992). C-heterochromatin of *G. barbipes* consists of two different C-bands: (1) dark C-bands at the periphery of the centromere of the chromosomes which correspond to satellite II DNA; and (2) pale C-bands in the region of the centromere area corresponding to the satellite I DNA in the middle of the centromere. Differentiation in the centromere region of the chromosomes of *G. salinus* is weakly expressed (Figure 7.5.5). It is quite possible that these insects share a common library of satellite sequences which are amplified to various degrees so that in each species one may see different amounts of heterochromatin. During species divergence (*G. barbipes* is phylogenetically the younger of the two species) mutations could be accumulated within the members of cluster repeats of this heterochromatin and could account for the observed heterogeneity of cytological appearance in the *G. barbipes* heterochromatin. Thus the heterochromatin can be considered a dynamic element during speciation.

There has been no record of interspecies hybridization within this group in nature. However, laboratory crossing experiments among some sibling species were possible using artificial insemination and these crosses produce variable degrees of reproductive isolation. Data from the hybridization experiments between *C. plumosus* and *C. muratensis* indicate that they are the most reproductively isolated of all the species (Michailova and Fischer, 1986). Crosses of *C. plumosus* ♀ × *C. balatonicus* ♂ produced only 50% fertile F_1 hybrids. From the reciprocal cross, no egg masses were obtained. Backcrossing is only possible with one of the parents. Almost the same result was obtained in the artificial hybridization *G. salinus* × *G. barbipes*. Hybridization was successful only in one direction (*G. salinus* ♀ × *G. barbipes* ♂). Egg hatchability of this cross was about 70%. Larval viability was reduced so that only ~27% of the eggs developed into adults. The females of the F_1 generation could be backcrossed against both parents. Although the banding patterns were very similar the

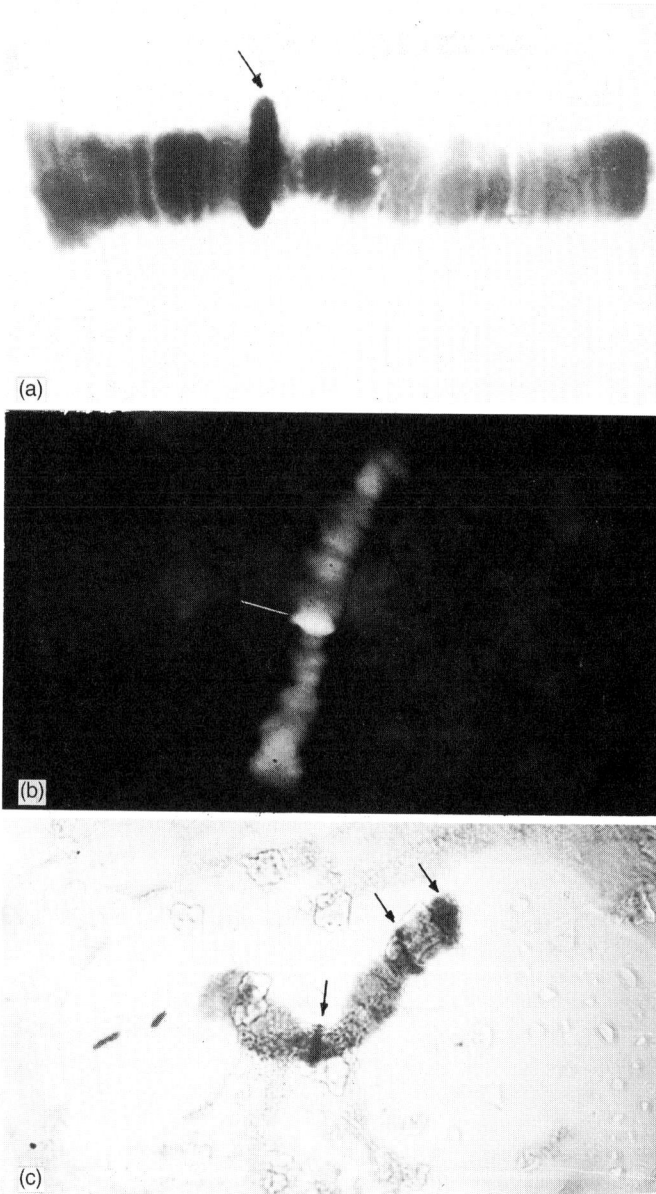

Figure 7.5.3 Differential staining of polytene chromosomes of *C. plumosus*. (a) C-banding of IIIrd (EF) chromosome of *C. plumosus*; (b) Q-banding of IInd (CD) chromosome of *C. plumosus*; (c) C-banding of IIIrd chromosome of *C. balatonicus*.

giant chromosomes of the hybrids showed almost complete asynapsis (Figure 7.5.4 (a, b)), a phenomenon which must be caused by genetic changes which may not be reflected in the banding pattern. Therefore, between these species there are postmating barriers preventing hybrid survival; the hybrid zygotes are non-viable or adaptively inferior to those of the parental species.

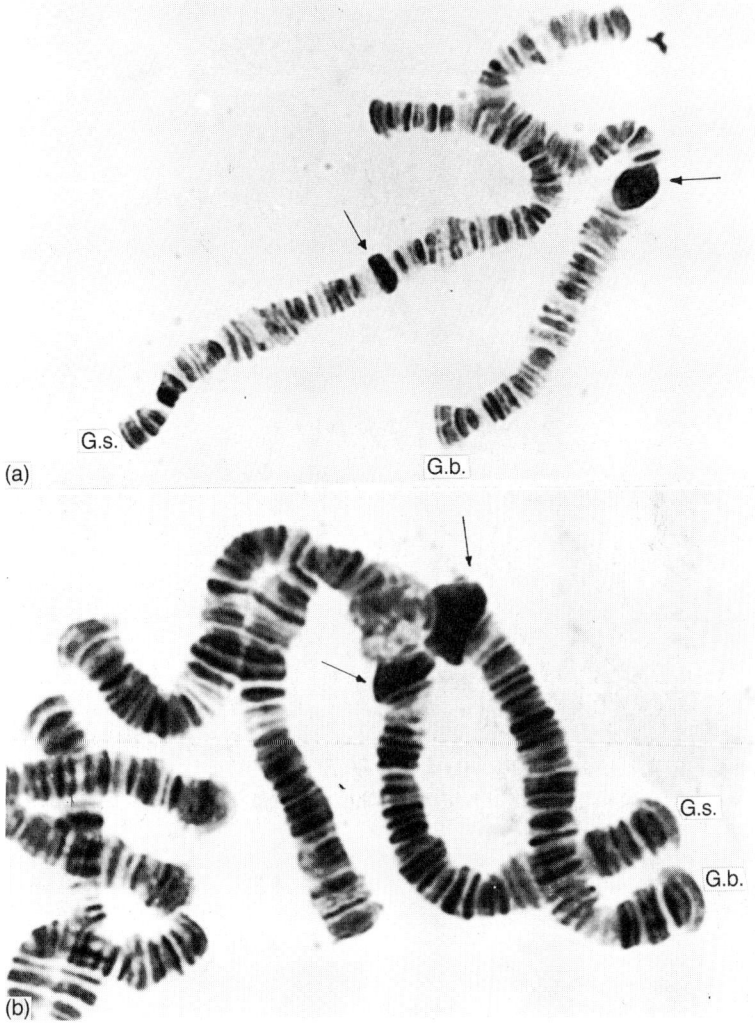

Figure 7.5.4 Chromosomes of a hybrid *G. salinus* ♀ × *G. barbipes* ♂. (a) IIIrd (EF) chromosome; (b) IInd chromosome.

Also, between some of these species there are effective premating isolating mechanisms. From reciprocal crosses of certain different sibling species no egg masses were obtained, although spermatozoa were found in the spermatheca of the females. It seems that spermatozoa are not able to enter the eggs. These data confirm the occurrence of gametic incompatibility, possibly due to a failed insemination reaction.

The differences in mating behaviour of sibling species in the *plumosus* group (Ryser, Scholl and Wülker, 1983) may be an additional mechanism for species differentiation. Furthermore, the larvae of sibling species of the *plumosus* group and those of the genus *Glyptotendipes* live in different parts of water pools or sometimes in different niches (Devai, Wülker and Scholl, 1983; Michailova, 1987a). Thus, *G. salinus* is a sibling species of *G. barbipes* which has reached a degree of reproductive isolation which enables the two

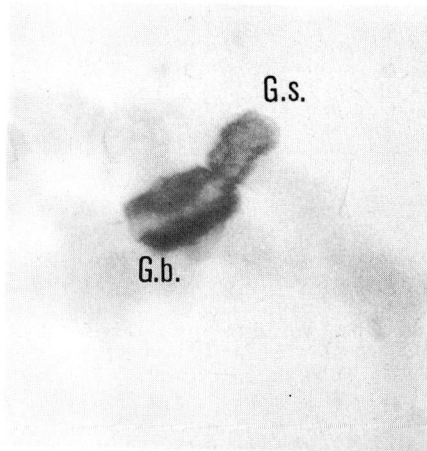

Figure 7.5.5 C-banding of a hybrid *G. salinus* ♀ × *G. barbipes* ♂.

species to coexist sympatrically in certain areas without merging. *G. salinus* occurs in brackish water, while *G. barbipes* prefers fresh water.

In conclusion, I propose the following:

1. Sympatric speciation in the family Chironomidae is accompanied by the development in the population of stable groups with a specific genetic structure. These groups may be considered the genetic reserve of the species.

2. Sympatric speciation takes place following complex rearrangements in the karyotype (homozygous inversions, translocations, changes in the quantity and quality of heterochromatin, changes in chromosomal functions, changes at a molecular level). This can produce reproductive barriers between two groups in a population and can cause the split of these two groups into new species. Karyotype transformation results in redistribution of the genetic material in the species formed by sympatric speciation. However, the initial genetic material is not lost and no new structures appear; the material is only redistributed.

Acknowledgements

This study was supported by the Ministry of Science, Culture and Education, Sofia, Grant B-27. I thank Andy Beaumont for assistance with the English.

References

Ayala, F. and Kiger, Ir. (1984) *Modern Genetics*, 2nd edn, Benjamin-Cummings, California.

Bush, G. (1975) Modes of animal speciation, *Annu. Rev. Ecol. Syst.*, **6**, 339–64.

Devai, G. Wülker, W. and Scholl, A. (1983) Revision der Gattung *Chironomus* Meigen (Diptera). IX. *C. balatonicus* sp. n. aus dem Flachsee Balaton (Ungarn). *Acta Zool. Acad, Sci. Hung.*, **29**, 357–74.

Fischer, J. (1969) Zur Fortpflanzungsbiologie von *Chironomus nuditarsis*. *Rev. Suisse Zool.*, **76**, 23–35.

Kerkis, I., Filippova, M., Shobanov, N. and Gunderina, L. (1988) Karyological and genetic-biochemical characterization of *C. borokensis* sp. n. from *plumosus* group. *Tsitologia*, **30**, 1304–12 (in Russian).

Keyl, H. and Keyl, I. (1959) Die cytologische Diagnostik der Chironomiden Bestimmungstabelle für die Gattung *Chironomus* Mg. auf Grund der Speicheldrüsen Chromosomen. *Arch. Hydrobiol.*, **56**, 43–57.

Maximova, F. (1979) On the cytodiagnosis of *Chironomus plumosus* L. larvae, in *Karyosystematics of the Invertebrate Animals* (ed. L. Chubareva), Institute of Zoology, Leningrad, pp. 51–5.

Mayr, E. (1970) *Populations, Species and Evolution*, Harvard University Press, Cambridge, MA.

Michailova, P. (1985) Method of breeding the species from family Chironomidae, Diptera in experimental conditions. *CR Acad. Bulg. Sci.*, **38**, 1179–81.

Michailova, P. (1987a) Comparative karyological studies of three species of the genus *Glyptotendipes* Kieff. (Diptera, Chironomidae) from Hungary and Bulgaria. *Glyptotendipes salinus* sp. n. from Bulgaria. *Folia Biol.*, **35**, 43–68.

Michailova, P. (1987b) 'C' banding in the polytene chromosomes of species of a group *plumosus* (Diptera, Chironomidae). *Genetica*, **74**, 41–51.

Michailova, P. (1987c) 'C' banding in polytene chromosomes of *Glyptotendipes* (Staeger) and *Glyptotendipes salinus* Michailova (Diptera, Chironomidae) and their experimental hybrid. *Acta Zool Bulg.*, **35**, 3–15.

Michailova, P. and Fischer, J. (1986) Speciation within the *plumosus* group of the genus *Chironomus* Mg. (Diptera, Chironomidae). *Z. Zool. Syst. Evol.*, **24**, 207–22.

Michailova, P. and Nikolov. H. (1992) Different DNA content in centromere regions of sibling species *Glyptotendipes barbipes* (Staeger) and *Glyptotendipes salinus* Michailova (Chironomidae, Diptera). *Cytobios*, **70**, 179–84.

Michailova, P. and Petrova, N. (1991) Chromosomal polymorphism in geographically isolated populations of *Chironomus plumosus* L. (Chironomidae, Diptera). *Cytobios*, **67**, 161–76.

Ryser, H., Scholl, A. and Wülker, W. (1983) Revision der Gattung *Chironomus* Meigen (Diptera). VII. *C. muratensis* n. sp. und *C. nudiventris* n. sp. Geschwisterarten aus der *plumosus* Gruppe. *Rev. Suisse Zool.*, **99**, 299–316.

Schlee, D. (1968) Zur Präparation von Chironomiden. II. Die Behandlung ausgeblichenen bzw getrockneten Materials und die Reparieren schadhaften Praparate. *Ann. Zool. Fenn.*, **5**, 127–9.

Shobanov, N. and Demin, S. (1989) *Chironomus agilis* sp. n., group *plumosus* (Diptera, Chironomidae). *Zh. Zool.*, **67**, 1483–97 (in Russian).

Stegnii, V. (1984) Evolution significance of chromosome inversions. *J. Gen. Biol.*, **XLV**, 3–14 (in Russian).

8

Genetics and pollution

8.1 GENETIC EFFECTS OF POLLUTANTS ON MARINE AND ESTUARINE INVERTEBRATES

Herman Hummel and Tomaso Patarnello

Abstract

A review of the state of the art is provided on research dealing with the effects of pollutants on the genetics of marine and estuarine invertebrates. Significant changes in the genetic structure often occur. Many contrasting trends are found. Possible causes for the differential fitness of genotypes are discussed and related to protein structure, metabolism and high contaminant concentrations used in experiments. Some recommendations are given concerning genetic monitoring.

8.1.1 Introduction

From historical times people have concentrated their settlements in estuarine and coastal regions, which resulted in a significant increase of both organic and inorganic pollution in these areas. In addition, estuaries receive pollutants from the hinterland through freshwater networks. In the mixing zone of fresh and salt water, sedimentation of a major part of the contaminants occurs (the turbidity zone), causing their concentration. This phenomenon suggests estuarine and coastal areas as environments where effects of pollution on the genetic structure of benthonic and planctonic organisms are more likely.

It has been widely demonstrated that several marine and estuarine organisms accumulate pollutants. Due to this biological characteristic, some of these organisms have been suggested as bio-indicators of the degree of environmental contamination (Goldberg, 1975; Beardmore *et al.*, 1980; Bayne *et al.*, 1985).

Pollutants can influence the genetic constitution in two ways (Dixon, 1985). Firstly, they can cause mutations by directly damaging the DNA molecule within the individual cell nucleus. This may result in gene mutations (changes at a single locus) or chromosomal aberrations (changes of (parts of) chromosomes with many loci). Although in laboratory experiments direct effects of pollutants on vertebrates have often been shown, not much is known about chromosomal abberrations in marine invertebrates (Dixon, 1985) or the

Genetics and Evolution of Aquatic Organisms. Edited by A.R. Beaumont.
Published in 1994 by Chapman & Hall, 2–6 Boundary Row, London SE1 8HN.
ISBN 0-412-49370-5

correlation between environmental pollution and the incidence of mutations (Bolognesi, 1989). For a further discussion of these cytogenetic effects the reader is referred to Dixon (1985), World Health Organization (1989) and section 8.2. The second way in which pollutants can affect the genetic constitution is by exerting a selective pressure on the genetic structure of the population by modifying the environment. This will be discussed in this section.

A genetic basis for tolerance to pollutants was first proposed more than 20 years ago for specimens of the marine fouling alga *Ectocarpus siliculosus* from boats with antifouling paint (Russell and Morris, 1970). Positive growth ceased in normal uncontaminated field populations at Cu concentrations of 10 μg/l. On the other hand, populations of this alga living under the keel of fishing boats were able to survive up to 100 μg/l of Cu. Similarly, populations of *Nereis diversicolor* from areas with metal (copper and zinc) concentrations above 2 mg/g dry sediment showed a higher resistance to copper pollution compared with populations collected in areas with a lower level of contamination (Bryan and Hummerstone, 1973).

With the introduction of allozyme research a genetic basis for adaptation and differences in tolerance became more evident. Gooch and Schopf (1970) first reported changes in allele frequencies, suggesting pollutants as a possible selecting agent. Since then, most efforts have been focused on understanding mechanisms of genetic adaptation and evolution. Research on the genetic effect of pollution remained a side-issue.

The aim of this section is to provide a general view of the state of the art of research addressing the problem of the genetic effects of pollution, with special reference to marine and estuarine organisms.

8.1.2 Scarcity of data

Field studies on the effects of pollutants are scarce; only six relevant publications with field observations and three with experimental manipulations in the field are reported in this study (Table 8.1.1). Laboratory studies are also available; they are more inclined to yield detailed indications concerning the influencing factor. However, the species studied in laboratory experiments were others than those used in the field studies, so that a validation of experimental data was not possible.

Only a single study has considered the genetic effects of pollution in relation to the age class (Patarnello, Guinez and Battaglia, 1991). Comparisons between juveniles (2 weeks old) and adults of the barnacle *Balanus amphitrite* were carried out in polluted and unpolluted areas in order to detect the action of some selective factor during the post-settlement period.

8.1.3 Differential fitness of genotypes

A summary of the main studies (Table 8.1.1) so far available shows that populations living in contaminated environments often exhibit significant changes in their genetic structure compared to control populations. However, the direction of these changes can be variable and contrasting trends emerge. Even within the same study for one species, results might show contrasting results over time (Nevo, Shiminy and Libni, 1977; Battaglia, Bisol and Rodino, 1980; Battaglia *et al.*, 1980; Fevolden and Garner, 1986). Fevolden and Garner (1986) proposed the occasional action of different selecting environmental agents. Contrasting results might be the expression of genetic, biochemical or metabolic differences in the ability of the species to face stressful conditions such as pollution. These differences in the pollutant–organism interaction are discussed below.

Table 8.1.1 Summary of the effects of pollutants on the genetic constitution of estuarine and marine organisms (after Hummel and Herman, 1993)

Type of stress	Species	Effect on Alleles[a]	Effect on Heter[b]	Type[c] of experiment	Pollutant concentration (ppb) and period of exposure	Loci studied and type (allele changed)[d]	Reference
Metals	Balanus amphitrite		–	FO	4 × higher than control	Mpi, Pgm, Pgi	Patarnello, Guinez and Battaglia (1991)
	Monodonta turbinata	+		LE	5000–55 000 Cu 2–5 days	Pgi (S. M; only 2 alleles)	Lavie and Nevo (1982)[e]
		0[f]			100 000–600 000 Zn 1–2 days	Pgi	
	Monodonta turbiformis	+			5000–100 000 Cu 2–5 days	Pgi (S, M; only 2 alleles)	
		0			500 000–600 000 Zn 1–3 days	Pgi	
	Palaemon elegans	0[g]		LE	200 Hg	Pgm	Ben-Shlomo and Nevo (1988)[e,i,j]
		0[h]			20 000 Cd	Pgm	
		0[h]			200 μg + 20 000 Cd 1–3 days	Pgm	
	Macoma balthica		+	LE	300–500 Cu 17–23 days	Pgi, Pgm	Kok (1990)
	Mytilus edulis	+	+	LE	50–300 Cu 5–56 days	Pgi, Pgm	
	Mytilus edulis		+	LE	150 Cu 14–28 days	Pgi	Hawkins et al. (1989b)
	Mytilus edulis	+		LE	100–500 Cu 6–15 days	Pgi, Pgm	Hvilsom (1983)[e]
Oil	Balanus amphitrite	+	0	FO	(10 000) sulfet 5 months	Pgi(F), Pgm(F), Me(m+), Ao-1(M), Ao-2(M), Ao-3(F), Acph-1(M), Est-1(F), Est-2(F), Est-3(M), Est-4(S), Gp(M) No change: Mdh-2, Est-7, Est-8	Nevo, Shimony and Libni (1978)
	Littorina littorea	0	0	FE	129 and 31 oil 2 years	31 loci studied; only 6Pgdh and Pgm-2 polymorph	Fevolden and Garner (1987)

Table 8.1.1 (contd)

Type of stress	Species	Effect on Alleles[a]	Effect on Heter[b]	Type[c] of experiment	Pollutant concentration (ppb) and period of exposure	Loci studied and type (allele changed)[d]	Reference
Oil (contd)	Mytilus edulis	0	—	FE	129 and 31 oil 2 years	31 loci studied	Fevolden and Garner (1986)
	Tishe holothuriae	+		LE	5000–18 000 oil 40 days (2 generations)	Pgi-1 (a,b; only 2 alleles), Ap-1	Battaglia, Bisol and Rodino (1980); Battaglia et al. (1980)
Thermal	Balanus amphitrite	0		FO	10 °C higher than than control	Mpi, Pgm, Fgi	Patarnello, Guinez and Battaglia (1991)
	Balanus amphitrite	+	—	FE	9–12 °C higher than control 5 months	Me(M+), Ac-3(F), Mdh-1(F), Acph-1(M), Est-3(F), Est-4(F), Est-8(F) No change: Pgi, Ao-1, Est-1, Est-7	Nevo, Shimony and Libni (1977)[k]
	Gammarus insensibilis	+	+	LE	27–37 °C 3 days (control at 10 °C)	Pgi	Patarnello, Bisol and Battaglia (1989)[e]
Several (undetermined) sources of pollution	Acartia clausi Centropagus typicus	+ +		FO	Eutrophication	Est, Me, Lap, Mdh	Kerambrun and Guérin (1984)
	Mytilus galloprov.	+	—	FO		Ap, Lap, 6-Pgd, Idhs, Idhm, Pgi, Pgm	Battaglia, Bisol and Rodino (1980); Battaglia et al. (1980)
	Schizoporella unicornis		+	FO	Oil (in harbour)	Lap-3	Goch and Schopf (1970)

[a] + = change of allele or genotype frequencies due to pollution (in live animals); 0 = no effect.

[b] -, + = decrease and increase in heterozygote frequencies (in live animals) because of exposure to pollutants (i.e. animals sensitive or resistant to pollutants); 0 = no effect.

[c] FO = field observations, FE = field experiments, LE = laboratory experiments.

[d] Changing allele as indicated by authors or abstracted from data.

[e] Statistics based on comparison between live and dead animals and no data on control or blank.

[f] Result was said to be significant, but is not.

[g] A non-significant ($p > 0.05$) trend was found.

[h] Results were hardly significant ($0.09 > p > 0.07$ with the χ^2 test and/or Wilcoxon signed rank sum test).

[i] Statistics only on individual data and not on overall results.

[j] Direction of changes is very inconsistent.

[k] Proportion of heterozygotes between parts of experiments or between dates differs strongly.

8.1.4 Pollutant–organism interaction

(a) Differences in energy balance

The exposure of an organism to any kind of environmental contamination may produce a significant change in its metabolism, requiring adaptation to stressful conditions (Hoffmann and Parsons, 1991; Calow, 1991). This adaptation is an energy-demanding process which involves protein turnover, excretion, synthesis of specific stress-induced proteins like heat-shock proteins, or detoxification proteins such as metallothioneins. All this contributes to an appreciable increase of the (basal) maintenance metabolism (Basha *et al.*, 1984; Gaudy, Guerin and Kerambrun, 1991). The increase of the metabolic rate is measurable by a raised O_2 consumption. It has been suggested that the energy cost for maintenance is genotype-dependent (Koehn and Bayne, 1989). Heterozygotes are supposed to require less energy for their maintenance, as inferred from their lower oxygen consumption.

Moreover, according to Clark and Koehn (1993), a differential rate in protein turnover (major component of maintenance) might be also genotype-dependent, resulting in a significant difference in available energy to be invested in production (growth rate, reproduction, allocation of food, escape). The decrease of food assimilation which has been described during exposure to pollutants is a further reduction of energy supply (Gaudy, Guerin and Kerambrun, 1991) and might play a relevant role when energy is a limiting factor. In such conditions it is likely that we will find differential fitness by genotypes (Calow, 1991). Therefore, in a polluted stressful environment the energy balance might favour the heterozygote genotypes because of their less energy-demanding maintenance.

Thus, differences in fitness are more easily detectable in stressful conditions (Koehn and Bayne, 1989) such as pollution, especially at loci involved in energy production (Blackstock, 1980) or in relevant metabolic pathways. A good example is given by the *Pgi* locus, which is considered in most of the studies listed in Table 8.1.1.

The enzyme phosphoglucose isomerase (PGI) has been widely analysed because of its high genetic polymorphism and because of its position in the glycolytic pathway. Several authors have suggested an adaptive nature of specific *Pgi* genotypes (Watt, 1985; Hawkins *et al.*, 1989a,b; Patarnello and Battaglia, 1992), based on the metabolic role of the PGI enzyme (Koehn, Diehl and Scott, 1988). However, no common trend emerged for the observed changes (Table 8.1.1). In this respect, it has to be considered that we are comparing *Pgi* in many different species, which means that we are dealing with the same protein in terms of function but not in terms of structure. Differences in enzyme structure might explain the different behaviour of *Pgi* variants in the presence of pollutants, irrespective of their homo- or heterozygote genotype condition. Therefore, a higher fitness of *Pgi* homozygotes *in Balanus amphitrite* (Patarnello, Guinez and Battaglia, 1991) does not mean that the homozygote genotypes are better *per se* but that, in this species, in the presence of chemical pollution, they may have a selective advantage.

(b) Protein structure

The high variability of the data reported in Table 8.1.1 might be related to the different protein structure of allelic products at a polymorphic locus which interact differently with pollutants.

At a two-allele locus coding for a monomeric enzyme (e.g. *Pgm* or *Mpi*) it is likely that enzyme activity reduction due to some chemical pollutants mostly affects one of the two homozygous genotypes, giving a selective advantage to the other one. The heterozygote is expected to have an intermediate activity between the two homozygotes.

On the other hand, heterozygotes at loci coding for non-monomeric enzymes might have higher (overdominant) enzyme activity compared to both homozygotes. This could be explained by considering that heterozygotes for non-monomeric enzymes have a quaternary structure characterized by heteroduplex chains which neither of the homozygotes possesses (Traut, Casiano and Zecherle, 1989). These heteroduplex chains may confer on heterozygotes biochemical properties which are not intermediate between homozygotes but which may include a higher resistance to chemical pollution.

It is therefore difficult to infer any general rule which may finally link genotypic fitness and metabolism/biochemistry. As mentioned before, it is not a question of 'heterozygotes (or homozygotes) are better *per se*'. It depends on the case in question.

(c)   Direct interactions

Conflicting results might be related to the interference of the pollutant directly with the biochemistry of the enzymes coded by the loci considered in the genetic analyses. In fact, several studies have been able to demonstrate significant effects of pollution on enzyme activity (Rivière *et al.*, 1983). It is especially true in the case of chemical pollution since heavy metals have long been known to interact with proteins, modifying their biochemical properties (Milstein, 1961; Webb, 1966). In most cases they reduce the enzyme activity (Berglind, 1985; George and Young, 1986), especially after long exposure of the organism to contamination (Hilmy, Shabana and Daabees, 1985a,b; Sastry and Subhadra, 1985). The lowered enzyme activity might be due to interactions between the enzyme and heavy metal at several levels. Metal ions may affect the catalytic efficiency of the enzyme, reducing its affinity for the substrate. It is also possible that heavy metals produce a breakdown of S–S bonds of the protein with a consequent modification of its secondary and tertiary structure. Moreover, interactions between SH groups and heavy metals may strongly decrease the activity of enzymes (Volkin and Klibanov, 1989).

The direct effect of pollutants by means of mutation of the genetic material, as hypothesized by Beardmore (1980) and Beardmore *et al.* (1980), is thought to be doubtful. The shifts in genetic composition in all the studies reviewed are caused by changes in the frequencies of existing alleles. Therefore, there is no evidence for the creation of new alleles due to pollutants.

(d)   High contaminant concentrations

Another source of conflicting results, especially for laboratory experiments, lies with the unrealistically high contaminant concentrations used. At high concentrations the metabolic functioning of the tested organisms might be blocked, whereas at low concentrations adaptation and selection mechanisms might work in such a way that the organisms best adapted (or most able to withstand the toxicological impact on an enzyme) will survive. So, high and low concentrations of pollutants interact at different functional levels of an organism, and thus give rise to differential selection. Adaptation at very high concentrations of pollutants, with the consequent short timespan (days), is unlikely to be the case. Hvilsom (1983) and Kok (1990) found that the shifts in allele frequencies were stronger at the lower concentrations (100 and 50 ppb copper respectively). However, Hvilsom (1983) probably used a mixture of two lineages from the *Mytilus edulis* group (*Mytilus edulis* and *Mytilus trossulus* and their hybrids). Therefore, her results could also be explained as selection (on different adaptation capacities) between different lineages or hybrids. Both studies showed also that at higher concentrations the mussels died too fast to achieve any selection between genotypes. Only selection at the lower pollutant concentrations for a longer period (weeks, months) might indeed reflect adaptation to the pollutant.

8.1.5 Field observations

An interesting observation on heterozygosity and pollution was made by Nevo *et al.* (1985, 1986). Species of the gastropod genera *Monodonta, Littorina* and *Cerithium* living at the higher levels of the intertidal zone showed a higher genotypic diversity (mean number of alleles per locus, proportion of polymorphic loci per population, proportion of heterozygous loci per individual), as well as higher survival during exposure to pollutants (metals, oil detergents), compared to other related species from lower levels in the tidal zone. Therefore, genotypic diversity and resistance to pollutants appear to be positively correlated. However, it has to be considered that the more resistant species generally came from the higher intertidal levels. Thus, the specimens analysed by Nevo and colleagues might have already been adapted to different levels of tidal stress and their relation between resistance to pollutants and levels of genetic heterogeneity in their tests may be coincidental.

Attempts to validate experimental data in the field are rare. In studies on *Littorina neritoides* and *L. punctata*, populations from mercury-polluted sites along the Israeli coast were compared with populations from unpolluted stations (Nevo, Lavie and Noy, 1987). It was observed that the allele which was negatively selected by mercury pollution in laboratory experiments was also significantly less frequent in the population from the polluted area. In similar studies carried out on the shrimp *Palaemon elegans* and the gastropod *Monodonta turbinata* (Nevo, Ben-Shlomo and Lavie, 1984) it was found that the allele frequency distributions in nature were consistent with the results on mercury-exposed animals in laboratory tests. However, the validation of some of these reported studies might be doubtful because of weak statistics (Hummel and Herman, 1993) (studies using Baileys' test on proportions were doubtful and omitted; χ^2 tests should have been used).

8.1.6 Biomonitoring

When marine or estuarine organisms are employed in biomonitoring programmes, it is important to carefully choose the species most suitable for the kind of study to be under-taken. A possible source of problems comes from sibling species, as demonstrated by the example of *Capitella* spp. (Grassle and Grassle, 1976). Several species of this polychaete are morphologically quite undistinguishable but have clear differences in their life-history, genetics and physiology. A similar confusion may arise for the bivalve mussel *Mytilus edulis*, which is commonly used as pollution indicator in worldwide mussel-watch programmes for water-quality control (Goldberg, 1975). Recently, on the basis of bio-chemical genetic analyses, it was proposed that the *Mytilus edulis* group might be separated into three evolutionary lineages, namely *Mytilus edulis*, *M. trossulus* and *M. galloprovincialis* (McDonald and Koehn, 1988; Varvio, Koehn and Vainola, 1988; McDonald, Seed and Koehn, 1991). *Mytilus trossulus* tends to accumulate elements up to 50% more than *M. edulis* of the same length (Lobel *et al.*, 1990) (*M. trossulus* specimens are most probably older). This may cause differences (elevated trace element concentrations) in mussel-watch programmes due to species differences rather than to differences in the level of contamination. Genetic research, accompanying pollution-monitoring studies, is thus indispensable.

8.1.7 Conclusion and recommendations

Genetic analyses have been considered an important tool for monitoring the quality of aquatic environments and the degree of environment contamination. However, it is not possible to draw any general rule with universal validity. Genetic changes do not necessarily follow a common trend in relation to the environmental modification. It will depend upon the species, the genes and the kind of pollution considered in each case. Hence, to successfully address the relation between genetics and pollution the approaches suggested are:

1. choice of a sessile species as bio-indicator (possibly with a filtration habit which enables the organism to accumulate pollutants);
2. comparison of alleles and genotype frequencies as well as multilocus structure between populations (of the same species) from polluted and unpolluted areas over time (seasons, years);
3. translocation of animals from non-polluted areas to polluted areas (and vice versa) and record of their eventual genetic changes;
4. *in vivo* and *in vitro* laboratory experiments at different concentrations of pollutants, only to be carried out at field-relevant low concentrations and for long periods (months);
5. characterization of the physiology of species and the biochemistry of enzymes in the presence of pollutants.

Acknowledgements

Centre for Estuarine and Coastal Ecology Communication No. 673.

References

Basha, S.M., Prasada Rao, K.S., Sambasivq Rao, K.R.S. and Ramano Rao, K.V. (1984) Respiratory potentials of fish (*Tilapia mossambica*) under Malathion, Cabaryl and Lindane intoxication. *Bull. Environ. Contam. Toxicol.*, **32**, 570–4.

Battaglia, B., Bisol, P.M. and Rodino, E. (1980) Experimental studies on some genetic effects of marine pollution. *Helg. Meeresunt.*, **33**, 587–95.

Battaglia, B., Bisol, P.M. Fossato, V.U. and Rodino, E. (1980). Studies on the genetic effects of pollution in the sea. *Rapp. P.-v. Reun. Cons. Int. Explor. Mer*, **179**, 267–74.

Bayne, B.L., Brown, D.A., Burns, K. *et al.* (1985) *The Effects of Stress and Pollution on Marine Animals*, Praeger, New York.

Beardmore, J.A. (1980) Genetical considerations in monitoring effects of pollution. *Rapp. P.-v. Reun. Cons. Int. Explor. Mer*, **179**, 258–66.

Beardmore, J.A., Barker, C.J., Battaglia, B. *et al.* (1980) The use of genetical approaches to monitoring biological effects of pollution. *Rapp. P.-v. Reun. Cons. Int. Explor. Mer*, **179**, 299–305.

Ben-Shlomo R. and Nevo, E. (1988) Isozyme polymorphism as monitoring of marine environments: the interactive effect of cadmium and mercury pollution on the shrimp, *Palaemon elegans. Mar. Pollut. Bull.*, **19**, 314–17.

Berglind, R. (1985) The effects of cadmium on ALA-D activity, growth and haemoglobin content in the water flea, *Daphnia magna. Comp. Biochem. Physiol.*, **80C**, 407–10.

Blackstock, J. (1980) Estimation of activities of some enzymes associated with energy-yielding metabolism in the polychaete, *Glycera alba* (Muller), and application of the methods to the study of the effect of organic pollution. *J. Exp. Mar. Biol. Ecol.*, **46**, 197–217.

Bolognesi, C. (1989) Carcinogenic and mutagenic effects of pollutants in marine organisms: a review, in *Carcinogenic, Mutagenic, and Teratogenic Marine Pollutants: Impact on Human Health and the Environment* (World Health Organization), Gulf Publishing Company, London, pp. 67–83.

Bryan, G.W. and Hummerstone, L.G. (1973) Adaptation of the polychaete *Nereis diversicolor* to estuarine sediments containing high concentrations of heavy metals. *J. Mar. Biol. Assoc. UK*, **51**, 845–63.

Calow, P. (1991) Physiological costs of combating chemical toxicants: ecological implications. *Comp. Biochem. Physiol.*, **100C**, 3–6.

Clark, A.G. and Koehn, R.K. (1993) Enzymes and adaptation, in *Genes in Ecology* (ed. R.J. Berry), Oxford University Press. (in press).

Dixon, D.R. (1985) Chromosomal aberrations, in *The Effects of Stress and Pollution on Marine Animals* (eds B.L. Bayne, D.A. Brown, K. Burns *et al.*), Praeger, New York, pp. 75–80.

Fevolden, S.E. and Garner, S.P. (1986) Population genetics of *Mytilus edulis* (L.) from Oslofjorden, Norway, in oil-polluted and non oil-polluted water. *Sarsia*, **71**, 247–57.

Fevolden, S.E. and Garner, S.P. (1987) Environmental stress and allozyme variation in *Littorina littorea* (Prosobranchia). *Mar. Ecol. Progr. Ser.*, **39**, 129–36.

Gaudy, R., Guérin, J.-P. and Kerambrun, P. (1991) Sublethal effects of cadmium on respiratory metabolism, nutrition, excretion and hydrolase activity in *Leptomysis lingvura* (Crustacea: Mysidacea). *Mar. Biol.*, **109**, 493–501.

George, S.G. and Young, P. (1986) The time course of effects of cadmium and 3-methylcholanthrene on activities of enzymes of xenobiotic metabolism and metallothionein levels in the plaice, *Pleuronectes platessa. Comp. Biochem. Physiol.*, **83C**, 37–44.

Goldberg, E.D. (1975) The mussel watch: a first step to global marine monitoring. *Mar. Pollut. Bull.*, **6**, 111.

Gooch, J.L. and Schopf, T.J.M. (1970) Population genetics of marine species of the phylum Ectoprocta. *Biol. Bull.*, **138**, 138–50.

Grassle, J.P. and Grassle, J.F. (1976) Sibling species in the marine pollution indicator *Capitella* (Polychaeta). *Science*, **192**, 567–9.

Hawkins, A.J.S., Bayne, B.L., Day, A.J., Rusin, J. and Worrall, C.M. (1989a) Genotype-dependent interrelations between energy metabolism, protein metabolism and fitness, in *Reproduction, Genetics and Distributions of Marine Organisms* (eds J.S. Ryland and P.A. Tyler), Olsen & Olsen, Fredensborg, pp. 283–92.

Hawkins, A.J.S., Rusin, J., Bayne, B.L. and Day, A.J. (1989b) The metabolic/physiological basis of genotype-dependent mortality during copper exposure in *Mytilus edulis. Mar. Environm. Res.*, **28**, 253–7.

Hilmy, A.M., Shabana, M.B. and Daabees, A.Y. (1985a) Bioaccumulation of cadmium: toxicity in *Mugil cephalus. Comp. Biochem. Physiol.*, **81C**, 139–43.

Hilmy, A.M., Shabana, M.B. and Daabees, A.Y. (1985b) Effects of cadmium toxicity upon the *in vivo* and *in vitro* activity of proteins and five enzymes in blood serum and tissue homogenates of *Mugil cephalus. Comp. Biochem. Physiol.*, **81C**, 145–53.

Hoffmann, A.A. and Parsons, P.A. (1991) *Evolutionary Genetics and Environmental Stress*, Oxford University Press, Oxford.

Hummel, H. and Herman, P.M.J. (1994) A reconsideration of the effects of pollutants on the genetic constitution of marine organisms (in prep.).

Hvilsom, M.M. (1983) Copper-induced differential mortality in the mussel *Mytilus edulis. Mar. Biol.*, **76**, 291–5.

Kerambrun, P. and Guérin, J.P. (1984) L'electrophorèse dans l'étude des stress chez les invertébrés marins. *Bull. Soc. Zool. France*, **109**, 333–41.

Koehn, R.K. and Bayne, B.L. (1989) Towards a physiological and genetical understanding of the energetics of the stress response. *Biol. J. Linn. Soc.*, **37**, 157–71.

Koehn, R.K., Diehl, W.J. and Scott, T.M. (1988) The differential contribution by individual enzymes of glycolysis and protein catabolism to the relationship between heterozygosity and growth rate in the coot clam, *Mulinia lateralis. Genetics*, **118**, 121–30.

Kok, G. (1990) Het effect van koper geïnduceerde mortaliteit op isoenzympatronen van *Mytilus edulis* L. en *Macoma balthica* (L.). *Delta Institute Student Report*, Yerseke, D3-1990.

Lavie, B. and Nevo, E. (1982) Heavy metal selection of phosphoglucose isomerase allozymes in marine gastropods. *Mar. Biol.*, **71**, 17–22.

Lobel, P.B., Belkhode, S.P., Jackson, S.E. and Longerich, H.P. (1990) Recent taxonomic discoveries concerning the mussel *Mytilus*: implications for biomonitoring. *Arch. Environ. Contam. Toxicol.*, **19**, 508–12.

McDonald, J.H. and Koehn, R.K. (1988) The mussels *Mytilus galloprovincialis* and *M. trossulus* on the Pacific coast of North America. *Mar. Biol.*, **99**, 111–18.

McDonald, J.H., Seed, R. and Koehn, R.K. (1991) Allozymes and morphometric characters of three species of *Mytilus* in the Northern and Southern hemispheres. *Mar. Biol.*, **111**, 323–33.

Milstein, C. (1961) Inhibition of phosphoglucomutase by trace metals. *Biochem. J.*, **79**, 591.

Nevo, E., Ben-Shlomo, R. and Lavie, B. (1984) Mercury selection of allozymes in marine organisms: prediction and verification in nature. *Proc. Natl. Acad. Sci. USA*, **81**, 1258–9.

Nevo, E., Lavie, B. and Noy, R. (1987) Mercury selection of allozymes in marine gastropods: prediction and verification in nature revisited. *Environm. Monit. Assessm.*, **9**, 233–8.

Nevo, E., Shimony, T. and Libni, M. (1977) Thermal selection of allozyme polymorphisms in barnacles. *Nature*, **267**, 699–701.

Nevo, E., Shimony, T. and Libni, M. (1978) Pollution selection of allozyme polymorphism in barnacles. *Experientia*, **34**, 1562–4.

Nevo, E., Noy, R., Lavie, B. and Muchtar, S. (1985) Levels of genetic diversity and resistance to pollution in marine organisms. *FAO FIRI/R352* (Suppl), 175–81.

Nevo, E., Noy, R., Lavie, B., Beiles, A. and Muchtar, S. (1986) Genetic diversity and resistance to marine pollution. *Biol. J. Linn. Soc.*, **29**, 139–44.

Patarnello, T. and Battaglia, B. (1992) Glucosephosphate isomerase and fitness: effects of temperature on genotype dependent mortality and enzyme activity in two species of the genus *Gammarus* (Crustacea: Amphipoda). *Evolution*, **46**, 1568–73.

Patarnello, T., Bisol, P.M. and Battaglia, B. (1989) Studies on the differential fitness of PGI genotypes with regard to temperature in *Gammarus insensibilis. Mar. Biol.*, **102**, 355–9.

Patarnello, T., Guinez, R. and Battaglia, B. (1991) Effect of pollution on heterozygosity in the barnacle *Balanus amphitrite* (Cirripedia: Thoracica). *Mar. Ecol. Progr. Ser.*, **70**, 237–43.

Rivière, D., Kerambrun, P. *et al.* (1983) Impact d'une pollution d'origine urbaine sur les activités enzymatiques de deux copépodes planctoniques (*Acartia clausi* et *Centropagus typicus*). *Mar. Biol.* **75**, 25–35.

Russell, G. and Morris, O.P. (1970) Copper tolerance in the marine fouling alga *Ectocarpus siliculosus*. *Nature*, **228**, 288–9.

Sastry, K.V. and Subhadra, K. (1985) *In vivo* effects of cadmium on some enzyme activities in tissues of the freshwater catfish, *Heteropneustes fossilis*. *Environm. Res.*, **36**, 32–45.

Traut, R.R., Casiano, C. and Zecherle, N. (1989) Cross-linking of protein subunits and ligands by the introduction of disulphide bonds, in *Protein Function: a Practical Approach* (ed. T.E. Creighton), IRL Press, Oxford.

Varvio, S.-L., Koehn, R.K. and Vainola, R. (1988) Evolutionary genetics of the *Mytilus edulis* complex in the North Atlantic region. *Mar. Biol.*, **98**, 52–60.

Volkin, D.B. and Klibanov, A.M. (1989) Minimizing protein activation, in *Protein Function: a Practical Approach* (ed. T.E. Creighton), IRL Press, Oxford.

Watt, W.B. (1985) Bioenergetics and evolutionary genetics: opportunities for new synthesis. *Am. Nat.*, **125**, 118–43.

Webb, J.L. (1966) *Enzyme and Metabolic Inhibitors*, Vol. 2, Academic Press, New York.

World Health Organization (1989) *Carcinogenic, Mutagenic, and Teratogenic Marine Pollutants: Impact on Human Health and the Environment*, Gulf Publishing Company, London.

8.2   CHROMOSOME SET CHANGES IN MOLLUSCS FROM HIGHLY POLLUTED HABITATS

Janina Baršienė

Abstract

Chromosome set abnormalities in the tissues of 14 species of snails from the polluted areas of the main Lithuanian hydrosystem and from three uncontaminated lakes have been studied. Modal chromosome numbers in 56.6–74.5% of nuclei were found in molluscs inhabiting unfavourable biotopes and in 88.9–92% in those from the unpolluted lakes. The existence of a high proportion (40–60%) of hypoploid, polyploid, mosaic and hermaphroditic bivalve specimens as well as the presence of tumour cells, mitotic supression and various aberrations of chromosomes were characteristic features of clams inhabiting areas strongly affected by industrial and domestic pollution and Chernobyl radionuclides. In 80% of bivalve specimens a positive correlation between heavy metal, aromatic hydrocarbon or radionuclide bioaccumulation and chromosome set disturbances was determined.

8.2.1   Introduction

There is extensive anthropogenic contamination of the main Lithuanian hydrosystem (Neris–Nemunas–Kuršių Bay–Baltic Sea). Data on the concentrations of heavy metals, hydrocarbons and radionuclides in the water, the bottom sediments or the biota have been presented by several authors (Taraškevičius, 1991; Jokšas, 1991; Milukaitė and Bukelskis, 1991; Zareckas, 1991; Dušauskiene-Duž, 1992; Lukšienė *et al.*, 1992; Špirkauskaitė *et al.*, 1992).

Certain mollusc species, due to their distribution, size, sedimentary life and filtration activity, are very suitable for studies of environmental pollutants, their bioaccumulation and biological effects. Snails are indispensable for ecological risk assessment using the analysis of genetic changes, and relationships between aneuploidy, polyploidy, mosaicism, development of tumour cells and other peculiarities of reproduction.

In the present study freshwater snails were collected from highly polluted and uncontaminated areas of the main Lithuanian hydrosystem. Chromosome set disturbances in gonadal and somatic cells were detected, including the presence of neoplastic lesions.

8.2.2 Materials and methods

Material for the karyological studies of molluscs was collected in the summers of 1985–1991. Fourteen species of molluscs (560 specimens) were taken from the polluted sites of the Vilnia, Nemunas and Šventoji Rivers and Kuršių Bay by the Ventė, Dreverna, Klaipėda, Juodkrantė and Preila. Twenty five specimens of *Valvata piscinalis*, 21 *Bithynia tentaculata* and 20 *Lymnaea palustris* were collected from the uncontaminated lakes Totoriškės, Dubingiai and Verkiai.

Pieces of somatic and gonadal tissue were dissected from snails and prepared according to modified methods previously used in karyological studies of fish and trematodes (Baršienė, 1978; Baršienė and Grabda-Kazubska, 1988). Blocking of cell division at metaphase was obtained by injection of a 0.1–0.2% aqueous solution of colchicine into soft tissue of snails using ~1 ml per 100 g of mollusc weight, 4–6 or 12–15 h before dissection of the snail. Tissues were placed in a hypotonic medium, distilled water for 40–50 min, and fixed in Carnoy's fixative, which was changed three times: after 30 min, after 1 h and after 24 h. Following 1 or more days of fixation, pieces of tissue were washed in 45% acetic acid, smeared on slides or prepared as a cell suspension, before being air-dried and stained with 4% Giemsa solution (30–50 min). Slides were examined under a Jena Med microscope.

8.2.3 Results

The karyotypes of six mollusc species, possessing different diploid chromosome numbers and inhabiting polluted and unpolluted areas, were studied. The most stable chromosome sets of snails were described from the clean lakes and for *Viviparus viviparus* occurring in contaminated sites. Up to 45% of polyploid and up to 23.1% of hypoploid cells, and various meiotic abnormalities, were observed in the tissues of other species of snails from unfavourable polluted biotopes (Table 8.2.1; Figures 8.2.1 and 8.2.2). It should be stressed that most *Lymnaea ovata* males (especially from the mouth of Kuršių Bay) possessed 30–50% of polyploid spermatids (Figure 8.2.3). Remarkable and extensive chromosome set irregularities were observed in tissues of snails from the Nemunas estuaries (Table 8.2.1).

Individual cytogenetic analyses performed on eight bivalve species, listed in Tables 8.2.2 and 8.2.3 (from Nemunas by the Smalininkai, Kuršių Bay by the Ventė and Dreverna, and Šventoji River populations) revealed the predominance in all studied species of a diploid number of 38 chromosomes (Figure 8.2.4). The percentages of such normal nuclei ($2n = 38$) in the tissues of individual bivalve specimens varied from 6.7% to 90.9% (Tables 8.2.2, 8.2.3, 8.2.4 and 8.2.5). Both aneuploid and polyploid cells were observed in their tissues (Figures 8.2.5 and 8.2.6). Furthermore, some bivalve specimens (from different species) possessed $2n = 32, 36, 37$ (e.g. Figure 8.2.7), $3n = 57$, $4n = 76$. Most of the polyploid individuals expressed neoplastic changes. There were also fragmented or split polyploid nuclei and other interphase nuclei with highly irregular shapes (Figure 8.2.8 (a,b)).

There was a strong tendency for genome polyploidization in the clams taken from the Nemunas River at Smalininkai (Table 8.2.2). In contrast, there was a predominance of hypoploid nuclei in the tissues of clams from the Šventoji River estuaries and Kuršių Bay by

Table 8.2.1 Percentage of hypoploid, modal, hyperploid and polyploid cells in molluscs collected from polluted and unpolluted parts of the Lithuanian hydrosystem

Mollusc species	Collection site	2n	Hypo- ploid cells	Modal cells	Hyper- ploid cells	Poly- ploid cells	Number of studied cells
Lymnaea ovata	Vilnia	36	4.9	66.9	2.1	26.1	174
	Nemunas estuaries	36	18.5	66.7	7.4	7.4	181
	Kauno Bay	36	9.6	71.8	3.2	15.4	98
Lymnaea palustris	Kuršių Bay mouth	36	5.2	48.0	1.8	45.0	195
	Nemunas estuaries	36	23.1	69.2	7.7	–	72
Valvata piscinalis	Vilnia	20	11.5	63.4	1.6	23.5	122
	Nemunas estuaries	20	14.5	56.6	4.4	25.5	180
Viviparus viviparus	Nemunas estuaries	18	9.3	88.9	1.8	–	162
	Vilnia	18	1.4	74.5	1.4	22.6	274
	Kauno Bay	18	4.9	86.9	1.6	6.6	124
Dreisena polymorpha	Kuršių Bay (Preila)	32	11.0	62.8	1.5	24.8	258
	Kuršių Bay (Juodkrantė)	32	14.6	66.6	2.1	16.7	192
	Kuršių Bay mouth	32	22.0	71.4	2.5	4.1	275
	Kuršių Bay (Ventė)	32	13.5	61.5	15.4	9.6	104
	Nemunas over Smalininkai	32	13.7	64.2	1.8	20.1	218
Bithynia tentaculata	Lake Totoriškės (unpolluted)	34	6.1	92.0	0.9	0.9	213
Valvata piscinalis	Lake Dubingiai (unpolluted)	20	8.3	88.9	2.8	–	76
Lymnaea palustris	Lake Verkiai (unpolluted)	36	5.0	90.0	2.5	2.5	118

Table 8.2.2 Percentage of hypoploid, modal, hyperploid and polyploid cells in the tissues of some bivalve specimens from the Nemunas Rivers at Smalininkai

Mollusc number and species	Hypoploid cells	Modal cells	Hyperploid cells	Polyploid cells
438 *Unio tumidus*	18.7	62.5	–	18.8
439 *Unio longirostris*	18.1	77.4	4.5	–
440 *Unio longirostris*	3.3	13.3	6.7	76.7
437 *Unio pictorum*	–	–	17.8	82.2
533 *Unio pictorum*	17.0	23.5	4.3	55.2
537 *Unio pictorum*	15.1	43.4	5.7	35.8
541 *Unio pictorum*	20.0	80.0	–	–
544 *Unio pictorum*	85.7	14.3	–	–
545 *Unio pictorum*	50.0	50.0	–	–
547 *Unio pictorum*	20.0	80.0	–	–
548 *Unio pictorum*	23.1	15.4	15.4	46.2
549 *Unio pictorum*	25.0	17.5	2.5	55.0

(a)

(b)

Figure 8.2.1 Polyploid cells of molluscs *Lymnaea ovata*. (a) Polyploid metaphase; (b) pachytene. Arrows indicate non-conjugating sites of bivalents. Scale bar = 10 μm.

(a)

(b)

Figure 8.2.2 Abnormal meiosis of snail *Lymnaea palustris*. (a) Pachytene, including 15 normal bivalents, one circular bivalent and polychromosomal complex consisting of three bivalents (arrows); (b) diakinesis with linear polychromosomal complex (tetravalent) and two univalents (arrows).

the Ventè and Dreverna (Tables 8.2.3, 8.2.4 and 8.2.5). About 25–30% of the studied bivalve specimens were characterized as mosaic. Two individuals – 442 *Unio tumidus* and 699 *Anadonta piscinalis* (Ventè population) – were hermaphroditic, with primary oocytes and spermatocytes being found in their gonads. This is a very unusual observation for these dioecious clams. Furthermore, in their tissues only 35% and 25% respectively of their cells had the modal chromosome number (Table 8.2.5). It is important to note that mitotic inhibition and other different chromosome aberrations were also recorded in bivalve tissues.

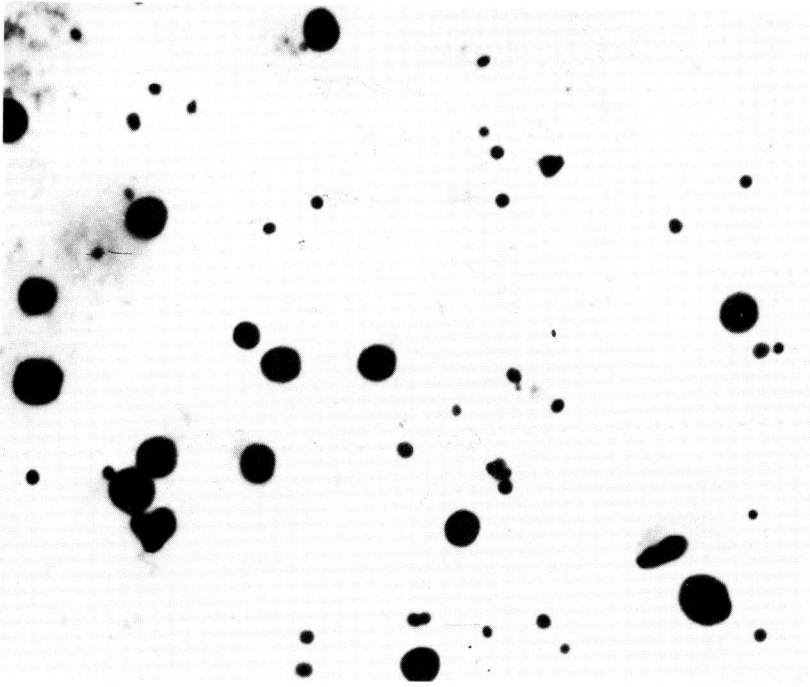

Figure 8.2.3 Typical ratio of polyploid and diploid spermatids in the gonads of *Lymnaea ovata* inhabiting highly polluted areas. Small nuclei = 2*n*; all larger nuclei are polyploid.

Table 8.2.3 Percentage of hypoploid, modal, hyperploid and polyploid cells in the tissues of some bivalve specimens from the Šventoji River estuary

Mollusc number and species	Hypoploid cells	Modal cells	Hyperploid cells	Polyploid cells
507 *Crassiana crassa*	28.6	71.4	–	–
509 *Crassiana crassa*	25.0	85.0	–	–
511 *Crassiana crassa*	29.2	66.6	–	4.2
518 *Crassiana crassa*	33.2	66.8	–	–
521 *Crassiana crassa*	75.0	25.0	–	–
670 *Pseudoanadonta anatina*	15.8	84.2	–	–
671 *Anadonta piscinalis*	24.0	73.8	–	2.2
678 *Anadonta piscinalis*	58.4	41.6	–	–
695 *Anadonta piscinalis*	18.6	65.6	7.9	7.9
672 *Anadonta cygnea*	30.8	69.2	–	–
673 *Anadonta cygnea*	69.8	30.2	–	–
676 *Anadonta cygnea*	23.8	76.2	–	–
677 *Anadonta cygnea*	56.0	41.6	2.4	–
684 *Anadonta cygnea*	50.0	50.0	–	–
685 *Anadonta cygnea*	6.6	13.3	–	80.1
686 *Anadonta cygnea*	86.6	6.7	–	6.7
679 *Anadonta subcircularis*	43.4	52.3	4.3	–
683 *Anadonta subcircularis*	80.2	19.8	–	–

Figure 8.2.4 Modal diploid chromosome set ($2n = 38$) of *Unio pictorum*, from the Nemunas River at Smalininkai.

Table 8.2.4 Percentage of hypoploid, modal, hyperploid and polyploid cells in the tissues of some bivalves from the Kuršių Bay by the Dreverna

Mollusc number and species	Hypoploid cells	Modal cells	Hyperploid cells	Polyploid cells
621 *A. cygnea*	27.0	72.0	1.0	–
622 *A. cygnea*	64.8	33.6	1.6	–
623 *A. cygnea*	15.9	72.7	11.4	–
624 *A. cygnea*	41.2	50.8	8.0	–
626 *A. cygnea*	21.0	73.8	2.6	2.6
628 *A. cygnea*	24.0	72.4	1.8	1.8
629 *A. cygnea*	18.5	71.6	5.4	4.4
630 *A. cygnea*	29.8	64.2	3.6	2.4
631 *A. cygnea*	40.6	56.3	3.1	–
632 *A. cygnea*	36.4	52.3	9.0	2.3
704 *A. cygnea*	42.9	57.1	–	–

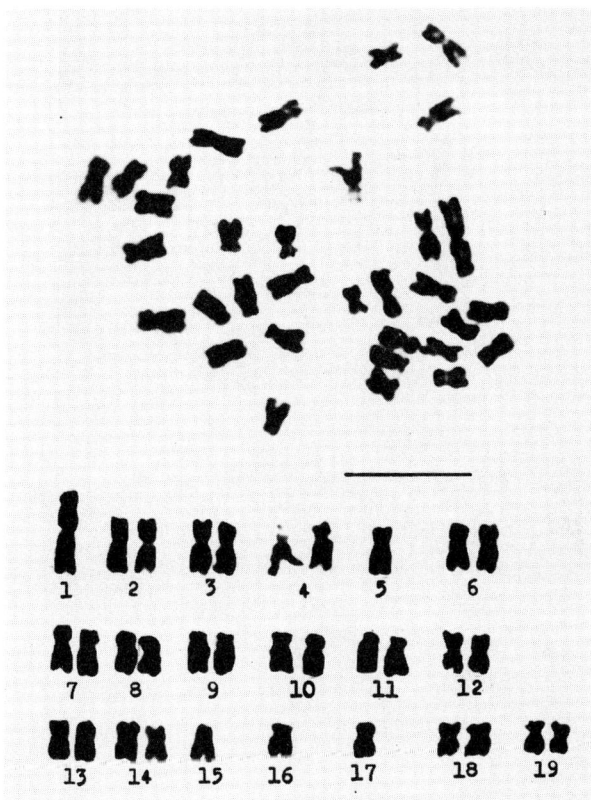

Figure 8.2.5 Hypoploid ($2n = 33$) chromosome set of gill cell of *Unio tumidus* No. 444, collected in Kuršių Bay by the Ventė. Scale bar = 10 μm.

Table 8.2.5 Percentage of hypoploid, modal, hyperploid and polyploid cells in the tissues of some bivalves from the Kuršių Bay by the Ventė ($\female$ = hermaphrodite)

Mollusc number and species	Hypoploid cells	Modal cells	Hyperploid cells	Polyploid cells
441 *U. tumidus*	25.6	68.6	5.8	–
442 $\female$ *U. tumidus*	16.2	35.1	5.4	43.3
443 *U. tumidus*	47.1	52.9	–	–
444 *U. tumidus*	48.8	37.4	1.3	12.5
445 *U. tumidus*	25.0	67.8	3.6	3.6
523 *U. tumidus*	20.0	64.0	16.0	–
524 *U. tumidus*	33.3	66.7	–	–
620 *A. piscinalis*	15.0	65.0	5.0	15.0
688 *A. piscinalis*	10.0	60.0	–	30.0
690 *A. piscinalis*	45.5	36.4	18.2	–
691 *A. piscinalis*	9.1	90.9	–	–
694 *A. piscinalis*	53.0	47.0	–	–
695 *A. piscinalis*	21.2	78.8	–	–
696 *A. piscinalis*	41.7	50.0	–	8.3
698 *A. piscinalis*	12.5	83.3	2.1	2.1
699 $\female$ *A. piscinalis*	75.0	25.0	–	–

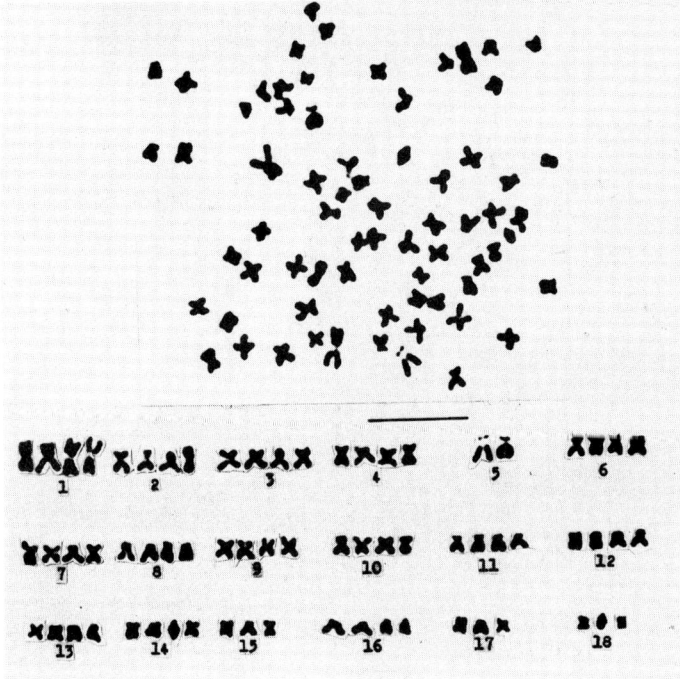

Figure 8.2.6 Chromosomes of polyploid spermatogonium of *Unio pictorum* No. 549 (Smalininkai locality). Scale bar = 10 μm.

8.2.4 Discussion

It is known that effects of pollutants are usually displayed first at the biochemical and molecular levels (Veldhuizen-Tsoerkan *et al.*, 1991). This then leads to genetic changes which become cytologically visible, especially in the tissues of organisms which are good pollutant bioaccumulators, e.g. molluscs. This study demonstrates the high level of mollusc chromosome set changes in populations inhabiting some of the most polluted aquatic areas in Lithuania: Kuršiu̯ Bay mouth (by the Klaipėda), the dam of Vilnia (Vilnius), and Kuršiu̯ Bay by the Ventė (near the Nemunas estuaries). Chromosome sets of *L. ovata* are very sensitive to the influence of environmental pollution. Mitotic suppression, nuclear polyploidization and degeneration were common in the tissues of most specimens inhabiting particularly the highly polluted northern part of Kuršiu̯ Bay, fed by the Nemunas River, which flows through the big cities of Gardinas, Alytus, Kaunas, and Sovietskas (Tilžė). In relatively unpolluted lakes the same mollusc species possessed significantly fewer chromosome set abnormalities.

There were remarkable differences in the karyotype instability between the various bivalve specimens. In clams, genome polyploidization was marked in Smalininkai and significant levels of hypoploidy were found in the Šventoji, Ventė and Dreverna populations. These last three populations are located in the zone of intensive Chernobyl radionuclide contamination. It should be stressed that the Dreverna population of clams disappeared in 1991. Furthermore, remarkable rarity of bivalves from the Ventė biotopes has also been noted. There is supporting literature on data concerning cytogenetic effects of radiation (Wong and Wagner, 1956; Narang, 1974), heavy metals and oil on molluscs (Dixon, 1982).

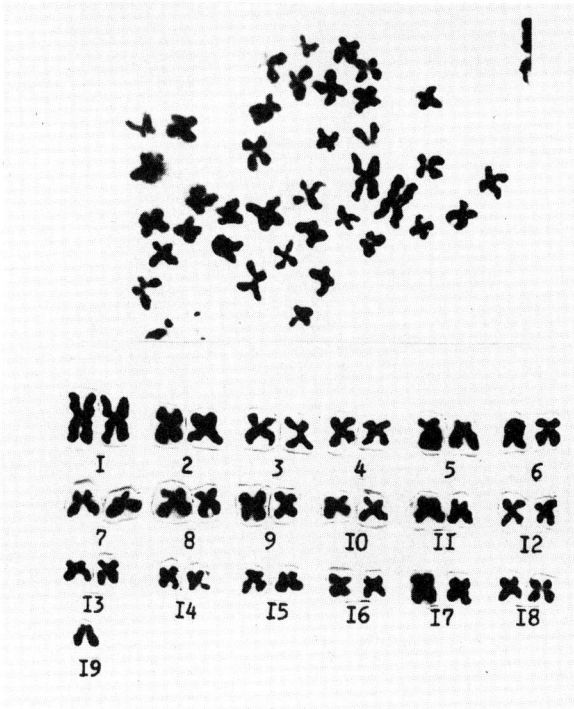

Figure 8.2.7 Hypoploid karyotype ($2n = 37$) of *Unio pictorum* No. 544 gill cell (Smalininkai locality).

Heavy metals are known to be able to disturb DNA (Choi and Simpkins, 1986). Different pollutants can retard development and physiological processes, and provoke mitotic suppression or sterility of aquatic organisms. In the Neris River below Vilnius, and the Nemunas River below Kaunas, Tilžè, these phenomena were significantly expressed in bivalve tissues.

Soft tissues of some of the bivalve specimens used for karyological examination were analysed for heavy metals (Fe, Mn, Cu, Zn, Ni, Cd, Pb, Cr) and aromatic hydrocarbons (naphthalene, anthracene, phenanthrene, fluoranthene, pyrene, triphenylene, chrizene, perylene, 3-4 benzopyrene) (K. Jokšas and S. Zareckas, unpublished data).

The shells were used for the determination of β–radioactivity (R. Dušauskienè-Duž, unpublished data). The data from some of these studies are presented in Table 8.2.6. In 80% of bivalve specimens a positive correlation was found between pollutant bioaccumulation and chromosome set changes. Aneuploid *Unio pictorum* No. 544 from the Smalininkai biotopes, exhibiting a high proportion of cells with only 32 chromosomes, had the highest level of Mn, Zn and Cd contamination. Increased hydrocarbon concentrations were detected in the tissues of polyploid *U. longirostris* No. 440 and *U. pictorum* No. 533 (Smalininkai). Polyploid *U. pictorum* No. 548 and *Anadonta cygnea* No. 685 possessed a high number of cancer cells even at relatively low tissue contaminations of aromatic hydrocarbons. A similar pattern was found for the tissues of hermaphroditic *A. piscinalis* No. 699. Especially high concentrations of hydrocarbons were found in *A. cygnea* (622, 621) and in *A. piscinalis* (695, 691) (from the Kuršiu Bay by the Dreverna and Ventè). In their tissues no extensive changes of chromosome sets were noticed. A high number of digestive tract micro-organisms –

(a)

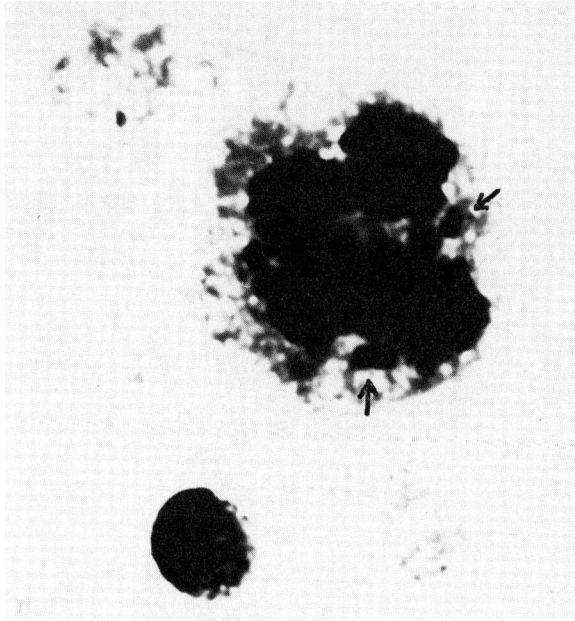

(b)

Figure 8.2.8 Neoplastic polyploid nuclei of *Anadonta cygnea* No. 685 (Šventoji River). (a) Nuclear fragmentation (arrows) of polyploid nucleus alongside a normal diploid nucleus; (b) split polyploid nucleus.

biodegraders of oil products – were present in these specimens and in other clams from the Ventè and Smalininkai localities (J. Šyvokienè, unpublished data).

Thus there was a wide variability in pollutant bioaccumulation and chromosome set changes in the tissues of various bivalve specimens. Biodegradation of pollutants in clams could partly explain this phenomenon as well as other physiological or genetical

Table 8.2.6 Tissue levels of aromatic hydrocarbons, heavy metals (10^{-4}% dry mass), β-radioactivity (Bq/kg shells mass: background < 25 Bq/kg) and chromosome set peculiarities in some bivalve specimens

Snail specimens: number, sex and species	Aromatic hydrocarbons ($\mu g/g$)	Benzopyrene (ng/g)	Mn (% dry mass)	Zn (% dry mass)	Cd (% dry mass)	Pb (% dry mass)	β-activity	Modal cells (%)	Cells with chromosomal abnormalities (%)
440 ♂ U. longirostris	52.5	67.0	0.23	261.3	2.1	0.1	250.0	13.3	76.7%, 3n
533 ♂ U. pictorum	109.9	28.2	0.07	46.5	0.1	0.2	275.0	23.5	55.2%, 4n—10n
544 ♀ U. pictorum	42.8	12.0	7.5	986.1	2.3	11.7	50.0	14.3	57%, 2n = 32
545 ♂ U. pictorum	51.5	11.7	0.45	398.0	2.4	9.0	150.0	50.0	25%, 2n = 32
547 ♀ U. pictorum	72.3	34.4	0.21	175.5	0.1	0.2	312.5	80.0	20%, 2n = 36
548 ♀ U. pictorum	52.1	11.2	0.16	142.0	0.1	0.3	225.0	15.4	46.2%, 4n; 15.4%, 8n
678 ♀ A. piscinalis	36.5	25.2	0.30	150.0	0.1	2.5	387.5	42.4	37.5%, 2n = 36
679 ♀ A. subcircularis	46.6	3.5	0.13	86.5	0.1	1.0	450.0	52.3	45.4%, 2n = 33–37
685 ♂ A. cygnea	35.3	11.6	0.04	69.5	0.1	0.5	225.0	13.3	69.3%, 4n; 10.8%, 8n
621 ♂ A. cygnea	835.7	318.9	0.06	64.4	0.1	0.1	Background	72.0	14%, 2n = 37
622 ♀ A. cygnea	1035.5	14.6	0.05	53.0	0.1	0.1	525.0	33.6	50%, 2n = 37
690 ♂ A. piscinalis	28.9	7.2	0.53	217.0	0.1	2.5	225.0	36.4	45.5%, 2n = 32, 36
691 ♂ A. piscinalis	125.1	29.9	0.25	186.0	0.1	9.6	275.0	90.9	
694 ♀ A. piscinalis	121.4	62.5	0.31	138.0	0.1	0.3	175.0	47.0	47%, 2n = 36
695 ♂ A. piscinalis	156.2	3.9	0.64	340.0	1.8	0.5	375.0	78.9	
696 ♂ A. piscinalis	66.6	2.0	0.78	359.0	2.2	13.0	512.5	50.0	41.7%, 2n = 33, 34, 36
698 ♀ A. piscinalis	7.2	2.0	0.51	476.5	0.1	8.5	62.5	83.3	
699 ♀ A. piscinalis	41.7	11.5	0.37	255.0	0.2	0.1	75.0	25.0	50%, 2n = 36

peculiarities. Nevertheless, the existence of polyploid, hypoploid, mosaic and herma-phroditic bivalve specimens, together with the presence of extensive evidence for polyploid tumour cells, mitotic supression and structural changes of chromosomes in diploid snails, indicates high levels of genetic risk to invertebrates living in the main Lithuanian hydrosystem below the biggest cities and in the Chernobyl radionuclide hot spots.

Acknowledgements

The author wishes to thank Dr R. Dušauskienė-Duž (Institute of Botany), J. Šyvokienė (Institute of Ecology), S. Zareckas and K. Jokšas (Institute of Geography) for scientific collaboration and performing the β-radioactivity, pollutant biodegradation, hydrocarbon and heavy metal analyses in the tissues of molluscs. I sincerely thank an anonymous reviewer for comments and especially Dr D.R. Dixon, who greatly improved the quality of the final work by a detailed and critical review of the manuscript.

References

Baršienė, J. (1978) Ontogenetic variability of the chromosome number of Atlantic salmon (*Salmo salar* L.). *Genetika*, **24**, 2029–36 (in Russian).

Baršienė, J. and Grabda-Kazubska, B. (1988) A comparative study on chromosomes in plagiorchiid trematodes. I. Karyotypes of *Opisthioglyphe ranae* (Frölich, 1791), *Haplometra cylindracea* (Zeder, 1800) and *Leptophallus nigrovenosus* (Bellingham, 1844). *Acta. Parasitol. Polonica*, **33**, 249–57.

Choi, B.H. and Simpkins, H. (1986) Changes in the molecular structure of mouse fetal astrocyte nucleosomes produced in vitro by methylmercuric chloride. *Environ. Res.*, **39**, 321–30.

Dixon, D.R. (1982) Aneuploidy in mussel embryos (*Mytilus edulis* L.) originating from a polluted dock. *Mar. Biol. Lett.* **3**, 155–61.

Dušauskienė-Duž R. (1992) Migration of ^{90}Sr in the cooling basin of the Ignalina atomic power plant and the Baltic Sea ecosystems, in *Radionuclide Pollution in Lithuania and its Effects*, (eds J. Baršiene *et. al.*), Academia, Vilnius, pp. 71–78 (in Lithuanian).

Jokšas, K. (1991) Peculiarities of heavy metals migration in the King Wilhelm canal in the Northern part of the Kuršiu Bay and in the Baltic near-shore in 1989, in *Nemunas Basin Water Pollution and Biological Effects in the Ecosystem*, (ed. R. Lekevičius), Academia, Vilnius, pp. 14–20 (in Lithuanian).

Lukšienė, B., Tarasiukas, N., Šalavėjus, S. *et al.*(1992) Investigations of migration and accumulation of radionuclides in the ecosystems of the Kuršiu Marios Lagoon, Kauno Marios and the Eastern part of the Baltic Sea, in *Radionuclide Pollution in Lithuania and its Effects* (eds J. Baršiene *et. al.*), Academia, Vilnius, pp. 100–6 (in Lithuanian).

Milukaitė, A. and Bukelskis, E. (1991) Investigations of polycyclic aromatic hydrocarbons in aquatic ecosystems of the Nemunas basin, in *Nemunas Basin Water Pollution and Biological Effects in the Ecosystem* (ed. R. Lekevičius) Academia, Vilnius, pp. 37–38 (in Lithuanian).

Narang, N. (1974) Cytogenetic effects of radiation on the planorbid snail, *Biomphalaria glabrata*. *Caryologia*, **27**, 387–93.

Špirkauskaitė, N., Tarasiukas, N., Lukšienė, B., Stelingis, K. and Šalavėjus, S. (1992) Investigation of the mechanisms of cleaning from radionuclides of the water way 'Nemunas–Curonian Lagoon–Baltic Sea', in *Radionuclide Pollution in Lithuania and its Effects* (eds J. Baršiene *et. al.*), Academia, Vilnius pp. 95–100 (in Lithuanian).

Taraškevičius, R. (1991) Important features of heavy metals geochemical behaviour in the Nemunas basin rivers, in *Nemunas Basin Water Pollution and Biological Effects in the Ecosystem*, (ed R. Lekevičius), Academia, Vilnius, pp. 6–12 (in Lithuanian).

Veldhuizen-Tsoerkan, M.B., Holwerda, D.A., de Bont, A.M.T. *et. al.* (1991) A field study on stress indices in the sea mussel, *Mytilus edulis*: application of the 'stress approach' in monitoring. *Arch. Environ. Contam. Toxicol.*, **21**, 297–304.

Wong, J.W. and Wagner, E.D. (1956) Some effects of ultra violet radiation on *Oncomelania nososphora* and *O. quadrasi* snail intermediate hosts of *Schistosoma japonicum*. *Trans. Am. Microsp. Soc.*, **75**. 204–10.

Zareckas, S. (1991) Technogenic and natural hydrocarbons in the water basins of Lithuania, in *Nemunas Basin Water Pollution and Biological Effects in the Ecosystem* (ed. R. Lekevičius), Academia, Vilnius, pp. 29–36 (in Lithuanian).

8.3 EFFECTS OF HEAVY METALS AND TEMPERATURE ON THE GENETIC STRUCTURE AND GPI ENZYME ACTIVITY OF THE BARNACLE *BALANUS AMPHITRITE* DARWIN (CIRRIPEDIA: THORACICA)

Consuelo Montero, Bruno Battaglia, Michele Stenico, Giancarlo Campesan and Tomaso Patarnello

Abstract

The effects of thermal and chemical pollution on genotype frequencies and enzyme activity were studied for about 1 year in three populations of the barnacle Balanus amphitrite *Darwin from the lagoon of Venice, Italy. In addition, the amount of specific heavy metals (Hg, Cd, Co, Ni, Cr, Cu, Pb, Zn, Mn, Fe) was measured in sediments and in the soft tissues of barnacles. Population A was subjected to high chemical pollution; population B was exposed to temperatures constantly higher than the rest of the lagoon (Δt up to +10 °C). The third population (C) was sampled in an unpolluted area of the lagoon of Venice and used in the experiments as control.*

Population A exhibited a deficiency of heterozygotes at the Gpi, Mpi *and* Pgm *loci. Such a deficiency was also found in populations B and C but significant departures from Hardy–Weinberg expectations were more frequent in population A.*

The sediments collected in the area of population A always exhibited a higher heavy metal concentration. On the contrary, the soft tissues of population B showed a higher concentration of Mg, Cd, Co, Cu, Zn, Mn and Fe compared to the other populations.

The effect of Cu, Zn and Cd on GPI enzyme activity was assayed at low and high metal concentration. The analyses were carried out at both 20 °C and 37 °C. Significant differences in GPI activity were found at 20 °C among the three populations. Mean overall activity was higher in homozygotes than in heterozygotes. Cd and Zn caused a decrease in activity which was more obvious at 20 °C. The GPI activity was not affected by Cd either at low or high concentration.

8.3.1 Introduction

The study of pollution in the marine environment should consider not only the measurements of chemical and physical diagnostic parameters but also the responses of biological systems. Over recent years many studies have been carried out with the aim of developing a genetic approach for studying the biological effects of pollution (Beardmore *et al.*, 1980). Genetic monitoring has proved to be a useful method for revealing changes in the genetic structure of populations inhabiting polluted environments. However, such an approach does not provide information concerning the mechanisms by which pollutants may affect specific genotypes differently. An integrated approach which combines field *(in vivo)* studies and *in vitro* experiments has been suggested (Eanes, Katona and Longine, 1990).

Balanus amphitrite is the most widespread barnacle species in the Lagoon of Venice. It typically settles on the low shore, requiring warm water for successful growth and breeding (Rainbow, 1984). *B. amphitrite* has been used in previous studies (Barbaro *et al.*, 1978; Nevo, Shimony and Libni, 1978) as a biological indicator of pollution. Barnacles, because of their feeding behaviour, are able to accumulate metals in tissues. Rainbow and White (1989) showed that the barnacle *Elminius modestus* concentrates Zn, Cd and Cu by filtering at a high rate (according to environmental levels) with no evidence of metal excretion. In *Balanus balanoides* heavy metals are incorporated into the cells of the liver-like organ 'stratum perintestinale'. They form special structures such as inert granules of zinc phosphate and insoluble deposits of copper and sulphur (Walker *et al.*, 1975; Walker, 1977).

Organisms can face the effects of chemical toxicants in different ways (Calow, 1991) which are in all cases energetically costly (Hawkins *et al.*, 1989; Koehn and Bayne, 1989; Calow, 1991). In a situation of long-term stress, and if all the proximate or physiological responses fail, pathological effects will occur, leading ultimately to death. In cyprid larvae of *B. amphitrite niveus,* exposure to Cu produced an increase in the oxygen uptake, and at high Cu concentrations drastically reduced survival rate (Bernard and Lane, 1963).

Heavy metals such as Zn, Cu, Cd and Hg have been suggested as selective agents in species of marine invertebrates under both laboratory (Nevo *et al.*, 1981; Nevo, Lavie and Ben-Shlomo, 1983; Lavie and Nevo, 1982, 1986, 1987) and field conditions (Nevo, Ben-Shlomo and Lavie, 1984; Nevo, Lavie and Noy, 1987). Heavy metals can interact with protein structure (Milstein, 1961; Webb, 1966) and might change the enzymatic properties of the different genotypes at a polymorphic locus. Less efficient genotypes might have an important bioenergetic impact on an organism's energy budget, conferring on them a lower fitness. This is more likely at loci which are involved in central metabolism such as enzymes in the glycolytic pathway or the Krebs cycle (Blackstock, 1980; Koehn and Bayne, 1989).

GPI (glucose phosphate isomerase, EC 5.3.1.9) is involved both in glycolysis and gluconeogenesis; it reversibly catalyses the isomerization between fructose 6-phosphate and glucose 6-phosphate. This protein may be considered as a coupling enzyme between rate-limiting reactions that distribute carbohydrates for different metabolic uses (Watt, 1985). Many studies have provided evidence that GPI is under natural selection, particularly under changing or unpredictable environments (Hoffmann, 1983; Watt, Cassin and Swan, 1983; Shiahab and Heath, 1987; Silva *et al.*, 1989; Patarnello and Battaglia, 1992).

The aim of this work was to evaluate heavy metal pollution and explore its possible genetic and biochemical effects on populations of *B. amphitrite* in the Lagoon of Venice.

8.3.2 Materials and methods

Samples of *B. amphitrite* were collected in three areas of the Lagoon of Venice, Italy (Figure 8.3.1). Populations A and B were sampled from the highly industrialized zone of Porto Marghera, population A from the 'South Industrial Channel', which has high levels of chemical pollution (Perin and Gabelli, 1983; Perin *et al.*, 1983), and population B from in front of the outflow of an electric plant cooling system (ENEL-CTE) where the temperature is consistently higher than the rest of the lagoon (Δt up to + 10 °C). At site B the level of chemical pollution is low. Population C, situated at Chioggia, a thermal and chemically unpolluted zone of the lagoon (Donazzolo *et al.*, 1984), was taken in this study as a control.

Sampling was carried out during May and November 1990, and January, April and June 1991. Samples were immediately frozen at −40 °C after collection. Barnacles were tested both electrophoretically and for heavy metal content. Electrophoretic analyses were carried out on cellulose acetate for the most polymorphic loci, mannose phosphate isomerase (*Mpi*, EC 5.3.1.8), phosphoglucomutase (*Pgm*, EC 2.7.5.1.) and *Gpi*, according to Patarnello *et al.* (1990). Genotype frequencies, mean heterozygosity and Hardy–Weinberg expectations were calculated using BIOSYS-1 (Swofford and Selander, 1981). Concentrations of Cd, Co, Ni, Cr, Cu, Pb, Zn, Mn, Fe and Hg were measured in the soft tissues according to Carmody, Pearce and Vasso (1973) and Agemian and Chau (1976). Metal concentration was expressed in terms of μg/g wet weight or ppm.

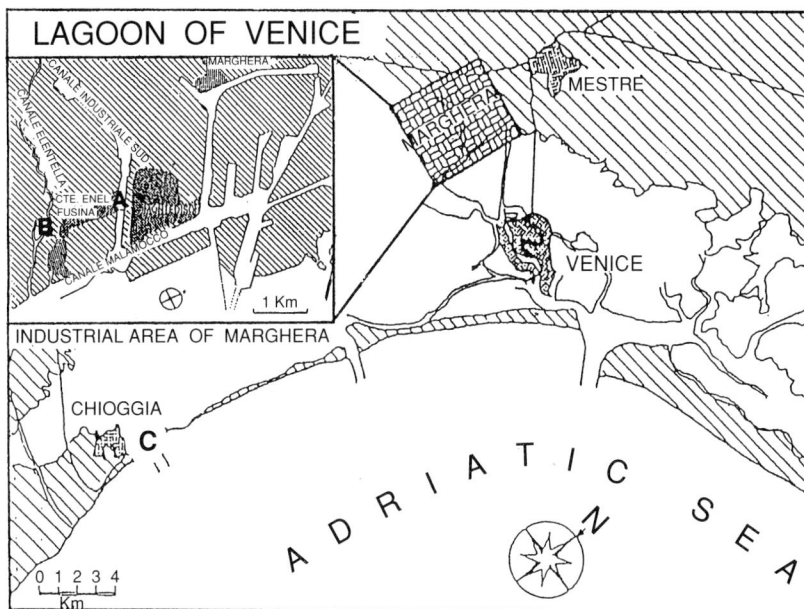

Figure 8.3.1 *B. amphitrite* sampling areas in the Lagoon of Venice, Italy. (A) Industrial channel; (B) electric power plant outflow; (C) Chioggia.

Heavy metal contamination in the upper layer of sediments was evaluated for the same metals listed above by seasonal collections from the same sampling sites as the barnacles (A, B and C).

(a) Enzyme activity experiments

GPI enzyme activity of the two most frequent genotypes, namely *Gpi* B/B and *Gpi* B/C, was assayed in samples of *B. amphitrite* (15 individuals from each site) collected in November 1991. Samples were prepared for the biochemical assays according to Patarnello and Battaglia (1992). Sigma kits containing fructose 6-phosphate 0.44 mM, glucose-6-phosphate dehydrogenase (G-6-PDH) (12 units) and NADP 6 μmol were used for the assays. The reaction was started by adding 100 μl of sample homogenate to 1 ml of kit solution. Only the linear phase of the reaction was considered for the calculation. The analyses were carried out both at 20 °C and 37 °C on single individuals scored for *Gpi* genotypes. To test the effect of Cu, Zn and Cd on GPI activity, low and high metal concentrations were added to the reaction mixture. Metal concentrations were chosen according to maximum and minimum metals content in the barnacles' tissues and were: 32 and 320 ppm of $CuSO_4$, 0.47 and 4.7 ppm $CdCl_2$ and 700 and 7000 ppm $ZnCl_2$. The enzyme activity of each sample was also measured without metal addition (blank) and has been expressed as units of activity/mg of total soluble protein. Total soluble proteins were measured for each individual according to Lowry *et al.* (1951).

8.3.3 Results

Tables 8.3.1, 8.3.2 and 8.3.3 report mean heterozygosity (observed and expected), *D* values ($D = \Sigma$ heterozygotes observed $- \Sigma$ heterozygotes expected/Σ heterozygotes expected) and

Table 8.3.1   Genotype frequencies at *Gpi*, *Mpi* and *Pgm* loci in population A (industrial channel population) of *B. amphitrite*

	$2N$	H_{obs}	H_{exp}	p(H–W)	D
Gpi					
May 1990	232	0.138	0.214	*	−0.356
November 1990	118	0.051	0.144	*	−0.646
January 1991	178	0.079	0.138	NS	−0.433
April 1991	100	0.080	0.243	*	−0.675
June 1991	110	0.091	0.088	NS	0.026
Pgm					
May 1990	200	0.200	0.331	**	−0.395
November 1990	120	0.317	0.584	**	−0.458
January 1991	174	0.230	0.422	***	−0.458
April 1991	98	0.143	0.188	NS	−0.246
June 1991	108	0.204	0.357	***	−0.434
Mpi					
May 1990	186	0.462	0.635	***	−0.272
November 1990	74	0.351	0.530	*	−0.337
January 1991	146	0.288	0.556	***	−0.486
April 1991	98	0.408	0.647	**	−0.376
June 1991	100	0.440	0.601	***	−0.275

$2N$ = number of genes scored; H_{obs} = observed heterozygosity; H_{exp} = expected heterozygosity; p(H–W) = evaluation of the Hardy–Weinberg equilibrium; $D = (H_{obs} - H_{exp})/H_{exp}$; $^{*}p < 0.05$, $^{**}p < 0.005$, $^{***}p < 0.001$; NS = not significant.

Table 8.3.2   Genotype frequencies of *Gpi*, *Mpi* and *Pgm* loci in population B, (power plant outflow)

	$2N$	H_{obs}	H_{exp}	p(H–W)	D
Gpi					
May 1990	228	0.096	0.093	NS	0.039
November 1990	152	0.145	0.136	NS	0.066
January 1991	142	0.056	0.133	NS	−0.058
April 1991	100	0.100	0.095	NS	0.042
June 1991	110	0.036	0.036	NS	0.000
Pgm					
May 1990	200	0.160	0.224	NS	−0.287
November 1990	142	0.268	0.373	NS	−0.282
January 1991	120	0.417	0.486	NS	−0.149
April 1991	100	0.200	0.266	NS	−0.256
June 1991	108	0.426	0.434	NS	−0.027
Mpi·					
May 1990	202	0.505	0.542	NS	0.021
November 1990	82	0.317	0.464	NS	−0.317
January 1991	128	0.453	0.625	*	−0.281
April 1991	100	0.600	0.647	NS	−0.082
June 1991	104	0.587	0.0596	NS	−0.005

$2N$ = number of genes scored; H_{obs} = observed heterozygosity; H_{exp} = expected heterozygosity; p (H–W) = evaluation of the Hardy–Weinberg equilibrium; $D = (H_{obs} - H_{exp})/H_{exp}$; $^{*}p < 0.05$; $^{**}p < 0.005$; $^{***}p < 0.001$; NS = not significant.

Table 8.3.3   Genotype frequencies of *Gpi*, *Mpi* and *Pgm* loci in population C (Chioggia)

	2N	H_{obs}	H_{exp}	p(H–W)	D
Gpi					
May 1990	216	0.057	0.203	NS	−0.361
November 1990	128	0.125	0.174	NS	−0.283
January 1991	140	0.057	0.083	NS	−0.315
April 1991	100	0.060	0.058	NS	0.021
June 1991	110	0.109	0.104	NS	0.040
Pgm					
May 1990	208	0.096	0.293	***	−0.671
November 1990	114	0.175	0.251	NS	−0.300
January 1991	140	0.271	0.502	***	−0.464
April 1991	100	0.300	0.430	NS	−0.310
June 1991	108	0.352	0.518	*	−0.327
Mpi					
May 1990	156	0.462	0.532	NS	−0.132
November 1990	80	0.300	0.519	**	−0.422
January 1991	132	0.364	0.597	***	−0.396
April 1991	98	0.510	0.665	NS	−0.241
June 1991	108	0.481	0.608	NS	−0.216

2N = number of genes scored; H_{obs} = observed heterozygosity; H_{exp} = expected heterozygosity; p(H–W) = evaluation of the Hardy–Weinberg equilibrium; $D = (H_{obs}-H_{exp})/H_{exp}$; $^*p < 0.05$; $^{**}p < 0.005$; $^{***}p < 0.001$; NS = not significant.

Hardy–Weinberg expectations for *Gpi*, *Mpi* and *Pgm* in populations A, B and C. Population A exhibited a significant heterozygote deficiency in many of the samples at the three analysed loci. Heterozygote deficiencies were also found in populations B and C but they were rarely significant. *Gpi* gene frequencies are reported in Tables 8.3.4, 8.3.5 and 8.3.6 for each of the five sampling times in the three populations.

Table 8.3.4   *Gpi* allele frequencies in population A (industrial channel) of *B. amphitrite*

Alleles	May 1990	November 1990	January 1991	April 1991	June 1991
a	0.058	0.042	0.045	0.130	0.018
b	0.884	0.924	0.927	0.860	0.955
c	0.058	0.034	0.028	0.010	0.027

Table 8.3.5   *Gpi* allele frequencies in population B (power plant outflow) of *B. amphitrite*

Alleles	May 1990	November 1990	January 1991	April 1991	June 1991
a	0.018	0.013	0.049	0.000	0.009
b	0.952	0.928	0.930	0.950	0.982
c	0.030	0.059	0.021	0.050	0.009

Table 8.3.6   *Gpi* allele frequencies in population C (Chioggia) of *B. amphitrite*

Alleles	May 1990	November 1990	January 1991	April 1991	June 1991
a	0.074	0.047	0.029	0.000	0.010
b	0.889	0.906	0.957	0.970	0.945
c	0.037	0.047	0.014	0.030	0.045

Table 8.3.7 Heavy metals in sediments of sampling sites; mean values and standard deviation (μg/g or ppm)

Population	Metal concentration									
	Hg	Cd	Co	Ni	Cr	Cu	Pb	Zn	Mn	Fe
Industrial channel (A)	1.13 ± 0.19	12.9 ± 6.7	16.8 ± 2.2	55.5 ± 9.1	46.7 ± 11.1	71 ± 12	142.1 ± 14.7	979 ± 212	424 ± 54	29491 ± 2725
Power plant outflow (B)	0.49 ± 0.23	2.6 ± 1.5	20.3 ± 17.4	27.2 ± 4.6	19.1 ± 5.2	46 ± 34	56.9 ± 20.9	381 ± 314	312 ± 59	15233 ± 3279
Chioggia (C)	0.42 ± 0.11	0.6 ± 0.2	10.9 ± 3.3	40.2 ± 6.6	41.9 ± 2.0	57 ± 13	38.4 ± 10.5	163 ± 22	285 ± 42	20222 ± 1084

Table 8.3.8 Heavy metals in soft tissues of *B. amphitrite*; mean values and standard deviation (μg/g or ppm)

Population	Metal concentration									
	Hg	Cd	Co	Ni	Cr	Cu	Pb	Zn	Mn	Fe
Industrial channel (A)	0.16 ± 0.09	1.90 ± 1.10	2.10 ± 0.80	6.1 ± 8.8	3.5 ± 4.7	47 ± 9	3.1 ± 1.4	2124 ± 830	35 ± 15	366 ± 239
Power plant outflow (B)	0.33 ± 0.24	2.60 ± 1.40	2.50 ± 1.10	2.1 ± 0.7	2.0 ± 0.7	111 ± 47	2.2 ± 0.9	3090 ± 1443	45 ± 29	346 ± 184
Chioggia (C)	0.10 ± 0.03	0.70 ± 0.30	0.50 ± 0.30	1.3 ± 1.2	1.7 ± 1.1	21 ± 9	1.1 ± 0.5	569 ± 225	13 ± 3	321 ± 108

Sediments collected in the industrial channel area (Table 8.3.7) showed a higher heavy metal concentration (with the exception of Co) compared with the other two sites. In the case of Cd, Pb and Zn the concentrations were at least three times higher in the industrial channel. In contrast heavy metal content in tissues (Table 8.3.8) showed considerably higher levels of Mg, Cd, Co, Cu, Zn and Mn for population B compared with the other two populations. The samples from Chioggia (C) exhibited the lowest metal concentrations both in tissues and sediments (with rare exceptions, see Tables 8.3.7 and 8.3.8) which confirms this population as a good control.

GPI activities of each of the three populations were compared by t-tests (significance levels adjusted by Bonferroni sequential method) (Rice, 1989). The results showed significant differences $(p < 0.05)$ only 20 °C between population C and population A (Figure 8.3.2). From 20 °C to 37 °C a significant $(p < 0.05)$ activity reduction was observed in all populations (Table 8.3.9). At 20 °C mean overall homozygote enzyme activity (calculated grouping the homozygotes from the three populations) was significantly higher $(p < 0.05)$ than the mean overall heterozygote activity; the same comparison at 37 °C revealed no significant differences between homozygotes and heterozygotes (Table 8.3.10). The total soluble protein assay showed differences by genotypes; in fact, heterozygotes were characterized by significantly lower protein concentration (t-test = 4.899, d.f. = 49, $p < 0.0001$).

ANOVA tests (adjusted by Bonferroni sequential method; $\alpha = 0.05$) (Rice, 1989) showed, in the presence of high concentrations of Zn, a significant decrease of GPI activity at both 20 °C and 37 °C (compared to the blank), whereas at low concentrations the Zn caused a significant reduction of activity only at 37 °C (Figure 8.3.3(a)). Also, Cu produced a decrease of GPI activity which was significant at 20 °C but not at 37 °C (Figure 8.3.3(b)). Enzyme activity was never affected by Cd, either at low or high concentration (Figure 8.3.3(c)). Comparing the two genotypes, GPI activity was, in most of the cases, higher in homozygotes. These differences are especially obvious at 20 °C, when the activity has its

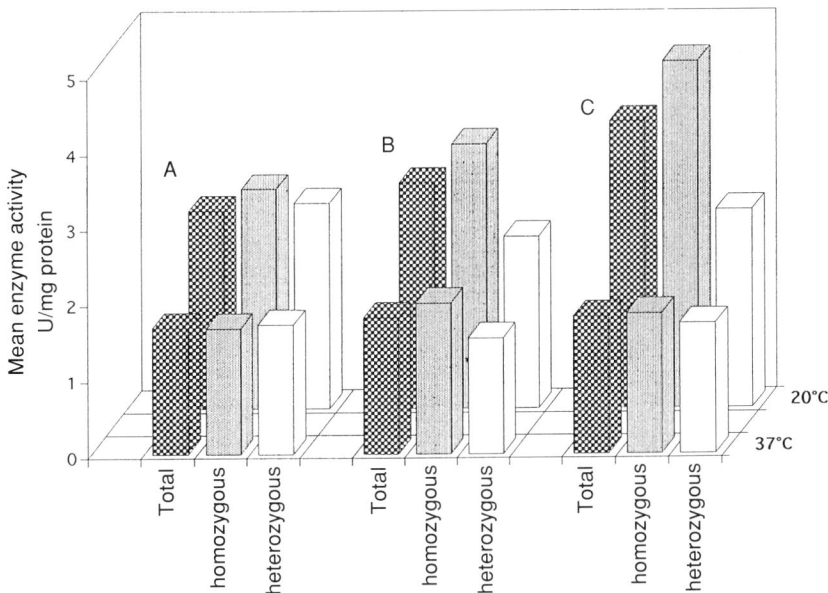

Figure 8.3.2 *GPI enzyme activity of homozygotes and heterozygotes of B. amphitrite*. Comparison between the three populations at 20° C and 37 °C. For abbreviations see Figure 8.3.1.

Table 8.3.9 *Gpi* enzyme activity in *B. amphitrite*: comparison between 20 °C and 37 °C

	Population					
	Industrial channel (A)		*Power plant outflow (B)*		*Chioggia (C)*	
	20 °C	*37 °C*	*20 °C*	*37 °C*	*20 °C*	*37 °C*
N	17	17	17	17	15	15
Mean	2.61 ± 1.59	1.66 ± 1.02	2.98 ± 1.43	1.79 ± 0.801	3.78 ± 1.61	1.80 ± 0.818
t-test		2.22		2.99		4.25
d.f.		32		32		28
p		*		**		***

N = number of tested individuals; $^{*}p < 0.05$; $^{**}p < 0.005$; $^{***}p < 0.001$. p values adjusted by Bonferroni sequential method.

Table 8.3.10 Mean overall GPI enzyme activity in *B. amphitrite*: comparison between homozygotes and heterozygotes

	Temperature			
	20 °C		*37 °C*	
	Homozygotes	*Heterozygotes*	*Homozygotes*	*Heterozygotes*
N	32	17	32	17
Mean	3.604 ± 1.625	2.496 ± 1.560	1.813 ± 0.878	1.631 ± 0.874
t-test		−2.30		−0.69
d.f.		47		47
p		*		NS

N = number of tested individuals; d.f. = degrees of freedom; NS = not significant; $^{*}p < 0.05$.

(a)

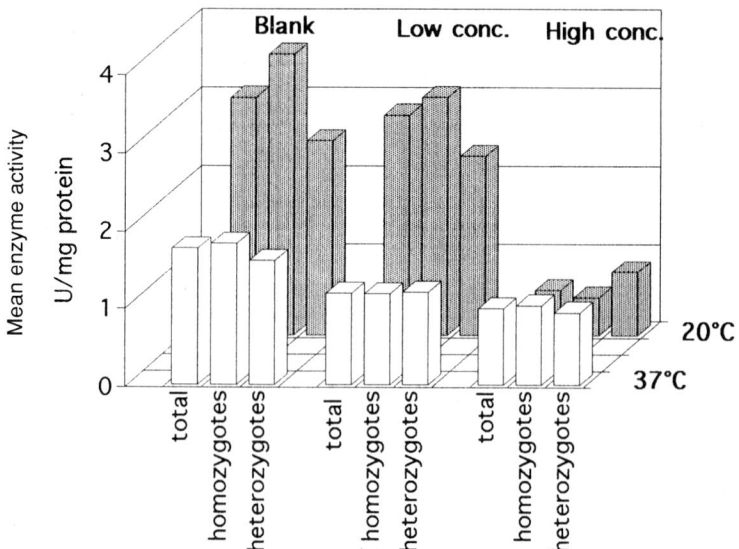

(b)

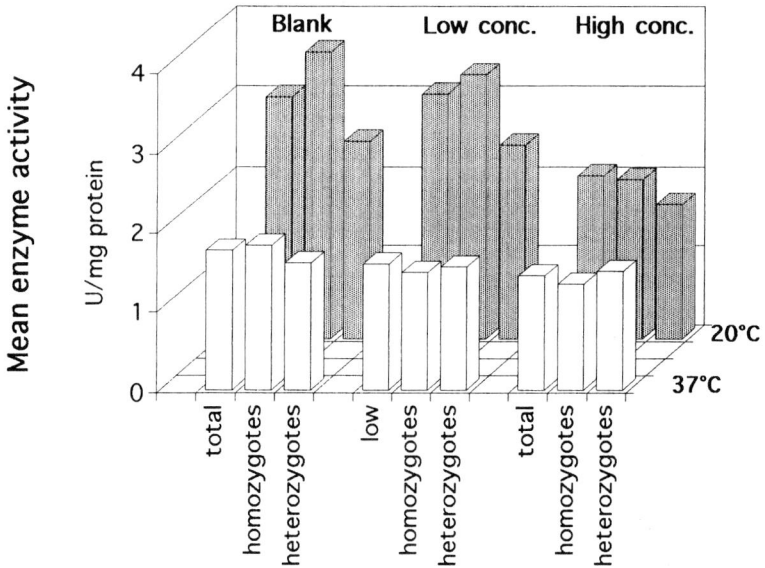

(c)

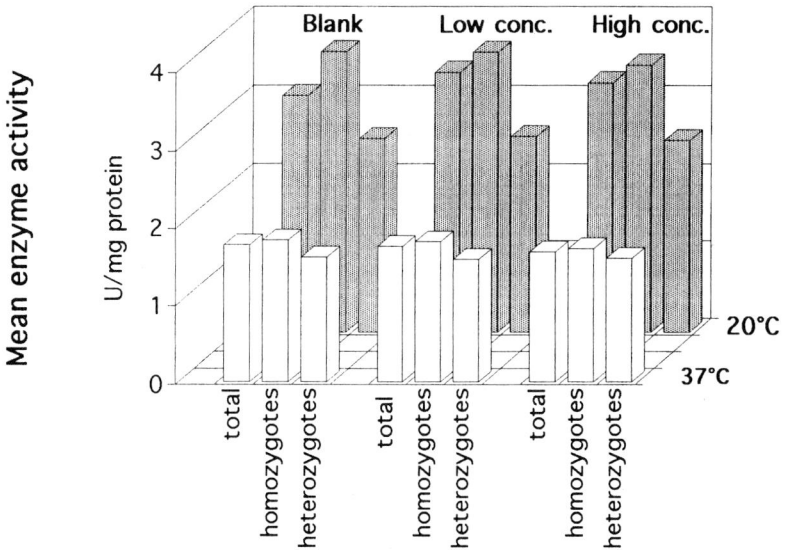

Figure 8.3.3 GPI enzyme activity of homozygotes and heterozygotes of *B. amphitrite*. (a) Zn effect (low and high concentration) at 20 °C and 37 °C; (b) Cu effect (low and high concentration) at 20 °C and 37 °C; (c) Cd effect (low and high concentration) at 20 °C and 37 °C.

optimum in all samples. Although it indicates a clear trend, the ANOVA analyses never reach significant values (often $0.10 < p < 0.05$), probably because of the low sample size.

8.3.4 Discussion

In the marine environment heavy metals are mainly bound to the sediment. They are partially remobilized into the overlying water by chemical and biological reactions. Industrial waste has been discharged for several decades directly in the industrial area (Porto Marghera) of the lagoon of Venice, causing high levels of chemical pollution. The heavy metal concentrations found in this study are similar to those reported in previous studies (Donazzolo *et al.*, 1984) and have to be considered as elevated, especially if compared to the open lagoon and to the control area of Chioggia (Barillari *et al.*, 1981; Campesan *et al.*, 1987; Donazzolo *et al.*, 1984). The metal content in tissues found in the present study (expressed in terms of wet weight) is similar to the values reported in literature for the same and other species (Stenner and Nickless, 1975; Barbaro *et al.*, 1978)

The general trend of heterozygote deficit in the industrial channel population (A) is in accordance with previous data from the same population which suggested the existence of some kind of post-settlement selection responsible for the significant increase of the multi-homozygote class from the young to the adult class age (Patarnello, Guinez and Battaglia, 1991).

The results of the present study concerning the levels of chemical pollution might indicate some of the considered heavy metals as possible selective agents. A high level of heavy metal contamination characterizes the sediment of the site A and might be related to the departure from the Hardy–Weinberg expectation observed in most of population A samples. In fact, chemical pollution might produce stressful conditions where mechanisms of genotype selection are more likely to operate (Koehn and Bayne, 1989). In such a situation an organism would shift part of the energy allocation from production to maintenance (Calow, 1991) in order to cope with a higher metabolic demand due to the activation of resistance mechanisms. Differential ability to cover the extra energy investment might then be responsible for genotype-dependent selection. Several studies have provided evidence for differential fitness of genotypes in changing environmental conditions, e.g. osmoregulatory loci like *Lap* in *Mytilus edulis* (Koehn and Hilbish, 1987) and *Gpt* in the copepod *Tigriopus californicus* (Burton and Feldman, 1983) or the energy-yielding genes *Ldh* in *Fundulus heteroclitus* (Powers *et al.*, 1991), and *Gpi* in *Colias* butterflies (Watt, Cassin and Swan, 1983; Watt, 1985) and in the amphipod *Gammarus insensibilis* (Patarnello, Bisol and Battaglia, 1989; Patarnello and Battaglia, 1992).

According to Hoffman and Parsons (1991), differences in biochemical traits are more likely to be detected in unfavourable environments, where fitness behaviour may change and normally neutral genotypes could determine differences in adaptation following environmental modification. Silva *et al.* (1989) suggest that because of their non-optimal biochemical properties, rare *Gpi* genotypes in *M. edulis* are selected against, particularly under stress-producing ecological circumstances. However, differences in electrophoretic mobility do not always result in functional differences. Very small differences in function among alleles, even though coding for enzymes in a central metabolic position, are unlikely to be individually visible to selection and could be nearly neutral.

The *in vitro* biochemical analysis carried out in the present study revealed a decrease of GPI activity at the higher temperature as well as in the presence of Zn and Cu. The general trend of a higher homozygote activity was found (with rare exceptions) both with and without the addition of Zn and Cu, particularly at 20 °C. Although it is difficult to decide whether the metal concentrations used in the biochemical assays exactly reproduce the metal concentrations available in the cell in a soluble state, the present results concerning the GPI activity might account for a higher homozygote efficiency in the interconversion of fructose 6-phosphate and glucose 6-phosphate, probably increasing the net flux through glycolysis (Watt, 1985). A constant heavy metal incorporation might produce a situation of physiological stress, coinciding with an increased metabolic requirement (Bernard and Lane, 1963) as a consequence of detoxification mechanisms. Chemical pollutants may also directly interact with the proteins, modifying their catalytic properties by blocking their active groups and destroying some specific molecular bonds (section 8.1).

In a long-term stress situation, if individuals heterozygous for bioenergetic central enzymes, such as GPI, have a lower efficiency, heterozygous individuals probably would not be able to cope with energetic demands. In this situation a differential glycolytic flux would contribute to determining a differential fitness.

The finding of a very high level of metal contamination in the samples of population B, not correlated with a significant heterozygote deficiency, apparently contrasts with the above considerations. The higher metal content in the organisms from the power plant outflow (probably due to the higher temperature, which causes an increase in the filtration and uptake rate) should produce a stronger excess of homozygotes than that observed in the industrial channel population. However, it has to be considered that very high levels of contamination might lead to a 'generalized' mortality (not genotype-dependent) which could possibly explain the lack of significant departure from the Hardy–Weinberg expectations in the power plant outflow population. It seems reasonable to assume that over a threshold amount of metals in tissues the fitness of all genotypes would be reduced, especially in the light of the direct effect of heavy metals on the function of the enzyme (as suggested by the present biochemical experiments). Evidence of a mass-mortality response to a high dose of metals has been presented by Diamond *et al.* (1991).

In contrast, the lower heavy metal content in the soft tissues of the industrial channel population indicates that these samples accumulate contaminants at a lower rate than the population from the power plant outflow. This lower accumulation might produce a situation of 'chronic' exposure of population A to the heavy metal action which might represent a situation where pollutants operate as selective agents (Hvilson, 1983). However, further *in vivo* and *in vitro* analyses are in progress to increase our understanding of the role of heavy metals in influencing the genetic structure of *B. amphitrite*.

Acknowledgements

We are grateful to Dr R. Cironi for his valuable suggestions and comments in designing the present study. This research was supported by ENEL (Ente Nazionale per l'Energia Elettrica) and by the 'Progetto Lagunare Veneziano (MURST)'.

References

Agemian, H. and Chau, A.S.Y. (1976) An improved digestion method for the extraction of mercury from environmental samples. *Analyst*, **101**, 91–5.

Barbaro, A., Francescon, A., Polo, B. and Bilio, M. (1978) *Balanus amphitrite* (Cirripedia: Thoracica) – A potential indicator of fluoride, copper, lead, chromium, and mercury in North Adriatic lagoons. *Mar. Biol.*, **46**, 247–57.

Barillari, A., Boldrin, A., Campesan, G. and Rabitti, S. (1981) Relazioni tra alcuni elementi in traccia e proprieta dei sedimenti in due aree compione della Laguna di Venezia. *Ist. Ven. Sci., Rapporti e Studi*, **8**, 87–103.

Beardmore, J.A., Barker, C.J., Battaglia, B. *et al.* (1980) The use of the genetic approaches to monitoring biological effects of pollution, *Rapp. P.-v Reun. Cons. Int. Explor. Mer*, **179**, 299–305.

Bernard, F.J. and Lane, C.E.(1963) Effects of copper ion on oxygen uptake by planktonic cyprids of the barnacle, *Balanus amphitrite niveus. Proc. Soc. Exp. Biol. Med.*, **113**, 418–20.

Blackstock, J. (1980) Estimation of activities of some enzymes associated with energy-metabolism in the polychaete *Glicera alba* (Muller), and application of the methods to the study of the effect of organic pollution. *J. Exp. Mar. Biol. Ecol.*, **46**, 197–217.

Burton, R.S. and Feldman, M.W. (1983) Physiological effect of an allozyme polymorphism: glutamate-pyruvate transaminase and response to hyperosmotic stress in the copepod *Tigriopus californicus. Biochem. Genet.*, **21**, 239–51.

Calow, P. (1991) Physiological costs of combating chemical toxicants: ecological implications. *Comp. Biochem. Physiol. C*, **100**, 3–6.

Campesan, G., Fossato, V.U., Barillari, A., Dolci, F. and Stocco, G. (1987) Metalli pesanti e idrocarburi clorurati nei sedimenti della valle di Brenta (Laguna di Venezia). *Ist. Ven. Sci., Rapporti e Studi.*, **11**, 21–8.

Carmody, D.J., Pearce, J.B. and Yasso, W.E. (1973) Trace metals in sediments of New York Bight. *Mar. Pollut. Bull.*, **4**, 132–5.

Diamond, S.A., Newman, M.C., Mulvey, M. and Guttman, S.I. (1991) Allozyme genotype and time-to-death of mosquitofish, *Gambusia holbrooki*, during acute inorganic mercury exposure: a comparison of populations. *Aquat. Toxicol.*, **21**, 119–34.

Donazzolo, R., Orio, A.A., Pavoni, B. and Perin, G. (1984) Heavy metals in sediments of the Venice Lagoon. *Oceanol. Acta*, **7**, 25–31.

Eanes, W.F., Katona, L. and Longtine, M. (1990) Comparison of *in vitro* and *in vivo* activities associated with the G6PD allozyme polymorphism in *Drosophila melanogaster. Genetics*, **125**, 845–53.

Hawkins, A.J.S., Bayne, B.L., Day, A.J., Rusin, J. and Worrall, C. (1989) Genotype-dependent inter-relation between energy metabolism, protein metabolism and fitness, in *Reproduction, Genetics and Distribution of Marine Organisms*, 23rd Eur. Mar. Biol. Symp (eds J.S. Ryland and P.A Tyler), Olsen & Olsen, Fredensborg, pp. 283–92.

Hoffman, A.A. and Parsons P.A. (1991) *Evolutionary Genetics and Environmental Stress*, Oxford University Press, New York.

Hoffman, R.J. (1983) Temperature modulation of the kinetics of phosphoglucose isomerase genetic variants from the sea anemone *Metridium senile. J. Exp. Zool.*, **227**, 361–70.

Hvilson, M.M. (1983) Copper-induced differential mortality in the mussel *Mytilus edulis. Mar. Biol.*, **76**, 291–5.

Koehn, R.K. and Bayne, B.L. (1989) Towards a physiological and genetical understanding of the energetics of the stress response. *Biol. J. Linn. Soc.*, **37**, 157–71.

Koehn, R.K. and Hilbish, T.J. (1987) The adaptive importance of genetic variation. *Am. Sci.*, **75**, 134–40.

Lavie, B. and Nevo, E. (1982) Heavy metal selection of phosphoglucose isomerase allozymes in marine gastropods. *Mar. Biol.*, **71**, 17–22.

Lavie, B. and Nevo, E. (1986) The interactive effects of cadmium and mercury pollution on allozyme polymorphisms in the marine gastropod *Cerithium scabridum. Mar. Pollut. Bull.*, **17**, 21–3.

Lavie, B. and Nevo, E. (1987) Differential fitness of allelic isozymes in the marine gastropods *Littorina punctata* and *Littorina neritoides*, exposed to environmental stress of the combined effects of cadmium and mercury pollution. *Environ. Manage.*, **11**, 345–9.

Lowry, O.H., Rosebrough, N.J., Farr, A.L. and Randal, R.J. (1951) Protein measurement with the Folin phenol reagent. *J. Biol. Chem.*, **193**, 265–75.

Milstein, C. (1961) Inhibition of phosphoglucomutase by trace metals. *Biochem. J.*, **79**, 591–6.

Nevo, E., Ben-Shlomo, R. and Lavie, B. (1984) Mercury selection of allozymes in marine organisms: prediction and verification in nature. *Proc. Natl. Acad. Sci. USA*, **81**, 1258–9.

Nevo, E., Lavie, B. and Ben-Shlomo, R. (1983) Selection of allelic isozyme polymorphism in marine organism: pattern, theory, and application, in *Isozymes: Current Topics in Biological and Medical Research*, Vol. 10 *Genetics and Evolution* (eds M.C. Rattazzi, J.G. Scandalios and G.S. Whitt), Alan R. Liss Inc., New York, pp. 69–92.

Nevo, E., Lavie, B. and Noy, R. (1987) Mercury selection of allozymes in marine gastropods: prediction and verification in nature revisited. *Environ. Monit. Assess.*, **9**, 233–8.

Nevo, E., Shimony, T. and Libni, M. (1978) Pollution selection of allozyme polymorphism in barnacles. *Experientia*, **34**, 1562–4.

Nevo, E., Perl, T., Beiles, A. and Wool, D. (1981) Mercury selection of allozyme genotypes in shrimps. *Experientia*, **37**, 1152–4.

Patarnello, T. and Battaglia, B. (1992) Glucosephosphate isomerase and fitness: effects of temperature on genotype dependent mortality and enzyme activity in two species of genus *Gammarus* (Crustacea: Amphipoda). *Evolution*, **46**, 1568–73.

Patarnello, T., Bisol, P.M. and Battaglia, B. (1989) Studies on differential fitness of PGI genotypes with regard to temperature in *Gammarus insensibilis* (Crustacea: Amphipoda). *Mar. Biol.*, **102**, 355–9.

Patarnello, T., Guinez, R. and Battaglia, B. (1991) Effects of pollution on heterozygosity in the barnacle *Balanus amphitrite* (Cirripedia: Thoracica). *Mar. Ecol. Prog. Ser.*, **70**, 237–43.

Patarnello, T., Guinez, R., Bressa, G. and Battaglia, B. (1990) Effetti genetici dell'inquinamento in popolazioni lagunari di *Balanus amphitrite* Darwin (Cirripedia: Thoracica). *Rend. Fis. Acc. Lincei s.* **9**, 193–202.

Perin, G. and Gabelli, A. (1983) Inquinamento chimico della laguna di Venezia. Contaminanti di origine urbana ed industriale nelle acque. *Acqua Aria*, **6**, 615–21.

Perin, G., Orio A.A., Pavoni B. *et al.* (1983) Nutrienti e metalli pesanti nei sedimenti. *Acqua Aria*, **6**, 623–32.

Powers, D.A., Laverman, T., Crawford, D and DiMichele, L. (1991) Genetic mechanisms for adapting to a changing environment. *Annu. Rev. Genet.*, **25**, 629–59.

Rainbow, P.S. (1984) An introduction to the biology of British littoral barnacles. *Field Studies*, **6**, 1–51.

Rainbow, P.S. and White, S.L. (1989) Comparative strategies of heavy metal accumulation by crustaceans: zinc, copper and cadmium in a decapod, an amphipod and a barnacle. *Hydrobiologia*, **174**, 245–62.

Rice, W.R. (1989) Analyzing tables of statistical tests. *Evolution*, **43**, 223–5.

Shiahab, A. and Heath, D.J. (1987) Components of fitness and Gpi polymorphism, in the freshwater isopod *Asellus aquaticus*. (L.). 2. Zygotic selection. *Heredity*, **58**, 289–95.

Silva, P.J.N., Koehn, R.J., Diehl, W.J. III *et al.* (1989) The effects of glucose-6-phosphate isomerase genotype on *in vitro* specific activity and *in vivo* flux in *Mytilus edulis*. *Biochem. Genet.*, **27**, 451–67.

Stenner, R.D. and Nickless, G. (1975) Heavy metals in organisms of the Atlantic Coast of S.W. Spain and Portugal. *Mar. Pollut. Bull.*, **6**, 89–92.

Swofford, D.L. and Selander, B.R. (1981) BIOSYS-1: a FORTRAN program for the comprehensive analysis of electrophoretic data in population genetics and systematics. *J. Hered.*, **72**, 282–3.

Walker, G. (1977) 'Copper' granules in the barnacle *Balanus balanoides*. *Mar. Biol.*, **39**, 343–9.

Walker, G., Rainbow, P.S., Foster, P. and Holland, D.L. (1975) Zinc phosphate granules in tissue surrounding the midgut of the barnacle *Balanus balanoides*. *Mar. Biol.*, **33**, 161–6.

Watt, W.B. (1985) Bioenergetics and evolutionary genetics. Opportunities for a new synthesis. *Am. Nat*, **125**, 118–43.

Watt, W.B., Cassin, R.C. and Swan, M.S. (1983) Adaptation at specific loci. III. Field behaviour and survivorship differences among *Colias* PGI genotypes are predictable from *in vitro* biochemistry. *Genetics*, **103**, 725–39.

Webb, J.L. (1966) *Enzyme and Metabolic Inhibitors*, Vol. 2, Academic Press, New York.

8.4 EFFECTS OF COPPER EXPOSURE DURING EARLY LIFE STAGES ON HETEROZYGOSITY IN LABORATORY-REARED MUSSEL (*MYTILUS EDULIS* L.) POPULATIONS

Kathrin Hoare, Andy R. Beaumont and John Davenport

Abstract

Mussels (Mytilus edulis L.) from the Menai Strait, UK, the Oosterschelde and the Westerschelde, the Netherlands, three sites with differing degrees of pollution stress, were spawned in the laboratory. Each of the three sets of offspring were divided into six different treatment groups, five of which were exposed to 8 ppb copper at one or more of the embryo, veliger and post-settlement stages. The sixth acted as an untreated control group. There were two replicate cultures per treatment. At the end of the experiments, which had an average duration of about 9 months, the largest mussels in each culture were scored using starch gel electrophoresis at the Gpi, Lap, Pgm, Odh, Gsr, 6Pgd and Icdh enzyme loci. No significant effects on single or multiple locus heterozygosities were found for copper exposure at any stage. No trends were seen in the effects of copper on heterozygosity at individual loci. There is a possible trend in multiple-locus heterozygosity towards reduction due to copper exposure at the embryo and veliger stages and increase due to exposure at the post-settlement stage.

8.4.1 Introduction

As Hummel and Patarnello (Section 8.1) point out, changes in heterozygosity caused by pollution selection have been demonstrated in a number of studies (Nevo, Shimony and Libni, 1977; Battaglia *et al.*, 1980; Hawkins *et al.*, 1989; Patarnello, Guinez and Battaglia, 1991). All studies so far published have used adult animals, but other developmental stages frequently show greater sensitivity to pollutant stress (e.g. Martin *et al.*, 1981). In addition, for practical reasons, previous studies have used pollutant concentrations which are rarely, if ever, encountered in the field. This section presents allozyme data from a study of the effects of a low, environmentally realistic, increased level of copper on the embryo, veliger (planktonic larva) and post-settlement stages of the mussel, *Mytilus edulis* L., over a period of months in the laboratory.

Most enzyme loci scored in this study were chosen because of their possible sensitivity to heavy metals. Hawkins *et al.* (1989) showed that adult *M. edulis* heterozygotes at the *Gpi* locus (glucose phosphate isomerase, EC 5.3.1.9) survive longer than homozygotes in 150 ppb copper. The enzymes PGM (phosphoglucomutase, EC 2.7.5.1) and LAP (leucine aminopeptidase, EC 3.4.11.-) are directly chemically affected by metal ions (Milstein, 1961; Young, Koehn and Arnheim, 1979). The *Gsr* locus (glutathione reductase, EC 1.6.4.2) was scored because Viarengo *et al.* (1988) demonstrated that under metal stress the glutathione reserves of *M. galloprovincialis* tissue become depleted, probably due to conjugation with metal ions. The enzyme 6PGD (6-phosphogluconate dehydrogenase, EC 1.1.1.44) is closely functionally linked to GPI in glycolysis, which may be affected by metal ions (Viarengo, 1985).

8.4.2 Materials and methods

Mussels (*M. edulis*) were collected during 1990 and 1991 from Kattendijke, Oosterschelde and Veerhaven, Westerschelde, The Netherlands, and from Beaumaris, Menai Strait, UK, just prior to natural spawning. The Westerschelde is an industrially polluted estuary of the river Scheldt, the Oosterschelde is a relatively unpolluted sea-arm in the same delta system and the Menai Strait can be regarded as a clean control site.

The mussels were maintained in the laboratory in flowing seawater at about 5 °C and fed about twice weekly with a mixture of cultured micro-algal species until used. Non-native adults and offspring were kept in quarantine conditions under Special Licence, in accordance with the Sea Fisheries (Shellfish) Act 1967 as varied by the Molluscan Shellfish (Control of Deposit)(Variation) Order 1983, S.I no. 159.

Separate experiments were carried out on mussels from each of the three sites. Adults were induced to spawn using injections of 0.5 M KCl into the mantle cavity combined with temperature shock (Loosanoff and Davis, 1963; Bayne, 1965) and the offspring cultured using standard techniques (Loosanoff and Davis, 1963; Bayne, 1965; Beaumont *et al.*, 1988). The Menai Strait spawning involved 16 parents, the Oosterschelde 9 and the Westerschelde 4. An embryo density of between 350/cm^2 and 700/cm^2 culture dish bottom area was used (Barnes, 1989) and veligers were maintained at a density of about 5/ml. *Pavlova lutheri* (Droop) Green and *Rhinomonas reticulata* (Lucas) Novarino were fed to the veliger larvae at a concentration of about 40 cells/μl and 10 cells/μl respectively until settlement began, whereupon rations of the two algal species were doubled to 80 cells/μl and 20 cells/μl. Settlement began within 1 month of fertilization and was completed by the time the cultures were 3 months old. Care was taken to maintain settled and unsettled veligers in the same culture vessels, using sieves, to avoid size selection (Beaumont *et al.*, 1988). All stages of the offspring were cultured in 0.2-μm-filtered and UV-sterilized seawater. Seawater and algal densities were renewed three times per week and during these changes copper chloride solution was added where appropriate. Beaumont, Tserpes and Budd (1987) demonstrated that under these conditions copper concentrations do not fall by more than 25%. The cultures were kept under identical conditions as far as possible, with copper concentration as the only known variable.

The copper concentration used was 8 ppb (8 μg Cu/l) above background, an environmentally realistic level that has been shown to have significant effects on both embryo survivorship (Barnes, 1989; Martin *et al.*, 1981) and adult growth (Strømgren, 1982; Redpath, 1985) in *M. edulis*.

The copper exposure regime (Table 8.4.1) was designed to identify any stage (embryo, veliger or post-settlement) at which the stressor might produce genetic changes.

There were two replicate cultures per treatment. These laboratory populations were maintained for between 7 and 12 months, depending upon the rate of growth. At the end of experiments the largest mussels from each replicate culture were genotyped using starch gel electrophoresis. For the Menai Strait and Westerschelde groups about 50 mussels were genotyped per replicate (200 per treatment), and for the Oosterschelde group all mussels which grew to the minimum size for reliable starch gel electrophoresis of the enzymes studied (a shell length of over 2.5 mm) within a year of fertilization were used for electrophoresis. Unfortunately the Oosterschelde veliger cultures suffered mortality associated with a *Vibrio* infection and survivorship to metamorphosis in some cultures was low. Because of the small numbers in some cultures, the treatment given during the veliger stage were maintained in each culture during the post-settlement stage.

Electrophoresis of the cultured juveniles and of 76 adult animals from each of the three parental sites was carried out on 12.5% horizontal starch gels. Homogenate from each animal was loaded onto two gels, one of 0.01 M citrate/phosphate pH 7.4 which was run on 0.1 M tris/EDTA/maleic acid electrode buffer pH 7.4 and one of 0.01 M tris/EDTA/maleic acid pH 6.0 which was run on 0.1 M tris/EDTA/maleic acid electrode buffer pH 6.0. The citrate/phosphate gels were stained for the enzymes GSR, LAP, 6PGD and isocitrate dehydrogenase (ICDH, EC 1.1.1.42) and the tris/EDTA/maleic acid gels were stained for GPI, PGM and octopine dehydrogenase (ODH, EC 1.5.1.11). Staining recipes were adapted from Harris and Hopkinson (1976), Beaumont and Beveridge (1983) and Beaumont, Day and Gade (1980).

Table 8.4.1 Copper exposure at three stages of development of six treatment groups (A–F) of laboratory-reared *Mytilus edulis*

Developmental stage	Treatment group					
	A	*B*	*C*	*D*	*E*	*F*
Embryo	−	−	−	+	+	+
Veliger	−	−	+	−	+	+
Post-settlement	−	+	+	−	−	+

+ = copper exposure.
− = control conditions.

8.4.3　Results

Mortality during the course of the experiments was over 95% in all cultures, giving sufficient scope for any selective forces to be expressed.

A null allele occurred at the *Odh* locus in the Westerschelde and Oosterschelde native populations at frequencies of about 0.2 and 0.1 respectively. This null also occurred in the Westerschelde laboratory-reared population, but, although ODH is monomeric in *M. edulis*, because of the restricted parentage of this group (one male, three females) it was possible to calculate the actual genotype of most individuals from the observed phenotype; there was only one phenotype for which there was any doubt as to whether the null was present and these were treated as being genotypically homozygous for the active allele. Since it was not possible to identify null heterozygotes in the native populations, all single-banded phenotypes were treated as homozygotes for active alleles. Null homozygotes were recorded as such. This has obvious effects on recorded heterozygosity at this locus, but the results are not significantly altered by exclusion of *Odh* from the analysis.

Table 8.4.2 reveals no obvious pattern of differences between treatment group heterozygosities, but shows that there are some large differences in single-locus heterozygosities between supposedly identical replicates.

The restricted parentage of the laboratory-reared populations precluded the use of any statistical comparisons based on the Hardy–Weinberg model. Instead, the results were analysed using heterogeneity χ^2 tests to compare the distribution of mussels between heterozygotes and homozygotes at each locus among treatments and among groups pooled according to treatment at each stage. Thus, treatment groups A, B and C were pooled and tested against groups D, E and F to assess the effects of embryo treatment; groups A, B and D were tested against groups C, E and F to assess the effects of veliger treatment; and groups A, D and E were tested against groups B, C and F to assess to effects of copper treatment post-settlement. Multiple-locus heterozygosity was investigated in the same way by summing the heterozygotes and homozygotes at each locus. Since this gives slightly different results from the usual estimate of overall heterozygosity, multiple-locus heterozygosities were also calculated in the normal way, using the unweighted heterozygote frequencies from each locus; no data were used from Oosterschelde cultures where the mean number of mussels

Table 8.4.2　Direct-count heterozygosities at seven loci for the Menai Strait laboratory-reared populations of *Mytilus edulis*

Locus	Treatment groups											
	A		B		C		D		E		F	
	a	b	a	b	a	b	a	b	a	b	a	b
Pgm	0.44	0.56	0.48	0.56	0.52	0.60	0.52	0.54	0.62	0.52	0.54	0.42
Lap	0.72	0.56	0.42	0.52	0.66	0.56	0.50	0.48	0.42	0.37	0.50	0.50
Odh	0.06	0.02	0.06	0.00	0.04	0.10	0.10	0.00	0.05	0.06	0.02	0.00
6Pgd	0.02	0.04	0.02	0.00	0.06	0.02	0.08	0.04	0.03	0.09	0.00	0.04
Gsr	0.02	0.06	0.10	0.02	0.06	0.10	0.12	0.06	0.06	0.06	0.02	0.10
Icdh	0.54	0.44	0.72	0.68	0.52	0.62	0.56	0.56	0.66	0.54	0.60	0.58
Gpi	0.76	0.68	0.72	0.86	0.58	0.72	0.74	0.80	0.72	0.77	0.81	0.77
N	50	50	50	50	50	50	50	50	65	35	48	52

For explanation of treatments A–F, see Table 8.4.1. a, b = Replicate cultures within treatments; *N* = number of mussels scored for each replicate.

Table 8.4.3 Results of χ^2 tests for heterogeneity in single- and multiple-locus heterozygosities among laboratory-reared mussel (*Mytilus edulis*) populations cultured under a range of copper exposure regimes

Locus	Grouping tested	Source population		
		Menai Strait	*Oosterschelde*	*Westerschelde*
Pgm	Treatment	NS	NS	NS
	Embryo	NS+	NS+	NS+
	Veliger	NS+	NS+	NS–
	Post-settlement	NS–		NS–
Lap	Treatment	$0.01 > p > 0.005$	NS	NS
	Embryo	$0.01 > p > 0.005+$	NS–	NS+
	Veliger	NS+	NS–	NS+
	Post-settlement	NS–		NS+
Odh	Treatment	[a]	NS	$0.05 > p > 0.025$
	Embryo	NS+	NS–	NS–
	Veliger	NS–	$0.05 > p > 0.025 -$	NS+
	Post-settlement	NS+		NS–
6Pgd	Treatment	[a]	[a]	[a]
	Embryo	NS–	[a]	[a]
	Veliger	NS–	$0.025 > p > 0.01+$	[a]
	Post-settlement	NS+		[a]
Gsr	Treatment	NS	$0.025 > p > 0.01$	NS
	Embryo	NS–	$0.005 > p > 0.001+$	NS+
	Veliger	NS–	NS+	NS–
	Post-settlement	NS		NS–
Icdh	Treatment	NS	$0.025 > p > 0.01$	NS
	Embryo	NS+	NS+	NS–
	Veliger	NS+	NS+	NS–
	Post-settlement	NS+		NS–
Gpi	Treatment	NS	[a]	$0.01 > p > 0.005$
	Embryo	NS+	NS+	NS+
	Veliger	NS–	NS+	NS–
	Post-settlement	NS=		NS–
All loci	Treatment	NS	$0.025 > p > 0.01$	NS
	Embryo	NS–	NS–	NS–
	Veliger	NS–	NS–	NS–
	Post-settlement	NS+		NS+

The experiment was performed three times, on the laboratory-reared offspring of mussels from Menai Strait, UK, and Oosterschelde and Westerschelde, the Netherlands, three sites with differing pollution exposure histories. Copper treatment regimes are shown in Table 8.4.1; tests were carried out on the data from individual treatment groups as well as on the data from groups pooled according to treatment at each of the embryo, veliger and post-settlement stages. The signs indicate whether there was a mean increase (+) or decrease (–) or no change (=) in heterozygosity with copper exposure at a particular stage.
[a] Tests were not done where expected values were < 5.
N.B. The Oosterschelde cultures received the same conditions post-settlement as at the veliger stage – see text.

scored per locus was less than 20. These estimates of multiple-locus heterozygosity were pooled according to treatment at each stage and the results analysed using the Mann-Whitney test (Siegel 1956). The results were very similar to those from the χ^2 tests. Table 8.4.3 gives the results of the χ^2 tests for all three laboratory-reared populations.

Table 8.4.4   Direct-count heterozygosities of the three wild source populations of *Mytilus edulis*

Locus	Menai Strait	Oosterschelde	Westerschelde
Pgm	0.43	0.32	0.38
Lap	0.49	0.62	0.55
Odh	0.24	0.13	0.16
6Pgd	0.08	0.09	0.09
Gsr	0.18	0.22	0.20
Icdh	0.59	0.61	0.51
Gpi	0.55	0.57	0.61
Mean	0.367	0.365	0.357

Seventy-six animals were scored for each population. χ^2 analysis of the distribution of mussels between homozygote and heterozygote classes indicates that there are no significant differences in single- or multiple-locus heterozygosities among the populations.

Table 8.4.5   The differences in mean weighted multiple-locus direct-count heterozygosity between control and copper-treated groups of laboratory-reared *Mytilus edulis* when pooled by treatment at three developmental stages

Parent population	Developmental stage			
	Embryo	*Veliger*	*Post-settlement*	*Sum*
Menai Strait	−0.006	−0.003	+0.005	−0.004
Oosterschelde	−0.034	−0.022		−0.026
Westerschelde	−0.007	−0.006	+0.010	−0.003

N.B. The Oosterschelde groups received the same conditions at the post-settlement stage as at the veliger stage – see text.

Any Bonferroni procedure to correct for type I error (Rice, 1989; Lessios, 1992) removes all of the single-test significances shown in Table 8.4.3. These data therefore show no significant effects of 8 ppb copper on single-locus or multiple-locus heterozygosity at any stage, for any of these three laboratory-reared populations.

Comparison of the differences between control and copper-treated groups (Table 8.4.3) among populations and among developmental stages indicates that there are no strong trends in the direction of heterozygosity change for any individual locus. Nor are any of these patterns of differences between control and copper-treated laboratory groups similar to the patterns of differences between the clean and polluted site natural populations shown in Table 8.4.4. However, comparison of the differences of multiple-locus heterozygosity between the control and copper-treated groups at each stage and for each population (Table 8.4.5) indicates that there might be a non-significant trend towards reduced overall heterozygosity due to copper treatment at the embryo and veliger larva stages. Copper exposure post-settlement seems to increase multiple-locus heterozygosity, though not enough to reverse earlier declines.

8.4.4   Discussion

The amount of variation in heterozygosities at single loci between supposedly identical replicates is surprising. There is no readily identifiable cause of these large differences, although similar variation within results in the literature has been described by Hummel and Patarnello (Section 8.1).

This study shows no evidence for an effect of copper on heterozygosity at any of the single loci examined. The difficulty of applying the results from such laboratory-based studies to the natural situation, where there is a much wider range of genetic backgrounds within a population, was pointed out by Beaumont *et al.* (1989).

The trends in the effects of copper on multiple-locus heterozygosity shown in Table 8.4.5 are not significant but are perhaps worth commenting on. If these trends are real, then the phenomenon of different heterozygosity effects at different life stages (biphasic selection) could explain why many studies on adult animals have demonstrated greater survivorship of heterozygotes under pollutant stress (Hawkins *et al.*, 1989; Patarnello, Bisol and Battaglia, 1989), while many field observations have recorded equal or lower heterozygosity in stressed compared with clean site populations (Nevo, Shimony and Libni, 1978; Battaglia *et al.*, 1980; Fevolden and Garner, 1987; Patarnello, Guinez and Battaglia, 1991).

It is not clear why copper should affect heterozygosity in different ways at different life stages, if indeed it does, but possible stress-induced trends in multiple-locus heterozygosity may well be worth further study. A characteristic feature of natural populations of bivalves is that small but consistent heterozygote deficiencies are generated during early life stages (Zouros and Foltz, 1984). Apparent differential mortality of homozygotes during later life tends to reduce these deficiencies. Although the causes of heterozygote deficiencies may be many and varied (Zouros and Foltz, 1984; Beaumont, 1991), ontogenetic changes in heterozygosity are clearly a feature of bivalve populations whether or not they are stressed by pollution.

Acknowledgements

We would like to thank Roelof Bogaards and the Centre for Estuarine and Coastal Fcology, Netherlands Institute of Ecology, for supplying mussels and pollution data from the Oosterschelde and Westerschelde. This study was carried out under the tenureship by K.H. of a Science and Engineering Research Council (UK) studentship.

References

Barnes, D.A. (1989) The effects of initial egg density and added copper concentration on the development of early larval stages of the mussel *Mytilus edulis*. MSc Thesis, University of Wales, Bangor.

Battaglia, B., Bisol, P.M., Fossato, V.U. and Rodino, E. (1980) Studies on the genetic effects of pollution in the sea. *Rapp. P.-v. Reun. Cons. Int. Explor. Mer*, **179**, 267–74.

Bayne, B.L. (1965) Growth and the delay of metamorphosis of the larvae of *Mytilus edulis* (L.). *Ophelia*, **2**, 1–47.

Beaumont, A.R. (1991) Genetic studies of laboratory reared mussels, *Mytilus edulis*: heterozygote deficiencies, heterozygosity and growth. *Biol. J. Linn. Soc.*, **44**, 273–85.

Beaumont, A.R., and Beveridge, C.M. (1983) Resolution of phosphoglucomutase isozymes in *Mytilus edulis* L. *Mar. Biol. Lett.*, **4**, 97–103.

Beaumont, A.R., Day, T.R. and Gade, G. (1980) Genetic variation at the octopine dehydrogenase locus in the adductor muscle of *Cerastoderma edule* (L.) and six other bivalve species. *Mar. Biol. Lett.*, **1**, 137–48.

Beaumont, A.R., Tserpes, G. and Budd, M.D. (1987) Some effects of copper on the veliger larvae of the mussel *Mytilus edulis* and the scallop *Pecten maximus* (Mollusca, Bivalvia). *Mar. Envir. Res.*, **21**, 299–309.

Beaumont, A.R., Beveridge, C.M., Barnet, E.A., Budd, M.D. and Smyth-Chamosa, M. (1988) Genetic studies of laboratory reared *Mytilus edulis* I. Genotype specific selection in relation to salinity. *Heredity*, **61**, 389–400.

Beaumont, A.R., Beveridge, C.M., Barnet, E.A. and Budd, M.D. (1989) Genetic studies of laboratory reared *Mytilus edulis* II. Selection at the leucine amino peptidase (*Lap*) locus. *Heredity*, **62**, 169–76.

Fevolden, S.E. and Garner, S.P. (1987) Environmental stress and allozyme variation in *Littorina littorea* (Prosobranchia). *Mar. Ecol. Progr. Ser.*, **39**, 129–36.

Harris, H. and Hopkinson, D.A. (1976) *Handbook of Enzyme Electrophoresis in Human Genetics*, North-Holland Publishing Company, London.

Hawkins, A.J.S., Rusin, J., Bayne, B.L. and Day, A.J. (1989) The metabolic/physiological basis of genotype-dependent mortality during copper exposure in *Mytilus edulis*. *Mar. Envir. Res.*, **28**, 253–7.

Lessios, H.A. (1992) Testing electrophoretic data for agreement with Hardy–Weinberg expectations. *Mar. Biol.*, **112**, 517–23.

Loosanoff, V.L. and Davis, H.C. (1963) Rearing of bivalve molluscs. *Adv. Mar. Biol.*, **1**, 1–136.

Martin, M., Osborn, K.E., Billig, P. and Glickstein, N. (1981) Toxicities of ten metals to *Crassostrea gigas* and *Mytilus edulis* embryos and *Cancer magister* larvae. *Mar. Pollut. Bull.*, **12**, 305–8.

Milstein, C. (1961) Inhibition of phosphoglucomutase by trace metals. *Biochem. J.*, **79**, 591–6.

Nevo, E., Shimony, T. and Libni, M. (1977) Thermal selection of allozyme polymorphisms in barnacles. *Nature*, **267**, 699–701.

Nevo, E., Shimony, T. and Libni, M. (1978) Pollution selection of allozyme polymorphisms in barnacles. *Experientia*, **34**, 1562–4.

Patarnello, T., Bisol, P.M. and Battaglia, B. (1989) Studies on the differential fitness of PGI genotypes with regard to temperature in *Gammarus insensibilis*. *Mar. Biol.*, **102**, 355–9.

Patarnello, T., Guinez, R. and Battaglia, B. (1991) Effect of pollution on heterozygosity in the barnacle *Balanus amphitrite* (Cirripedia: Thoracica). *Mar. Ecol. Progr. Ser.*, **70**, 237–43.

Redpath, K.J. (1985) Sub-lethal effects of trace metals on mussels. PhD Thesis, University of Wales, Bangor.

Rice, W.R. (1989) Analyzing tables of statistical tests. *Evolution*, **43**, 223–5.

Siegel, S. (1956) *Non-parametric Statistics for the Behavioral Sciences*, McGraw-Hill, New York.

Strømgren, T. (1982) Effect of heavy metals (Zn, Hg, Cu, Cd, Pb, Ni) on the length growth of *Mytilus edulis*. *Mar. Biol.*, **72**, 69–72.

Viarengo, A. (1985) Biochemical effects of trace metals. *Mar. Pollut. Bull.*, **16**, 153–8.

Viarengo, A., Pertica, M., Canesi, L. *et al.* (1988) Effects of heavy metals on lipid peroxidation in mussel tissues. *Mar. Envir. Res.*, **24**, 354.

Young, P.W., Koehn, R.K. and Arnheim, N. (1979) Biochemical characterization of 'LAP', a polymorphic aminopeptidase from the blue mussel, *Mytilus edulis*. *Biochem. Gent.*, **17**, 305–23.

Zouros, E. and Foltz, D.W. (1984) Possible explanations of heterozygote deficiency in bivalve molluscs. *Malacologia*, **25**, 583–91.

9

Genetics and aquaculture

9.1 THE APPLICATION AND RELEVANCE OF GENETICS IN AQUACULTURE

Andy R. Beaumont

Abstract

This section is divided into four main parts: (1) quantitative genetics, (2) chromosome ploidy manipulation, (3) allozyme genetics and (4) transgenic studies.

Quantitative genetics involves the measurement and correlation of phenotypic traits and the division of their variance into genetic and environmental components. In comparison with animals and plants used in agriculture, quantitative data are scarce for the majority of aquaculture species. Heritabilities of economically important traits in fish and shellfish are presented and discussed in relation to the potential for genetic improvement.

The development of techniques for producing triploid, tetraploid and gynogenetic diploid fish and shellfish is a recent innovation of considerable importance to aquaculture. The methods involved in ploidy manipulation and the various advantages of triploids and diploid gynogens are described and discussed.

Electrophoretic detection of allozymes has enabled: (1) the quantification of the changes in allelic diversity and heterozygosity in hatchery broodstocks; (2) the clarification of genetic differentiation in natural populations of fish in contrast to the founder effects in hatchery fish; and (3) the development of genetic markers for tagging hatchery stocks.

At present, there are few studies using modern DNA genetic techniques to study genetics in relation to aquaculture. However, the development of transgenic aquatic organisms has begun. Studies on the success of transfection techniques which are being used at a number of laboratories to incorporate novel DNA into the genome of fish are reviewed.

9.1.1 Introduction

Following two articles reviewing the potential for genetics in aquaculture by Newkirk (1980) and Wilkins (1981), a series of international triennial meetings on genetics in aquaculture were held during the 1980s at which all aspects of this wide subject have been explored (Wilkins and Gosling, 1983; Gall and Busack, 1986; Gjedrem, 1990).

Genetics and Evolution of Aquatic Organisms. Edited by A.R. Beaumont.
Published in 1994 by Chapman & Hall, 2–6 Boundary Row, London SE1 8HN.
ISBN 0-412-49370-5

The genetics of fish and shellfish interacts with their culture in four main areas of study. In this section I shall approach the subject first at the organismal or phenotypic level in the study of quantitative genetics, hybridization and inbreeding, and then move to the chromosomal level to explore the advances made using the techniques of ploidy manipulation. The third section will deal with single-locus allozyme studies which have increased our understanding of the genetic consequences of bringing wild populations under domestic control and of releases of cultured stocks into the wild. Finally, I will briefly cover the application of recent DNA techniques to the development of gene transfer in fish.

This section is not intended as an exhaustive review of all the papers published in this field; rather, the major principles involved in each section are outlined and salient features are illustrated using appropriate examples from the literature. In some areas of research covered in this section, reviews have been published within the last few years to which readers are directed for a more complete literature on the subject. In other areas, where recent reviews are not available, the literature is covered more extensively.

9.1.2 Quantitative genetics

Traits of importance of aquaculture are generally controlled by a number of genes and are therefore best studied by quantitative genetic techniques. From the commercial point of view, artificial selection needs to be carried out to enhance those features or traits which increase commercial potential or to instigate the process of domestication. The phenotypic variance (V_P) of a trait is the sum of the variance due to genes (V_G), to the environment (V_E) and to interaction between the two ($V_{G \times E}$):

$$V_P = V_G + V_E + V_{G \times E}$$

Variance due to genes can be further subdivided into that derived from the additive effects of alleles (V_A) and that resulting from non-additive effects – dominance effects within loci (V_D) and interaction, epistatic effects between loci (V_I).

$$V_G = V_A + V_D + V_I$$

Additive genetic variance is that part of the variance due to the sum of the effects of the alleles at all the loci that contribute to the phenotype and it is this variance which is amenable to artificial selection.

Non-additive genetic variation (V_D and V_I) is not amenable to controlled artificial selection because genotypic combinations both within and between loci are taken apart at meiosis and not reassembled in the same way at syngamy. Parents pass on their alleles, but not their genotypes, to the next generation. The heritability of a trait (denoted by the symbol h^2) is that proportion of the total phenotypic variation of the trait which is under additive genetic control:

$$h^2 = \frac{V_A}{V_P}$$

The heritability of a trait may be estimated experimentally using controlled matings to compare the variance of the trait among offspring of different relatedness (full sibs, half sibs, parent/offspring) (Falconer, 1989; Becker, 1984). Alternatively, the 'realized' heritability can be calculated following actual selection where the shift of the mean of the trait from the parent population to the offspring can be measured.

Measurement of heritability is complicated by a number of features. First, estimated or realized heritabilities may only be appropriate to the environment in which the experiments were conducted. If there are strong genotype – environment interactions, then selection programmes will not be applicable across different environments such as different hatcheries or fish farms. Second, different traits are frequently correlated such that an increase in value for one trait may be paralleled by a decrease in value for another. In such circumstances the degree of genetic correlation and the relative heritabilities of the two traits will be important factors in the success of selection for one or other trait. Third, 'standard errors on heritability estimates are uncomfortably large unless...based on very large numbers' (Falconer, 1989). The precision of h^2 estimates is notoriously weak because the scale of experiments is often constrained by physical limitations such as, for example, the number of crosses that can be performed, the number of families which can be reared, the number of replicates within families, or the number of individuals which can be scored or measured. Finally, if heritability is estimated from full sib families then this value is derived from the female (dam) component of genetic variation, which includes both additive (V_A) and dominance (V_D) variance together with maternal effects. Maternal effects can include, for example, the contribution to the egg cytoplasm, and therefore larval development, made by the female. Estimates based on half sibs relate to the male (sire) component and give a more realistic measure of additive genetic variation (Falconer, 1989).

Selection programmes, or breeding plans, for aquaculture species are based on heritabilities for important traits and, depending on the value of h^2, different types of selection can be instigated. Individuals may be selected for breeding based either on their own phenotypic value (individual or mass selection) or on the basis of the average phenotypic value of the family to which they belong (family selection). Alternatively, it is possible to select the best individuals from within a number of families (within-family selection). The relative merits of these three selection methods for altering a trait depend upon the value of h^2, the degree of potential inbreeding and the availability of relevant information on the trait for various families (Falconer, 1989). According to Gjedrem (1992), individual and family selection are the most efficient methods for selection in fish farming. More extended discussions of breeding plans and selection programmes in aquaculture are given by Shultz (1986), Gall (1990) and Tave (1993).

Published heritabilities are available for many important traits in salmonids and other fish (see Tave (1993) for detailed lists). In a recent review, Gjedrem (1992) has described the selection experiments which have been undertaken with rainbow trout (*Oncorhynchus mykiss*) and has tabulated estimates of heritability for many important traits. For example, heritability estimates for body weight in juveniles range from 0.06 to 0.55 (mean $h^2 = 0.21$) and in fish older than 1 year from 0.17 to 0.38 (mean $h^2 = 0.23$). Estimates of heritability from the dam component are generally higher than those from the sire component, indicating considerable non-additive genetic and/or maternal variance for this trait.

Besides the more obvious aquaculturally relevant traits of growth and meat quality, resistance to diseases has been investigated in several fish species (Chevassus and Dorson, 1990). For example, Gjedrem, Salte and Gjoen (1991) reported a high heritability of mortality from furunculosis in *Salmo salar* ($h^2 = 0.48 \pm 0.17$, sire component; $h^2 = 0.32 \pm 0.1$, dam component) and concluded that resistance to this disease can be effectively improved by selective breeding. In this study, as in others (Rye, Lillevik and Gjerde, 1990; Standal and Gjerde, 1987), there was a positive genetic correlation between growth rate and survival, indicating that selection for growth rate may give correlated responses in resistance to diseases. On the other hand, the generally restricted nature of the response to disease (death or survival), the possible negative genetic correlation between resistance to different diseases, and the high genetic variability of pathogens led Chevassus and Dorson (1990) to warn against expecting major advances in disease resistance through the quantitative genetics approach.

In contrast to fish, there are considerably fewer studies on the quantitative genetics of molluscan shellfish (Table 9.1.1). Nevertheless, selection programmes for various bivalves are now in place in several institutes worldwide and it is clear that the process of domestication, particularly of oysters, clams and scallops, has now begun (D. Hedgecock, personal communication; P.M. Gaffney and S.K. Allen, personal communication; Humphrey and Crenshaw, 1989; Crenshaw, Heffernan and Walker, 1991; Heffernan, Walker and Crenshaw, 1991).

Studies on the quantitative genetics of crustacean shellfish are rare and include work on the American lobster, *Homarus americanus* (Hedgecock, Schlester and Nelson, 1976; Hedgecock and Nelson, 1978), crayfish, *Procambarus clarkii* (Lutz and Wolters, 1989), *Penaeus vannamei* (Lester, 1988, 1989), and *Macrobrachium rosenbergii* (Malecha, Masuno and Onizuko, 1984; Suryono, 1990).

The number of studies on the quantitative genetics of particular aquaculture species is, as expected, clearly related to the length of time that controlled hatchery facilities have been available to carry out experimental crosses for those species. As more molluscs and crustacea are brought under hatchery production, so will the process of domestication and the study of quantitative genetics and selective breeding accelerate for these commercially important groups. Relevant books covering this field are Becker (1984), Kapuscinski and Jacobson (1987), Falconer (1989) and Tave (1993).

(a) Hybridization and inbreeding

The existence of hybrid vigour, or 'heterosis' (but see Zouros and Foltz, 1987; Zouros and Pogson, section 3.1) and inbreeding depression have been recognized for a long time. Indeed, Darwin (1859) was well aware of the phenomenon: 'I have collected so large a body of facts, and made so many experiments, showing, in accordance with the almost universal belief of breeders, that with animals and plants a cross between different varieties, or...strains, gives vigour and fertility to the offspring; and on the other hand, that *close* interbreeding diminishes vigour and fertility.' Amongst aquaculture organisms, the salmonid fish seem to have been the most studied and the most amenable to interspecific and interstrain crosses, but the potential exists for genetic improvement in all commercial species.

The degree of differentiation between strains required to produce measurable hybrid vigour is an important criterion. Very closely related strains may produce none while more distantly related groups (e.g. sibling species) may suffer from outbreeding depression, possibly brought about by the break-up of coadapted gene complexes or other changes in genetic architecture. Hybrids between salmonid species tend not to show measurable hybrid vigour and have little advantage in fish farming when compared with *S. salar* (Chevassus, 1979; Refstie, 1983). Nevertheless, according to Chevassus (1983), it is still important to explore interspecies hybridization in any new fish species with aquaculture potential.

In certain circumstances it may be advantageous to breed for ease of capture and this is the case with catfish which are grown for angling. Dunham *et al.* (1990) demonstrated that the hybrid cross between channel catfish, *Ictalurus punctatus*, and the blue catfish, *I. furcatus*, was more vulnerable to angling than either parental species. Whether this should be called 'heterosis for vulnerability to angling', as Dunham *et al.* (1986) suggest, or 'inbreeding depression' producing less cautious fish with reduced olfactory skills, is debatable. Nevertheless, channel and blue catfish hybrids do show heterosis for growth rate when stocked at high density (Dunham *et al.*, 1990).

In the case of crossing of strains of salmonids, results are equivocal: most studies fail to show evidence of significant enhancement of growth amongst hybrid offspring (e.g.

Table 9.1.1 *Quantitative genetic studies on commercially important bivalves*

Species	Comment	Author
Crassostrea virginica	h^2 larval growth	Longwell and Stiles (1973)
	Selection of strains resistant to MSX disease	Haskin and Ford (1978)
	h^2 larval growth rate	Newkirk *et al.* (1977)
	h^2 larval growth rate and spat length at 6 weeks	Losee (1979)
	Size selection after one generation	Haley and Newkirk (1982)
	Selection for growth rate	Newkirk (1980)
Crassostrea gigas	h^2 larval survival and setting success; spat traits	Lannan (1972)
	Selection for resistance to summer mortality	Beattie, Chew and Hershberger (1980)
		Hershberger, Perdue and Beattie (1984)
	Additive genetic variation for size at harvest	Hedgecock, Cooper and Hershberger (1991)
Ostrea edulis	Selection for growth rate	Newkirk (1980)
		Haley and Newkirk (1982)
		Newkirk and Haley (1982, 1983)
Ostrea chilensis	Realized h^2 shell height at 30 months old	Toro and Newkirk (1991)
Mytilus edulis	Genetic × salinity interaction in larval growth	Innes and Haley (1977)
	h^2 larval growth	Newkirk (1980)
	h^2 growth and survival of larvae, juvenile and adult	Mallet, Freeman and Dickie (1986)
	h^2 larval growth rate	Stromgren and Neilsen (1989)
	Larval growth; non-additive genetic variation	Newkirk, Haley and Dingle (1981)
Mulinia lateralis	Larval growth; maternal effects, non-additive genetic variation, G-E interaction	Ludwig and Gaffney (1991)
Mercenaria mercenaria	Selection for size	Chanley (1961)
	Realized h^2 for growth rate	Hadley (1988)
	h^2 juvenile growth	Rawson and Hilbish (1990)
	G–E interaction, juvenile growth	Rawson and Hilbish (1991)
	Negative response of larval growth to adult size	Heffernan, Walker and Crenshaw (1991)
Argopecten irradians concentricus	Realized h^2 growth rate	Crenshaw, Heffernan and Walker (1991)
	Larval phase independent of selection for adult size	Heffernan, Walker and Crenshaw (1992)
Pinctada fucata martensii	h^2 of shell shape	Wada (1986)

Horstgen-Schwark, Fricke and Langholtz, 1986; Gjedrem, 1992). Nevertheless, hybridization between different strains of fish species such as carp and catfish is reported to produce hybrids which are heterotic for growth and resistance to disease (Wohlfarth, Moav and Hulata, 1983; Dunham *et al.*, 1990; Eknath, 1991).

Interspecies hybridization can be used in tilapine fish to produce all-male offspring, a valuable feature because a major constraint on tilapia production is their uncontrolled reproduction in grow-out ponds (Hulata, Wohlfarth and Rothbard, 1983). Interstrain crosses in *Tilapia nilotica* can produce significant heterosis in both F_1 and F_2 generations (Tave *et al.*, 1990).

Crosses between stocks of the clam *Mercenaria mercenaria* do not appear to produce heterosis (Manzi, Hagley and Dillon, 1991), but, surprisingly, interspecies crosses between *M. mercenaria* and *M. campechiensis* apparently do (Menzel, 1962, 1989).

In contrast to the rather varied success in the search for hybrid vigour in interstrain or interspecies crosses in aquaculture species, the deleterious effects of inbreeding are usually strongly expressed even in the first inbred generation (Gjedrem, 1992; Kincaid, 1983; Mallet and Haley, 1983; Beaumont and Budd, 1983; Beaumont, 1986; Gall, 1987; Tave, 1993). Exceptions to this generalization may include certain apparently healthy populations of prawn and carp species which have been intensively farmed in the Far East for so many years that genetic bottlenecks and subsequent inbreeding must have been a common feature of their history.

9.1.3 Ploidy manipulation

The ability to manipulate the ploidy of aquaculture organisms was initially brought to prominence by Purdom (1969) working on fish, and the technology was only later taken up for use on molluscan shellfish (e.g. Stanley, Allen and Hidu, 1981). During the 1980s many more fish and shellfish species have been the subject of ploidy manipulation and developments have been reviewed by Thorgaard and Allen (1987), Beaumont and Fairbrother (1991), Thorgaard (1992) and Tave (1993). Triploid and gynogenetic diploid fish and shellfish are valuable for a number for reasons which are detailed later.

(a) Methodology

The eggs of many aquatic organisms are released directly into the sea before the maturation divisions have been completed; most molluscan eggs are released at metaphase of meiosis I, while most fish eggs are spawned at metaphase of meiosis II.

Induction of ploidy changes involves chromosome replication without cell division, effectively doubling the chromosome number, during meiosis I or II to induce triploids, or at first cleavage to produce tetraploids (Figure 9.1.1). If spermatozoa are subjected to appropriate levels of UV light or gamma irradiation to denature their DNA but retain sperm motility, then such sperm can be used, together with suppression of meiosis I, meiosis II or first cleavage, to produce diploid gynogenetic offspring (Figure 9.1.2). Similarly, androgenetic offspring can be produced by irradiation of eggs and doubling of (male) chromosomes at first cleavage. Because of the stage of the maturation divisions at the time of release, either the meiosis I or meiosis II division can be targeted in molluscs but only the meiosis II division is easily amenable to treatment in fish.

An important feature is that for both fish and molluscs the egg maturation divisions are usually halted upon release into water and require activation by sperm to proceed further. This allows accurate timing and targeting of treatment for particular divisions. For practical reasons, aquaculture organisms which brood their eggs (crustaceans, some bivalves) are not generally amenable to ploidy manipulation, although polyploid *Ostrea edulis* have now been produced (Gendreau and Grizel, 1990).

The maturation divisions and first cleavage can be suppressed by physical shock or chemical agents. Physical methods include heat or cold shock, which involve an instantaneous

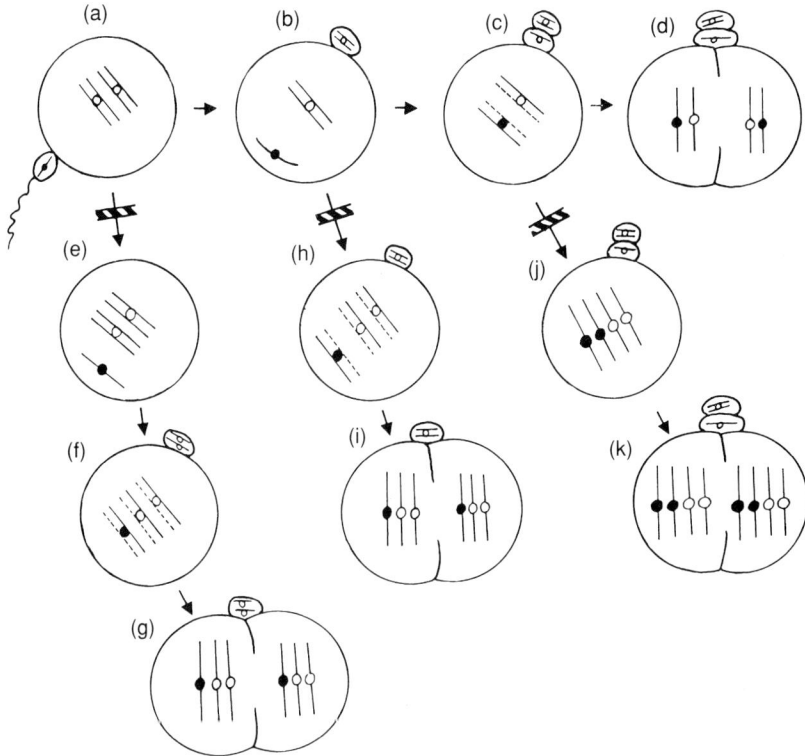

Figure 9.1.1 (a–d) Diagrammatic representation of the maturation divisions, syngamy and first cleavage in bivalve molluscs. For simplicity only one pair of chromosomes is shown. (a) Egg at release at metaphase of meiosis I, activation by sperm; (b) meiosis I complete, first polar body extruded, sperm nucleus (male pronucleus) has entered egg; (c) meiosis II completed, second polar body extruded, male and female pronuclei unite on first cleavage spindle; (d) first cleavage.
(e–k) Consequences of ploidy manipulation. (e) Shock administered at meiosis I, first polar body suppressed; (f) normal meiosis II, second polar body extruded, female pronucleus (2N) unites with male pronucleus (N); (g) first cleavage (meiosis I triploid); (h) shock administered during meiosis II, no second polar body produced; (i) first cleavage (meiosis II triploid); (j) shock administered at first cleavage; (k) tetraploid.

increase or decrease in temperature of between 5 and 10 °C, or the application of high pressure (41–62 MPa; 400–600 atm). Physical or chemical treatments are applied to eggs at appropriate times, and for a specific duration, to target meiosis I, meiosis II or first cleavage. In view of the success and ease of temperature and pressure shock treatment in fish (e.g. Quillet *et al.*, 1991), chemical methods are only extensively used in molluscs (Beaumont and Fairbrother, 1991). The most effective chemical is cytochalasin B, a fungal metabolite, and the northwestern USA oyster industry now routinely uses this treatment following the publication of Allen, Downing and Chew's (1989) hatchery manual. Further improvements to the technique have recently been made by Barber, Mann and Allen (1992) and 6-dimethylaminopurine (6-DMAP), a cheaper and less toxic chemical, has now been successfully used to induce triploidy in molluscs (Desrosiers *et al.*, 1993).

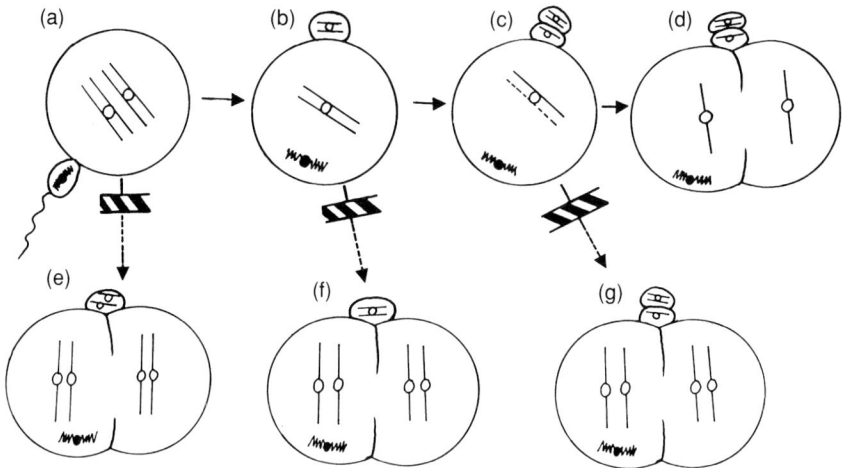

Figure 9.1.2 (a–d) Diagrammatic representation of bivalve mollusc egg maturation divisions and first cleavage following activation by UV-treated sperm. Haploid embryo produced (d). Diploid gynogenomes produced by shock administered at meiosis I (e), meiosis II (f) or first cleavage (g).

Accurate methods for assessing ploidy are important because treatments are seldom 100% effective and offspring from treated eggs are usually composed of mixtures of polyploids and diploids. Ploidy changes can be verified by a number of techniques, including chromosome counts, nuclear DNA staining using fluorescent dye with subsequent flow cytometry or microfluorimetry, direct nuclear or cellular sizing using light microscopy, counting of nucleoli and protein electrophoresis (Beaumont and Fairbrother, 1991). Identifying gynogenetic diploids is less easy but the use of control groups which are activated by treated sperm but which undergo no chromosome doubling (confirming production of only non-viable haploids) can provide indirect evidence that any diploids produced following chromosome doubling treatment will be gynogenetic. It is also possible to use a sperm donor which is homozygous for a dominant genetic marker, thus allowing identification of any non-gynogenetic diploid progeny from a homozygous recessive female (Thorgaard and Allen, 1987).

(b)   Gynogens and androgens

Gynogens and androgens have two important features of relevance to aquaculture. First, in fish which exhibit female homogamety, such as carp and many salmonids (XY system), all gynogenetic offspring should be female (XX). Similarly, androgenetic offspring should all be male (ZZ) in species with male homogamety (WZ system) (Tave, 1993). Production of monosex fish by androgenesis or gynogenesis, however, suffers from the disadvantage of a measure of inbreeding and, for most purposes, according to Thorgaard (1983), it is still probably better to use hormonal sex reversal. Ideally, a gynogenetic all-female (XX) population is sex-reversed to produce functional XX males which are then mated to unrelated normal females to produce outbred all-female offspring. Similarly, sex reversal can be used to produce outbred all-male lines in male homogametic species.

It is the element of inbreeding which offers the second important aquaculture potential for gynogenesis. It allows the possibility of rapid development of inbred lines for domesticated hatchery broodstock. However, that potential is difficult to realize because mitogynes (gynogens produced by suppression of first cleavage), though 100% homozygous, suffer high abnormalities and mortalities, and meiogynes (gynogens from suppression of meiosis I or II in molluscs, meiosis II in fish) may be less than 70% inbred, depending upon the amount of crossing-over at meiosis. Nevertheless, a useful feature of gynogens from a purely genetic viewpoint is that electrophoretic analysis can provide information on recombination frequencies at gene loci (e.g. Thompson and Scott, 1984; Allendorf and Leary, 1984; Thorgaard and Allen, 1987).

Although gynogens should, in principle, contain only maternal DNA, a recent study by Carter *et al.* (1991) using DNA fingerprinting on tilapia has demonstrated unexpected trans-mission of some paternal DNA during mitotic gynogenesis in *Oreochromis aureus*, although complete homozygosity was recorded at all allozyme loci scored. They suggested that this was possibly due to the incorporation of fragments of paternal chromosomal material into the female chromosomes before or during first cleavage.

(c)   Triploids and tetraploids

The most important feature of triploids is their sterility. Homologous chromosomes in the germ cells of triploids cannot synapse during early meiosis I and normal gametogenesis does not occur. Thus energy usually diverted to gamete production in mature animals is available for somatic growth in triploids. In fish the undesirable side-effects of sexual maturation, such as high mortality, reduced meat quality and slower growth, are not always eliminated in triploids though they are functionally sterile. In bivalves such as the oyster *Crassostrea gigas*, meat quality of triploids is sufficiently improved compared with diploids during the breeding season that year-round marketability is achieved (Allen, 1988; Allen and Downing, 1991). Generally, in both fish and molluscs, triploid males show more gonadal development than triploid females and occasionally active spermatozoa may be produced (Allen and Downing, 1990). However, where, for example, sperm from triploid plaice *Pleuronectes platessa* have been used to fertilize normal eggs, only non-viable aneuploid embryos develop (Lincoln, 1981). In general the physiological efficiency and performance of triploid rainbow trout has been found to be similar or slightly inferior to that of normal diploids at least until they are over 2 years old (Thorgaard, 1992).

Several trials have been carried out to produce triploid interspecific or intergeneric hybrid fish. Although some studies have demonstrated that triploid hybrids exhibit better survival than diploid hybrids (Thorgaard and Allen, 1987) this is not always the case (e.g. Quillet *et al.*, 1988; Blanc, Poisson and Vallee, 1992; McKay, Ihssen and McMillan, 1992).

A further feature of triploids which has potential relevance to aquaculture is their increased overall heterozygosity compared with diploids. At a locus with three alleles (*A, B, C*), triploids may be triple homozygotes (*AAA, BBB, CCC*), triple heterozygotes (*ABC*) or a range of double heterozygotes (e.g. *AAB, ABB, ACC* etc). Following traditional selectionist theory, the possession of three rather than two alleles at a locus might be expected to have the effect of increasing the fitness of an organism, but this remains an open question. Nevertheless, triploids should have a general increase in two-allele heterozygosity at a locus compared to diploids, and that increase will depend on which meiotic division is targeted and the frequency of recombination at the locus (Allendorf and Leary, 1984; Beaumont and Fairbrother, 1991). Such potential increased heterozygosity, which has indeed been demonstrated electrophoretically, would be expected to lead to the equivalent of 'hybrid vigour' (Zouros and Foltz, 1987). However, published results comparing growth rates of diploids and triploids before they reach maturity are equivocal; some studies show significant larval or juvenile growth improvement in triploids, while many others do not (Beaumont and Fairbrother, 1991).

The potential to produce sterile triploids of non-native fish and shellfish species enables their use for aquaculture in regions where introduced exotic species might challenge and compete with indigenous species. Also, as suggested by Thorgaard (1992), they may be useful as safe experimental animals for transgenic studies (see later). However, triploidy induction techniques are seldom 100% effective and regulatory bodies concerned with the introduction of non-native species for aquaculture are clearly aware that viable diploids are therefore likely to be included in any introduction of triploids. One way around this difficulty is to produce tetraploids by targeting first cleavage and using these tetraploids as adult broodstock to mate with diploid individuals of the other sex. In this way, all-triploid offspring can be guaranteed. Tetraploid salmonids have been successfully reared to maturity and triploid rainbow trout have been produced by crossing tetraploid and diploid parents (Blanc, Chourrout and Krieg, 1987; Myers and Hershberger, 1991) but little success has been achieved in other fish species. Furthermore, there are no reports of the successful production of viable tetraploid molluscs by targeting first cleavage. Surprisingly, a high proportion of tetraploid oyster (*C. gigas*) and clam (*Tapes philippinarum*) embryos resulted from chemical treatment of eggs immediately after release, at early meiosis I, a treatment normally expected to induce triploidy rather than tetraploidy (Stephens and Downing, 1988; Diter and Dufy, 1990). However, none of the tetraploid embryos survived beyond the spat stage.

Partly because of these difficulties in getting viable tetraploids in molluscs by suppressing first cleavage, attempts have recently been made to induce tetraploidy by cell fusion at the two-cell embryo stage using ethylene glycol (Guo *et al.*, 1990) or electrofusion (Cadoret, 1992), and this latter approach certainly looks promising.

9.1.4 Allozyme genetics

The development of the techniques of allozyme electrophoresis over the last 25 years has enable allozyme data to be used to address a number of questions of importance in aquaculture. There are three important aspects of fish or shellfish hatchery practice which are likely to have a significant impact on the genetics of cultured species, and all can be investigated using allozyme data. First, if the number of progenitors is small then genetic drift will lead to loss of rare alleles and a reduction in heterozygosity. Second, inbreeding, and consequent inbreeding depression, will occur if the offspring of the original broodstock are retained as new broodstock. Finally, cultured organisms are protected from some aspects of natural selection but are subjected to artificial selection for size or other aspects of culture as an unintentional result of normal husbandry procedures (Allendorf and Ryman, 1987; Tave, 1993).

(a)   Loss of genetic diversity in hatchery broodstocks

A number of studies have clearly demonstrated significant loss of genetic diversity following hatchery culture of salmonid fish, including *S. salar* (Cross and King, 1983; Stahl, 1983; Verspoor, 1988; Koljonen, 1989), *S. trutta* (Ryman and Stahl, 1980; Vourinen, 1984), *O. mykiss* (Hershberger, 1992), cutthroat trout, *O. clarki* (Allendorf and Phelps, 1980); other fish, such as black sea bream, *Acanthopagrus schlegeli* (Taniguchi, Sumantadinata and Iyama, 1983) and red sea bream, *Pagrus major* (Sugama, Taniguchi and Umeda, 1988); crustaceans, including *Penaeus japonicus* (Sbordoni *et al.*, 1986) and *P. vannamei* (Sunden and Davis, 1991); and molluscs, such as *C. gigas* (Gosling, 1982; Hedgecock and Sly, 1990), *C. virginica* (Vrijenhoek, Ford and Haskin, 1990; Gaffney, Davis and Hawes, 1992) (see also Brown and Paynter (1991) for genetic diversity based on a study of mitochondrial DNA), *M. mercenaria* (Dillon and Manzi, 1987) and *Haliotis iris* (Smith and Conroy, 1992).

An important conclusion drawn in several of these studies (e.g. Sbordoni *et al.*, 1986; Hedgecock and Sly, 1990; Gaffney, Davis and Hawes, 1992; see also Gaffney and Scott,

1984; Gile and Ferguson, 1990) is that the effective population size (N_e) involved in hatchery 'mass spawnings' can be very much smaller than the apparent number of progenitors used. In the particular case of the prawn *P. japonicus* (Sbordoni *et al.*, 1986), N_e was calculated to lie between two and four individuals per generation in broodstock tanks with actually held several hundred adult prawns! Heterozygosity and allelic diversity were dramatically reduced over seven generations. By contrast, only small reductions in heterozygosity and allelic diversity, and no evidence for inbreeding, were reported for a hatchery population of *P. vannamei* (Sunden and Davis, 1991). While differences in husbandry practice are evident between these two aquaculture stocks, it is nevertheless important to bear in mind that they are in fact different species and other factors such as mating behaviour, physiology and genetic architecture could play an important role in the reported differences in N_e. Hedgecock, Chow and Waples (1992) have estimated N_e in hatchery broodstocks of various shellfish using temporal variance in allele frequencies and conclude that in at least half of these broodstocks the true N_e is significantly less than the apparent number of progenitors.

Effective population size is strongly affected by the sex ratio of the progenitors (Allendorf and Ryman, 1987) so it is important to ensure that approximately equal numbers of males and females are used in spawnings. Allendorf and Ryman (1987) recommend a minimum of 100 males and 100 females be used to maintain genetic variability in a hatchery population of fish. Gosling (1982) has estimated that at least 45 individuals (sex ratio about 1 : 1) are needed to provide a 99% chance of retaining alleles at a frequency of 0.01 at a locus. Tave (1993) provides a tabulation of the probabilities of losing alleles of increasing rarity in relation to the effective number of progenitors. Gaffney, Davis and Hawes' (1992) data prompted them to recommend a simple practical approach to ensure a high N_e in shellfish culture. Rather than carrying out a single mass spawning involving many ripe adults, multiple small spawnings with subsequent pooling of larvae should be attempted.

(b)   Impact of release of cultured fish

The allozyme approach has recently been employed to assess the effect and impact of released cultured fish on natural wild stocks (Hindar, Ryman and Utter, 1991; Saunders, 1991). Cultured Atlantic salmon have been and are being released in quite large numbers either intentionally, in the context of sea ranching and stock enhancement, or accidentally by escape from cage culture operations. In some spawning populations in Norway, released fish actually outnumber native ones (Gausen and Moen, 1991).

A particular population genetic consideration with Atlantic salmon is the fact that wild populations are generally subdivided into small enclaves associated with specific rivers, tributaries or even short sections within tributaries. It is generally agreed that this population genetic subdivision is a result of restricted gene flow and natural selection operating in subdivided environments such that individual subpopulations are uniquely fitted to live where they do. Clearly, introduced cultured fish from any different source population are likely to be less fit than native fish, even ignoring the potential loss of genetic diversity associated with their culture. Indeed, reduced fitness in cultured fish has been recorded in many ways, including poor juvenile stamina and survival, weaker territorial behaviour, poor concealment, lower oceanic survival, increased straying rate, lower disease resistance and less active spawning behaviour (Hindar, Ryman and Utter, 1991; Fleming and Gross, 1992).

Studies on the effects of cultured fish on wild populations have demonstrated that a range of effects are possible. In many cases coadapted gene complexes may be broken down and unique alleles lost in wild fish populations following extensive hybridization and introgression of genes from cultured fish. However, in other cases introduced fish seem to have little effect. For example, following stocking using hatchery-reared anadromous sea trout *S. trutta* within a single river system, stocked fish hybridized with resident fish in some tributaries but not in others (Hauser *et al.*, 1991). For a more extensive review of this topic, see Hindar, Ryman and Utter (1991).

In certain cases where released fish can be genetically tagged they can be used to investigate their impact on native populations. One such example involving the release of large numbers of genetically tagged cod into a Norwegian fjord system is given elsewhere in Section 9.5.

9.1.5 Transgenic studies

Modern techniques for DNA analysis have been used in a few instances to address questions of importance to aquaculture (e.g. Brown and Paynter, 1991; Carter *et al.*, 1991), but the principal involvement of DNA technology is currently in transfection, the introduction of novel genes into aquaculture organisms. There are many potential benefits of this approach, including, for example, the development of disease-resistant stocks, faster-growing animals or freeze resistance in salmonids.

An important paper by Palmiter *et al.* (1982) first demonstrated the feasibility of the transfer of the rat growth hormone gene into mice, producing a dramatic increase in growth rate. However, the success of this experiment was due to expression of the growth hormone in all the cells of the liver rather than in just the pituitary gland, the normal site of expression. This was because the gene was spliced to a mouse metallothionein promoter which caused the gene to be switched on in the liver.

Fish are currently a valuable group of animals for the application of gene technology and transgenic induction because ethical considerations which tend to constrain the use of genetic manipulation in mammals and other 'higher' animals are less likely to apply to fish involved in aquaculture. A variety of fish species have been studied using a number of gene constructs and with varying degrees of success (Table 9.1.2). Chen, Agellon and Van Beneden (1987) have investigated a number of genes which might be appropriate for genetic engineering in fish and several fish growth hormone genes have now been identified and cloned (Schneider *et al.*, 1992; see below) and their chemical identification explored (e.g. Watanabe *et al.*, 1992).

Gene constructs which have been used in transfection experiments include Palmiter *et al.*'s (1982) rat growth hormone with mouse metallothionein promoter (MTrGH) (e.g. Maclean, Penman and Talwar, 1987; Penman *et al.*, 1991), human growth hormone with SV40 promoter (Chourrout, Guyomard and Houdebine, 1986), salmon growth hormone (sGH) with carp β-actin promoter (Liu *et al.*, 1990; Rahman and Maclean, 1992), bovine or chinook salmon growth hormone (bGH, csGH) with carp β-actin promoter (Gross *et al.*, 1992), salmonid growth hormone with Rouse sarcoma virus (RSV) (Hayat *et al.*, 1991), neomycin resistance gene with RSV (Yoon *et al.*, 1990) and a chinook salmon growth hormone with an ocean pout antifreeze protein (AFP) promoter (Du *et al.*, 1992).

Using standard techniques (Maniatis, Fritsch and Sambrook, 1982), the gene sequence or construct is cloned into a plasmid, grown up in a suitable bacterial strain and then harvested to provide a high copy number of the DNA. Either the whole plasmid (circularized) or just the gene construct, isolated using restriction endonucleases (linearized), can be used. Although circularized DNA is more resistant to denaturation within the cell, it is also less available for chromosome integration than linearized DNA (Maclean and Penman, 1990).

The principal methodology for introduction of novel DNA into fish is by microinjection of a solution containing around 10^6 copies of the DNA into the egg following insemination but before first cleavage. Injection after first cleavage is likely to produced mosaicism, with the transgene occurring in some cells or tissues but not in others. Indeed, such chimaeric fish are

Table 9.1.2   Transgenic studies on fish

Species	Comment	Authors
Oncorhynchus	MI, MT-rGH, int	Maclean and Talwar (1984)
mykiss	MI, SV40-hGH, int	Chourrout, Guyomard and Houdebine (1986)
	MI, MT-rGH, int	Maclean, Penman and Talwar (1987)
	MI, MT-hGH, int, F_1	Guyomard *et al.* (1989)
	MI, MT-rGH, int	Penman *et al.* (1988)
	MI, MT-rGH, int	Penman *et al.* (1990)
	MI, MT-rGH, int, F_1	Penman *et al.* (1991)
Salmo salar	MI, wfAFP, int	Fletcher *et al.* (1988)
	MI, MT-CAT, exp	McEvoy *et al.* (1988)
	MI, hGH, int, exp	Rokkones *et al.* (1988)
	MI, opAFP-csGH, int, exp	Du *et al.* (1992)
Oreochromis	MI, MT-hGH, int	Brem *et al.* (1988)
niloticus	MI, MT-rGH-CAT, int, exp	Rahman and Maclean, 1992
Oryzias	MI, cCR, int, exp	Ozato *et al.* (1986)
latipes	MI, USV-CAT, int, exp	Chong and Vielkind (1989)
	EL, MT-rGH, int, F_1	Inoue *et al.* (1990)
Ictalurus	MI, MT-hGH, int	Dunham *et al.* (1987)
punctatus	MI, MT-rGH, RSV-rtGH, RSV-csGH, RSV-rtV, int	Hayat *et al.* (1991)
Brachydanio	MI, USV-CAT, int, F_1	Stuart, McMurray and Westerfield (1988)
rerio	MI, USV-CAT, int, F_1	Stuart *et al.* (1990)
	SP, USV-CAT, int, F_1, F_2	Khoo *et al.* (1992)
Cassarius	MI, MT-hGH, int	Zhu *et al.* (1985)
auratus	MI, MT-hGH, int	Maclean, Penman and Zhu (1987)
	MI, RSV-NEO, int, exp	Yoon *et al.* (1990)
Misgurnus anguillicaudatus	MI, MT-hGH, int	Zhu *et al.* (1986)
Cyprinus	MI, RSV-rtGH, int, exp, F_1	Zhang *et al.* (1990)
carpio	MI, MT-rGH, RSV-rtGH, RSV-csGH, RSV-rtV, int	Hayat *et al.* (1991)
Esox lucius	MI, RSV-bGH, FV-2-csGH, int, exp	Gross *et al.* (1992)

Methods; MI = microinjection; EL = electroporation; SP = sperm vector. Promoters: MT = mammalian metallothionein; RSV = Rouse sarcoma virus; SV40 = SV40 promoter; USV = Rouse sarcoma virus long terminal repeat (RSV-LTR) + SV40 promoter; FV-2 = (fish expression vector) carp β-actin. Genes; GH = growth hormone; rGH = rat GH; hGH = human GH; bGH = bovine GH; rtGH = rainbow trout GH; csGH = chinook salmon GH; cCR = chiken crystallin; CAT = (bacterial) chloramphenicol acetyltransferase; opAFP = ocean pout antifreeze protein; wfAFP = winter flounder antifreeze protein; rtV = rainbow trout vitellogenin; NEO = neomycin resistance. Results; int = integration; exp = expression; F_1 = transgene detected in F_1 offspring of founder transgenics; F_2 = transgene detected in F_2 generation.

commonly produced in transgenic experiments even when microinjection only takes place before the first cleavage division is complete. In addition to the critical nature of the timing of microinjection (e.g. Hayat *et al.*, 1991), the use of linear versus circularized DNA and the type of buffer used (Penman *et al.*, 1990), the positioning close to the nucleus of the injected multiple-copy DNA is a further important criterion. In some species, such as salmon, eggs are best injected via the micropyle, and in others direct injection through the chorion is possible, but in all cases precise location of the DNA solution near to the nucleus is required.

An alternative method for getting the novel DNA into the egg is electroporation (Inoue *et al.*, 1990) whereby oocytes or fertilized eggs are subjected to electrical pulses which increase cell membrane porosity, allowing direct infusion of DNA into the egg. Eggs are held in a small volume of buffered saline containing a high copy number of the DNA to be introduced. The advantages of electroporation are that large numbers of eggs can be treated at a time and, unlike microinjection, this method requires little technical skill.

At the time of writing there are no published accounts of attempts to produce transgenic crustaceans or molluscs. The delay in transfer of the technology from fish to invertebrate aquaculture organisms is probably partly due to the greater commercial importance of fish but must also be due to the fact that crustacean eggs are generally brooded and molluscan eggs, though released before fertilization, are far too small for microinjection. Nevertheless, recent preliminary studies on electroporation of *C. gigas* eggs and sperm have indicated that this method has great potential as a technique for transgenic induction in bivalves (S. Watson and A.R. Beaumont, unpublished data). Another possibility resides in the apparent ability of spermatozoa of zebrafish and *Mytilus edulis* to absorb DNA directly from solution and might provide a simple and effective method of getting novel DNA into the egg (Khoo *et al.*, 1992; A.R. Beaumont and K. Hoare, unpublished data).

The usual technique for confirming success of transgenism is to sample putative transgenic fish, either destructively or by biopsy, and to purify DNA from the individual fish. Dot or slot blots are carried out by hybridizing a radiolabelled probe of the transgene DNA to the DNA from individual fish (Hallerman and Beckmann, 1988). Autoradiographic identification of fish containing the transgene is then possible. The next question is whether the DNA is incorporated into the genome. Southern blotting is required to confirm this (Maclean and Penman, 1990) and works on the principle that if the introduced DNA sequence is incorporated into the genome of the fish it will produce longer (higher molecular weight) fragments of DNA than non-integrated sequences following digestion with an appropriate restriction endonuclease.

Confirmation of the integration of the transgene into the genome of the host organism can be followed by a search for its expression by probing for the appropriate mRNA, by using protein immunoassay, or by direct testing for enzyme activity in the case of reporter genes such as chloramphenicol acetyltransferase (CAT). If the promoter spliced into the gene construct is not operative in the host then the gene will not be expressed. The search for gene expression in transgenic fish injected with early constructs using mouse metallothionein (MTrGH) was generally unsuccessful (Penman *et al.*, 1991). More recently there has been evidence of successful expression of an 'all-fish' growth hormone/ocean pout AFP promoter construct which produced a significant increase in growth rate of transgenic salmon compared to non-transgenic siblings (Du *et al.*, 1992). Furthermore, Gross *et al.* (1992) have reported expression of a chinook salmon growth hormone (csGH) in transgenic northern pike, *Esox lucius*, which had been microinjected with a gene construct containing csGH and the carp β-actin gene promoter.

The final goal, following demonstration of gene integration and expression, is to establish that the gene is incorporated into the germ-line. Several studies have provided confirmation that there is often widespread mosaicism in the first-generation transgenics and that the transgene seldom appears to be present in 50% of their gametes: the ratio which would be expected if the fish are transgenic heterozygotes with integration at a single locus (Guyomard *et al.*, 1989; Stuart *et al.*, 1990; Zhang *et al.*, 1990; Penman *et al.*, 1991).

It seems most likely that the development of techniques for producing transgenic fish, and possibly also shellfish, will continue to accelerate, making this the most rapidly growing area of genetic research in aquaculture for the immediate future. However, it is still too soon to tell whether any of the immediately obvious potential benefits such as disease resistance or faster growth will accrue and become commercially viable from this approach. Apart from the purely technical difficulties inherent in producing transgenics, there remain important ethical considerations and considerable barriers to consumer acceptance of genetically modified organisms.

Acknowledgements

I am very grateful to Norman Maclean, Stan Allen, Kate Hoare and Craig Wilding for helpful and constructive criticism of the manuscript.

References

Allen, S.K. (1988) Triploid oysters ensure year-round supply. *Oceanus*, **31**, 58–63.

Allen, S.K. and Downing, S.L. (1990) Performance of triploid Pacific oysters *Crassostrea gigas*: gametogenesis. *Can. J. Fish. Aquat. Sci.*, **47**, 1213–22.

Allen, S.K. and Downing, S.L. (1991) Consumers and 'experts' alike prefer the taste of sterile triploid over gravid diploid oysters *(Crassostrea gigas). J. Shellfish Res.*, **10**, 19–22.

Allen, S.K., Downing, S.L. and Chew, K.K. (1989) *Hatchery Manual for Producing Triploid Oysters*, University of Washington Press, Seattle.

Allendorf, F.W. and Leary, R.F. (1984) Heterozygosity in gynogenetic diploids and triploids estimated by gene-centromere recombination rates. *Aquaculture*, **43**, 413–20.

Allendorf, F.W. and Phelps, S.R. (1980) Loss of genetic variation in hatchery stock of cutthroat trout. *Trans. Am. Fish. Soc.*, **109**, 537–43.

Allendorf, F.W. and Ryman, N. (1987) Genetic management of hatchery stocks, in *Population Genetics and Fishery Management* (eds N. Ryman and F. Utter), University of Washington Press, Seattle, pp. 141–59.

Barber, B.J. Mann, R. and Allen, S.K. (1992) Optimization of triploid induction for the oyster *Crassostrea virginica* (Gmelin). *Aquaculture*, **106**, 21–6.

Beattie, J.H., Chew, K.K and Hershberger, W.H. (1980) Differential survival of selected strains of Pacific oysters (*Crassostrea gigas*) during summer mortality. *Proc. Natl. Shellfish Assoc.*, **70**, 184–9.

Beaumont, A.R. (1986) Genetic aspects of hatchery rearing of the scallop, *Pecen maximus* (L.). *Aquaculture*, **57**, 99–110.

Beaumont, A.R. and Budd, M.D. (1983) Effects of self-fertilisation and other factors on the early development of the scallop *Pecten maximus. Mar. Biol.*, **76**, 285–9.

Beaumont, A.R. and Fairbrother, J.E. (1991) Ploidy manipulation in molluscan shellfish: a review. *J. Shellfish Res.*, **10**, 1–18.

Becker, W.A. (1984) *Manual of Quantitative Genetics*, 4th edn, Academic Enterprises, Pullman.

Blanc, J-M., Chourrout, D. and Krieg, F. (1987) Evaluation of juvenile rainbow trout survival and growth in half-sib families from diploid and tetraploid sires. *Aquaculture*, **65**, 215–20.

Blanc, J-M., Poisson, H. and Vallee, F. (1992) Survival, growth and sexual maturation of the triploid hybrid between rainbow trout and arctic char. *Aquat. Living Resour.*, **5**, 15–21.

Brem, G., Brenig, B., Horstgen-Schwark, G. and Winnacker, E.L. (1988) Gene transfer in tilapia (*Oreochromis niloticus*). *Aquaculture*, **68**, 209–19.

Brown, B.L. and Paynter, K.T. (1991) Mitochondrial DNA analysis of native and selectively inbred Chesapeake Bay oysters, *Crassostrea virginica. Mar. Biol.*, **110**, 343–52.

Cadoret, J-P. (1992) Electric field-induced polyploidy in mollusc embryos. *Aquaculture*, **106**, 127–39.

Carter, R.E., Mair, G.C., Skibinski, D.O.F., Parkin, D.T. and Beardmore, J.A. (1991) The application of DNA fingerprinting in the analysis of gynogenesis in tilapia. *Aquaculture*, **95**, 41–52.

Chanley, P.E. (1961) Inheritance of shell marking and growth in the hard clam *Mercenaria mercenaria. Proc. Natl. Shellfish Assoc.*, **50**, 163–9.

Chen, T.T., Agellon, L.B. and Van Beneden, R.J. (1987) Genetic engineering of fish, in *Selection, Hybridization and Genetic Engineering in Fish* (ed. K. Tiews), vol. II, Heenemann, Berlin, pp. 347–60.

Chevassus, B. (1979) Hybridization in salmonids: results and perspectives. *Aquaculture*, **17**, 113–28.

Chevassus, B. (1983) Hybridization in fish. *Aquaculture*, **33**, 245–62.

Chevassus, B. and Dorson, M. (1990) Genetics of resistance to disease in fish. *Aquaculture*, **85**, 83–107.

Chong, S.S.C. and Vielkind, J.R. (1989) Expression and fate of CAT reporter gene microinjected into fertilized medaka (*Oryzias latipes*) eggs in the form of plasmid DNA, recombinant phage particles and its DNA. *Theoret. Appl. Genet*, **78**, 369–80.

Chourrout, D., Guyomard, R. and Houdebine, L. (1986) High efficiency gene transfer in rainbow trout (*Salmo gairdneri*) by microinjection into egg cytoplasm. *Aquaculture*, **51**, 143–50.

Crenshaw, J.W. Jr, Heffernan, P.B. and Walker R.L. (1991) Heritability of growth rate in the southern bay scallop, *Argopecten irradians concentricus* (Say, 1822). *J. Shellfish Res.*, **10**, 55–63.

Cross. T.F. and King, J. (1983) Genetic effects of hatchery rearing in Atlantic salmon. *Aquaculture*, **33**, 33–40.

Darwin, C. (1859) *The Origin of Species*, John Murray, London.

Desrosiers, R.R., Gérard, A., Peignon, J.-M. *et al.* (1993) A novel method to produce triploids in bivalve molluscs by the use of 6-dimethylaminopurine. *J. Exp. Mar. Biol. Ecol.*, **170**, 29–43.

Dillon, R.T. and Manzi, J.J. (1987) Hard clam, *Mercenaria mercenaria*, broodstocks: genetic drift and loss of rare alleles without reduction in heterozygosity. *Aquaculture*, **60**, 99–105.

Diter, A. and Dufy, C. (1990) Polyploidy in the Manila clam, *Ruditapes philippinarum*. II. Chemical induction of tetraploid embryos. *Aquat. Living Resour.*, **3**, 107–12.

Du, S.J., Gong, Z., Fletcher, G.L. *et. al.* (1992) Growth enhancement in transgenic Atlantic salmon by the use of an 'all fish' chimeric growth hormone gene construct. *BioTechnology*, **10**, 176–81.

Dunham, R.A., Smitherman, R.O., Goodman, R.K. and Kemp, P (1986) Comparison of strains, crossbreeds and hybrids of channel catfish for vulnerability to angling. *Aquaculture*, **57**, 193–201.

Dunham, R.A., Eash, J., Askins, J. and Townes, T.M. (1987) Transfer of the metallothionein-human growth hormone fusion gene into channel catfish. *Trans. Am. Fish. Soc.*, **116**, 87–91.

Dunham, R.A., Brummet, R.E., Ella, M.O. and Smitherman, R.O. (1990) Genotype–environment interactions for growth of blue, channel and hybrid catfish in ponds and cages at varying densities. *Aquaculture*, **85**, 143–51.

Eknath, A.E. (1991) A review of carp genetic research and possible approaches to genetic improvement of Asian carps, in *The Second Asian Fisheries Forum*, Tokyo, Japan, April, 1989 (eds R. Hirano and I. Hanyu), Asian Fisheries Soc., Manila, Philippines, pp. 473–6.

Falconer, D.S. (1989) *Introduction to Quantitative Genetics*, 3rd edn, Longman, New York.

Fleming, I.A. and Gross, M.R. (1992) Reproductive behaviour of hatchery and wild coho salmon (*Oncorhynchus kisutch*): does it differ? *Aquaculture*, **103**, 101–21.

Fletcher, G.L., Shears, M.A., King, K.J., Davies, P.L. and Hew, C.L. (1988) Evidence for antifreeze protein gene transfer in Atlantic salmon (*Salmo salar*). *Can. J. Fish. Aquat. Sci.*, **45**, 352–7.

Gaffney, P.M., Davis, C.V. and Hawes, R.O. (1992) Assessment of drift and selection in hatchery populations of oysters (*Crassostrea virginica*). *Aquaculture*, **105**, 1–20.

Gaffney, P.M. and Scott, T.M. (1984) Genetic heterozygosity and production traits in natural and hatchery populations of bivalves. *Aquaculture*, **42**, 289–302.

Gall, G.A.E. (1987) Inbreeding, in *Population Genetics and Fishery Management* (eds N. Ryman and F. Utter), University of Washington Press, Seattle, pp. 47–87.

Gall, G.A.E. (1990) Basis for evaluation breeding plans. *Aquaculture*, **85**, 125–42.

Gall, G.A.E. and Busack, C.A. (eds) (1986) *Genetics in Aquaculture II*, Elsevier Press, Amsterdam.

Gausen, D. and Moen, V. (1991) Large-scale escapes of farmed Atlantic salmon (*Salmo salar*) into Norwegian rivers threaten natural populations. *Can. J. Fish. Aquat. Sci.*, **48**, 426–8.

Gendreau, S. and Grizel, H. (1990) Induced triploidy and tetraploidy in the European flat oyster *Ostrea edulis* L. *Aquaculture*, **90**, 229–38.

Gile, S.R. and Ferguson, M.M. (1990) Crossing methodology and genotypic diversity in a hatchery strain of rainbow trout (*Oncorhynchus mykiss*). *Can. J. Fish. Aquat. Sci.*, **47**, 719–24.

Gjedrem, T. (ed.) (1990) *Genetics in Aquaculture III*, Elsevier Press, Amsterdam.

Gjedrem, T. (1992) Breeding plans for rainbow trout. *Aquaculture*, **100**, 73–83.

Gjedrem, T., Salte, R. and Gjoen, H.M. (1991) Genetic variability in susceptibility of Atlantic salmon to furunculosis. *Aquaculture*, **97**, 1–6.

Gosling, E.M. (1982) Genetic variability in hatchery produced Pacific oysters (*Crassostrea gigas* Thunberg). *Aquaculture*, **26**, 273–87.

Gross, M.L., Schneider, J.F., Moav, N. *et al.* (1992) Molecular analysis and growth evaluation of northern pike (*Esox lucius*) microinjected with growth hormone genes. *Aquaculture*, **103**, 253–73.

Guo, X., Cooper, K., Hershberger, W.K. and Chew, K. (1990) Production of tetraploid embryos in the Pacific oyster, *Crassostrea gigas*: comparison among different approaches. *J. Shellfish Res.*, **10**, 236 (abstract only).

Guyomard, R., Chourrout, D., Leroux, C., Houdebine, L.M. and Pourrain, F. (1989) Integration and germ line transmission of foreign genes microinjected into fertilized trout eggs. *Biochimie*, **71**, 857–63.

Hadley, N. (1988) Improving growth rates of hard clams through genetic manipulation. *World Aquacul.*, **19**, 65–6.

Haley, L.E. and Newkirk, G.F. (1982) The genetics of growth rate of *Crassostrea virginica* and *Ostrea edulis*. *Malacologia*, **22**, 399–401.

Hallerman, E.M. and Beckmann, J.S. (1988) DNA-level polymorphisms as a tool in fisheries science. *Can. J. Fish. Aquat. Sci.*, **45**, 1075–87.

Haskin, H.H. and Ford, S.E. (1978) Mortality patterns and disease resistance in Delaware Bay oysters. *Proc. Natl. Shellfish Assoc.*, **68**, 80.

Hauser, L., Beaumont, A.R., Marshall, G.T.H. and Wyatt, R.J. (1991) Effects of sea trout stocking on the population genetics of landlocked brown trout, *Salmo trutta* L., in the Conwy River system, North Wales, UK. *J. Fish Biol.*, **39**, (Suppl. A), 109–16.

Hayat, M., Joyce, C.P., Townes, T.M. *et al.* (1991) Survival and integration rate of channel catfish and common carp embryos microinjected with DNA at various developmental stages. *Aquaculture*, **99**, 249–56.

Hedgecock, D., Chow, V. and Waples, R.S. (1992) Effective population numbers of shellfish broodstocks estimated from temporal variance in allele frequencies. *Aquaculture*, **108**, 215–32.

Hedgecock, D., Cooper, K. and Hershberger, W. (1991) Genetic and environmental components of variance in harvest body size among pedigreed Pacific oysters *Crassostrea gigas* from controlled crosses. *J. Shellfish Res.*, **10**, 516 (abstract only).

Hedgecock, D. and Nelson, K. (1978) Components of growth rate variation among families of lobster *(Homarus)*. *Proc. World Maricult. Soc.*, **9**, 125–37.

Hedgecock, D., Schlester, R.A. and Nelson, K. (1976) Application of biochemical genetics to aquaculture. *J. Fish. Res. Bd. Can.*, **33**, 1108–19.

Hedgecock, D. and Sly, F. (1990) Genetic drift and effective sizes of hatchery-propagated stocks of the Pacific oyster, *Crassostrea gigas*. *Aquaculture*, **88**, 21–38.

Heffernan, P.B., Walker, R.L. and Crenshaw, J.W. Jr. (1991) Negative larval response to selection for increased growth rate in northern quahogs *Mercenaria mercenaria* (Linnaeus, 1758) *J. Shellfish Res.*, **10**, 199–202.

Heffernan, P.B., Walker, R.L. and Crenshaw, J.W. Jr. (1992) Embryonic and larval responses to selection for increased rate of growth in adult bay scallops, *Argopecten irradians concentricus* Say. *J. Shellfish Res.*, **11**, 21–5.

Hershberger, W.K. (1992) Genetic variability in rainbow trout populations. *Aquaculture*, **100**, 51 71.

Hershberger, W.K., Perdue, J.A. and Beattie, J.H. (1984) Genetic selection in Pacific oyster culture. *Aquaculture*, **39**, 237–45.

Hindar, K., Ryman, N. and Utter, F. (1991) Genetic effects of cultured fish on natural fish populations. *Can. J. Fish. Aquat. Sci.*, **48**, 945–57.

Horstgen-Schwark, G., Fricke, H. and Langholtz, H.-J. (1986) The effect of strain crossing on the production performance in rainbow trout. *Aquaculture*, **57**, 141–52.

Hulata, G., Wohlfarth, G. and Rothbard, S. (1983) Progeny-testing selection of tilapia broodstocks producing all-male hybrid progenies–preliminary results. *Aquaculture*, **33**, 263–8.

Humphrey, C.M. and Crenshaw, J.W. Jr. (1989) Clam genetics, in *Clam Mariculture in North America* (eds J.J. Manzi and M. Castagna), Elsevier, Amsterdam, pp. 323–56.

Innes, D.J. and Haley, L.E. (1977) Genetic aspects of larval growth under reduced salinity in *Mytilus edulis*. *Biol. Bull.*, **153**, 312–21.

Inoue, K., Yamashita, S., Hata, J-I. *et al.* (1990) Electroporation as a new technique for producing transgenic fish. *Cell Diff. Dev.*, **29**, 123–8.

Kapuscinski, A.R. and Jacobson, L.D. (1987) *Genetic Guidelines for Fisheries Management*, Minnesota Sea Grant, University of Minnesota, Duluth.

Khoo, H.-W., Ang, L.-H, Lim, H.-B. and Wong, K.-Y. (1992) Sperm cells as vectors for introducing foreign DNA into zebrafish. *Aquaculture*, **107**, 1–19.

Kincaid, H.L. (1983) Inbreeding in fish populations used for aquaculture. *Aquaculture*, **33**, 215–27.

Koljonen, M-L. (1989) Electrophoretically detectable genetic variation in natural and hatchery stocks of Atlantic salmon in Finland. *Hereditas*, **110**, 23–35.

Lannan, J.E. (1972) Estimating heritability and predicting response to selection for the Pacific oyster, *Crassostrea gigas*. *Proc. Natl. Shellfish Assoc.*, **62**, 62–6.

Lester, L.J. (1988) Differences in larval growth among families of *Penaeus stylirostris* Stimpson, and *Penaeus vannamei* Boone. *Aquacult. Fish. Manag.*, **19**, 243–51.

Lester, L.J. (1989) Inheritance of size in *Penaeus vannamei*. *J. World. Aquat. Soc.*, **20**, 50A (abstract only).

Lincoln, R.F. (1981) Sexual maturation in triploid male plaice (*Pleuronectes platessa*) and plaice × flounder (*Platichthys flesus*) hybrids. *J. Fish. Biol.*, **19**, 415–26.

Liu, Z., Moav, B., Faras, A.J. *et al.* (1990) Developments of expression vectors for transgenic fish. *BioTechnology*, **8**, 1268–71.

Longwell, A.C. and Stiles, S.S. (1973) Oyster genetics and the probable future role of genetics in aquaculture. *Malacol. Rev.*, **6**, 151–77.

Losee, E. (1979) Influences of heredity on larval and spat growth in *Crassostrea virginica* in *Proceedings of the Nineteenth Annual Meeting, World Mariculture Society* (ed. J.W. Avault), World Mariculture Society, Baton Rouge, pp. 101–8.

Ludwig, A.M. and Gaffney, P.M. (1991) Quantitative genetics of growth in the dwarf surf clam *Mulinia lateralis. J. Shellfish Res.*, **10**, 451–4.

Lutz, C.G. and Wolters, W.R. (1989) Estimation of heritabilities for growth, body size, and processing traits in red swamp crawfish *Procambarus clarkii* (Girard). *Aquaculture*, **78**, 21–33.

Maclean, N. and Penman, D.J. (1990) The application of gene manipulation to aquaculture. *Aquaculture*, **85**, 1–20.

Maclean, N., Penman, D. and Talwar, S. (1987) Introduction of novel genes into the rainbow trout, in *Selection, Hybridization and Genetic Engineering in Fish* (ed. K. Tiews), Vol. II, Heenemann, Berlin, pp. 325–33.

Maclean, N., Penman, D. and Zhu, Z. (1987) Introduction of novel genes into fish. *BioTechnology*, **5**, 257–61.

Maclean, N. and Talwar, S. (1984) Injection of cloned genes into rainbow trout eggs. *J. Embryol. Exp. Morphol.*, **82**, 187.

Malecha, S.R., Masuno, S. and Onizuko, D. (1984) The feasibility of measuring the heritability of growth pattern variation in juvenile freshwater prawn, *Macrobrachium rosenbergii* (De Man). *Aquaculture* **38**, 347–63.

Mallet, A.L., Freeman, K.R. and Dickie, L.M. (1986) The genetics of production characters in the blue mussel *Mytilus edulis.* I. Preliminary analysis. *Aquaculture, 57,* 133–40.

Mallet, A.L. and Haley, L.E. (1983) Effects of inbreeding on larval and spat performance in the American oyster. *Aquaculture*, **33**, 229–35.

Maniatis, T., Fritsch, E.T. and Sambrook, J. (1982) *Molecular Cloning. A Laboratory Manual*, Cold Spring Harbor Laboratory.

Manzi, J.J., Hagley, N.H. and Dillon, R.T. (1991) Hard clam, *Mercenaria mercenaria*, broodstocks: growth of selected hatchery stocks and their reciprocal crosses. *Aquaculture*, **94**, 17–26.

McEvoy, T., Stack, M., Keane, B. *et al.* (1988) The expression of a foreign gene in salmon embryos. *Aquaculture*, **68**, 27–37.

McKay, L.R., Ihssen, P.E. and McMillan, I. (1992) Early mortality of tiger trout (*Salvelinus fontinalis* × *Salmo trutta*) and the effects of triploidy. *Aquaculture*, **102**, 43–54.

Menzel, W. (1962) Seasonal growth of the northern and southern quahogs and their hybrids in Florida. *Proc. Natl. Shellfish Assoc.*, **53**, 111–18.

Menzel, W. (1989) The biology, fishery and culture of quahog clams, *Mercenaria*, in *Clam Mariculture in North America* (eds J.J. Manzi and M. Castagna), Developments in Aquaculture and Fisheries Science, Vol. 19, Elsevier, Amsterdam, pp. 201–42.

Myers, J.M. and Hershberger, W.K. (1991) Early growth and survival of heat-shocked and tetraploid-derived triploid rainbow trout. *Aquaculture*, **96**, 97–107.

Newkirk, G.H. (1980) Review of the genetics and the potential for selective breeding of commercially important bivalves. *Aquaculture*, **19**, 209–28.

Newkirk, G.F. and Haley, L.E. (1982) Progress in selection for growth rate in the European oyster *Ostrea edulis. Mar. Ecol. Progr. Ser.*, **10**, 77–9.

Newkirk, G.F. and Haley, L.E. (1983) Selection for growth rate in the European oyster, *Ostrea edulis*: response of second generation groups. *Aquaculture*, **33**, 149–55.

Newkirk, G.F., Haley, L.E. and Dingle, J. (1981) Genetics of the blue mussel *Mytilus edulis* (L.): nonadditive genetic variation in larval growth rate. *Can. J. Genet. Cytol.*, **23**, 349–54.

Newkirk, G.F., Haley, L.E., Waugh, D.L. and Doyle, R. (1977) Genetics of larvae and spat growth rate in the oyster *Crassostrea virginica. Mar. Biol.*, **41**, 49–52.

Ozato, K., Kondoh, H., Inohara, H. *et al.* (1986) Production of transgenic fish: introduction and expression of chicken crystallin gene in medaka embryos. *Cell Differ.*, **19**, 237–44.

Palmiter, R.D., Brinster, R.L., Hammer, R.E. *et al.* (1982) Dramatic growth of mice that develop from eggs microinjected with metallothionein-growth hormone fusion genes. *Nature*, **300**, 611–15.

Penman, D.J., Beeching, A.J., Iyengar, A. and Maclean, N. (1988) Introduction of metallothionein-somatotropin fusion gene into rainbow trout: analysis of adult transgenics. *Bull. Aquacult. Assoc. Can.*, **88**, 137–9.

Penman, D.J., Beeching, A.J., Penn, S. and Maclean, N. (1990) Factors affecting survival and integration following microinjection of novel DNA into rainbow trout eggs. *Aquaculture*, **85**, 35–50.

Penman, D.J., Iyengar, A., Beeching, A.J. *et al.* (1991) Patterns of transgene inheritance in rainbow trout (*Oncorhynchus mykiss*). *Molec. Reprod. Dev.*, **30**, 201–6.

Purdom, C.E. (1969) Radiation-induced gynogenesis and androgenesis in fish. *Heredity*, **24**, 431–44.

Quillet, E., Chevassus, B., Blanc, J-M., Kreig, F. and Chourrout, D. (1988) Performance of auto and allotriploids in salmonids I. Survival and growth in fresh water farming. *Aquat. Living Resour.*, **1**, 29–43.

Quillet, E., Foisil, B., Chevassus, B., Chourrout, D. and Liu, F.G. (1991) Production of all-triploid and all-female brown trout for aquaculture. *Aquat. Living Resour.*, **4**, 27–32.

Rahman, M.A. and Maclean, N. (1992) Production of transgenic tilapia (*Oreochromis niloticus*) by one-cell-stage microinjection. *Aquaculture*, **105**, 219–32.

Rawson, P.D. and Hilbish, T.J. (1990) Heritability of juvenile growth for the hard clam *Mercenaria mercenaria*. *Mar. Biol.*, **105**, 429–36.

Rawson, P.D. and Hilbish, T.J. (1991) Genotype-environment interaction for juvenile growth in the hard clam *Mercenaria mercenaria* (L.). *Evolution*, **45**, 1924–35.

Refstie, T. (1983) Hybrids between salmonid species. Growth rate and survival in seawater. *Aquaculture*, **33**, 281–5.

Rokkones, E., Alestrom, P., Skjervold, H. and Gautvik, K.M. (1988) Microinjection and expression of a mouse metallothionein human growth hormone fusion gene in fertilized salmon eggs. *J. Comp. Physiol. B*, **158**, 751–8.

Rye, M., Lillevik, K.M. and Gjerde, B. (1990) Survival in early life of Atlantic salmon and rainbow trout: estimates of heritabilities and genetic correlations. *Aquaculture*, **89**, 209–16.

Ryman, N. and Stahl, G. (1980) Genetic changes in hatchery stocks of brown trout (*Salmo trutta*). *Can. J. Fish. Aquat. Sci.*, **37**, 82–7.

Saunders, R.L. (1991) Potential interaction between cultured and wild Atlantic salmon. *Aquaculture*, **98**, 51–60.

Sbordoni, V., De Matthaeis, E., Cobolli-Sbordoni, M., La Rosa, G. and Mattoccia, M. (1986) Bottleneck effects and the depression of genetic variability in hatchery stocks of *Penaeus japonicus* (Crustacea, Decapoda). *Aquaculture*, **57**, 239–51.

Schneider, J.F., Myster, S.H., Hackett, P.B. *et al.* (1992) Molecular cloning and sequence analysis of the cDNA for northern pike (*Esox lucius*) growth hormone. *Mol. Mar. Biol. Biotech.*, **1**, 106–12.

Shultz. F.T. (1986) Developing a commercial breeding programme. *Aquaculture*, **57**, 65–76.

Smith, P.J. and Conroy, A.M. (1992) Loss of genetic variation in hatchery-produced abalone, *Haliotis iris N Z. J. Mar. Freshwat Res.*, **26**, 81–5.

Stahl, G. (1983) Differences in the amount and distribution of genetic variation between natural populations and hatchery stocks of Atlantic salmon. *Aquaculture*, **33**, 23–32.

Standal, M. and Gjerde, B. (1987) Genetic variation in survival of Atlantic salmon during the sea-rearing period. *Aquaculture*, **66**, 197–207.

Stanley, J.G., Allen, S.K. and Hidu, H. (1981) Polyploidy induced in the American oyster *Crassostrea virginica* with cytochalasin B. *Aquaculture*, **12**, 1–10.

Stephens, L.B. and Downing, S.L. (1988) Inhibiting first polar body formation in *Crassostrea gigas* produces tetraploids, not meiosis I triploids. *J. Shellfish Res.*, **7**, 550–1 (abstract only).

Stromgren, T. and Neilsen, M.V. (1989) Heritability of growth in larvae and juveniles of *Mytilus edulis*. *Aquaculture*, **80**, 1–6.

Stuart, G.W., McMurray, J.V. and Westerfield, M. (1988) Replication, integration and stable germ line transmission of foreign sequences injected into early zebrafish embryos. *Development*, **103**, 403–12.

Stuart, G.W., Vielkind, J.R., McMurray, J.V. and Westerfield, M. (1990) Stable lines of transgenic zebrafish exhibit reproducible patterns of transgene expression. *Development*, **109**, 577–84.

Sugama, K., Taniguchi, N. and Umeda, S. (1988) An experimental study on genetic drift in hatchery population of red sea bream. *Nippon Suisan Gakkaishi*, **54**, 739–44.

Sunden, S.L.F. and Davis, S.K. (1991) Evaluation of genetic variation in a domestic population of *Penaeus vannamei* (Boone): a comparison with three natural populations. *Aquaculture*, **97**, 131–42.

Suryono (1990) Heritability estimates of larval growth from inbred broodstocks of freshwater prawn *Macrobrachium rosenbergii* (De Man), MSc thesis, University of Wales.

Taniguchi, N., Sumantadinata, K. and Iyama, S. (1983) Genetic changes in the first and second generations of hatchery stocks of black sea bream. *Aquaculture*, **35**, 309–20.

Tave, D. (1993) *Genetics for Fish Hatchery Managers*, 2nd edn, Van Nostrand Reinhold, New York.

Tave, D., Smitherman, R.O., Jayaprakas, V. and Kuhlers, D.L. (1990) Estimates of additive genetic effects, maternal genetic effects, individual heterosis, maternal heterosis, and egg cytoplasmic effects for growth in *Tilapia nilotica*. *J. World Aquat. Soc.*, **21**, 263–70.

Thompson, D. and Scott, A.P. (1984) An analysis of recombination data in gynogenetic diploid rainbow trout. *Heredity*, **53**, 441–52.

Thorgaard, G.H. (1983) Chromosome set manipulation and sex control in fish, in *Fish Physiology* (eds W.S. Hoar, D.J. Randall and E.M. Donaldson), Vol. 9(B) Academic Press, London, pp. 405–34.

Thorgaard, G.H. (1992) Application of genetic technologies to rainbow trout. *Aquaculture*, **100**, 85–97.

Thorgaard, G.H. and Allen, S.K. (1987) Chromosome manipulation and markers in fishery management, in *Population Genetics and Fishery Management* (eds N. Ryman and F. Utter), University of Washington Press, Seattle, pp. 319–31.

Toro, J.E. and Newkirk, G.F. (1991) Responses to artificial selection and realised heritability estimate for shell height in the Chilean oyster *Ostrea chilensis. Aquat. Living Resour.*, **4**, 101–8.

Verspoor, E. (1988) Reduced genetic variability in first-generation hatchery populations of Atlantic salmon (*Salmo salar*). *Can. J. Fish. Aquat. Sci.*, **45**, 1686–90.

Vrijenhoek, R.C., Ford, S.E. and Haskin, H.H. (1990) Maintenance of heterozygosity during selective breeding of oysters for resistance to MSX disease. *J. Hered.*, **81**, 418–23.

Vourinen, J. (1984) Reduction in genetic variability in a hatchery stock of brown trout, *Salmo trutta. J. Fish Biol.*, **24**, 339–48.

Wada, K.T. (1986) Genetic selection for shell traits in the Japanese pearl oyster, *Pinctada fucata martensii. Aquaculture*, **57**, 171–6.

Watanabe, K., Igarashi, A., Noso, T. *et al.* (1992) Chemical identification of catfish growth hormone and prolactin. *Mol. Mar. Biol. Biotechnol.*, **1**, 239–49.

Wilkins, N.P. (1981) The rationale and relevance of genetics in aquaculture: an overview. *Aquaculture*, **22**, 200–28.

Wilkins, N.P. and Gosling, E.M. (eds) (1983) *Genetics in Aquaculture*, Developments in Aquaculture and Fisheries Science, Vol. 12, Elsevier Press, Amsterdam.

Wohlfarth, G.W., Moav, R. and Hulata, G. (1983) A genotype-environment interaction for growth rate in the common carp, growing in intensively manured ponds. *Aquaculture*, **33**, 187–95.

Yoon, S.J., Hallerman, E.M., Gross, M.L. *et al.* (1990) Transfer of the gene for neomycin resistance into goldfish, *Carassius auratus. Aquaculture*, **85**, 21–33.

Zhang, P., Hayat, M., Joyce, C. *et al.* (1990) Gene transfer, expression and inheritance of pRSV-rainbow trout-GH cDNA in the common carp, *Cyprinus carpio* (Linnaeus). *Molec. Reprod. Dev.*, **25**, 3–13.

Zhu, Z., Li, G., He, L. and Chen, S. (1985) Novel gene transfer into the fertilised eggs of goldfish (*Carassius auratus*). *Z. Angew. Ichthyol.*, **1**, 32–4.

Zhu, Z., Xu, K., Li, G., Xei, Y. and He, L. (1986) Biological effects of human growth hormone gene microinjected into the fertilized egg of the loach *Misgurnis anguillicaudatus. Kex. Tong. Acad. Sin.*, **31**, 988–90.

Zouros, E. and Foltz, D.W. (1987) The use of allelic isozyme variation for the study of heterosis, in *Isozymes: Current Topics in Biological and Medical Research* (eds M.C. Rattazzi, J.C. Scandalios and G.S. Witt), Vol. 13, Liss, New York, pp. 1–59.

9.2 ACCLIMATION TO FRESH WATER OF THE SEA BASS: EVIDENCE OF SELECTIVE MORTALITY OF ALLOZYME GENOTYPES

Giuliana Allegrucci, Carlo Fortunato, Stefano Cataudella and Valerio Sbordoni

Abstract

This section reports data on allozyme variation in samples of the sea bass, Dicentrarchus labrax, *before and after acclimation to freshwater.*

Study samples were reared in seawater conditions up to a weight of 70 g. At this size, a subsample was acclimated to fresh water. The acclimation trial was repeated for two generations (1989 and 1990) from the same broodstock.

Genetic variation at 28 gene loci coding for 24 enzymes was studied in each sample, before and after acclimation to fresh water. Six out of seven polymorphic loci showed statistically significant heterozygote deficiencies, probably due to the genetic heterogeneity of the broodstock.

The overall mortality at sampling, most of which occurred during the acclimation phase, averaged 94% and 75% in the generations 1989 and 1990, respectively.

Multivariate analyses of individual allozymic profiles indicate that survival was not at random in respect to genotype. Survival rates and relative fitnesses per genotype per locus were estimated and in both yearly samples the same genotypes showed the highest probability of survival. This result strongly suggests a selective response to the changed environmental regime. Even if the genetic markers involved do not necessarily represent the target of selection, these results open a promising field of investigation related to selection and management of broodstock for sea bass aquaculture.

9.2.1 Introduction

The sea bass *Dicentrarchus labrax* is a carnivorous marine fish commonly found in estuaries and lagoons of the European Atlantic and Mediterranean sea. It is one of the most important commercial species of finfish in Europe. From 1970 this euryhaline species has been considered for aquaculture, and artificial spawning, larval rearing and experimental growth-out in ponds, cages and lagoons have been investigated (Barnabé,1980; Barnabé and Billard, 1984).

The performance characteristics, such as growth and survival, of sea bass within a particular environment could be genetically determined and the possibility of characterizing such different genotypes is highly desirable. Isozyme electrophoresis has been used successfully to determine the stock composition of intermixed fish populations (Grant *et al.*, 1980; Murphy, Nielesen and Turner, 1983; Macaranas and Fujio, 1990). Moreover, there is a tendency for species with multiple alleles at several biochemical loci to be highly differentiated in performance traits during artificial selection (Wohlfarth, Moav and Hulata, 1975; Brody *et al.*, 1979; Hallerman, Dunham and Smitherman, 1986). The influence of selection and/or drift on loss or fixation of alleles at some loci argues strongly for the use of isozyme markers in fish culture, particularly in the estimation of genetic variation in a species, the estimation of selection pressure in hatchery stocks and the marking of performance traits.

Little is known about the adaptability of *D. labrax* to fresh water. It has been demonstrated that this species is able to tolerate a wide range of salinity (Chervinski, 1979; Cataudella *et al.*, 1991). This capability is certainly of interest when considering it for aquaculture development.

In this section we analyse allozyme electrophoretic variation in samples of *D. labrax* before and after acclimation to fresh water. The aims of this study are (1) to assess genetic structure in the starting samples, reared in salt water, (2) to test the possible occurrence of changes in genotypic frequencies after acclimation, and (3) to investigate the possible role of selection in the changed environmental regime.

9.2.2 Material and methods

Study samples were from the ENEL fish farms 'Torrevaldaliga' (Civitavecchia, Latium) and 'La Casella' (Piacenza, Lombardy). Samples from 'Torrevaldaliga' were obtained through artificial reproduction in 1989 and 1990. In both years, the same broodstock, consisting of individuals coming from different localities on the Italian coasts, contributed to artificial reproduction, sex ratio being 2:1 males to females. Fingerlings (about 60 000 per year) were reared in seawater conditions up to 70 g in weight. Average age was 10 months (1989 sample) and 1 year (1990 sample). Yearly subsamples (about 7000 sea bass in 1989 and 2600 in 1990) were taken from this stock and transferred into the ENEL fish farm 'La Casella', where they were subjected to different salinity conditions according to two acclimation protocols:

1. fast adaptation to fresh water (48 h, on average);
2. slow adaptation to fresh water (17 days).

In the fast adaptation, dechlorinated tapwater (salinity = 0 ppt) was slowly added to the tank in order to replace seawater completely over a 48-h period. In the slow adaptation, seawater was replaced in 17 days by adding water at proportionally decreasing salinities each day. Individuals born in 1989 were partitioned into both kinds of trials while those born in 1990 were subjected only to slow acclimation. No statistically significant allele differences were revealed between the fast and slow adaptation to fresh water, and as a consequence acclimated individuals born in 1989 were pooled into a single sample.

Starting samples from 'Torrevaldaliga', collected before fresh water acclimation, were named CVB for 1989 and CVC for 1990. Samples from 'La Casella', collected after accli-mation trials, were named CST for 1989 and CSD for 1990.

At least 50 individuals from each sample were assayed electrophoretically for genetic variation at 28 gene loci coding for 24 enzymes. Liver tissue from previously frozen indi-viduals was homogenized and centrifuged to remove debris. Aliquots of the supernatant were electrophoresed on cellulose acetate strip gel using procedures from Richardson, Baverstock and Adams (1986) Allele frequencies, heterozygosities and other genetic parameters were calculated using the BIOSYS-1 program of Swofford and Selander (1981).

Gametic disequilibrium between polymorphic loci was estimated and tested for the hypo-thesis that $D = 0$ using the likelihood ratio statistic Q (Hill, 1974). Pairwise estimates of D between alleles were calculated using formulae for two codominant alleles at each of two polymorphic loci.

Overall mortality rate was estimated in each yearly sample as the proportion of dead indi-viduals over the initial number subjected to acclimation experiment. Relative mortality rate of any genotype at a given locus was estimated as:

$$G_1 - G_2/M$$

where M = number of dead fish at sampling

G_1 = number of individuals carrying a given genotype before acclimation to fresh water
G_2 = number of individuals of the same genotype after acclimation to fresh water

Survival rate for each genotype per locus was computed as:

$$G_2/G_1$$

Relative fitness per genotype (per locus) was estimated by dividing its frequency after accli-mation by its starting frequency. This index was normalized and corrected by the starting genotypic frequency.

The mean fitness, $\overline{W}$, per locus was computed as the sum of the relative contributions of the different genotypes.

Multivariate ordination of individual fish based on the polymorphic set of loci was studied by means of correspondence analysis, utilizing multilocus individual profiles. Each indi-vidual was characterized for each variable as 1 if homozygous for that allele, 0.5 if hetero-zygous and 0 in the absence of the allele. This analysis was carried out on the yearly samples separately (CVB-CST and CVC-CSD respectively) because at some loci, the starting samples showed different gene frequencies.

Correspondence analysis was also carried out utilizing multilocus genotypic profiles on the basis of the relative mortality rate per genotype per locus. Each individual was character-ized for each variable (locus) by a mortality coefficient depending on its genotype at that locus.

9.2.3 Results

Acclimation to fresh water involved high mortality rates. After 30 days, mortality reached about 80% in CST (1989) and 40% in CSD (1990). However, at the sampling time (i.e. after 15 months in CST and 7 months in CSD) mortality rates were 94.4% and 74.7% respectively.

Of the 28 loci examined for electrophoretic variability (Table 9.2.1), 21 showed little variation, with the most common allele at a frequency of 90% or greater in the starting samples CVB and CVC. No significant variation occurred at these loci in the acclimated samples, CST and CSD, where in at least six loci the rare allele was lost.

Seven additional loci, however, showed interesting patterns (Table 9.2.1). At one locus, *FDP-2**, which was moderately polymorphic, the starting samples in the two years (CVB and CVC) were slightly different ($F_{st} = 0.009$, $P < 0.25$) but the acclimated ones showed the same gene frequencies.

A second locus, *NP**, showed a 60 : 40 two-allele situation in both starting samples. Acclimated samples (CSD and CST) were fixed or nearly fixed for the more common allele.

At the *CK-4** locus there is evidence that the starting samples in the two years were different ($F_{st} = 0.04$, $P < 0.001$) but the acclimated samples, CST and CSD, showed similar gene frequencies.

The *ADK-3** locus was characterized as being moderately polymorphic in the starting samples and was out of Hardy–Weinberg proportions, but both acclimated samples went near to fixation for the common allele.

The *EST-2** locus showed clearly different allele frequencies in the CVB and CVC samples, one being 30 : 70 and the other 70 : 30 for the two alleles ($F_{st} = 0.297$, $P < 0.001$). Despite this difference in starting frequencies, the acclimated samples, CST and CSD, were fixed for the same allele. Both starting samples were out of Hardy–Weinberg proportions, showing in all cases a deficiency of heterozygotes.

The *G6PD** locus showed a similar situation to *EST-2**. Also in this case the frequencies of starting samples were drastically different, being 66 : 34 in CVB and 36 : 64 in CVC ($F_{st} = 0.085$, $P < 0.001$). On the other hand, the acclimated samples showed frequencies very similar to each other, being 20 : 80 in CST and 34 : 66 in CSD ($F_{st} = 0.025$, $P < 0.05$). All samples had genotype frequencies not in agreement with Hardy–Weinberg expectations, showing in all cases a deficiency of heterozygotes.

At the *ME** locus the starting frequencies were only slightly different in years 1989 and 1990 ($F_{st} = 0.02$, $P < 0.025$) and the acclimated samples drastically changed in the same direction. In both these experiments F_{st} values calculated between starting and acclimated samples were significant (χ^2 test, $P < 0.001$). Deviations from Hardy–Weinberg expectations were observed in all samples.

Linkage disequilibrium (D) was estimated between polymorphic loci, and Table 9.2.2 reports the Q values. Three loci, *CK-4**, *EST-2** and *G6PD**, showed significant ($P < 0.05$) association with each other in the starting sample CVB (1989 trial). The situation was completely different for the starting sample CVC (1990 trial), where the same loci did not show association with each other, while there was association of *G6PD** with *ME** ($P < 0.05$).

Table 9.2.3 reports genetic variability estimates at 28 loci in all study samples. A considerable reduction in heterozygosity levels from the starting samples (CVB and CVC) to the acclimated ones (CST and CSD) was observed. Mean observed heterozygosity varied from 0.09 to 0.045 and the percentage of polymorphic loci from 35.7% to 17.9%.

Table 9.2.1 Allele frequencies at 28 loci in the two yearly samples of *D. labrax* before and after acclimation to fresh water

Locus	1989		1990	
	CVB	CST	CVC	CSD
AAT-2*				
(*N*)	61	80	37	36
A	0.926	0.956	0.946	0.986
B	0.074	0.044	0.054	0.014
ACON*				
(*N*)	75	102	48	49
A	0.987	0.975	0.979	1.000
B	0.013	0.025	0.000	0.000
C	0.000	0.000	0.021	0.000
ADA*				
(*N*)	76	82	47	50
A	0.928	0.976	0.979	0.970
B	0.072	0.024	0.021	0.030
ADH*				
(*N*)	68	85	43	43
A	0.007	0.012	0.000	0.000
B	0.993	0.988	1.000	1.000
ADK-3*				
(*N*)	42	74	43	49
A	0.155	0.007	0.105	0.010
B	0.845	0.993	0.895	0.990
ALDO*				
(*N*)	73	87	46	46
A	0.027	0.000	0.022	0.000
B	0.925	0.989	0.935	1.000
C	0.048	0.011	0.043	0.000
CK-4*				
(*N*)	75	96	41	50
A	0.020	0.005	0.195	0.000
B	0.820	0.865	0.659	0.870
C	0.160	0.130	0.146	0.130
EST-2*				
(*N*)	75	96	50	50
A	0.773	0.000	0.220	0.000
B	0.227	1.000	0.780	1.000
FDP-2*				
(*N*)	64	84	32	31
A	0.937	0.940	0.875	0.952
B	0.062	0.060	0.109	0.048
C	0.000	0.000	0.016	0.000
FUM*				
(*N*)	75	95	50	49
A	0.013	0.021	0.000	0.000
B	0.987	0.979	0.990	1.000
C	0.000	0.000	0.010	0.000

Table 9.2.1 *(cont)*

	1989		1990	
Locus	CVB	CST	CVC	CSD
GDA*				
(N)	74	86	50	47
A	0.034	0.012	0.040	0.064
B	0.966	0.988	0.960	0.936
GLUD*				
(N)	70	94	49	48
A	0.007	0.021	0.000	0.021
B	0.986	0.973	1.000	0.979
C	0.007	0.005	0.000	0.000
G6PD*				
(N)	74	96	46	49
A	0.007	0.000	0.000	0.000
B	0.655	0.193	0.359	0.337
C	0.338	0.807	0.641	0.663
GPI*				
(N)	79	97	50	50
A	1.000	0.995	1.000	1.000
B	0.000	0.005	0.000	0.000
IDH*				
(N)	78	97	48	50
A	0.006	0.026	0.000	0.020
B	0.981	0.974	1.000	0.980
C	0.013	0.000	0.000	0.000
ME*				
(N)	73	94	48	47
A	0.021	0.000	0.000	0.000
B	0.630	0.335	0.490	0.117
C	0.349	0.665	0.510	0.883
MPI*				
(N)	70	97	50	50
A	0.986	1.000	0.970	0.990
B	0.014	0.000	0.030	0.010
NP*				
(N)	74	95	50	50
A	0.399	0.047	0.420	0.000
B	0.601	0.953	0.580	1.000
PGM*				
(N)	76	102	50	50
A	0.007	0.029	0.060	0.070
B	0.961	0.941	0.920	0.930
C	0.033	0.029	0.020	0.000

AAT-3*, *ADK-1**, *ADK-2**, *EST-1**, *GAPD**, *LAP-1**, *MDH**, *6PGD** and *PK-3** were monomorphic.

AAT* = aspartate aminotransferase; *ACON**= aconitase; *ADA** = adenosine deaminase; *ADH** = alcohol dehydrogenase; *ADK** = adenylate kinase; *ALDO** = aldolase; *CK** = creatine kinase; *EST** = esterase; *FDP** = fructose-1 6-diphosphatase; *FUM** = fumarase; *GDA** = guanine deaminase; *GLUD** = glutamate deaminase; *G6PD** = glucose-6-phosphate dehydrogenase; *GPI** = glucose phosphate isomerase; *IDH** = isocitrate dehydrogenase; *ME** = malic enzyme; *MPI** = mannose phosphate isomerase; *NP** = nucleoside phosphorylase; *PGM** = phosphoglucomutase; *GAPD** = glyceraldehyde phosphate dehydrogenase; *LAP** = leucine amino peptidase; *MDH** = malate dehydrogenase; *6PGD** = 6-phosphogluconate dehydrogenase; *PK** = pyruvate kinase.

Table 9.2.2 *D. labrax*: values of likelihood ratio statistics Q, testing the hypothesis that D (estimate of linkage disequilibrium) = 0, in the two starting samples CVB (A) and CVC (B)

	A					B			
	EST-2*	G6PD*	ME*	NP*		EST-2*	G6PD*	ME*	NP*
G6PD*	10.6**				G6PD*	1.3			
ME*	1.9	2.9			ME*	0.0	3.9*		
NP*	0.3	1.8	0.2		NP*	0.0	0.1	0.6	
CK-4*	7.9*	8.6*	0.5	0.1	CK-4*	0.5	0.0	1.1	0.2

*$P < 0.05 > 0.01$; **$P < 0.01 > 0.01$.

Table 9.2.3 Genetic variability estimates at 28 loci in the two yearly samples of *D. labrax* before and after acclimation to fresh water

				Mean heterozygosity	
Population	Mean sample size per locus	Mean no. of alleles per locus	Percentage of polymorphic loci	Observed	Expected
1989					
CVB	68.4	1.9	35.7	0.094	0.114
	(2.0)	(0.1)		(0.024)	(0.030)
CST	85.3	1.7	17.9	0.047	0.062
	(2.6)	(0.1)		(0.013)	(0.020)
1990					
CVC	40.8	1.7	35.7	0.094	0.120
	(2.2)	(0.1)		(0.025)	(0.033)
CSD	45.9	1.4	17.9	0.045	0.052
	(1.4)	(0.1)		(0.014)	(0.019)

A locus is considered polymorphic if the frequency of the most common allele does not exceed 0.95. Standard errors in parentheses.

9.2.4 Discussion

Unexpectedly high levels of polymorphism were revealed in starting samples (CVB and CVC) with a percentage of polymorphic loci equal to 35.7% and mean observed heterozygosity equal to 0.09 (Table 9.2.3). Previous research on the same species, *D. labrax* (Cervelli, 1985), and on the close American species, *Morone saxatilis* (Otto, 1975; Grove, Berggren and Powers, 1976; Sidell *et al.*, 1980; Rogier, Ney and Turner, 1985) reported much less genetic variability, with a percentage of polymorphic loci ranging between 10% and 4% and mean heterozygosity estimates ranging between 0.009 and 0.016. Our values are also well above average among fish species that have been electrophoretically surveyed, where heterozygosity averages 0.05 and the mean percentage of polymorphic loci is about 17% (Nevo, 1978; Kirpichnikov, 1981).

Starting samples (CVB and CVC) differed from each other to some extent, indicating that the gene pools were different at the start of the 1989 and 1990 experiments. This was especially clear at the *EST-2*, *G6PD*, *CK-4* and *FDP-2* loci (Table 9.2.1), suggesting that these samples originated from genetically different parents within the same broodstock. This broodstock was indeed heterogeneous, as it was composed of individuals from different geographic populations from Italian coastal waters.

In the starting samples, almost all polymorphic loci deviated from a Hardy–Weinberg expected distribution due to heterozygote deficiencies, the only exception being the NP^* locus (Tables 9.2.4 to 9.2.7). Interestingly, this deviation carries over from the starting

Table 9.2.4 χ^2 test for deviation from Hardy–Weinberg equilibrium in population CVB (1989 starting sample) of *D. labrax*

Locus	Class	Observed frequency	Expected frequency	χ^2	d.f.	P	Fixation index F
$ADK\text{-}3^*$							
	AA	2	0.940				
	AB	9	11.120				
	BB	31	29.940				
				1.638	1	0.200	0.181
$CK\text{-}4^*$							
	AA	0	0.020				
	AB	3	2.477				
	AC	0	0.483				
	BB	51	50.356				
	BC	18	19.812				
	CC	3	1.852				
				1.499	3	0.682	0.072
$EST\text{-}2^*$							
	AA	50	44.765				
	AB	16	26.470				
	BB	9	3.765				
				12.032	1	0.001	0.391
$FDP\text{-}2^*$							
	AA	56	56.220				
	AB	8	7.559				
	BB	0	0.220				
				0.247	1	0.619	− 0.067
$G6PD^*$							
	AA	0	0.000				
	AB	1	0.660				
	AC	0	0.340				
	BB	37	31.673				
	BC	22	32.993				
	CC	14	8.333				
				8.927	3	0.030	0.319
ME^*							
	AA	0	0.021				
	AB	3	1.903				
	AC	0	1.055				
	BB	35	28.869				
	BC	19	32.359				
	CC	16	8.793				
				14.431	3	0.002	0.373
NP^*							
	AA	11	11.639				
	AB	37	35.721				
	BB	26	26.639				
				0.096	1	0.756	− 0.043

Table 9.2.5 χ^2 test for deviation from Hardy–Weinberg equilibrium in population CVC (1990 starting sample) of *D. labrax*

Locus	Class	Observed frequency	Expected frequency	χ^2	d.f.	P	Fixation index F
ADK-3[*]							
	AA	2	0.424				
	AB	5	8.153				
	BB	36	34.424				
				7.159	1	0.007	0.380
CK-4[*]							
	AA	2	1.481				
	AB	7	10.667				
	AC	5	2.370				
	BB	20	17.667				
	BC	7	8.000				
	CC	0	0.815				
				5.607	3	0.132	0.245
EST-2[*]							
	AA	5	2.333				
	AB	12	17.333				
	BB	33	30.333				
				4.923	1	0.027	0.301
FDP-2[*]							
	AA	25	24.444				
	AB	5	6.222				
	AC	1	0.889				
	BB	1	0.333				
	BC	0	0.111				
	CC	0	0.000				
				1.711	3	0.634	0.156
G6PD[*]							
	BB	10	5.802				
	BC	13	21.396				
	CC	23	18.802				
				7.269	1	0.007	0.386
ME[*]							
	BB	17	11.379				
	BC	13	24.242				
	CC	18	12.379				
				10.543	1	0.001	0.458
NP[*]							
	AA	10	8.697				
	AB	22	24.606				
	BB	18	16.697				
				0.573	1	0.449	0.097

populations to the acclimated ones, at least at those loci that maintain polymorphism (*ME*[*] and *G6PD*[*], Tables 9.2.4 to 9.2.7). Deviations from Hardy–Weinberg proportions at individual loci might have contributed to the deviation from expected di-locus gene combinations (Table 9.2.2). The results of tests for linkage disequilibrium are markedly different

Table 9.2.6 χ^2 test for deviation from Hardy–Weinberg equilibrium in population CST (1989 acclimated sample) of *D. labrax*

Locus	Class	Observed frequency	Expected frequency	χ^2	d.f.	P	Fixation index F
ADK-3*							
	AA	0	0.000				
	AB	1	1.000				
	BB	73	73.000				
				0.000	1	1.000	0.000
CK-4*							
	AA	0	0.000				
	AB	1	0.869				
	AC	0	0.131				
	BB	70	71.702				
	BC	25	21.728				
	CC	0	1.571				
				2.254	3	0.521	− 0.150
FDP-2*							
	AA	74	74.269				
	AB	10	9.461				
	BB	0	0.269				
				0.301	1	0.583	− 0.063
G6PD*							
	BB	15	3.524				
	BC	7	29.952				
	CC	73	61.524				
				57 104	1	0.000	0.766
ME*							
	BB	20	10.444				
	BC	23	42.112				
	CC	51	41.444				
				19.621	1	0.000	0.451
NP*							
	AA	1	0.190				
	AB	7	8.619				
	BB	87	86.190				
				3.752	1	0.053	0.184

between the 1989 and 1990 samples; in CVB (1989) three loci (*EST-2**, *G6PD** and *CK-4**) showed significant association while in CVC (1990) two other loci (*G6PD** and *ME**) revealed significant association with each other. This finding, together with the observed heterozygote deficiency, suggests that the hypothesized linkage disequilibrium may only be apparent, and could be the result of a sample raised by non-random union of gametes that caused an interlocus dependency among the genotypic combinations. This was possibly caused by the genetically heterogeneous broodstock.

Figures 9.2.1 and 9.2.2 show the scatter diagrams derived from correspondence analysis of the two yearly samples carried out on individual genotype profiles, for the most informative loci, before and after acclimation to fresh water. These plots represent the ordination of both

Table 9.2.7　χ^2 test for deviation from Hardy–Weinberg equilibrium in population CSD (1990 acclimated sample) of *D. labrax*

Locus	Class	Observed frequency	Expected frequency	χ^2	d.f.	P	Fixation index F
ADK-3[*]							
	AA	0	0.000				
	AB	1	1.000				
	BB	48	48.000				
				0.000	1	1.000	0.000
CK-4[*]							
	BB	37	37.788				
	BC	13	11.424				
	CC	0	0.788				
				1.022	1	0.312	− 0.149
FDP-2[*]							
	AA	28	28.049				
	AB	3	2.902				
	BB	0	0.049				
				0.053	1	0.819	− 0.051
G6PD							
	BB	10	5.443				
	BC	13	22.113				
	CC	26	21.443				
				8.539	1	0.003	0.406
ME[*]							
	BB	2	0.591				
	BC	7	9.817				
	CC	38	36.591				
				4.218	1	0.040	0.279

starting and acclimated genotypes on the plane described by the first two axes, which together explain about 60% (1989) and 50% (1990) of the total variance. Each point in the diagrams represents a given genotype, corresponding to one or more individuals. It is immediately clear from this analysis that, in both yearly samples, genotypes sampled after acclimation are displaced over only a reduced and definite portion of the spread of the starting genotypes, indicating that survival of individuals to the acclimation regime was not at random in respect of their genotype. Moreover, in both analyses (Figures 9.2.1 and 9.2.2) the same alleles (i.e. *G6PD*C, *EST*A, *ME*B, *NP*A and *CK*C, Table 9.2.8) are responsible for the observed pattern. This accounts for changes in gene frequency always occurring in the same direction in both yearly samples acclimated to fresh water, even if the starting samples showed markedly different gene frequencies at some loci.

However, all genotypes, even those carrying the favoured alleles, contributed to the overall observed mortality rate, as seen in Figures 9.2.3 and 9.2.4, where the relative survival rates of genotypes are plotted against their relative fitnesses. In both yearly samples the same genotypes were found to be better adapted to fresh water. Particularly, *EST*BB and *NP*BB showed a greater survival rate than the others, even if their relative fitnesses were rather different in the two years.

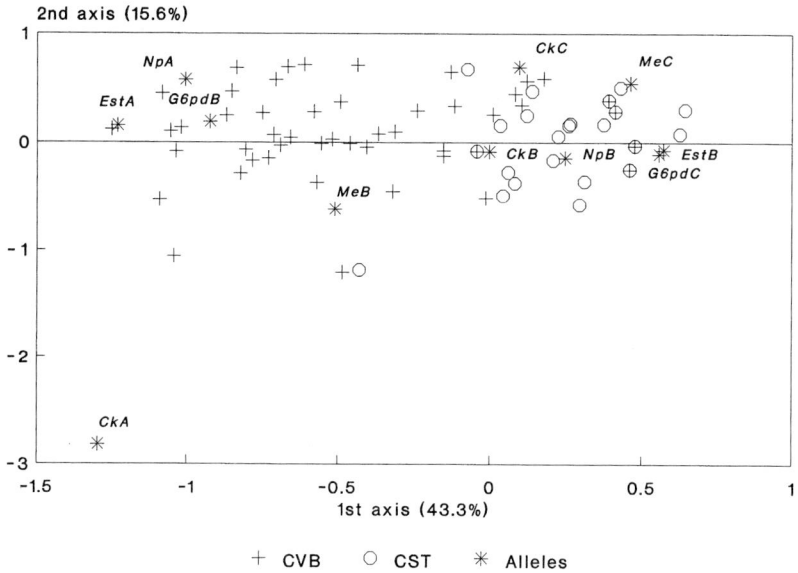

Figure 9.2.1 *D. labrax*, 1989 sample. Correspondence analysis on multilocus individual profiles. Ordination of both starting (CVB) and acclimated (CST) genotypes on the plane described by the first two axes is represented.

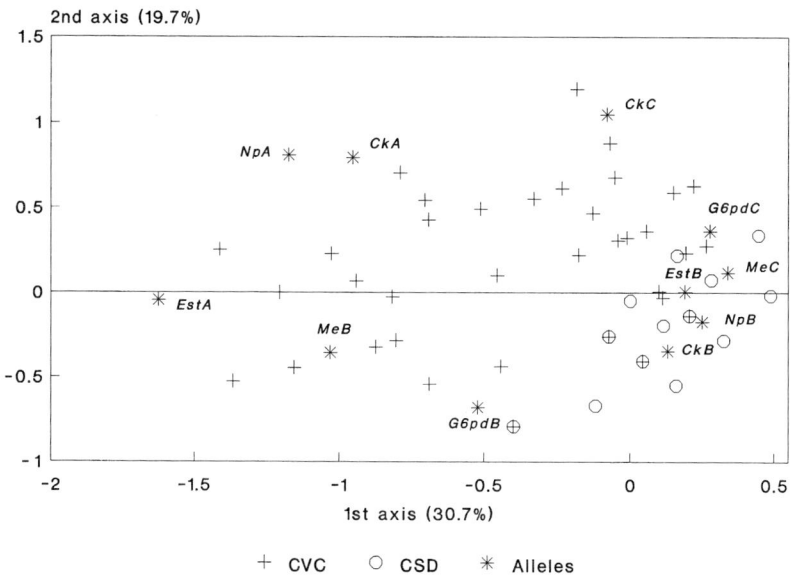

Figure 9.2.2 *D. labrax*, 1990 sample. Correspondence analysis on multilocus individual profiles. Ordination of both starting (CVC) and acclimated (CSD) genotypes on the plane described by the first two axes is represented.

Table 9.2.8 *D. labrax*: relative contributions of variables (alleles) to the variance explained by the first two axes from correspondence analyses of multilocus genotypic profiles (see also Figures 9.2.1 and 9.2.2)

	1989 Sample		1990 Sample	
	1st axis	*2nd axis*	*1st axis*	*2nd axis*
*CK*A*	0.012	0.160	0.067	0.073
*CK*B*	0.000	0.009	0.011	0.115
*CK*C*	0.001	0.117	0.001	0.252
*EST*A*	0.280	0.012	0.234	0.000
*EST*B*	0.129	0.006	0.028	0.000
*G6PD*B*	0.185	0.022	0.079	0.210
*G6PD*C*	0.112	0.012	0.042	0.112
*ME*B*	0.070	0.285	0.220	0.041
*ME*C*	0.065	0.246	0.072	0.014
*NP*A*	0.117	0.105	0.203	0.150
*NP*B*	0.028	0.026	0.044	0.032

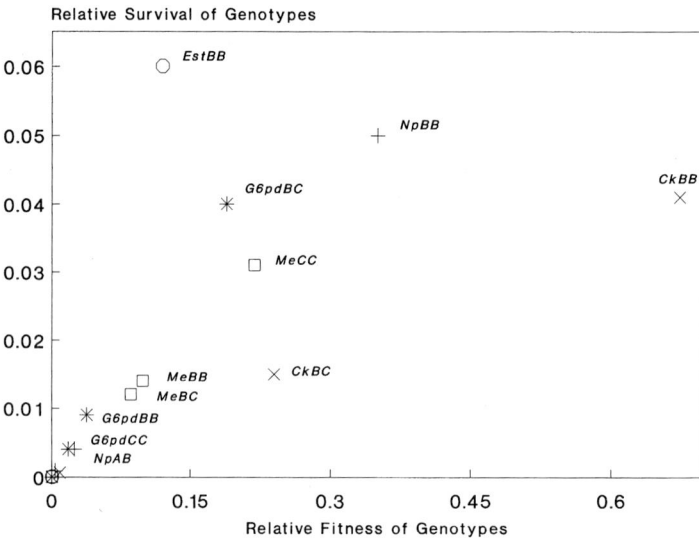

Figure 9.2.3 *D. labrax*, 1989 sample. Relationship between relative survival of genotypes at each of five loci and their estimated relative fitness.

From these results it appears that only those individuals carrying given combinations of allozyme genotypes survived acclimation to fresh water. Ideally, from plots in Figures 9.2.3 and 9.2.4 one could extract the most favoured genotypic combinations which, in both replicated trials, include *EST*BB*, *NP*BB*, *CK*BB*, *ME*CC* and the *C* allele of *G6PD** either in the homozygous condition (1990) or in a heterozygous one, paired with the *B* allele (1989). Notably, the *BB* homozygotes are the only genotypes at *EST-2** and *NP** loci actually surviving in both years; yet these two loci were not associated with each other (Table 9.2.2).

The statistical significance of differences in genotypic frequencies between starting and acclimated samples was tested by χ^2 tests applied to F_{st} values, by considering starting and acclimated samples as two subsets of a single gene pool.

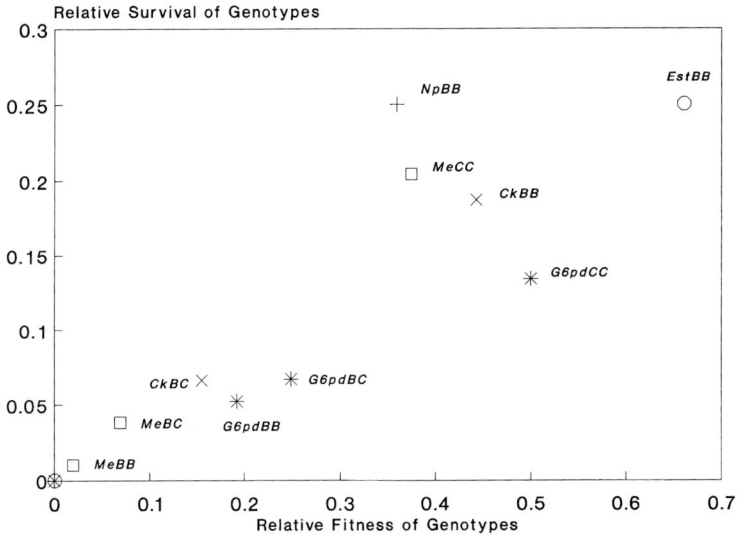

Figure 9.2.4 *D. labrax*, 1990 sample. Relationship between relative survival of genotypes at each of five loci and their estimated relative fitness.

Moreover, to estimate the probability of obtaining differing genotypic distributions as measured by F_{st} between starting and acclimated samples, we used resampling simulation tests. In each randomization we considered a gene pool built up from pooling starting and acclimated samples in each of two trials. We subdivided each yearly sample into two subsamples of the size of the experimental ones (starting and acclimated). Genotypes at each locus were randomly partitioned between subsamples. Then F_{st} values were calculated for each of 250 000 randomizations per locus per yearly trial. From these simulations we estimated the resampling probabilities that F_{st} values obtained were greater than the experimental ones.

Results from this analysis are given in Table 9.2.9 and clearly confirm that survival of individuals was not at random, but dependent on their genotype at the ADK-3^*, EST-2^*, $G6PD^*$, ME^* and NP^* loci in the 1989 sample and the CK-4^*, EST-2^*, ME^* and NP^* loci in the 1990 sample.

Table 9.2.9 *D. labrax* F_{st} values in the two yearly samples tested by χ^2 test ($^{***}P < 0.001$)

Locus	1989 Sample		1990 Sample	
	F_{st}	P	F_{st}	P
ADK-3^*	0.066^{***}	0.000	0.026	0.028
CK-4^*	0.005	0.215	0.085^{***}	0.000
EST-2^*	0.633^{***}	0.000	0.134^{***}	0.000
FDP-2^*	0.002	0.491	0.019	0.044
FUM^*	0.0007	0.703	0.067	0.000
$G6PD^*$	0.252^{***}	0.000	0.001	0.583
ME^*	0.095^{***}	0.000	0.120^{***}	0.000
MPI^*	0.009	0.179	0.008	0.335
NP^*	0.193^{***}	0.000	0.253^{***}	0.000
PGM^*	0.0003	1.000	0.002	0.576

P represents the probability that F_{st} values from 250 000 randomizations are greater than observed values.

In order to investigate the relative importance of different loci in determining overall genotype viability we carried out a correspondence analysis of multilocus individual genotypes described by the relative mortality rate per genotype per locus. Only the five most informative loci were utilized, for both yearly samples, before and after acclimation to fresh water. The plots in Figure 9.2.5 and 9.2.6 represent the ordination of genotype viabilities on

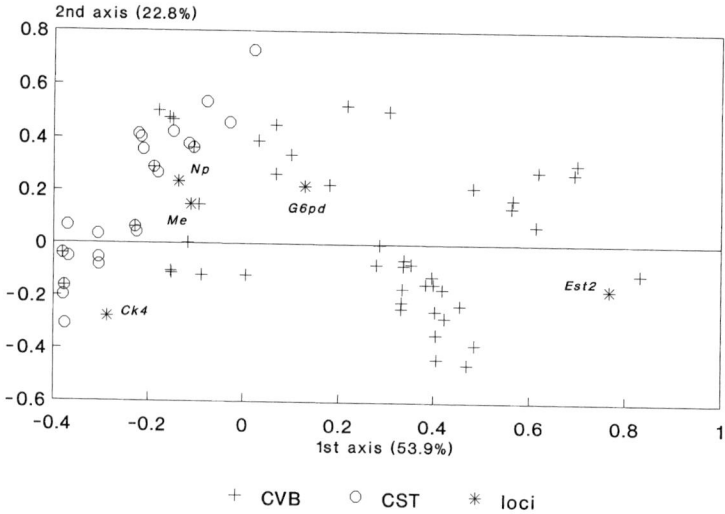

Figure 9.2.5 *D. labrax*, 1989 sample. Correspondence analysis of multilocus individual genotypes described by their relative mortality rates. This plot represents ordination of genotype viabilities on the plane described by the first two axes and the relative contribution of five marker loci to mortality.

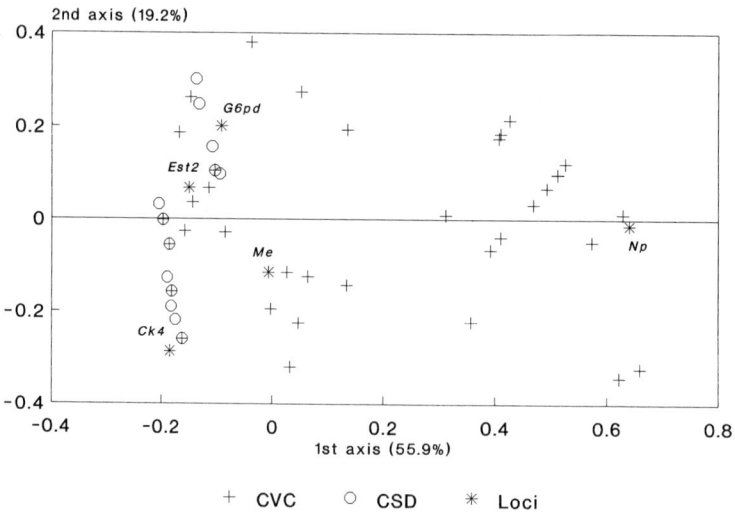

Figure 9.2.6 *D. labrax*, 1990 sample. Correspondence analysis of multilocus individual genotypes described by their relative mortality rates. This plot represents ordination of genotype viabilities on the plane described by the first two axes and the relative contribution of five marker loci to mortality.

the plane described by the first two axes, which together explain about 75% of the total variance, in both analyses. The role played by $EST\text{-}2^*$ and NP^* loci in survival to acclimation to fresh water is clear, but the most important variable, responsible for the observed pattern, was the $EST\text{-}2^*$ locus in the 1989 sample and the NP^* locus in the 1990 sample. The difference between the two years is probably due to the different allele frequencies in the starting samples CVB and CVC. In both replicates, ordination of loci on the first principal axis is related to the mean fitness per locus (Table 9.2.10).

Table 9.2.10   *D. labrax*: mean fitness per locus in acclimation to fresh water

	1989 Sample	*1990 Sample*
$CK4^*$	0.922	0.598
$EST\text{-}2^*$	0.120	0.660
$G6PD^*$	0.245	0.941
ME^*	0.403	0.464
NP^*	0.383	0.360

The conclusion from these experiments is that acclimation regimes have caused selective, non-random genetic changes, which occurred in the same direction in both experiments, even if starting from different allele and genotype frequencies. In our opinion such clear-cut results stem from the high-stress selective regime to which fish have been subjected, leading to extremely high mortality rates. The question arises whether any of these allozyme loci are directly involved as the target of selection or simply represent markers for other unknown linked genes.

A likely hypothesis is that the observed differential mortality of genotypes occurred within a diverse array of coadapted gene pools, representing 'races' of sea bass more or less adapted to fresh water, brought together into the 'Torrevaldaliga' broodstock. Preliminary electrophoretic analysis of some wild populations of the sea bass from different geographic origins within the Tyrrhenian sea have revealed the occurrence of geographic variation in some of the most informative loci assayed in this work.

Testing of these hypotheses, however, requires further investigation and opens a promising field of research connected with selection and management of broodstock for sea bass aquaculture.

Acknowledgements

We are grateful to Paolo Bronzi for providing fish samples and hatchery data, and to Donatella Cesaroni for her help in some statistical analyses. Massimo Regoli and Antonio Sansotta helped us in the simulation analysis and Jeffrey R. Powell provided useful hints on this work. Acclimation trials of sea bass were carried out with an EEC grant to Stefano Cataudella. This research was supported by a National Research Council grant to Valerio Sbordoni, Special Project RAISA, Sub-project N.3 Paper N.730.

References

Barnabé, G., (1980) Exposé synoptique des données biologiques sur le loup ou bar, *Dicentrarchus labrax* (Linne', 1758). *Synop. FAO Peches*, **126**.
Barnabé, G. and Billard, R., (eds) (1984) *L'Aquaculture du Bar et des Sparidés*, INRA Publ., Paris.
Brody, T., Kirsht, D., Parag, G. *et al.* (1979) Biochemical genetic comparison of the Chinese and European races of the common carp. *Anim. Blood Groups Biochem. Genet.*, **10**, 141–9.

Cataudella, S., Allegrucci, G., Bronzi, P. *et al.* (1991) Multidisciplinary approach to the optimization of sea bass (*Dicentrarchus labrax*) rearing in fresh water. *EAS Special Publ.*, **14**, 56–61.

Cervelli, M. (1985) Primi risultati dell'analisi genetica di alcuni sistemi enzimatici di *Dicentrarchus labrax* L. *Oebalia*, (New Series), **11**, 45–8.

Chervinski, J. (1979) Preliminary experiments on the adaptability of juvenile European sea bass (*Dicentrarchus labrax* (L.)) and gilthead sea bream (*Sparus aurata* (L.)) to brackish water. *Bamidgeh*, **31**, 14–17.

Grant, W.S. Milner, G.B., Krasnowski, P. and Utter, F.M. (1980) Use of biochemical genetic variants for identification of sockeye salmon (*Oncorynchus nerka*) stocks in Cook Inlet, Alaska. *Can. J. Fish. Aquat. Sci.*, **37**, 1236–47.

Grove, T.L., Berggren, T.J. and Powers, D.A. (1976) The use of innate tags to segregate spawning stocks of striped bass *Morone saxatilis*. *Est. Process.*, **1**, 166–76.

Hallerman, E.M., Dunham, R.A. and Smitherman, R.O. (1986) Selection or drift-isozyme allele frequency changes among channel catfish selected for rapid growth. *Trans. Am. Fish. Soc.*, **115**, 60–8.

Hill, W.G. (1974) Estimation of linkage disequilibrium in randomly mating populations. *Heredity*, **33**, 229–39.

Kirpichnikov, V.S. (1981) *Genetic Bases of Fish Selection*, Springer-Verlag, Berlin.

Macaranas, J. and Fujio, Y. (1990) Strain differences in cultured fish – isozymes and performance traits as indicators. *Aquaculture*, **85**, 69–92.

Murphy, B.R., Nielesen, L.A. and Turner, B.J. (1983) Use of genetic tags to evaluate stocking success for reservoir walleyes. *Trans. Am. Fish. Soc.*, **112**, 457–63.

Nevo, E.(1978) Genetic variation in natural populations: patterns and theory. *Theoret. Pop. Biol.*, **13**, 121–77.

Otto, R.S. (1975) Isozyme systems of the striped bass and congeneric percichthyid fishes. PhD thesis, University of Maine, Orono, MA.

Richardson, B.J., Baverstock, P.R. and Adams, M. (1986) *Allozyme Electrophoresis. A Handbook for Animal Systematics and Population Studies*, Academic Press, New York.

Rogier, C.G., Ney, J.J. and Turner, B.J. (1985) Electrophoretic analysis of genetic variability in a landlocked striped bass population. *Trans. Am. Fish. Soc.*, **114**, 244–9.

Sidell, B.D., Otto, R.G., Powers, D.A., Karweit, M. and Smith, J. (1980) Apparent genetic homogeneity of spawning striped bass in the upper Chesapeake bay. *Trans. Am. Fish. Soc.*, **109**, 99–107.

Swofford, D.L. and Selander, R.K. (1981) BIOSYS-1: a FORTRAN program for the comprehensive analysis of eletrophoretic data in population genetics and systematics. *J. Hered.*, **72**, 281–283.

Wohlfarth, G., Moav, R. and Hulata, G. (1975) Genetic differences between the Chinese and European races of the common carp. II. Multi-character variation – a response to the diverse methods of fish cultivation in Europe and China. *Heredity*, **34**, 341–50.

9.3 ESTIMATING GENE–CENTROMERE RECOMBINATION FREQUENCIES IN GYNOGENETIC DIPLOIDS OF *OREOCHROMIS NILOTICUS* L., USING ALLOZYMES, SKIN COLOUR AND A PUTATIVE SEX-DETERMINATION LOCUS (*SDL-2*).

Mohammad G. Hussain, Brendan J. McAndrew, David J. Penman and Panom Sodsuk

Abstract

The proportions of heterozygotes (y) resulting from recombination between non-sister chromatids in meiotic gynogenetic offspring of females heterozygous at a number of allozyme loci, a red skin colour locus and a new putative sex-determination locus (SDL-2) were analysed. The frequency of heterozygotes ranged from y = 0.0 at the EST-2* locus to y = 1.00 at the ADA* locus; other allozyme loci were intermediate (AAT-2*, y – 0.53; AH*, y = 0.46; FH-1*, y = 0.20; MEP 2*, y – 0.26). The proportion of heterozygotes observed at the red gene ranged from 0.02 to 1.00 but gave a mean frequency of y = 0.45. The estimated recombination frequency at SDL-2 was y = 0.85 based on the frequency of non-recombinant spontaneously sex-inverted males. When these data were combined with other published *results this gave an overall mean frequency of* y *= 0.47 in this species. These results confirm that meiotic gynogenetic individuals are only partially inbred. The fixation index (*F *= (1–ȳ)) for a single generation is* F *= 0.53, a value equivalent to selfing or three generations of full sib mating. The loci used also show that there is no evidence of paternal inheritance in any gynogenetic offspring in this study, confirming the effectiveness of the milt UV irradiation treatment (300–310 µW/cm² for 2 min). Mitotic gynogenetics were also produced and were always homozygous at all loci studied. These results are compared to those obtained in studies on other species of fish.*

9.3.1 Introduction

There has been a rapid increase in the application of ploidy manipulation techniques to various finfish in recent years, either for the improvement of cultured species or for research purposes. The production of gynogenetic individuals is of particular interest to fish breeders because of the high levels of inbreeding that can be induced in populations in a single generation. Meiotic gynogenetics may be used in the development of inbred strains (Nagy and Csányi, 1982, 1984) or in the analysis of traits such as sex-determination mechanisms in fish (Stanley, 1976; Penman *et al.*, 1987; Mirza and Shelton, 1988; Mair *et al.*, 1991 a,b). Mitotic gynogenetics can be used to produce completely homozygous individuals which can then be used to establish inbred or outbred clonal lines (Streisinger *et al.*, 1981; Quillet, Garcia and Guyomard, 1991).

Gynogenetics can be produced by fertilizing an egg with irradiated milt so it will not contain any viable paternal DNA. With no further treatment this would result in a haploid embryo. The second meiotic division occurs shortly after penetration of the sperm head and if the egg is given an appropriate physical or chemical shock this results in the retention, rather than the loss, of the second polar body within the nucleus. As these two sets of chromosomes originated from sister chromatids, such diploid 'meiotic' gynogenetics would be homozygous if no recombination happened at the first meiotic division. If recombination does occur, heterozygosity will be generated in the meiotic gynogenetic offspring of a heterozygous female. The frequency of heterozygotes in the offspring will reflect the effective crossover rate (single or uneven number of crossovers) between non-sister chromatids. A number of studies have shown that in many fish species there is a high level of chiasma interference (Allendorf *et al.*, 1986): thus for most chromosomes there appears to be only one chiasma event per meiosis and therefore the probability of any recombinant event is proportional to the proximity of a locus to its centromere.

In diploid 'mitotic' gynogenesis the egg is diploidized by the suppression of first cleavage. In this case the gynogenetics should all be completely homozygous at all loci as this is simply the doubling by replication of a haploid chromosome complement. Thus for any individual locus, a heterozygous female will give rise to both kinds of homozygotes but no heterozygotes.

The absence of visible marker genes in the majority of aquatic organisms has resulted in a reliance on the use of allozyme variation for assessing parental contribution and the genetic status of the progeny in all forms of ploidy manipulations. The ability to study a large number of variable allozyme loci has also been particularly useful in estimating the frequency of heterozygotes in meiotic gynogenetics and thereby the level of recombination that has occurred at the first meiotic division (review: Seeb and Miller, 1990).

This section describes the way data from allozyme and other marker genes have assisted in a study to produce meiotic and mitotic gynogenetics in the cichlid *Oreochromis niloticus*. The various loci were used to confirm the lack of paternal inheritance when using UV-irradiated milt and to determine the gynogenetic status of the fish. The proportion of heterozygotes in mitotic gynogenetics was used to estimate gene–centromere recombination rates. No heterozygotes were observed in any of the mitotic gynogenetics, confirming the differing levels of inbreeding in these two types of gynogenetics.

9.3.2 Materials and methods

The strain of *O. niloticus* broodstock used in this study came from a pure stock maintained in the Tilapia Reference Collection at the Institute of Aquaculture. Tissue samples (fin, muscle and blood) were collected under anaesthesia (Benzocaine: used at 1 : 10 000 w/v) from a total of 70 female and 35 male mature broodstock. The fish were then tagged (ICES suture tags) for future reference. The samples were analysed using horizontal starch gel electrophoresis to visualize the known polymorphic enzyme loci in *O. niloticus* (aspartate aminotransferase, AAT, EC. 2.6.1.1; adenosine deaminase, ADA, EC. 3.5.4.4; aconitate hydratase, AH, EC. 4.2.1.3; esterase, EST, EC. 1.1-; fumarate hydratase, FH, EC. 4.2.1.2; malic enzyme (NADP$^+$), MEP, EC. 1.1.1.40) using the methods of McAndrew and Majumdar (1983) and Sodsuk and McAndrew (1991). This enabled fish of known genotype to be used in subsequent breeding experiments. The genotypes of the offspring were determined in a number of ways; if only *ADA** was used to check the gynogenetic status then fin biopsies were taken from the juveniles at 8–12 weeks of age; otherwise the fish were killed at 33 weeks and the appropriate tissues removed (liver, kidney, muscle and blood). The results are expressed as the proportion of heterozygotes (*y*) in the meiotic gynogenetic batches:

$$y = 1 - \frac{\text{NR}}{\text{Total}}$$

where NR = number of non-recombinants

The origin of the dominant red colour variant and a description of the homozygote and heterozygote phenotypes are given in McAndrew *et al.* (1988). Sibling heterozygous red females (*Rr*) were produced for this study by crossing a homozygous pure red male (*RR*) with a wild-type female (*rr*). The phenotypic appearance of the *Rr* genotype can be variable (generally red body colour with varying degrees of blotching covering up to 30% of the body but some unblotched reds also) and in batches containing both *RR* and *Rr* genotypes some misidentification is possible. Non-recombinant females could not be identified for the putative *SDL-2* locus (see below). For the red and the *SDL-2* loci the proportion of hetero-zygotes was estimated using the number of unequivocally identifiable non-recombinants (wild-type colour or male fish respectively: the total number of non-recombinants, NR, is equal to twice this).

The techniques used to produce the gynogenetic diploids using both early heat and pressure shocks are described in Hussain *et al.* (1991). The techniques for the UV irradiation of milt and late heat and pressure shocks are described in Hussain *et al.* (1993).

9.3.3 Results

The variable loci enabled confirmation of the identity of offspring from various ploidy manipulations within this species. The most useful allozyme marker in this respect was *ADA**, which always gave *y* = 1.0 (100%) recombinants in six different meiotic gynogenetic progeny groups, a total of 150 meiotic gynogenetic offspring from heterozygous females (Table 9.3.1). All mitotic gynogenetics in these groups were homozygous for one or other of the maternal alleles. No evidence of any paternal inheritance was observed in any meiotic or mitotic offspring at this or any other locus, confirming the adequacy of the milt irradiation technique. This was checked by using males which were homozygous for one allele and

Table 9.3.1 Examples of heterozygosity in meiotic gynogenetic fish arising from recombination at meiosis

Species	Locus (proportion of heterozygotes, y)				Reference
Oncorhynchus mykiss	AAT-3*	(0.66)	GLP-1*	(0.50)	Allendorf *et al.* (1986)
	ACO-2*	(0.64)	MEP-2*	(0.96)	
	CK-1*	(0.46)	PEPA-1*	(0.44)	
	CKCI*	(0.86)	PEPD*	(0.41)	
	EST-1*	(0.78)	PMI*	(0.09)	
	EST-2*	(0.84)	SDH-1*	(0.40)	
	G3P-1*	(1.00)	TF*	(0.83)	
	HEX*	(0.03)	MDHM1*	(0.62)	
	IDH-2*	(0.69)	IDH-3*	(0.58)	
	LDH-4*	(0.02)	MDH-1*	(0.24)	
	MDH-3,4*	(0.98)	PGM-2*	(0.14)	
	SOD-1*	(1.00)	PGM-1*	(0.03)	
	MDH-3,4*	(0.75)	PGM*	(0.16)	Thompson and Scott
	SOD*	(0.95)			(1984)
	Mean = 0.56 (24 loci)				
Salmo trutta	AAT-1*	(1.00)	AAT-4*	(1.00)	Guyomard (1986)
	CPK-1*	(0.84)	FDP-1*	(0.80)	
	FUM-1*	(1.00)	IDH-3*	(0.98)	
	MDH-2*	(0.60)	MDH-3*	(1.00)	
	6PGDH-2*	(0.60)	PMI-2*	(0.76)	
	PGI-2*	(1.00)	SDH-1*	(1.00)	
	Mean = 0.88 (12 loci)				
Oreochromis niloticus	AAT-2*	(0.53)	ADA*	(1.00)	This section
	MEP-2*	(0.26)	AH*	(0.46)	
	RED+	(0.45)	EST-2*	(0.00)	
	SDL-2+1	(0.85)	FH-1*	(0.20)	
	SDL-1+	(0.69)	BI+	(0.33)	Mair *et al.* (1991),
	AP*	(0.57)	MEP-2*	(0.60)	Mair *et al.* (1993)
	SDL-2+1	(0.88)	CDS+	(0.07)	
	Mean = 0.47 (12 loci)				
Cyprinus carpio	EST-'S'*	(0.09)	EST-'F'*	(0.28)	Reviewed by Komen
	Tf+	(0.08)	Sc(S)+	(0.08)	(1990)
	Sc(N)+	(0.97)	Tp+	(0.01)	
	Pat+	(0.74)	Pig+	(0.70)	
	YE+	(> 0.9)	Mal+	(> 0.9)	
	BII+	(0.27)	Bl2+	(0.38)	
	Mean = 0.45 (12 loci)				

All loci except those marked + are isozymes. *SDL-1* and *SDL-2* = putative sex determining loci; *RED* = dominant red body coloration; *Tf* = transferrin; *Sc* = scalation, two loci *S* and *N*; *Tp* = transparent; *Pat* = pattern, *Pig* = pigmentation; *YE* = yellow eggs; *Mal* = maleness; *Bl* = blond (two loci, *1* and *2*, in carp); CDS = caudal deformity syndrome. Mean proportion of heterozygotes in *C. carpio* is conservative (uses $y = 0.90$ for loci *YE* and *Mal*).
[1] Data recalculated from original paper.

females which were either heterozygous or homozygous for a different allele. Control crosses using non-irradiated milt gave no significant deviation from the expected Mendelian genotype frequencies (data not included). The results for the *ADA** locus suggest that it is placed distally on a chromosome arm.

The *EST-2** locus was observed to be at the other extreme, as there was no evidence of any recombinant genotypes ($y = 0.0$) in any of the 150 meiotic gynogenetic offspring from six different sets of progeny (Table 9.3.1). This would suggest that this locus is very close to its respective centromere. The other allozyme loci gave a range of intermediate values; *AAT-2**, $y = 0.53$; *AH**, $y = 0.46$; *MEP-2**, $y = 0.26$; and *FH-1**, $y = 0.20$. The above results were from a single female which was heterozygous at all the allozyme loci (Table 9.3.2). In all cases the mitotic offspring were fixed for one or other of the maternal alleles. The control crosses (24 individuals) did not differ significantly from the expected Mendelian frequencies (data not included).

The meiotic gynogenetics produced from females heterozygous for the red skin colour gene gave a wide range of recombination frequencies ($y = 0.02–1.00$: Table 9.3.3). These values showed significant heterogeneity ($p < 0.005$) with a mean of $y = 0.45$. The range of recombination rates observed for this trait is unusually large, particularly considering that all six females used to produce the gynogenetics were siblings.

Other evidence of crossing over is seen in the sex ratios obtained in some meiotic gynogenetic progeny where small numbers of males were observed (Table 9.3.4). Most broods of gynogenetic offspring in *O. niloticus* were all-female. Table 9.3.4 shows sex ratios from control, meiotic and mitotic gynogenetic offspring from a single female which did not fit this pattern. These sex ratios could have arisen from an epistatic locus with a recessive allele causing female to male sex-reversal in the homozygous state. Thus, from a female heterozygous at this locus (XX *SRsr*), meiotic gynogenesis would be expected to yield non-recombinant homozygous males (XX *srsr*) and both non-recombinant and recombinant females (XX *SRSR* and XX *SRsr* respectively). Mitotic gynogenesis would give only XX *srsr* males and XX *SRSR* females in the offspring. On this basis, the proportion of

Table 9.3.2 Distribution of genotypes and proportion of heterozygotes (y) at six loci in meiotic and mitotic gynogenetic progeny derived from heterozygous (at all loci) female parent of *O. niloticus*

Enzyme loci	Female parent genotype	Progeny		Progeny genotypes			Proportion heterozygotes (y)
		Type	No.	F/F	F/S	S/S	
*ADA**	*135/*113	Meio.	60	0	60	0	1.00
		Mito.	20	10	0	10	0.00
*EST-2**	*105/*100	Meio.	30	8	0	22	0.00
		Mito.	20	12	0	8	0.00
*AAT-2**	*100/*80	Meio.	70	21	37	17	0.53
		Mito.	7	4	0	3	0.00
*MEP-2**	*110/*100	Meio.	70	25	18	27	0.26
		Mito.	7	3	0	4	0.00
*AH**	*110/*85	Meio.	70	13	32	25	0.46
		Mito.	7	3	0	4	0.00
*FH-1**	*107/*100	Meio.	15	0	3	12	0.20
		Mito.	7	4	0	3	0.00

F/F = homozygous for faster allele of maternal genotype;
F/S = heterozygous for both faster and slower alleles of maternal genotype;
S/S = homozygous for slower allele of maternal genotype.
Meio = Meiotic gynogenetic progeny; Mito = Mitotic gynogenetic progeny.

Table 9.3.3 Distribution of body colour pattern and frequency of heterozygotes in meiotic gynogenetic progeny derived from six red F_1 (Red *O. niloticus* × wild-type *O. niloticus*) heterozygous females (*Rr*)

| Female parent number | Number of meiotic gynogenetic progeny | Progeny phenotypes and genotypes | | | Proportion of heterozygotes (*y*) |
| | | Red | | Wild type | |
		Full Presumed *RR*	Blotch Presumed *Rr*	*rr*	
1	43	22	0	21	0.02
2	25	13	0	12	0.04
3	9	0	9	0	1.00
4	16	3	9	4	0.50
5	26	8	12	6	0.54
6	32	8	18	6	0.63
Mean					0.45

N.B. Observed frequencies of body colour pattern in progenies derived from all control crosses related to the above experiments were not significantly different from the expected Mendelian segregation ratio of 3 red to 1 wild type. Proportion of heterozygotes (*y*) = (No. of reds – No. of wild types)/total.

Table 9.3.4 Number of each sex observed in control, meiotic and mitotic gynogenetic *O. niloticus* groups at 33 weeks of age (percentages in parentheses)

| Treatment | Number | Sex | | |
		Female	Male	Other
Control[a]	40	22 (55.0)	18 (45.0)	0
Meiotic	40	37 (92.5)	3 (7.5)	0
Mitotic[a]	40	20 (50.0)	19 (47.5)	1 (2.5)

[a]The observed sex ratio in these two treatments were not significantly different ($p > 0.05$) from a 1 : 1 ratio (χ^2 goodness-of-fit).

recombinants was calculated to be $y = 0.85$. Offspring from three such gynogenetic males in crosses to non-related females were all female (50 fry per cross).

9.3.4 Discussion

Mair *et al.* (1991a), using the same strain of *O. niloticus*, also observed small numbers of males in meiotic gynogenetic progeny from four out of nine females tested (4.0–17.2% males). Mitotic gynogenesis from three of these females gave one male and four females. Progeny from this male in crosses to control females were all female or nearly so (96–100% female). These results are explicable on the basis of the same locus proposed above. The sex ratios from the meiotic gynogenetics from these four females give a recombination frequency of $y = 0.88$, a value not significantly different from the results obtained here. Pooling the two sets of data gives a value of $y = 0.87$. We use the name *SDL-2* for this putative locus, with alleles *SR* and *sr*. A further test of this hypothesis would be to cross male gynogenetics back to their dam, a cross predicted to give a 1 : 1 sex ratio (female XX *SRsr* × male XX *srsr* → 1 : 1 female XX *SRsr*/male XX *srsr*).

The mean heterozygote frequency estimated for all the loci in meiotic gynogenetic offspring produced in this study is $y = 0.47$. There have been few other studies which report on recombination rates in gynogenetic tilapia (review: Mair, 1993). If the mean values for these studies and the value for *SDL-2* derived from Mair *et al.* (1991a) are included with the present data, the mean proportion of heterozygotes observed in this species is still $y = 0.47$ (12 loci). Data from other large studies (> 9 loci) have been compiled for comparison (Table 9.3.4). The tilapia value is very similar to that observed for a number of other fish species, $y = 0.56$ (25 loci) for the rainbow trout (Allendorf *et al.*, 1986; Thompson and Scott, 1984), $y = 0.45$ (12 loci) for the common carp (Komen, 1990) and $y = 0.42$ (5 loci) for the plaice (Thompson, 1983). The highest value observed so far, $y = 0.88$ (12 loci), has been in the brown trout (Guyomard, 1986). Guyomard (1986) stated that comparison between species was difficult because of the large standard error associated with these estimates. It should be remembered that the use of a large number of loci randomly placed on their respective chromosome arms is likely to result in a large interlocus variation in recombination rates. Intralocus variation in recombination rates between individuals within a species does, however, appear to be relatively small, as exemplified by the work of Allendorf *et al.* (1986) on four strains of rainbow trout. The high figure for the brown trout results from the large number of loci (10 out of 12 studied) with $y > 0.66$ (maximum value of y if there is no chiasma interference). The large differences in the y value obtained for the red locus in *O. niloticus* are very unusual considering the low intralocus variation that is normally observed and could be related to this locus being situated near a recombination hot spot.

The combined estimate (Table 9.3.4) for y in *O. niloticus* gives a fixation estimate (Allendorf and Leary, 1984) of $F = 0.53$, which is approximately equal to a single generation of selfing ($F = 0.50$) or about three generations of full sib mating. The rate of inbreeding in the tilapia will, however, not be equal over the entire genome. Loci such as *ADA**, which are distally placed on their respective chromosome, will never become homozygous in eggs from heterozygous females. In contrast, *EST-2** will be fixed in the first generation. It can be seen that meiotic gynogenetic individuals will have a level of inbreeding equivalent to selfing but the level at any particular locus will be dependent on the distance from that locus to its respective centromere. Nagy and Csányi (1982, 1984) utilized a combination of meiotic gynogenesis and sib mating to overcome this problem in the production of inbred strains of common carp.

In contrast, the data on the mitotic gynogenetics shows that all these fish appear to be homozygous at all the loci studied. This results in complete fixation in a single generation but with a very low survival rate (Hussain *et al.*, 1993). Chourrout (1984) suggested that the survival of mitotic gynogenetics would be increased if they were produced from meiotic gynogenetics as this would have selected against many deleterious alleles.

Work is continuing to increase the number of females analysed to determine the variability in recombination rates and to identify more polymorphic loci which may be used as markers and may enable us to undertake gene–centromere mapping and linkage studies.

References

Allendorf, F.W. and Leary, R.F. (1984) Heterozygosity in gynogenetic diploids and triploids estimated by gene–centromere recombination rates. *Aquaculture*, **43**, 413–20.

Allendorf, F.W., Seeb, J.E., Knudsen, K.L., Thorgaard, G.H. and Leary, R.F. (1986) Gene centromere mapping of 25 loci in rainbow trout. *J. Hered.*, **77**, 307–12.

Chourrout, D. (1984) Pressure-induced retention of second polar body and suppression of first cleavage in rainbow trout: production of all-triploid, all-tetraploid, and heterozygous and homozygous diploid gynogenetics. *Aquaculture*, **36**, 111–26.

Guyomard, R. (1986) Gene segregation in gynogenetic brown trout (*Salmo trutta* L.): systematically high frequencies of post-reduction. *Genet. Sel. Evol.*, **18**, 385–92.

Hussain, M.G., Chatterji, A., McAndrew, B.J. and Johnstone, R. (1991) Triploidy induction in Nile tilapia, *Oreochromis niloticus* L. using pressure, heat, and cold shocks. *Theoret. Appl. Genet.,* **81,** 6–12.

Hussain, M.G., Penman, D.J., McAndrew, B.J. and Johnstone, R. (1993) Suppression of first cleavage in the Nile tilapia *Oreochromis niloticus* L. – a comparison of the relative effectiveness of pressure and heat shocks. *Aquaculture,* **111,** 263–70.

Komen, J. (1990) Clones of common carp: new perspectives in fish research. PhD Thesis, Agricultural University of Wageningen, The Netherlands.

Mair, G.C., Scott, A.G., Penman, D.J., Beardmore, J.A. and Skibinski, D.O.F. (1991a) Sex determination in genus *Oreochromis.* 1. Sex reversal, gynogenesis and triploidy in *O. niloticus* (L.) *Theoret. Appl. Genet.,* **82,** 144–52.

Mair, G.C., Scott, A.G., Penman, D.J., Skibinski, D.O.F. and Beardmore, J.A. (1991b) Sex determination in genus *Oreochromis.* 2. Sex reversal, hybridization, gynogenesis and triploidy in *O. aureus* Steindachner. *Theoret. Appl. Genet.,* **82,** 153–60.

Mair, G.C. (1993) Chromosome-set manipulation in tilapia: techniques, problems and prospects. *Aquaculture,* **111,** 227–44.

McAndrew, B.J. and Majumdar, K.C. (1983) Tilapia stock identification using electrophoretic markers. *Aquaculture,* **30,** 240–61.

McAndrew, B.J., Roubal, F.R., Roberts, R.J., Bullock, A.M. and McEwen, I.M. (1988) The genetics and histology of red, blond and associated colour variants in *Oreochromis niloticus. Genetica,* **76,** 127–37.

Mirza, J.A. and Shelton, W.L. (1988) Induction of gynogenesis and sex reversal in silver carp. *Aquaculture,* **68,** 1–14.

Nagy, A. and Csányi, V. (1982) Change of genetic parameters in successive gynogenetic generations and some calculations for carp gynogenesis. *Theoret. Appl. Genet.,* **63,** 105–10.

Nagy, A. and Csányi, V. (1984) A new breeding system using gynogenesis and sex-reversal for fast inbreeding in carp. *Theoret. Appl. Genet.,* **67,** 485–90.

Penman, D.J., Shah, M.S., Beardmore, J.A. and Skibinski, D.O.F. (1987) Sex ratios of gynogenetic and triploid tilapia, in *Proceedings of the World Symposium on Selection, Hybridization and Genetic Engineering in Aquaculture,* Bordeaux May, 1986, Vol. II (ed K. Tiews) Heeneman, Berlin, pp. 267–76.

Quillet, E., Garcia, P. and Guyomard, R. (1991) Analysis of the production of all homozygous lines of rainbow trout by gynogenesis. *J. Exp. Zool.,* **257,** 361–14.

Seeb, J.E. and Miller, G.D. (1990) The integration of allozyme analyses and genomic manipulations for fish culture and management, in *Electrophoretic and Isoelectric Focusing Techniques in Fisheries Management* (ed. D.H. Whitmore), CRC Press, Boca Raton, pp. 265–80.

Sodsuk, P. and McAndrew, B.J. (1991) Molecular systematics of three tilapiine genera *Tilapia, Sarotherodon* and *Oreochromis* using allozyme data. *J. Fish Biol.,* **39,** (Suppl. A), 301–8.

Stanley, J.G. (1976) Female homogamety in the grass carp (*Ctenopharyngodon idella*) determined by gynogenesis. *J. Fish. Res. Bd. Can.,* **33,** 1372–4.

Streisinger, G.C., Walker, N., Dower, D., Knauber, D. and Singer, F. (1981) Production of clones of homozygous diploid zebra fish (*Branchydanio rerio*). *Nature,* **291,** 293–6.

Thompson, D. (1983) The efficiency of induced diploid gynogenesis in inbreeding. *Aquaculture,* **33,** 237–44.

Thompson, D. and Scott, A.P. (1984) An analysis of recombination data in gynogenetic rainbow trout. *Heredity,* **53,** 441–52.

9.4 STUDY ON SEX DETERMINATION IN THE COMMON BARBEL (*BARBUS BARBUS* L.) (PISCES, CYPRINIDAE) USING GYNOGENESIS

Manola Castelli

Abstract

The sex-determining mechanism XX–XY is the general rule in cyprinids, with female homogamety. To test the hypothesis of female homogamety in common barbel Barbus barbus, *the sex ratio of gynogenetic offspring was analysed. Almost 50% of males were observed in all gynogenetic lines produced whereas we expected 100% females in the case of female homogamety. When gynogenetic males were crossed with control females their progenies showed sex ratios close to 1:1. When gynogenetic females were crossed with control males their progenies were 100% female. These results demonstrate female heterogamety (ZW) and male homogamety (ZZ) in barbel.*

9.4.1 Introduction

The common barbel is a European gonochoric cyprinid. No sex chromosomes have been detected in the subspecies *Barbus barbus plebejus* Vall. (Cataudella *et al.*, 1977). Sex-determining mechanisms can be studied using various technical approaches such as karyology, induced gynogenesis and steroid-induced sex-inversion. Female homogamety (XX) and male heterogamety (XY) seems to be the general rule in cyprinids (Table 9.4.1). But in the goldfish *Carassius auratus*, which is characterized by female homogamety XX (Yamamoto and Kajishima, 1968), some gynogenetic offspring showed a high percentage of males (Oshiro, 1987). Oshiro (1987) mentioned the possible selection of goldfish varieties with female heterogamety or with other complex genetic systems involving multilocus and multiallelic sex-determining genes. Komen and Richter (1990) and Komen *et al.* (1990) discovered a minor female sex-determining gene in the common carp *Cyprinus carpio* and demonstrated the existence of a complex system of genes apart from those involved in the XX–XY mechanism demonstrated by Nagy *et al.* (1978) and Nagy, Bercsenyi and Csányi (1981). Nearly 50% of males and intersexes were produced in homozygous gynogenetic offspring. The genotype of the original female was assumed to be heterozygous *mas⁺/mas-1* segregating into *mas-1/mas-1* males and intersexes and *mas⁺/mas⁺* females. Moreover, the low percentage of males in the heterozygous gynogenetic progeny (6.6%) indicated possible gene–centromere recombination at the sex-determining locus.

Our objectives were to test for presumed female homogamety in common barbel through gynogenesis and subsequent progeny testing for expected Mendelian sex-ratios.

9.4.2 Material and methods

(a) Gynogenesis, triploidy and normal fertilization

Eggs were stripped from naturally ripened common barbel into a dry container. Milt was stripped from naturally ripened common carp *C. carpio* and was irradiated with UV (253.7 nm, at 7.02 W cm^2 for 5 min.). A 0.2-ml volume of sperm was diluted in 1.8 ml of

Table 9.4.1 Sex-determining mechanisms (SDM) and methods of analysis for sex determination in cyprinids

Species	SDM	Analysis method	Reference
Cyprinus carpio	XX–XY	Gynogenesis	Nagy *et al.* (1978)
		Sex inversion	Nagy, Bercsényi and Csányi (1981)
Carassius auratus	XX–XY	Sex inversion	Yamamoto and Kajishima (1968)
Ctenopharyngodon idella	XX–XY	Gynogenesis	Stanley (1976)
		Sex inversion	Shelton (1986a)
Hypophthalmichthys molitrix	XX–XY	Gynogenesis	Mirza and Shelton (1988)
Aristichthys nobilis	XX–XY	Gynogenesis	Shelton (1986b)

saline (glycine 3.75 g/l, Tris, 2.42 g/l, NaCl 5.52 g/l) (Billard, 1974). The 2-ml volume of diluted sperm was magnetically stirred (120 rev/min) during the irradiation in a 5-cm-diameter dish. The barbel eggs were inseminated with the irradiated milt of carp. Milt was activated by the addition of a small quantity of water.

As the diploid and triploid hybrids *B. barbus* × *C. carpio* are non-viable, individuals produced by accidental fertilization by non-irradiated sperm should not survive (Table 9.4.2). This method allows the production of 100% gynogenetic progeny. Thermal shock was administered 3 min after incubation of eggs at 21 °C by immersing at 37 °C for 2 min. After treatment, eggs were returned to 21 °C until hatching. The progenies were raised at 21 °C ± 1 °C.

Triploid and control diploid offspring were produced according to the same procedure except with no milt irradiation for triploids and no milt irradiation or heat shock for diploid controls. Eggs and sperm were obtained from barbel broodstock and milt from carp broodstock reared in the Tihange experimental fish culture station of the University of Liège (Philippart, 1982).

The cytological differentiation (sex cell differentiation) of the gonads was completed at an average weight of 155 mg. After this time, sex reversal was impossible to achieve (Castelli, unpublished data). To test the effect of the temperature on secondary sex determination (differentiation of phenotypic sex) three samples of eggs obtained from a mating of one female with five males maintained at 19–20 °C were incubated respectively at 18 °C (control batch), 26 °C and 29 °C. Fry ($N = 300$) were kept under the same temperature regime (18 °C for 79 days and 26 °C for 62 days) in 66-l aquaria with a semi-flow-through system. At approximately 450 mg, when the gonads were visible to the naked eye, fry were removed from 18 °C and 26 °C and were reared under the same temperature regime (22.6 °C ± 1.3 °C) to a size at which they could be sexed by gonad squash (Guerrero and Shelton, 1974).

(b) Sex identification

An aceto-carmine squash method (Guerrero and Shelton, 1974) was used for sexing juvenile fish when they had reached a length of 6–7 cm. This method was used for sexing gynogenetic, triploid and control offspring (Table 9.4.3) and the overall progeny of gynogenetic males and females (Table 9.4.4 and 9.4.5). For some gynogenetic and diploid control offspring not sacrificed for gonad squash (Table 9.4.3), the sex of the fish was determined when they began to mature. In this case, we identified the gametes produced at the age of 1 year for the males and at the age of 2 years for the females.

Table 9.4.2 Use of common carp sperm (C) as a marker for gynogenesis in common barbel (B)

	Hatching survival (%)	Treated eggs (N)	Ploidy[a] (n = 20)
(B × B)	34	109	2n
(B × B) + heat shock	31	85	3n
(B × C)	0	237	2n
(B × C irradiated)	6	198	n
(B × C) + heat shock	0	236	3n
(B × C irradiated) + heat shock	14	344	2n

[a] Chromosome preparation.

Table 9.4.3 *B. barbus* sex ratio in gynogenetic, triploid and control biparental progenies

	Number of eggs	Hatching survival (%)	Number of observations	Sex ratio (M : F)	Method of sex identification	χ^2 test (d.f. = 1)
Gyn 1	716	59.8	49 : 72 (121)	1 : 1.47	Stripped gametes examination	4.37 (0.0365)*
Control 1	No data	No data	No data	No data	–	–
Gyn 2	315	59.7	41 : 51 (92)	1 : 1.24	Histology	1.08 (0.2971) NS
Control 2	156	82.0	32 : 32 (64)	1 : 1	Histology	0 (1.0000) NS
Gyn 3	1880	52.7	27 : 30 (57)	1 : 1.11	Stripped gametes examination	0.16 (0.6911) NS
Tripl 3	948	78.4	42 : 28 (70)	1.5 : 1	Histology	2.8 (0.0943) NS
Control 3	1053	89.4	71 : 26 (97)	2.73 : 1	Stripped gametes examination	20.87 (0.0001)***
Gyn 4	830	43.2	58 : 72 (130)	1 : 1.24	Histology	1.50 (0.2195) NS
Tripl 4	1006	72.2	70 : 60 (130)	1.16 : 1	Histology	0.769 (0.3805) NS
Control 4	660	77.6	67 : 63 (130)	1.06 : 1	Histology	0.12 (0.7257) NS
Gyn 5	959	47	10 : 8[a] (18)	1.25 : 1	Stripped gametes examination	0.22 (0.6374) NS
Tripl 5	1196	74.8	35 : 33 (68)	1.06 : 1	Histology	0.06 (0.8084) NS
Control 5	582	83.1	37 : 37 (74)	1 : 1	Stripped gametes examination	0 (1.0000) NS
Control 6	911	90.8	30 : 25 (55)	1.2 : 1	Histology	0.45 (0.5002) NS
Gyn 6	No data	No data	No data	No data	–	–

[a] Surviving offspring after disease.

Gyn = gynogens. Tripl = triploids; NS = not significant; * $0.05 > p > 0.01$; ** $0.01 > p > 0.001$; *** $p < 0.001$.

Table 9.4.4 *B. barbus*: overall and mean sex ratio of controls, gynogenetics and triploids expressed as the percentage of male offspring

Progeny	Number of progenies (total number of fish)	Total % male	Mean % male ± SE	Homogeneity χ^2 (p)	Total χ^2 (p)
Control	10 (760)	58.2	57.2 ± 2.9	27.80 (0.0010)***	20.23 (0.0001)***
Gynogenetic	5 (418)	44.3	46.5 ± 2.5	1.86 (0.7616) NS	5.512 (0.0189)*
Gynogenetic (except Gyn 1)	4 (297)	45.8	48.0 ± 2.6	0.88 (0.8311) NS	2.10 (0.1469) NS
Triploid	3 (268)	54.8	55.1 ± 2.5	1.12 (0.5723) NS	2.522 (0.1122) NS

Homogeneity χ^2 values are for comparison between sex ratios in each group respectively. Total χ^2 values are for comparison of the overall sex ratio of each group with the expected sex ratio 1 : 1. For abbreviations see Table 9.4.3.

Table 9.4.5 *B. barbus*: sex ratio expressed as the percentage of females

Comparison of progeny pair	Sex ratio difference (% female)	Homogeneity χ^2	p
Control 3–Gyn 3	− 25.8	10.35	0.0013**
Control 3–Tripl 3	− 13.2	3.24	0.0720 NS
Tripl 3–Gyn 3	− 12.6	2.02	0.1552 NS
Control (2 + 4 + 5)–Gyn (2 + 4 + 5)	− 5.3	1.44	0.2301 NS
Control (2 + 3 + 4 + 5)–Gyn (2 + 3 + 4 + 5)	− 10.9	7.82	0.0052**
Control (3 + 4 + 5)–Gyn (3 + 4 + 5)	− 11.8	6.82	0.0090**
Control (3 + 4 + 5)–Tripl (3 + 4 + 5)	− 3.2	0.62	0.4295 NS
Tripl (3 + 4 + 5)–Gyn (3 + 4 + 5)	− 8.6	3.37	0.0666 NS

Homogeneity χ^2 values are for comparison between control and corresponding gynogenetic or triploid progenies (pooled or not pooled) and for comparison between triploid and corresponding gynogenetic progenies (pooled or not pooled). For abbreviations see Table 9.4.3.

9.4.3 Results

(a) Sex identification in control offspring

Sex ratios recorded in 10 control progenies (Table 9.4.3 and Tables 9.4.6, 9.4.7 and 9.4.8) were not significantly different from 1:1 apart from the control offspring 3 (Table 9.4.3) and the control offspring raised at 18 °C (Table 9.4.8) showing an excess of males (nearly 75%). These two skewed sex ratios were not significantly different from a 3 : 1 ratio ($p = 0.752$, $p = 0.817$). The sex ratios in the control diploid group were significantly heterogeneous ($p = 0.001$) and the overall sex ratio was significantly different from 1 : 1 ($p = 0.0001$) with a greater proportion of males (58.2%) (Table 9.4.4).

Table 9.4.6 *B. barbus*: sex ratio (M : F) in progeny of four gynogenetic males and one control male crossed with control females

Male × Female	Control 1	Control 2	Control 3
Gyn 1	1 : 1 0.00[a] (1.0000) NS	−	−
Gyn 2	1.04 : 1 0.02[a] (0.8886) NS	−	−
Gyn 3	1 : 1 0.00[a] (1.0000) NS	−	−
Gyn 4	1.21 : 1 0.49[a] (0.4838) NS	1.08 : 1 0.08[a] (0.7772) NS	1.68 : 1 3.31[a] (0.0687) NS
Control	1.08 : 1 0.08[a] (0.7772) NS	1.17 : 1 0.32[a] (0.5716) NS	1.50 : 1 2.00[a] (0.1573) NS

Sex ratios are determined from 50–51 observations of 150 individuals per progeny.
[a] χ^2 test (d.f. = 1). For abbreviations, see Table 9.4.3.

Table 9.4.7 *B. barbus*: sex ratio (M : F) in progeny of three gynogenetic females and one control female respectively crossed with males from an initial pool of eight individuals

Female	Number of eggs	Number of males used	Hatching survival (%)	Number of observations M : F	Sex ratio (M : F)	χ^2 test (d.f. = 1)
Gyn 1	1377	8	50.8	0 : 80 (80)	0 : 1	80 (0.0001)***
Gyn 2	2698	6	75.8	0 : 80 (80)	0 : 1	80 (0.0001)***
Gyn 3	1988	5	88.2	3 : 447 (450)	0.007 : 1	438.1 (0.0001)***
Control	559	8	20.4	48 : 42 (90)	1.14 : 1	0.4 (0.5271) NS

For abbreviations see Table 9.4.3.

(b)   Sex identification in gynogenetic and triploid offspring

Sex ratios of gynogenetics were found to be homogeneous ($p = 0.7616$; Table 9.4.4) and were not significantly different from 1 : 1 apart from the gynogenetic offspring 1 (Table 9.4.3) showing more females than males. We examined each of these progeny at the beginning of the breeding season when all the males had probably not yet ripened. A re-examination of these offspring later in the breeding season was not possible. In this case, the number of females has probably been slightly overestimated.

Table 9.4.8 The effect of incubation and feeding temperature on sex ratio in common barbel

Incubation and feeding temperature (°C)	Number of eggs	Hatching survival (%)	At the end of treatment		Sex ratio (number of observations M : F)	χ^2 test (d.f. = 1)
			Weight (mg)	Age from fertilization (days)		
13.4	213	0.0	–	–	–	–
18	832	93.1	474	79	2.84 : 1 (74 : 26)	23.04 (0.0001)***
26	1816	46.4	450	62	2.57 : 1 (72 : 28)	19.36 (0.0001)***
29	2066	0.0	–	–	–	–

One female mated to five males except for results at 13.4 °C which come from a mating of one female with four males. Control batch from this mating: incubation temperature = 22 °C, N = 213, survival = 71.5% (Absil, 1989).

Based on the general rule of female homogamety in cyprinids, 100% females should be produced. As we observed more or less 50% males in each gynogenetic line, the *ZW–ZZ* hypothesis of sex determination can be proposed to explain these results.

Moreover, sex ratios of gynogenetic and diploid control fish were significantly different from each other (p = 0.0052) (Table 9.4.5). Gynogenetic progenies showed a deficit of males (44.3%) unlike diploid control progenies (58.2%). Sex ratios of triploids were homogeneous (p = 0.5723) and not significantly different from 1 : 1 (p = 0.1122) (Table 9.4.4) and from sex ratios of diploid controls (p = 0.4295). But sex ratios of triploids were found to be almost significantly different from sex ratios of gynogenetics (p = 0.0666) (Table 9.4.5).

(c) Progeny of gynogenetic males

Table 9.4.6 summarizes the results of nine different single pair matings between yearling gynogenetic and control males with control females. All sex ratios were close to 1 : 1 and those from gynogenetic male progenies were not significantly different from control ratios.

Moreover, control female 3 mated with two different males gave similar progeny sex ratios (1.68 : 1; 1.5 : 1) whereas control females 1 and 2 gave progeny sex ratios closer to 1 : 1 than control female 3. These results could indicate a female influence on progeny sex ratio which might involve autosomal sex-determining factors.

(d) Progeny of gynogenetic females

Sex ratios of progenies from three gynogenetic females and a control female crossed respectively with eight, six, five and eight males out of an initial pool of eight individuals are given in Table 9.4.7. To avoid any male influence on sex ratio (Wohlfarth and Wedekind, 1991), we fertilized the eggs of each female with the mixed milt of the males. The progenies of the first two gynogenetic females were 100% females, whereas the sex ratio of the control progeny was close to 1 : 1. Very few males (0.7%) were produced in the progeny of the third gynogenetic female. This confirms the hypothesis of female heterogamety (*ZW*) and male homogamety (*ZZ*) in barbel even if we suspect the existence of minor sex-determining factors because of the presence of some males in the progeny from the third gynogenetic female.

(e)   Sex ratio under different water temperature conditions

No difference was found between sex ratios of progeny raised at 18 °C (from fertilization to average weight of 474 mg) and progeny raised at 26 °C (from fertilization to average weight of 450 mg) (Table 9.4.8). Below 14 °C (Absil, 1989) (breeding occurs above 15 °C in nature) and above 26 °C (abnormal conditions) no eggs hatch. However, the sex ratios of these progenies are significantly different from the expected sex ratio 1:1 (nearly 75% of males) although the female has been crossed with five males. This again suggests a possible maternal influence on progeny sex ratio.

9.4.4   Discussion

Cataudella *et al.* (1977) detected no heteromorphic sex chromosomes in barbel *Barbus barbus plebejus* (Vall.). But banding techniques (G, C, Q, restriction endonuclease banding) could be suitable tool to discover possible cytological sex differences (Section 7.1).

In cases where the sex chromosomes cannot be detected gynogenesis and sex reversal are helpful techniques. In this study, gynogenesis in barbel produced about 50% gynogenetic males, whereas we expected all females according to the general rule of female homogamety in cyprinids. Sex ratios recorded in progenies of gynogenetic males (close to 1:1) and gynogenetic females (100% or nearly 100% of females) prove female heterogamety (ZW) and male homogamety (ZZ) in this species.

The results recorded by gynogenesis and by mating gynogenetic males and females respectively with control females and males confirm a sex determining mechanism ZW–ZZ in barbel, but the observation of diploid control sex ratios of 3:1 in favour of the males could suggest the presence of a minor sex-determining gene (*M, m*) epistatic to the major sex-determining gene (*Z,W*). The recessive allele '*m*' in the homozygous condition would induce maleness in fish which are ZW. A cross between a heterozygous female ZWMm and a recessive homozygous male ZZmm would produce a 3:1 sex ratio (Table 9.4.9). Other data could attest to the existence of autosomal sex-determining factors, such as the production of 0.7% males in one of the progenies from gynogenetic females or the observation of spontaneous hermaphrodites in a population obtained from wild barbels, originating from the

Table 9.4.9   *B. barbus*: atypical sex ratio 3:1 in diploid control progeny. A model involving a minor male sex-determining gene (*M, m*). Genotypes and sex ratios in diploid control, gynogenetic and triploid progenies

Progeny	*Male ZZmm*	*Female ZWMm*	*Ratio*
Diploid	ZZMm	ZWMm	3:1
	ZZmm		
	ZWmm		
Gynogenetic	ZZMM	WWMM	1:1.25
	ZZmm	WWmm	
	ZZMm	WWMm	
	ZWmm	ZWMM	
		ZWMm	
Triploid	ZZZMMM	ZWWMMM	1.25:1
	ZZZmmm	ZWWMmm	
	ZZZMmm	ZZWMMM	
	ZWWmmm	ZZWMmm	
	ZZWmmm		

River Brouffe (Belgium) (Philippart, personal communication), although a possible environmental influence on sex differentiation cannot be excluded in the latter case (Conover and Heins, 1987; Mair, Beardmore and Skibinski, 1990). On the other hand, these atypical sex ratios could simply be due to a differential segregation of the sex-determining genes to ova and polar bodies (Bulmer and Bull, 1982). In this case, the sex ratio of gynogenetic and triploid progenies would be nevertheless close to 1:1 because of retention of the second polar body.

Gynogenetic sex ratios were not significantly different from a 1:1 sex ratio but were significantly different from diploid control sex ratios, with a larger proportion of females. This result could point towards crossing over between the sex-determining gene (Z, W) and the centromere during meiosis. Consequently, gynogenetic ZW and triploid ZZW genotypes would be produced in addition to the expected gynogenetic (ZZ, WW) and triploid (ZZZ, ZWW) genotypes (Table 9.4.9). Moreover, similar sex ratios should be expected in gynogenetic and triploid groups (Mair *et al.*, 1991). The comparison of the triploid sex ratio 3 (40% females) with the corresponding diploid control (26.8% females) and gynogenetic (52.6% females) sex ratios showed no significant difference (Table 9.4.5). But these percentages are quite different enough from each other to suggest the hypothesis of crossing over between the minor sex-determining gene (M, m) and the centromere. Furthermore, we could postulate inequalities between gonosomal pairs and autosomal pairs –$WW > mm > ZW$ – to be consistent with the experimental data (Table 9.4.9). The analysis of the pooled data (3 + 4 + 5) (Table 9.4.5) showed similar results, with a significant difference between the gynogenetic (53.7% females) and the diploid control (41.9% females) sex ratios ($p = 0.009$) and a nearly significant difference between the gynogenetic and triploid (45.1% females) sex ratios ($p = 0.0666$), while there is no significant difference between diploid control and triploid sex ratios ($p = 0.4295$). These results probably indicate that there would be not only recombination between the sex-determining genes and their centromere but also complex interactions between these genes as postulated above.

The results concerning the sex ratios of progenies raised at different temperatures point towards no influence of water temperature on sex determination but more data are needed to confirm or invalidate this hypothesis. Indeed, in *Clarias lazera* (van den Hurk and Lambert, 1982), a temporary rise in temperature from 30 to 39 °C for 3 h at day 9 of development resulted in an all-male brood. A similar shock (above 26 °C or below 15 °C) could have an effect on the sex ratio in barbel which might not be evident in long-term rearing at 26 °C.

Until now the cyprinid species studied have been characterized by female homogamety. This work on barbel proves that the Cyprinidae show both female homogamety and heterogamety, as do the Anostomidae (Galetti *et al.*, 1981), the Cyprinodontidae (Chen and Ruddle, 1970; Khuda-Bukhsh, 1979), the Gasterosteidae (Ebeling and Chen, 1970) and the Cichlidae (Jalabert *et al.*, 1974; Guerrero, 1975). Moreover, minor sex-determining genes could be involved in sex determination in barbel as has been reported in the other cyprinid species, *Carassius auratus* (Oshiro, 1987) and *Cyprinus carpio* (Komen and Richter, 1990; Komen *et al.*, 1990). However, more investigations into possible maternal/paternal influences and temperature effects on the progeny sex ratio are necessary to confirm the model of sex determination proposed above for *B. barbus*.

In addition, this work could initiate a study of the evolution of sex-determining mechanisms in *Barbus*. The origin and the phylogeny of the genus *Barbus sensu lato* (subgenera *Barbus* and *Luciobarbus*) are still debated (Almaça, 1981; Doadrio, 1990; Agnèse *et al.*, 1990). Diploid and tetraploid species of *Barbus* are found in Africa and Asia whereas only tetraploid species are present in Europe. This genus was probably dispersed to Europe from Africa (Novacek and Marshall, 1976) or Asia (Banarescu, 1973) after the tetraploidization event at the origin of the European species of *Barbus*. Sex determination in African and

Asian diploid and tetraploid species of *Barbus* in relation to European tetraploid species would be interesting to study in regard to its evolution from the diploid species to the tetraploid species. Indeed polyploidization would only have been possible at a time before the chromosomal sex-determining mechanism was firmly established (Ohno and Atkin, 1966). In this case, sex determination could have evolved differently in diploid and tetraploid species of *Barbus*.

Acknowledgements

This work has been supported by the 'Centre d'Etude pour la Récupération des Energies résiduelles (CERER-Pisciculture, Université de Liège – Electrabel; Directeur Dr J.C. Philippart). During this time, M. Castelli was a fellow of IRSIA (Institut pour l'Encouragement de la Recherche Scientifique dans l'Industrie et l'Agriculture).

References

Absil, P. (1989) *Biologie des stades précoses du barbeau fluviatile, Barbus barbus (Linnaeus, 1758). Effets de la température sur le développement et la mortalité.* Mémoire de Licence en Sciences Zoologiques, Université de Liège, Belgium.

Agnèse, J.F., Berrebi, P., Lévêque C. and Guégan, J.F. (1990) Two lineages, diploid and tetraploid, demonstrated in African species of *Barbus* (Osteichthyes, Cyprinidae): on the coding of differential gene expression. *Aquat. Living Resour.*, **3**, 305–11.

Almaça, C. (1981) La collection de Barbus d'Europe du Muséum national d'Histoire naturelle (Cyprinidae, Pisces). *Bull. Mus. Natn. Hist. Nat., Paris*, **3**, 277–307.

Banarescu, P. (1973) Origin and affinities of the freshwater fish fauna of Europe. *Ichthyologia*, **5**, 1–8.

Billard, R. (1974) L'insémination artificielle de la truite *Salmo gairdneri* Richardson. IV–Effets des ions K et Na sur la conservation de la fertilité des gamètes. *Bull. Fr. Pisc.*, **256**, 88–100.

Bulmer, M.G. and Bull, J.J. (1982) Models of polygenic sex determination and sex ratio control. *Evolution*, **36**, 13–26.

Cataudella, S., Sola, L., Accame Muratori, R. and Capanna, E. (1977) The chromosomes of 11 species of Cyprinidae and one Cobitidae from Italy, with some remarks on the problem of polyploidy in the cypriniformes. *Genetica*, **47**, 161–71.

Chen, T.R. and Ruddle, F.M. (1970) A chromosome study of four species and a hybrid of the killifish genus *Fundulus* (Cyprinodontidae). *Chromosoma*, **29**, 255–67.

Conover, D.O. and Heins, S.W. (1987) Adaptative variation in environmental and genetic sex determination in a fish. *Nature*, **326**, 496–8.

Doadrio, I. (1990) Phylogenetic relationships and classification of west palaearctic species of the genus *Barbus* (Osteichthyes, Cyprinidae). *Aquat. Living Resour.*, **3**, 265–82.

Ebeling, A.W. and Chen, T.R. (1970) Heterogamety in teleostean fishes. *Trans. Am. Fish. Soc.*, **99**, 131–8.

Galetti, P.M., Foresti, F., Bertollo, L.A.C. and Moreira-Filho, O. (1981) Heteromorphic sex chromosomes in 3 species of the genus *Leporinus* Pisces Anastomidae. *Cytogenet. Cell. Genet.*, **29**, 138–42.

Guerrero R.D. III (1975) Use of androgens for the production of all-male *Tilapia aurea* (Steindachner). *Trans. Am. Fish. Soc.*, **104**, 342–8.

Guerrero R.D. III and Shelton, W.L. (1974) An aceto-carmine squash method for sexing juvenile fishes. *Prog. Fish-Cult.*, **36**, 56.

Jalabert, B., Moreau, J., Planquette, P. and Billard, R. (1974) Déterminisme du sexe chez *Tilapia macrochir* et *Tilapia nilotica*: action de la méthyltestostérone dans l'alimentation des alevins sur la différentiation sexuelle; proportion des sexes dans la descendance des mâles 'inversés'. *Ann. Biol. Anim. Bioch. Biophys.*, **14**, 729–39.

Khuda-Bukhsh, A.R. (1979) Chromosomes of 3 species of fishes *Aplocheilus panchax*, Cyprinodontidae, *Lates calcerifer*, Percidae, and *Gadusia chapra*, Clupeidae. *Caryologia*, **32**, 161–70.

Komen, J., and Richter, C.J.J. (1990) Clones of common carp, *Cyprinus carpio*. New perspectives in fish research, PhD Thesis, Agricultural University Wageningen, pp. 141–59.

Komen, J., Wiegertjes, G.F., van Ginneken, V., Eding, E.H. and Richter, C.J.J. (1990) Clones of common carp, *Cyprinus carpio*. New perspectives in fish research, PhD Thesis, Agricultural University Wageningen, pp. 101–22.

Mair, G.C., Beardmore, J.A. and Skibinski, D.O.F (1990) Experimental evidence for environmental sex determination in *Oreochromis* species, in *The Second Asian Fisheries Forum* (eds R. Hirano and I. Hanyu), Asian Fisheries Society, Manila, Philippines, pp. 555–8.

Mair, G.C., Scott, A.G., Penman, D.J., Skibinski, D.O.F. and Beardmore, J.A. (1991) Sex determination in the Genus *Oreochromis*. 2. Sex reversal, hybridisation, gynogenesis and triploidy in *O. aureus* Steindachner. *Theoret. Appl. Genet.*, **82**, 153–60.

Mirza, J.A. and Shelton, W.L. (1988) Induction of gynogenesis and sex reversal in silver carp. *Aquaculture*, **68**, 1–14.

Nagy, A., Bercsényi, M. and Csányi, V. (1981) Sex reversal in carp (*Cyprinus carpio*) by oral administration of methyltestosterone. *Can. J. Fish. Aquat. Sci.*, **38**, 725–8.

Nagy, A., Rajki, K., Horvath, L. and Csanyi, V. (1978) Investigation in carp, *Cyprinus carpio* L. Gynogenesis. *J. Fish Biol.*, **13**, 215–24.

Novacek, M.J. and Marshall, L.G. (1976) Early biogeographic history of Ostariophysan fishes. *Copeia*, **1976** (1), 1–12.

Ohno, S. and Atkin, N.B. (1966) Comparative DNA values and chromosome complements of eight species of fishes. *Chromosoma*, **18**, 455–66.

Oshiro, T. (1987) Sex ratios of diploid gynogenetic progeny derived from five different females of goldfish. *Bull. Japan. Soc. Sci. Fish.*, **53**, 1899.

Philippart, J.C. (1982) Mise au point de l'alevinage contrôlé du barbeau, *Barbus barbus* (L.), en Belgique. Perspectives pour le rempoissonnement des rivières. *Cah. Ethol. Appl.*, **2**, 173–202.

Shelton, W.L. (1986a) Broodstock development for monosex production of grass carp. *Aquaculture*, **57**, 311–19.

Shelton, W.L. (1986b) Control of sex in cyprinids for aquaculture, in *Aquaculture of Cyprinids* (eds R. Billard and J. Marcel), INRA, Paris, pp. 179–94.

Stanley, J.G. (1976) Female homogamety in grass carp (*Ctenopharyngodon idella*) determined by gynogenesis. *J. Fish. Res. Bd Can.*, **33**, 1372–4.

van den Hurk, R. and Lambert, J.G.D. (1983) Temperature and steroid effects on gonadal sex differentiation in rainbow trout, in *Proceedings of the International Symposium on Reproduction and Physiology of Fish* (eds C.J.J. Richter and H.J. Th. Goos), Pudoc, Wageningen, pp. 69–72.

Wohlfarth, G.W. and Wedekind, H. (1991) The heredity of sex determination in tilapias. *Aquaculture*, **92**, 143–56.

Yamamoto, T. and Kajishima, T. (1968) Sex hormone induction of sex reversal in the goldfish and evidence for male heterogamety. *J. Exp. Zool.*, **168**, 215–22.

9.5 RELEASE AND RECAPTURE OF GENETICALLY TAGGED COD FRY IN A NORWEGIAN FJORD SYSTEM

Knut Eirik Jørstad, Gunnar Nævdal, Ole I. Paulsen and Solveig Thorkildsen

Abstract

Genetic studies were performed in connection with a large-scale cod enhancement programme in western Norway. In 1990 and 1991, artificially produced cod fry were partially based on genetically tagged cod broodstock being homozygous for the rare allele PGI-1*30. In these years, about 215 000 cod fry were released and 56% of the individuals were genetically tagged. Analyses of cod catches in 1991 in the release area demonstrate that no change in the frequency of the genetically tagged cod had occurred under natural conditions, and these fish constitute about 30% of the total 1990 year class of cod in the area. So far, the genetic tagging approach seems to be a reliable method for identifying released fish, and further, reproductive success and gene introgression with wild cod in the area may be studied when the released fish reach maturity.

9.5.1 Introduction

Artificial production of cod (*Gadus mortua* L.) for enhancement of wild stocks was initiated in Norway more than 100 years ago. This activity includes mass rearing of cod eggs and large-scale (in billions) release of yolksac larvae in coastal areas. The hatching and releases continued for many decades, in spite of great controversies, mainly due to difficulties in the evaluation of the effects of this programme (Solemdal *et al.*, 1984).

The breakthrough of mass rearing of cod fry in a saltwater pond in western Norway in 1983 (Øiestad, Kvenseth and Folkvord, 1984) caused an increased interest in cod farming and enhancement of local cod populations. Several new release projects were initiated in different geographic areas in order to obtain scientific information for evaluation of the potential for commercial ranching.

During the last decade much attention has been paid to the potential interaction between cultured and wild stocks of the same species, especially focusing on preservation of native gene pools (Ryman, 1981). For this reason, genetic aspects and genetic analyses were incorporated in the cod sea ranching programmes in Norway.

The evaluation of any release project is dependent on a reliable tagging of the fry to be released and identification of the recaptured individuals. This is usually carried out by mechanical and/or chemical tagging methods which have limitations with respect to the numbers and cost of large-scale tagging programmes as well as reliable identification being dependent on time of recapture or age of the fish.

Several workers have discussed the use of genetic markers in connection with genetic studies of wild fish (Allendorf and Utter, 1979; Shaklee, 1983) and aquaculture (Moav *et al.*, 1976; Wilkins, 1981). Recently, potential uses of genetic tagging have been reviewed (Utter and Seeb, 1990; Gharrett and Seeb, 1990). There are, however, very few cases (see Skaala *et al.* (1989) and Hindar, Ryman and Utter (1991) for references) in which specific genetically marked strains have been developed under controlled conditions.

In Norway, artificially produced cod fry of the 1983 year class were raised under farming conditions, and therefore screening of rare heterozygotes as basis for developing a genetically tagged strain was easily carried out. The establishment of a cod strain homozygous for a rare allele at the phospho-glucose isomerase locus (*PGI-1*30*) has been described elsewhere (Jørstad *et al.*, 1987; Jørstad, Skaala and Dahle, 1991), and these fish were available to be used as broodstock in connection with the cod enhancement project carried out in Masfjord, western Norway.

In this section we describe the release experiments in 1990 and 1991 including the use of genetically tagged fry. In addition, the first data on recapture and survival of the release fry compared to wild fry in the area are presented. The preliminary data indicate that the genetic tagging approach can be used successfully as a reliable tagging method in enhancement programmes.

9.5.2 Materials and methods

The development of a genetically tagged strain of cod was carried out at the Aquaculture Station Austevoll, Institute of Marine Research, in the period from 1983 to 1989–1990. Initial screening of rare heterozygotes in tissue enzymes in farmed cod of wild origin suggested the development of a strain homozygous for the rare allele *PGI-1*30*. Matured cod heterozygous for this allele were allowed to spawn naturally in a net pen. Eggs were collected and yolksac larvae were released and reared in an artificial saltwater pond (Jørstad *et al.*, 1987). Among the offspring, the frequency of *PGI-1*30/30* genotypes was 25%, and these fish were identified by enzyme electrophoresis and raised as a genetically tagged broodstock (Jørstad, Skaala and Dahle, 1991).

In each of the years 1990 and 1991, two different spawning pens were set up at the Aquaculture Station Austevoll. In the first year, 203 mature cod (all *PGI-1*30/30* homozygotes) were used in one pen for production of genetically tagged yolksac larvae. In the other spawning pen, 116 individuals of an unselected cod broodstock originating from wild cod in western Norway were used for egg production. In both pens the parent fish spawned naturally, and the eggs were kept separately until release as 5-day-old yolksac larvae in an enclosed saltwater pond called Parisvatnet in Øygarden northwest of Bergen. The larvae from the two different strains were mixed in the pond and fed on natural zooplankton until metamorphosis. After that they were given formulated feed for supplement (Blom *et al.*, 1990). The surviving fish were caught in midsummer by sinking nets and transferred to net pens for feeding; the feed includes the chemical tag oxytetracycline (Nordeide *et al.*, 1992). In this way offspring from both strains were chemically tagged and a mixture of the two different strains was released in Masfjorden (Figure 9.5.1) in autumn 1990. A similar production approach was used in 1991.

The total release of cod fry in the Masfjord area in 1990 was about 30 000 individuals (Table 9.5.1) and more than 12 000 of them were genetically tagged. The same production design was used in 1991 based on the same genetically tagged broodstock. This year the total production of cod fry in the pond was more than 315 000 individuals and of these nearly 60% were offspring from the genetically tagged broodstock. More than 180 000 individuals were released in the Masfjord area, mainly the outer part, and over 100 000 of these fish were genetically tagged.

Samples of the fish produced were collected and genetically analysed before release. Cod sampling by net fisheries in the Masfjorden area was carried out regularly each month and blood and white muscle samples were taken for genetic analyses. Biological sampling, including age determination,was always carried out and available for all individuals caught. The released and recovered fish were identified by the chemical tagging, and the genetically tagged fish were scored by their *PGI-1** genotypes (Figure 9.5.2).

Blood and tissue samples were analysed by the methods described by Jørstad (1984). Agar electrophoresis was used for scoring of haemoglobin (*HB*) genotypes and horizontal starch gel electrophoresis for variation in white tissue enzymes. Polymorphic enzymes in cod have been described earlier (Mork *et al.*, 1982; Jørstad, 1984), and the enzymes identified as being most informative, including lactate dehydrogenase (LDH), phosphoglucose isomerase (PGI), phosphoglucomutase (PGM) and glucose phosphate dehydrogenase (GPD), were routinely stained for (Jørstad and Nævdal, 1989).

Tests of accordance between observed distribution of genotypes and expected Hardy–Weinberg distributions as well as sample collection heterogeneity were carried out using standard statistical tests (χ^2) in the programme package (BIOSYS-1) developed by Swofford and Selander (1981).

9.5.3 Results

In both years, genetic analyses of the artificially produced fry were carried out before release, and the results for the three most important loci are summarized in Table 9.5.2. Here the two genetically tagged strains are also given as separate samples. In 1990 the frequency of the marker allele in the totally released fry was as high as 0.54, compared to the more normal frequencies of 0.02–0.05 in natural populations of cod. The high frequency is mainly due to the presence of the genetically tagged offspring, but the frequency of the marker allele was also higher than expected among the rest of the released cod fry. This was due to a higher frequency of *PGI-1** heterozygotes in the general cod broodstock at the aquaculture station.

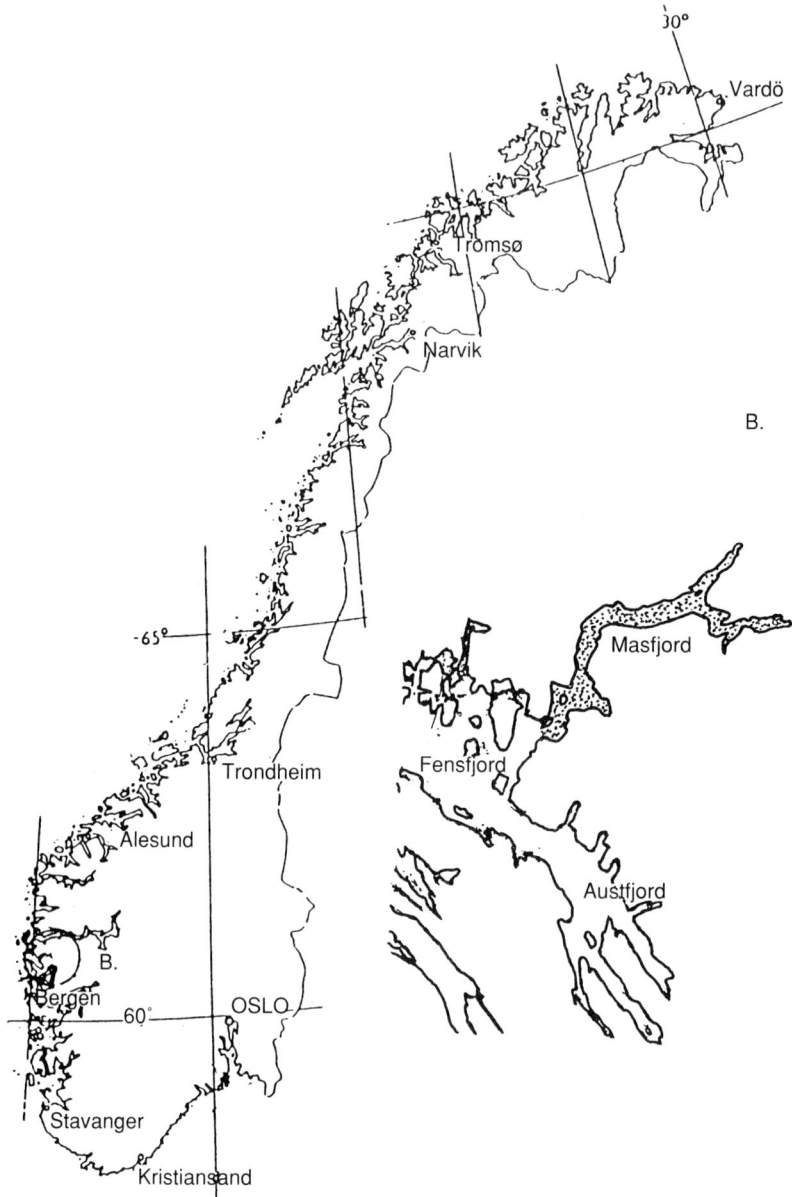

Figure 9.5.1 Map showing the release and recapture area in Masfjord (stippled).

In 1991 precautions were taken and the excess of marker heterozygotes (offspring from the genetically tagged broodstock development) were removed from the general cod broodstock; the offspring had a normal frequency (Table 9.5.2) of the marker allele (0.02). The overall frequency of the marker allele in the 1991 year class (including the pond-produced offspring from both strains) was 0.5. For both year classes the genetically tagged

Table 9.5.1   The number of cod fry produced in Parisvatnet and released in Masfjord in 1990 and 1991

Year class	Total no. fry released	No. genetically tagged fry
1990	30 109	12 645
1991	184 598	107 066
Total	214 707	119 711

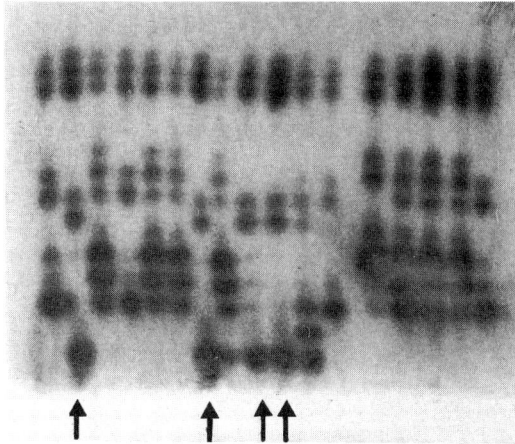

Figure 9.5.2 Starch gel stained for phosphoglucose isomerase after electrophoresis of white muscle samples taken from cod captured in Masfjord. The genetically tagged fish have a single, slow moving band which is indicated by the arrows, and these fish have a genotype of *PGI-1*30/30*.

Table 9.5.2   Allele frequencies at three loci in artificially produced cod fry in Parisvatnet in 1990 and 1991

	*HB1**			*LDH-3**			*PGI-1**		
Year class	N	*1	*2	N	*70	*100	N	*30	*100	*150
1990	160	0.620	0.380	160	0.320	0.680	160	0.130	0.720	0.160
1990 (GM)	145	0.570	0.430	145	0.250	0.740	145	1.000	–	–
1991	162	0.550	0.450	162	0.430	0.570	162	0.020	0.700	0.280
1991 (GM)	243	0.580	0.420	243	0.320	0.680	243	1.000	–	–

GM = genetically marked; *N* = number of individuals scored.

strain was fixed (frequency = 1.00) for the *PGI-1*30* allele. No differences were, however, observed for the other polymorphic loci, indicating no association between the marker loci and the other loci analysed.

The genetic analyses are incorporated in the measurements taken when sampling cod in the Masfjorden area. Thus all individuals are measured, their ages are estimated and they are classified as wild or released according to the presence of oxytetracycline rings in the otoliths or vertebrae. By combining this information with the individual results from the genetic analyses, allele frequencies for the different polymorphic enzymes are obtained for the different groups and year classes of cod in the fjord area. The allele frequencies at five protein loci are given in Table 9.5.3 for released and wild cod which were captured during 1991. With respect to the wild cod population in the area, no deviation from the Hardy–Weinberg distribution was found at any of the loci investigated. Further, there were no differences between the wild cod and the released/recaptured cod from the 1988 and 1989 year classes. As expected, the wild cod population in the area have a very low frequency of the *PGI-1*30* allele (0.03–0.05). The sample of recaptured released cod of the 1990 year class, on the other hand, has a frequency of 0.49 of the marker allele, which represents a significant gene pulse into the 1990 year class of the cod population in the area.

A comparison of the allele frequencies of the released and the recaptured individuals of the 1990 year class is given in Table 9.5.4. During 1991, 76 of the released fish were captured at 1 year old, and no changes in allele frequencies are indicated. The fraction of *PGI-1*30/30* homozygotes in the 1990 year class produced in Parisvatnet varied significantly through the production period and values as low as 38% were estimated (Blom *et al.*, 1990). The frequency of homozygotes in the actual population that were released in the autumn was, however, estimated to be 47%, which is used in this comparison. During the first year after release, a total of 76 cod of this year class were recaptured and 33 individuals (or 43%) were easily identified as *PGE-1*30/30* homozygotes (Figure 9.5.2). Thus the data indicate that these homozygous fish have the same survival rate under natural conditions as the offspring from the unselected broodstock of cod, at least for the period measured.

The genotype distributions detected at the *PGI-1** locus for the total cod material captured in Masfjord during 1991 are shown in Table 9.5.5. Homozygotes for the marker allele are only found in the released 1990 year class and this allele is present in the wild population at low frequencies of heterozygotes. The released fish of the 1988 and 1989 year class that were captured in 1991 have the same distribution as the wild fish in the area and no deviation from the Hardy–Weinberg distribution was detected. On the other hand, a significant deviation was found, as expected, at the marker locus for the 1990 year class.

Based on the number of fish in the different groups and year classes shown in Table 9.5.5, the proportion of released fish in the Masfjord area in 1991 was about 30% in total. Considering the year classes where releases have been carried out, as a whole, 52% of the captured fish were identified as released. With respect to the 1990 year class, the released fish constitute the major part (68%, where the genetically tagged fish were 30%). Thus three different components or cod strains of this year class (wild, general broodstock and genetically tagged fish) can be identified and compared with respect to survival, growth and future recruitment to the natural spawning stock in the area.

9.5.4 Discussion

The rarity of the *GPI-1*30* marker allele in natural populations of cod suggests that this could be associated with some disadvantage under natural conditions and any experiment aiming at artificial development of genetically tagged populations should be carefully considered (Allendorf and Utter, 1979). Recently, the principles and potential of genetic tagging have been reviewed (Utter and Seeb, 1990) and guidelines for genetically marking populations have been proposed.

Intentional breeding of a broodstock which is homozygous for a rare allele takes several years. The experiment performed for cod has been described elsewhere (Jørstad *et al.*, 1987) and an evaluation of survival and growth for the genetically tagged individuals has been carried out (Jørstad, Skaala and Dahle, 1991). The data obtained so far in several experi-

Table 9.5.3 Allele frequencies at five loci in wild and released cod groups caught in Masfjord during 1991

Year class/Age	N	HB-1*		LDH-3*			PGI-1*				GPD*			PGM*		
		*1	*2	*70	*100	*150	*30	*70	*100	*150	*90	*100	*120	*30	*100	*150
Released																
1988/3	32	0.59	0.40	0.44	0.56	–	0.02	0.22	0.75	0.21	–	0.95	0.05	–	1.00	–
1989/2	40	0.49	0.51	0.26	0.74	–	0.04	0.37	0.67	0.22	–	0.97	0.03	–	0.99	0.01
1990/1	76	0.52	0.48	0.31	0.69	–	0.49	0.32	0.43	0.06	–	0.87	0.13	–	1.00	–
Wild																
1987/4	196	0.52	0.48	0.35	0.65	–	0.03	–	0.71	0.26	0.01	0.95	0.04	0.01	0.99	–
1988/3	69	0.44	0.56	0.30	0.70	–	0.03	–	0.71	0.26	0.01	0.98	0.01	0.03	0.92	0.05
1989/2	37	0.53	0.47	0.28	0.72	–	0.03	0.03	0.66	0.28	–	0.99	0.01	–	0.99	0.01
1990/1	36	0.53	0.47	0.28	0.72	0.01	0.05	0.01	0.75	0.19	–	1.00	–	–	0.95	0.05

The ages were estimated from otolith reading, and the released fish were identified by chemical tagging. N = number of individuals scored.

Table 9.5.4 Allele frequencies at five loci in cod released in 1990 (0+) and recovered (1 year old) in 1991

Sample/Age	N	HB-1*		LDH-3*			PGI-1*				GPD*			PGM*		
		*1	*2	*70	*100	*150	*30	*70	*100	*150	*90	*100	*120	*30	*100	*150
Released/0+	305	0.59	0.41	0.29	0.71	–	0.54	0.01	0.38	0.01	–	0.88	0.12	0.06	0.94	–
Recaptured/1	76	0.52	0.48	0.31	0.69	0.01	0.49	0.01	0.43	0.06	–	0.87	0.13	0.02	0.98	–

N = Number of individuals scored.

Table 9.5.5 Distribution of *PGI-1** genotypes of released and wild cod of different year classes captured in Masfjord in 1991

		Genotypes					
Year class/Age	N	*30/30	*30/100	*30/150	*100/100	*100/150	*150/150
Released							
1988/3	33	0	0	1	23	6	3
1989/2	44	0	4	0	25	13	2
1990/1	76	33	10	0	25	7	1
Wild							
1987/4	196	0	9	3	95	77	12
1988/3	69	0	3	1	33	29	3
1989/2	37	0	2	0	19	12	4
1990/1	36	0	1	3	23	9	0

N = Number of individuals scored.

ments do not indicate that the individuals possessing the genetic tag have any disadvantage compared with other individuals. Lack of significant variation in allele frequencies does not, however, exclude small selection forces associated with the allele in question. Similar data have been reported for brown trout (Taggart and Ferguson, 1986), pink salmon (Lane *et al.*, 1990) and chum salmon (Seeb *et al.*, 1990).

The application of genetic tagging in fish has mainly been connected to freshwater species and especially salmonids. The establishment of a genetically marked broodstock of cod is the first example of this approach for a marine fish species. As discussed by Shaklee (1983), such stocks can be used in many different experiments, and so far the cod broodstock has been incorporated in early-stage studies (Svåsand *et al.*, 1991; Blom *et al.*, 1990) and in a cod enhancement programme.

As reported here, about 215 000 artificially produced cod fry were released in the Masfjord area in 1990 and 1991, and 56% of these are homozygous for the rare *PGI-1*30* allele present in white muscle. Each genetically tagged individual is easily detected by enzyme electrophoresis, and analyses of the recaptured fish show no indication of differential survival under natural conditions after 1 year in the fjord. Recapture data will be collected in the future, and survival and growth data will be obtained for both year classes (1990 and 1991) of released fish as well as the wild year classes in the area. The recapture data and the genetic analyses so far suggest that the genetic tagging approach has been very promising for the identification of released fish. Thus genetic tagging could be a very valuable identification technique in large-scale ranching programmes where hundreds of thousands or millions of individuals are being released.

In the 1990 year class of cod that were captured in the Masfjorden area and classified to cod group, about 30% were of the genetically tagged strain. This resulted in an increase of the frequency of the *PGI-1*30* allele in the year class from 0.08 (wild stock and unselected stock) to 0.29. The number of genetically tagged cod fry was even higher in 1991 and these fish will probably be coming into the fisheries in 1992–1993. In this fjord system we therefore have a situation where three different stocks of cod (two of cultured origin) are present. These can be identified by chemical and genetic tags and survival and growth can be compared at different ages. Further, and more interesting, the different stock contribution to the spawning population of local cod can be measured and the spawning success of the released fish can be estimated through the genetic marker and its contribution in the new generation. Thus the extent of the gene flow from the released fish can be estimated and its persistence in the future can be monitored.

This situation offers a unique possibility to study genetic interaction between wild, local and cultured stocks. The potential harmful effects on native gene pools due to escape of farmed fish and large-scale ranching programmes have been discussed recently (Saunders, 1991; Hindar, Ryman and Utter, 1991; Jørstad, Skaala and Dahle, 1991), and these questions are highly controversial (Kapuscinski and Lannan, 1986; Bentsen, 1991; Ryman, 1991). This is mainly due to lack of empirical information about gene flow between populations which can only be obtained through controlled experiments including some kind of genetic marker.

Acknowledgement

We are greatly indebted to the staff of the Institute of Marine Research for collecting the samples and providing information on chemical tagging and age of individual fish. Likewise we also want to thank the Norwegian Fisheries Research Council for financial support.

References

Allendorf, F.W. and Utter, F. (1979) Population genetics, in *Fish Physiology* (eds W.S. Hoar, D. Randall and J.R. Brett), Vol. 8, Academic Press, London, pp. 407–54.

Bentsen, H.B. (1991) Quantitative genetics and management of wild populations. *Aquaculture*, **98**, 263–6.

Blom, G., Svåsand, T., Jørstad, K.E. *et al.* (1990) Comparative growth and survival of two genetic strains of Atlantic cod (*Gadus morhua* L.) reared through the early life stages in a marine pond in western Norway. *ICES* C.M. 1990/F: 48.

Gharrett, A.J. and Seeb, J.E. (1990) Practical and theoretical guidelines for genetically marking fish populations, *Am. Fish. Soc. Symp.*, **9**, 407–17.

Hindar, K., Ryman, N. and Utter, F. (1991) Genetic effects of cultured fish on natural fish populations. *Can. J. Fish. Aquat. Sci.*, **48**, 945–57.

Jørstad, K.E. (1984) Genetic analyses of cod in northern Norway, in *The Propagation of Cod* Gadus morhua *L.* (eds E. Dahl, D.S. Danielsen, E. Moksness and P. Solemdal), *Flødevigen rapportser.*, **1**, 734–60.

Jørstad, K.E. and Nævdal, G. (1989) Genetic variation and population structure of cod, *Gadus morhua* L., in some fjords in northern Norway. *J. Fish. Biol.*, **35** (Suppl. A), 245–52.

Jørstad, K.E., Skaala, Ø. and Dahle, G. (1991) The development of biochemical and visible genetic markers and their potential use in evaluating interaction between cultured and wild fish populations. *ICES Mar. Sci. Symp.*, **192**, 200–5.

Jørstad, K.E., Øiestad, V., Paulsen, O.I., Naas, K. and Skaala, Ø. (1987) A genetic marker for artificially reared cod (*Gadus morhua* L.) *ICES* C.M. 1987/F:22.

Kapuscinski, A.R.D. and Lannan, J.E. (1986) A conceptual genetic fitness model for fisheries management. *Can. J. Fish. Aquat. Sci.*, **43**, 1606–16.

Lane, S., McGregor, A.J., Taylor, S.G. and Gharret, A.J. (1990) Genetic marking of an Alaskan pink salmon population, with an evaluation of the mark and the marking process. *Am. Fish. Soc. Symp.*, **7**, 395–406.

Moav, R., Brody, T., Wholfart, G. and Hulata, G. (1976) Applications of electrophoretic markers to fish breeding. I. Advantages and methods. *Aquaculture*, **9**, 217–28.

Mork, J., Reuterwall, C., Ryman, N. and Ståhl, G. (1982) Genetic variation in Atlantic cod (*Gadus morhua* L.): a quantitative estimate from a Norwegian coastal population. *Hereditas*, **96**, 55–62.

Nordeide, J.T., Holm, J.C., Otterå, H., Blom, G. and Borge, A.(1992) The use of oxytetracycline as a marker for juvenile cod (*Gadus morhua* L.). *J. Fish. Biol.*, **41**, 21–30.

Øiestad, V., Kvenseth, P.G. and Folkvord, A. (1985) Mass production of Atlantic cod juveniles (*Gadus morhua*) in a Norwegian saltwater pond. *Trans. Am. Fish. Soc.*, **114**, 590–5.

Ryman, N. (ed.) (1981) Fish gene pools. *Ecol. Bull.* (Stockholm), **34**.

Ryman, N. (1991) Conservation genetics considerations in fishery management. *J. Fish Biol.*, **39**, (Suppl. A), 211–24.

Saunders, R.L. (1991) Potential interaction between cultured and wild Atlantic salmon. *Aquaculture*, **98**, 51–60.

Seeb, L.W., Seeb, J.E., Allen, R.L. and Hershberger, W.K. (1990) Evaluation of adult returns of genetically marked Chum salmon, with suggested future applications. *Am. Fish. Soc. Symp.*, **9**, 418–25.

Shaklee, J. (1983) The utilization of isozymes as gene markers in fisheries management and conservation, in *Isozymes: Current Topics in Biological and Medical Research.* (eds M.C. Ratazzi, J.G. Schandalios and G.S. Whitt), Alan R. Liss, New York, pp. 213–47.

Skaala, Ø., Dahle, G., Jørstad, K.E. and Nævdal, G. (1989) Interaction between natural and farmed fish populations: information from genetic markers. *J. Fish. Biol.*, **36**, 449–60.

Solemdal, P., Dahl, E., Danielssen, D.S. and Moksness, E. (1984) The cod hatchery in Flødevigen – background and realities, in *The Propagation of Cod* Gadus morhua *L.* (eds E. Dahl, D.S. Danielssen, E. Moksness and P. Solemdal) *Flødevigen Rapportser*, Vol. 1, pp. 17–45.

Svåsand, T., Jørstad, K.E., Blom, G. and Kristiansen, T.S. (1991) Application of genetic markers for early life history investigations on Atlantic cod (*Gadus morhua* L.). *ICES Mar. Sci. Symp.*, **192**, 5–12.

Swofford, D.L. and Selander, R.B. (1981) BIOSYS-1: a FORTRAN program for the comprehensive analyses of electrophoretic data in population genetics and systematics. *J. Hered.*, **72**, 281–3.

Taggart, J. and Ferguson, A. (1986) An electrophoretically detectable genetic tag for hatchery reared brown trout, *Salmo trutta* L. *Aquaculture*, **41**, 119–30.

Utter, F.M. and Seeb, J.E. (1990) Genetic marking of fishes: overview focusing on protein variation. *Am. Fish. Soc. Symp.*, **7**, 426–38.

Wilkins, N.P. (1981) The rationale and relevance of genetics in aquaculture: an overview. *Aquaculture*, **22**, 209–28.

Index

Page numbers in *italic* refer to tables and page numbers in **bold** refer to figures.